SOUTH WESTERN
Algebra 1
AN INTEGRATED APPROACH

GERVER, SGROI, CARTER, HANSEN
MOLINA & WESTEGAARD

Editor-in-Chief	Peter McBride
Managing Editor	Eve Lewis
Project Manager	Enid Nagel
Developmental Editor	Janet Heller
Production Coordinator	Patricia M. Boies
Production Consultant	Tamara S. Jones
Marketing Manager	Colleen J. Thomas
Art Director	John Robb
Photographic Consultant	Devore M. Nixon
Design Consultant	Elaine St. John Lagenaur
Editorial Assistant	Mary Schwarz
Marketing Assistant	Dawn Zimmer
Editorial Development and Production	Gramercy Book Services, Inc.
Cover Design	Photonics Graphics
About the Cover	The cover design is a collage of real world images from the chapter themes. How many can you find?

ISBN: 0-538-64417-6
1 2 3 4 5 6 7 8 VH 03 02 01 00 99 98 97 96
Printed in the United States of America

I(T)P
International Thomson Publishing
South-Western Educational Publishing is an ITP Company. The ITP trademark is used under license.

Robert Gerver, Ph.D., has been a mathematics instructor at North Shore High School in Glen Head, New York for more than 15 years. During that time, he has also taught at several other institutions, grade 4 through graduate level. He has served on the New York Regents Competency Test Committee. He received his B.A. and M.S. degrees from Queens College of the City University of New York where he was elected to Phi Beta Kappa. He received his Ph.D. from New York University. In 1988, Dr. Gerver received the Presidential Award for Excellence in Mathematics Teaching for New York State.

Claudia Carter, mathematics teacher at the Mississippi School for Mathematics and Science, Columbus, Mississippi, has taught all levels of secondary mathematics and is an active advisor for Mu Alpha Theta. Ms. Carter is a Woodrow Wilson National Foundation Master Teacher and a consultant for IBM as a demonstrator of *Mathematics Exploration Toolkit*. She is past president of Mississippi Council of Teachers of Mathematics. She was the recipient of the Presidential Award for Excellence in Mathematics Teaching in 1989 and received Mississippi's Top Twenty Star Teacher Award in 1990, 1991, and 1992. Ms. Carter received her B.A. and M.A. degrees from the University of New Orleans.

David Molina, Ph.D., is the Glenadine Gibb Teaching Fellow in Mathematics Education and assistant professor in the Department of Curriculum and Instruction at the University of Texas in Austin. He has taught secondary mathematics and at present is researching curricular and instructional changes through technology. Dr. Molina is a consultant to school districts which are implementing systemic change. Dr. Molina received his B.S. degree from the University of Notre Dame and his M.A. and Ph.D. from the University of Texas, Austin.

Richard Sgroi, Ph.D., is mathematics director of Newburgh Enlarged City School District, Newburgh, New York. He has taught mathematics at all levels for more than 20 years. While teaching at North Shore High School, Glen Head, New York, he and his colleague Robert Gerver developed both *Dollars and Sense: Problem Solving Strategies in Consumer Mathematics* and *Sound Foundations: A Mathematics Simulation*. He and Dr. Gerver received recognition from the U.S. Department of Education's Program Effectiveness Panel for *Sound Foundations* to be included in Educational Programs That Work. Dr. Sgroi received his B.A. and M.S. degrees from Queens College of the City University of New York, and his Ph.D. from New York University.

Mary Hansen, a mathematics teacher at Independence High School, Independence, Kansas, teaches algebra and geometry. While teaching in Texas and North Carolina, Ms. Hansen was active in mathematics teachers staff development. She received her B.A. and M.S. degrees from Trinity University, San Antonio, Texas, where she was elected to Phi Beta Kappa.

Susanne Westegaard is a mathematics department chair of Montgomery-Lonsdale Public School, Montgomery, Minnesota where she has taught mathematics for more than 20 years. Ms. Westegaard has been on the Woodrow Wilson National Fellowship Traveling Team for several summers. She received the Distinguished Educator Award from the Minnesota Mutual Foundation and MAEF in May 1994 and the Presidential Award for Excellence in Teaching Mathematics in 1991. Ms. Westegaard received her B.S. degree from Yankton College, Yankton, South Dakota and her M.Ed. from the University of Minnesota.

ALGEBRA 1: AN INTEGRATED APPROACH

Reviewers

George Bratton
Associate Professor of Mathematics
University of Central Arkansas
Conway, Arkansas

Gail Brenner
Mathematics Teacher
McCord Junior High School
Sylvania, Ohio

Rosemary Garmann
Mathematics Consultant
Hamilton County Office of Education
Cincinnati, Ohio

Judith Gerwe
Mathematics Department Chair
Notre Dame Academy
Park Hills, Kentucky

Jacqueline Brannon Giles
Multicultural Consultant
Central College
Houston Community College System
Houston, Texas

Sue Hunsinger
Mathematics Department Chair
Leon County Schools
Tallahassee, Florida

Dale Johnson
Professor of Research
University of Tulsa
Tulsa, Oklahoma

Alicia Jones
Algebra Teacher
Athens Drive High School
Raleigh, North Carolina

Carl Kalota
Assistant Professor
Division of Professional Practice
University of Cincinnati
Cincinnati, Ohio

Bonnie Knowles
Mathematics Department Chair
Deep Creek Middle School
Chesapeake, Virginia

Joan Lamborne
Supervisor of Mathematics
Egg Harbor Township High School
Pleasantville, New Jersey

James Mayes
Algebra Teacher
Elkhart Memorial High School
Elkhart, Indiana

Bob Mora
Coordinator, Mathematics/Technology
Carrollton-Farmers ISD
Carrollton, Texas

Edward Okuniewski
Mathematics Teacher
Andover High School
Bloomfield Hills, Michigan

Robert Pacyga
Mathematics Supervisor
Argo Community High School
Summit, Illinois

Justine Replinger
Mathematics Department Chair
East Leyden High School
Franklin Park, Illinois

James Rudolph
Mathematics Curriculum Specialist K–12
Riverside Unified School District
Riverside, California

Paul Tisdel
Mathematics Department
Robert E. Lee High School
San Antonio, Texas

Connecting the World to Algebra

To the Student

You are about to embark on a most exciting algebra program!

South-Western Algebra bridges your skills and knowledge of today to what you will be doing in the future. In *South-Western Algebra*, you will connect algebra to topics such as geometry, earth science, biology, travel, ecology, entertainment, and personal finance.

Chapter themes help you to see how mathematics applies to daily life.

The *Chapter Project* helps you develop mathematical understanding and connect algebra to the real world. You can use the Internet to research your projects by using the address given in the Internet Connection.

The *Data Activity* and *Explore Statistics* lessons help you use data in various ways, including reading and interpreting charts and graphs.

AlgebraWorks showcases a specific job or career which requires the algebra you are learning. You will see the importance algebra has in everyday life!

The *Problem Solving File* prepares you with problem solving strategies that will help you on standardized tests. It will also sharpen your critical thinking and problem solving skills!

Algebra Workshops and *Explore Activities* help you understand ideas so you can become comfortable with algebra.

You will use *calculators* and *computers* as tools for learning and doing mathematics. Graphing calculators and computer spreadsheet applications show you how graphs communicate concepts about data.

South-Western Algebra will make you a confident problem solver who can clearly communicate in mathematics. You will feel empowered as you realize the endless opportunities that algebra brings to you.

1 Data and Graphs

THEME: Entertainment

Applications and Connections

Technology and Other Tools

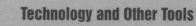

2 Variables, Expressions, and Real Numbers

THEME: Meteorology: The Science of Weather

3 Linear Equations

THEME: Health and Fitness

4 Functions and Graphs

THEME: Sports

Applications and Connections

Technology and Other Tools

5 Linear Inequalities

THEME: Graphic Arts and Advertising

6 Linear Functions and Graphs

THEME: Save the Planet

Applications and Connections

Technology and Other Tools

7 Systems of Linear Equations

THEME: Cities and Municipalities

Technology and Other Tools

8 Systems of Linear Inequalities

THEME: Be a Smart Shopper!

9 Absolute Value and the Real Number System

THEME: Traffic

10 Quadratic Functions and Equations

THEME: Business and Industry

11 Polynomials and Exponents

THEME: Exploring Flight

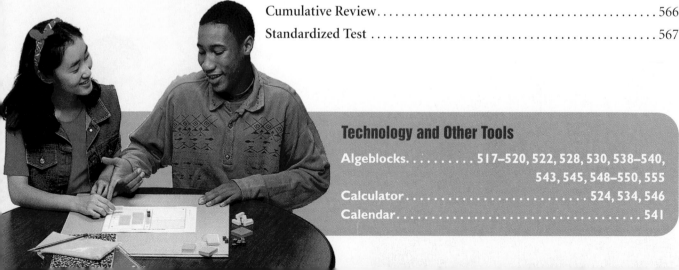

Technology and Other Tools

12 Polynomials and Factoring

THEME: U.S. Rivers

Technology and Other Tools

13 Geometry and Radical Expressions

THEME: Travel and Transportation

14 Rational Expressions

THEME: Agribusiness

Also...

1 Data and Graphs

Take a Look
AHEAD

Make notes about things that look familiar.

- List the different topics covered in this chapter. Put a check mark next to the topics you have studied in previous mathematics courses.
- What do you think algebra is about? Why do you think the topics you listed above are included in an algebra book? List what the topics have in common.

Make notes about things that look new.

- Look for new terms. Find an example of "variable," "spreadsheet," "statistic," "probability," and "matrix."
- Locate the problem solving guide. List the steps you can use to solve a problem.

DATA Activity

How Does it Rate?

A television show's popularity is rated by the percent of TV households that tune in to that program. The table on the next page shows the most popular programs for the week of June 12 through June 18, 1995, according to Nielsen Media Research. As the agency that tracks this information, Nielsen Media Research surveys a representative sample of households and uses the results to describe the entire population.

SKILL FOCUS

- Add, subtract, multiply, and divide whole numbers and decimals.
- Round numbers to different places.
- Solve two-step problems.
- Determine percents.

ALGEBRAWORKS

HELP WANTED

Entertainment!

In this chapter, you will see how:

- **MEDIA RESEARCHERS** use data to interpret and predict TV viewing patterns. (Lesson 1.5, page 26)

- **INVENTORY CONTROL MANAGERS** use algebra to keep track of shipments going into and out of warehouses. (Lesson 1.7, page 37)

TELEVISION RATINGS		
Title	Network	Percent of TV Households
1. Prime Time Live	ABC	25.9
2. NBA Final Game 4	NBC	14.9
3. Friends	NBC	14.2
4. Roseanne	ABC	13.9
5. E.R. TIE	NBC	13.4
Seinfeld TIE	NBC	13.4
7. Grace Under Fire	ABC	12.3
8. Home Improvement	ABC	12.2
9. 20/20	ABC	11.5
10. Coach	ABC	11.4

Source: Nielsen Media Research

Use the table to answer the following questions.

1. How many rating points separated the first-place show from the tenth-place show?

2. To the nearest tenth of a rating point, what was the average rating of the top ten shows?

3. If each rating point equals 954,000 households, how many households tuned in to the NBA Final Game 4?

4. Show the number you answered for Question 3 rounded to the nearest thousand, the nearest ten thousand, and the nearest million.

5. How many more households tuned in to *ER* than to *20/20*?

6. What percent of the top ten shows did each of ABC, NBC, and CBS have?

7. **WORKING TOGETHER** Work with a group to research and report on the methods Nielsen uses to collect and analyze data. If you know of a household that is a member of the sample population, try to arrange an interview with a participant.

What's Really on TV?

More than 98% of U.S. households own at least one TV set, and about 60% of them have cable hook-ups. People in such a huge audience are likely to have different preferences about what types of shows they like to see. Just how much variety does current programming actually offer? What types of shows seem most popular? Aside from specialized cable stations, are the types of programs similar on most stations or not? How many minutes of commercials can you expect each hour? In this project, you will investigate these and other questions about "what's on TV today."

PROJECT GOAL

To collect, analyze, and display data about television programming.

Getting Started

Work in small groups. Obtain several TV programming guides to use as references.

1. Discuss the different types of programs that are on television. Plan a method to catagorize the various kinds of programs. Too few categories might not be meaningful, while too many may be confusing.

2. Decide how you will treat movies or special programs.

3. Select the channels you will research. Will you study each one in the same way?

4. Think about how to summarize information about weekly programming, channel-by-channel offerings, or the variety on different days or time periods. Design and use data sheets.

PROJECT Connections

Lesson 1.2, page 12:
Take a survey to determine how student preferences compare to actual TV programming.

Lesson 1.5, page 25:
Investigate how much television time is devoted to commercials and the kinds of products promoted on different programs.

Lesson 1.8, page 41:
Explore the future of television programming guides.

Chapter Assessment, page 49:
Report on the present state of television programming and make recommendations for change.

Internet Connection

www.swpco.com/
swpco/algebra1.html

1.1 Algebra Workshop
Work with Data

Think Back

Many times in your life you have drawn conclusions based on information you have. Work with a partner to take a look at a conclusion someone drew from given information.

A music store chain was interested in knowing the most popular type of music in one city. A regional manager for the chain gathered this information from the stores in that city.

CDs and Tapes Most Frequently Sold, July 1995	
Store	**Type of CDs and Tapes**
A	Country
B	Rock
C	Country
D	Country
E	Rock

The manager concluded that country music was the most popular type of music in this city.

1. *Data* and *statistics* are terms often used in discussing information and conclusions. Define these terms in your own words.

2. The conclusion made by the manager seems reasonable based on the data in the table, but this table is misleading. Why do you think it could be misleading?

Algebra Workshop

Explore

● As shown in the table in Think Back, each music store named the types of CDs and tapes most frequently sold in July 1995, based on these data.

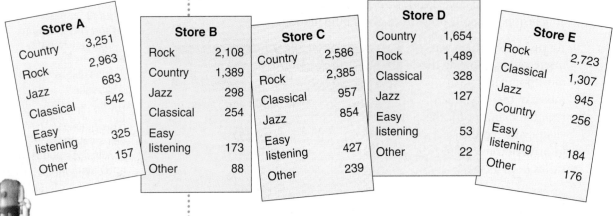

Store A

Country	3,251
Rock	2,963
Jazz	683
Classical	542
Easy listening	325
Other	157

Store B

Rock	2,108
Country	1,389
Jazz	298
Classical	254
Easy listening	173
Other	88

Store C

Country	2,586
Rock	2,385
Classical	957
Jazz	854
Easy listening	427
Other	239

Store D

Country	1,654
Rock	1,489
Classical	328
Jazz	127
Easy listening	53
Other	22

Store E

Rock	2,723
Classical	1,307
Jazz	945
Country	256
Easy listening	184
Other	176

The manager could have drawn a more accurate conclusion had she analyzed all these data. The first step is to organize the data.

3. Create a table like the one at the right. Use the data above to complete the table.

4. Using the table, draw a conclusion about the most popular type of music in this city.

CDs and Tapes Sold, July 1995	
Music Type	**Number of CDs and Tapes**

Depending upon how data are organized, they can often be analyzed to answer several different questions.

5. Write two new questions that could be answered by analyzing the data above. Then use the data to answer your questions. If necessary, reorganize the data in another form of table.

Make Connections

● Work in a group of four to six students to collect, organize, and analyze data that will enable you to draw a conclusion about your class' favorite TV show.

6. Collect data from each member of your group to determine your group's favorite show.

7. Give a copy of your data, both for the group as a whole and for individual group members, to each of the other groups in your class.

8. Analyze the data collected from your class. Draw a conclusion about your class' favorite TV show based on the favorite show of each group. Then draw a conclusion based on each student's favorite show. Organize the data in tables to help you.

9. Compare the conclusions you drew in Step 8. Are they the same or different? Explain the reason for this.

10. What other types of conclusions could you draw from the class data? Use the data to draw one other conclusion.

Summarize

11. **WRITING MATHEMATICS** What decisions did your group make to organize the class data for Steps 8 and 10? Was it helpful to organize the data in a table? Why or why not?

12. **THINKING CRITICALLY** Analyzing correct data can sometimes lead to incorrect or misleading conclusions. Explain.

13. **GOING FURTHER** Was analyzing sales of CDs and tapes the best method of determining the most popular type of music in that city? Explain. If not, suggest a better method.

14. **GOING FURTHER** During a three-day holiday weekend, Cineplex Five sold the following numbers of movie tickets.

> **Saturday** *The Little Penguin:* 1105 adults, 2161 children; *Mark's Misadventure:* 1397 adults, 680 children; *The Graduation:* 370 adults, 73 children; *Speedway Dreams:* 975 adults, 239 children; *Blue Water:* 1223 adults, 410 children

> **Sunday** *Speedway Dreams:* 801 adults, 55 children; *Blue Water:* 1088 adults, 294 children; *The Graduation:* 364 adults, 779 children; *The Little Penguin:* 537 adults, 725 children; *Mark's Misadventure:* 1450 adults, 172 children

> **Monday** *Mark's Misadventure:* 1025 adults, 308 children; *The Graduation:* 221 adults, 56 children; *Blue Water:* 1299 adults, 498 children; *Speedway Dreams:* 1675 adults, 561 children; *The Little Penguin:* 344 adults, 390 children

Organize the Cineplex Five data to answer these questions.

a. Which film had the best attendance?

b. Approximately what percent of the weekend's ticket sales were for *The Little Penguin*?

c. On which day were the most tickets sold?

> **THINK BACK**
>
> To find what percent one number is of another, divide the part by the whole.
>
> What percent of 150 is 76?
>
> part $\rightarrow \dfrac{76}{150} \approx 0.51$
> whole
>
> So, 76 is about 51% of 150.

Explore

The director of a community center asked some teens to name their favorite activity at the center. He received these responses.

chess	art	swimming	basketball	swimming
drama	dance	art	basketball	swimming
chess	basketball	dance	dance	basketball
art	basketball	chess	dance	basketball
drama	drama	art	art	dance
basketball	drama	art	basketball	drama
basketball	dance	swimming	dance	swimming

1. How many teens named each activity?

2. The director has to cut some programs due to limited funds. Which programs would you recommend he cut? Which would you recommend he not cut? Explain.

CHECK UNDERSTANDING

Would organizing the data in Explore have made your work in Questions 1 and 2 easier? Explain.

Build Understanding

When a set of data is not organized, it is difficult to analyze the data for the purpose of drawing conclusions. For example, by simply looking at the responses as they are listed above, it is hard to make recommendations to the community center director. In Lesson 1.1 you learned that it is helpful to organize data in a table. One type of table that you can use to organize data is a **frequency table**.

To organize data in a frequency table, use tally marks to record each time a piece of data occurs. By looking at the tally marks for each item, you can record the **frequency**, or number of times, that each item occurs.

EXAMPLE 1

Make a frequency table for these data. Refer to the frequency table to find the number of household members that occurs most often.

Number of Household Members per Apartment at 1232 Crestwood Ave.									
4	3	1	2	2	3	4	3	2	6
4	3	4	3	3	5	4	4	3	3
3	4	2	3	1	3	4	3	5	4

Solution

List each different number that occurs in the set of data. Then make a tally mark for each item, as it occurs, in the set of data. For every fifth tally mark, draw a diagonal mark across four vertical marks (卌). Finally, record the frequency of each different number.

The number of household members that occurs most often is 3. ◄

Number of Household Members

Number	Tally	Frequency
1	II	2
2	IIII	4
3	卌 卌 II	12
4	卌 IIII	9
5	II	2
6	I	1

If there are many different items in a set of data, sometimes it is more helpful to group the data in *intervals*.

EXAMPLE 2

PUBLIC UTILITIES Make a frequency table for this set of data. Group the data in intervals of 10. Then refer to the frequency table to answer these questions.

a. On how many days were fewer than 30 calls made?

b. On what percent of the days were at least 40 calls made?

Number of Service Calls (Daily) by Town and Country Heating in November

20	1	20	45	31
41	24	3	20	37
48	51	37	36	6
29	54	44	28	38
20	8	38	27	30
32	32	4	3	29

Solution

The least number of service calls is 1; the greatest number is 54. Show intervals of 10 from 1 to 59 in the frequency table.

a. Add the frequencies shown for 0–9, 10–19, and 20–29.

$$6 + 0 + 9 = 15$$

On 15 days, fewer than 30 calls were made.

b. Days with at least 40 calls:

$$4 + 2 = 6$$

Total days: 30

$$\frac{6}{30} = \frac{1}{5} = 20\%$$

On 20% of the days, at least 40 calls were made. ◄

Number of Service Calls Made Daily

Number	Tally	Frequency
0–9	卌 I	6
10–19		0
20–29	卌 IIII	9
30–39	卌 IIII	9
40–49	IIII	4
50–59	II	2

COMMUNICATING ABOUT ALGEBRA

In Example 2, there are no items of data in the interval 10–19. Why do you think this interval was included in the frequency table?

TRY THESE

PHYSICAL FITNESS Use the health club data shown at the right.

1. Make a frequency table for the data.

2. What percent of members exercise at least 5 h per week?

3. Find the number of hours per week most club members exercise.

4. Can you use the data to determine how many times a week each member visits the club?

Number of Hours Health Club Members Exercise per Week									
3	0	2	1	5	2	7	2	3	4
4	1	6	7	3	5	2	1	2	0

LITERATURE Use the reading data shown below.

5. Make a frequency table with the data grouped into intervals of 5.

6. How many participants read at least 10 books?

7. What percent of the participants read fewer than 20 books?

Number of Books Read During Summer Reading Program				
6	3	10	15	12
7	2	9	14	21
8	4	20	12	10
4	5	22	16	18
9	10	8	11	13

8. **WRITING MATHEMATICS** Write one question you can answer using the reading data as shown above, but not as shown in your frequency table.

PRACTICE

For Problems 1–6, make a frequency table for each set of data. Group the data by intervals if appropriate. Then refer to the appropriate frequency table to answer the questions.

ENTERTAINMENT Use the theater data shown below.

1. How many theaters have fewer than 5 screens?

2. What fraction of the theaters have more than 10 screens?

Number of Movie Screens per Theater in Indianapolis, 1995									
2	12	8	8	12	7	3	4	3	2
4	4	12	4	10	10	8	6	2	2

EDUCATION Use the exam score data shown below.

3. A passing score on the exam is 60. To the nearest whole percent, what percent of the class did not pass the exam?

Exam Scores of Students in Ms. Rivera's Algebra Class													
98	83	83	87	79	81	47	82	95	48	78	88	70	68
78	86	76	60	90	65	89	80	82	92	84	53	88	72

4. What fraction of students scored in the 80s?

PUBLISHING Use the circulation data shown below.

5. How many more magazines had a circulation of under 10 million than had a circulation of over 20 million?

6. What percent of the magazines had a circulation of 6 million to 15 million?

Circulation of Top 14 U.S. Magazines, 1993			
Magazine	**Circulation**	**Magazine**	**Circulation**
Better Homes and Gardens	7,600,960	National Geographic	9,390,787
Family Circle	5,114,030	NRTA/AARP Bulletin	22,064,262
Good Housekeeping	5,162,597	People Weekly	3,446,569
Ladies' Home Journal	5,153,565	Reader's Digest	16,261,968
McCall's	4,605,441	Time	4,103,772
Modern Maturity	22,226,063	TV Guide	14,122,915
National Enquirer	3,403,330	Women's Day	4,858,625

7. WRITING MATHEMATICS In Problems 1–6, how did you decide whether to group the data by intervals? If you grouped the data by intervals, how did you decide which intervals to use?

EXTEND

MEASUREMENT Some students were asked to measure the thickness of a steel plate using calipers that give measurements to thousandths of an inch. Here are their results, in inches.

0.373	0.369	0.367	0.372
0.369	0.366	0.370	0.371
0.368	0.369	0.371	0.372
0.370	0.371	0.368	0.369
0.371	0.373	0.365	0.374

8. Make a frequency table showing the number of students for each measurement.

9. Find the measurement that has an equal number of measurements greater than and less than it.

10. What percent of the measurements are between 0.369 in. and 0.371 in. inclusive?

11. What do you think is the best estimate of the actual thickness? Explain.

BUSINESS Adrianne is planning to start a pet care service for people who are away on business or vacation but who want to leave their pets at home, rather than at the kennel. She surveyed others in her city who were already performing the same service and made the frequency table below at the left. Refer to this table to answer the following questions.

12. How many people did Adrianne survey?

13. Which price-range interval contains the greatest number of charges? How many people charge less than this? How many charge more than this?

14. What would you recommend that Adrianne charge per visit? Explain your decision.

Charge per Visit at Home With 1 or 2 Pets		
Charge	Tally	Frequency
$4.00–$4.99	ЖЖ	5
$5.00–$5.99	ЖЖ I	6
$6.00–$6.99	ЖЖ III	8
$7.00–$7.99		0
$8.00–$8.99	IIII	4
$9.00–$9.99	II	2
$10.00–$10.99	I	1

THINK CRITICALLY

SPORTS Following basketball tryouts, Coach Fox made a frequency table of the number of free throws the boys made. Dante knew how many free throws he had made, and he looked at the table to see how many other boys had made the same number.

15. If Coach Fox had made a frequency table such as the table in the solution part of Example 1, could Dante have gotten this information? Explain.

16. If Coach Fox had made a frequency table such as the table in the solution part of Example 2, could Dante have gotten this information? Explain.

PROJECT *Connection*   In this activity you will compare student preferences with actual television programming.

1. Use the same program categories as in the opening project activity. Prepare a survey form that lists each category and includes space for students to rank each of the categories in terms of their own preferences. As a group, decide on the ranking method you will use.

2. Survey at least 30 students. Tally the results and display them in a table or graph.

3. Write a report about whether student preferences are well represented by actual programming.

1.3 Use Variables to Represent Data

Explore

● A football coach created the following symbols.

$A @ B$ means "passing average if you have A completions in B attempts."

$$A @ B = \frac{A}{B}$$

$A \# B$ means "percent of games won if you have A wins and B losses."

$$A \# B = \frac{A}{A + B} \cdot 100\%$$

Give the meaning of each of the following expressions. Then evaluate the expression.

1. $15 @ 25$ **2.** $15 \# 25$ **3.** $21 @ 84$ **4.** $21 \# 84$

Write a mathematical expression for $A \, \mathcal{c} \, B$ if $A \, \mathcal{c} \, B$ means:

5. percent of games lost if you have A wins and B losses

6. number of points scored on A touchdowns and B field goals (touchdown = 6 points; field goal = 3 points)

Build Understanding

● One of the principal uses of computers is to analyze data. Computer programmers use symbols to tell a computer what to do with a set of data. In a computer language such as BASIC, capital letters are used to represent numbers. These letters are called **variables**, because their values *vary* depending on the numbers that they represent. A computer uses variables to mark locations in the computer's memory where data represented by the variables are stored.

In the following computer programs, the letters A and B are variables. (10 and 20 are line numbers. The next line would be numbered 30.)

```
10 INPUT A, B
20 PRINT A + B
```

If you input 6 for A and 9 for B, the computer would print 15.

An expression containing one or more variables is a **variable expression** or an **algebraic expression.**

BASIC uses the familiar symbols + and – to represent addition and subtraction. The symbol ∗ represents multiplication and the symbol / represents division. Many graphing calculators use a version of BASIC as a programming language for the calculator.

CHECK UNDERSTANDING

How would you change the program so that if you input 6 for A and 9 for B, the computer would print 54?

ALGEBRA: WHO, WHERE, WHEN

BASIC stands for "Beginner's All-purpose Symbolic Instruction Code." It was invented by John Kemeny and Thomas Kurtz at Dartmouth College in 1964.

CHECK UNDERSTANDING

In Example 2, what do the numbers in the printout represent?

EXAMPLE 1

Evaluate each expression if A = 8 and B = 3. Expressions are in BASIC.

a. A − B   **b.** 9 ∗ B   **c.** 44 / (A + B)

Solution

a. A − B = 8 − 3 = 5

b. 9 ∗ B = 9 · 3 = 27

c. 44 / (A + B)
  A + B = 8 + 3 = 11   Evaluate expressions in parentheses first.
  44 / (A + B) = 44 ÷ 11 = 4   ◄

When data have been collected, a program can be written to analyze the data. The design of the program will depend on what the programmer wants to know about the data.

EXAMPLE 2

ACCOUNTING The following program calculates the net daily receipts for one weekend at the Einstein Science Museum. The variable A represents adult attendance at $5 each. C represents child attendance at $3 each.

```
10  INPUT A, C
20  PRINT 5 ∗ A + 3 ∗ C    Perform multiplication before addition.
```

Weekend Attendance		
	Adult	**Child**
Friday	106	255
Saturday	348	491
Sunday	196	304

If you input data from the table, what will the computer print?

Solution

A = 106, C = 255: 5 ∗ A + 3 ∗ C = 5 · 106 + 3 · 255 = 1295

A = 348, C = 491: 5 ∗ A + 3 ∗ C = 5 · 348 + 3 · 491 = 3213

A = 196, C = 304: 5 ∗ A + 3 ∗ C = 5 · 196 + 3 · 304 = 1892

The computer will print 1295, 3213, and 1892.   ◄

Computer spreadsheets are designed to operate using formulas and data. A spreadsheet contains **cells** designated by column and row. Columns are identified by letters in alphabetical order from left to right, and rows are identified by numbers in order from top to bottom.

In any cell, you can enter data or a formula. The data can be a **value** (numerical data) or a **label** (text). Values, but not labels, can be used in calculations. Labels are often used as titles for, or explanations of, the data in nearby cells. Entering a formula in a cell tells the spreadsheet program to perform a calculation, usually by referring to other cells that contain the required values.

In the spreadsheet below, row 1 and column A have been reserved for labels. The number 9, a value, appears in cell C2.

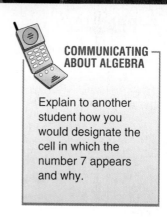

Column A			
	A	B	C	D
1			Cell C2		Row 1
2		5	9	45	=B2*C2
3		7	4	28	=B3*C3

The formula entered in cell D2 tells the computer what operation to perform, using the data in cells B2 and C2. When you enter a formula in a cell, instead of seeing the formula, you will see the result of the calculation. For example, in cell D2, you see 45 (the result of B2 $*$ C2).

EXAMPLE 3

ACCOUNTING Use the data in Example 2. Design a spreadsheet to find the total daily attendance on Friday through Sunday at the science museum. Then find the totals the computer will calculate.

Solution

The total each day is the sum of the numbers of adults and children. Therefore, the formulas entered in the cells of the last column tell the computer to add the data from the appropriate cells.

	A	B	C	D	
1		Adult	Child	Totals	Label
2	Friday	106	255	361	=B2+C2
3	Saturday	348	491	839	=B3+C3
4	Sunday	196	304	500	=B4+C4

The totals are 361, 839, and 500.   ◄

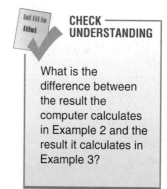

COMMUNICATING ABOUT ALGEBRA

Explain to another student how you would designate the cell in which the number 7 appears and why.

CHECK UNDERSTANDING

What is the difference between the result the computer calculates in Example 2 and the result it calculates in Example 3?

TRY THESE

Evaluate each expression if L = 2, M = 6, and N = 10. Expressions are in BASIC.

1. N – M

2. L $*$ M

3. M / L

4. 6 $*$ M – N

5. 5 $*$ 2 / N

6. 4 $*$ M / (L + N)

7. WRITING MATHEMATICS Give an example of a formula with which you are familiar. Are the expressions on each side of the formula algebraic expressions? Explain.

ELECTRONICS The table at the right gives data on sales of three television models at TV King stores on the plaza and at the mall. The Premier sells for $699, the Mark IV for $425, and the Entertainer for $249.

	Premier	Mark IV	Entertainer
Plaza	9	21	16
Mall	15	30	34

8. What would the program at the right be used to find?

```
10 INPUT A, B, C
20 PRINT A + B + C
```

9. If you input these data, what would the computer print?

10. Design a spreadsheet to find the net receipts on the sale of television sets at each store. Then find the totals that the computer will calculate.

PRACTICE

Evaluate each expression if R = 9, M = 5, and N = 16. Expressions are in BASIC.

1. R + N
2. N − M
3. 7 * M
4. R / 3
5. N + M − R
6. 15 + N + R
7. N / (M − 3)
8. 10 * R / M
9. (4 * N) / (2 * M)

VIDEO RENTALS This table gives the number of videos rented at three Video Extra stores, grouped by MPAA rating.

	G	PG	PG13	R
Store 1	73	55	38	71
Store 2	28	30	41	51
Store 3	107	88	95	120

10. Design a spreadsheet to find the total number of videos rented at each store.

11. Determine the totals the computer will calculate.

GEOMETRY The program at the right finds the area of a triangle with base B and height H. Find each area.

```
10 INPUT B, H
20 PRINT 0.5 * B * H
```

12. B = 24
    H = 9

13. B = 41
    H = 35

14. B = 0.62
    H = 0.8

15. How would you change the program to print the areas of rectangles with base B and height H?

METEOROLOGY The program at the right finds the Fahrenheit temperature that corresponds to a Celsius temperature of C degrees. Find each Fahrenheit temperature.

```
10 INPUT C
20 PRINT 1.8 * C + 32
```

16. C = 45

17. C = 100

18. C = 0

19. **WRITING MATHEMATICS** Write a few sentences explaining why variables are useful in computer programs.

OCEANOGRAPHY At sea level, the *air pressure*—the weight of the atmosphere pressing on you—is 14.7 lb/in.^{2} Divers descending into the ocean feel an additional 0.44 lb/in.^{2} of *water pressure* for each foot they descend.

20. Explain how you could find the *total pressure* in pounds per square inch at a depth of *d* ft in the ocean.

21. Design a spreadsheet showing the water pressure and the total pressure at each of these points in the ocean (depths in feet in parentheses): Puerto Rico Trench (28,232), Japan Trench (27,599), Brazil Basin (20,076), Peru-Chile Trench (26,457), Eurasia Basin (17,881), Mariana Trench (35,840).

EXTEND

VIDEO RENTALS Use the data for Exercise 10.

22. Design a spreadsheet to find the total number of films of each rating that were rented at the three stores.

23. Find the totals the computer will calculate.

GEOMETRY Write a BASIC program for finding each of the following.

24. the perimeters of rectangles with base B and height H

25. the areas of trapezoids with height H and bases M and N

26. the surface areas of rectangular prisms with length L, width W, and height H

THINK CRITICALLY

Let A = 4, B = 10, and C = 5. Write two expressions in BASIC that are equal to the given number.

27. 8   **28.** 24   **29.** 50   **30.** 90   **31.** 200

32. Is X + Y * Z equal to Y * Z + X? Give examples to illustrate your answer.

33. Is X * Y + Z equal to X * (Y + Z)? Give examples to illustrate your answer.

MIXED REVIEW

34. Make a frequency table using these daily high temperatures (°F) for February.

26, 32, 31, 25, 19, 16, 27, 26, 15, 16, 23, 28, 33, 35,
34, 30, 35, 40, 41, 31, 30, 29, 25, 25, 26, 30, 31, 33

To the nearest whole percent, on what percent of the days in February was the high temperature freezing (32°) or greater?

35. STANDARDIZED TESTS For K = 5, L = 8, and M = 3, what is the value of 7 * (K + M) / (4 * L)?

   A. 0.57   **B.** 2.8   **C.** 1.75   **D.** 2.84

Think Back

● Work with the same group you worked with in Lesson 1.1.

1. Compile a table of information like the one below. Base the information in your table on the data you collected about favorite television shows. This table shows the results for three shows chosen by 25 students polled.

Name of Show	Number of Students Choosing This Show	Fractional Portion of All Students	Decimal Portion of All Students	Percent of All Students
Favorite Algebra Teachers	6	$\frac{6}{25}$	0.24	24%
Unsolved Math Problems	5	$\frac{1}{5}$	0.2	20%
House of Variables	2	$\frac{2}{25}$	0.08	8%

Explore

● One way to display data is with a **pictograph**. The pictograph shown at the right displays data on the favorite types of music of the members of one class.

Favorite Type of Music

👤 = 2 people selecting as favorite

(pictograph with categories: R & B, Rock, Reggae, Latin, Country, Jazz, Rap)

2. How many students are represented on the pictograph? How can you tell?

3. How many students chose rap as their favorite type of music?

4. Suppose that each symbol represented 3 people. How many symbols would appear in the "Rock" column? in the "Country" column?

5. Would the symbol used in this pictograph be a good symbol to represent 3 people? Justify your answer.

6. In a much larger poll, 1000 people chose Latin music as their favorite type of music. If you were drawing a pictograph of this data, would you let each symbol represent 2 people? Explain.

7. The information depicted on the previous page in the pictograph is shown here in a **bar graph**. Describe similarities and differences between pictographs and bar graphs.

8. When drawing a bar graph, how can you decide which scale to use on the vertical axis?

9. In a **line plot**, each item of data is represented by an **X** and the **X**'s are stacked as shown at the right. There may or may not be a vertical axis for measuring the number of **X**'s. How could you convert a line plot into a bar graph?

10. What advantages do a bar graph and a line plot have over a pictograph? What advantages does a pictograph have?

11. In a **line graph**, which is closely related to a line plot, the tops of the stacks are connected by line segments. A line graph is appropriate for displaying certain types of data, though not "Favorite Type of Music." What type of data might be appropriately displayed in a line graph? Give an example.

12. In a **circle graph**, each category of data is represented by a **sector**, a slice-of-pie-shaped portion of a circle. Each sector is drawn to show the relative size of the category. In the circle graph shown, what is the measure of the central angle of the largest sector? Explain how you found the measure.

13. What information does the circle graph tell you that the other types of graphs do not?

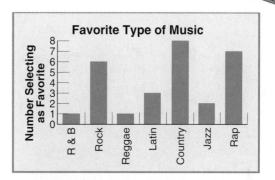

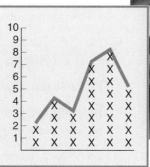

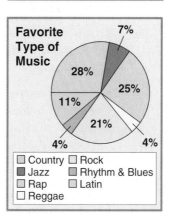

THINK BACK

Angles can be measured in degrees (°). A right angle measures 90°. A straight angle measures 180°. There are 360° in a circle. A central angle is an angle at the center of a circle formed by two radii.

Algebra Workshop

Make Connections

14. Refer to the data you collected on favorite television shows in Lesson 1.1 and the table of information you made at the beginning of this workshop. How many categories should you include in a display of your data?

15. The **range** is the difference between the greatest and least values in a set of data. What is the range of your data?

16. Draw a pictograph, a bar graph, a line plot, and a circle graph to display your data. Most spreadsheet programs can display data in various types of graphs. If you have a computer with a spreadsheet program, use it to draw the graphs.

Summarize

17. MODELING Describe how you drew each of your displays. If you ran into any problems, describe them and explain how you solved them.

18. THINK CRITICALLY Which one of your graphs best displays the data? Explain your reasoning.

19. How did you use the range of your data in making your displays?

20. Compare your graphs with those of other students. If there are differences between the graphs, what might be the reason?

21. What kind of data about television shows could be displayed on a line graph?

22. WRITING MATHEMATICS When a person looks at the graphs, what information can be obtained from each of the four types of graphs you drew?

23. GOING FURTHER You can use a graphing utility to display data. The graph shown at the left displays the data on "Favorite Type of Music" which you investigated in the Explore section of this workshop. Most graphing utilities can display data in several different ways. Experiment with the graphing utility that you use. Find out how to enter data, how to set the range, and how to choose a particular type of display. Then use the utility to display your data on favorite television shows.

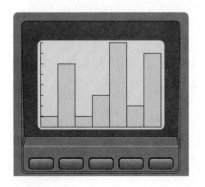

Explore/Working Together

- Your school has been selected to send ten students to a national convention. Delegates to the convention will discuss and make recommendations on issues of importance to teenagers. The ten students from your school are to be chosen so that they "truly represent the entire student body."

  Work with your group. Decide how you would choose the ten students. Explain why you believe that the students you select would be truly representative of all the students in your school.

Build Understanding

- A **survey** is a study of the opinions or behavior of a **population**, or large group of people. The person conducting the survey designs a set of questions, conducts the survey or poll, and records and analyzes the results.

  In order to obtain completely accurate results, the surveyor must poll the entire population. Since that is rarely practical, surveyors usually direct their questions to representative **samples**, or portions, of the population. They then extend the conclusions that they draw from the sample group to the entire population. The accuracy of the results of such a poll depends on whether or not the sample population is truly representative of the total population.

SPOTLIGHT ON LEARNING

WHAT? In this lesson you will learn
- to design a survey, sample a population, interpret your results, and make predictions.

WHY? Surveys and sampling help you solve problems about politics, snack foods, and media research.

Findings based on a sample that is not truly representative are said to be **biased**. The size and method of choosing the sample population, how the questions are formulated, and whether people answer honestly are factors that affect survey results.

EXAMPLE 1

ATHLETICS The Lincoln High School student council decided to sample student opinion about high school athletics. The council designed a questionnaire and distributed it to the first 50 students to arrive at a varsity basketball game. Were the findings biased? Explain.

Solution

The findings were biased. Students who attend a varsity basketball game are likely to be strong supporters of high school athletics. Therefore, they may not be representative of the entire school population, which probably contains some students who do not support athletics. Many students at a basketball game will not want to complete a questionnaire. Mostly students with strong opinions about athletics will participate in the poll, which will further bias the results.

Several standard methods for choosing a representative sample of 25 people are listed below.

Random Sampling
Each member of the population has an equal chance of being selected. To sample your entire school population randomly, for example, you could write the name of each student on an index card, place the cards in a box, draw 25 cards, and survey only those students whose names were drawn.

Cluster Sampling
A specifically defined portion of the population is chosen at random, and everyone in that portion is interviewed. To cluster sample your school population, a letter of the alphabet could be chosen at random. Every student whose last name begins with that letter would then be surveyed.

Convenience Sampling
A readily available group is chosen and all members of the group are surveyed. The first 25 students to arrive at school in the morning might represent a convenience sample.

Systematic Sampling
Members are chosen by a rule or pattern that applies to the entire population. Choosing every twentieth student on an alphabetical list of all the students in your school would be a systematic sampling.

COMMUNICATING ABOUT ALGEBRA

Describe a method you could use to choose an unbiased sample of students at your school for a poll on athletics.

COMMUNICATING ABOUT ALGEBRA

Is it possible for any of the sampling methods described here to be biased? Give an example to illustrate your answer.

EXAMPLE 2

PUBLISHING A publisher wants to know which types of books are most popular with high school students. Name the sampling method represented by each of the following.

a. Choose twenty high schools at random around the country and survey every student in each school.

b. Give questionnaires one afternoon to all high school students who visit the White House on class trips to Washington, D.C.

c. Dial phone numbers at random around the country and ask to speak to a teenager.

Solution

a. This method represents cluster sampling.

b. This method represents convenience sampling.

c. This method represents random sampling. ◀

Once unbiased findings have been assembled for a representative sample of the population, a proportion can be written to extend the findings to the entire population.

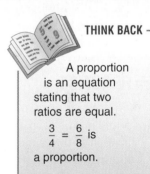

THINK BACK

A proportion is an equation stating that two ratios are equal.
$\frac{3}{4} = \frac{6}{8}$ is a proportion.

EXAMPLE 3

POLITICS Of 240 Lockwood citizens who were surveyed, 107 stated that they supported a proposed airport expansion. If the survey was unbiased, how many of Lockwood's 23,488 citizens can be expected to support the expansion? Round to the nearest hundred.

Solution

The ratio $\frac{107}{240}$ expresses the portion of the people surveyed who support the expansion. Let n represent the number of all the citizens of Lockwood who support the expansion. Write a proportion.

number who support → $\dfrac{107}{240} = \dfrac{n}{23{,}488}$ ← number who support
total number → $\phantom{\dfrac{107}{240}}$ ← total number

$107 \cdot 23{,}488 = 240 \cdot n$ In a proportion, **cross products** are equal.

$2{,}513{,}216 = 240 \cdot n$ 240 *times n* equals 2,513,216. Therefore, find *n* by *dividing* 2,513,216 by 240.

$2513216 \boxed{\div} 240 \boxed{=} 10471.733$ Use a calculator.

To the nearest hundred, 10,500 people in Lockwood support the expansion. ◀

POLITICS During her re-election campaign, the mayor of Lockwood decided to conduct a popularity poll. Name the sampling method represented by each of the following. For each method, give one reason why the findings obtained by that method might be biased.

1. Give a questionnaire to everyone stopping at the mayor's campaign headquarters.

2. Choose a city building and survey everyone going into that building on a particular day.

3. Give a questionnaire to every tenth commuter driving to work across the Lockwood Bridge in the morning.

Solve each proportion for n.

4. $\dfrac{1}{4} = \dfrac{n}{8}$

5. $\dfrac{2}{3} = \dfrac{n}{12}$

6. $\dfrac{n}{20} = \dfrac{3}{4}$

7. $\dfrac{2}{7} = \dfrac{16}{n}$

8. POLITICS A random sample of 160 registered voters showed that 75 supported the mayor in her bid for re-election. If the survey was unbiased, how many of the 8240 registered voters in Lockwood can be expected to vote for the mayor? Round your answer to the nearest hundred.

9. WRITING MATHEMATICS Does everyone who conducts a survey try to be unbiased? Are there reasons a surveyor might want to conduct a biased poll? Give reasons to support your answer.

PRACTICE

SNACK FOODS A popcorn manufacturer wants to identify the most popular type of snack food. Name the sampling method represented by each of the following. For each method, give one reason why the findings obtained by that method might be biased.

1. Place a questionnaire in every fifth package of popcorn produced.

2. Place questionnaires in every package of popcorn sold in a state that produces popcorn.

3. Give questionnaires to all of the people who tour the plant where the company's popcorn is processed and packaged.

Solve each proportion for n.

4. $\dfrac{1}{2} = \dfrac{n}{10}$

5. $\dfrac{1}{3} = \dfrac{n}{6}$

6. $\dfrac{n}{10} = \dfrac{2}{5}$

7. $\dfrac{n}{8} = \dfrac{12}{32}$

8. $\dfrac{7}{10} = \dfrac{49}{n}$

9. $\dfrac{18}{81} = \dfrac{n}{117}$

10. $\dfrac{200}{750} = \dfrac{96}{n}$

11. $\dfrac{0.75}{n} = \dfrac{0.375}{0.5}$

12. PUBLIC OPINION An unbiased survey of 640 adults in one city found that 415 of them favored smoke-free public buildings. How many of the 32,700 adults in the city might be expected to favor smoke-free public buildings? Round to the nearest hundred.

13. WRITING MATHEMATICS Describe ways you can use to judge whether the results of a survey that you hear or read about are accurate.

EXTEND

SPORTS The circle graph at the right shows the results of a survey of 300 people asked to name their favorite spectator sport.

14. How many of those surveyed chose baseball as their favorite sport?

15. If the survey was unbiased, how many of 250 million Americans might be expected to choose soccer as their favorite spectator sport?

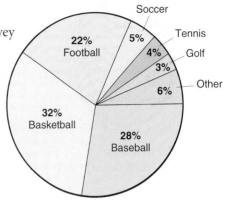

OPINION POLLS Reporters from the *Belltown Daily Dispatch* asked a random sample of the city's 684,000 registered voters which of the three candidates for mayor they favored. The results are shown in the table at the right.

16. What percent of registered voters were sampled?

17. How many voters would you expect to be undecided?

18. If 50% of the undecided voters decide to vote for Rivera, how many votes would you expect Rivera to receive if 90% of registered voters go to the polls?

Candidate	Number of Votes
Rivera	922
Dayton	845
Lazio	1047
Undecided	606

THINK CRITICALLY

GOVERNMENT A *margin of error* is sometimes reported along with the results of a survey. An unbiased poll of 850 voters in one state found that 527 favored term limits for senators. The margin of error was ± 3%.

19. Give the maximum and minimum percents of those surveyed who favored term limits.

20. Of 2 million voters in the state, what is the maximum number who might be expected to favor term limits? What is the minimum number?

PROJECT *Connection*   In this activity you will investigate how much television time is devoted to commercials.

1. With your group, select a variety of programs to watch for a week. Include programs that are shown at different times of the day and week, are of different length, are aimed at different audiences, and so on.

2. Agree on a uniform method of timing the commercials and recording data. A stopwatch may be helpful. Also record the types of products being advertised.

3. Compile and analyze your data. Determine the ratio of commercial time to total program length. Does it appear to be consistent, or does it vary by program type, channel, time of day, and so on? Do there appear to be any relationships between program type and products advertised?

4. Make graphs to display some of the similarities and differences supported by your data.

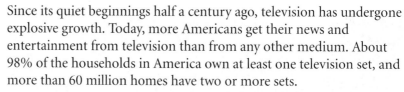

Career
Media Researcher

Since its quiet beginnings half a century ago, television has undergone explosive growth. Today, more Americans get their news and entertainment from television than from any other medium. About 98% of the households in America own at least one television set, and more than 60 million homes have two or more sets.

Each day, media researchers sample the public's opinion of network, cable, and local TV shows. The Nielsen Media Research Corporation is the nation's largest television ratings service. For their major polls, Nielsen researchers randomly sample about 4,000 households nationwide—about 10,000 viewers. For local shows, Nielsen's samples are much smaller—about 500 viewers.

Decision Making

1. There are 95 million households in the United States. In how many of them is there at least one TV set?

2. The last episode of "M*A*S*H," broadcast on February 28, 1983, received the highest Nielsen rating up to that time. In a major poll, Nielsen estimated that 125 million people watched the show, 77% of those who were watching television at the time. How many people were watching television that evening?

3. Of the viewers polled by Nielsen researchers, how many watched the last episode of "M*A*S*H"?

4. Nielsen researchers estimate that their results are subject to possible errors of ± 2%. Of all the viewers in America, what is the greatest number who might have watched the final "M*A*S*H" episode? What is the least?

5. In another survey, a Nielsen sample of viewers in a local area found that 86 of the 500 polled were tuned to "News at 5." If 240,000 people were watching television in the area at the time, about how many were watching "News at 5"?

1.6 Use Matrices for Data

Explore

The following data give the numbers of medals won by the top four countries that competed in the 1992 Summer Olympics. The numbers in parentheses represent gold, silver, and bronze medals, and total number of medals won by the countries: Unified Team (formerly USSR) (45, 38, 29, 112); United States (37, 34, 37, 108); Germany (33, 21, 28, 82); China (16, 22, 16, 54).

1. Organize the data in a convenient form that allows you to easily find the number and type of medal won by any country.

2. Explain why you chose the method you used to organize the data.

3. Create a system for assigning a unique variable to each item of data in your display. The system should allow another student to find an item of data knowing only the variable.

SPOTLIGHT ON LEARNING

WHAT? In this lesson you will learn
- to organize and display data in matrices and to operate on matrices.
- to perform basic operations using matrices.

WHY? Matrices help you solve problems about appliance sales, video equipment, and car prices.

Build Understanding

● As you know, data can be displayed in a table or in a spreadsheet.

If you remove the labels from a table or spreadsheet, you have a rectangular arrangement of numbers. By placing brackets around the numbers, you create a **matrix** like the one below. (The plural of *matrix* is *matrices*.)

$$\begin{bmatrix} 87 & 35 & 105 \\ 44 & 40 & 51 \end{bmatrix} \leftarrow \text{row}$$
column

Each number in a matrix is called an **element** or entry. The sample matrix above has elements 87, 35, 105, 44, 40, and 51.

The number of rows and columns in a matrix specifies its **dimensions**. Since the matrix has 2 rows and 3 columns, its dimensions are 2 × 3, read "2 by 3." Remember that rows are horizontal and columns are vertical. If a matrix has the same number of rows and columns, it is called a **square matrix**.

	Store 1	Store 2	Store 3
Camcorder Plus	87	35	105
Easy-Cam	44	40	51

Table

	A	B	C
1	87	35	105
2	44	40	51

Spreadsheet

$$\begin{bmatrix} 87 & 35 & 105 \\ 44 & 40 & 51 \end{bmatrix}$$

Matrix

Write a matrix that
uses the data in
Example 1 and has
dimensions 5 × 3.
(*Hint*: Write the rows
as columns.)

EXAMPLE 1

Give the dimensions of the matrix
shown at the right.

$$\begin{bmatrix} 3 & 1 & 5 & 16 & 4 \\ 0 & 1 & 9 & 3 & 2 \\ 21 & 3 & 3 & 6 & 0 \end{bmatrix}$$

Solution

The matrix has 3 rows and 5 columns, so its dimensions are 3×5. ◀

Capital letters such as A, B, . . . are often used to name matrices so that
they can be referred to easily.

If two matrices have the same dimensions, you can add or subtract the
corresponding elements to find the sum or difference. **Corresponding
elements** are the elements in the same position of each matrix.

EXAMPLE 2

$$A = \begin{bmatrix} 14 & 7 \\ 21 & 19 \\ 35 & 12 \end{bmatrix} \qquad B = \begin{bmatrix} 11 & 0 \\ 9 & 18 \\ 25 & 5 \end{bmatrix}$$

Find $A + B$ and $A - B$.

Solution

$$A + B = \begin{bmatrix} 14 + 11 & 7 + 0 \\ 21 + 9 & 19 + 18 \\ 35 + 25 & 12 + 5 \end{bmatrix} = \begin{bmatrix} 25 & 7 \\ 30 & 37 \\ 60 & 17 \end{bmatrix}$$

$$A - B = \begin{bmatrix} 14 - 11 & 7 - 0 \\ 21 - 9 & 19 - 18 \\ 35 - 25 & 12 - 5 \end{bmatrix} = \begin{bmatrix} 3 & 7 \\ 12 & 1 \\ 10 & 7 \end{bmatrix} \qquad ◀$$

You can use a graphing calculator or a
computer program to add or subtract
matrices. The calculator screens below show
the addition and subtraction of two matrices
C and D.

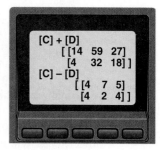

Adding or subtracting matrices often proves useful. In business, for example, matrices are useful for inventory control.

EXAMPLE 3

APPLIANCE SALES This table gives the inventory of two types of microwave ovens at three appliance stores on March 1. During March, the manufacturer delivered the following

Microwave Inventory, March 1		
	Regal	Ovenmaster
Campus	38	52
Eastwood	49	70
Central	25	41

quantities of Regals and Ovenmasters to the stores: Campus (25, 32); Eastwood (30, 35); Central (18, 12). During the month, customers purchased Regals and Ovenmasters in these numbers: Campus (31, 19); Eastwood (39, 43); Central (11, 17).

a. Write inventory matrix A.

b. Write delivery matrix B.

c. Write purchase matrix C.

d. Calculate $A + B - C$ to find the store inventories at the end of March.

ALGEBRA: WHO, WHERE, WHEN

In 1850, British mathematician James Sylvester (1814–1897) coined the word *matrix* from the Hebrew word *gematria*, an ancient system that assigned numbers to the letters in Hebrew words. He then added the numbers and interpreted the results. Sylvester and his lifelong friend Arthur Cayley (1821–1895) developed the theory of matrices.

Solution

a.

[A]
$$\begin{bmatrix} 38 & 52 \\ 49 & 70 \\ 25 & 41 \end{bmatrix}$$

b.

[B]
$$\begin{bmatrix} 25 & 32 \\ 30 & 35 \\ 18 & 12 \end{bmatrix}$$

c.

[C]
$$\begin{bmatrix} 31 & 19 \\ 39 & 43 \\ 11 & 17 \end{bmatrix}$$

d.
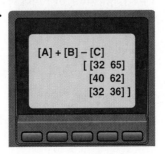
[A] + [B] − [C]
$$\begin{bmatrix} 32 & 65 \\ 40 & 62 \\ 32 & 36 \end{bmatrix}$$

Use matrices *A*, *B*, *C*, and *D* for Exercises 1–4.

$$A = \begin{bmatrix} 9 & 15 & 7 & 2 & 6 \end{bmatrix}$$

$$B = \begin{bmatrix} 3 & 5 & 9 & 19 \\ 13 & 6 & 20 & 2 \end{bmatrix}$$

$$C = \begin{bmatrix} 7 & 16 & 20 & 5 \\ 31 & 15 & 19 & 40 \end{bmatrix}$$

$$D = \begin{bmatrix} 5 & 9 & 13 & 5 \\ 55 & 11 & 4 & 19 \end{bmatrix}$$

1. Give the elements and dimensions of *A*.

2. Find *B* + *C*.   3. Find *C* − *D*.   4. Find *B* + *C* + *D*.

RECREATION The table at the right shows membership in the Central City Astronomy Club by age and gender at the beginning of 1995. During the year, males and females joined in these numbers: senior (38, 35); junior (25, 40). Members who stopped attending occurred in these numbers: senior (13, 9); junior (6, 10).

CC Astronomy Club	Male	Female
Senior	117	88
Junior	91	95

5. Write initial membership matrix *A*, new member matrix *B*, and stopped attending matrix *C*.

6. Calculate *A* + *B* − *C* to find the club membership at the end of 1995.

7. WRITING MATHEMATICS Describe a practical use for matrices and for adding or subtracting matrices.

Use matrices *J* through *N* for Exercises 1–6.

$$J = \begin{bmatrix} 4 & 5 \\ 8 & 3 \end{bmatrix} \qquad K = \begin{bmatrix} 12 & 9 \\ 8 & 6 \\ 20 & 13 \end{bmatrix} \qquad L = \begin{bmatrix} 9 & 5 \\ 8 & 0 \\ 16 & 7 \end{bmatrix} \qquad M = \begin{bmatrix} 26 & 14 \\ 9 & 12 \\ 5 & 25 \end{bmatrix} \qquad N = \begin{bmatrix} -1 & 6 \\ 2 & -7 \end{bmatrix}$$

1. Give the elements of *J*.   2. Give the dimensions of *K*.

3. Find *J* + *N*.   4. Find *K* + *M*.   5. Find *K* − *L*.   6. Find *L* + *M* − *K*.

VIDEO EQUIPMENT The table at the right lists inventories of three models of an overhead projector in two warehouses at the beginning of June. During the month, new projectors were received in the following numbers: W-North (52, 70, 48); W-West (88, 86, 66). Projectors were shipped out in these numbers: W-North (61, 90, 77); W-West (114, 98, 50).

OP Inventory	OP-8	Sunlight	Clarity
W-North	317	490	166
W-West	555	207	181

7. Write initial inventory matrix *I*, new inventory matrix *N*, and shipping matrix *S*.

8. Calculate *I* + *N* − *S* to find the final inventories at the end of June.

CAR PRICES The matrix at the right summarizes the sticker prices of four car models at three automobile dealerships. Each dealer will add 8% in taxes to the sticker price.

$$\begin{bmatrix} 12{,}400 & 12{,}600 & 12{,}000 \\ 11{,}100 & 11{,}400 & 11{,}500 \\ 14{,}800 & 14{,}100 & 14{,}400 \\ 10{,}200 & 10{,}900 & 10{,}300 \end{bmatrix}$$

9. Write a matrix giving the tax on each car.

10. Add the matrices to show the total prices.

11. WRITING MATHEMATICS Write and solve a real world problem that requires adding or subtracting two or more matrices.

EXTEND

Find matrix M that makes each equation true.

12. $M + \begin{bmatrix} 29 & 63 \\ 77 & 49 \end{bmatrix} = \begin{bmatrix} 81 & 83 \\ 115 & 62 \end{bmatrix}$

13. $\begin{bmatrix} 19 & 44 & 30 \\ 62 & 91 & 55 \end{bmatrix} - M = \begin{bmatrix} 8 & 29 & 6 \\ 43 & 60 & 27 \end{bmatrix}$

THINK CRITICALLY

14. Add: $\begin{bmatrix} 6 & 3 \\ 5 & 9 \end{bmatrix} + \begin{bmatrix} 6 & 3 \\ 5 & 9 \end{bmatrix}$

15. Write a rule for doubling a matrix.

16. Write a rule for multiplying a matrix by the number N.

17. How could the total prices of the cars in Exercises 9–10 be written as a matrix multiplied by a number?

Use matrices A, B, and C for Exercises 18–20.

$$A = \begin{bmatrix} a & b \\ c & d \end{bmatrix} \quad B = \begin{bmatrix} e & f \\ g & h \end{bmatrix} \quad C = \begin{bmatrix} i & j \\ k & l \end{bmatrix}$$

18. Does $A + B = B + A$? Justify your answer.

19. Does $A + (B + C) = (A + B) + C$? Justify your answer.

20. Find a matrix D such that $A + D = A$, $B + D = B$, and $C + D = C$.

MIXED REVIEW

21. STANDARDIZED TESTS An unbiased poll of 315 voters showed that 183 favored Mason for mayor, 102 favored Dunn, and 30 favored Green. How many of 16,400 voters can be expected to vote for Green for mayor? Round to the nearest hundred.

   A. about 1,400   **B.** about 1,600   **C.** about 2,700   **D.** about 172,200

Use matrices R and S for Exercises 22–24.

$$R = \begin{bmatrix} 114 & 91 & 172 \\ 88 & 65 & 80 \end{bmatrix} \quad S = \begin{bmatrix} 151 & 114 & 198 \\ 156 & 79 & 80 \end{bmatrix}$$

22. Give the dimensions of R.   23. Find $R + S$.   24. Find $S - R$.

Explore

Live Math National Tour	
City	**Attendance**
Somewhere, Maine	100
Middle City, Iowa	100
Potato Park, Idaho	100
Heavenly, California	99,700

After playing with 20 bands at the Heavenly Rock Festival, rock group Live Math issued a press release with the headline "Live Math National Tour Huge Success! Average Concert Attendance 25,000!"

1. Was the band's claim regarding average attendance accurate?

2. Explain why the claim was misleading.

3. Write a headline that gives a more truthful picture of attendance at the group's concerts.

4. Why is the "average" of a set of data not always a good measure of the data as a whole?

5. In what other ways was the headline misleading?

Build Understanding

A set of data may contain hundreds of items. In order to understand the data, it is often useful to have a single number that represents all the items. Such a number is called a **measure of central tendency.** One commonly used measure of central tendency is the *mean,* or *average.*

> The *mean,* or *average,* of a set of data is the sum of the data divided by the number of items of data.

You can use a calculator to find the mean of a set of data.

SPORTS In five games, Mei bowled these scores: 112, 108, 106, 197, 102. Find her mean score.

Solution

Mean: $\dfrac{112 + 108 + 106 + 197 + 102}{5} = 125$

Mei's mean score was 125.

The mean is an appropriate measure of central tendency when all the data are approximately equal. However, when the **range** of values—the difference between the greatest and least values—is large compared to the values themselves, the mean may give an unrealistic view of the data. The mean may also be inappropriate when the data include a few extremely high or low values. In such cases, the *median* may better represent the data.

CHECK UNDERSTANDING

Is the mean an appropriate measure of central tendency in Example 1? Explain.

> The *median* of a set of data is the middle value when the data are arranged in numerical order. When there are two middle values, the median is the average of the two.

EXAMPLE 2

BIRTH WEIGHTS The birth weights, in pounds, of eight babies born at Memorial Hospital were 7.8, 8.8, 7.7, 6.6, 8.2, 8.9, 6.2, and 9.1. Find the median weight.

Solution

Write the data in numerical order.

6.2, 6.6, 7.7, 7.8 , 8.2 , 8.8, 8.9, 9.1
         ↑   ↑
       middle

There are two middle values, 7.8 and 8.2. The median is the average of the values, $\frac{7.8 + 8.2}{2} = 8.0$.

The median birth weight was 8.0 pounds.

The *mode* is another measure of central tendency that, like the median, may be appropriate when the mean is not.

COMMUNICATING ABOUT ALGEBRA

A set of data has a mean of 5, a median of 6, and a mode of 7. Which, if any, of the numbers 5, 6, and 7 must be in the set? Explain your answer to another student.

> The *mode* of a set of data is the element that occurs most often in the set. A set may have no mode, one mode, or several modes. If it has two modes, the set is **bimodal**.

The mode is especially useful when several items of data are the same. This may indicate that elements which are the same have special importance in the set and should be granted extra emphasis. The mode can be used for numerical data or nonnumerical data, such as eye color, since the mode is the most frequently occurring element.

EXAMPLE 3

MONEY A cashier wanted to know which denomination of bill customers used most often. In one half-hour period, the cashier took in bills in these denominations ($): 1, 1, 1, 2, 5, 5, 5, 5, 5, 10, 10, 10, 10, 20, 20, 20, 50. Find the mode.

Solution

The number 5 appears most often in the set, so the mode is $5. ◀

In Example 3, the mean of $11.125 and the median of $7.50 are useless for the cashier's purpose. These denominations do not exist. In this case, the mode is the most appropriate measure of central tendency.

You can find measures of central tendency from a frequency table or graph.

CHECK —
UNDERSTANDING

Why are 44, 45, and 49 multiplied by 2, 2, and 4 to find the mean in Example 4?

EXAMPLE 4

RADIO ADVERTISING The table shows the number of minutes of music aired in one hour on nine radio stations. Find the mean, median, mode, and range.

Music Minutes per Hour	
Minutes	**Number of Stations**
41	1
44	2
45	2
49	4

Solution

Mean: $\dfrac{41 + 2(44) + 2(45) + 4(49)}{9} \approx 46.1$

The mean amount of music air time is about 46.1 min.

Median: 41, 44, 44, 45, $\boxed{45}$, 49, 49, 49, 49

The middle value is 45, so the median amount is 45 min.

Mode: The value 49 occurs most often in the set, so the mode is 49 min. It is very easy to find the mode from a frequency table.

Range: The greatest and least values in the set are 49 and 41, so the range is 49 − 41 = 8 min. ◀

TRY THESE

Find the mean, median, mode, and range of each set of data.

1. 98, 77, 89, 93, 75, 81, 77, 88, 78

2. 5, 9, 3, 1, 0, 4, 5, 2, 1, 8, 7, 6

3. 188, 260, 174, 206

4. 3.8, 2.6, 4.1, 4.8, 5.9, 2.7, 6.9

SPORTS The heights, in centimeters, of five basketball players are 180, 186, 185, 180, and 210.

5. Find the mean, median, mode, and range.

6. Which measure of central tendency best represents the data? Give reasons for your answer.

READING The table gives the number of seconds needed by twelve students to read one page of a history book.

Time, s	Number of Students
68	1
73	2
75	1
80	3
85	3
91	2

7. Find the mean, median, mode, and range.

8. Which measure of central tendency best represents the data? Give reasons for your answer.

9. **WRITING MATHEMATICS** In your own words, define the mean, median, and mode of a set of data. Make up a set of data and give examples.

10. **ENTERTAINMENT** Find the mode of the set of responses to a favorite music category survey.

country	rock	classical	rock	jazz	show music
jazz	folk	rock	classical	rap	blues
blues	rock	rap	country		

PRACTICE

Find the mean, median, mode, and range of each set of data.

1. 23, 36, 25, 28, 21, 30, 32, 23

2. 9.5, 9.9, 10.3, 8.4, 7.7, 10.0

3. 280, 295, 235, 210, 230, 235, 195

4. 5, 5, 7, 6, 7, 5, 7, 7, 8, 5

5. 4237, 4516, 4444, 4379

6. $3\frac{1}{2}, 5\frac{3}{4}, 4\frac{3}{8}, 3\frac{7}{8}$

7. **EDUCATION** The tuition costs at six universities are $12,560, $14,300, $13,750, $12,400, $13,680, and $15,420. Find the mean, median, mode, and range of the costs.

SALARIES Two employees of City Cleaners earn $18,000, another earns $22,000, and a manager earns $45,000.

8. Find the mean, median, mode, and range of the salaries.

9. Which measure of central tendency best represents the data? Give reasons for your answer.

10. A new employee is hired at a salary of $17,000. Without calculating, tell how the mean, median, and mode of the salaries will be affected.

11. **SPORTS** The table gives the distance from home plate to the center field fence in the 14 American League baseball stadiums. Find the mean, median, and mode of the data. Round to the nearest tenth.

Center Field Home Run Distance, ft			
Distance	No. of Parks	Distance	No. of Parks
390	1	404	3
400	5	410	3
402	1	440	1

12. **WRITING MATHEMATICS** Give an example of a data set you might encounter in your daily life. Which measure of central tendency best represents the data? Explain.

13. **CHEMISTRY** Althea was working on a chemistry experiment and was measuring the volume of water in beakers. She recorded these measurements: 210.5 mL, 212.4 mL, 210.0 mL, 209.8 mL, 213.5 mL, 212.6 mL, 210.5 mL, 211.5 mL, and 210.8 mL. Find the mean, median, mode, and range of the volumes.

14. Michelle's mean score on eight math quizzes is 89. What must she score on her next quiz to raise her mean score to exactly 90?

15. Construct a set of data that has at least three different numbers and that has 20 as its mean, median, and mode.

Tell whether the mean, median, mode, or range is being used.

16. Half of the homes were priced at $150,000 or less.

17. The most popular shoe size is $7\frac{1}{2}$.

18. The average bill for a customer at Big Burg's is $3.38.

19. The interest rates for different types of accounts varied by 2.25%.

20. The best-selling T-shirt colors were white and navy.

DINING OUT The prices for different entrées at Casa Mexicana are $12, $14, $15, $x, $15, and $14. Find a possible value of x such that

21. there is one mode   22. there are two modes   23. there is no mode

24. Ms. Shannon had graded all the tests and determined that the mean of the scores was 78. She forgot to enter one score in her grade book. The tests she had entered were 90, 95, 75, 80, 85, 65, 70, 80, and 60. Find the missing score.

THINK CRITICALLY

A circle graph of the ages of 240 seniors at McKinley High School was drawn. The table gives the measure, in degrees, of the central angle of the sector representing each age.

25. Find the number of students of each age.

26. Find the mean, median, and mode of the ages.

Age	Angle	Age	Angle
15	9°	17	180°
16	144°	18	27°

27. What must be true about a new item of data if it causes the mode of the set to change?

28. What must be true about a new item of data if it causes the mean to decrease?

29. Find two sets of five scores that have a mean of 30, the same range, and different medians. Explain your work.

MIXED REVIEW

Solve each proportion.

30. $\frac{13}{26} = \frac{1}{m}$ 31. $\frac{x}{3} = \frac{36}{27}$ 32. $\frac{10}{15} = \frac{p}{12}$ 33. $\frac{35}{y} = \frac{25}{30}$

34. STANDARDIZED TESTS What is the median of the following values: 56, 77, 75, 44, 81, 66, 56, and 80?

    A. 70.5   B. 66.8   C. 56   D. 37

Career
Inventory Control Manager

In 1982, long-playing records (LPs) were the most popular recording medium in the United States, with 244 million records sold. Only 182 million cassettes were sold that year, and not a single CD!

Just nine years later, sales of LPs had plunged almost to zero. Meanwhile, CD sales had skyrocketed. With 434 million sold in 1991, the CD surpassed the cassette as the most popular recording medium. The gap between the two has been widening ever since.

An inventory control manager for a CD distributor keeps track of all shipments into and out of the warehouse. At any time, the manager must know exactly how many copies of every CD are on hand.

Decision Making

1. If CD sales increased at the same rate after 1991 as they did from 1982 to 1991, how many would you expect to sell in 1999?

2. An inventory control manager for a CD distributor stores data in matrices. The three matrices below give data on four different CDs. A lists present inventory, B lists incoming shipments, and C gives numbers of CDs to be shipped out.

$$A = [88{,}000 \quad 195{,}000 \quad 52{,}000 \quad 436{,}000]$$
$$B = [46{,}000 \quad 123{,}000 \quad 18{,}000 \quad 257{,}000]$$
$$C = [35{,}000 \quad 98{,}000 \quad 21{,}000 \quad 209{,}000]$$

Write a matrix showing inventory of each CD after the completion of incoming and outgoing shipments.

3. Each row of the spreadsheet below represents one CD. Each column from A through D gives the number of that CD shipped during one of the past four months. Column E gives the total shipped during the four months. Column F gives the mean number shipped each month.

	A	B	C	D	E	F
1						
2	45,000	62,000	75,000	106,000		
3	91,000	77,000	42,000	13,000		

   a. Write the formulas for cells E2, F2, E3, and F3.

   b. Find the values for those four cells.

1.8 Experimental Probability

Explore/Working Together

1. Work with a partner. Count the total number of letters in two lines from a book, newspaper, or magazine. If the total is less than 100, continue to a third line. Then count the number of E's and T's that appear in the lines.

2. Write the ratio $\dfrac{\text{number of E's and T's}}{\text{number of letters}}$ as a percent. Is your result the same as that of other students? Explain.

3. The mean number of letters in the 150,000 words in a novel is 5. Based on your results, how many E's and T's are in the novel?

Build Understanding

The **probability** that something will happen is how often you can expect the event to occur. When you toss a fair coin, for example, you are equally likely to toss a head or a tail. Therefore, the mathematical or *theoretical* probability that you will toss a head is 50%. (You can also express the probability as the decimal 0.5 or the fraction $\frac{1}{2}$.)

However, the theoretical probability that something will happen is often not the same as what actually happens in the real world. If you tossed a penny 10,000 times, you would probably not get exactly 5,000 heads and 5,000 tails. The British mathematician John Kerrich tossed a coin 10,000 times. He obtained 5,067 heads and 4,933 tails. Using Kerrich's results, you can say that the *experimental probability* of tossing a head is $\frac{5,067}{10,000}$, or 0.5067, or 50.67%.

> **EXPERIMENTAL PROBABILITY**
>
> The experimental probability $P(E)$ of an event E is given by
>
> $$P(E) = \frac{\text{number of times } E \text{ occurs}}{\text{total number of trials}}$$

Probabilities range from 0 to 1. A probability of 0 means that an event cannot happen. A probability of 1 means that an event must happen. Most probabilities are between 0 and 1.

Real world situations often involve possible outcomes that are not equally likely. You can base experimental probabilities on data gathered about past events.

CHECK UNDERSTANDING

In Kerrich's coin experiment, what was the event E? the number of times E occurred? the number of trials?

COMMUNICATING ABOUT ALGEBRA

Suppose you tossed a coin with two heads 100 times. Explain to a classmate why your results would show that the experimental probability of tossing a head is 1 and a tail is 0.

EXAMPLE 1

BOOK/MOVIE SURVEY The table shows the results of interviewing 1000 high school students chosen at random to find out if they had read a certain novel or seen the movie based on it.

Consider another high school student, not among the 1000 students interviewed.

a. Estimate to the nearest tenth the probability that the student will have read the book.

b. Estimate to the nearest tenth the probability that the student will not have seen the movie.

Survey of High School Students			
	Movie (yes)	Movie (no)	Total
Book (yes)	276	205	481
Book (no)	373	146	519
Total	649	351	1000

c. Consider a group of 500 high school students, none of whom are among the 1000 students interviewed. How many would you expect neither to have seen the movie nor read the book?

Solution

a. $\dfrac{481}{1000} \approx 0.5$ **The experimental probability is about 0.5.**

b. $\dfrac{351}{1000} \approx 0.4$ **The experimental probability is about 0.4.**

c. $\dfrac{146}{1000} = \dfrac{x}{500} = 73$ **Out of 500 students, you would expect that 73 will have neither seen the movie nor read the book.** ◄

You can use the results of experiments or surveys to calculate probabilities and to make predictions. Probabilities are often written as percents.

EXAMPLE 2

QUALITY CONTROL A quality control engineer at Everglow Bulbs tested 400 bulbs and found 6 of them to be defective.

a. What is the experimental probability that an Everglow bulb will be defective?

b. In a shipment of 75,000 bulbs, how many are likely to be defective?

Solution

a. P (bulb will be defective) $= \dfrac{\text{number of defective bulbs}}{\text{number of bulbs tested}}$

$= \dfrac{6}{400} = 0.015$, or 1.5%

b. $1.5\% \cdot 75{,}000 = 1{,}125$

About 1,125 bulbs will be defective. ◄

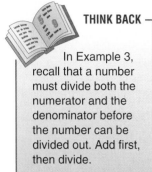

THINK BACK

In Example 3, recall that a number must divide both the numerator and the denominator before the number can be divided out. Add first, then divide.

You may have to combine items of data in order to find an experimental probability.

EXAMPLE 3

ADVERTISING An advertising firm surveyed consumers to find which one of three commercials (A, B, or C) they preferred. The table gives the results by age category. What is the probability that a person 18 or over will prefer commercial A or B?

Age	A	B	C
Under 18	15	72	44
18–40	91	60	29
Over 40	17	19	88

Solution

Let E represent the event of a person 18 or over choosing A or B.

$$P(E) = \frac{\text{number of persons 18 or over who chose A or B}}{\text{number of persons 18 or over}}$$

$$= \frac{91 + 60 + 17 + 19}{91 + 60 + 29 + 17 + 19 + 88} \approx 0.615$$

The probability is approximately 0.615, or 61.5%.

TRY THESE

METEOROLOGY In the town of Wilcox, snow has fallen in 80 of the past 80 Januarys, none of the past 80 Julys, and 30 of the past 80 Octobers. Find the experimental probability of the event.

1. It will snow next January. **2.** It will snow next July. **3.** It will snow next October.

HEALTH A public health survey revealed that 56 out of 175 adults had high blood pressure.

4. What is the experimental probability that a person will have high blood pressure?

5. In a city of 250,000 adults, how many can be expected to have high blood pressure?

The sides of a number cube are numbered from 1 to 6. You roll the cube 50 times and obtain the number 3 eight times. What is the experimental probability that your next roll will:

6. be 3 **7.** be 1, 2, 4, 5, or 6 **8.** be less than 7 **9.** be 7

10. POLITICS A news magazine surveyed public opinion on imposing term limits for legislators. The table gives the results of the survey by party affiliation. What is the probability that a Democrat or Independent will have a favorable opinion or no opinion on the issue?

	Favorable	Unfavorable	No Opinion
Republican	125	144	36
Democrat	151	130	21
Independent	31	19	2

11. WRITING MATHEMATICS Suppose you wanted to determine the probability of tossing an index card folded in half so that it lands in a tent position. Explain the procedure you would use. You may wish to carry out your procedure and report the results.

1. On average, thunder is heard in Tororo, Uganda, 251 days each year. What is the experimental probability that there will be thunder in Tororo tomorrow?

A spinner's 8 equal sections are lettered from A to H. You spin 75 times, getting C 12 times and D 6 times. What is the experimental probability that your next spin will:

2. be C   3. be C or D   4. not be D   5. be T

PUBLIC HEALTH A doctor's records showed that 63 of 248 adult patients and 11 out of 90 child patients had cholesterol readings over 240.

6. Give the experimental probability that a patient will have a reading over 240.

7. In a city of 800,000, how many residents would you expect to have readings over 240?

8. WRITING MATHEMATICS Explain the following statement by a TV weather forecaster: "The probability of rain today is 75%." How do you think the 75% is determined?

SCHOOL ENROLLMENT The table shows a high school's enrollment by gender and class. If students are chosen randomly, give each probability.

9. A sophomore or junior is a girl.

10. A student is a boy.   11. A student is a senior.

	Boys	Girls
Freshman	142	165
Sophomore	136	145
Junior	109	114
Senior	122	102

EXTEND

12. CAR PURCHASES Based on the results of one survey, the probability that a family has purchased a car in the last two years is 28%. Of the families surveyed, 105 bought cars in the last two years. How many families were surveyed?

13. WRITING MATHEMATICS Write a problem that has as its solution, "The experimental probability of the event is 45%."

THINK CRITICALLY

14. A fair penny tossed 20 times came up heads every time. Therefore, the experimental probability of tossing a head on the next throw is 100%. Estimate the actual likelihood of tossing a head. Explain your answer.

PROJECT Connection   As the cable television industry develops, hundreds of new channels are on the horizon. How will viewers keep track of what's on?

1. Make a chart showing the number of channels included and length of current TV guides.

2. If 100 channels are available in the future, how long would similar printed guides have to be? What if there are 500 channels? Explain how you arrived at your conclusions.

3. Will printed guides be practical in the future? Brainstorm other types of guides. What new problems might need to be solved? Design a sample of your best idea.

Use Data and Four Problem Solving Steps

Because problem solving is the main goal of mathematics, you will find opportunities throughout this book to develop your problem solving skills. You will learn how algebraic strategies can be used with other familiar approaches, such as drawing a diagram or finding a pattern.

The general process you can use to solve problems involves four main steps.

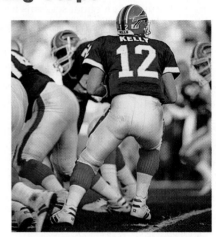

Super Bowls		
Year	**Winner**	**Loser**
1995	San Francisco (NFC) 49	San Diego (AFC) 26
1994	Dallas (NFC) 30	Buffalo (AFC) 13
1993	Dallas (NFC) 52	Buffalo (AFC) 17
1992	Washington (NFC) 37	Buffalo (AFC) 24
1991	New York Giants (NFC) 20	Buffalo (AFC) 19
1990	San Francisco (NFC) 55	Denver (AFC) 10
1989	San Francisco (NFC) 20	Cincinnati (AFC) 16
1988	Washington (NFC) 42	Denver (AFC) 10
1987	New York Giants (NFC) 39	Denver (AFC) 20
1986	Chicago (NFC) 46	New England (AFC) 10
1985	San Francisco (NFC) 38	Miami (AFC) 16
1984	Los Angeles Raiders (AFC) 38	Washington (NFC) 9
1983	Washington (NFC) 27	Miami (AFC) 17
1982	San Francisco (NFC) 26	Cincinnati (AFC) 21
1981	Oakland (AFC) 27	Philadelphia (NFC) 10
1980	Pittsburgh (AFC) 31	Los Angeles Rams (NFC) 19
1979	Pittsburgh (AFC) 35	Dallas (NFC) 31
1978	Dallas (NFC) 27	Denver (AFC) 10
1977	Oakland (AFC) 32	Minnesota (NFC) 14
1976	Pittsburgh (AFC) 21	Dallas (NFC) 17
1975	Pittsburgh (AFC) 16	Minnesota (NFC) 6
1974	Miami (AFC) 24	Minnesota (NFC) 7
1973	Miami (AFC) 14	Washington (NFC) 7
1972	Dallas (NFC) 24	Miami (AFC) 3
1971	Baltimore (AFC) 16	Dallas (NFC) 13
1970	Kansas City (AFL) 23	Minnesota (NFL) 7
1969	New York Jets (AFL) 16	Baltimore (NFL) 7
1968	Green Bay (NFL) 33	Oakland (AFL) 14
1967	Green Bay (NFL) 35	Kansas City (AFL) 10

Step 1 **UNDERSTAND** the problem. Analyze the information.

Step 2 **PLAN** an approach. Decide on the strategies, computations, and technology you want to use.

Step 3 **SOLVE** the problem. Apply the strategy. Carry out the work.

Step 4 **EXAMINE** the result. Check your answer. Look back at your work and ask yourself if the answer is reasonable, whether you could have solved the problem another way, and if you notice any general patterns that might apply to other problems.

To explore how the steps can help you, consider the following problem.

Problem

Many sportscasters believe an "exciting" football game ends with a close score, say a winning margin of 7 points (equal to a touchdown plus a 1-point conversion) or less. Using these guidelines, what percent of Super Bowl games through 1995 provided exciting entertainment?

Explore the Problem

● **UNDERSTAND** the problem. Analyze the information by answering questions such as these.

┌─ PROBLEM ─
SOLVING PLAN

• Understand
• Plan
• Solve
• Examine

1. What information is given?

2. What is meant by the term "winning margin"?

3. How would you restate the question in your own words?

4. What are you asked to find?

PLAN an approach. Again, use questions to guide your thinking.

5. Do you need to reorganize the data?

6. Do you know the winning margin for each game? If not, what should you do first?

┌─ PROBLEM ─
SOLVING TIP

Remember to record carefully the results of any preliminary computations.

7. What will you do next?

8. Can you use technology?

SOLVE the problem. Carry out the work you planned above.

9. What percent of Super Bowls were "worth watching"? (Round to the nearest percent.)

EXAMINE the result. Check your answer. Look back at your work.

10. How can you use estimation to decide if your answer is reasonable?

Investigate Further

● From 1985 to 1995, the National Football Conference (NFC) had an unbroken series of wins over the American Football Conference (AFC). What is the difference between the average winning margin during that winning streak and the average winning margin for the period from the first Super Bowl, in 1967, to 1995?

┌─ COMMUNICATING ─
ABOUT ALGEBRA

Suppose you were an advertiser buying commercial time on the Super Bowl. Discuss how analyzing the winning margins might influence what time during the game you would want your commercial shown and how much you would be willing to spend.

11. After you read the problem and analyze the information, describe your plan for finding a solution.

12. Carry out your plan. Write your results to the nearest tenth.

13. Look back at your work. Do you see any ways in which you could have made the work easier?

14. **WRITING MATHEMATICS** Write your own problem based on the Super Bowl data. Your problem should require at least two operations for the solution. Then show how to solve and check your problem.

Apply the Strategy

● Use the best-seller list below for Problems 15–21.

Best Sellers			
This Week	Fiction	Last Week	Weeks on List
1	The Augustine Prophecy	1	58
2	Fair Game	2	4
3	Nursery Rhyme Politics	4	?
4	The Dark Lake	8	7
5	We Are the Jury	9	5
6	The Caves of Cook County	3	139
7	A Cat's Tale	7	8
8	Don't Blink	6	9
9	Racer	12	2
10	Sincerely, Joe	11	12
11	Nightfall	5	30
12	The Plush Pillow	10	21

15. *Nursery Rhyme Politics* has been on the list twice as long as *The Plush Pillow*. Find the number of weeks for *Nursery Rhyme Politics*.

16. Use the completed list. What is the average number of weeks books have been on the list? Round to the nearest whole number.

17. What percent of books on the list for the current week rose from their position the previous week? Round to the nearest percent.

18. During the time *Nightfall* has been on the best-seller list, Betsy's Bookstore has sold an average of 150 copies of that book per week at a discount price of $16.94. What is the dollar total of *Nightfall* sales at Betsy's during this period?

19. During the time *The Caves of Cook County* has been on the best-seller list, it has sold 4.5 million hardcover copies. To the nearest whole number, what are its average weekly sales?

20. *The Caves of Cook County* is $2\frac{3}{8}$ in. thick. If all the copies sold thus far were stacked one on top of the other, how many miles high would the stack be? Round to the nearest tenth of a mile.

21. *The Caves of Cook County* is $7\frac{1}{2}$ in. wide and 10 in. long. If all the copies sold thus far were laid out to form a (very) large rectangle, what would the area of the rectangle be in square feet?

REVIEW PROBLEM SOLVING STRATEGIES

On the Road

1. Leon, Yvonne, and Keith are road maintenance workers. They are repainting the yellow center lines along a 7900-ft stretch of highway. Working together steadily, they can complete the job in 3 days if each of them works at the same rate. At the end of the first day, Keith is assigned to a different job, so Leon and Yvonne must complete the painting themselves. How many more days will it take Leon and Yvonne to finish?

There are several ways to solve this problem. The questions below will help you with one approach. If you think of a different way, share it with the class.

   a. What part of the job can the three workers complete each day?

   b. What part of the job does each worker complete each day?

   c. At the end of the first day, what part of the job remains to be done?

   d. How much of the job will Leon and Yvonne complete together each day?

   e. How many more days will it take Yvonne and Leon to finish?

   f. What information was not needed?

2. NUMBER Logic

Fill in the empty boxes with one of the digits from 1 to 9. Use each digit just once.

$$\square - \square = \square$$
$$\times$$
$$\square \div \square = \square$$
$$=$$
$$\square + \square = \square$$

Answer these questions to help guide your thinking.

   a. Is it possible for any number in the column at the right or in the middle row to be a 1? Why or why not?

   b. What is the greatest number that can occur in the upper right corner? Why?

   c. What are the only two numbers that can occur in the lower right corner? How do you know?

   d. Fill in the missing numbers.

Ups and Downs

3. Start with this arrangement of arrows. ↑ ↑ ↑ ↓ ↓ ↓

By switching the direction of two adjacent arrows at a time, what is the least number of steps in which you can create this arrangement of arrows?

• • • CHAPTER REVIEW • • •

VOCABULARY

Choose the word from the list that correctly completes each statement.

1. The average of a set of data is its ___?___ .

2. ___?___ describes how often you can expect an event to occur.

3. The number of times an item occurs is its ___?___ .

4. A ___?___ is a portion of a population.

5. The ___?___ is the value that occurs most often in a set of data.

a. mean

b. sample

c. mode

d. probability

e. frequency

Lessons 1.1, 1.2, and 1.4 USE DATA; TABLES AND GRAPHS   pages 5–12, 18–20

- To organize data in a **frequency table**, list each different item individually or in intervals, make a tally mark each time an item occurs, and record the number of tally marks.

Use the following data for Exercises 6–7.

Summer Concert Series ticket prices: $25, $38, $32, $17, $25, $32, $35, $20, $24, $25, $32, $20, $18, $32, $32, $24, $25, $38, $18, $24

6. Make a frequency table and then display the data in a graph.

7. For what percent of the concerts do tickets cost at least $30?

Lesson 1.3 USE VARIABLES TO REPRESENT DATA   pages 13–17

- Data can be analyzed on computers through the use of **programs** or **spreadsheets**. **Variables** can be used to represent data in programs. Different rows and columns are designated to hold specific types of data in spreadsheets.

Use the table on page 29 for Exercises 8–9.

8. Design a spreadsheet that shows the total inventory of microwave ovens at each of the three stores.

9. The program at the right finds the income from the sale of Regal microwaves A and Ovenmaster microwaves B. If A = 5 and B = 9, what would the computer print?

```
10  INPUT  A, B
20  PRINT 179 * A + 229 * B
```

Lessons 1.5 and 1.9 PREDICTION AND PROBLEM SOLVING   pages 21–26, 42–45

- You can use the findings from an unbiased survey to write and solve a proportion and make a prediction about an entire population.

All adults attending the Boone County Fair were given a questionnaire to survey their knowledge of the warning signs of cancer.

10. What sampling method does this represent?

11. Of the 3,278 questionnaires that were completed, 650 indicated knowledge of the warning signs. Based on these findings, how many of Boone County's 90,546 adult residents could be expected to know the warning signs? Round to the nearest hundred.

Lesson 1.6 USE MATRICES FOR DATA	pages 27–31

- Numerical data can be displayed in a **matrix**.
- To add or subtract matrices having the same dimensions, add or subtract corresponding elements.

Use matrices *D*, *E*, and *F* for Exercises 12–14.

$$D = \begin{bmatrix} 7 & 4 & 9 & 6 \\ 2 & 9 & 1 & 5 \end{bmatrix} \quad E = \begin{bmatrix} 3 & 3 & 8 & 4 \\ 2 & 0 & 0 & 5 \end{bmatrix} \quad F = \begin{bmatrix} 16 & 9 & 18 & 7 \\ 25 & 30 & 2 & 15 \end{bmatrix}$$

12. Find $D + E$.   **13.** Find $F - E$.   **14.** Find $D - E + F$.

Lesson 1.7 MEAN, MEDIAN, AND MODE	pages 32–37

- **Measures of central tendency** are single numbers that represent all data in a set. The **mean**, the **median**, and the **mode** are measures of central tendency. The **range** of a set of data is the difference between the greatest and least values.

Find the mean, median, mode, and range of each set of data.

15. Heights of basketball players, in.: 84, 80, 78, 80, 79, 81, 84, 86, 82, 83

16. Number of household members shown in the frequency table on page 9.

Lesson 1.8 EXPERIMENTAL PROBABILITY	pages 38–41

- **Experimental probability** is based on the results of actual trials or past events. It is equal to

$$\frac{\text{number of times event occurs}}{\text{total number of trials}}$$

This table displays students' participation in fall and winter sports at Martin Luther King Jr. High School. If a student's name is chosen at random, give each probability.

17. The student participates in both a fall and a winter sport.

18. The student participates in a fall sport.

	In Fall Sport	Not In Fall Sport
In Winter Sport	152	124
Not In Winter Sport	96	578

CHAPTER ASSESSMENT

CHAPTER TEST

Evaluate each expression if $P = 12$, $Q = 10$, and $R = 5$. Expressions are in BASIC.

1. $Q - R$ **2.** $\dfrac{P * R}{6}$ **3.** $3 * Q + R$

Use matrices S, T, and U for Questions 4–6.

$$S = \begin{bmatrix} 3 & 7 \\ 8 & 2 \\ 5 & 6 \end{bmatrix} \quad T = \begin{bmatrix} 10 & 9 \\ 4 & 12 \\ 5 & 15 \end{bmatrix} \quad U = \begin{bmatrix} 12 & 9 \\ 7 & 14 \\ 8 & 20 \end{bmatrix}$$

4. $U - T$ **5.** $S + U$ **6.** $S + T - U$

STANDARDIZED TESTS Use statements I–IV about car prices for Question 7.

Prices of six new cars: $16,999, $14,490, $17,249, $24,309, $18,160, $17,399

   I. Mean is $18,101. **II.** Median is $20,779.
   III. Mode is $0. **IV.** Range is $9,819.

7. Which of the following is true?

   A. I and III only **B.** I and IV only
   C. I, III, and IV **D.** I, II, III, and IV

The sides of a number cube are numbered 2, 4, 6, 8, 10, and 12. You roll 30 times, getting 4 six times and 8 twelve times. What is the experimental probability that on the next roll you will get:

8. 8 **9.** 2, 6, 10, or 12

A candidate wants to see how many people will vote in an election. For Questions 10–12, match each survey with the sampling method it represents.

   A. random **B.** cluster
   C. convenience **D.** systematic

10. Ask all adults who visit the library one day.

11. Call 100 registered voters chosen at random.

12. Call every registered voter whose last name begins with a letter selected at random.

13. WRITING MATHEMATICS Write a paragraph that explains how the candidate could use the findings from any one of the surveys described in Questions 10–12 to find how many people can be expected to vote in the next election.

Use the following data for Questions 14–15.

Highway fuel efficiency (mi/gal) of cars: 27, 24, 39, 27, 30, 25, 38, 25, 27, 30, 25, 30, 32, 24, 30, 32, 32, 30, 25, 27

14. Make a frequency table for the set of data.

15. What percent of the cars got more than 30 miles to a gallon of gas?

The table shows the number of posters and T-shirts sold at three stands on the first day of a basketball tournament. Use this data for Questions 16–18.

	Posters	T-shirts
Stand 1	235	258
Stand 2	456	379
Stand 3	329	402

16. Design a spreadsheet that shows the total of each item sold.

17. Before the tournament began, each stand had these numbers of items: posters (400, 500, 500) and T-shirts (400, 600, 600). At the end of the day, more of each were delivered in these numbers: posters (300, 600, 500) and T-shirts (200, 400, 400). Write initial inventory matrix I, sales matrix S, and delivery matrix D. Calculate $I - S + D$ to find the inventory at each stand at the end of the day.

18. What is the probability that the first item sold at Stand 2 the next day will be a poster?

PERFORMANCE ASSESSMENT

DESIGN AN EXPERIMENT Design an experiment that you can conduct to find the probability of getting more heads than tails when you toss three of the same type of coin. Carry out the experiment and find the probability. Compare your results with those of other students. Refine your experiment and repeat it if you think it is necessary.

MAKE A PREDICTION Gather data on one or more aspects of your local weather for one week. On the basis of this data, make a prediction about the next day's weather. State the probability in your prediction. Then compare the actual weather with your prediction. Explain why it is or is not what you predicted.

ANALYZE A GRAPH Find an example of a pictograph, bar graph, line plot, line graph, or circle graph in a newspaper, magazine, or book other than your math textbook. Present the information displayed in the graph in a table. Then find the mean, median, mode, and range of the data.

ANALYZE DATA Use the data given (from page 6) for at least three of the stores to do the following.

Store A	
Country	3,251
Rock	2,963
Jazz	683
Classical	542
Easy listening	325
Other	157

Store B	
Rock	2,108
Country	1,389
Jazz	298
Classical	254
Easy listening	173
Other	88

Store C	
Country	2,586
Rock	2,385
Classical	957
Jazz	854
Easy listening	427
Other	239

Store D	
Country	1,654
Rock	1,489
Classical	328
Jazz	127
Easy listening	53
Other	22

Store E	
Rock	2,723
Classical	1,307
Jazz	945
Country	256
Easy listening	184
Other	176

a. Write matrix S of the CDs and tapes sold. Make up two other sets of data to accompany this matrix (such as inventory on July 1, 1995) and a problem that could be solved using these data and matrices. Have another student solve the problem. Then check his or her solution.

b. Design a spreadsheet that uses the data in some way. Find the amounts the computer will calculate.

c. Write a program to analyze the data in some way. Briefly describe what your program is to accomplish. Then find what the computer will print. Make up any additional data, such as prices, if it is necessary.

PROJECT ASSESSMENT

PROJECT *Connection*   Your group has been commissioned by a nonprofit agency to write a report titled *What's Really on TV* and to make recommendations for how programming needs to change. Work with your group to prepare a comprehensive, accurate, and interesting report.

1. In the report provide a careful description of the methods used to collect data and draw conclusions. Include a copy of your survey form.

2. Make suggestions for additional surveys or types of data that could be useful for decision making about future programming.

3. Imagine a new channel called The Student Channel. Plan a programming schedule for this channel.

• • • CUMULATIVE REVIEW • • •

Solve each proportion.

1. $\dfrac{3}{5} = \dfrac{n}{35}$

2. $\dfrac{4}{t} = \dfrac{10}{15}$

Compare. Use <, =, or >.

3. $\dfrac{7}{9} \, \blacksquare \, \dfrac{3}{4}$

4. 5% of 80 $\blacksquare$ 80% of 5

5. four and sixty-three hundredths $\blacksquare$ four and seven tenths.

Use the picture graph to answer Questions 6–8.

Seasonal Birthdays

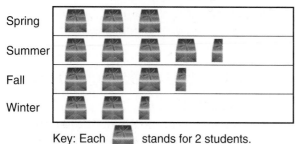

Key: Each $\blacksquare$ stands for 2 students.

6. Which season has the fewest birthdays?

7. How many students have a fall birthday?

8. What percent of the birthdays are in the summer? Round to the nearest percent.

Evaluate each expression if $A = 6$, $B = 5$, and $C = 8$. Expressions are in BASIC.

9. $C * B$

10. $\dfrac{A}{C - B}$

11. $4 * A + B - C$

Find $A + B - C$.

12. $A = \begin{bmatrix} 3 & 7 \\ 4 & 2 \end{bmatrix}$ $B = \begin{bmatrix} 6 & 3 \\ 1 & 9 \end{bmatrix}$ $C = \begin{bmatrix} 5 & 8 \\ 2 & 3 \end{bmatrix}$

13. **WRITING MATHEMATICS** Can a 2×3 matrix be added to a 3×2 matrix? Explain.

The graph below shows the number of heads that resulted from 10 tosses of a coin for each student. Use the data to answer Questions 14–17.

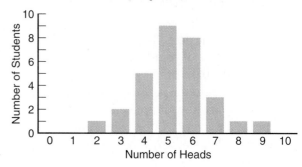

14. What is the mode of the data?

15. What fraction of the students tossed 7 heads or more?

16. **STANDARDIZED TESTS** Given the following statements about the number of occurrences of heads in 10 tosses, which is true?

   I. More than half the students tossed 5 or 6 heads.

   II. It is impossible to get either 0 heads or 10 heads.

   III. Theoretically, the number of students tossing 3 heads should be the same as the number that tossed 7 heads.

   A. I only   **B.** II only
   C. I and III   **D.** I, II, and III

17. **WRITING MATHEMATICS** Joey was the one student who tossed 9 heads. Would Joey be correct in concluding that the probability of tossing a head is 0.9 or 90%? Explain.

18. In a random sample of 50 students, 38 said that they would order pizza on pizza day. In a school of 475 students, how many might be expected to order pizza?

··· STANDARDIZED TEST ···

STANDARD FIVE-CHOICE Select the best choice for each question.

1. After 36 people responded that they usually voted for liberal candidates, a pollster stated that there was a 75% probability that a person would vote for a liberal candidate. How many people must have been surveyed?

   A. 24   **B.** 27   **C.** 48
   D. 63   **E.** 75

2. What is the least common multiple of the first four prime numbers?

   A. 1   **B.** 24   **C.** 30
   D. 120   **E.** 210

3. If a telemarketer's goal is to make 60 calls per shift and he has reached $\frac{3}{5}$ of his goal, how many calls does he have left to make?

   A. 12   **B.** 20   **C.** 24
   D. 36   **E.** 100

4. In the table, A represents the number of people surveyed who reported they consumed an average of 20–39 g of fat daily. If A represents 15% of those surveyed, then $A =$

Average Number of Fat Grams Consumed Daily	
Number of Grams	Frequency
0 –19	1
20 – 39	A
40 – 59	4
60 – 79	7
80 or more	5

   A. 3   **B.** 5   **C.** 15
   D. 17   **E.** 20

5. How many cubes that are 2 cm on a side will completely fill a cube that is 8 cm on a side?

   A. 4   **B.** 16   **C.** 64
   D. 128   **E.** 512

6. How much greater than 3.5 is the sum of 0.32, 2.4, and 6.85?

   A. 5.71   **B.** 6.07   **C.** 6.74
   D. 7.03   **E.** 9.57

7. The base of a triangle is twice as long as its height. If the area of the triangle is 16 in.^{2}, what is the length of its base in inches?

   A. 2   **B.** 3   **C.** 4
   D. 6   **E.** 8

8. $A = \begin{bmatrix} 9 & X \\ 7 & 3 \end{bmatrix}$ $B = \begin{bmatrix} 3 & 2 \\ X & 0 \end{bmatrix}$ $A - B = \begin{bmatrix} 6 & 4 \\ Y & 3 \end{bmatrix}$

   The problem above shows the subtraction of two matrices. If X and Y represent two different nonzero digits, then $Y =$

   A. 1   **B.** 2   **C.** 5
   D. 6   **E.** 13

9. If the graph displays data about the 800 people who attended the game, which of the following statements must be true?

   High School Football Game Attendance

   5% Sophomores — Freshman 15%
   25% Juniors — Others 30%
   25% Seniors — Faculty/Staff 10%

   I. More than 35% of those at the game were students.
   II. The central angle of the largest sector is 90°.
   III. The largest sector represents 240 people.

   A. I only   **B.** II only   **C.** III only
   D. I and II   **E.** I and III

10. The mean of a set of numbers is 16. If the sum of the numbers is 96, how many numbers must be in the set?

    A. 5   **B.** 6   **C.** 12
    D. 16   **E.** 80

2 Variables, Expressions, and Real Numbers

DATA Activity

Tornado Alley

Many people have seen the dark funnel cloud formed by the strong, whirling winds of a tornado. A tornado can develop in less than an hour and quickly destroy whatever lies in its path. Because so many tornadoes occur in the region extending from northern Texas through Oklahoma, Kansas, and Missouri, this area is called "Tornado Alley." The table on the next page gives data about the number of tornadoes reported in the United States.

METEOROLOGY
The Science *of* Weather

In this chapter, you will see how:

- **METEOROLOGISTS** use a barometer to collect data about atmospheric conditions. (Lesson 2.3, page 72)

- **SERVICE STATION ATTENDANTS** use temperature data and product information to determine the proper antifreeze mixture to add to a car's radiator. (Lesson 2.7, page 94)

- **WEATHER FORECASTERS** use a wind-chill table to make their reports more useful to people. (Lesson 2.8, page 101)

Number of Tornadoes Reported in the United States			
Year	**Number**	**Year**	**Number**
1980	866	1987	656
1981	783	1988	702
1982	1046	1989	856
1983	931	1990	
1984	907	1991	1132
1985	684	1992	1297
1986	764	1993	1173

1. The number of tornadoes reported in 1990 was 235 less than twice the number reported in 1985. Write a numerical expression you can use to find how many tornadoes were reported in 1990. Then find the number.

Use your completed table to answer Questions 2–5.

2. What was the average yearly number of tornadoes in the United States from 1980 to 1989. (Round to the nearest whole number.) What is the range of the numbers reported?

3. Between which two years was there the sharpest increase in the number of tornadoes reported?

4. **a.** How would you describe the trend in the number of tornadoes reported from 1990 to 1993?

   b. If this trend were to continue, about how many tornadoes would you have predicted for 1994?

5. **WORKING TOGETHER** Make a pictograph of the data shown in your completed table. How will you round the numbers? How many tornadoes will each symbol represent? Compare your graph with those of other students and discuss similarities and differences.

PROJECT Taking the Nation's Temperature

When you woke up this morning, did you wonder "what's the weather going to be today?" Weather is an important topic because it affects you individually. Weather also can affect the economy and well-being of the entire country.

PROJECT GOAL

To use algebraic symbols and expressions to analyze temperature data and make predictions about weather conditions.

Balloons carry radar, computers, and sophisticated instruments into the sky to collect data about temperature, air pressure, winds, humidity, and precipitation at locations all around the globe. Meteorologists analyze changes over periods of time and look for patterns that will help them predict future weather more accurately.

Getting Started

Work in groups of four or five students. For an eight-day period beginning today, collect data on temperatures throughout the United States using a newspaper that prints national weather reports.

1. Decide on the newspaper to use. You must be able to get the newspaper for eight consecutive days.

2. Choose 20 cities listed in the newspaper's weather report section. Try to choose cities from all over the United States. Include your own city if you wish.

3. Divide the cities among the group members. Assign each member cities that are located in the same geographical region. (For example, New York City, Boston, Philadelphia, and Baltimore might be one member's assignment.)

4. Create a form that can be used to record the high and low temperatures for the eight-day period. Each member will use a copy of the form and fill in data for his/her cities.

5. Decide how your group will divide the work of collecting and organizing data.

Internet Connection

www.swpco.com/
swpco/algebra1.html

PROJECT *Connections*

Lesson 2.2, page 64:
Perform an experiment to determine the dew point and use it to calculate the altitude of nearby clouds.

Lesson 2.5, page 83:
Determine average temperatures and explore the relationship between the average and the data set.

Lesson 2.7, page 93:
Construct a high/low temperature chart.

Chapter Assessment, page 109:
Prepare final project results in a presentation.

2.1 Algebra Workshop
Represent Numbers, Variables, and Expressions

Think Back

Recall that cells of a spreadsheet can include algebraic expressions. Suppose cell A2 contains the number of people aged 14–17 registered to play soccer in Wellsville.

1. Write an expression for cell B2 that will compute the total cost of uniforms at $26 per uniform.

2. Write an expression for cell C2 that will compute the number of teams that can be formed if each team must include at least 15 people.

Suppose the value in cell A2 is 145.

3. Find the total cost of uniforms.

4. Find the number of teams that can be formed. Will there be exactly 15 members on each team? How can you use all the people on teams?

	A	B	C	D
1	no. of players	uniform costs	no. of teams	
2				
3		=26*A2	=A2/15	
4				
5				

Explore

A **model** is a representation. A weather map is a model used by TV weather forecasters to display information about present weather conditions and future weather conditions.

On the weather map model, symbols pointing from left to right represent a weather system moving from west to east. The model is not only a display but also a tool for simulating how the weather system changes.

Algebra Workshop

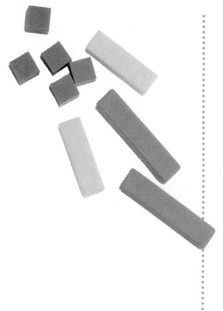

Algeblocks are models that give a physical representation of important algebraic concepts. These blocks can be placed and moved according to the rules of algebra. In this way, Algeblocks simulate how the system of algebra works.

The Algeblocks that you will use in this chapter include a green units block that represents 1, a yellow rectangular block that represents any number and is called the x-block, and an orange rectangular block that represents any number different from x and is called the y-block.

The numbers in the set $\{\ldots -3, -2, -1, 0, 1, 2, 3, \ldots\}$ are called **integers**. The numbers 1, 2, 3, $\ldots$ are the **positive** integers and the numbers $-1, -2, -3, \ldots$ are the **negative** integers. Positive and negative integers can be modeled using the unit blocks of the Algeblocks.

The **Basic Mat** for Algeblocks is divided into two equal parts, one positive and one negative.

To model $+2$, show two unit blocks on the positive side of the mat.

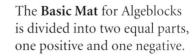

To model -3, show three unit blocks on the negative side of the mat.

A letter can be used to represent any number. The letter is called a **variable**. The x-block and the y-block represent two different variables. Look at these mats.

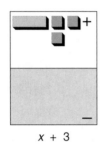

$x + 3$ $y + 4$ $x + x$ or $2x$ $-y + -y + -y$ or $-3y$

Work in a group. Write the number or expression modeled on each mat.

5.

6.

7.

8.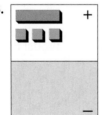

9.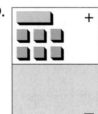

10.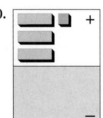

11.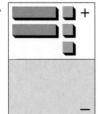

12.

13.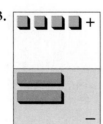

Algebra Workshop

Represent each of these expressions on the Basic Mat using Algeblocks. Make sketches to record your work.

14. −5   **15.** $x + 4$ **16.** $x + 8$ **17.** $x + y$

18. $2y$ **19.** $4y + 2$ **20.** $-2x$ **21.** $-x + 1$

Make Connections

- Integers, variables, and algebraic expressions can model real world relationships.

  Use Algeblocks to model each description. Then write algebraic symbols for the blocks.

  22. 7° below zero

  23. five dollars more than x, the money you have saved

  24. the total cost of x adult tickets and y children's tickets if adult tickets cost $5 each and children's tickets cost $2 each

  25. five dollars an hour for y hours of work plus $2 for transportation

Summarize

- **26.** WRITING MATHEMATICS Write a paragraph explaining how you can show positive and negative numbers, variables, and variable expressions using Algeblocks. Sketch some examples.

  27. MODELING Use Algeblocks to model these expressions. Sketch your model.

      a. $3x + 4$

      b. $x + 2y - 3$

      c. $-5y + 7$

  28. GOING FURTHER The mat at the right shows $y + 6$. Suppose the value of y is 3. How would you find the value of the expression on the mat using Algeblocks? What is the value of $y + 6$ if $y = 3$?

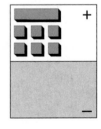

2.2 Variables and Expressions

Explore/Working Together

Maria and her friends were on a hike in the woods. They stopped to listen to some crickets. Maria said she knew a way to estimate the temperature from the number of cricket chirps in one minute. She uses this formula.

$$T = 40 + c \div 4$$

They counted 88 chirps in one minute. Maria wrote the expression $40 + 88 \div 4$ on a sheet of paper and said, "The value of this expression is the number of degrees Fahrenheit!"

Joan said the temperature was 32°F. Leah said the temperature was 62°F.

Work with a partner to answer these questions.

1. How do you think Joan got her answer?

2. How do you think Leah got her answer?

3. What did Joan and Leah do differently from one another?

4. Who do you think is correct? Explain.

5. Try it on a calculator. What is the result?

Build Understanding

A **numerical expression** is a combination of numbers and operations. Expressions use the operations of addition, subtraction, multiplication, or division. Grouping symbols, such as parentheses, may be used in expressions, such as

$$4 + 7 - 3 \qquad 4 \cdot 7 + 3 \qquad 2(4 + 3) - 7$$

Expressions may also contain exponents. In the expression 2^3, the number 2 is the **base**, and the number 3 is the **exponent**. The expression 2^3 represents the number you obtain when 2 is used as a factor 3 times.

$$\text{base} \rightarrow 2^{3 \; \leftarrow \text{exponent}} = 2 \cdot 2 \cdot 2 = 8$$

The number 2^3 is a **power**. It is read "two to the third power" or "two cubed." Most calculators have a special key for exponents. It is usually labeled $\boxed{\wedge}$ or $\boxed{y^x}$ or $\boxed{x^y}$.

THINK BACK

Review the different ways to show multiplication:

$3 \times 4 \quad 3 \cdot 4 \quad 3(4)$

Review different ways to show division:

$16 \div 4 \quad 4\overline{)16} \quad \dfrac{16}{4}$

To **evaluate** an expression, find the number that the expression represents. For example, the value of $4 + 5$ is 9. When evaluating expressions with more than one operation, you must follow the **order of operations**. As you saw in Explore, the value of a numerical expression depends on the order in which you perform the operations.

> **ORDER OF OPERATIONS**
> 1. **Perform operations within grouping symbols first.**
> 2. **Perform all calculations involving exponents.**
> 3. **Multiply or divide in order from left to right.**
> 4. **Add or subtract in order from left to right.**

EXAMPLE 1

Evaluate each expression.

a. $(4 + 3) \cdot 5$ **b.** $8 \cdot 3 - 6 \div 3 + 7$

c. $55 - 7^2 - 12 \div 4$ **d.** $(5 - 3)^4 - 7 \cdot 2$

Solution

Use the order of operations.

a. $(4 + 3) \cdot 5$
$= 7 \cdot 5$
$= 35$

b. $8 \cdot 3 - 6 \div 3 + 7$
$= 24 - 2 + 7$
$= 22 + 7$
$= 29$

c. $55 - 7^2 - 12 \div 4$
$= 55 - 49 - 12 \div 4$
$= 55 - 49 - 3$
$= 6 - 3$
$= 3$

d. $(5 - 3)^4 - 7 \cdot 2$
$= 2^4 - 7 \cdot 2$
$= 2^4 - 14$
$= 16 - 14$
$= 2$ ◄

CHECK UNDERSTANDING

Look back at Explore. Who is correct? How do you know?

THINK BACK

Remember a **factor** is a number that is multiplied; 5 and 2 are both factors of 10 since $5 \cdot 2 = 10$.

Expressions with exponents indicate repeated multiplication.

Recall from Lessons 1.3 and 2.1 that a **variable** is a symbol, usually a letter, used to represent a number. A **variable expression** or **algebraic expression** is a combination of numbers, operations, and one or more variables. Some examples are $-3x$, $4y + 1$, and $5z^2$.

The value of an expression depends on the value of the variables in it. To evaluate an expression, substitute a value for the variable.

EXAMPLE 2

Evaluate each expression for $x = 2$.

a. $3x + 1$
b. $\dfrac{3}{4} - \dfrac{x}{5}$
c. $5x^2$

Solution

In each expression, substitute 2 for x. Then use the order of operations to find the value of the expression.

a. $3x + 1 = 3 \cdot 2 + 1$ $3x$ means 3 times x.
$\qquad\qquad\; = 6 + 1$
$\qquad\qquad\; = 7$

b. $\dfrac{3}{4} - \dfrac{x}{5} = \dfrac{3}{4} - \dfrac{2}{5}$

$\qquad\qquad = \dfrac{15}{20} - \dfrac{8}{20}$ 20 is the LCM (Least Common Multiple) of 4 and 5.

$\qquad\qquad = \dfrac{7}{20}$

c. $5x^2 = 5 \cdot 2^2$
$\qquad\;\; = 5 \cdot 4$ Find 2^2 before multiplying by 5.
$\qquad\;\; = 20$ ◀

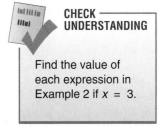

CHECK UNDERSTANDING

Find the value of each expression in Example 2 if $x = 3$.

The **replacement set** is the set of numbers that can be substituted for the variable.

EXAMPLE 3

Find all values of the expression $4y - 3$ if the replacement set of y is $\{1, 3, 6\}$.

Solution

Make a table. Substitute a value from the replacement set for y. Then evaluate the expression. Do this for each member of the replacement set.

y	$4y - 3$
1	$4(1) - 3 = 1$
3	$4(3) - 3 = 9$
6	$4(6) - 3 = 21$

◀

PROBLEM SOLVING TIP

Making a table can help you organize information when there are several possibilities.

The parts of a variable expression separated by addition or subtraction signs are called **terms** of the expression.

$\quad 5x^2$ one term $2x + 6x^2 + 4$ three terms

In the term $5x^2$, x is the variable, 2 is the exponent, and 5 is the **coefficient**. In the expression $3x^4 - 2x^3 + 7x + 20$, the first three terms are variable terms. The last term, 20, is a constant. A **constant** is a number not multiplied or divided by a variable.

The **degree** of an expression is the greatest exponent of the variables in the expression in simplest form. In the expression $3x^4 - 2x^3 + 7x + 20$, the degree is 4.

EXAMPLE 4

Give the degree of each expression. Tell how many terms it has, name any constant terms, and name the coefficient of the first term.

  a. $4x^5 + 2x$ **b.** $y^3 + 6y^2 - 3y - 11$ **c.** $-3z^2 + 2z + 8$

Solution

  a. $4x^5 + 2x$ Degree is 5; two terms; no constant term; the coefficient of the first term is 4.

  b. $y^3 + 6y^2 - 3y - 11$ Degree is 3; four terms; constant term is -11; the coefficient of the first term is 1 ($y^3 = 1y^3$).

  c. $-3z^2 + 2z + 8$ Degree is 2; three terms; constant term is 8; the coefficient of the first term is -3.   ◄

TRY THESE

Evaluate each expression. Explain the order of your evaluation process.

1. $9 \cdot 4 - 3 \cdot 8$ **2.** $8 + 12 \div 4$ **3.** $7 + 2^3$ **4.** $2 + 3 \cdot 5^2 - 6 \cdot 2$

5. $10 + 6 \div 4 \cdot 2$ **6.** $(10 + 6) \div 4 \cdot 2$ **7.** $(10 + 6) \div (4 \cdot 2)$ **8.** $10 + [6 \div (4 \cdot 2)]$

Evaluate each expression for $y = 10$.

9. $y - 1$ **10.** $2y^2$ **11.** $3y + 10$ **12.** $\dfrac{y}{2}$

13. Make a table for the values of $5x + 2$, given the replacement set $\{3, 5, 8, 10\}$.

Give the degree of each expression, tell how many terms it has, and name its constant term.

14. $y^4 + 3y^6 + 7 + 8y$ **15.** $b^2 + 3b - 1$ **16.** $c + 24$

Name the coefficient of each term.

17. $4x^3$ **18.** $-7y$ **19.** a **20.** $-b$ **21.** $-\dfrac{3x}{4}$

22. MODELING Use Algeblocks to model the expression $2x + 5$. Draw a sketch of your model. Then, replace each x-block with 3 unit blocks. Draw another sketch to show your work. Use your model to show the value of $2x + 5$ when $x = 3$.

23. GEOMETRY A gazebo in the shape of a regular octagon has a railing all around, except for an opening that is 4 ft wide. Write an expression for the number of feet of railing that will be needed if the length of a side is s feet, where s is greater than or equal to 4 ft.

24. WRITING MATHEMATICS Write a paragraph explaining how to evaluate an algebraic expression. Give one or two examples.

PRACTICE

Evaluate each expression.

1. $8 + 3 \cdot 9$ **2.** $34 - 10 \div 2$ **3.** $9 \cdot 2 + 5 \cdot 3$

4. $9 \cdot (2 + 5) \cdot 3$ **5.** $42 - 5 \cdot 5$ **6.** $79 - 3 \cdot 8$

7. $7 + 81 \div 9 + 3$ **8.** $(12 + 15) \div 3 + 20$ **9.** $200 \div 40 \cdot 8$

10. $(25 - 5) \cdot 5 + 5$ **11.** $36 \div 18 \div 2$ **12.** $\frac{1}{4}(6 \cdot 8) + 2 \cdot 12$

13. $7^2 - 2 \cdot 8$ **14.** $8^2 - 5 \cdot 4$ **15.** $3 \cdot (4^2 + 2)$

16. $2 \cdot 5^3 + 10$ **17.** $3^4 - 2^3$ **18.** $4^3 - 4^2$

19. $10 \cdot 2^5 + 4 \cdot 2 + 36 \div 4$ **20.** $2 \cdot 10^3 + 2 \cdot 30 + 200 \div 5$

21. MAIL ORDER A mail-order company charges \$3.50 handling on all orders. Write an expression for the total to be paid if the amount of the mail order is x dollars.

Evaluate each expression for $x = 5$.

22. $3x - 1$ **23.** $x^2 + 23$ **24.** $4x^3$ **25.** $5x - 2$ **26.** $\frac{3x}{10}$

Copy and complete the table for each expression and its replacement set.

27.

y	$3y - 2$
3	?
4	?
5	?
6	?

28.

a	$5a^2 + 6$
4	?
5	?
6	?
10	?

29.

g	$\frac{g}{4}$
4	?
8	?
12	?
20	?

30.

k	$\frac{k}{3}$
3	?
6	?
18	?
24	?

Give the degree of each expression, tell how many terms it has, and name its constant term, if any.

31. $a^3 + 4a^4 - 3 - 9a$ **32.** $12 + x + 3x^2 + 8x^3$

33. $y^6 + y^4 + 2y^5 + 5y^2 - 4y + 10$ **34.** $21g - 2g^8$

Name the coefficient of each term.

35. $9a^3$ **36.** x^2 **37.** $-3x$ **38.** $-v$

39. PLANT NURSERY Tomato plants are sold in flats of 8 plants each. Write an expression that will tell the number of flats that must be bought if a customer wants y tomato plants.

40. WRITING MATHEMATICS Suppose you are planning to take a few friends to the movies to celebrate your birthday. Write an expression for the total amount of money you will spend. Explain your answer.

EXTEND

Evaluate each expression for $x = 3$ and $y = 4$.

41. $2x + 3y$ **42.** $x^2 + y^2$ **43.** $3x^2 - y$ **44.** $5x + y^4$

Use a calculator to evaluate these expressions. Round answers to the nearest tenth.

45. $12^4 + 7^5$ **46.** $125 + 28 \cdot 45$ **47.** $(2.5)^3$

48. $(3.04)^4$ **49.** $452 + 1360 \div 8$ **50.** $674 + 2943 \div 4$

51. SALES REPRESENTATIVE Mrs. Wallace sells computer chips. She gets paid $450 per week plus 2% commission on sales. Write an expression for the amount of money she will earn in a week if she sells x dollars worth of computer chips.

52. SALES TAX Tia plans to buy a new dress. Sales tax is 4%. Write an expression for the amount she will pay, including sales tax, for a dress that costs y dollars.

53. FORENSIC SCIENCE When part of a skeleton is found, police detectives are interested in the height of the victim, for identification purposes. If the victim is male and the length of the femur (large thigh bone) is known, this expression can be used to determine the victim's height in centimeters: $69.089 + 2.238F$, where F is the length in centimeters of the femur. Find the height of a male victim whose femur measures 42 cm.

THINK CRITICALLY

54. NUMBER THEORY A number a is divisible by b if $a \div b$ has no remainder. In the expression $3x$, what must be true of x so that $3x$ is divisible by 6?

55. It is true that 2^4 equals 4^2. Is it always true that $a^b = b^a$? Justify your answer.

56. ESTIMATE/TECHNOLOGY Predict how many decimal places $(1.2)^5$ will have. Explain your answer. Calculate to verify your prediction.

PROJECT *Connection*   The **dew point** is the temperature at which condensation (change from gas to liquid state) first takes place. If you know the outside air temperature and the dew point, it is possible to find the approximate altitude of nearby clouds.

1. You should go outside to perform this experiment. First, determine the outside air temperature.

2. Remove the paper label from an empty metal soup or vegetable can. Fill the can with ice water. Insert a Fahrenheit thermometer into the water and stir. At the moment when water droplets appear on the sides of the can, remove the thermometer and take a reading. This is the dew point.

3. Let T represent the outside air temperature in degrees Fahrenheit. Let D represent the dew point. The approximate altitude A in feet of nearby clouds is given by the formula $A = 1000\left(\dfrac{T - D}{5.5}\right)$.

4. Find the approximate altitude of nearby clouds using the air temperature you recorded.

2.3 Real Numbers and the Number Line

Explore

● This thermometer shows the temperature outside Liu's house on a winter day. Notice the thermometer has long and short scale lines.

1. What does each long line on the scale of the thermometer represent?

2. What does each short line on the scale of the thermometer represent?

3. How would an increase in temperature of 10° be shown on this thermometer?

4. How would a decrease in temperature of 8° be shown on this thermometer?

5. By how many degrees must the temperature increase to get to 0°?

6. What other temperature is the same distance from 0° as the one shown on the thermometer?

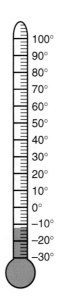

Build Understanding

● Numbers like -3, 5, $\frac{1}{2}$, 0.51, and $0.333\ldots$ are all examples of rational numbers. A **rational number** is a number that can be written in the form $\frac{a}{b}$ where a and b are integers and b is not equal to 0. The fraction $\frac{1}{2}$ is already written in this form. Each of the others can be.

$$-3 = -\frac{3}{1} \qquad 5 = \frac{5}{1} \qquad 0.51 = \frac{51}{100} \qquad 0.333\ldots = \frac{1}{3}$$

Rational numbers include all integers and fractions. They also include decimals which either terminate, such as 0.51, or are nonterminating but repeating, such as $0.333\ldots$.

Decimals that are nonterminating and nonrepeating are called **irrational numbers**. Examples of irrational numbers are π, $0.12112111211112\ldots$, and $\sqrt{2}$.

Most calculators have a $\sqrt{}$ key. The expression $\sqrt{x}$ means "the square root of x." This is the number that, when multiplied by itself, is x. Numbers such as 1, 4, and 9 are called **perfect squares** since $1 = 1 \cdot 1$, $4 = 2 \cdot 2$, and $9 = 3 \cdot 3$. Their square roots will always be integers: $\sqrt{1} = 1$, $\sqrt{4} = 2$, $\sqrt{9} = 3$. The square root of any number that is not a perfect square is an irrational number.

COMMUNICATING ABOUT ALGEBRA

Write a sentence explaining why you think the word *square* is used to describe a number like 9.

EXAMPLE 1

Tell whether each is a rational number. Give a reason for your answers.

a. −3.4   **b.** 0.121212...   **c.** $\sqrt{8}$

d. 0.141724213...   **e.** $\sqrt{36}$

Solution

a. −3.4 is rational since it is a terminating decimal: $-3.4 = -\dfrac{34}{10}$.

b. 0.121212... is rational since it is a nonterminating, repeating decimal.

c. $\sqrt{8}$ is not rational since 8 is not a perfect square.

d. 0.141724213... is not rational since it is a nonterminating, nonrepeating decimal.

e. $\sqrt{36}$ is rational since 36 is a perfect square: $\sqrt{36} = 6$. ◄

The **real numbers** consist of the rational and irrational numbers. Every real number can be matched with a point on a number line. The graph below represents the **real number line**.

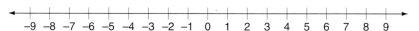

The point that corresponds to 0 is called the **origin**. Positive real numbers are located to the right of 0; negative real numbers are located to the left of 0. Numbers increase in value from left to right.

EXAMPLE 2

Graph each point on the real number line.

a. $-2\dfrac{1}{3}$ **b.** 4.5   **c.** $\sqrt{7}$ **d.** $\dfrac{3}{4}$ **e.** −0.8

Solution

a. $-2\dfrac{1}{3}$ is one third of the way between −2 and −3.

b. 4.5 is halfway between 4 and 5.

c. A calculator shows that $\sqrt{7} = 2.6457513\ldots$. Therefore, it is located a little more than halfway between 2 and 3.

d. $\dfrac{3}{4}$ is three-fourths of the way between 0 and 1.

e. −0.8 is between 0 and −1, closer to −1. ◄

Graphing real numbers on a number line shows the order of the numbers from least to greatest. The number line in Example 2 shows the order of the numbers.

$$-2\frac{1}{3} \qquad -0.8 \qquad \frac{3}{4} \qquad \sqrt{7} \qquad 4.5$$

You can use **inequality symbols** to **order**, or **compare**, real numbers.

The symbol < means "is less than." $\qquad -2\frac{1}{3} < -0.8$

The symbol > means "is greater than." $\qquad 4.5 > \sqrt{7}$

A sentence containing an inequality symbol is called an **inequality**.

PROBLEM SOLVING TIP

To remember the meaning of < and >, try this device. The small "point" of the symbol always points to the smaller value.

EXAMPLE 3

Compare each using < or >.

a. −12 ▢ −4   **b.** −3 ▢ −3.5   **c.** 2 + 3 • 4 ▢ 19

Solution

Think of a number line.

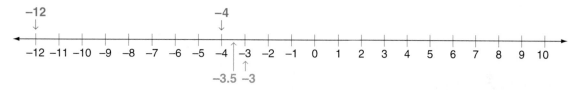

a. On the number line, −12 is farther to the left than −4. Therefore, −12 < −4.

b. On the number line, −3 is farther to the right than −3.5. Therefore, −3 > −3.5.

c. First, evaluate the expression on the left.

$$2 + 3 \cdot 4 = 2 + 12$$
$$= 14$$

Since 14 < 19, 2 + 3 • 4 < 19. ◄

CHECK UNDERSTANDING

Suppose a number line has −2.3 and −2.2 already graphed on it. Would you place a point for $-2\frac{1}{4}$ to the left of −2.3, to the right of −2.2, or in between? How did you decide?

An important idea in algebra is the idea of an opposite. The **opposite** of any real number x is $-x$. For example, the opposite of 5 is −5 and the opposite of −2.1 is 2.1. On a number line, opposites are the same distance from 0, but in different directions.

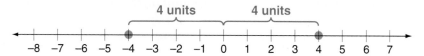

As shown above, 4 and −4 are each 4 units from 0.

COMMUNICATING ABOUT ALGEBRA

Explain why a number and its opposite have the same absolute value.

The **absolute value** of a number x is its distance from 0 on a number line. Distance is always a positive number. The absolute value of x is written as $|x|$.

EXAMPLE 4

Find the value of each.

a. $|-1.3|$　　　　b. $\left|2\frac{3}{4}\right|$　　　　c. $|-0.5|$

Solution

Find the distance of the number from 0 on a number line.

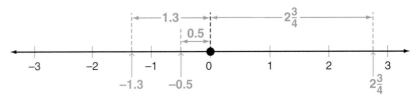

a. $|-1.3| = 1.3$　　　b. $\left|2\frac{3}{4}\right| = 2\frac{3}{4}$　　　c. $|-0.5| = 0.5$ ◄

Absolute value can be defined algebraically, as follows.

ABSOLUTE VALUE

For any real number x,

$$|x| = x \text{ if } x > 0 \text{ or } x = 0 \quad \text{and} \quad |x| = -x \text{ if } x < 0.$$

EXAMPLE 5

Use the definition of absolute value to find the value of each.

a. $|5.1|$　　　　b. $\left|-\frac{2}{3}\right|$　　　　c. $-|-8|$

Solution

a. $|5.1| = 5.1$　　　b. $\left|-\frac{2}{3}\right| = -\left(-\frac{2}{3}\right) = \frac{2}{3}$

c. $-|-8| = -(8) = -8$ ◄

TRY THESE

Tell whether each is a rational number. Give a reason for each answer.

1. $-\frac{2}{3}$　　　2. 0.9　　　3. $\sqrt{3}$　　　4. $0.252525\ldots$　　　5. $\sqrt{11}$

For each number below, write the letter of the corresponding point on the real number line.

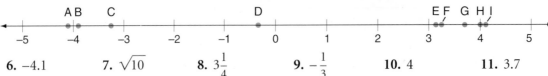

6. -4.1　　　7. $\sqrt{10}$　　　8. $3\frac{1}{4}$　　　9. $-\frac{1}{3}$　　　10. 4　　　11. 3.7

Compare each using < or >.

12. −4 [ ] −4.8 **13.** −2 [ ] 0 **14.** −120 [ ] −40 **15.** 3 + 12 ÷ 3 [ ] 6

Find the value of each.

16. |−3| **17.** |9| **18.** $\left|-\frac{1}{2}\right|$ **19.** |4.9| **20.** −|−101|

21. MODELING How can you describe the absolute value of a number using Algeblocks?

22. WRITING MATHEMATICS A real world example of the idea of opposites is that walking four steps forward is the opposite of walking four steps backward. Write three more real world examples of opposites.

23. METEOROLOGY The average temperature in Barrow, Alaska, in the month of February is −20°F. The average temperature in Fairbanks, Alaska, in the month of February is −4°F. Which place is warmer in February? How do you know?

PRACTICE

Tell whether each is a rational number. Give a reason for your answers.

1. $5\frac{1}{4}$ **2.** $3\frac{1}{9}$ **3.** 2.44948 . . . **4.** −21 **5.** $\sqrt{5}$ **6.** $\sqrt{17}$

7. 6.2 **8.** −9.1 **9.** $\sqrt{49}$ **10.** $\sqrt{81}$ **11.** 0.444 . . . **12.** 0.9191 . . .

For each, name the letter of the corresponding point on the real number line that follows.

13. −3 **14.** −0.9 **15.** $\sqrt{11}$ **16.** $\sqrt{20}$ **17.** −2.3

18. −5.2 **19.** $-5\frac{7}{8}$ **20.** $-2\frac{3}{4}$ **21.** $\frac{2}{3}$ **22.** $3\frac{1}{2}$

Compare. Use < or >.

23. −12 [ ] −4 **24.** −20 [ ] −21 **25.** −16 [ ] 2 **26.** −11 [ ] 7

27. −3 [ ] 4 **28.** −1 [ ] 0 **29.** 5.1 [ ] −6.22 **30.** 10.2 [ ] −4.3

Find the value of each.

31. |−2| **32.** |−11| **33.** $\left|7\frac{2}{3}\right|$ **34.** $\left|-4\frac{3}{4}\right|$ **35.** −|−2.3| **36.** −|4.7|

37. WRITING MATHEMATICS Explain what is meant by the statement, "Every real number can be matched with a point on a number line."

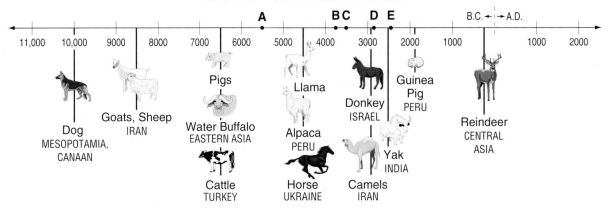

Time Line – Domesticated Animals

A   B C D E   B.C. ←│→ A.D.

11,000 10,000 9000 8000 7000 6000 5000 4000 3000 2000 1000   1000 2000

Pigs

Goats, Sheep
IRAN

Dog
MESOPOTAMIA,
CANAAN

Water Buffalo
EASTERN ASIA

Cattle
TURKEY

Llama

Alpaca
PERU

Horse
UKRAINE

Donkey
ISRAEL

Camels
IRAN

Guinea
Pig
PERU

Yak
INDIA

Reindeer
CENTRAL
ASIA

ANCIENT HISTORY The time line above shows when and where certain animals were domesticated by humans. At which point would you place each of the following events?

38. Zebu cattle were domesticated in Thailand between 4000 and 3500 B.C.

39. The cat was domesticated in Egypt between 2500 and 2400 B.C.

40. The elephant was domesticated in India between 3000 and 2900 B.C.

41. GROCERY SHOPPING Tom wants to buy meat for a stew. There are several packages of prewrapped meat, marked 2.12 lb, 2.63 lb, 2.39 lb, 2.78 lb, and 3.21 lb. Which package should he buy to get at least $2\frac{1}{2}$ lb of meat, but not much more than that?

EXTEND

42. Write these numbers in order, from least to greatest.

$$-3.6, 3.2, -3.31, \sqrt{11}, \sqrt{9}, -\sqrt{13}, -3\frac{1}{3}, 3\frac{3}{8}$$

43. Write these numbers in order, from greatest to least.

$$0.5, -\frac{3}{10}, \frac{9}{20}, \sqrt{3}, -1.1, 0.51, -0.98, -\sqrt{4}$$

Compare. Use <, > or =.

44. $2 \cdot 5 + 36 \div 4$ ▨ $4 \cdot 2 + 16 \div 8$

45. $-4 |{-3}|$ ▨ $-|{-5.2}|$

46. $3 + 6 \cdot 7$ ▨ $5^2 + 2 \cdot 10$

47. $6 + 3 \cdot 9$ ▨ $4 + 8^2$

48. $-|{-14}|$ ▨ $2 |{-7}|$

49. $|{-4}| \cdot |6|$ ▨ $-|4| \cdot |6|$

50. CHEMISTRY On the Celsius scale, the melting point of water is 0°C. This is the point at which water in the solid state will change to the liquid state. This table gives the melting points of some other chemical solvents (solutions). Draw a number line and graph each of the solvents at its melting point.

Solvent	Melting Point, °C
Acetic acid	16.6
Ammonia	−77.7
Bromine	−7
Carbon tetrachloride	−23
Methanol	−97.5
Nitric acid	−41.6
Sulfur dioxide	−75.5

51. **HISTORY** Draw a timeline to illustrate these landmarks in the development of the modern automobile.

A.D. 540   Book appears in China describing wind-driven land vehicles actually in use.

300 B.C.   Chinese invent an improved harness for horses.

A.D. 1887   Gottlieb Daimler uses the internal combustion engine to power a four-wheeled vehicle.

80 B.C.   Greek engineers invent the differential gear.

A.D. 1913   Henry Ford sets up the first assembly line to manufacture automobiles.

A.D. 1564   Horse-drawn coach introduced in England from Holland.

A.D. 1769   Joseph Cugnot builds a steam carriage that can carry four people at 2.25 mi/h, the first true automobile.

1500 B.C.   Light carts with two spoked wheels are used in warfare.

2000 B.C.   People in Asia Minor and Persia develop wheels with spokes.

3500 B.C.   Wheeled vehicles used in Sumeria.

THINK CRITICALLY

52. True or false: The absolute value of a number is always the opposite of the number. Explain your answer.

53. True or false: If $a < 0$, $b < 0$, and $a < b$, then $|a| > |b|$. Explain your answer.

MIXED REVIEW

A company has *x* widgets on hand. Write a variable expression for each description.

54. the value of the widgets if one is worth $0.15

55. the number of defective widgets if one third are defective

56. **STANDARDIZED TESTS** A number cube numbered from 1 to 6 was rolled 75 times. It came up 5 nine times and 6 thirteen times. What is the experimental probability that on the next roll a 1, 2, 3, or 4 will come up?

    A. $29.\overline{3}\%$ **B.** $70.\overline{6}\%$ **C.** $83.\overline{3}\%$ **D.** none of these

Compare using < or >.

57. −33 ▨ −12   58. −2 ▨ −4   59. −6 ▨ −10   60. −5 ▨ 3

Career
Meteorologist

Meteorology is the science of weather and climate. In making weather forecasts, meteorologists use measures of temperature, air pressure, humidity, and wind speed and direction.

The barometer was invented in 1643 by Evangelista Torricelli. A tube closed at one end is filled with mercury and placed, open end down, into a container of mercury. An attached ruler measures the height of the mercury in the tube. The space above the mercury in the tube is a vacuum; it has no air. Air outside the tube exerts pressure on the mercury in the container. As the air pressure changes, the height of the mercury in the tube changes.

Decision Making

1. In the barometer shown above, what is the air pressure reading in inches?

2. As the air pressure increases, what will happen to the mercury in the tube? Explain.

3. As the air pressure decreases, how do you think the mercury in the tube will change? Explain your thinking.

4. A high pressure area is defined as an area in which the air pressure is higher than that of its surroundings. The weather map below shows some barometric readings for several locations. Name some locations that appear to be in a high pressure area.

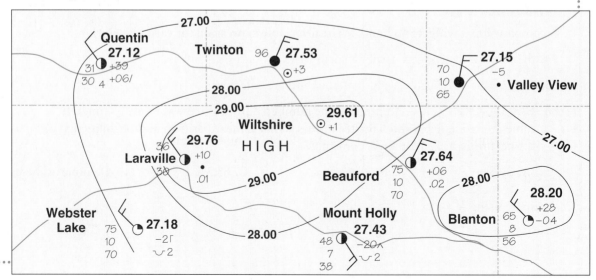

Algebra Workshop
Addition, Subtraction, and Zero Pairs

Think Back

- On a basic mat, drop a handful of unit blocks. Sort out which fall on the positive side of the mat and which fall on the negative side.

  1. Write an integer for the number of unit blocks on the positive side of the mat.

  2. Write an integer for the number of unit blocks on the negative side of the mat.

Explore

- **3.** Suppose that the blocks on the positive side of the mat at the right represent points you win in a game, and that the blocks on the negative side represent points you lose in a game. How would you describe your overall game score?

  4. Suppose on the next round of the game, you gain 1 point and lose 1 point.
    a. How would you show the new points on the mat?
    b. How is your overall score affected?
    c. Write an integer to represent your overall score. How would you change the mat to show only this overall score?

To add two integers using Algeblocks, represent each integer with unit blocks on the same mat, then write an integer for the sum.

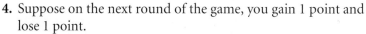

For example, to add $-3 + (-4)$, place 3 unit blocks on the negative side of the mat, and then place 4 more unit blocks on the negative side of the mat. The sum is -7.

  5. Add using Algeblocks. Write an integer for the sum.
    a. $-4 + (-2)$ **b.** $3 + 5$
    c. $-1 + (-3)$ **d.** $-2 + (-3)$

  6. How are the additions you did in Question 5 alike?

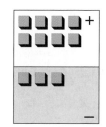

$$-3 + (-4) = -7$$

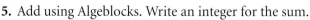

Algebra Workshop

Each number in a pair such as −4 and 4 is called the **opposite** of the other number. To use Algeblocks to find the opposite of an integer, begin by representing the integer on the appropriate side of the mat. Then model the opposite by moving the blocks to the opposite side of the mat. Read the answer from the mat.

The opposite of 5 is −5.

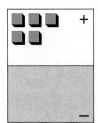 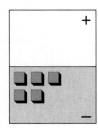

The opposite of −5 is 5.

Use Algeblocks to find each of the following.

7. the opposite of 6   **8.** the opposite of −3   **9.** the opposite of 0

Add using Algeblocks. Write an integer for the result.

10. −4 + 4   **11.** 3 + (−3)

12. 5 + (−5)   **13.** −2 + 2

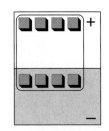

Pairs of integers such as −4 and 4 or 5 and −5 are called **zero pairs**. The mat at the right shows the zero pair 4 and −4.

14. Why do you think pairs of opposites are called zero pairs?

15. Predict the result of adding −17 + 17. How did you determine your prediction?

zero pair

The mat at the right shows the first step involved in adding +5 and −7.

16. Can you make some zero pairs from the blocks shown on the mat? What is the largest zero pair you can make?

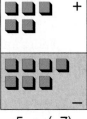

17. Remove the zero pair from the mat. Have you changed the total value shown on the mat? Explain.

5 + (−7)

18. Write an integer for the blocks that remain on the mat.

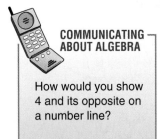

COMMUNICATING
ABOUT ALGEBRA

How would you show
4 and its opposite on
a number line?

Show each addition using Algeblocks. Record your result as an integer.

19. −3 + 4   **20.** 3 + (−4)   **21.** 5 + (−8)

22. −5 + 8   **23.** −6 + 3   **24.** 6 + (−3)

25. −1 + 7   **26.** 1 + (−7)   **27.** 6 + (−5)

Write an addition sentence for each mat below.

28. 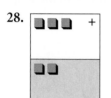   **29.**  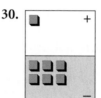   **30.** 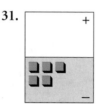   **31.**

To show addition with Algeblocks, you add blocks to the mat. To model subtracting one integer from another using Algeblocks, you take blocks away from the mat. To subtract −5 − (−3), place 5 unit blocks on the negative side of the mat. Then take 3 unit blocks away from the negative side of the mat. The result is −2. So, −5 − (−3) = −2.

Subtract using Algeblocks.

32. 6 − 4   **33.** −7 − (−3)   **34.** 9 − 8

35. 4 − 3   **36.** −8 − (−5)   **37.** −12 − (−8)

Sometimes you cannot take away the number of blocks that would model the subtraction. For example, to subtract −3 − (−4) using Algeblocks, begin by placing 3 unit blocks on the negative side of the mat.

You want to be able to take away 4 unit blocks from the negative side. You can use zero pairs to help you get enough unit blocks on the negative side.

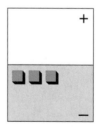

Add the zero pair +1 and −1 to the mat. Did the value of the mat change?

Now take away 4 unit blocks from the negative side of the mat. One unit is left on the positive side of the mat.

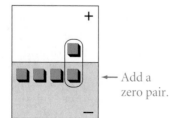
← Add a zero pair.

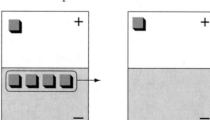

So, −3 − (−4) = 1.

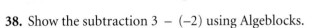

38. Show the subtraction $3 - (-2)$ using Algeblocks.
  a. Show 3 unit blocks on the positive side of the mat. Draw a diagram to show your work.
  b. How can you use zero pairs to help you get enough unit blocks on the negative side of the mat?
  c. Take away 2 unit blocks from the negative side of the mat. Draw a diagram to show your work.
  d. What integer represents the integer shown on the map?
  e. Complete: $3 - (-2) =$ ▢

Show each subtraction using Algeblocks.

39. $-4 - (-6)$ **40.** $-8 - (-9)$ **41.** $-3 - (-6)$

42. $3 - (-3)$ **43.** $3 - (-4)$ **44.** $6 - (-4)$

Make Connections

Use Algeblocks to compute each of the following.

45. **a.** $5 - 3$ **46.** **a.** $-2 - 6$ **47.** **a.** $-4 + 3$
  b. $5 + (-3)$ **b.** $-2 + (-6)$ **b.** $-4 - (-3)$

48. Compare your answers in Exercises 45–47. What do you notice?

49. Complete: Subtracting an integer yields the same result as ___?___.

Summarize

50. WRITING MATHEMATICS In your own words, tell how to add two integers when they have the same sign and when they have different signs. Give examples of each type of addition.

51. MODELING Draw pictures that show how to add −11 and 9 using Algeblocks. Then, draw pictures to show $-7 - (-9)$.

52. WRITING MATHEMATICS In your own words, tell how to subtract two integers when they have the same sign and when they have different signs. Give examples of each type of subtraction.

53. THINKING CRITICALLY Consider this statement: "When you add two integers, you always get an integer greater than the two integers you are adding." Give examples that show if this is true or not.

54. GOING FURTHER How would you use Algeblocks to compute $3 + (-4) + (-2)$? Draw a picture of your work. Compare your work to that of another student.

2.5 Addition and Subtraction of Real Numbers

Explore/Working Together

Play this game with a partner. Take turns spinning this spinner. If you spin a positive integer, you get a score of that many points. A negative integer indicates that you lose that many points. Write down your score. Indicate points lost with negative integers. Continue taking turns for ten spins each. Combine each new score with your previous score. The winner is the person with the highest score after ten rounds.

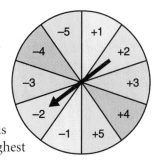

SPOTLIGHT ON LEARNING

WHAT? In this lesson you will learn
• to add and subtract real numbers.
• to identify and apply properties of real numbers.

WHY? Addition and subtraction of real numbers can help you solve problems about weather records, personal finance, and geography.

1. Each player records his or her spins and scores in a table like this.

Turn Number	1	2	3	4	5	6	7	8	9	10
Spinner Says										
Score										

2. What would you record as a new score if the previous score was 12 and you spin –3? Explain or draw a picture to justify your answer.

3. What would you record as a new score if the previous score was –4 and you spin –5? Explain or draw a picture to justify your answer.

Build Understanding

In Lesson 2.4, you used Algeblocks to add real numbers. Adding real numbers can also be shown on a number line. A positive number is shown by moving to the right, and a negative number is shown by moving to the left.

EXAMPLE 1

Add on a number line.

a. $-3 + (-4)$ **b.** $-5 + 2\frac{1}{2}$ **c.** $4 + (-1)$

Solution

For each addition, draw a number line and start at 0.

a. Move left to –3. Then, from –3, move 4 units to the left. The result is –7.

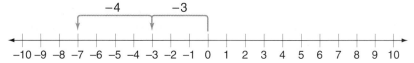

b. Move left to –5. Then, from –5, move $1\frac{1}{2}$ units to the right. The result is $-3\frac{1}{2}$.

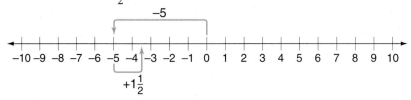

c. Move right to 4. Then, from 4, move 1 unit to the left. The result is 3.

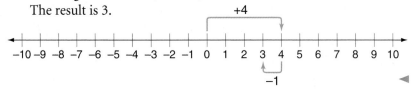

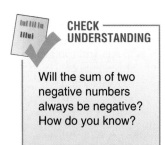

CHECK UNDERSTANDING

Will the sum of two negative numbers always be negative? How do you know?

From your work adding on a number line and adding with Algeblocks, you can see a pattern that suggests rules for adding real numbers.

ADDITION OF REAL NUMBERS

1. **When the signs of the numbers are the same, add the absolute values of the numbers. The sum has the same sign as the sign of the numbers.**

2. **When the signs of the numbers are different, subtract their absolute values. The difference has the sign of the number with the greater absolute value.**

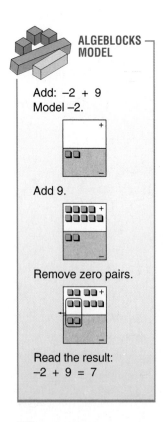

ALGEBLOCKS MODEL

Add: $-2 + 9$
Model -2.

Add 9.

Remove zero pairs.

Read the result:
$-2 + 9 = 7$

EXAMPLE 2

Add using the rules for adding real numbers.

a. $-4 + (-5)$ **b.** $-2 + 9$ **c.** $\frac{7}{8} + \left(-\frac{3}{4}\right)$ **d.** $-9.1 + 7.08$

Solution

a. $-4 + (-5) = $ ▨
The numbers have the same sign.
Add the absolute values.

$$|-4| + |-5| = 4 + 5 = 9$$

Both numbers are negative, so the sum is negative.

$$-4 + (-5) = -9$$

b. $-2 + 9 = $ ▨
The numbers have different signs.
The difference between their absolute values is 7.
The number with the greater absolute value, 9, is positive.

$$-2 + 9 = 7$$

c. $\dfrac{7}{8} + \left(-\dfrac{3}{4}\right) = \blacksquare$

The numbers have different signs.
The difference between their absolute values is $\dfrac{1}{8}$.

The number with the greater absolute value, $\dfrac{7}{8}$, is positive.

$$\dfrac{7}{8} + \left(-\dfrac{3}{4}\right) = \dfrac{1}{8}$$

d. $-9.1 + 7.08 = \blacksquare$

The numbers have different signs.
The difference between their absolute values is 2.02.
The number with the greater absolute value, -9.1, is negative.

$$-9.1 + 7.08 = -2.02 \qquad \blacktriangleleft$$

In Lesson 2.4 you saw that to subtract a real number, you add the opposite of the number. Thus, subtraction of real numbers may be defined in terms of addition.

SUBTRACTION OF REAL NUMBERS

For all real numbers a and b,
$$a - b = a + (-b)$$

EXAMPLE 3

Find.

a. $-13 - (-8)$ **b.** $2 - (-4)$

c. $7 - 10$ **d.** $-15 - 9$

Solution

Rewrite each subtraction as an addition.

a. $-13 - (-8)$
     $= -13 + 8$ The opposite of -8 is 8.
     $= -5$ Subtract the absolute values.

b. $2 - (-4)$
     $= 2 + 4$ The opposite of -4 is 4.
     $= 6$ Add.

c. $7 - 10$
     $= 7 + (-10)$ The opposite of 10 is -10.
     $= -3$ Subtract the absolute values.

d. $-15 - 9$
     $= -15 + (-9)$ The opposite of 9 is -9.
     $= -24$ Add the absolute values. $\blacktriangleleft$

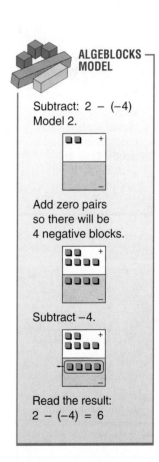

ALGEBLOCKS MODEL

Subtract: $2 - (-4)$
Model 2.

Add zero pairs so there will be 4 negative blocks.

Subtract -4.

Read the result:
$2 - (-4) = 6$

Addition of real numbers has the following properties.

┌─ **COMMUTATIVE PROPERTY OF ADDITION** ─┐
For any real numbers a and b,
$$a + b = b + a$$

$2 + (-3) = -3 + 2$

PROBLEM SOLVING TIP

The associative property of addition aids mental math by letting you group numbers in ways that make them easier to add.

┌─ **ASSOCIATIVE PROPERTY OF ADDITION** ─┐
For any real numbers a and b, and c,
$$(a + b) + c = a + (b + c)$$

$-5 + [(-4) + 7]$
$= [-5 + (-4)] + 7$

┌─ **ADDITIVE IDENTITY PROPERTY** ─┐
For any real number a,
$$a + 0 = a$$

$-8 + 0 = -8$

┌─ **ADDITIVE INVERSE PROPERTY** ─┐
For any real number a,
$$a + (-a) = 0$$

$6 + (-6) = 0$

CHECK UNDERSTANDING

Write a sentence that describes each property in your own words. Give an additional example of each property.

EXAMPLE 4

Evaluate each expression.
   a. $-4 + 3 - (-5)$ **b.** $-7 + 8 + (-2) + 4$

Solution

Use the properties of addition and the definition of subtraction.

a. $-4 + 3 - (-5)$

     $= -4 + 3 + 5$ Rewrite the subtraction as an addition of the opposite.

     $= -4 + (3 + 5)$ Use the associative property to add the two numbers with the same sign first.

     $= -4 + 8$ Subtract absolute values.

     $= 4$

COMMUNICATING ABOUT ALGEBRA

Would you get the same result for Example 4a if you added $-4 + 3$ first? Explain.

b. $-7 + 8 + (-2) + 4$

     $= -7 + (-2) + 8 + 4$ Use the commutative property to group numbers with the same sign.

     $= [-7 + (-2)] + (8 + 4)$ Use the associative property and add.

     $= -9 + 12$ Subtract absolute values.

     $= 3$ ◄

Addition and subtraction of real numbers can be used to solve a variety of problems.

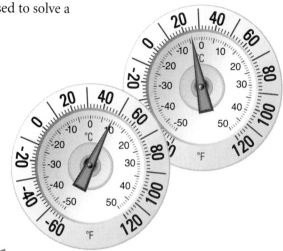

EXAMPLE 5

WEATHER The temperature at 6:15 A.M. was −4°C. By noon, the temperature was 9°C. By how much did the temperature rise in that time?

Solution

To find the difference between 9°C and −4°C, add the opposite.

$$9 - (-4) = 9 + 4 = 13$$

The temperature rose 13°C between 6:15 A.M. and noon.

◄

TRY THESE

Add.

1. $6 + (-7)$

2. $10 + (-3)$

3. $-\dfrac{7}{11} + \left(-\dfrac{2}{11}\right)$

4. $-\dfrac{4}{5} + \left(-\dfrac{9}{10}\right)$

5. $-0.1 + 0.8$

6. $-1.5 + 0.7$

7. $-\dfrac{1}{4} + \dfrac{5}{6}$

8. $-0.6 + 2$

9. **WRITING MATHEMATICS** Write the rules for adding real numbers. Include some examples with answers.

Subtract.

10. $5 - 2$

11. $11 - 4$

12. $-\dfrac{8}{9} - \left(-\dfrac{1}{3}\right)$

13. $-\dfrac{7}{8} - \left(-\dfrac{3}{4}\right)$

14. $-6 - 2$

15. $-8 - 1$

16. $0.6 - (-1.1)$

17. $4 - (-0.5)$

18. **WEATHER RECORDS** The record high temperature in the world is 136°F, which occurred on September 13, 1922, at Azizia, Libya. The record low temperature is −126.9°F, observed on August 24, 1960, at Vostok, Antarctica. What is the difference between these extreme temperatures?

19. **PERSONAL FINANCE** Ms. Twinning's checking account had a balance of $45 last week. Since then, she deposited $75 and wrote checks for $80, $32, and $16. What is her account balance now?

Evaluate.

20. $-4 + 2 - (-7)$

21. $8 - 10 + (-8)$

22. $1.4 - 2 + 0.7$

23. $7 - 1.1 + (-0.3)$

24. $-\dfrac{1}{2} + \left(-\dfrac{2}{3}\right) + \left(-\dfrac{3}{4}\right)$

25. $-\dfrac{1}{4} + \dfrac{2}{5} - \left(-\dfrac{7}{10}\right)$

PRACTICE

1. **WRITING MATHEMATICS** Write a paragraph describing how you would explain to a friend why subtracting a number is the same as adding its opposite.

Add.

2. $12 + (-6)$

3. $14 + (-9)$

4. $-6 + (-8)$

5. $-9 + (-8)$

6. $-1.5 + 0.2$

7. $-1 + 0.4$

8. $-1 + 3.2$

9. $-0.5 + 2$

10. $-2.1 + (-3.9)$

11. $-4 + (-2.3)$

12. $-3 + 1.7$

13. $-31 + 19$

14. $\frac{5}{6} + \left(-\frac{2}{3}\right)$

15. $\frac{2}{5} + \left(-\frac{7}{10}\right)$

16. $3\frac{1}{6} + \left(-5\frac{1}{2}\right)$

17. $5 + \left(-2\frac{1}{3}\right)$

18. $-\frac{3}{7} + \left(-\frac{5}{7}\right)$

19. $-\frac{2}{3} + \left(-\frac{1}{2}\right)$

20. $-\frac{1}{4} + \frac{7}{10}$

21. $-\frac{2}{9} + \frac{2}{3}$

22. **GEOGRAPHY** The highest point on Earth's surface is Mount Everest, which is 29,028 ft above sea level. The lowest point above water is at the shores of the Dead Sea, which is 1299 ft below sea level. What is the difference in feet between the two points?

Subtract.

23. $-5 - (-1)$

24. $-9 - (-3)$

25. $-3 - 8$

26. $7 - (-3)$

27. $12 - 18$

28. $17 - (-9)$

29. $-8 - 1.3$

30. $-5.1 - (-9)$

31. $-0.6 - (-0.7)$

32. $6.2 - (-4.5)$

33. $0.9 - (-0.03)$

34. $-0.4 - 0.7$

35. $\frac{5}{6} - \left(-\frac{3}{8}\right)$

36. $2 - \left(-1\frac{1}{2}\right)$

37. $-\frac{1}{2} - \frac{3}{4}$

38. $\frac{1}{2} - \frac{7}{8}$

39. $\frac{4}{5} - \frac{9}{10}$

40. $-\frac{3}{5} - \left(-\frac{1}{5}\right)$

41. **HISTORY** The earliest known sundial dates from about 800 B.C. in Egypt. The most accurate clock in the world is the atomic clock, built in Washington, D.C., in 1948. What is the difference in years between these two dates?

Evaluate.

42. $-3 + 5 - 8$

43. $-1 + 4 - 10$

44. $36 - 40 - (-2)$

45. $51 - 12 - (-5)$

46. $-\frac{8}{9} + \frac{4}{9} - \frac{1}{3} - \left(-\frac{2}{3}\right)$

47. $\frac{7}{10} + \left(-\frac{3}{5}\right) - \frac{1}{10} - \left(-\frac{2}{5}\right)$

48. $\frac{1}{4} + \frac{5}{6} - \frac{1}{3} - \frac{3}{4}$

49. $-1.2 + 0.83 - 0.3$

50. $-42.9 + 2.3 - 11.05$

51. **SPORTS** During the last possession of the football in a football game, the Panthers gained 36 yards, lost 21 yards, gained 3 yards, gained 5 yards, and lost 15 yards. If they began the possession at the 20-yard line, where was the ball at the end of these plays?

Evaluate.

52. $16 \cdot 2 - 5^2 + (-13)$ 53. $12 \cdot 3 - 7^2 - (-3)$ 54. $-14 + 2^3 - 11$

55. $-20 + 3^3 - 5$ 56. $120 \div 40 - 6^3 + 1$ 57. $-12 + 36 \div 6 - 8$

58. $|4 - 9|$ 59. $|6| - |13|$ 60. $|-(3 - 7)|$ 61. $-|5 - 8|$

62. **STOCK MARKET** A stock that closed yesterday at $58\frac{1}{4}$ closed today at $53\frac{1}{2}$. Write a number for the change in value from yesterday to today.

63. **STOCK MARKET** A stock closed at $12\frac{3}{8}$ today, a change of $-1\frac{1}{2}$ from yesterday. What was yesterday's closing price?

WEATHER DATA Answer the following questions about the table below.

64. Which state had the greatest difference in record extreme temperatures?

65. What is the difference between the lowest recorded temperatures in Alabama and in Arizona?

State	Record High	Record Low
Alabama	112°F	−24°F
Alaska	100°F	−80°F
Arizona	127°F	−40°F
Arkansas	120°F	−29°F

THINK CRITICALLY

66. What is the opposite of $a + b$? Justify your answer.

67. If $a \neq b$, what is the opposite of $a - b$? Justify your answer.

68. The equation $(a - b) - c = (a - c) - b$ is (*sometimes, always, never*) true. Justify your answer.

69. Give a positive number and a negative number such that the sum of the two numbers is positive. Give another positive number and another negative number such that the sum of the numbers is negative.

PROJECT *Connection*   The average (mean) is often used to analyze temperatures collected over a period of time. Use the daily high temperatures (degrees F) collected over the eight-day period for two of your cities (City A and City B).

1. Determine the average high temperature for City A and City B for the eight days. (Do not round.)

2. Subtract the City A average from each of the daily high temperatures in City A. Record these positive and negative differences.

3. Find the sum of the differences recorded for City A.

4. Repeat Steps 2 and 3 for City B.

5. Compare the two sums of the differences. What do you notice?

6. Explain how the sums of the differences relate to the average and why you would expect the outcome to always be the same.

Algebra Workshop
Multiplication and Division

Think Back

● When you see a pattern in a set of numbers, you can use the pattern to predict other numbers in the set. Write the next three numbers in each set below.

1. 2, 7, 12, 17, . . .   **2.** 4, 8, 12, 16, . . .   **3.** 5, 3, 1, –1, . . .

Explore

● Copy the table at the right. It has four sections, or quadrants. They are numbered counterclockwise, starting from the top right quadrant.

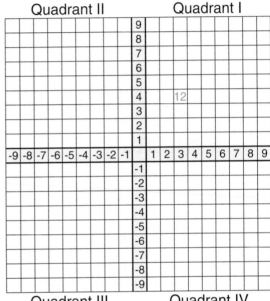

Quadrant II   Quadrant I

Quadrant III   Quadrant IV

4. Fill in Quadrant I by multiplying each of the horizontal numbers, 1 through 9, by the vertical numbers, 1 through 9. For example, write the product of 3(4) in the box that corresponds to 3 on the horizontal scale and 4 on the vertical scale, as shown in the table.

5. Look at the first row in the completed part of your table. What pattern do you see in the numbers as you read to the left?

6. If you continue this pattern, what number would you put in the center of the first row?

7. For each row in Quadrant I, find a pattern as you read to the left. Then use that pattern to fill the boxes in Quadrant II (top left). When you finish, Quadrants I and II will be filled in.

Use your table to find each product.

8. 3(5)   **9.** –3(5)   **10.** –8(6)   **11.** –4(9)

12. What is true about the sign of all the products in Quadrant I?

13. What is true about the sign of all the products in Quadrant II?

14. Look at the third column in Quadrant I. What pattern do you see as you read down the column?

15. Continue the pattern from Quadrant I into Quadrant IV (bottom right). When you finish, three quadrants will be filled in.

Use your table to find each product.

16. 4 (–6)   **17.** 5(–2)   **18.** 8(–9)   **19.** 7(–3)

20. What is true about the sign of all the products in Quadrant IV?

21. Look at the column in Quadrant II that contains (–5)(9). What pattern do you notice as you read down the column?

22. Continue this pattern into Quadrant III (bottom left). Do this for each column in Quadrant II and Quadrant III.

23. Look at the row in Quadrant IV that contains 9(–2). What pattern do you observe as you read along the row to the left?

24. Is the pattern you observed in Question 23 continued into Quadrant III? What does this tell you?

COMMUNICATING ABOUT ALGEBRA

How are the longitude and latitude lines on a globe like the lines on a coordinate plane? How are they different?

Use your table to find each product.

25. –4(–5)   **26.** –2(–8)   **27.** –8(–7)   **28.** –6(–3)

29. What is true about the sign of all the products in Quadrant III?

30. What is the sign of the product of two positive numbers? of two negative numbers? of one positive and one negative number?

31. Is multiplication of integers commutative? How do you know?

32. Is multiplication of integers associative? Give an example.

33. Describe how to find a quotient like –36 ÷ 9 by using the multiplication table you completed.

Use the table to find each quotient.

34. –12 ÷ 4   **35.** –45 ÷ 5   **36.** 16 ÷ (–4)   **37.** 8 ÷ (–4)

38. **WRITING MATHEMATICS** In your own words, write the rules for finding the sign of a quotient of integers.

Algebra Workshop

Make Connections

● Work with a partner. Use Algeblocks and the Quadrant Mat.

39. What do you notice about the signs on the Quadrant Mat?

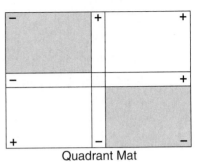

Quadrant Mat

To show multiplication with Algeblocks on the Quadrant Mat, use unit blocks to make a rectangle whose sides are the factors. The product is the number of blocks with the sign of the quadrant.

For example, to multiply 2 times 4, place 2 unit blocks in the positive part of the horizontal axis. Place 4 unit blocks in the positive part of the vertical axis. Make a rectangle.

Count the unit blocks in the rectangle. The blocks are in Quadrant I, which is positive. Since there are 8 blocks in a positive quadrant, $2(4) = 8$.

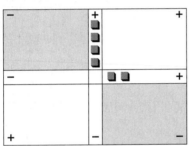
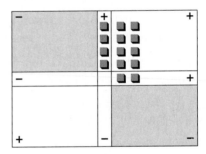

40. What multiplication is modeled on the mat at the right?

41. Write the product of this multiplication as an integer.

42. Model the multiplication $3(-2)$ on a Quadrant Mat. Draw a picture to show your work.

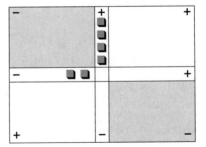

43. How does your work on the mat show you the product of $3(-2)$?

Use Algeblocks to model each of the following. Draw a picture to show your work.

44. $4(-5)$ **45.** $-3(-4)$

46. $-5(2)$ **47.** $-2(-4)$

48. Describe how you can use Algeblocks to model division. Draw a picture using the example $-12 \div 3 = $ ▢.

Use Algeblocks to model each of the following. Draw a picture to show your work.

49. $-14 \div (-7)$ **50.** $-18 \div 9$ **51.** $-35 \div 7$ **52.** $32 \div (-8)$

The following problem shows another way to think about the rules for multiplying integers.

The population of Signtown is decreasing at a rate of 50 citizens per year. If this rate continues, how many fewer citizens will there be in Signtown 3 years from now?

$$3 \cdot (-50) = -150$$

years
from now (+)

decreasing
rate (−)

decrease
in population (−)

There will be 150 fewer citizens 3 years from now.

53. What rule for multiplying integers does the problem above model?

Next, consider this problem. How many more citizens lived in Signtown 4 years ago?

54. Copy and complete this model.

$$\boxed{} \times \boxed{} = \boxed{}$$

years
ago (−)

decreasing
rate (−)

increase
in population (+)

55. What rule for multiplying integers does your model in Question 54 show?

Summarize

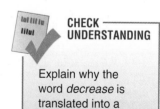

CHECK UNDERSTANDING

Explain why the word *decrease* is translated into a negative value.

- **56.** WRITING MATHEMATICS Write a paragraph describing how to find the product of two integers using patterns in a multiplication table.

- **57.** WRITING MATHEMATICS Write a paragraph describing how you find the product of two integers using Algeblocks.

- **58.** MODELING Model the division $-24 \div 3$. Draw a picture, describe how you set up the model, and read the quotient from it.

- **59.** THINKING CRITICALLY Is division of integers associative? Explain.

- **60.** GOING FURTHER Evaluate $(-1)^2$, $(-1)^3$, $(-1)^4$, and $(-1)^5$. Can you predict the value of $(-1)^{201}$? Explain.

2.7 Multiplication and Division of Real Numbers

Explore

Use a calculator.

1. **a.** Start with zero and subtract 7 six times. How could you write this computation as a multiplication problem?

   b. What is the result?

2. **a.** Start with zero. How many times must 6.5 be added on the calculator to get 52?

   b. Based on this computation, what division problem and answer could you write?

3. **a.** Start with zero. How many times must –6.5 be added to get –52?

   b. What division problem and answer could you write?

Build Understanding

You can use patterns to investigate the product of two negative numbers. Remember that in the multiplication $3(4) = 12$; 3 and 4 are **factors** and 12 is their **product**.

CHECK UNDERSTANDING

Find the product, $(-4)(2)$.

EXAMPLE 1

Describe and continue the pattern you observe in these products through $-5(-3)$.

$$-5(3) = -15$$
$$-5(2) = -10$$
$$-5(1) = -5$$
$$-5(0) = 0$$

Solution

As the second factor decreases by 1, the product increases by 5. Therefore, the pattern continues in this way.

$$-5(-1) = 5$$
$$-5(-2) = 10$$
$$-5(-3) = 15$$

This example suggests the following rule.

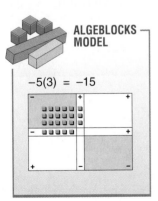

ALGEBLOCKS MODEL

$-5(3) = -15$

MULTIPLICATION OF REAL NUMBERS

1. **When two factors have different signs, the product is negative.**
2. **When two factors have the same sign, the product is positive.**

EXAMPLE 2

Multiply using the rules for multiplying real numbers.

a. $-3(21)$ **b.** $-\dfrac{1}{2}\left(-\dfrac{2}{5}\right)$ **c.** $0.6(-1.3)$

Solution

a. $-3(21) = $

The factors have different signs. When two factors have different signs, the product is negative.

$$-3(21) = -63.$$

b. $-\dfrac{1}{2}\left(-\dfrac{2}{5}\right) = $

Both factors are negative. When two factors have the same sign, the product is positive.

$$-\frac{1}{2}\left(-\frac{2}{5}\right) = \frac{-1(-2)}{2(5)} = \frac{2}{10} = \frac{1}{5}$$

c. $0.6(-1.3) = $

The factors have different signs. When two factors have different signs, the product is negative.

$$0.6(-1.3) = -0.78.$$ ◀

COMMUNICATING ABOUT ALGEBRA

Use a pattern, starting from 3(4), to find 3(–2). Explain your answer.

Multiplication of real numbers is associative and commutative. This means that you can group factors in any order, and that the order of factors does not affect the product.

> **ASSOCIATIVE PROPERTY OF MULTIPLICATION**
>
> For any real numbers a, b, and c,
> $(a \cdot b) \cdot c = a \cdot (b \cdot c)$

$(8 \cdot 2) \cdot 3$
$= 8 \cdot (2 \cdot 3)$

> **COMMUTATIVE PROPERTY OF MULTIPLICATION**
>
> For any real numbers a and b,
> $a \cdot b = b \cdot a$

$9 \cdot 3 = 3 \cdot 9$

The numbers 0 and 1 have special properties for multiplication. The product of any number and 0 is 0. The number 1 is called the **identity element for multiplication**, because the product of any number and 1 is the number itself. The product of –1 and any number except 0 is the opposite of the number.

> **MULTIPLICATIVE PROPERTY OF ZERO**
>
> For any real number a,
> $a \cdot 0 = 0 \cdot a = 0$

$-7 \cdot 0$
$= 0 \cdot -7$
$= 0$

CHECK UNDERSTANDING

Show how the definition of division can be used to find $24 \div 10$.

MULTIPLICATIVE IDENTITY PROPERTY

For any real number a,
$$a \cdot 1 = 1 \cdot a = a$$

$$\begin{aligned} 8 &\cdot 1 \\ &= 1 \cdot 8 \\ &= 8 \end{aligned}$$

PROPERTY OF −1 FOR MULTIPLICATION

For any real number a, where $a \neq 0$,
$$a \cdot -1 = -1 \cdot a = -a$$

$$-1(4) = -4$$

Every real number b, except 0, has a **reciprocal** or **multiplicative inverse**. The product of a number and its reciprocal is 1.

MULTIPLICATIVE INVERSE PROPERTY

For any nonzero real number b, there exists a real number $\dfrac{1}{b}$ such that $b \cdot \dfrac{1}{b} = 1$.

$$9 \cdot \frac{1}{9} = 1$$

EXAMPLE 3

Find the reciprocal of each of the following.

a. −9 **b.** $\dfrac{2}{3}$ **c.** $-\dfrac{4}{5}$

Solution

	Number	Reciprocal	Check
a.	-9	$-\dfrac{1}{9}$	$(-9)\left(-\dfrac{1}{9}\right) = 1$
b.	$\dfrac{2}{3}$	$\dfrac{3}{2}$	$\left(\dfrac{2}{2}\right)\left(\dfrac{3}{2}\right) = 1$
c.	$-\dfrac{4}{5}$	$-\dfrac{5}{4}$	$\left(-\dfrac{4}{5}\right)\left(-\dfrac{5}{4}\right) = 1$

◀

Multiplication and division are **inverse operations**. That is, if $a \cdot b = c$, then $c \div b = a$. Thus, division can be defined in terms of multiplication.

DEFINITION OF DIVISION

For any real numbers a and b, $b \neq 0$,
$$a \div b = a \cdot \frac{1}{b}$$

You can use multiplication to investigate division of real numbers.

Division	Related Multiplication	Result
$-32 \div 4$	$-32\left(\dfrac{1}{4}\right)$	-8
$-45 \div (-5)$	$-45\left(-\dfrac{1}{5}\right)$	9

The rules for division are similar to the rules for multiplication.

DIVISION OF REAL NUMBERS

1. When two numbers have different signs, their quotient is negative.
2. When two numbers have the same sign, their quotient is positive.

EXAMPLE 4

Divide.

  a. $5.1 \div -0.03$ **b.** $\left(-\dfrac{1}{5}\right) \div \left(-\dfrac{2}{15}\right)$ **c.** $1\dfrac{1}{3} \div -\dfrac{2}{7}$

Solution

  a. $5.1 \div -0.03 = -170$

  b. $-\dfrac{1}{5} \div -\dfrac{2}{15} = -\dfrac{1}{5} \cdot -\dfrac{15}{2} = \dfrac{3}{2}$ or $1\dfrac{1}{2}$

  c. $1\dfrac{1}{3} \div -\dfrac{2}{7} = \dfrac{4}{3} \div -\dfrac{2}{7} = \dfrac{4}{3} \cdot -\dfrac{7}{2} = -\dfrac{14}{3}$ or $-4\dfrac{2}{3}$ ◄

You can use addition and division to find the mean of a set of numbers.

EXAMPLE 5

WEATHER The high temperatures for four winter days were 2°C, −1°C, −4°C, and −2°C. Find the mean high temperature for those days.

Solution

Add the temperatures, then divide by 4.

$$\frac{2 + (-1) + (-4) + (-2)}{4} = \frac{-5}{4} = -1.25°C$$

The mean high temperature was −1.25°C.   ◄

TRY THESE

1. WRITING MATHEMATICS Describe the rules for multiplying integers.

Multiply.

  2. $-6(3)$ **3.** $8(-0.4)$ **4.** $\left(-\dfrac{1}{2}\right)\left(-\dfrac{5}{9}\right)$ **5.** $\left(-6\dfrac{1}{3}\right)(-4)$

Find the reciprocal of each.

  6. -11 **7.** $-\dfrac{3}{4}$ **8.** 0.8 **9.** $-3\dfrac{2}{5}$

Divide.

  10. $72 \div (-8)$ **11.** $-9.3 \div (-6)$ **12.** $-\dfrac{2}{3} \div \dfrac{3}{4}$ **13.** $-4 \div \left(-1\dfrac{1}{7}\right)$

14. BUSINESS The monthly profits of a shoe store for the last four months were $2000, $1500, −$3600, and $2200. What is the mean profit for those four months?

15. MODELING What multiplication is represented by the picture of Algeblocks shown at the right? Write the product.

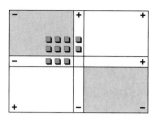

PRACTICE

Multiply.

1. 9(−11)

2. −3(14)

3. −5(−21)

4. −25(−4)

5. 0.6(−3.4)

6. 1.5(−9)

7. (−2.7)(−0.001)

8. −3.2(−5)

9. $-\dfrac{3}{4} \cdot \dfrac{7}{12}$

10. $-\dfrac{1}{9} \cdot -\dfrac{3}{8}$

11. $1\dfrac{1}{2} \cdot -2\dfrac{3}{4}$

12. $-15 \cdot -3\dfrac{1}{5}$

Find the reciprocal.

13. −21

14. $\dfrac{8}{9}$

15. $-3\dfrac{1}{6}$

16. −0.02

17. Find four pairs of numbers whose product is
    a. negative   **b.** positive   **c.** one

Divide.

18. 144 ÷ (−12)

19. −150 ÷ 6

20. −201 ÷ (−3)

21. −289 ÷ (−17)

22. −2.4 ÷ 0.5

23. 2.8 ÷ (−70)

24. $-\dfrac{8}{1.6}$

25. $\dfrac{-200}{-2.5}$

26. $\dfrac{5}{8} \div -\dfrac{3}{4}$

27. $-\dfrac{6}{7} \div \dfrac{2}{5}$

28. $2\dfrac{1}{2} \div -\dfrac{1}{10}$

29. $-6\dfrac{1}{4} \div (-5)$

30. WEATHER For four consecutive years, the average temperature in February in Anchorage, Alaska, was −5.7°C, −3.7°C, −8°C, and −15.7°C. Find the mean temperature for those four years.

31. BANKING Mrs. Williams deposits $1125 every month in her checking account. At the same time, $100 is automatically withdrawn and placed in a savings account. The bank automatically deducts $5 per month for checking fees.

  a. Write an algebraic expression for the net amount for these three transactions after m months.

  b. Find the value of the expression you wrote for $m = 4$.

32. WRITING MATHEMATICS Write and solve one problem that involves multiplication of negative real numbers and another that involves division of negative real numbers.

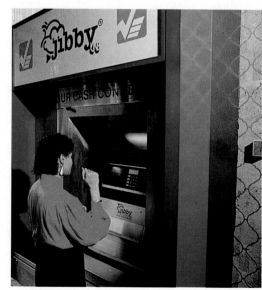

EXTEND

Simplify.

33. $4 - 3(-5)$ **34.** $10 + (-7)(-3)$ **35.** $-8 + (-2)(-9)$

36. $(-7)^2 + (-9)(-2)$ **37.** $(-2)^3 - (3)(-7)$ **38.** $3(-4) + 5(-3)$

39. $-7(-4) + 9(-6)$ **40.** $-6(2^2) - 11(-3)$ **41.** $-4(3^2) - 2(-8)$

42. POPULATION The population of Detroit, Michigan is estimated to be 2,735,000 by the year 2000. At an annual growth rate of -0.3%, what is the estimated population for 2001?

43. PATTERNS Find the next number in this pattern. Describe the general rule for the next number. $2, -5, 16, -47, \ldots$

THINK CRITICALLY

44. If a and b are negative and c is positive, determine whether each of the following is positive or negative.
    a. $(b - c)(a - c)$ **b.** $a(c - b)$

45. Is the quotient of two integers always an integer? Justify your answer.

Use absolute value notation to complete Exercises 46–49.

46. If $a > 0$ and $b < 0$, then $ab = $ ▢ . **47.** If $a < 0$ and $b < 0$, then $ab = $ ▢ .

48. If $a > 0$ and $b < 0$, then $a \div b = $ ▢ . **49.** If $a < 0$ and $b < 0$, then $a \div b = $ ▢ .

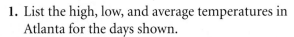 A **high/low chart** can be used to display a span of temperatures for each day during a time period. A high/low chart for a six-day period in Atlanta is shown at the right. Each day a vertical line segment is drawn connecting the high and low temperatures. That segment is intersected by a horizontal segment showing the average temperature for that date.

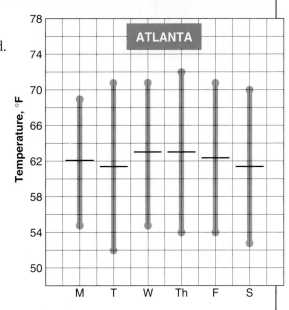

1. List the high, low, and average temperatures in Atlanta for the days shown.

2. The average Atlanta temperature for Sunday of the week shown was 2.5°F lower than the highest average temperature for the previous six days.
    a. What was Sunday's average temperature?
    b. Study the trends in the chart. Using your result for Sunday's average, what might have been the day's high and low? Explain.

3. Construct high/low charts for at least two cities for which you have data. Choose cities with different weather patterns.

Water in a car radiator is used to cool the engine. The freezing point of water is 32°F. The air temperature can dip below this point in many parts of the world during the winter months.

Service station attendants know that the addition of certain chemicals to water decreases the freezing point. These types of mixtures are called antifreeze and are added to the water in a car radiator. The antifreeze keeps the solution in the radiator from freezing even when the air temperature is below the freezing point of water.

The graph at the right can be used to compare three commonly used antifreeze solutions. It shows the relationship between the percent of total volume of antifreeze mixture in the radiator (the horizontal axis) and the freezing point of that mixture (the vertical axis).

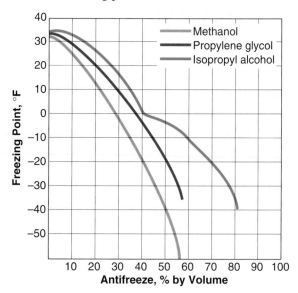

Decision Making

1. What is the freezing point of a mixture that is 60% isopropyl alcohol?

2. What percent mixture of propylene glycol is needed to protect against a freezing point of −20°F?

3. Which antifreeze mixture produces the greatest decrease in freezing temperature for the least percent of volume?

4. What factors would affect the type of antifreeze you would buy?

5. How would a service station attendant decide what volume of antifreeze mixture would be enough to protect a customer's car?

6. Cars need antifreeze in the summer. Do research to find out why this is true.

2.8 The Distributive Property

Explore

- The midpoint of a line segment is the point exactly halfway between the endpoints.

  1. Draw a number line from −10 to 10. Use it to find the midpoint of the segment with each pair of endpoints.
     a. 1 and 5   **b.** 2 and 7   **c.** −3 and 5   **d.** −2 and 9

  2. Compare the midpoint of each of the segments in Question 1 with the mean of the coordinates of their endpoints.

  3. Write an algebraic expression for the midpoint m of the line segment with endpoints a and b.

SPOTLIGHT ON LEARNING

WHAT? In this lesson you will learn
- to use order of operations for expressions having other grouping symbols.
- to use the distributive property.

WHY? The order of operations and distributive property can help you solve problems about meteorology, geometry, and accounting.

Build Understanding

- In Lesson 2.2, you learned the order of operations. You can use the order of operations to simplify expressions. You **simplify** an expression by changing it to an equivalent expression that has fewer terms.

  > **ORDER OF OPERATIONS**
  > 1. **Perform operations within grouping symbols first.**
  > 2. **Perform all calculations involving exponents.**
  > 3. **Multiply or divide in order from left to right.**
  > 4. **Add or subtract in order from left to right.**

The fraction bar is also a grouping symbol. It shows that operations in the numerator or denominator should be performed first. When there is more than one set of grouping symbols in an expression, perform operations inside the innermost grouping symbols first.

EXAMPLE 1

Simplify each expression.

a. $\dfrac{-9(6 + 14)}{5}$

b. $11 - 3[10 - (5 - 2)^2]$

Solution

a.
$$\dfrac{-9(6 + 14)}{5}$$
$$= \dfrac{-9(20)}{5} = \dfrac{-180}{5}$$
Simplify the numerator first.
$$= -36$$

b. $11 - 3[10 - (5 - 2)^2]$ Simplify $(5 - 2)$ first.
$= 11 - 3[10 - (3)^2]$
$= 11 - 3[10 - 9]$
$= 11 - 3[1] = 11 - 3$
$= 8$ ◀

One of the most important properties of real numbers combines the operations of addition and multiplication. This property is called the distributive property.

The rectangular plot shown at the right is subdivided into two parts. The total area of the plot is equal to the sum of the areas of both parts.

$$A = 10(5) + 10(7)$$
$$= 50 + 70$$
$$= 120$$

The area can also be expressed as the product of the width, 10, and the entire length, $5 + 7$.

$$A = 10(5 + 7)$$
$$= 10(12)$$
$$= 120$$

No matter which way you compute the area, the result is the same.

$$10(5) + (10)7 = 10(5 + 7)$$

This illustrates the distributive property.

DISTRIBUTIVE PROPERTY OF MULTIPLICATION OVER ADDITION

For all real numbers a, b, and c,
$$a(b + c) = ab + ac \quad \text{and} \quad (b + c)a = ba + ca$$

EXAMPLE 2

Use the distributive property to simplify each expression.

a. $3(x + 7)$ **b.** $(y + 5)2$ **c.** $-4(x + 6)$ **d.** $(c - 3)(-2)$

Solution

a. $3(x + 7) = 3(x) + 3(7) = 3x + 21$

b. $(y + 5)2 = y(2) + 5(2) = 2y + 10$

c. $-4(x + 6) = -4(x) + (-4)(6) = -4x - 24$

d. $(c - 3)(-2) = c(-2) - 3(-2) = -2c + 6$ ◀

COMMUNICATING ABOUT ALGEBRA

Is it always true that $a(b - c) = ab - ac$? Explain.

Expressions can be simplified by combining like terms. **Like terms** are terms that have the same variable base and exponent. They may have different coefficients.

EXAMPLE 3

Which are pairs of like terms?

a. $3x$ and $-15x$ **b.** $-5ab$ and $9ab$ **c.** $7x$ and $8x^2$

d. $8x^2$ and x^2 **e.** $3x$ and $3y$ **f.** $5xy$ and $3xz$

Solution

Like terms have the same variable base and the same exponent. These pairs are like terms.

a. $3x$ and $-15x$ **b.** $-5ab$ and $9ab$ **d.** $8x^2$ and x^2

These pairs are *not* like terms.

c. $7x$ and $8x^2$ Variable bases are the same but exponents are different.

e. $3x$ and $3y$ Variable bases are not the same.

f. $5xy$ and $3xz$ Variable base x is in both, but y is in $5xy$ and z is in $3xz$. ◀

Like terms can be combined by applying the distributive property in reverse order to combine their coefficients.

EXAMPLE 4

Simplify each expression.

a. $4x - 2y - 9y + 6x$ **b.** $x^2 + 3(2x^2 + 7x)$

Solution

The commutative and associative properties can be used to rearrange the terms so that like terms are together. Then use the distributive property to collect like terms. Then combine the coefficients of the variables.

a. $4x - 2y - 9y + 6x$
$= 4x + 6x - 2y - 9y$
$= (4 + 6)x + (-2 - 9)y$ Collect like terms.
$= 10x + 11y$ Combine coefficients.

b. By the distributive property, $3(2x^2 + 7x) = 3 \cdot 2x^2 + 3 \cdot 7x$. Therefore,

     $x^2 + 3(2x^2 + 7x)$
$= x^2 + 6x^2 + 21x$
$= 1 \cdot x^2 + 6x^2 + 21x$ $x^2 = 1 \cdot x^2$
$= (1 + 6)x^2 + 21x$ Collect like terms.
$= 7x^2 + 21x$ Combine coefficients. ◀

When simplifying algebraic expressions, it is often helpful to rewrite subtraction as the addition of the opposite.

EXAMPLE 5

Simplify $5x - 4(3x - 2)$.

Solution

$$5x - 4(3x - 2)$$
$$= 5x + (-4)[3x + (-2)] \quad \text{Rewrite the subtractions.}$$
$$= 5x + (-12x) + 8 \quad \text{Use the distributive property.}$$
$$= x[5 + (-12)] + 8 \quad \text{Collect like terms.}$$
$$= -7x + 8 \quad \text{Do the operation inside brackets.}$$

The expression cannot be simplified further since $-7x$ and 8 are not like terms. ◀

TRY THESE

Simplify each expression.

1. $5 + 4(9 - 11)$

2. $12 \div (-4) + 2(3 + 4)$

3. $5 \cdot 4^2 - (3 - 6)^2$

4. $2.3 + 5(1.4 + 8.3)$

5. $0.2(1 - 0.7) + (-0.1)^3$

6. $\frac{2}{3}\left(\frac{3(4 + 8)}{5}\right)$

7. $3 + [1 - 2(8 + 7)]$

8. $4 - [(5 - 2) - 12]$

9. $100 \div [-4 + 3(2 - 9]$

Use the distributive property to simplify.

10. $9(x + 2)$

11. $(b - 3)(4)$

12. $-6(y + 2)$

13. $(m - 8)(-2)$

Determine which pairs are like terms. Write *yes* or *no*.

14. $4x, -4x$

15. $15y, 15z$

16. $3ab, ab$

17. $7y, 8y^3$

Simplify each expression.

18. $3x + 2(5x - 1)$

19. $9 - 4(x - 8)$

20. $7x - 2(8 + 3x)$

21. MODELING Write an algebraic expression for the area of each figure at the right. Tell what property is shown.

Figure 1

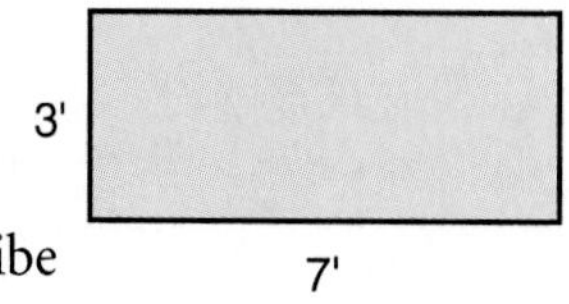

3'

7'

Figure 2

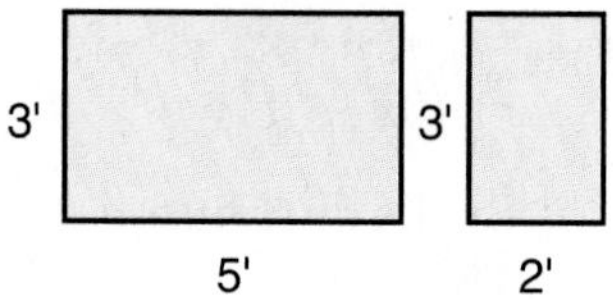

3'

5'

3'

2'

22. WRITING MATHEMATICS Describe what it means to combine like terms. Give at least two examples.

23. METEOROLOGY The formula that relates Celsius to Fahrenheit temperature is $C = \frac{5}{9}(F - 32)$, where C is the temperature in degrees Celsius and F is in degrees Fahrenheit. Find the number of degrees Celsius equivalent to a temperature of 23°F.

Simplify each expression.

1. $3 - 6(7 - 11)$

2. $5 + 2(9 - 16)$

3. $(9 - 4)^2 - 7(8)$

4. $(2 + 1)^3 - 5(-4)$

5. $4.3 + (9 - 3.02)$

6. $7.7 - (5.12 + 2.4)$

7. $(0.3 - 0.5)^2 - 2(0.9)$

8. $(3.9 - 2.5)^2 + 5(0.01)$

9. $0.1(4.5 - 1.2) - 0.01(6 - 2.2)$

10. WRITING MATHEMATICS Some calculators follow the order of operations. Some do not. Describe how you would determine whether or not a calculator followed the order of operations. Tell how you would enter $9 - 3(5 + 2)$ in a calculator that did *not* follow the order of operations.

Simplify each expression.

11. $\dfrac{3}{10}\left(\dfrac{1 + 5}{2 \cdot 5}\right) + \dfrac{1}{5}$

12. $\dfrac{4}{5}\left(\dfrac{3 - 5}{2 \cdot 4}\right) + 1\dfrac{1}{2}$

13. $15 - [10 - (6 - 11)]$

14. $-9 + [16 - (3 - 5)]$

15. $7 - [4(1 - 8) + 2]$

16. $-8 - [3(2 - 7) + 6]$

17. $14 - [6 - (2 + 1)]^3$

18. $1 + [3 - 2(1 - 3)]^2$

19. $5x + 4 - 3x - 11$

20. $-6a + 9b - 3a - 2b$

21. $15 - 3(8c + 2)$

22. $-5(3x + 2) - 19$

23. $7a - 3b + 2c + 10a + b$

24. $x - y + 3z + 5y - 4x$

25. $3(r + 2) - (r + s)$

26. GEOMETRY Write an expression for the area of the figure at the right.

27. GEOMETRY Recall that the perimeter of a rectangle is represented by the formula $P = 2l + 2w$. Use the distributive property to write another representation of the formula.

28. ACCOUNTING Ms. Franco set up a spreadsheet to compute the cost per item of producing different quantities of mugs. If A represents the unit cost, B represents the number of mugs produced, and C represents the fixed costs, write a formula for column D that will compute the total production cost per mug.

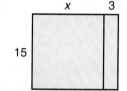

EXTEND

For Exercises 29–32, evaluate each expression for $x = -1$.

29. $-5x^2$

30. $(-5x)^2$

31. $2x^2 - 3(x + 2)$

32. $4x^3 - 5(2x + 1)$

For Exercises 33–36, evaluate for $a = 3$ and $b = -4$.

33. $-6ab$

34. $4a^2b$

35. $5a + 2b - 3a - 8b$ **36.** $9a^2 - 3ab + 2b^2$

37. If $x + y = x$, what is the numerical value of $y(3x + 5)$?

38. If $xy = y$, what is the numerical value of x^{10}?

Find a value for n that makes each of the expressions in Exercises 39–40 true.

39. $8(3 + 8) = 8 \cdot 3 + 8n$

40. $(-6 + 13)4 = -6n + 13 \cdot 4$

Place parentheses in each expression to make its value 20.

41. $7 + 1 \cdot 2 + 4$

42. $15 - 20 \div -3 - 1$

43. BIOPHYSICS The way a person feels outside on a cold day is related not only to the temperature but also to the speed of the wind. The formula for the heat loss K in kilocalories, by 1 m² of skin in 1 h, is $(10.45 + 10\sqrt{v} - v)(33 - t)$ where v is the wind speed in meters per second and t is the air temperature in degrees Celsius. Find K to the nearest tenth when $v = 11$ m/s (40 km/h) and $t = 10°C$.

44. BUSINESS MANAGEMENT A business that makes and sells a product is interested in knowing the break-even number of units. This is the number of units that must be sold for the income from sales to equal the cost of producing those units.

  a. The formula for the number of items Q needed to break even is $Q = \dfrac{F}{S - v}$ where F is the fixed costs of production, S is the selling price of the item, and v is the variable cost of producing one unit. Find Q if F is \$40,000, S is \$2.00, and v is \$1.20.

  b. What conditions, other than a decrease in Q, would give the company profit?

THINK CRITICALLY

45. Is there a distributive property of division over addition? Justify your answer.

46. NUMBER THEORY Is $(2n + 1)^2$ always an odd number if n is an integer? Justify your answer.

47. Is addition distributive over multiplication; that is, does $a + (bc) = (ab) + (ac)$? Give an example or a counterexample.

MIXED REVIEW

Use matrices A and B to answer Exercises 48–50.

48. Which are square matrices?

49. Find the matrix $A + B$.

50. Find the matrix $A - B$.

$$A = \begin{bmatrix} 3 & 5 & -6 \\ 8 & -2 & 3 \\ -1 & 4 & 4 \end{bmatrix} \qquad B = \begin{bmatrix} 2 & 1 & 0 \\ 5 & 2 & 2 \\ 1 & 2 & 1 \end{bmatrix}$$

Name the letter of the corresponding point on the real number line.

51. -0.4

52. 1.2

53. $-\dfrac{3}{4}$

54. $\sqrt{3}$

55. STANDARDIZED TESTS The daily low temperatures for four consecutive winter days were $-3°F$, $0°F$, $5°F$, and $-2°F$. What was the mean low temperature for those days?

  A. $-5°F$ **B.** $0°F$ **C.** $5°F$ **D.** none of these

Weather forecasters use the term *wind chill* to describe how cold it feels outside. The **wind-chill factor** measures the heat loss from exposed human skin due to air temperature and wind speed. It is based on experimental observations by Dr. Paul Siple in Antarctica. Dr. Siple was the youngest member of Admiral Byrd's first expedition to Antarctica, 1928–1930. Calculations based on Dr. Siple's observations have been used to make a table that gives equivalent temperatures for combinations of air temperature and wind speed.

Wind speed (mi/h)	Wind Chill Factors Thermometer reading (degrees Fahrenheit)																
	35	30	25	20	15	10	5	0	−5	−10	−15	−20	−25	−30	−35	−40	−45
5	33	27	21	19	12	7	0	−5	−10	−15	−21	−26	−31	−36	−42	−47	−52
10	22	16	10	3	−3	−9	−15	−22	−27	−34	−40	−46	−52	−58	−64	−71	−77
15	16	9	2	−5	−11	−18	−25	−31	−38	−45	−51	−58	−65	−72	−78	−85	−92
20	12	4	−3	−10	−17	−24	−31	−39	−46	−53	−60	−67	−74	−81	−88	−95	−103
25	8	1	−7	−15	−22	−29	−36	−44	−51	−59	−66	−74	−81	−88	−96	−103	−110
30	6	−2	−10	−18	−25	−33	−41	−49	−56	−64	−71	−79	−86	−93	−101	−109	−116
35	4	−4	−12	−20	−27	−35	−43	−52	−58	−67	−74	−82	−89	−97	−105	−113	−120
40	3	−5	−13	−21	−29	−37	−45	−53	−60	−69	−76	−84	−92	−100	−107	−115	−123
45	2	−6	−14	−22	−30	−38	−46	−54	−62	−70	−78	−85	−93	−102	−109	−117	−125

Decision Making

1. What is the equivalent wind-chill temperature if the air temperature is 30°F and the wind speed is 35 mi/h?

2. Between which two wind speeds does the wind-chill temperature seem to drop the most?

3. At an air temperature of 25°F, what is the difference between the wind-chill temperatures for speeds of 20 and 40 mi/h?

4. How is the wind-chill table useful?

Use Algebra to Generalize

You can use variable expressions and properties of numbers to construct general arguments to explain or justify many different number facts and relationships.

Problem

Look at the following multiplication problems. Describe the pattern that seems to exist and predict the missing products.

$$23(101) = 2323$$
$$47(101) = 4747$$
$$12(101) = 1212$$
$$85(101) = 8585$$
$$59(101) = \blacksquare$$
$$78(101) = \blacksquare$$
$$94(101) = \blacksquare$$

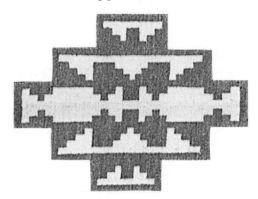

Explore the Problem

1. Write a description of the pattern that seems to exist. Your description should give information about each factor and the resulting product.

2. What are your predictions for the missing products? Verify using a calculator.

3. Show that your pattern always works. Let *ba* represent any two-digit number. Complete the multiplication of *ba* and 101 in vertical format, as shown to the right.

$$\begin{array}{r} 101 \\ \times\ ba \\ \hline \end{array}$$

4. Multiply. Look for a pattern.

$$315(101) = \blacksquare$$
$$283(101) = \blacksquare$$
$$452(101) = \blacksquare$$
$$426(101) = \blacksquare$$
$$715(101) = \blacksquare$$
$$832(101) = \blacksquare$$

5. State and justify a pattern that can be used to find the product of any three-digit number and 101.

6. State and justify a pattern that can be used to find the product of any three-digit number and 1001.

┌─ PROBLEM
  SOLVING PLAN

• Understand
• Plan
• Solve
• Examine

Investigate Further

• The algebraic representation you choose may depend on what fact you are trying to establish. Suppose you wanted to show that "the sum of two even (whole) numbers is an even number."

7. Let $2m$ and $2n$ represent the two even numbers. Represent their sum.

8. Rewrite the sum you wrote in Question 7 using the distributive property.

9. Explain why the expression you now have represents a whole number that is even.

10. How could you represent two odd numbers?

11. Use your representations of odd numbers to justify the statement, "The sum of two odd numbers is an even number."

12. Write and justify a statement about the sum of an odd number and an even number.

13. WRITING MATHEMATICS Write a paragraph explaining why it is useful to formulate and justify general results and how algebra helps you do this.

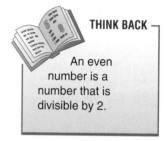

THINK BACK

An even number is a number that is divisible by 2.

CHECK UNDERSTANDING

In Question 7, why must $2m$ and $2n$ be even numbers for any integer values of m and n?

Apply the Strategy

• Remember, the representation you choose for a number may vary according to the problem.

14. Study the following sums. What do you notice about the addends? What do all the sums have in common?

$$32 + 23 = 55$$
$$16 + 61 = 77$$
$$52 + 25 = 77$$
$$45 + 54 = 99$$
$$93 + 39 = 132$$
$$78 + 87 = 165$$

THINK BACK

Addend is the name given to numbers which are added.

15. Recall that two-digit numbers can be written in expanded form: $10b + a$, where b is a digit in the tens place and a is a digit in the ones place. For example, $38 = 10(3) + 8$.

   a. Use this form to represent the general addends in the sums above.

   b. Write an expression for the sum.

   c. Simplify the expression and interpret your result. Remember to name the property you use for each step.

16. **a.** Try this several times: Take a two-digit whole number and form another number by reversing the digits.

   b. Subtract the smaller number from the greater. What do you notice about each result?

   c. Write and justify a statement about the nonnegative difference between any two-digit whole number and the number formed by reversing its digits.

17. For any integers A, B, C, consider the product

   $$(A - B)(B - C)(C - A)$$

   Without performing any multiplication, show that the product is divisible by 2. (*Hint:* Consider the cases where at least two of the integers A, B, and C are even and where at least two of them are odd. Write expressions to represent these even or odd numbers. Show what must be true of the factor that represents their difference, and relate this to the original expression for the product.)

18. WRITING MATHEMATICS Look through some mathematics or puzzle books. Find a numerical pattern that interests you. Show examples and describe the pattern. Then try to justify the pattern using algebra.

REVIEW PROBLEM SOLVING STRATEGIES

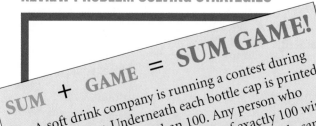

SUM + GAME = SUM GAME!

1. A soft drink company is running a contest during the summer. Underneath each bottle cap is printed a whole number less than 100. Any person who collects a set of caps with a sum of exactly 100 wins a motorcycle. Below are several typical bottle caps.

6 15 27 33 48 60 75 93

 a. Can you find a winning set of caps?

 b. Study the numbers on the caps for a common property. Explain how the property relates to forming a sum of 100.

 c. Suppose the company wants to advertise truthfully, "Up to 2000 possible winners nationwide." Explain a simple strategy for controlling the number of potential winners.

EASY AS XYZ

2.

+	x	y	z
x	y	z	x
y	z	x	y
z	x	y	z

 a. Use the addition table. Find $x + y$.

 b. What is the additive identity element in this system? Explain.

 c. What is the opposite of x in this system? Why?

 d. Find $z - (y - x)$. Explain your thinking.

PATH PUZZLE

3. Moving only in the direction of the arrows in the figure below, how many paths are there from A to B?

Work in a small group so you can share the work, discuss patterns, and check results.

 a. Start with a simpler problem. How many paths are there from A to B in the figure at the right?

 b. Draw the figure you will analyze next. Decide on a systematic way of keeping track of the paths. How many did you find?

 c. Draw the figure you will analyze next. How many paths from A to B are there this time?

 d. Now try to find a pattern in the numbers of paths you found above. Describe the pattern and use it to solve the original problem.

· · · CHAPTER REVIEW · · ·

VOCABULARY

Choose the word from the list that completes each statement.

1. A(n) __?__ is a symbol that represents a number.

2. The distance of a number from zero on the number line is the __?__ of the number.

3. Decimals that neither terminate nor repeat are __?__.

4. Terms that have the same variable base and exponent are __?__ terms.

5. The set of __?__ includes all the whole numbers and their opposites.

6. A number whose square root is an integer is a(n) __?__.

a. perfect square

b. absolute value

c. like

d. irrational numbers

e. integers

f. variable

Lesson 2.1 REPRESENT NUMBERS, VARIABLES, AND EXPRESSIONS · · · · pages 55–58

- Algeblocks may be used to model integers, variables, and algebraic expressions.

Write the expression that is modeled.

7.
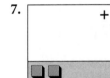

8.
 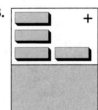

9.
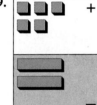

Lesson 2.2 VARIABLES AND EXPRESSIONS · · · · pages 59–64

- To **evaluate an expression** with more than one operation, substitute a value for the variable. Then use the **order of operations**.

Evaluate each expression for $x = 6$.

10. $2x - 5$

11. $3x^2 + 4$

12. $\dfrac{5x}{3}$

Lesson 2.3 REAL NUMBERS AND THE NUMBER LINE · · · · pages 65–72

- **Real numbers** include all **rational** and **irrational** numbers. Every real number can be represented by a point on a **real number line**.

Draw a number line and graph each point on it.

13. $\sqrt{18}$

14. $3\dfrac{1}{2}$

15. -0.6

Lessons 2.4 and 2.6   ADDITION, SUBTRACTION, MULTIPLICATION, DIVISION   pages 73–76, 84–87

- Algeblocks may be used to model addition, subtraction, multiplication, and division of integers.

Use Algeblocks to model each of the following. Draw a picture to show your work.

16. $4 - (-3)$ **17.** $7 + (-3)$ **18.** $5(-2)$ **19.** $-8 \div (-4)$

Lesson 2.5   ADDITION AND SUBTRACTION OF REAL NUMBERS   pages 77–83

- To add real numbers having the same sign, add their absolute values. The sum has the same sign as the numbers.
- To add numbers having different signs, subtract their absolute values. The difference has the same sign as the number with the greater absolute value.
- To subtract a real number, add its opposite.

Simplify.

20. $4\frac{1}{2} + \left(-6\frac{1}{4}\right)$ **21.** $7 - (-2.8)$ **22.** $-8 - 5 + (-4)$

23. $3\frac{1}{3} - \left(-9\frac{1}{6}\right)$ **24.** $8 + (-3.4) - 5$ **25.** $-47 + 8.7 + (-5.7)$

Lesson 2.7   MULTIPLICATION AND DIVISION OF REAL NUMBERS   pages 88–94

- The product or quotient of two real numbers having the same signs is positive.
- The product or quotient of two real numbers having different signs is negative.

Simplify.

26. $(-7)(20)$ **27.** $\frac{2}{3} \div \left(-\frac{1}{4}\right)$ **28.** $-400 \div (-25)$

29. $(-5)(-8)(-4)$ **30.** $\left(-\frac{8}{9}\right) \div \left(5\frac{1}{3}\right)$ **31.** $-5780 \div 17$

Lesson 2.8   THE DISTRIBUTIVE PROPERTY   pages 95–101

- Use the **distributive property** to combine like terms or to simplify an expression with addition or subtraction and multiplication.

Simplify.

32. $(3 - 5)^2 + 4(-2 + 1)$ **33.** $-2(4c - 7) - 10$ **34.** $4x + 2y - 5z - 3x + 2z$

Lesson 2.9   USE ALGEBRA TO GENERALIZE   pages 102–105

- Algebraic expressions are useful for describing and justifying numerical patterns.

Use algebra to justify this statement.

35. The sum of two consecutive whole numbers is an odd number.

CHAPTER ASSESSMENT

CHAPTER TEST

Simplify the expressions.

1. $20 - 15 \div 5$

2. $(4^2 + 8) \div 4 \cdot 2$

3. $-1 - (-5)$

4. $(6)(-0.4)$

5. $-\dfrac{1}{2} \div \left(-\dfrac{2}{3}\right)$

6. $\dfrac{3}{4}\left(\dfrac{2(5-7)}{5}\right)$

7. $6\left(\dfrac{1}{8} - \dfrac{1}{3}\right)$

8. $\dfrac{7}{8}\left(\dfrac{2(4-9)}{15}\right)$

9. $-7.3 + 2.5$

10. $-4(b + 2)$

11. $-3 + (-2) - (-4) + 8$

12. $4.5 + (0.2)(0.5 - 5)$

13. $10 \div (-5) - (3 - 7)^2$

14. $-20 \div 5 \div 4 - 3^2$

15. $3x + 2(x - 7)$

16. $7s + 5t - 3u - s + 4u$

Evaluate each expression.

17. Evaluate for $x = 2$: $\dfrac{x^2}{20}$

18. Evaluate for $y = -7$: $4y - 12$

19. Evaluate for $z = -3$: $2(7 - z) + 4$

20. Evaluate for $x = 0.8$: $\dfrac{x^2}{4} - \dfrac{x}{5}$

21. Evaluate for $a = -4$: $\dfrac{28a}{3a - 2}$

22. WRITING MATHEMATICS Write a paragraph to explain how you could use $<$ or $>$ to compare $-2 \cdot 3 + 5$ and -3.

23. Choose A, B, C, or D. Which of the following has the value -2.5?

   I. $-|2.5|$ **II.** $|-2.5|$ **III.** $-|-2.5|$

  A. I and II   **B.** I and III
  C. II and III   **D.** I, II, and III

For Questions 24–27, use estimation to name the letters of the corresponding points on the number line below.

24. $-3\dfrac{1}{3}$ **25.** 3.4 **26.** $\sqrt{5}$ **27.** -0.4

28. Choose A, B, C, or D. Which of the following is *not* a rational number?

  A. $2\dfrac{1}{6}$ **B.** -0.3

  C. $\sqrt{9}$ **D.** $\sqrt{27}$

Solve.

29. The temperature t dropped 15° during the past hour as a cold front moved through. Write an expression that represents the current temperature in degrees.

30. An automotive service station charges c for a new tire plus \$4 for mounting and balancing. Write an expression for the total cost in dollars of 4 new tires.

31. An electrician charges a fee of \$50 per service call plus \$18 per hour h. Write an expression for the cost of these services.

32. The width of a family room is 5 ft less than its length in feet, l. Write an expression for the area of the room in square feet.

33. The daily changes in value of a share of stock for a week are $-\dfrac{5}{8}$, $1\dfrac{1}{2}$, 1, $-\dfrac{3}{4}$, and -2. Find the mean daily change for the week.

34. WRITING MATHEMATICS Explain how you can use algebra to justify this statement: The product of two even numbers is an even number.

PERFORMANCE ASSESSMENT

USE ALGEBLOCKS Choose a variety of types of numerical and algebraic expressions from this chapter. State each expression to a partner. Then ask your partner to model each with Algeblocks and simplify it, if possible. Check that each is modeled correctly.

IDENTIFY MODELS Decide whether an elevator could be used as a model for operating with integers. Use examples to justify your decision and describe any limitations. Work with a partner to identify at least one other model that you believe would be more appropriate or that would have fewer limitations.

RECREATIONAL EXPRESSIONS Use expressions to give information about the annual cost for you or your family to use recreational facilities in your area. Research two or more facilities that interest you, such as state parks, swimming pools, fitness centers, or bowling alleys. Ask other students in your class to evaluate each expression by substituting a value for the variable that is appropriate for them.

USE FLOWCHARTS You can describe how to use the order of operations to simplify $3 \cdot 4 - 2 + 7$ with the flowchart below.

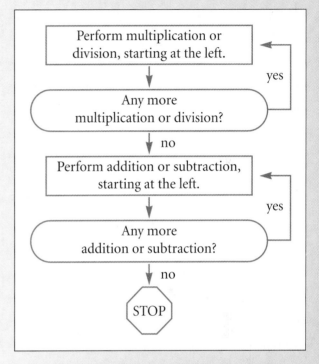

Draw flowcharts for different types of numerical and algebraic expressions from this chapter. Try to include flowcharts for expressions having exponents and grouping symbols.

PROJECT ASSESSMENT

PROJECT *Connection* Work together with your group. Organize all the project data that you have collected and the graphs and tables you have made and decide if additional displays would be helpful. Design a presentation poster which shows the different uses of mathematics in the science of meteorology. Include the following.

1. Explain how an average daily temperature is computed and how this statistic is used to give information about the weather.
2. Discuss different ways a forecaster might study temperatures in order to identify weekly weather patterns or trends.
3. Give examples of regional patterns that your group observed during the project.
4. Show illustrations of different types of weather symbols, graphs, and maps from various publications.

CUMULATIVE REVIEW

Simplify each expression.

1. $5 + (-8)$

2. $-4 - 9$

3. $-13 + 24$

4. $-8 - (-2)$

5. $-7 \cdot 8$

6. $-32 \div (-4)$

7. $20 \div (-10)$

8. $(-3)(-6)$

A number cube is tossed 15 times. Each result is recorded. Use the data for Questions 9–12.

$$2, 4, 1, 2, 5, 3, 4, 2, 6, 2, 3, 1, 5, 2, 3$$

9. Make a frequency table for the data.

10. What is the mode?

11. What is the median?

12. In what fraction of the tosses did 2 occur?

Evaluate each expression.

13. $3 + 4 \cdot 5$

14. $7 - 12 \div 4 + 2$

15. $4 + 2^3 \div 2$

16. $12 \div (4 + 2) - 7$

Name the property illustrated.

17. $5 + 7 = 7 + 5$

18. $7 \cdot \frac{1}{7} = 1$

19. $10(2x) = (10 \cdot 2)x$

20. $-8 + 0 = -8$

21. $5(2x + 7) = 5 \cdot 2x + 5 \cdot 7$

Compare. Use $<$, $=$, or $>$.

22. $-14 \ \blacksquare \ -5$

23. $\sqrt{17} \ \blacksquare \ |-5|$

24. $3^2 \ \blacksquare \ 2^3$

25. $-6 + 8 \ \blacksquare \ 8 - 6$

26. $-4.5 \ \blacksquare \ -4\frac{1}{3}$

27. $\frac{-12}{-2} \ \blacksquare \ \sqrt{25}$

28. **STANDARDIZED TESTS** Which of the following expressions is positive?

   I. $-|4 \cdot (-3)|$ **II.** $(-2)(-3)(-4)$

   III. $-(-1)^{17}$ **IV.** $\dfrac{-8 + (-4)}{-2}$

   A. I only   **B.** IV only

   C. II and III   **D.** III and IV

Perform the matrix operations.

29. $\begin{bmatrix} 6 & 12 \\ 9 & 4 \end{bmatrix} + \begin{bmatrix} 10 & 4 \\ 6 & 7 \end{bmatrix} - \begin{bmatrix} 5 & 13 \\ 7 & 6 \end{bmatrix}$

30. $\begin{bmatrix} 3 & -1 \\ -2 & 8 \end{bmatrix} + \begin{bmatrix} -2 & -9 \\ 6 & -8 \end{bmatrix}$

31. $\begin{bmatrix} -3 & 12 \\ 2 & 0 \end{bmatrix} + \begin{bmatrix} -4 & -4 \\ 3 & -2 \end{bmatrix} - \begin{bmatrix} 2 & 8 \\ -2 & -6 \end{bmatrix}$

32. List the coefficients in the expression $3x^3 - 2x^2 - x + 5$.

Evaluate each expression.

33. $\dfrac{-2}{5} + \dfrac{7}{10}$ 34. $12.6 - 24.25$

35. $\dfrac{3}{7} \div \left(\dfrac{-9}{14} \right)$ 36. $-9.2 \cdot (-8.43)$

37. **WRITING MATHEMATICS** For any integer n, explain why the expression $2n + 1$ must be an odd number.

• • • STANDARDIZED TEST • • •

STUDENT PRODUCED ANSWERS Solve each question and on the answer grid write your answer at the top and fill in the ovals.

Notes: Mixed numbers such as $1\frac{1}{2}$ must be gridded as 1.5 or 3/2. Grid only one answer per question. If your answer is a decimal, enter the most accurate value the grid will accommodate.

1. Grid the sum of $-|3.2|$ and 5.8.

2. Each letter of the alphabet is written on a card once and placed in a bag. Suppose that, as you draw one card and replace it 50 times, you draw a vowel 8 times. What decimal represents the probability that you will get a consonant for your next draw?

3. If $x = -3$, what is the value of $3x^2 - 2x + 4$?

4. Grid the value that is the opposite of $-3 - (-2.46) + (-0.7)$.

5. If $-3(h + t) = -3h - 21$, what is the value of t?

6. Suppose $p = -0.2$. Grid the absolute value of the reciprocal of p.

7. The record low for one midwestern U.S. city in January is $-29°F$. How many degrees difference are there between this and that city's record high for January, $64°F$?

8. If $b = \left(\frac{-3}{4}\right)\left(\frac{2}{5}\right)$, what is the value of $-\frac{-b}{6}$?

9. What element is in row 3, column 2 of $P - Q + R$?

$$P = \begin{bmatrix} 12 & 15 & 35 \\ 24 & 18 & 30 \\ 14 & 27 & 27 \end{bmatrix} \quad Q = \begin{bmatrix} 9 & 8 & 10 \\ 12 & 0 & 15 \\ 14 & 10 & 6 \end{bmatrix}$$

$$R = \begin{bmatrix} 4 & 10 & 2 \\ 3 & 18 & 15 \\ 2 & 2 & 4 \end{bmatrix}$$

10. The least value in a set of four items of data is 12. The mean of the data is 19, and its range is 20. If the set of data has a mode, what must it be?

11. In the table below, c represents the number of long distance calls that lasted 9 min. If 36% of the calls lasted less than 10 min, what is the value of c?

Long Distance Phone Calls	
Length of Calls (min)	Frequency
2	2
5	1
9	c
12	4
17	3
21	5
28	3
54	1

12. According to the graph below, what was the greatest increase in enrollment between two consecutive years?

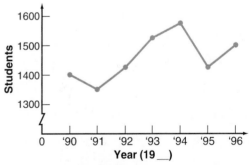

13. If $\frac{3}{7} = \frac{n}{12}$, what is the value of n?

14. The program below finds the inventory at the end of the month for starting inventory A, inventory received during the month B, and inventory sold C. If 238 was input for A and 576 for B and the computer printed 198, what value must have been input for C?

```
10 INPUT A, B, C
20 PRINT A + B - C
```

3 Linear Equations

Take a Look
AHEAD

Make notes about things that look new.

- If you read an English sentence from right to left, it probably won't make sense. Does the meaning of a mathematical sentence change if you read left to right or right to left?

- Which careers highlighted in the chapter interest you?

Make notes about things that look familiar.

- Can you identify all the symbols used in this chapter? What different ways are there to show multiplication and division?

- Find some problems in this chapter you can solve using any method you know.

DATA Activity

Health Costs

Each year, Americans spend billions of dollars on health services and supplies including fees for doctors and dentists, home health care, medicines and optical products, nursing facilities, health insurance, research, and government public health activities. The total expenditure for all these categories is shown in the table on the next page.

SKILL FOCUS

▶ Add, subtract, multiply, and divide real numbers.

▶ Find the mean and range of data items.

▶ Calculate percent increase.

▶ Determine trends and make predictions on the basis of data.

▶ Construct a line graph and a bar graph and compare alternate forms of data display.

HEALTH *and* FITNESS

In this chapter, you will see how:

- **NUTRITION THERAPISTS** use algebra to determine a client's body type. (Lesson 3.4, page 134)

- **PERSONAL TRAINERS** use equations to design individual fitness programs. (Lesson 3.5, page 139)

- **AEROBICS INSTRUCTORS** use formulas to determine a range of acceptable heartbeat rates during exercise. (Lesson 3.7, page 151)

U.S. Expenditures for Health, 1985–1991 (in billions of dollars)						
1985	1986	1987	1988	1989	1990	1991
$422.6	$454.8	$494.1	$546.0	$604.3	$675.0	$751.8

Use the table to answer each question.

1. Find the mean and range of the health care expenditures from 1985 to 1991. Round to one decimal place.

2. For each of the years from 1985 to 1991, find the dollar amount of the increase in expenditures from the previous year. Which two-year period showed the greatest increase?

3. For each of the years from 1986 to 1991, find the percent increase from the previous year. Round to the nearest tenth of a percent.

4. Are health expenditures increasing at a steady rate? Would your conclusion be different depending on whether you used the results from Question 2 or Question 3? Explain.

5. Predict the total expenditures for 1992, 1993, and 1994. Explain your method. Use an almanac or other reference to check your predictions.

6. **WORK TOGETHER** Work with a partner. One partner will display the data from the table using a bar graph and the other partner will use a line graph. Compare and discuss your work with other pairs of students. Which method makes it easiest to identify exact numbers? Which method makes it easiest to determine patterns of change?

113

PROJECT

Is "LITE" Right?

As part of their fitness program, many people keep track of the total calories, fat, sodium, and cholesterol they consume daily. In May of 1994, Food and Drug Administration (FDA) rules went into effect defining exactly what criteria must be met for food manufacturers to claim their products are "Light" or "Lite," "Lo-Cal," "Reduced Sodium," and so on. All but the smallest companies are required to provide standard nutritional information on their labels.

In this project you will compare "Lite" and regular products to discover how much information you need to eat right.

PROJECT GOAL

To use algebra and statistics to analyze data about nutrition and prices of "Lite" and regular food items.

Getting Started

Work in groups of four or five students. Use these ideas to get started.

1. Research which terms FDA rules govern and what criteria must be met in order to use the terms on food labels or in advertisements.

2. Decide on at least twenty different food products that can be compared in "Lite" and regular versions.

3. Decide what nutritional and price information you will need to collect for each product.

4. Create a form that can be used to record data for later use.

Internet Connection

www.swpco.com/
swpco/algebra1.html

PROJECT Connections

Lesson 3.3, page 128:
Determine the number of calories in a food that come from fat and decide if the food meets dietary guidelines.

Lesson 3.6, page 145:
Compare the percent decrease of fat or salt in "lite" foods with the percent increase in price.

Lesson 3.7, page 150:
Compare fat content and price of different supermarket varieties of ground beef and home mixtures.

Chapter Assessment, page 163:
Plan a presentation to communicate final project results.

Solve Equations with Mental Math and Estimation

Explore/Working Together

● Work with a partner. You will need two number cubes.

1. One partner rolls both cubes. Add, subtract, multiply, or divide the numbers showing to form a mystery number. For example, with 6 and 3 the set of possible whole numbers is 9 (by adding), 3 (by subtracting), 18 (by multiplying), and 2 (by dividing). Give one clue so the other partner can guess the number. Possible clues might be "Three times my number is 27" or "The sum of my number and 11 is 13."

2. Play several rounds. Then discuss the thinking process you used to solve each puzzle.

SPOTLIGHT ON LEARNING

WHAT? In this lesson you will learn
• to solve equations by using mental math and estimation.

WHY? Mental math and estimation can help you solve equations that model problems about tipping, amount of markup, and budgeting.

Build Understanding

● An **equation** is a statement that two numbers or expressions are equal. Examples are:

$$-7 + 3 = -4 \qquad 2x - 1 = 9 \qquad 5 - p = 8 + p$$

An equation that contains one or more variables is an **open sentence**. An open sentence may be *true* or *false*, depending on what **values of the variable** are substituted. A value of the variable that makes the equation true is a **solution** of the equation.

Some equations can be solved by mental math, using number facts and properties you already know.

EXAMPLE 1

Your partner rolls a 5 and a 4, and you know that 7 less than the mystery number is 13. What is the number?

Solution
Think: "Some number minus 7 equals 13." Write an equation. Let x represent the mystery number.

$$x - 7 = 13$$

To solve this equation mentally, ask yourself, "What number must I have so that if 7 is subtracted, 13 will remain?" Mental math gives you the solution: $20 - 7 = 13$. The mystery number was 20.

If the solution to an equation is not obvious, you can try guessing the solution. Test your guess by substituting it into the equation. If the guessed value makes the equation true, your guess was correct. If not, you can use what you have learned to improve your next guess.

COMMUNICATING ABOUT ALGEBRA

How might you change your guessing strategy if the equation in Example 2 were changed to
$70 = (3 - y)5$?

EXAMPLE 2

Find the value of y in this equation.

$$70 = (y - 3)5$$

Solution

First Guess: Try $y = 20$ as the value of the variable. Replace y with 20 and simplify the right side of the equation.

$70 \stackrel{?}{=} (20 - 3)5$
$70 \stackrel{?}{=} (17)5$
$70 < 85$ **Since 70 < 85, 20 is not a solution. It is too large.**

Second Guess: Try $y = 15$ since 20 is too large.

$70 \stackrel{?}{=} (15 - 3)5$
$70 \stackrel{?}{=} (12)5$
$70 > 60$ **Since 70 > 60, 15 is not a solution. It is too small.**

Third Guess: Try $y = 17$ since 17 is between 15 and 20.

$70 \stackrel{?}{=} (17 - 3)5$
$70 \stackrel{?}{=} (14)5$
$70 = 70$ ✓   **Since both sides are the same number, 17 is a solution.**

The first two guesses help you see that the solution must be greater than 15 and less than 20. ◀

A different guessing technique works for other types of equations.

EXAMPLE 3

Solve: $58 - 3w = 40$

CHECK UNDERSTANDING

In Example 3, how did covering the term $3w$ help you find that the value of $3w$ must be 18?

Solution

Cover the term $3w$. You see that 58 minus some number must be equal to 40. Since $58 - 18 = 40$, this tells you that $3w = 18$. What number multiplied by 3 equals 18? You know that $3 \cdot 6 = 18$, so the solution of the equation $58 - 3w = 40$ must be 6.

Check

$58 - 3w = 40$
$58 - 3(6) \stackrel{?}{=} 40$ Replace w with 6.
$58 - 18 \stackrel{?}{=} 40$
$40 = 40$ ✓   **The solution is 6.** ◀

A number line can also be helpful in solving some equations.

EXAMPLE 4

MODELING An elevator rose from 3 floors below street level to the fourth floor. How many floors did it climb?

Solution

Think of an elevator's path as a number line. Let x represent the number of floors the elevator climbed. The equation that describes the situation is $-3 + x = 4$. Represent the equation on the number line.

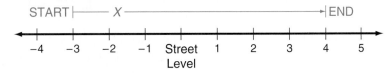

Count the number of floors along the line. The elevator rose 7 floors. ◄

TRY THESE

Use mental math to solve each equation. Check your answer.

1. $x + 5 = 21$ **2.** $y \div 3 = 9$ **3.** $6m = 30$ **4.** $32 - r = 16$

5. $t - 11 = 17$ **6.** $\frac{1}{4}n = 12$ **7.** $\frac{e}{7} = 7$ **8.** $8 + g = 19$

9. $9 = 36 \div p$ **10.** $10 = x - 4$ **11.** $-2 = h + 3$ **12.** $a - 5 = -2$

Use the guess-and-check method to solve each equation. Check your answer.

13. $2y + 6 = 22$ **14.** $45 + 3p = 81$ **15.** $60 \div 4g = 5$

16. $7(m + 9) = 105$ **17.** $81 \div 3r = 3$ **18.** $64 \div (12 - x) = 8$

Cover the term with the variable to solve each equation. Check your answer.

19. $9u + 42 = 96$ **20.** $3e \div 4 = 12$ **21.** $13.7 - 2m = 9.5$

22. WRITING MATHEMATICS Explain how you would choose between the guess-and-check method and the cover-up method to solve an equation.

Use the number line to solve each equation.

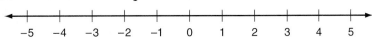

23. $-5 + t = 3$ **24.** $4 + b = -2$ **25.** $-2 = x + (-5)$

26. WRITING MATHEMATICS How did you use the number line differently in Exercises 23 and 24? How did that affect your solution?

27. GEOMETRY The formula for the area of a triangle is $A = \dfrac{bh}{2}$, where A represents the area, b represents the base, and h represents the height. A triangle has a base 14 units long and an area of 56 square units. What is the height? What method did you use to solve?

PRACTICE

Solve each equation. Check your answer.

1. $q - 17 = 23$ **2.** $38 + y = 65$ **3.** $\frac{t}{4} = 120$ **4.** $61 - p = 14$

5. $\left(\frac{1}{3}\right)b = 25$ **6.** $\frac{81}{e} = 27$ **7.** $59 = 73 - d$ **8.** $51 = 3z$

9. $23.9 - c = 14.8$ **10.** $37 = y - 57$ **11.** $\frac{156}{r} = 3$ **12.** $\left(\frac{3}{4}\right)w = 75$

13. $2x - 1 = 19$ **14.** $12 + y = -3$ **15.** $-17 + t = -12$ **16.** $2.1g = 8.4$

17. $-14 + r = -6$ **18.** $2(d - 9) = 22$ **19.** $\frac{y + 4}{6} = 7$ **20.** $\frac{45.5}{n} = 9.1$

21. NUTRITION A frankfurter has 13 g of fat. There is 12.5 g less fat in a serving of roast turkey. Write and solve an equation to find how much fat is in a serving of roast turkey.

22. BUSINESS A buyer for a chain of electronics stores paid $75.25 apiece for new model CD players. She marked them up to sell at $125.99. Write and solve an equation to find the dollar amount of markup per player.

23. WRITING MATHEMATICS Write a paragraph explaining the different ways you know to solve an equation. Use examples to illustrate.

EXTEND

Solve each equation. Use technology where appropriate.

24. $23.2 - y = 15.5$ **25.** $-4 + m = 34$ **26.** $-6h = 96$

27. $6.2t = 31$ **28.** $r + 27 + 2r = 42$ **29.** $w - 7 + w = 5$

30. $\frac{b}{-4} = -3$ **31.** $\frac{r}{-3} = 27$ **32.** $x + 19.4 = 54$

33. WRITING MATHEMATICS Describe how you might use your calculator to solve Exercise 24.

Estimate to determine whether each solution is greater than or less than zero.

34. $x - 9 = 6$ **35.** $-16 + m = -28$ **36.** $36 - p = -10$

37. $7 + 2a = 1$ **38.** $18 - x = -4$ **39.** $\frac{r}{-3} = 12$

40. WRITING MATHEMATICS Describe how you determined your answers to Exercises 34–39.

ECOLOGY In 1962 Rachel Carson wrote the book *Silent Spring*, in which she predicted that 40 species of birds were likely to become extinct. None of the species has become extinct. Solve Exercises 41–43 to find out what did happen to those species.

41. Solve $y + 11 = 40$ to find the number of species whose populations have remained stable.

42. Solve $40 = 47 - z$ to find the number of species whose populations decreased.

43. Use Exercises 41 and 42 to find the number of species whose populations increased.

CONSUMERISM Ray bought a container of orange juice and three cans of tomatoes. The receipt was blurred when he looked at it. He knew that the total amount he spent was $6.16 and that the orange juice cost $2.59. He could not read the cost of each can of tomatoes.

44. Write an equation that Ray could use to determine the cost of one can of tomatoes.

45. If you use guess-and-check to solve Ray's equation, would $2.00 be a reasonable first guess for the cost of one can? Why or why not?

46. Use any method you choose to solve Ray's equation.

47. **BUDGETING** An after-school job pays $5.75 per hour. Keiko is trying to save enough to buy a gift that costs $89.95. She wrote the equation $5.75h = 89.95$ to estimate the number of hours she will have to work, where h represents hours. If she saves all her income, Keiko estimates she will have enough money after working about 15 or 16 hours. Describe the reasoning that Keiko might have used.

48. **WRITING MATHEMATICS** Explain how whole number estimating skills help you estimate the solution to an equation.

THINK CRITICALLY

49. If the sum of a variable and a positive number is a negative number, does the variable represent a positive or a negative number?

50. If the difference between a variable and a positive number is a negative number, does the variable represent a positive or a negative number?

Use mental math to solve Exercises 51–53.

51. If $x + 7 = 29$, what is the value of $x + 16$?   **52.** If $2y - 23 = 45$, what is the value of $2y - 35$?

53. If $\frac{4w}{6} + 13 = 50$, what is the value of $\frac{4w}{6} + 40$?

MIXED REVIEW

54. Calculate the mean, median, mode, and range for the set of data in the table.

Number of Points Scored	23	17	9	6	5	2
Number of Players	1	1	2	2	3	4

55. **STANDARDIZED TESTS** Which is the value of the expression when $a = 4$, $b = 6$, and $c = 8$?

$c^2 - ab$

A. 32   **B.** −16   **C.** 40   **D.** 92

Think Back

● Work with a partner. Use Algeblocks. Recall that when you add a number and its opposite, the result is zero.

1. Which of the mats below show a model of zero? If the mat does not show a model of zero, add blocks so that it does. In what other way could you change it to model zero?

a.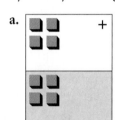

b.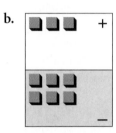

c.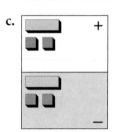

d.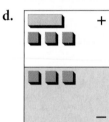

2. Show how to simplify the expression modeled at the right.

Explore

You know that a scale is in balance when both sides hold the same amount. An equation is like a scale in balance. The expression on one side of the equal symbol is equivalent to the expression on the other side.

You can use Algeblocks to model an equation. On the Sentence Mat at the right, the equation $x - 3 = 8$ is modeled. (A Sentence Mat is two Basic Mats with an equal symbol between them.)

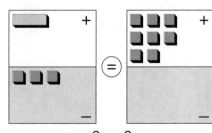

$$x - 3 = 8$$

3. What equation is modeled on each Sentence Mat?

a.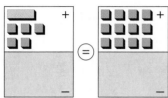

b.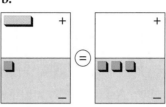

c.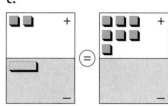

4. Model some equations. Have your partner identify them. Then you and your partner switch roles.

COMMUNICATING ABOUT ALGEBRA

In 3a, you could either add unit blocks to the negative part of both sides or remove unit blocks from the positive part of both sides. Explain why both methods work.

You can use Algeblocks to solve an equation by getting the *x*-block alone on one side of the Sentence Mat. Whatever is on the other side of the Sentence Mat is the solution, the value of the variable that makes the equation true.

Examine these Sentence Mats.

Show $x - 4 = 5$.

Add 4 unit blocks to each side of the Sentence Mat.

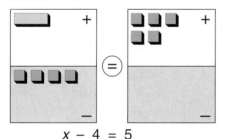

$$x - 4 = 5$$

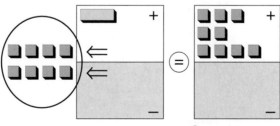

$$x - 4 + 4 = 5 + 4$$

Opposites on the same side of the equal symbol make sums of zero. These are called **zero pairs**. You can simplify this equation by removing zero pairs.

Read the solution: $x = 9$

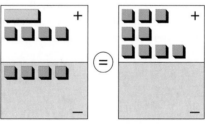

$$x = 9$$

Remember, opposites on the same side of the Sentence Mat make zero, and you can remove those blocks. You can add or remove blocks, but you must add or remove the same number of blocks from *each* side of the Sentence Mat.

5. Identify the equation being modeled. Then solve. Sketch each step.

a.

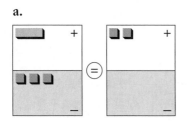

b.

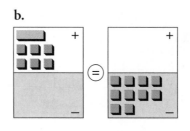

c.

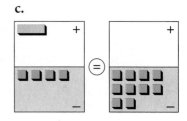

6. Make up two equations of your own. Use Algeblocks to solve each equation. Sketch each step.

7. Adam started with the model below.

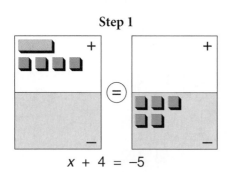

This is the Sentence Mat when Adam was finished.

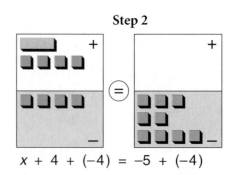

Do you think his work was correct? Explain. If you disagree with Adam's work, show what you think is the correct result.

Make Connections

- You know how to write an equation for a given model. You can also write an equation for the other steps in solving the equation. Here is an example.

Step 1

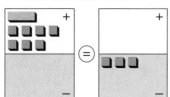

$x + 4 = -5$

Step 2

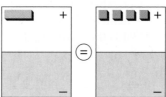

$x + 4 + (-4) = -5 + (-4)$

Step 3

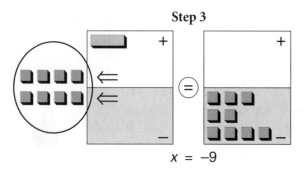

$x = -9$

8. Write an equation for each modeled step.

Step 1

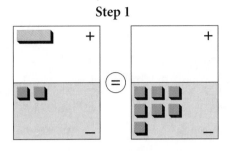

Step 2

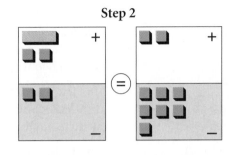

Step 3

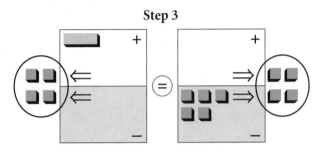

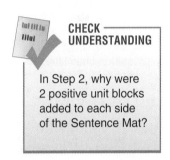

CHECK UNDERSTANDING

In Step 2, why were 2 positive unit blocks added to each side of the Sentence Mat?

Sketch an Algeblocks model for solving each equation.

9. $x + 5 = -7$ **10.** $x - 4 = -3$ **11.** $x - (-3) = 8$

12. $x + 2 = 6$ **13.** $x - 3 = 2$ **14.** $x - (-4) = 6$

Summarize

15. WRITING MATHEMATICS Write a paragraph explaining how you decide what your first step is for solving an equation that is modeled with Algeblocks. How can you tell when you have found a solution to the equation?

16. MODELING Model the equation $x - 5 = -3$ and model the steps in solving the equation. Sketch how the mat should look for each step. Write the equation that corresponds to each step shown in your model.

17. THINKING CRITICALLY Consider this statement: "When you add Algeblocks to both sides of an equation, it will always be necessary to simplify both sides." Give an example to show this is not always true.

18. GOING FURTHER How could you solve the equation modeled at the right? Compare and discuss your results with those of another student.

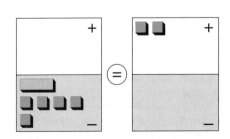

Explore

Figure A shows a balance with an unknown weight on one side and a 12-lb weight on the other.

In Figure B, a balloon with a lifting force of 3 lb has been attached to the left side.

1. Is the unknown weight x greater or less than 12 lb?

2. What is the effect of attaching the balloon to the left side?

3. What equation can you write to describe the relation in Figure B?

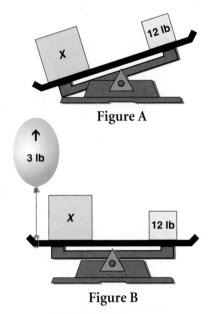

Figure A

Figure B

4. If a 3-lb weight were placed on each side of the scale in Figure B, would the scale remain balanced?

5. How much weight would be on each side of the scale after the addition of the 3-lb weights?

Build Understanding

COMMUNICATING ABOUT ALGEBRA

Are the equations $x - 11 = 7$ and $x - 8 = 10$ equivalent? Explain why or why not. Compare your explanation with those of some classmates.

When you add the same weight to both sides of a scale that is balanced, the scale remains balanced. When you add the same blocks to both sides of a sentence mat, the equation remains balanced. This is called the *addition property of equality*.

ADDITION PROPERTY OF EQUALITY

For all real numbers a, b, and c,
  if $a = b$, then $a + c = b + c$.

When you add the same number to both sides of an equation, you create an equation that is *equivalent* to the original equation. **Equivalent equations** have the same solution. Look at these equations.

$$x - 9 = 15 \qquad \text{In this equation, } x = 24.$$
$$x - 9 + 3 = 15 + 3 \qquad \text{3 is added to each side.}$$
$$x - 6 = 18 \qquad \text{In this equation, } x = 24.$$

The equations $x - 9 = 15$ and $x - 6 = 18$ are equivalent.

You solve an equation by getting the variable alone on one side. When one side of the equation is a variable having a coefficient of 1, the value of the expression on the other side is the solution of the equation.

To solve an equation such as $r - 2 = 21$, for example, you use the addition property of equality to add 2 to both sides.

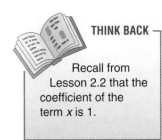

THINK BACK

Recall from Lesson 2.2 that the coefficient of the term x is 1.

EXAMPLE 1

Solve: $r - 2 = 21$

Solution

$$r - 2 = 21$$
$$r - 2 + 2 = 21 + 2$$
$$r = 23$$

The opposite of -2 is 2.
Add 2 to both sides.

Check

$$r - 2 = 21$$
$$23 - 2 \stackrel{?}{=} 21$$
$$21 = 21 \checkmark$$

Substitute 23 for r.
The solution is 23. ◄

To solve an equation such as $y + 3 = 8$, first isolate the variable y on one side of the equation. You can do this by adding -3, the opposite of 3, to both sides. Since subtracting 3 has the same result as adding -3, you can also solve the equation by subtracting 3 from both sides.

Subtracting the same number from both sides of an equation produces an equivalent equation. This is called the *subtraction property of equality*.

> **SUBTRACTION PROPERTY OF EQUALITY**
>
> **For all real numbers a, b, and c,**
> **if $a = b$, then $a - c = b - c$.**

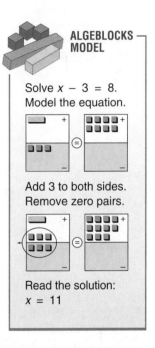

ALGEBLOCKS MODEL

Solve $x - 3 = 8$. Model the equation.

Add 3 to both sides. Remove zero pairs.

Read the solution: $x = 11$

EXAMPLE 2

Solve: $m + 17 = 3$

Solution

$$m + 17 = 3$$
$$m + 17 - 17 = 3 - 17$$
$$m = -14$$

The opposite of 17 is -17.
Subtract 17 from both sides.

Check

$$m + 17 = 3$$
$$-14 + 17 \stackrel{?}{=} 3$$
$$3 = 3 \checkmark$$

Substitute -14 for m.
The solution is -14. ◄

CHECK UNDERSTANDING

In Example 2, how would you use addition instead of subtraction to solve the problem?

To solve an equation such as $b - (-4) = 9$, remember that $-(-4) = 4$.

EXAMPLE 3

Solve: $b - (-4) = 9$

Solution

$$b - (-4) = 9$$
$$b + 4 = 9 \qquad \text{The opposite of 4 is } -4.$$
$$b + 4 - 4 = 9 - 4 \qquad \text{Subtract 4 from both sides.}$$
$$b = 5 \qquad \textbf{The solution is 5.}$$

Check by substituting in the original equation. ◀

EXAMPLE 4

WEATHER The record temperature increase (recorded in January 1943, in South Dakota) is 49°F in 2 min, after which the temperature was 45°F. What was the original temperature?

Solution

Let x be the original temperature. Write and solve an equation.

Original temperature	+	Increase	=	New temperature

$$x + 49 = 45$$
$$x + 49 - 49 = 45 - 49$$
$$x = -4$$

The temperature before the rise was -4°F. Check by substituting -4 into the original equation. ◀

TRY THESE

Name the number you would add to each side of the equation to isolate the variable.

1. $x + 3 = 19$ **2.** $r - 4 = -2$ **3.** $-17 + x = -4$ **4.** $-5 + m = 12$

5. $-6 + a = 15$ **6.** $-20 + d = 32$ **7.** $c - (-4) = 3$ **8.** $w - (-17) = 14$

9. **MODELING** What equation is modeled by these Algeblocks? Use Algeblocks to solve. Sketch each step. Write the appropriate equation below each sketch.

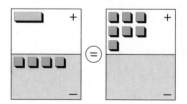

Solve and check each equation.

10. $r - 8 = 10$ **11.** $a - 11 = -16$ **12.** $z + 14 = -2$ **13.** $g + 19 = 12$

14. $x - (-16) = 8$ **15.** $p - (-3) = 19$ **16.** $y + 3 = 0$ **17.** $k - 47 = 0$

18. $n + \dfrac{1}{2} = 3$ **19.** $f + \dfrac{3}{4} = 5$ **20.** $(c - 7) + 15 = 26$ **21.** $(b + 1) - 12 = 9$

22. **WRITING MATHEMATICS** Describe how to solve equations with the addition property of equality.

PRACTICE

Solve and check each equation.

1. $b + 15 = 40$ **2.** $z + 28 = 79$ **3.** $c + 10 = -10$ **4.** $j + 8 = -5$

5. $x - 9 = 9$ **6.** $b - 41 = 42$ **7.** $w - (-2) = -1$ **8.** $k - (-18) = -3$

9. $31 + x = 17$ **10.** $50 + p = 13$ **11.** $31 + x = -17$ **12.** $37 + g = -25$

13. EARTH SCIENCE The temperature at noon was 71°F. This was 14°F more than the temperature at 11 A.M. Write and solve an equation to find the temperature at 11 A.M.

14. $e + (-7) = 12$ **15.** $f + (-15) = -8$ **16.** $-4 + n = -7$ **17.** $-9 + h = -9$

18. $r + 3.5 = -6.5$ **19.** $m + 2.6 = -1$ **20.** $x + \dfrac{5}{8} = \dfrac{7}{8}$ **21.** $g + \dfrac{7}{12} = 2$

22. $5 + (2 + y) = 8$ **23.** $7 + (4 + a) = 13$ **24.** $3 = -2 - (5 - c)$ **25.** $16 = 9 - (1 - b)$

26. CHEMISTRY At the start of an experiment, the temperature of a solution was 168°F. After some ice was added, the temperature dropped to 105°F. Find the change in temperature.

27. WRITING MATHEMATICS Write a problem you can solve using the equation $k - 32 = 48$.

EXTEND

28. TECHNOLOGY How would you use a calculator to solve and check $x - 2394 = 339$?

ESTIMATION/TECHNOLOGY Estimate. Then use a calculator to solve and check.

29. $t + 3.024 = -9.74$ **30.** $-2.6015 = a + 83.3$

31. $x - 0.1008 = 4.992$ **32.** $-18.9 + n = 5.01$

HEALTH The table shows Recommended Daily Dietary Allowances (RDAs) of calories and protein. Use the table and write equations to answer Exercises 33–34.

	Age	Height, in.	Calories Needed	Protein Needed, g
Male	11–14	63	2800	44
	15–18	69	3000	54
Female	11–14	62	2400	44
	15–18	65	2100	48

33. A 16-year-old male who is 69 in. tall requires 300 more calories than a 35-year-old male. What is the RDA of calories for the 35-year-old male?

34. The RDA of calories for a 55-year-old woman who is 65 in. tall is 1000 less than that for a 12-year-old boy. What is the RDA of calories for the 55-year-old woman?

35. HEALTH For a person who weighs approximately 150 lb, one hour of gardening uses 220 Cal. This is 130 Cal less than is used in an hour of roller-skating. How many calories does an hour of roller-skating consume? Write an equation to solve the problem.

THINK CRITICALLY

36. Suppose $x + a = b$.

    a. What happens to x if a increases and b remains the same?

    b. What happens to b if x decreases and a remains the same?

37. Suppose $x - a = b$.

    a. What happens to x if a decreases and b remains the same?

    b. What happens to b if x increases and a remains the same?

38. If $z + 9 = -5$, find the value of $z - 3$.

39. If $c - 1\frac{1}{2} = 2\frac{1}{4}$, find the value of $2c$.

Write and solve an equation for each situation. Tell what question(s) you have answered.

40. An exercise machine is on sale for $59 less than its regular price. The sale price is $177.

41. The length of the pool is 25.5 ft more than its width. The length is 53 ft.

PROJECT *Connection*

Health professionals recommend adults eat foods low in fat. One guideline for judging if a food is low fat is whether less than 30% of the food's caloric value is derived from fat. A gram of fat has 9 Cal. Suppose a 180-Cal serving of yogurt has 4 g of fat.

- **Multiply to find the number of calories from fat.** $4 \cdot 9 = 36$

- **Divide by total calories per serving.** $\dfrac{36}{180} = 0.2$

- **Write the answer as a percent.** $0.2 = 20\%$

1. A serving of soup has 140 Cal and 8 g of fat. What percent of the soup's caloric value comes from fat?

2. If a food product has 370 Cal per serving, what is the maximum number of grams of fat the product can have to fall within the above limits?

3. Gather food labels from a variety of food products, including some so-called junk foods such as cookies and potato chips. Rank the foods you investigated from lowest to highest percentage of calories from fat. Create a visual display of your results.

Nutrition Facts
Serving Size 1 Container (210 g)

Amount Per Serving

Calories 180	Calories from Fat 36

	% Daily Value
Total Fat 4g	7%
Saturated Fat 1.5g	8%
Cholesterol 15mg	5%
Sodium 135mg	6%
Potassium 500mg	14%
...hydrates 46g	15%
...y Fiber 1g	4%
...s 44g	

3.4 Solve Equations with Multiplication and Division

Explore

1. At the health club, you and 11 people in your exercise class are to form four equal teams. How many people should be on each team?

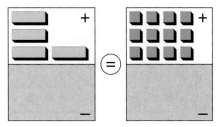

SPOTLIGHT ON LEARNING

WHAT? In this lesson you will learn
- to use the multiplication property of equality to solve equations.

WHY? The multiplication property of equality helps you solve equations that model problems about consumerism, earth and life science, and astronomy.

   a. What expression is shown on each side of the mat? How do the Algeblocks model the problem? You can solve the problem by dividing the blocks on each side into 4 equal groups. Sketch how you would arrange the blocks to solve the problem.

   b. What equation can represent the problem? What equation can represent the solution?

2. Suppose it takes you 18 min to run 3 mi around the track at the health club. At that rate, how many minutes does it take to run 1 mi? Use Algeblocks to model and solve the problem. Sketch each step.

Build Understanding

In the previous lesson you used the addition property of equality and the subtraction property of equality to solve certain equations. You will use the *multiplication property of equality* to solve other types of equations. This property allows you to multiply both sides of an equation by the same number.

COMMUNICATING ABOUT ALGEBRA

How would you solve the equation $6x = 30$? How would you solve the equation $6 + x = 30$? How are the methods alike? How are they different?

> **MULTIPLICATION PROPERTY OF EQUALITY**
>
> For all real numbers a, b, and c,
> if $a = b$, then $ca = cb$.

EXAMPLE 1

Solve: $\dfrac{p}{16} = -4$

Solution

$$\frac{p}{16} = -4$$

$$16\left(\frac{p}{16}\right) = 16(-4) \quad \text{Multiply both sides by 16.}$$

$$p = -64$$

Check

$$\frac{p}{16} = -4$$

$$\frac{-64}{16} \overset{?}{=} -4 \quad \text{Substitute } -64 \text{ for } p.$$

$$-4 = -4 \checkmark \quad \textbf{The solution is } -64. \blacktriangleleft$$

THINK BACK

Recall that when a number is multiplied by its **reciprocal**, the product is 1.

To solve an equation such as $3c = 15$, multiply both sides by $\frac{1}{3}$, the reciprocal of 3. Since multiplying by $\frac{1}{3}$ has the same result as dividing by 3, you can also solve the equation by dividing both sides by 3. Dividing both sides of an equation by any nonzero number produces an equivalent equation. This is called the *division property of equality*.

DIVISION PROPERTY OF EQUALITY

For all real numbers a, b, and c,
if $a = b$, and $c \neq 0$, then $\dfrac{a}{c} = \dfrac{b}{c}$.

EXAMPLE 2

Solve: $-7m = 91$

Solution

$-7m = 91$

$\dfrac{-7m}{-7} = \dfrac{91}{-7}$ Divide both sides by -7.

$m = -13$

Check

$-7m = 91$

$-7(-13) \overset{?}{=} 91$ Substitute -13 for m.

$91 = 91$ ✓ **The solution is -13.** ◄

CHECK UNDERSTANDING

In Example 2, how could you solve the equation using multiplication instead of division?

When you know the value of $-x$, you can find the value of x by using the property of -1 for multiplication.

EXAMPLE 3

Solve: $14 = -x$

Solution

$14 = -x$

$(-1)14 = (-1)(-1x)$

$-14 = x$

Check

$14 = -x$

$14 \overset{?}{=} -(-14)$ Substitute -14 for x.

$14 = 14$ ✓   **The solution is -14.** ◄

EXAMPLE 4

CONSUMERISM Sport! sells exercise books for $8.95 each. In one day, they took in $1816.85. How many exercise books were sold?

Solution

Let x represent the number of exercise books sold. Write and solve an equation that describes the information.

Price per book	·	Number of books sold	=	Total amount taken in

$8.95x = 1816.85$

$\dfrac{8.95x}{8.95} = \dfrac{1816.85}{8.95}$

$x = 203$

There were 203 exercise books sold.

Check

Is the solution reasonable? The amount of money taken in from the book sale is about $1800 and the cost of each book is about $9.00. Since 200 • $9 is $1800, the solution is reasonable. Complete the check by substituting the solution into the original equation. ◄

You may wish to use a calculator in solving equations like the one given in Example 4. If so, it will be more efficient to write in a step that shows the variable alone on one side. This allows you to see clearly what calculations are to be performed.

$$x = \frac{1816.85}{8.95}$$
$$x = 203$$

TRY THESE

Indicate whether you would multiply or divide, and by what number, to solve each equation. *Do not solve the equation.*

1. $7a = -35$ **2.** $-5p = 40$ **3.** $\frac{c}{3} = 15$ **4.** $\frac{b}{-4} = 12$

5. $\frac{2}{3}n = 60$ **6.** $\frac{3}{4}x = -48$ **7.** $9z = 36$ **8.** $-6k = -36$

What equation do the Algeblocks model?

9. 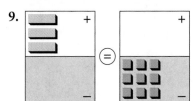 **10.**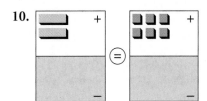

Solve and check.

11. $\frac{a}{3} = 5$ **12.** $12 = \frac{x}{5}$ **13.** $5b = 65$ **14.** $36 = 9c$

Assume you will use a calculator to solve the following equations. What equation should you write just before you perform your calculations?

15. $9.2b = 32.2$ **16.** $22.1 = 3.4c$ **17.** $\frac{d}{28.5} = 7.5$ **18.** $\frac{f}{16} = 6.4$

PRACTICE

Indicate whether you would multiply or divide, and by what number, to solve each equation. *Do not solve the equation.*

1. $6v = 36$ **2.** $84 = 12d$ **3.** $-4m = 28$ **4.** $-32 = 8h$

5. $-7q = -56$ **6.** $-108 = -9s$ **7.** $\frac{t}{5} = 25$ **8.** $15 = \frac{2}{3}x$

9. $\frac{-y}{12} = 5$ **10.** $\frac{z}{-10} = 2$ **11.** $\frac{3}{4}c = -8$ **12.** $-\frac{2}{5}k = -13$

Solve and check. Use a calculator where appropriate.

13. $-4f = -480$ **14.** $80c = 560$ **15.** $\frac{2}{3}x = 8$ **16.** $-\frac{1}{4}y = 7$

17. $-5x = 90$ **18.** $\frac{3}{4}h = -6$ **19.** $6y = -102$ **20.** $3.2p = 14.4$

21. $\frac{z}{6} = 13$ **22.** $2.6y = 24.7$ **23.** $\frac{2}{5}z = 1.2$ **24.** $\frac{2}{7}w = 3.4$

25. $\frac{-c}{13.6} = 3.5$ **26.** $\frac{-j}{3.4} = 5.5$ **27.** $5.4g = 13.5$ **28.** $3.8q = -9.5$

29. GAMES At a games tournament, 1008 people are divided into teams of 6. How many teams can be formed? Write and solve an equation. Answer the question.

30. MONEY Game Boosters' most popular board game sells for $19.95. One afternoon the store took in $1017.45 on the sales of this game. Write an equation to find the number of games sold. Estimate the solution, then use a calculator to check your estimate.

31. WRITING MATHEMATICS In solving the equation $-3x = 27$, your friend gets the solution $x = 9$. Describe how you would show your friend the error and how to correct it.

32. WRITING MATHEMATICS Write two equations, one that requires multiplication to solve and one that requires division to solve.

EXTEND

Solve and check. Round the answers to the nearest hundredth.

33. $-4(-5) = 8n$ **34.** $\frac{k}{3} = 2(6 - 8 + 12)$ **35.** $\frac{p}{4 - 10} = \frac{4 + 5}{8 - 5}$

36. $\frac{64.8}{3} = -2.4x$ **37.** $6 + 18 \div 9 = (9 - 29)y$ **38.** $4(-2.1) = \frac{n}{12(-3.5)}$

39. $56.8g = 1209.84$ **40.** $13.6 = \frac{e}{-9.7}$ **41.** $\frac{h}{123} = 0.88$

42. $8.036 = \frac{y}{-0.45}$ **43.** $2142 = -2520p$ **44.** $2.7x = 8.1(-9.3)$

45. $-1 = \frac{3}{4}x$ **46.** $-\frac{x}{2} = 1\frac{1}{4}$ **47.** $-\frac{3}{4}x = -\frac{2}{3}$

Which equation correctly models the problem situation? Identify and solve the appropriate equation.

48. SPENDING MONEY Your budget allows you to put $\frac{1}{3}$ of your weekly earnings into a savings account, spend $\frac{1}{3}$ on clothing, and spend $\frac{1}{3}$ on all other expenses. Last week you deposited $39.15 in your savings account. Let x represent last week's earnings. How much did you earn last week?

    a. $\frac{1}{3x} = 39.15$ **b.** $\frac{x}{3} = 39.15$ **c.** $3x = 39.15$

49. SAVING MONEY Of the money in your savings account, $\frac{3}{4}$ is for future college expenses and $\frac{1}{4}$ is for a summer trip. Your total savings account balance is $4968.80. Let s represent the amount of trip money saved. How much trip money has been saved?

    a. $s = \frac{3}{4}(4968.80)$ **b.** $s = 4968.80 - \frac{3}{4}(4968.80)$ **c.** $s = 4968.80 - \frac{3}{4}$

Write and solve an equation. Answer the question.

50. GEOMETRY A house is built on a rectangular lot that has an area of 7200 ft². The lot is 120 ft long. Let *w* represent the width of the lot. How wide is the lot?

51. GEOMETRY A square game board has a perimeter of 72 in. Let *x* represent the length of one side. Write and solve an equation to find this length.

52. EARTH SCIENCE The highest point in the United States is Mt. Whitney in California, which is 14,494 ft above sea level. The lowest point, Death Valley, California, is 282 ft below sea level. Find *d*, the difference in elevation between the two points.

53. EARTH SCIENCE The Marianas Trench in the Pacific Ocean is almost 11 km deep. It is 6.875 times as deep as the Grand Canyon. Find *g*, the depth of the Grand Canyon.

54. LIFE SCIENCE A researcher conducted experiments with the ocean plant kelp. She found the plant grew at a steady rate of 0.45 m per day. At the end of her experiment the plant had grown 12.6 m. Find *d*, the number of days the experiment lasted.

55. ASTRONOMY The planet Mercury has an average distance from the sun of 36,000,000 mi. This distance is about 11,612 times greater than Mercury's diameter. Let *m* represent Mercury's diameter. Find the approximate length of *m*.

THINK CRITICALLY

56. Look at the following equations.

    I. $7x = 42$ **II.** $0 \cdot 7x = 0 \cdot 42$ **III.** $(0 \cdot 7)x = 0 \cdot 42$ **IV.** $0 \cdot x = 0$

    a. How many solutions does equation I have?
    b. How many solutions does equation IV have?
    c. Are equation I and equation IV equivalent? Why or why not?
    d. Should you multiply both sides of an equation by zero? Explain.

MIXED REVIEW

57. STANDARDIZED TESTS Which is the simplified form of the expression

    $8 - 3r + 17 + 5r - 2$

    A. $8r - 23$ **B.** $-2r - 23$ **C.** $2r + 27$ **D.** $2r + 23$

Solve and check.

58. $0.4c = 2.6$ **59.** $\dfrac{d}{3} = 18$ **60.** $f - 10 = 3.5$ **61.** $28.7 = 16.2 + c$

Write and solve an equation for each.

62. The square of −4 is the product of 8 and what number?

63. Half of 72 is equal to adding 15 to what number?

Career
Nutrition Therapist

People have different body types. To plan an individual's nutrition and exercise program, a health professional will determine whether the person has a small, medium, or large body frame. You can use the method below to approximate body frame size. Work with a partner and take turns measuring and recording data.

- Extend your arm. Bend the forearm up to form a 90° angle. Keep your fingers straight and the inside of your wrist toward your body.

- Have your partner use a caliper to measure the distance between the two protruding bones on either side of the elbow. Measure to the nearest $\frac{1}{8}$". If a caliper is not available, your partner can place the thumb and index finger of their other hand on the two bones. Then use a ruler to measure the distance between the fingers.

- Compare the measurement with this table that gives information for *medium-framed* people. Measurements less than those given indicate a small frame; greater measurements, a large frame.

MALES		FEMALES	
Height*	Elbow Measurement	Height*	Elbow Measurement
5'2"–5'3"	$2\frac{1}{2}$"–$2\frac{7}{8}$"	4'10"–4'11"	$2\frac{1}{4}$"–$2\frac{1}{2}$"
5'4"–5'7"	$2\frac{5}{8}$"–$2\frac{7}{8}$"	5'0"–5'3"	$2\frac{1}{4}$"–$2\frac{1}{2}$"
5'8"–5'11"	$2\frac{3}{4}$"–3"	5'4"–5'7"	$2\frac{3}{8}$"–$2\frac{5}{8}$"
6'0"–6'3"	$2\frac{3}{4}$"–$3\frac{1}{8}$"	5'8"–5'11"	$2\frac{3}{8}$"–$2\frac{5}{8}$"
6'4"	$2\frac{7}{8}$"–$3\frac{1}{4}$"	6'0"	$2\frac{1}{2}$"–$2\frac{3}{4}$"

* Heights include shoes with 1-inch heels.

Decision Making

1. If you know a male's elbow measurement is 3", can you decide what size frame the person has? Explain.

2. Write an equation that shows how your elbow measurement compares to a person with a measurement of $2\frac{1}{2}$". Solve your equation and interpret the result.

3. Miguel's elbow measurement is $2\frac{5}{8}$". Theo's measurement is $\frac{1}{4}$" greater than Miguel's. Write an equation that gives Theo's measurement. Solve your equation.

3.5 Proportion, Percent, and Equations

Explore/Working Together

Suppose you want to buy the calculator shown. Work with a group. Rank the six prices in the order you believe is least to most expensive.

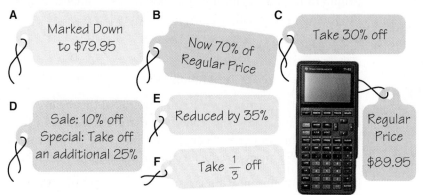

A Marked Down to $79.95

B Now 70% of Regular Price

C Take 30% off

D Sale: 10% off Special: Take off an additional 25%

E Reduced by 35%

F Take $\frac{1}{3}$ off

Regular Price $89.95

1. How can you estimate to see that $79.95 is the highest price?

2. Why are prices D and E not identical? Can you substitute a different price for $89.95 that might make the difference easier to see?

3. What percent is equivalent to $\frac{1}{3}$?

4. Must you know each actual price to know where to shop? Explain.

Build Understanding

There are several types of problems that involve percents. You can write an equation to solve each type of problem.

EXAMPLE 1

a. What number is 35% of 60?
b. What percent of 120 is 90?
c. 27 is 45% of what number?

Solution

a.
What number	is	35%	of	60?
x	=	0.35	·	60
x	=	21		

Let x be the unknown number.
Write 35% as a decimal.
21 is 35% of 60.

b.
What percent	of	120	is	90?
x	·	120	=	90

$$x = \frac{90}{120}$$

$$x = 75\%$$

Let x be the unknown percent.
Write an equation.
Divide both sides by 120.
Write x in percent form.
90 is 75% of 120.

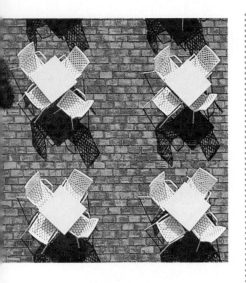

c. 27 is 45% of what number?

Let x represent the unknown number.

$$27 = 0.45 \cdot x$$

Write 45% as a decimal.

$$\frac{27}{0.45} = x$$

Divide both sides by 0.45.

$$60 = x$$

27 is 45% of 60. ◄

You can also use proportions to solve percent problems. Recall that a **proportion** is an equation that states that two ratios are equal. Proportions are usually written in one of two ways:

$$a:b = c:d \quad \text{or} \quad \frac{a}{b} = \frac{c}{d}$$

where a, b, c, and d are real numbers and $b \neq 0$ and $d \neq 0$.

In either of the forms above, b and c are called the **means** of the proportion and a and d are called the **extremes**. You can use the multiplication property of equality to show that in any proportion the product of the means is equal to the product of the extremes: $ad = bc$.

If one term of a proportion is a variable, you can **solve the proportion** by finding the value of the variable that makes the equation true.

In Explore, the questions dealt with **percent discount**. You estimated sale prices using any means you wished. You can find percent discount using either the equation method or the proportion method.

EXAMPLE 2

RETAILING The manager of a clothing store was having no success selling a particular shirt at $35.99. She decided to lower the price to $23.39. What percent discount did she use to determine the sale price?

Solution

Find the dollar amount of the discount.

$$\$35.99 - \$23.39 = \$12.60$$

You want to find what percent $12.60 is of $35.99, the original price.

Use an equation

Let x represent the percent.

$$12.60 = x \cdot 35.99$$

$$\frac{12.60}{35.99} = x$$

$$0.35 \approx x \quad \text{Round.}$$

Use a proportion

Let $\dfrac{x}{100}$ represent the percent.

$$\frac{12.60}{35.99} = \frac{x}{100}$$

$$\frac{100(12.60)}{35.99} = x$$

$$x \approx 35$$

$$\text{So, } \frac{x}{100} \approx 0.35$$

The manager used a discount of 35% from the original price. ◄

COMMUNICATING ABOUT ALGEBRA

Work together to prepare a list of fractional equivalents for some often-used percents. Discuss how the fraction form can help you solve some percent equations mentally.

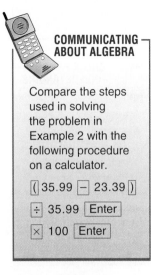

COMMUNICATING ABOUT ALGEBRA

Compare the steps used in solving the problem in Example 2 with the following procedure on a calculator.

[(35.99 − 23.39)]

[÷] 35.99 [Enter]

[×] 100 [Enter]

Solve. If the answer is not a whole percent, round to the nearest tenth.

1. 40% of 120

2. 15% of 260

3. 9% of 90

4. 25% of 56

5. 84% of 155

6. 3% of 30

7. What percent of 70 is 20?

8. What percent of 134 is 45?

9. What percent of 18 is 30?

10. What percent of 200 is 32?

11. 64 is 80% of what number?

12. 78 is 16% of what number?

13. 130 is 125% of what number?

14. 90 is 200% of what number?

15. BANKING On July 1, Sari deposited $540 in a bank account paying simple interest. Exactly one year later, he closed the account and withdrew $564.30. At what percent was interest paid on his account during the year?

16. RETAILING During a pre-holiday sale, a bookstore marked down all mysteries 25%. One book was on sale for $11.96. What was the original price of the book?

17. WRITING MATHEMATICS Explain how you could find 3% of 158,000,000 using a calculator with a display of fewer than 9 digits.

PRACTICE

Solve. If the answer is not a whole percent, round to the nearest tenth.

1. What is 23% of 45?

2. 35 is 80% of what number?

3. What percent of 150 is 250?

4. Find 125% of 36.

5. 75 is what percent of 120?

6. What percent of 7 is 28?

7. What is 230% of 230?

8. 14 is what percent of 50?

9. What percent of 300 is 190?

10. What is 56% of 150?

11. What is 8% of 32?

12. What percent of 12 is 40?

13. RETAILING Percent markup is the difference between the price a store pays for an item and the price it charges customers. A store pays $79 for a CD player, which it then sells for $135. To the nearest percent, what is the percent markup from the original price?

14. ENVIRONMENT On a typical day, sanitation workers collect about 170 million tons of household garbage. It is expected that the amount collected will grow approximately 2% annually over the next several years. If that forecast is correct, about how many tons of household garbage will be collected on a typical day next year?

15. ANTHROPOLOGY In 1994, anthropologists made a startling discovery. They found what they believe was a system of standard weights in use by inhabitants of the Andes Mountains in Peru as long as 700 years ago. This was considerably earlier than scientists previously thought this skill was in use. What they found were three weights: 190 g, 380 g, and 1140 g. These were, in order, 5000%, 10,000%, and 30,000% of what appears to have been a standard weight, which was not found. What was the weight of the standard?

EXTEND

Solve. Round to the nearest tenth.

16. What is 0.5% of 120?

17. What percent of 12.6 is 3?

18. 45.8 is 20% of what number?

19. Find 13.9% of 15.

20. What percent of 49 is 3.8?

21. What is 150.5% of 34?

22. Find 12.5% of 246.

23. 11.6 is $15\frac{1}{2}$% of what number?

FINANCE The table gives the sales tax rate for several states. Use the table to answer Exercises 24–26.

24. In which state will a car advertised at $11,000 actually cost $11,440 when the sales tax is included?

25. What is the difference in cost between a $23,500 car bought in Mississippi and the same car bought in Colorado? Explain two different ways of solving the problem.

26. In which states would it be possible to purchase a car with a sticker price of $17,990 with no more than $19,000?

Sales Tax Rates	
Alabama	4%
Colorado	3%
Connecticut	6%
Maine	5%
Mississippi	7%

THINK CRITICALLY

For Exercises 27–30, write *true* or *false*. If false, use an example to show why.

27. 100% of 50 is equivalent to 50% of 100.

28. 10% of 30 plus 5% of 30 equals 15% of 30.

29. A discount of 20% followed by another discount of 20% is equivalent to a discount of 40%.

30. The price you pay for a $25 shirt is the same in both of the following cases: **a.** A discount of 30% is figured and then sales tax of 6% is added, and **b.** sales tax of 6% is added and then a discount of 30% is figured.

31. NUTRITION Wheetos cereal includes the following nutritional information on its box: one serving $\left(\frac{3}{4}\ \text{cup}\right)$ of the dry cereal provides 3 g of protein, or 4% of the RDA. One serving with skim milk $\left(\frac{1}{2}\ \text{cup}\right)$ provides 7 g, or 15% of the RDA. What do you think of the information provided? Explain.

MIXED REVIEW

Make a table of possible sums when two number cubes, each labeled 1 to 6, are rolled. What is the probability that, on one roll of the two cubes, the sum will be:

32. an even number

33. a two-digit number

34. a prime number

35. a number less than 7

36. a number no greater than 7

37. a number divisible by 3

38. STANDARDIZED TESTS Which is the value of the expression $|23.5| + |-14|$?

    A. 9.5   **B.** 37.5   **C.** 13.5   **D.** 24.9

Career
Personal Trainer

A personal trainer helps his or her clients create a fitness program. The trainer will often assess the client's eating habits and design an exercise program to burn a certain number of calories. Different activities burn different numbers of calories. Each of the activities below burns a total of 2000 Cal during a week, and might be suggested by a trainer.

Walking—4 mi in 1 h, 5 days per week

Swimming—30 min per day, 6 days per week

Tennis—1 h per day, 5 days per week

Jogging—3 mi in 30 min, 6 days per week

A client may become bored with the same exercise routine from day to day and ask for a program that has varied activities. To design a program, a trainer needs to know how many calories each activity burns per day.

Decision Making

1. Write and solve an equation to determine the number of calories burned by each of the following activities.

    a. walking 1 mi in 15 min

    b. swimming for 30 min

    c. playing tennis for 1 h

    d. jogging 1 mi in 10 min

2. A client has requested a more varied routine. Design a weekly exercise program that meets the following restrictions.

    a. There are no more than 5 days of exercise.

    b. There are no fewer than 3 days of exercise.

    c. No consecutive exercise days have the same activity.

    d. No activity is done more than twice in the week.

    e. Exercise cannot last longer than 1 h per day.

    f. Between 2000 and 2100 Cal must be burned per week.

    g. No more than 35% of total calories are burned on any one day.

3.6 Solve Equations Using Two or More Operations

Explore

In Figure A, each bag contains the same unknown weight of nails. The bags together with the 5 additional nails, each of which weighs one ounce, exactly balance 41 oz.

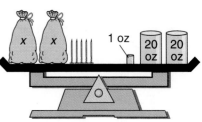

Figure A

1. Use the guess-and-check strategy to estimate the weight of each bag. Check your guess. Continue until you have found the value of x.

2. What does the value of x represent in this situation?

3. If the two bags were combined into one larger bag, weighing the same as the two bags, would the balance be maintained?

Figure B shows the bags combined.

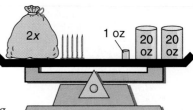

4. Write an equation to describe the relation you see in Figure B.

5. Test the solution you found using guess-and-check in this equation. Does it satisfy the equation you wrote?

Figure B

Build Understanding

To solve equations requiring more than one operation, you will first need to isolate the term containing the variable. Usually this means you should add or subtract before multiplying or dividing.

EXAMPLE 1

Solve and check: $7x - 2 = -37$

Solution

$$7x - 2 = -37$$
$$7x - 2 + 2 = -37 + 2 \qquad \text{Add 2 to both sides.}$$
$$\frac{7x}{7} = \frac{-35}{7} \qquad \text{Divide both sides by 7.}$$
$$x = -5$$

SPOTLIGHT ON LEARNING

WHAT? In this lesson you will learn
- to solve equations requiring two or more steps.

WHY? Solving two-step equations helps you solve problems about travel mileage, automobile depreciation, and electrical usage.

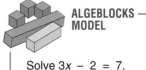

ALGEBLOCKS MODEL

Solve $3x - 2 = 7$. Model the equation.

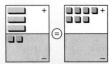

Add 2 to both sides. Remove zero pairs.

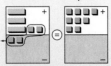

Divide both sides by 3.

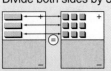

Read the solution: $x = 3$.

Check

$$7(-5) - 2 \stackrel{?}{=} -37$$
$$-35 - 2 \stackrel{?}{=} -37$$
$$-37 = -37 \checkmark \qquad \textbf{The solution is -5.} \qquad \blacktriangleleft$$

To solve certain linear equations, it may be necessary to combine like terms first.

THINK BACK

Recall that like terms are terms that have the same variable base with the same exponent.

EXAMPLE 2

Solve and check: $10b - 6b - 3 = 9$

Solution

Combine like terms first.

$$10b - 6b - 3 = 9$$
$$4b - 3 = 9 \qquad \text{Combine like terms.}$$
$$4b - 3 + 3 = 9 + 3 \qquad \text{Add 3 to both sides.}$$
$$\frac{4b}{4} = \frac{12}{4} \qquad \text{Divide both sides by 4.}$$
$$b = 3$$

Check

$$10(3) - 6(3) - 3 \stackrel{?}{=} 9 \qquad \text{Substitute 3 for } b.$$
$$30 - 18 - 3 \stackrel{?}{=} 9$$
$$9 = 9 \checkmark \qquad \textbf{The solution is 3.} \qquad \blacktriangleleft$$

ALGEBRA: WHO, WHERE, WHEN

Problems in the Egyptian documentary remains of the period 2000–1500 B.C. include the sorts of problems we now solve by a linear equation such as $2x + 5 = 15$.

When linear equations contain parentheses, you will need to remove the parentheses first. In order to clear the parentheses from the equation, you may need to use the distributive property.

EXAMPLE 3

Solve and check: $5(x + 3) = -30$

Solution

Remove parentheses first.

$$5(x + 3) = -30 \qquad \text{Use the distributive property to}$$
$$5x + 15 = -30 \qquad \text{eliminate parentheses.}$$
$$5x + 15 - 15 = -30 - 15 \qquad \text{Subtract 15 from both sides.}$$
$$\frac{5x}{5} = \frac{-45}{5} \qquad \text{Divide both sides by 5.}$$
$$x = -9$$

Check

$$5((-9) + 3) \stackrel{?}{=} -30 \qquad \text{Substitute -9 for } x.$$
$$5(-6) \stackrel{?}{=} -30$$
$$-30 = -30 \checkmark \qquad \textbf{The solution is -9.} \qquad \blacktriangleleft$$

CHECK UNDERSTANDING

When you apply the distributive property to an expression such as $-(24 - 3m)$, why does the sign preceding the term $3m$ change from negative to positive?

In some problems, you can describe the relationship between several pieces of information using one variable.

EXAMPLE 4

BASKETBALL Two teams scored a total of 122 points during a basketball game. The winning team outscored the losing team by 24 points. What was the final score of the game?

Solution

Define the variable. Let x represent the number of points scored by the losing team. Then $x + 24$ represents the number of points scored by the winning team.

$x + (x + 24) = 122$	Write an equation that describes the problem conditions.
$2x + 24 = 122$	Use the distributive property and combine like terms.
$2x + 24 - 24 = 122 - 24$	Subtract 24 from both sides.
$2x = 98$	Simplify.
$\dfrac{2x}{2} = \dfrac{98}{2}$	Divide both sides by 2.
$x = 49$	

The losing team scored 49 points. Since the winning team outscored the losing team by 24, the winning team scored $49 + 24$, or 73, points.

Check

$$49 + 73 = 122$$
$$122 \stackrel{?}{=} 122 \checkmark \qquad \textbf{The solution is 49.} \qquad \blacktriangleleft$$

142

Describe the first step you would take to solve each equation for the underlined variable.

1. $\underline{a} - b = c$ **2.** $m\underline{n} + 4 = p$ **3.** $6\underline{x} - y = s$ **4.** $\dfrac{n + p}{7} = r$

Solve and check each equation.

5. $3m - 12 = 15$ **6.** $\dfrac{r}{8} - 7 = -1$ **7.** $\dfrac{z}{3} + 70 = 98$

8. $9w + 6.25 = 123.25$ **9.** $\dfrac{a}{2.5} - 3.8 = 8.2$ **10.** $\dfrac{-6f}{2} + 1.9 = -8.6$

11. $-3x + 6 = 30$ **12.** $-8y - 15 = -87$ **13.** $4(x - 8) = -4$

14. $5(15 + h) = 65$ **15.** $-2(3 - x) = -14$ **16.** $(c - 3) - 2(8 - c) = -37$

17. MODELING Use Algeblocks to solve $3x + 4 = -2$. Make sketches to record each step.

18. SPORTS The length of a football field is 67 yd greater than the width. The perimeter of the field is 346 yd. Find the length and width of the football field.

19. AUTOMOBILE DEPRECIATION A new car purchased for $14,000 loses an average of $1800 value per year for each of the first six years of ownership. After a certain number of years the value of the car has dropped to $6800. Write and solve an equation to determine the number of years for the car's value to drop to $6800.

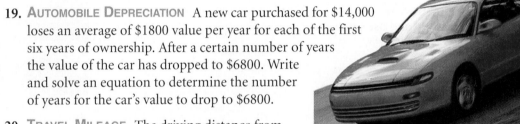

20. TRAVEL MILEAGE The driving distance from Dallas, Texas, to Cleveland, Ohio, is m mi. The distance from Cleveland to Seattle, Washington, is $2m$. The total distance from Dallas to Cleveland to Seattle is 3600 mi. What is the distance between Dallas and Cleveland? between Cleveland and Seattle?

21. WRITING MATHEMATICS Explain the steps you use to write and solve an equation from the data in a word problem.

PRACTICE

Decide whether 3 or –3 is the solution to each equation.

1. $3x - 9 = 0$ **2.** $\dfrac{2y}{2} + 5 = 2$ **3.** $-17 + 2x = -11$

4. $2.5r + 6 = -1.5$ **5.** $-4e - 8 = -20$ **6.** $24 - 8.5c = 50.5$

Solve and check.

7. $14 = 8r - 58$ **8.** $3.9w - 46.8 = 0$ **9.** $\dfrac{1}{3}x - 15 = 10$

10. $\dfrac{a}{11} + 16 = 25$ **11.** $-35 - 4y = 5$ **12.** $6(3 - e) = 42$

13. $13x - 7x = 24$ **14.** $-8e - 3e = 22$ **15.** $23 - \dfrac{1}{2}h = 17.5$

16. $15(2 + x) - 3x = 114$ **17.** $-x + (9 - 2x) = 39$ **18.** $-6 = -\dfrac{1}{4}(a + 4)$

19. $4c - 5(c - 3) = 12.5$ **20.** $19 - 14x = 6.4$ **21.** $2(x - 8) - 3(x + 2) = -24$

22. $\frac{5y}{4} - 9 = -49$ **23.** $\frac{x}{7} - 3 = 5$ **24.** $\frac{2}{3}(z + 1) = 4$

25. $6m - 3(m + 4) = 6$ **26.** $8.9c + 2.3 = -16.39$ **27.** $1.7a - (a + 3) = 3.3$

28. TAXI FARE A taxi company uses the following formula to determine the cost of a ride: $F = \$1.25 + \$0.75(m - 1)$ where m represents the total number of miles traveled and F represents the fare. At the conclusion of a ride, a passenger pays the driver \$10.25. How far did the passenger travel?

29. TYPING SPEED An estimate of a person's accurate typing speed is given by the formula $S = \frac{w - 5e}{10}$ where S is the accurate speed in words per minute, w is the actual number of words typed in 10 min, and e is the number of errors made during the test. An applicant for a job requiring typing was told that he had an accurate speed of 36 words per minute. During the 10-min test, he typed a total of 400 words. How many errors did he make?

30. WRITING MATHEMATICS Think about how you solve equations that involve more than one operation. Which operation do you tend to perform first? Explain your reasoning and use some equations as examples.

EXTEND

Solve and check each equation.

31. $(4.5 - w)3 + \frac{w}{2} = 21$ **32.** $-87 = \frac{7}{2}(x - 6) + 2x$ **33.** $-14\frac{2}{3} = \frac{4}{9}(-9y - 6)$

34. GRADE AVERAGES A student's grades on three math tests were x, $x + 5$, and $x - 2$. If the total number of points scored on the three tests was 273, what was the student's average? What was the score on each test?

35. PROBABILITY A spinner can be designed so that the probability of landing on an even number is three times as great as the probability of landing on an odd number. Draw a circular spinner and construct it to meet the above requirements. Write and solve an equation to determine the probability of landing on an even or an odd number.

36. ELECTRICAL USAGE An electric company charges \$0.13 per kilowatt-hour for the first 250 kilowatt-hours used and \$0.09 for each kilowatt-hour beyond 250. An appliance store advertises an efficient refrigerator that uses 400 kilowatt-hours of electricity per month. If you don't want the electricity cost to exceed \$40.00 per month, should you buy the refrigerator?

37. GEOMETRY What is the measure of each angle in parallelogram *QRST*? (*Hint*: The sum of the measures of the angles is 360°.)

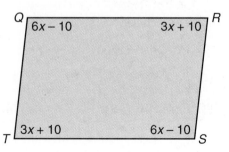

38. WRITING MATHEMATICS Write a two-operation equation involving addition and division that has -8 as its solution. Then write a two-operation equation involving subtraction and multiplication that has 7 as its solution.

THINK CRITICALLY

39. BUSINESS The equation $P = 19x - 1300$ represents the calculation done by the owner of the All Sweaters $19.00 Shop to determine her monthly profit. Explain what each term of the equation represents. If the profit one month was $1778, how many sweaters were sold?

40. Consecutive integers are integers that differ by one. You can represent consecutive integers as $n, n + 1, n + 2$, and so on. Find three consecutive integers with a sum of 111.

41. If you begin with an odd integer n and count by two's, then $n, n + 2, n + 4$ are consecutive odd integers. Find three consecutive odd integers with a sum of 87.

42. Can you find three consecutive even integers whose sum is 100? Explain.

Solve each problem using mental math.

43. If $8m - 3m = 18$, find the value of $6m + 4m$.

44. If $\dfrac{9 - r}{2} = 14$, find the value of $2(9 - r)$.

45. If $5(a + 3) - 2(a + 3) = 7$, find the value of $12(a + 3)$.

46. Write an equation similar to Exercise 45 whose solution is –1.

PROJECT *Connection*

Often foods marketed as healthful, low-fat, or low-salt are more expensive than the standard form of the food. To assess the advantages and disadvantages of eating these modified foods, you can compare the percent of decrease in the substance being reduced with the percent of increase in the price.

Recall that to find the percent of increase or decrease, subtract the new value from the original value and divide the result by the original value.

1. Choose one food advertised as low fat. Compare its nutrition label with the label of the standard version of the product. Determine the percent decrease in fat and the percent increase in price, if any. How do the two compare?

2. Repeat the activity for several other items. Do you notice any pattern in either the amount of reduction or the amount of price increase?

3. For the items that you compared, prepare a bar graph comparing the percent decrease of the substance with the percent increase in the price.

Solve Equations with Variables on Both Sides

Explore

• The figure shows a balanced scale with the unknown weight on both sides.

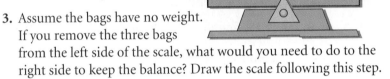

1. Write a short paragraph to describe what is shown.

2. Write an equation that reflects the diagram.

3. Assume the bags have no weight. If you remove the three bags from the left side of the scale, what would you need to do to the right side to keep the balance? Draw the scale following this step.

4. What is the effect of cutting the string holding down the balloon?

5. What must you do to the left side to keep the balance after cutting the balloon?

6. Write the equation following this step.

7. What is the value of x?

Build Understanding

• Solving equations with the variable on both sides of the equal symbol is an extension of solving equations requiring two or more operations. Since the variable represents a number, you can apply the addition, subtraction, multiplication, and division properties to a variable.

EXAMPLE 1

Solve and check: $5x - 21 = -2x + 28$

Solution

Eliminate the variable from the right side first.

$$5x - 21 = -2x + 28$$
$$5x + 2x - 21 = -2x + 2x + 28 \quad \text{Add } 2x \text{ to both sides.}$$
$$7x - 21 = 28$$
$$7x - 21 + 21 = 28 + 21 \quad \text{Add 21 to both sides.}$$
$$7x = 49$$
$$\frac{7x}{7} = \frac{49}{7} \quad \text{Divide both sides by 7.}$$
$$x = 7 \quad \textbf{The solution is 7.}$$

Check the solution by substituting 7 in the original equation. ◄

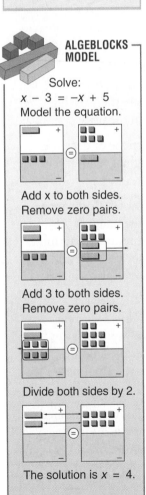

ALGEBLOCKS MODEL

Solve:
$x - 3 = -x + 5$
Model the equation.

Add x to both sides.
Remove zero pairs.

Add 3 to both sides.
Remove zero pairs.

Divide both sides by 2.

The solution is $x = 4$.

To solve equations with the variable on both sides, you will often follow these steps.

1. Use the distributive property to clear the equation of parentheses.
2. Combine like terms.
3. Use the addition or subtraction property to isolate the variable.
4. Use the multiplication or division property to solve for the variable.

EXAMPLE 2

Solve and check: $-3n - 2(n + 4) = -6n - 5$

Solution

$$\begin{aligned}
-3n - 2(n + 4) &= -6n - 5 \\
-3n - 2n - 8 &= -6n - 5 \quad &\text{Use the distributive property.} \\
-5n - 8 &= -6n - 5 \quad &\text{Combine like terms.} \\
-5n + 6n - 8 &= -6n + 6n - 5 \quad &\text{Add } 6n \text{ to both sides.} \\
n - 8 &= -5 \\
n - 8 + 8 &= -5 + 8 \quad &\text{Add 8 to both sides.} \\
n &= 3 \quad &\textbf{The solution is 3.}
\end{aligned}$$

Check the solution in the original equation. ◄

Solving equations of this type can help you make choices.

EXAMPLE 3

COMMUNICATIONS A cellular phone company offers two payment plans. Under the first, there is a monthly fee of $40 and a charge of $0.30 per minute. The second charges a $25 monthly fee and a $0.50 per-minute charge.

a. Write an expression for the cost under each plan if you speak m min per month.

b. For what number of minutes are the costs equal?

Solution

a. First plan: Cost is $40 + $0.30 • number of minutes
$$C = 40 + 0.30m$$
Second plan: Cost is $25 + $0.50 • number of minutes
$$C = 25 + 0.50m$$

b.
$$\begin{aligned}
40 + 0.30m &= 25 + 0.50m \quad &\text{The costs are equal.} \\
40 + 0.30m - 0.30m &= 25 + 0.50m - 0.30m \quad &\text{Subtract } 0.30m \\
40 &= 25 + 0.2m \quad &\text{from both sides.} \\
40 - 25 &= 25 - 25 + 0.2m \quad &\text{Subtract 25} \\
15 &= 0.2m \quad &\text{from both sides.} \\
75 &= m \quad &\textbf{The solution is 75.}
\end{aligned}$$

The costs are equal if you speak for 75 min per month. ◄

CHECK UNDERSTANDING

Rewrite the solution to Example 2 by adding $5n$ to both sides of the equation. What is accomplished by collecting the variables on the side of the equation with the greater variable coefficient, as was done in Example 2?

PROBLEM SOLVING TIP

Note that to find m, it is easier to multiply both sides of $15 = 0.2m$ by 5 than to divide by 0.2. Plan your steps so you can use mental math when possible.

Use the steps to solve Exercises 1–3.

1. $8a + 12 = 11a$
  a. Subtract $8a$ from both sides.
  b. Divide both sides by 3.

2. $-3c - 4 = -5c + 6$
  a. Add $5c$ to both sides.
  b. Add 4 to both sides.
  c. Divide both sides by 2.

3. $6(r - 3) = 4r - 4$
  a. Use the distributive property.
  b. Subtract $4r$ from both sides.
  c. Add 18 to both sides.
  d. Divide both sides by 2.

Solve and check each equation.

4. $9b + 8 = 10b$

5. $6f + 24 = -2f$

6. $-8 + 3h = 2(h - 5)$

7. $-3(z + 4) = -4z$

8. $24 - 9t = -13t + 8$

9. $7x = 10(x - 1.5)$

10. $61 - 13r = -3r + 1$

11. $-n = 5n - 72$

12. $\frac{1}{2}(3c - 4) = c + 8$

13. $3(5.5 - m) = 8m$

14. $e - 17 = -4e + 24.25$

15. $\frac{1}{4}x = 2x + 17.5$

BANKING Pedro and Moira each opened savings accounts on the same day, intending to withdraw the money for weekly expenses. Pedro opened his account with $710 and withdrew $35 per week. Moira opened her account with $570 and withdrew $25 weekly. Let w be the number of weeks. In how many weeks will their accounts be equal?

16. Select the equation that models the problem.
  a. $710 + 35w = 60w - 570$ **b.** $710 - 35w = 570 - 25w$ **c.** $570 + 25w = 710 + 35w$

17. Solve the equation in Exercise 16 that models the problem. What does the solution represent in terms of this problem?

18. How much will be in each account when they are equal?

19. **GEOMETRY** Write and solve an equation to find the length and area of the rectangle at the right.

20. **WRITING MATHEMATICS** Choose one of the equations from Exercises 4 through 15. Write the steps you used to solve the equation and what each step accomplished.

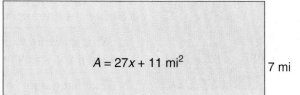

$A = 27x + 11$ mi^2 7 mi

$(3x + 5)$mi

PRACTICE

Solve and check each equation.

1. $m - 7 = -13 - m$

2. $2r = 35 - 3r$

3. $x + 6 = -6x + 13$

4. $23 - 4e = -7e + 2$

5. $5y = 32 - 3y$

6. $5x - 7 + 2x = 3x - 2 + 5x$

7. $-3a + 1 = -4a + 8$

8. $17 + 7n = 8 + 10n$

9. $-8c + 3(c - 2) = -3c + 2$

10. $(3y + 4) - y = 10y$

11. $2(-3h + 5) = -9h - 17$

12. $-17w - 4(w - 1) = -20w + 1$

13. **WRITING MATHEMATICS** Write a word problem leading to an equation with variables on both sides of the equal symbol. Be sure your problem has a solution that makes sense.

14. $6(p + 5) - 3(p - 2) = 12p + 18$

15. $12(x - 7) + 3(2x + 2) = 50x - 62$

16. $\frac{1}{2}(a + 15) = 3a - 1.25$

17. $\frac{2}{3}(6d + 3) + (d - 8) = -4d + 12$

18. $-3 + 2(4k - 13) = -40k + 7$

19. $-5(6 - w) = 9(w + 2) - 16$

20. $\frac{3}{5}(15c + 10) = 12c - 9$

21. $2(-3m + 4) = 4(-2m + 6)$

22. **BUSINESS** To raise money to fund a recycling program, students agreed to sell T-shirts. The company from which they planned to purchase the shirts sent them the following table, which includes the price for which the students could expect to sell each shirt.

Description	Color	Cost	Price
50% cotton, 50% polyester	White	$48/doz	$7.50 ea
50% cotton, 50% polyester	Green, Blue	$54/doz	$9.50 ea
100% cotton	White	$60/doz	$10.00 ea
100% cotton	Green, Blue	$66/doz	$12.00 ea

If the students buy only one type of shirt, which type will enable them to make at least $700 profit by selling the fewest shirts? Write and solve equations to show your work.

23. **CONSUMERISM** The Magic Carpet charges $90 for installation and $9 per square yard of carpeting. The Carpeteria's installation price is $50, but the store charges $13 for each square yard. Write and solve an equation to find x, the number of square yards of carpeting for which the cost, including installation, is the same for both stores.

24. In Problem 23, at the point where the cost at both stores is the same, how much must the buyer pay the store?

NUMBERS The sum of 6 times a number and 3 is equal to 17 less than the number.

25. Which equation models the data in the problem?
   a. $6x + 3 = 17 - x$ **b.** $6x + 3 = x - 17$
   c. $6x - 3 = x - 17$ **d.** $3 - 6x = x - 17$

26. Find the number.

EXTEND

Solve each equation.

27. $\dfrac{x + 6}{3} = \dfrac{5x}{9}$

28. $\dfrac{2m + 1}{9} = \dfrac{12 - 7m}{6}$

29. $\dfrac{2(3a - 4)}{7} = \dfrac{-4(10 - a)}{-4}$

30. Find a value of n such that -4 is a solution of the equation $2x + 3 - 4n = x + 7$.

31. Find a value of n such that -3 is a solution of the equation $2n + w - 10 = 3w + 1 - 5n$.

32. Of three numbers, the second number is 12 less than 4 times the first. The third number is 25 more than $\frac{3}{4}$ the second. The second and third numbers are equal. What are the three numbers?

33. CARPENTRY A carpenter needs some boards for a construction project. If he buys 19 boards, he has 8 ft extra. If he buys 18 boards, he is short 4 ft. If the boards are each the same length, how long is each board? What is the exact number of feet of lumber the carpenter needs?

34. PROBABILITY In the two-section spinner shown at the right, the probability of landing on red in a single spin is $\frac{x-3}{20}$. The probability of landing on green is $\frac{x+4}{16}$. What is the probability of landing in each section? Find the value of x.

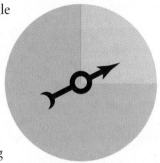

THINK CRITICALLY

35. Describe the similarities in each of the following equations.
  a. $10 - 3x = -4x + 6$
  b. $5a + 7 = 6a - 2$
  c. $8 - 2m = 32 - m$

36. How do the similarities allow each of the equations in Exercise 35 to be solved using the same method?

37. Solve each of the following equations. Then write a general statement explaining what you found.
  a. $3y + 2 = 3y + 7$ **b.** $6 - 8m = -12m + 9 + 4m$ **c.** $3(5x - 2) - 6x = 7 + 9x$

PROJECT *Connection* Most food markets are now providing several varieties of ground beef. One difference is the percent of fat. Research the choices of ground beef at local stores, noting the price per pound as well as the percent of fat in each type of ground beef sold.

Because the leaner meats are more expensive, some shoppers may purchase some beef with the higher fat content and some with the lower fat content. Then they mix the two varieties, creating a ground beef with a new fat content. In doing this, people try to save money yet eat food with less fat.

  1. Using the data you collected, determine the fat content of ground beef that uses 3 lb of the lowest fat content meat offered and 2 lb of the highest fat content meat.

  2. Find the cost of the 5-lb meat mixture.

  3. Find the cost per pound of the meat mixture. Compare price and fat content with the store's varieties.

  4. What other considerations might make the low-fat beef a better buy?

Career
Aerobics Instructor

In designing an aerobic fitness program, an aerobics instructor will determine a client's desirable exercise heart rate range. Heart rate is measured in beats per minute. To obtain aerobic benefits, a person must exercise at least three times per week, keeping his or her heart rate above a minimum level for between 20 and 30 min. Trainers use these steps to determine the range of heart beats during exercise.

a. Subtract your age from 220.

b. To find the minimum rate during exercise, find 60% of the result from Step 1.

c. To find the maximum rate, find 75% of the result in Step 1.

1. Write expressions that can be used to find the minimum and maximum heart rates.

2. Determine your own minimum and maximum heart rates during exercise.

Decision Making

3. If a 35-year-old client told you that during his workouts he kept his heart rate at 150 beats per minute for 25 min, what would your advice be?

4. Describe two ways that you can use the data in Problem 3 to reach your conclusion.

3.8 Use Literal Equations and Formulas

Explore

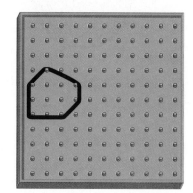

The geoboard polygon has pegs that touch the boundary, called *boundary pegs*, and pegs inside the polygon, called *interior pegs*.

1. How many of each type of peg are in the figure?

2. If the area of the smallest square on the geoboard is 1 square unit, what is the area A of the figure?

The formula $A = \dfrac{B}{2} + I - 1$ relates the area of a geoboard polygon to the number of boundary pegs B and interior pegs I.

3. Test the formula for the figure shown above.

4. Use a geoboard to create other polygons. Count the boundary and interior pegs for each. Use the formula to calculate the area. Find the area another way to check the answer found with the formula.

Build Understanding

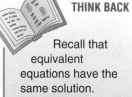

THINK BACK

Recall that equivalent equations have the same solution.

The *formula* for the area of a geoboard polygon is an example of a *literal equation*. A **literal equation** contains at least two different variables. Common literal equations are often called **formulas**.

You can write an equivalent literal equation by using the same properties and operations you used to solve other equations. Solving a literal equation often makes it easier for you to find the value of a particular variable when values are substituted for the other variables.

EXAMPLE 1

How many boundary pegs are on a geoboard polygon with an area of 7 square units and 2 interior pegs?

Solution

$$A = \frac{B}{2} + I - 1 \qquad \text{Solve for } B.$$

$$A - I + 1 = \frac{B}{2} + I - 1 - I + 1 \qquad \text{Subtract } I \text{ from both sides and add 1 to both sides.}$$

$$A - I + 1 = \frac{B}{2} \qquad \text{Simplify.}$$

$$2(A - I + 1) = B \qquad \text{Multiply both sides by 2.}$$

$$2(A - I + 1) = B$$
$$2(7 - 2 + 1) = B$$
$$2(6) = B$$
$$12 = B$$

Use the equation.
Substitute 7 for A and 2 for I.

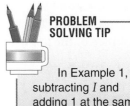

PROBLEM SOLVING TIP

In Example 1, subtracting I and adding 1 at the same time saves a step.

A geoboard polygon with an area of 7 units^2 and 2 interior pegs has 12 boundary pegs.

Solving a formula for a particular variable can often help you see a relationship from a different perspective.

EXAMPLE 2

UTILITY BILLS Julissa and her father were discussing how much using her hair dryer cost each month. Her father thought that she must be adding at least $10 per month to the utility bill. Julissa knew the formula that relates cost, time, and amount of electricity used.

$$W = \frac{1000C}{tc}$$

where W is the number of watts the appliance uses, C is the cost to use the appliance per month, t is the time in hours, and c is the cost of electricity per kilowatt hour. What is the cost of operating Julissa's hair dryer per month?

Solution

Julissa first solved the formula for C, the cost per month.

$$W = \frac{1000C}{tc}$$

Write the formula.

$$W(tc) = \left(\frac{1000C}{tc}\right)tc$$

Multiply both sides by tc.

$$Wtc = 1000c$$

Simplify.

$$\frac{Wtc}{1000} = \frac{1000C}{1000}$$

Divide both sides by 1000.

$$\frac{Wtc}{1000} = C$$

From the electric bill, Julissa found the cost c per kilowatt hour was $0.13. Her hair dryer uses 1500 watts. She uses her hair dryer 15 min or 0.25 h per day for 30 days a month. To find the monthly cost of using her hair dryer, Julissa substituted the values she knew.

$$C = \frac{Wtc}{1000}$$

$$C = \frac{1500(0.25 \cdot 30)(0.13)}{1000}$$

Substitute 1500 for W, $0.25 \cdot 30$ for t, and 0.13 for c.

$$C = 1.46$$

Julissa demonstrated that the cost of using her hair dryer was approximately $1.46 per month, not $10 per month.

CHECK UNDERSTANDING

In Example 2, how do you know the cost is in dollars?

Which of the following are literal equations? Write *yes* or *no*.

1. $3x + 7 = -2 + 5x$ **2.** $6(4 + 2a) = 8a$ **3.** $9c = 6d$ **4.** $-p(1 + 3a) = 2p$

Describe the first step you would take to solve each equation for the underlined variable.

5. $a\underline{b} + 7 = c$ **6.** $x - \underline{z} = y$ **7.** $3\underline{w} - m = r$

8. $\dfrac{\underline{m} + 5}{9} = t$ **9.** $9\underline{a} = C$ **10.** $d + 2 = 12 - \underline{e}$

Solve each equation for the underlined variable.

11. perimeter of a rectangle
$$P = 2(l + \underline{w})$$

12. consumer price index
$$I = \dfrac{100P}{\underline{p}}$$

13. hours and sunblock number
$$H = \dfrac{15\underline{N}}{60}$$

14. **HOME INSULATION** The net R-value of a substance used for insulation describes how much resistance it offers to the flow of heat. The net R-value is $R = rT$ where r is the value of the substance per inch of thickness T. Determine which substances give a homebuilder a minimum R-value of 19 using 6.5 in. of insulation. (Data are from the U.S. Department of Energy.)

Substance	R-value
Fiberglass blanket	3.25
Loosefill fiberglass	2.2
Loosefill cellulite	3.7

15. **WRITING MATHEMATICS** Describe the similarities and differences between solving literal equations and solving those containing only one variable.

PRACTICE

Solve each equation for the underlined variable.

1. gravity
$$g = \dfrac{Gm}{s^2}$$

2. kinetic energy
$$m = \dfrac{2E}{v^2}$$

3. volume of a cone
$$V = \dfrac{1}{3}\pi r^2 \underline{h}$$

4. simple interest
$$I = Pr\underline{t}$$

5. cost including tax
$$C = c + r\underline{c}$$

6. electrical charge
$$W = VI\underline{t}$$

7. matter and energy
$$E = \underline{m}c^2$$

8. electrical power
$$P = \underline{I}V$$

Solve each equation for the underlined variable. Then find the value of the underlined variable for the given values of the other variables.

9. $3\underline{x} + 7 = y$ when $y = 19$

10. $9 - 4m + \underline{b} = 15$ when $m = 2$

11. $V = lw\underline{h}$ when $V = 48, l = 8, w = 2$

12. $C = 2\pi\underline{r}$ when $C = 94.2$ (Use 3.14 for π.)

13. $\underline{V}I = \dfrac{E}{t}$ when $I = 0.5, E = 900, t = 300$

14. $P = \dfrac{k\underline{w}}{H}$ when $P = 18.8, H = 3, k = 1.2$

TRAVEL To solve Exercises 15–17, use the formula: distance = rate · time ($d = rt$).

15. At an average speed of 55 mi/h, how many hours of driving will a 715-mi trip require?

16. By sharing the driving, a group of friends was able to complete a 988-mi trip in 19 h. What was their average speed?

17. At an average speed of 61 mi/h, how many miles can be covered in 11.5 h of driving?

GEOMETRY Use the formula for the area of a trapezoid, $A = \frac{1}{2}h(b_1 + b_2)$, where b_1 and b_2 represent the two parallel bases, to solve Exercises 18–19.

18. A trapezoid has area 112 in.^{2} and height 8 in. One base of the figure is 12 in. Find the length of the other base.

19. The lengths of the bases of a trapezoid are 9 cm and 17 cm. If the area of the trapezoid is 39 cm^2, what is the height?

INVESTMENTS Use the simple interest rate formula, $A = P + Prt$, to solve Exercises 20–21. In this formula, A is the amount returned, P is the amount invested, r is the annual interest rate, and t is the length of time in years.

20. Sonia deposited some money in a savings account earning 6% simple interest. At the end of four years, she had $434. How much had she deposited?

21. Brian borrowed $8000 to buy a car and took five years to repay the loan. At the end of five years, he paid a total of $12,200. What interest rate was he charged?

22. **WRITING MATHEMATICS** In a circle, $A = \pi r^2$. Describe a situation where you might need to solve for r instead of A.

EXTEND

NUMBER THEORY You can write any three-digit number ABC as $100A + 10B + C$. If there is a whole number Q such that $Q(A + B + C) = 100A + 10B + C$, then ABC is divisible by the sum of its digits, $A + B + C$. Use this information for Exercises 23–25.

23. Solve $Q(A + B + C) = 100A + 10B + C$ for Q.

24. For $A = 3$, $B = 9$, and $C = 6$, find Q.

25. Find Q if $A = 5$, $B = 0$, and $C = 4$.

THINK CRITICALLY

Describe the outcome if the value of the underlined variable is zero.

26. cost including tax
$C = c + \underline{r}c$

27. simple interest
$I = Pr\underline{t}$

28. consumer price index
$I = 100\underline{P} \div p$

For Exercises 29–30, solve each equation for the underlined variable. Then examine the incorrect solution given. Explain the error in reasoning that was most likely made.

29. $3a - 4\underline{b} = 8 \quad b = \dfrac{8 - 3a}{4}$

30. $9c + 6\underline{d} = 12 \quad d = 12 - \dfrac{9c}{6}$

MIXED REVIEW

Solve each equation.

31. $-9 - 7m = 12$

32. $15 + 5x = -25$

33. $-13r + 28 = -89$

34. $4b + 8 = 6b$

35. **STANDARDIZED TESTS** What number is 19% of 80?

    A. 152   **B.** 1.52   **C.** 0.152   **D.** 15.2

Solve Mixture Problems

Many real world problems are about mixtures, where two or more parts are combined into a whole. You can solve these problems by using what you know about the parts and the combined mixture to write equations. Sometimes, organizing the information in a table will help you understand the relationships among the quantities.

Problem

At Jenna's Juice Café, the Cranberry Cooler is supposed to be 50% pure juice. Jenna discovers that her helpers have mixed two batches incorrectly; one helper made 20 gal that was 60% juice, and the other made 20 gal that was only 10% juice. How many gallons of the 10% juice should Jenna add to the whole batch of 60% juice to obtain the correct 50% mixture?

Explore the Problem

PROBLEM SOLVING TIP

Remember that a percent is a rate or a part of a whole.

1. Write an expression for the number of gallons of pure cranberry juice that is in the 60% batch. Explain your thinking.

2. Let n represent the number of gallons of 10% juice to be added. How many gallons of pure juice are in these n gallons?

3. When the two mixtures are combined, how many gallons will there be in all? If this combined mixture is 50% juice, how many gallons of pure juice will be in it?

CHECK UNDERSTANDING

Suppose instead of a 10% juice mixture, Jenna had a 30% mixture. Do you think the number of gallons she would have to add to the 20 gal of 60% would be more or less than the n gallons you found in answering Question 5? Why?

4. Use your results above. Write the equation that represents this verbal model.

Pure juice in 10% mixture	+	Pure juice in 60% mixture	=	Pure juice in 50% mixture

5. Solve for n. (*Hint:* Multiply by 100 to get integer coefficients.) Check and explain your answer.

6. How many gallons of 50% juice will there be?

7. You could use a table to show the relationship in the problem. Review your thinking in Problems 1–4. Complete the table on the next page.

Mixture	Pure Juice in Mixture	Amount of Mixture, gal	Amount of Pure Juice
10%	0.10		
60%	0.60	20	0.60(20)
50%			

PROBLEM SOLVING PLAN

- Understand
- Plan
- Solve
- Examine

Investigate Further

- Think about how this next problem about investing is similar to the juice problem on the previous page.

COMMUNICATING ABOUT ALGEBRA

Discuss why it is helpful in solving problems to organize the information in a table with labeled headings.

Problem
Jenna invested $2000, part at 6% simple interest and the remainder at 8%. Her yearly income from the two investments is $132. Find the amount invested at each rate.

Solution
In the case of simple interest, the amount of interest I earned during a single year is equal to the amount of principal P times the interest rate r: $I = Pr$. A verbal model is:

$$\boxed{\text{Amount of interest from 6\% part}} + \boxed{\text{Amount of interest from 8\% part}} = \boxed{\text{Total amount of interest}}$$

8. Let a represent the amount invested at 6%. How can you represent the amount invested at 8% in terms of a?

9. Represent the amount of interest earned from the 6% investment.

10. Complete the table.

Investment	Rate	Amount	Interest
6%	0.06	a	
8%			

11. Write an equation relating the interest from each investment and the total interest income.

12. Solve and check the equation you wrote. How much was invested at each rate?

Apply the Strategy

13. **CHEMISTRY** A chemist has 40 mL of 35% alcohol solution. How many milliliters of 80% solution should she add to obtain a 60% solution?

14. **FINANCE** Mr. Rosario invested $11,000 in bank certificates and bonds. The bank certificates paid 4% and the bonds paid 6%, giving him an annual income of $530. How much did he invest in bonds?

15. **CONSUMER MATH** At Pop's Pantry, ground chuck is labeled 80% lean, and ground sirloin is labeled 95% lean. How many pounds of ground sirloin would have to be mixed with 30 lb of ground chuck to make a hamburger mixture labeled 90% lean?

16. **CHEMISTRY** How much pure alcohol should be added to 120 mL of a 45% solution so that the resulting solution is 60% alcohol?

17. **METALLURGY** To the nearest gram, how much of an alloy containing 33% silver must be melted with an alloy containing 75% silver to obtain 300 g of an alloy containing 58% silver?

18. **MANUFACTURING** A manufacturer blends 1200 L of olive oil worth $10.99 per liter with 800 L of vegetable oil worth $4.99 per liter. What is the resulting blended oil worth per liter?

19. **AUTOMOTIVE** A car radiator contains 8 qt of a mixture of water and antifreeze.

PROBLEM SOLVING TIP

Use 100% = 1.00 to represent the pure substance.

   a. If only 20% of the mixture is antifreeze, how much should a filling-station attendant drain and replace with pure antifreeze to raise the concentration to 60% antifreeze?

   b. How many quarts of antifreeze were in the original mixture?

   c. In draining the number of quarts you answered in part a, how much antifreeze is lost?

   d. When the amount drained is replaced with pure antifreeze, what is the amount of antifreeze in the final mixture?

20. **BIOLOGY** A biologist is testing the effects of acid rain on plant growth. She has 500 mL of a 12% acid solution, but wants to use a 7% solution. To the nearest mL, how much water should she add to obtain the desired concentration?

REVIEW PROBLEM SOLVING STRATEGIES

Quilt QUESTIONS

1. Copy and color this quilt.

Row 1 has a yellow, a blue, and a pink.
Column 2 has two yellows and a pink.
Row 2 has two whites and a pink.
Column 3 has a pink, a white, and a blue.
One of the diagonals has two pinks and a
yellow.

TRIANGLE COUNT

2. A *cevian*—named for the
seventeenth-century Italian
mathematician and
engineer Giovanni Ceva—
is a line segment that
joins a vertex of a
triangle to a point on
the opposite side.

a. How many triangles are there if one cevian
   is drawn from one vertex of a triangle?
b. How many triangles are there if seven
   cevians are drawn from one vertex of a
   triangle?
c. How many cevians were drawn from one

STRATEGIC sub-traction

3. Play this game with a partner several times. Alternate who goes first. Record
the moves and the winner each time.

   Start at 100. In turn, each player subtracts a single digit number.
   The player who ends at 0 is the loser.

Can you find a winning strategy in this game? These questions may help you
think about the problem.

a. Suppose it is almost the end of the game. The number that remains forces
   the next player to become the loser. What is the number?
b. Suppose the number that remains is 18. You go next. Describe how you
   can be sure to win.
c. Now work out a general winning strategy from the beginning of the
   game. If both players know the strategy, who wins?

··· CHAPTER REVIEW ···

Choose the word from the list that completes each statement.

1. Two numerical or variable expressions joined by an equal symbol form a(n) __?__.

2. A(n) __?__ is an equation containing more than one letter as a variable.

3. Any value of a variable that turns an open sentence into a true statement is called a(n) __?__ of the sentence.

4. A(n) __?__ is an equation that states that two ratios are equal.

5. Equations with the same solutions are __?__.

a. proportion

b. equivalent equations

c. equation

d. literal equation

e. solution

Lesson 3.1   SOLVE EQUATIONS WITH MENTAL MATH AND ESTIMATION   pages 115–119

- Mental math, guessing, and number lines may be used to solve some equations.

Use mental math, guess-and-check, or a number line to solve.

6. $b - 7 = 11$ **7.** $26 + 3c = 71$ **8.** $-2 + d = -8$

Lesson 3.2   SOLVE LINEAR EQUATIONS   pages 120–123

- Algeblocks may be used to represent and to solve linear equations using addition and subtraction.

Write the equation that is modeled. Then solve.

9. 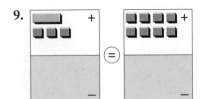   **10.** 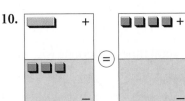   **11.**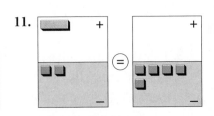

Lessons 3.3 and 3.4   SOLVE EQUATIONS USING PROPERTIES OF EQUALITY   pages 124–134

- To solve an equation, use the **addition property of equality**, the **subtraction property of equality**, the **multiplication property of equality**, or the **division property of equality**.

Solve and check.

12. $x - 4 = 5$ **13.** $x + 12 = 4$ **14.** $-5 + x = 14$

15. $-3x = 15$ **16.** $10x = 40$ **17.** $\frac{n}{3} = 6$

18. $4y = 56 - 12$ **19.** $w - \frac{2}{3} = 3$ **20.** $5.2 = p - 1.3$

Lesson 3.5 PROPORTION, PERCENT, AND EQUATIONS pages 135–139

- To solve a percent problem, write an equation or a proportion in the form

$$\frac{\text{part}}{\text{whole}} = \frac{\text{part}}{\text{whole}}$$

Solve and check.

21. Find 46% of 75. **22.** What percent of 8 is 36? **23.** 12 is what percent of 50?

Lesson 3.6 SOLVE EQUATIONS USING TWO OR MORE OPERATIONS pages 140–145

- To solve equations requiring more than one operation, first use the addition or subtraction properties of equality to isolate the term containing the variable. Then use the multiplication or division properties of equality to complete the solution.

Solve and check.

24. $3x - 6 = 12$ **25.** $\frac{x}{4} + 2 = -3$ **26.** $\frac{2}{3}y - 3 = -13$

27. $2(2y + 3) = 18$ **28.** $12(2 + p) - 4p = 96$ **29.** $1.3m + 4 = -48$

Lesson 3.7 SOLVE EQUATIONS WITH VARIABLES ON BOTH SIDES pages 146–151

- To solve equations having variables on both sides, collect like variables on the side on which the variable has the greater coefficient.

Solve and check.

30. $3x + 12 = -2x + 42$ **31.** $15 - n = 2n$ **32.** $4(3 - y) = 2y$

33. $2(-5y + 3) = 3(-2y + 10)$ **34.** $\frac{1}{2}(3c - 1) = c$ **35.** $\frac{1}{2}x + 2 = \frac{5}{2} - x$

Lesson 3.8 USE LITERAL EQUATIONS AND FORMULAS pages 152–155

- A **literal equation** is one that contains more than one letter used as a variable.
- **Formulas** are literal equations.

Solve for the indicated variable.

36. Solve for b: $a - 4 = 2b - 7$ **37.** Solve for r: $8r + 7s = 12$ **38.** Solve for l: $P = 2l + 2w$

Lesson 3.9 SOLVE MIXTURE PROBLEMS pages 156–159

- Organizing information in a table is a useful strategy for solving problems.

Use a table to solve.

39. Luis invested $2100, part at 5% interest and the remainder at 7%. His annual income from the two investments is $123. Find the amount he invested at each rate.

CHAPTER ASSESSMENT

CHAPTER TEST

1. **WRITING MATHEMATICS** Write a paragraph to explain how the equation $6x = 18$ can be solved by multiplication and how it can be solved by division.

Solve each equation.

2. $x + 14 = 2$

3. $26 = t - (-4)$

4. $9 + b = 0$

5. $w - 3.2 = 17.6$

6. $13a = 39$

7. $-6x = -72$

8. $\dfrac{c}{2} = 2.5$

9. $-\dfrac{2b}{3} = -6$

10. $97 + 4j = 17$

11. $\dfrac{1}{2}x + 5 = -9$

12. $2(x + 2) = 46$

13. $8m - (8 - m) = 28$

14. $6x - 7 = 13 - x$

15. $3x - 21 = -3x + 15$

16. $\dfrac{5}{21} = \dfrac{h}{84}$

17. $\dfrac{16}{64} = \dfrac{a}{12}$

18. $2(x - 4) - 3(x + 4) = -50$

19. **WRITING MATHEMATICS** Write a paragraph to explain how you would use an equation and how you would use a proportion to answer the question: 35 is what percent of 40?

Solve.

20. 80 is what percent of 50?

21. What is 15% of 24?

22. 30 is 12% of what number?

23. **STANDARDIZED TESTS** Which of the following equations is *not* equivalent to $3v - 16 = -6v + 2$?

   A. $6s = 12$ **B.** $-9t = 18$

   C. $r - 4 = -2$ **D.** $25x = 50$

24. Estimate the solution of each equation.

   a. $6.0219f = -178.92$ **b.** $c - 24.89 = 49.6$

   c. $p + 3.67 = -16.25$ **d.** $-11.908 = \dfrac{35.899}{m}$

25. **WRITING MATHEMATICS** Write a paragraph explaining how you would solve the equation $3y - 5 = 16$. Do not solve the equation.

26. Solve for t: $d = rt$

27. Solve for b_1: $A = \dfrac{1}{2}h(b_1 + b_2)$

28. Solve for r: $A = P + Prt$

29. Solve for m: $wb = \dfrac{nm}{a}$

30. **STANDARDIZED TESTS** For which of the following equations is -2 a solution?

   I. $1.5 - g = 3.5$ **II.** $\dfrac{y}{4} = -8$ **III.** $c + 4 = 2$

   A. I and II   **B.** I and III

   C. II and III   **D.** I, II, and III

Write and solve an equation for each problem.

31. Flavia earns \$1575 every two weeks. She works 35 h per week. Find her hourly rate.

32. The perimeter of a rectangular lot is 400 ft. If the lot is 80 ft wide, what is the length?

33. A mixture of dried apricots and prunes costs \$5.50 per kilogram. The apricots cost \$6.00 per kilogram and the prunes cost \$4.00 per kilogram. How many kilograms of apricots are in 14 kg of the mixture?

34. Sun-hee invested savings in two certificates of deposit, each at a different rate of interest. She invested \$1500 more at 7% than at 5%. Her yearly income from the two investments is \$465. How much did she invest at each rate?

PERFORMANCE ASSESSMENT

USE ALGEBLOCKS Choose a variety of types of equations from this chapter. Show how to model each equation with Algeblocks. Then ask a partner to show the steps for solving each equation. Check that each step is correct.

DESIGN A DEMONSTRATION Use a balance scale and known and unknown weights to design a demonstration that shows:

a. how to model different equations.

b. the steps needed to solve the equations.

USE FLOWCHARTS You can describe $3x + 4 = 19$ with the flowchart below at the left. The reverse flowchart at the right shows how the solution is obtained, working backward with inverse operations.

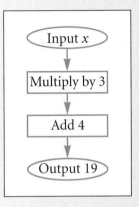
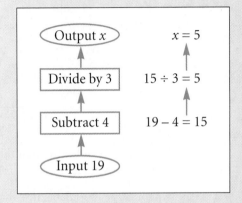

Draw both types of flowcharts for different equations. Try to show both one-step and two-step equations as well as equations with variables on both sides.

STATE EQUATIONS Use equations to give information and make comparisons about different American states. Research two or more states that interest you. Make note of information such as the current population, the population change from the last census, the land area of the state, the land area covered by forests, and so on. Express information algebraically. Ask other students in your class to find the missing facts by solving the equations.

PROJECT ASSESSMENT

PROJECT *Connection* Work with your group. Agree on how to present your findings about "lite" and regular versions of food items to other students. For example, you might want to make a collage including product advertisements, design a brochure, or videotape a panel discussion. When you have made all your decisions, complete the work and make a final presentation.

1. What general conclusions can you draw from your investigation about how "lite" and regular foods compare? Do results vary for different food categories?

2. Are there other pieces of information that you need to complete the project? How can you obtain these items?

3. How could you best display your data visually?

4. Can you use algebraic expressions or equations to explain relationships among food items?

5. Why might the result of your project be interesting or important to your audience?

6. What will each member of your group do to help complete the project?

··· CUMULATIVE REVIEW ···

The graph at the right shows the attendance at one high school's home football games. Use the data to answer Questions 1–4.

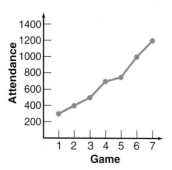

1. How many more people attended the sixth game than the second game?

2. From which game to which game did the attendance increase the least?

3. **STANDARDIZED TESTS** Which of the following seems to be a logical inference from the graph?

   A. The team probably had a losing record.

   B. The attendance was miscounted the last four games.

   C. The team played surprisingly well and won most of their games.

   D. Concessions were too expensive at the beginning of the season.

4. **STANDARDIZED TESTS** Which of the following is true?

   A. The mean attendance is greater than the median attendance.

   B. The median attendance is greater than the mean attendance.

   C. The mean attendance is equal to the median attendance.

   D. There is not enough information provided.

Solve each equation.

5. $x - 6 = -2$

6. $n + 21 = 15$

Solve each equation.

7. $5y = -20$

8. $\frac{2}{3}x = 12$

9. $4(c - 2) - c = -11$

10. $2(5k + 3) = k - 12$

Evaluate each expression.

11. $\left[\left(\frac{2}{3}\right)2 - \frac{2}{3}\right] \cdot (-9)$

12. $-7 \cdot 3 - 6 \div 3 + 12 \cdot (-2)$

13. **TECHNOLOGY** Juan tried to enter $4 - (-2)$ on his calculator by pressing $\boxed{4}\ \boxed{-}\ \boxed{-}\ \boxed{2}$, but could not get the correct answer of 6. Why didn't Juan get the correct answer?

14. 24 is 75% of what number?

15. What percent of 5 is 7?

16. What is 28% of 150?

17. Find the largest of three consecutive integers whose sum is -24.

18. Martina invested $1200, part at 5% simple interest and the remainder at 7% interest. She earned a total of $74.40 from both investments. How much did Martina invest at 7%?

19. Patrick had 22 answers correct on a test. This earned him a score of 88%. How many questions were on the test?

20. **WRITING MATHEMATICS** To solve the equation $2x + 9 = 5x + 3$, Ralph subtracted $2x$ from both sides and then subtracted 3 from both sides. Alice subtracted $5x$ from both sides and then subtracted 9 from both sides. Both obtained the correct answer of 2. Which method would you have chosen? Why?

STANDARDIZED TEST

STANDARD FIVE-CHOICE Select the best choice for each question.

1. Mariko bought a blouse that regularly costs $20 for $14. What percent was the discount?

   A. 6%   **B.** 14%   **C.** 20%
   D. 30%   **E.** 70%

2. Which point corresponds to $\sqrt{8}$ on the number line?

   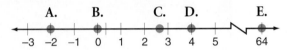

3. For which operations is the associative property true?

   A. addition and subtraction
   B. multiplication and division
   C. addition and multiplication
   D. subtraction and division
   E. all four operations

4. In an analysis of ten data values, which measure of central tendency is most likely to be affected by one exceptionally high value?

   A. mean
   B. median
   C. mode
   D. all will be affected
   E. not enough information

5. What is the degree of the expression $5x^3 + 7x + x^4 - 8x^2 - 10$?

   A. 10   **B.** 7   **C.** 5
   D. 4   **E.** 3

6. The perimeter of a rectangle is 60. The width of the rectangle is 10. What is the length of the rectangle?

   A. 6   **B.** 20   **C.** 50
   D. 600   **E.** cannot be determined

Use the following information to answer Questions 7 and 8.

Ed asked the 20 students in his class their favorite number from 1 through 10. Eight people responded that 7 was their favorite.

7. Which type of sampling did Ed employ?

   A. random
   B. cluster
   C. convenience
   D. systematic
   E. not enough people for a sample

8. What is the probability that the next person that Ed asks will say that 7 is their favorite number?

   A. 8%   **B.** 20%   **C.** 40%
   D. 60%   **E.** 80%

9. An equation for *two less than the product of 5 and a number is 10* is

   A. $(x + 5) - 2 = 10$
   B. $2 - 5x = 10$
   C. $5(x - 2) = 10$
   D. $\dfrac{5}{x} - 2 = 10$
   E. $5x - 2 = 10$

10. Evaluate: $(-1)^{23} \cdot (-2)^3$

    A. 138   **B.** 8   **C.** 6
    D. −6   **E.** 184

11. The range of a set of three numbers is 26. The lowest number is 32. Which of the following statements could be true?

    I. The mode is 58.

    II. The median is 32.

    III. The mean is 58.

    A. I only   **B.** II only   **C.** III only
    D. I and II   **E.** II and III

4 Functions and Graphs

Take a Look
AHEAD

Make notes about things that look new.

- There are many graphs shown in this chapter. In what ways are the graphs similar? In what ways are they different? Can you suggest reasons for these differences?
- What can you learn about a set of data by making a scatter plot?

Make notes about things that look familiar.

- The word "function" has several different meanings in everyday language. Explain and give examples of these meanings. Which do you think is closest to the mathematical meaning?
- What units of measurement are used in this chapter? What physical quantities do they measure? How are the units abbreviated?

DATA Activity

Section 204 Row 10 Seat 1

Where do you go to listen to a rock concert? It may be the same place you watch your favorite teams play: a baseball stadium. Because of their size, these arenas are sometimes used for other large-attendance events. Baseball stadiums and their seating capacities are listed in the chart on the next page.

SKILL FOCUS

- ▶ Use estimation.
- ▶ Determine percents.
- ▶ Find the mean and range of a set of data.
- ▶ Explore methods for analyzing data relationships.

SPORTS

In this chapter, you will see how:

- **GREENSKEEPERS** estimate areas on golf courses. (Lesson 4.2, page 176)

- **SPORTS REFEREES** use functions and equations to interpret the rules. (Lesson 4.4, page 188)

- **SPORTS STATISTICIANS** analyze the large amounts of data generated by different sports. (Lesson 4.9, page 211)

	Team	Stadium (built)	Seating Capacity
National League	Atlanta Braves	Atlanta-Fulton County Stadium (1966)	52,003
	Chicago Cubs	Wrigley Field (1916)	38,710
	Cincinnati Reds	Riverfront Stadium (1970)	52,952
	Colorado Rockies	Coors Field (1995)	50,000
	Houston Astros	Astrodome (1965)	54,816
	Los Angeles Dodgers	Dodger Stadium (1962)	56,000
	Montreal Expos	Olympic Stadium (1977)	43,739
	New York Mets	Shea Stadium (1964)	55,601
	Philadelphia Phillies	Veterans Stadium (1971)	62,382
	Pittsburgh Pirates	Three Rivers Stadium (1970)	58,727
	St. Louis Cardinals	Busch Stadium (1966)	56,227
	San Diego Padres	Jack Murphy Stadium (1969)	59,022
	San Francisco Giants	Candlestick Park (1960)	58,000
American League	Baltimore Orioles	Camden Yards (1992)	48,041
	Boston Red Sox	Fenway Park (1912)	34,142
	California Angels	Anaheim Stadium (1966)	64,593
	Chicago White Sox	Comiskey Park (1991)	43,500
	Cleveland Indians	Cleveland Stadium (1932)	74,483
	Detroit Tigers	Tiger Stadium (1912)	52,416
	Kansas City Royals	Royals Stadium (1973)	40,625
	Milwaukee Brewers	Milwaukee County Stadium (1953)	53,192
	Minnesota Twins	Hubert H. Humphrey Metrodome (1982)	55,883
	New York Yankees	Yankee Stadium (1923)	57,545
	Oakland A's	Oakland Alameda County Coliseum (1968)	47,313
	Seattle Mariners	Kingdome (1977)	57,748
	Texas Rangers	Arlington Stadium (1972)	43,521
	Toronto Blue Jays	Skydome (1989)	50,516

Use the table to answer each question.

1. Estimate the total seating capacity of the two New York stadiums. Explain if the actual total is less than or greater than your answer.

2. What percent of National League stadiums seat over 55,000? What percent of American League stadiums seat fewer than 55,000? Round answers to the nearest percent.

3. Find the mean and range of seating capacities for National League stadiums. Round answers to the nearest whole number.

4. **WORKING TOGETHER** How can you determine if there is a relationship between the seating capacity of a stadium and the year it was built?

5. Try the method you agreed on in Question 4. Explain your conclusions. Is there a difference between the two leagues?

School Sports Showcase

Think about all the people who are involved with sports but don't actually play the game. Professionals such as television and radio commentators, journalists, and public relations consultants focus on providing information about a particular sport, team, or player. Keeping sports in the news maintains the public's enthusiasm. You can help gain support for your school's teams by creating a Sports Showcase full of interesting information.

PROJECT GOAL

To collect, organize, and analyze school sports data and create a Sports Showcase.

Getting Started

Work in small groups. Discuss the group's interests and current knowledge about school sports. Plan how to best work together to accomplish the project.

1. Make a list of all school teams or sports clubs. Find out where and when each team practices or each club meets. Determine whether any of the teams has an official spokesperson. Make arrangements for any meeting or interviews you wish to conduct.

2. Find out where information about past years is available. Sources may include school records, team coaches, past issues of school and local newspapers, and oral histories from former students.

PROJECT Connections

Lesson 4.3, page 182:
Determine if there is a relationship between hand size and foot size for football players and, if so, how this relationship compares with the general student population.

Lesson 4.4, page 187:
Experiment to determine how distances and practice affect a basketball player's shooting accuracy.

Lesson 4.6, page 196:
Collect, organize, and display data about school teams.

Chapter Assessment, page 215:
Assemble Sports Showcase and plan related school events.

✉ Internet Connection

www.swpco.com/
swpco/algebra1.html

4.1 Algebra Workshop
Paired Data

Think Back

● Work with a partner. Use a graphing utility. Recall that when you collect data, you can organize it in a table and display it on a graph.

A baseball pitcher has just returned to active play after recovering from an injury. The team trainer has limited the pitcher to throwing only 15 pitches in each practice session on the first day. The pitcher throws curveballs and fastballs. The trainer keeps a tally chart for each type of pitch and will stop the pitcher after 15 pitches.

SPOTLIGHT ON LEARNING

WHAT? In this lesson you will learn
- to represent paired data as points on a graph.
- to represent paired data as ordered pairs.

WHY? Representing paired data as ordered pairs can help you solve problems in which analysis of data is necessary, such as reporting sports data.

1. Make a table of all possible pairs of fastballs and curveballs the pitcher can throw.

Fastballs	Curveballs
0	15

Explore

● You can use the data in your table to write **ordered pairs** of numbers.

0 fastballs → (0, 15) ← 15 curveballs

2. How can you graph all the ordered pairs from the table?

3. Enter the data from your table into a graphing utility. Then use your graphing utility to draw a graph. Make a sketch of this graph.

Your graph should consist of **data points** and a **horizontal axis** and a **vertical axis** that intersect at a point called the **origin**.

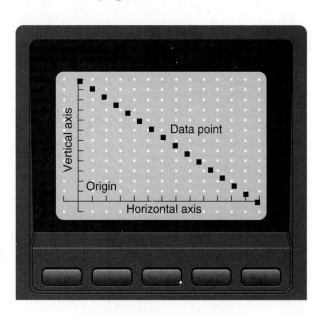

4.1   Algebra Workshop: Paired Data   **169**

4. What do the marks on the horizontal axis of the graph represent? What do the marks on the vertical axis represent?

5. Look at the graph displayed on your graphing utility screen. The first data point represents 0 fastballs and 15 curveballs. Use the TRACE feature of your graphing utility to move the cursor to this point and check that the numbers displayed at the bottom of the screen match the data. Try several other points and match the point to the line in your data table.

6. What happens to the number of curveballs as the number of fastballs increases? Do you think it is easier to determine this relationship from the table or the graph?

7. What do you notice about the pattern made by the points on your graph?

CHECK UNDERSTANDING

Explain how you choose where to write the number of fastballs and the number of curveballs in each ordered pair.

Make Connections

● Let *f* represent the number of fastballs and *c* the number of curveballs.

8. Write an algebraic expression using *f* that shows the number of curveballs the pitcher can throw.

9. Write an equation using *c* and *f* that shows how *c* and *f* are related.

10. Write an equation using *c* and *f* that shows what the number of curveballs equals. Substitute several ordered pairs of fastballs and curveballs in your equation to check that you wrote it correctly.

PROBLEM SOLVING TIP

Are the number of possible fastball and curveball combinations for the problem in this Workshop finite or infinite? Are the number of real number combinations for Question 13 finite or infinite?

Summarize

● **11.** WRITING MATHEMATICS Briefly describe the different ways in which you can represent paired data. Name at least one advantage and disadvantage to using each way.

12. THINK CRITICALLY Would it have made sense to draw a line through the points on your graph of fastballs and curveballs? Why or why not?

13. GOING FURTHER Suppose the sum of two real numbers is 15. Make a table of some possible values for this relation. Graph this paired data and write ordered pairs. How are the representations of the possible values for this relation the same as the representation of the paired data for fastballs and curveballs? How are they different?

4.2 Relations and Functions

Explore

From 1949 to 1953, the New York Yankees won five straight World Series championships. The first team to win four games in a World Series is the winner. Table 1 compares the number of games the Yankees won to the number their opponents won.

From 1984 to 1988, the Chicago Bears won the Central Division title of the National Football Conference. Except for 15 games during 1987, there are 16 games in a regular season of NFL football. Table 2 compares the number of games the Bears won to the number they lost.

SPOTLIGHT ON LEARNING

WHAT? In this lesson you will learn
- to identify tables and ordered pairs that represent functions.

WHY? Identifying tables and ordered pairs for functions can help you solve problems about geometry, physics, and health.

Table 1 World Series, 1949–1953	
Yankee Wins	Opponent Wins
4	1
4	0
4	2
4	3
4	2

Table 2 Regular Season, 1984–1988	
Bears Won	Bears Lost
10	6
15	1
14	2
11	4
12	4

1. How are Tables 1 and 2 the same? How are they different?

2. If you covered up Table 1 and were given a number in the first column, could you name the corresponding number in the second column? Why or why not? Could you do this for Table 2? Why or why not?

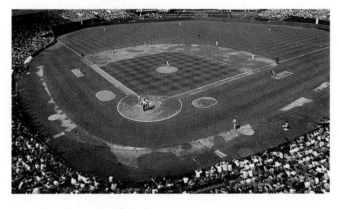

Build Understanding

In Tables 1 and 2 above, each value in the first column is paired with a value in the second column. The values in the first column of such tables are often referred to as *input values*, while the values in the second column are called *output values*. A set of paired data, or values, is a **relation**. When there is only one output value for each different input value in a set of paired values, the relation is a **function**. However, more than one input value may have the same output value.

Relations and functions can be shown in several ways. Two ways are in tables and in sets of ordered pairs. In an ordered pair, the first number is considered the input and the second number, the output.

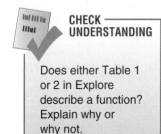

CHECK UNDERSTANDING

Does either Table 1 or 2 in Explore describe a function? Explain why or why not.

EXAMPLE 1

Tell whether each table or set of ordered pairs shows a function.

a.

Input	3	4	5	6	7
Output	5	8	10	8	9

b. {(1, −1), (5, −4), (3, −2), (2, −4), (1, −3)}

Solution

a. This table shows a function because each different input value has only one output value. Notice that both input values 4 and 6 have the same output value, 8.

b. This set of ordered pairs does not show a function because each different input value does not have only one output value. The input value 1, shown twice, is paired with an output value of −1 and an output value of −3. ◀

In a function, the input values make up the **domain** of the function and the output values make up the **range** of the function.

EXAMPLE 2

For the following function, name the values in
a. the domain   **b.** the range

{(1, 1.5), (2, 2.5), (3, 3.5), (4, 4.5)}

Solution

a. 1, 2, 3, 4   **b.** 1.5, 2.5, 3.5, 4.5 ◀

In Example 2, the output value in each ordered pair is 0.5 greater than the corresponding input value. Let x represent an input value. The corresponding output value can then be represented by $x + 0.5$. A description of a function is called a **function rule**. You can express the rule $x + 0.5$ using **function notation** by naming the function f.

$$f(x) = x + 0.5$$

The symbol $f(x)$ means the value of the function f at x, and is read, "f of x." The domain of the function f is the set of all possible values of x, and the range of the function is the set of all possible values of $f(x)$. For $f(x) = x + 0.5$, both the domain and the range are the real numbers. To **evaluate a function**, find the output, or $f(x)$, for a given input x. To evaluate $f(x) = x + 0.5$ when $x = 1$, substitute 1 for x.

$$f(x) = x + 0.5$$
$$f(1) = 1 + 0.5 \quad \text{Substitute 1 for } x.$$
$$f(1) = 1.5$$

The input value 1 yields the output value 1.5 in Example 2.

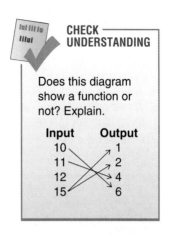

CHECK UNDERSTANDING

Does this diagram show a function or not? Explain.

Input	Output
10	1
11	2
12	4
15	6

EXAMPLE 3

Evaluate $f(x) = x^2 - 3$ at each given value of x.

a. $x = 0$ **b.** $x = -4$

Solution

a. $f(x) = x^2 - 3$
$f(0) = 0^2 - 3$ Substitute 0 for x.
$f(0) = 0 - 3$
$f(0) = -3$

b. $f(x) = x^2 - 3$
$f(-4) = (-4)^2 - 3$ Substitute
$f(-4) = 16 - 3$ -4 for x.
$f(-4) = 13$ ◀

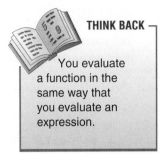

THINK BACK

You evaluate a function in the same way that you evaluate an expression.

EXAMPLE 4

GEOMETRY The area of a square is a function of the length of a side of the square. The area function is $f(x) = x^2$ where x is the length of a side. Find the area of a square with sides 3 cm long.

Solution

$f(x) = x^2$
$f(3) = 3^2$ Substitute 3 for x.
$f(3) = 9$

The area of a square with sides 3 cm long is 9 cm². ◀

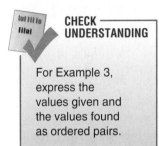

CHECK UNDERSTANDING

For Example 3, express the values given and the values found as ordered pairs.

TRY THESE

Determine whether each table or set shows a function. For each function, name the values in the domain and the values in the range. If not a function, explain why.

1.

Input	-1	-2	-3	-4	-5
Output	2	2	2	2	2

2.

Input	$\frac{1}{2}$	$\frac{3}{4}$	$\frac{7}{8}$	$\frac{1}{2}$	$\frac{2}{3}$
Output	$\frac{1}{4}$	$\frac{3}{8}$	$\frac{1}{2}$	$\frac{1}{3}$	$\frac{3}{4}$

3. $\{(6, 3), (5, 4), (4, 5), (3, 6)\}$

4. $\{(-1, 1), (-2, 4), (-5, 25), (-3, 9), (-4, 16)\}$

Evaluate each function at the given values of x.

5. $f(x) = \dfrac{x}{4}$ at $x = -4$, $x = 0$, and $x = 8$

6. $f(x) = 3x + 1$ at $x = \dfrac{1}{3}$, $x = \dfrac{1}{2}$, and $x = \dfrac{3}{4}$

7. **MODELING** What function rule is modeled by these Algeblocks? Use Algeblocks to evaluate this function when $x = 3$. Make a sketch of each step.

8. **PHYSICS** The distance in feet an object freely falls through space is a function of the time in seconds that it falls. This is shown by the equation $f(t) = 16t^2$. Evaluate this function to find how far a ball will fall in 10 seconds.

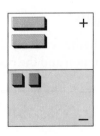

Tell whether each table or set shows a function. For each function, name the values in the domain and the values in the range. If not a function, explain why.

1.

Input	Output
$0.50	$0.02
$0.60	$0.02
$0.70	$0.02
$0.80	$0.02
$0.90	$0.02

2.

Input	Output
2	−4
3	−2
2	0
3	2
2	4

3. $\{(2, 0), (4, 0), (6, 0), (8, 0), (10, 0)\}$

4. $\{(-1, -1), (-3, -3), (-5, -5), (-7, -7)\}$

5. $\{(3, \$1.50), (2, \$1.29), (3, \$1.80), (4, \$2.29), (3, \$2.49), (5, \$2.50)\}$

6.

Speed, mi/h	55	55	55	55	55
Distance, mi	55	110	165	220	275

7.

Time, h	1	2	3	4	5
Distance, mi	50	100	150	200	250

8. WRITING MATHEMATICS Write a brief explanation of how you determined whether each table or set in Exercises 1–8 shows a function.

Evaluate each function at the given values of x.

9. $f(x) = 0.5x$ at $x = -4$, $x = 1$, and $x = 20$ **10.** $f(x) = 3(x + 4)$ at $x = 0$, $x = 3$, and $x = 10$

11. $f(x) = x^2 - 2$ at $x = -3$, $x = -1$, and $x = 3$ **12.** $f(x) = 2x - \dfrac{1}{2}$ at $x = \dfrac{1}{4}$, $x = \dfrac{1}{2}$, and $x = \dfrac{3}{4}$

13. HEALTH The relationship of Calories from fat to the amount of fat in any food is $f(x) = 9x$, where x is the amount of fat in grams. Evaluate this function to find the Calories from fat in one serving of these foods.

    a. bread, 0.5 g   **b.** macaroni and cheese, 2 g   **c.** pizza, 17 g

14. Describe the domain and the range of the function in Exercise 13.

EXTEND

Tell whether each set shows a function. For each function, name the domain and the range. If not a function, explain why.

15. $\{(a, 1), (a, 2), (a, 3), (a, 4)\}$

16. $\{(a, -1), (b, -3), (c, -5), (b, -3)\}$

17. $\{(a, d), (b, e), (c, f), (d, g)\}$

18. TECHNOLOGY You can use the program at the right to evaluate the function $f(x) = x^2 - 2$ on some graphing calculators. Ifthe program does not work on your calculator, use your calculator manual to modify the program. Then change the program to evaluate the function $f(x) = x^2 + 3$ when $x = -2$, $x = 4$, and $x = 15$.

```
Prgm: EVALUATE
:Lbl 1
:Disp "X VALUE"
:Input X
:X²–2→A
:Disp A
:Goto 1
```

19. BUSINESS An appliance store pays its full-time employees a weekly base salary plus a commission, as shown by these functions where x is the dollar amount of sales per week.

$$f(x) = 200 + 0.05x \text{ when } x < \$2,000 \qquad f(x) = 100 + 0.1x \text{ when } x \geq \$2,000$$

Evaluate the appropriate function to find the earnings for salespeople who made the following dollar amounts in sales last week.

a. $3,500   **b.** $2,000   **c.** $1,900

20. What is the domain and range of each function in Exercise 19?

21. WRITING MATHEMATICS Use function notation to write a function for which the domain is your age from 0 to 18 years. Tell what the function is and describe its range.

THINK CRITICALLY

Study each table of input and output values. Look for a pattern in each. Then use function notation to describe the function shown.

22.

Input	8	9	12	13	18
Output	5	6	9	10	15

23.

Input	−1	−3	−5	−7	−9
Output	−1	−5	−9	−13	−17

Describe a situation for each equation. Be sure to tell what x represents.

24. $f(x) = 4x$

25. $f(x) = 7 + 0.75x$

26. Given $f(x) = \frac{x}{3}$ and $f(x) = 12$, find all values of x.

27. Given $f(x) = x^2 + 2$ and $f(x) = 18$, find all values of x.

MIXED REVIEW

Find the following measures of central tendency for these data: 35 cm, 27 cm, 56 cm, 72 cm, and 41 cm.

28. mean   **29.** median   **30.** mode   **31.** range

Solve and check.

32. $\frac{d}{7} = 42$

33. $3t + 2 = -13$

34. STANDARDIZED TESTS The value of $f(x) = 4x^2 - 0.2x$ at $x = -3$ is

A. 35.4   **B.** −36.6   **C.** 36.6   **D.** 12.6

Career
Greenskeeper

Keeping golf courses beautiful is as much a science as it is an art. Greenskeepers must understand the role of earth science, biology, chemistry, and mathematics as they care for the special grasses used. The amounts of fertilizers, herbicides, and pesticides applied to lawns are related to the area of each lawn.

A pin or flag stands in the hole. The area surrounding the hole is called a putting green. It is an irregularly shaped region with a curved border. The area of a circle is a function of the radius: $f(r) = \pi r^2$, where r is the length of the radius. Follow these steps to use an adaptation of this function to find the area of a putting green such as the one shown.

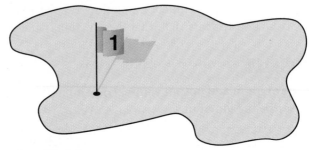

Decision Making

1. Trace the outline of the putting green above. On your sketch, draw a point that approximates the center of the green. From this point, draw ten equally spaced radii. Draw the radii 36° apart because $360° \div 10 = 36$.

2. Measure and record the length of each radius in centimeters. Find the average length of a radius by adding the lengths of all the radii and dividing by the number of radii.

3. Evaluate the function $f(r) = \pi r^2$, using the average radius length as the value for r.

4. Compare your finding of the area of the golf green to the areas found by other students in your class.

5. Write an equation that can be used to find the average length r of n radii that are labelled $r_1, r_2, r_3, \ldots r_n$. Use the equation you wrote to help you find the area of a golf green having the following radius measurements: 12 m, 10 m, 14 m, 15 m, 11 m, 9 m, 11 m, 12 m, 13 m, 15 m, 11 m, and 14 m.

4.3 The Coordinate Plane and Graphs of Functions

Explore

Julianne is a hockey fan. She wants to paint an enlarged copy of the logo of her favorite hockey team on her bedroom wall. If she tries to draw it freehand, she knows the drawing will not be in correct proportion. She needs a way to get the exact shape correct, but in a much larger size.

Julianne decided to trace the logo onto a piece of graph paper, as shown at the right. Then she drew a larger grid on her wall. She will draw points on the larger grid that correspond with points on the graph paper and then connect these points.

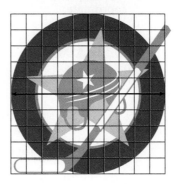

1. What does the intersection of the horizontal and vertical axes on the graph paper represent? Why might Julianne want to know this location?

2. To make it easier to draw points on the wall grid that correspond with points on the graph paper, how could Julianne use numbers to describe the points on the graph paper? Give an example to describe one point in each of the four sections of the graph paper.

3. Use Julianne's method to make an enlarged copy of the logo of your favorite sports team.

Build Understanding

A grid such as the one Julianne used is a **coordinate plane**. The horizontal axis and the vertical axis intersect at zero, called the *origin*. The axes divide the coordinate plane into four **quadrants**.

You can write an ordered pair for any point on a coordinate plane. Look straight up or straight down from the point to the horizontal axis to find the first number, or **coordinate**. Then look straight across left or right from the point to the vertical axis to find the second number, or coordinate.

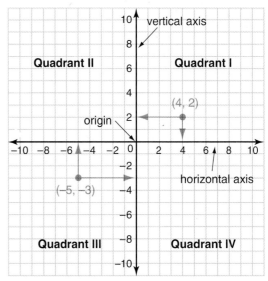

SPOTLIGHT ON LEARNING

WHAT? In this lesson you will learn
• to identify tables, ordered pairs, and graphs that are functions.
• to graph ordered pairs on the coordinate plane.

WHY? Graphs of functions on the coordinate plane can help you solve problems about interior decoration, postage, sports, and other applications where location is important.

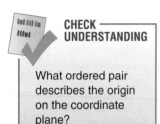

CHECK UNDERSTANDING

What ordered pair describes the origin on the coordinate plane?

EXAMPLE 1

Write an ordered pair to describe each point.

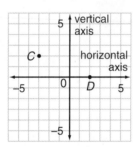

Solution

$A(1, -2)$
A is 1 unit to the right of the origin and 2 units below the origin.

$B(-1, 3)$
B is 1 unit to the left of the origin and 3 units above the origin. ◄

To graph, or plot, an ordered pair on a coordinate plane, start at the origin and move right or left along the horizontal axis the number of units represented by the first coordinate. Then move up or down parallel to the vertical axis the number of units indicated by the second coordinate.

EXAMPLE 2

Graph each ordered pair on a coordinate plane.
   a. $C(-3, 2)$ **b.** $D(2, 0)$

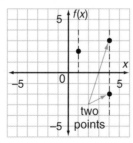

Solution

a. The first coordinate of C is –3, so move 3 units left along the horizontal axis. The second coordinate is 2, so move 2 units up.

b. The first coordinate of D is 2, so move 2 units right. The second coordinate is 0, so do not move any units up or down. ◄

A function can be shown in a table and as a set of ordered pairs $(x, f(x))$. A function can also be shown as a graph on a coordinate plane. In this case, the horizontal axis is the **x-axis** and the vertical axis is the **$f(x)$-axis**. The first number of the ordered pair is the **x-coordinate** and the second number is the **$f(x)$-coordinate**.

You can use the **vertical line test** to determine whether a graph of a relation is also a graph of a function. Visualize a vertical line drawn through each point plotted. If there is more than one plotted point on any vertical line, the graph does not show a function. At the left is a graph of the following.

$$\{(1, 2), (4, 3), (4, -2)\}$$

There is only one point on the vertical line through $(1, 2)$. That is because the input value 1 has only one output value: 2.

On the other vertical line, there are two points, (4, 3) and (4, –2). That is because the input value 4 has two output values: 3 and –2. Because there are two outputs for this input, the graph does not show a function.

EXAMPLE 3

Determine whether each graph represents a function. For each function, name or describe the domain and the range.

a.

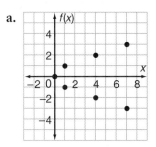

Solution

a. This graph does not show a function. You could draw four vertical lines to include all the points. Three of the lines would each pass through two points.

✓ **CHECK UNDERSTANDING**

Use ordered pairs to explain why the graph in Example 3a is not a function.

b.

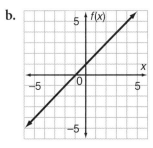

b. This graph shows a function. A vertical line drawn through any point on the graph will not go through any other point on the graph. Both the domain and range of this graph are all the real numbers. ◄

EXAMPLE 4

SPORTS The only event in the first Olympic games in 776 B.C. is said to have been a 200-yard foot race. This graph shows the winning times for the men's 200-meter run in recent Olympic games. Does this graph represent a function? If so, what are the domain and range?

Solution

This graph shows a function. Although it appears that you could draw a horizontal line through the points for 1972 and 1992, you could not draw any vertical lines that pass through more than one point.

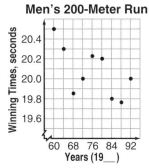

Men's 200-Meter Run

The domain is the years from 1960 to 1992 in which the Olympic games were held. The range is the winning times, which are approximately 20.5, 20.3, 19.85, 20.0, 20.24, 20.2, 19.8, 19.76, and 20.0. ◄

Write an ordered pair to describe each point in the graph at the right.

1. A 2. B 3. C 4. D

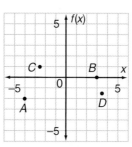

Graph each ordered pair on a coordinate plane.

5. $E(0, -1)$ 6. $F\left(-1\frac{1}{2}, 4\right)$

7. $G(3, 5)$ 8. $H(-3.75, -4.5)$

Determine whether each graph shows a function. For each function, name or describe the domain and the range.

9.

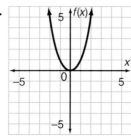

10.

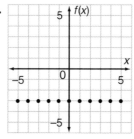

11.

Men's Jean Sizes

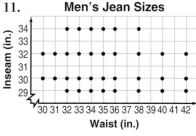

Name the quadrant or axis on which each point is located.

1. $M(-3.2, 0)$ 2. $N(-8, 7)$ 3. $P\left(4\frac{1}{2}, -4\frac{1}{2}\right)$ 4. $Q(5, 1)$

Write an ordered pair to describe each point in the graph at the right.

5. R 6. S 7. T

8. U 9. V 10. W

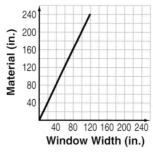

Graph each ordered pair on a coordinate plane.

11. $J(1, -3)$ 12. $K\left(-2\frac{1}{4}, -1\frac{1}{2}\right)$

13. $X(4, 0)$ 14. $Y(0.8, 2)$

Determine whether each graph shows a function. For each function, name or describe the domain and the range.

15. **INTERIOR DECORATOR** For pleated draperies, an interior decorator orders twice the amount of material as the width of a window. This graph shows the relation between the width of the window and the material ordered.

Drapery Material Needed

16. POSTAGE In 1995, first class mail cost 32¢ for the first ounce or fraction thereof and 23¢ for each additional ounce or fraction thereof. This graph shows the relation between the cost of mailing a first class letter and the weight of the letter.

17. WRITING MATHEMATICS How did you determine whether each graph in Exercises 15 and 16 is a function?

TECHNOLOGY/WEATHER For Exercises 18–19, graph each set of data. Use a graphing utility if appropriate. Tell whether each set represents a function.

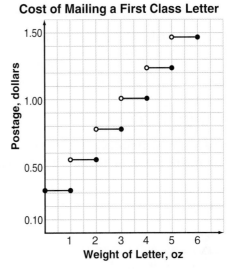

Cost of Mailing a First Class Letter

18. Wind makes cold temperatures feel even colder. The table below shows the wind chill effect when the actual temperature is 30°F.

Wind Speed, mi/h	5	10	15	20	25	30	35	40	45
Apparent Temperature, °F	27	16	9	4	1	−2	−4	−5	−6

19. The normal maximum and minimum temperatures for locations in the United States are based on records kept from 1951–1980. For each city, the table below shows the maximum and minimum temperatures for January.

City	Juneau, AK	Phoenix, AZ	Los Angeles, CA	Jacksonville, FL	New York, NY
Maximum Temperature, °F	27	65	67	65	37
Minimum Temperature, °F	16	39	48	42	26

EXTEND

SPORTS An ice hockey game consists of 60 minutes of play divided into 3 equal periods that are separated by 10-minute intermissions. Part way through one period, a spectator wondered how many more minutes of play until the next intermission.

20. Make a table of at least four pairs of data that could help the spectator solve his problem. Define the input and output values.

21. Look for a pattern in the input and output values shown in your table. Then use function notation to describe the function.

22. Show the function as a graph.

23. WRITING MATHEMATICS Find how many minutes there are until the next intermission if 8 minutes of the period have been played. To what did you refer? Why?

Use the graph at the right for Exercises 24 and 25.

24. Write a story that this graph might tell.

25. Write another story this graph might tell. Change the title and labels for the axes, as needed.

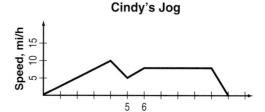

Cindy's Jog

THINK CRITICALLY

Let *a* and *b* represent positive real numbers. Use *a*, *b*, and zero to represent a point in each of the following places on the coordinate planes.

26. Quadrant III

27. Quadrant I

28. the *x*-axis

29. Quadrant IV

30. the *f(x)*-axis

31. Quadrant II

Match each table of input and output values with the graph of the same function. Explain how you made each match.

32.

Input	Output
10	20
15	25
20	30
25	30
30	25

33.

Input	Output
20	40
40	80
60	100
80	90
100	80

a.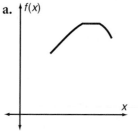

b.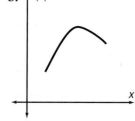

PROJECT *Connection*

The essence of softball is running, hitting, and catching. "Good hands" and "good feet" are hallmarks of good softball players. For this activity, you will need to measure the hand length and shoe length of your school's softball players, to the nearest centimeter.

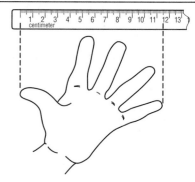

1. Record the data as an ordered pair (hand length, shoe length) for each player.

2. Graph the ordered pairs on an appropriately labeled coordinate plane.

3. Make a conjecture about the relationship between the hand length and shoe length of softball players.

4. Gather the hand length and shoe length data for the students in your math class and, using a different color, graph these results onto the graph you made above for softball players.

5. Compare the softball team's data to that of the math class. Make a conjecture based on the data.

6. Make a conjecture about the relationship between the hand length and shoe length of the general student population.

4.4 Linear Functions

Explore

1. Write the equation represented by these Algeblocks.

2. Use Algeblocks to find some solutions for the equation you wrote. Copy the table at the right. Substitute the values for *x* shown in the table. Record the corresponding values for *y* that you find.

3. Is the relation between *x* and *y* in the equation you wrote in Question 1 a function? Explain your answer.

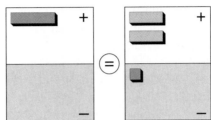

Input (x)	Output (y)
−2	
0	
1	
2	

Build Understanding

In Lessons 4.2 and 4.3, you learned that there are two types of mathematical relations: those that are functions and those that are not functions. Likewise, there are two types of functions: linear functions and nonlinear functions. You will learn about linear functions in this lesson and nonlinear functions in Lesson 4.6.

A **linear function** is a function that can be represented by a straight line. It can also be represented by a **linear equation in two variables** of the form

$$y = ax + b$$

in which *x* and *y* are variables and *a* and *b* are constants. This equation is similar to function notation, except *y* is written in place of $f(x)$. Both *y* and $f(x)$ represent the same thing—the output values for a function. Therefore, another name for the vertical axis of the coordinate plane is the **y-axis**, and another name for the $f(x)$-coordinate of an ordered pair is the **y-coordinate**.

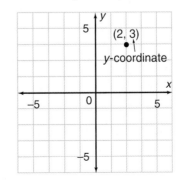

To graph any linear function, you can make a table of *x*- and *y*-values, plot those ordered pairs, and connect the points with a line.

EXAMPLE 1

Graph the linear equation $y = 3x + 2$. Use a table of values.

Solution

Make a table of values. In the table below, –1, 0, 2, and 3 are substituted for x, but you can use any real number for x.

x	y
−1	−1
0	2
2	8
3	11

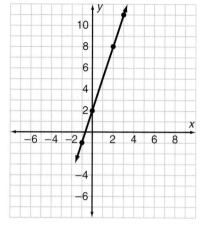

Plot the ordered pairs in your table of values and connect them with a line.   ◄

The graph of a linear equation shows solutions of the equation. By looking at the graph, you can tell whether an ordered pair is a solution.

EXAMPLE 2

Determine whether each ordered pair is a solution of the equation $y = 3x + 2$, which was graphed in Example 1.

   a. $(2, 0)$ **b.** $(1, 5)$ **c.** $(−3, −7)$

Solution

  a. $(2, 0)$ is not a solution. It does not lie on the graph of the equation.

  b. $(1, 5)$ is a solution of the equation. It lies on the graph.

  c. $(−3, −7)$ is also a solution of the equation. If you extend the line drawn in Example 1, you see the point lies on the graph.   ◄

If a function is linear, equal changes in the x-values cause equal changes in the y-values. Look again at the graph of the equation from Example 1. Each time the x-value increases by 1, the y-value increases by 3.

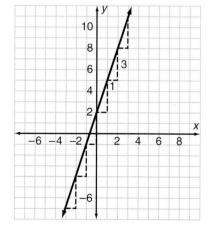

EXAMPLE 3

GEOMETRY Both the perimeter and the area of a square are functions of the length of a side of the square. Some input and output values for each function are given in these tables. Determine whether each function is linear or nonlinear.

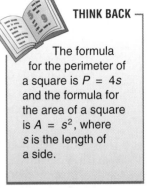

THINK BACK

The formula for the perimeter of a square is $P = 4s$ and the formula for the area of a square is $A = s^2$, where s is the length of a side.

a.

length of side, units	1	2	3	4	5
perimeter, units	4	8	12	16	20

b.

length of side, units	1	2	3	4	5
area, square units	1	4	9	16	25

Solution

a. The perimeter function is linear. Each time the length of a side increases by 1 unit, the perimeter increases by 4 units.

b. The area function is nonlinear. Each time the length of a side increases by 1 unit, the area increases by a different number of square units. ◄

TRY THESE

1. **MODELING** Write the equation represented by the Algeblocks shown at the right. Then make and use a table of values to help you graph the equation.

For each equation, make a table of at least four pairs of values. Use the table to graph the linear function.

2. $y = x + 3$ 3. $y = \dfrac{x}{2} - 4$ 4. $y = 2x + 4$

Determine whether each ordered pair is a solution of the equation graphed at the right. Explain why or why not.

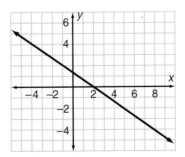

5. $(1, -2)$ 6. $(-1, 2)$

7. $(-8, 8)$ 8. $(2, 0)$

9. $(-4, 3)$ 10. $(0, 3)$

11. **SPORTS** The distance Marco rides on his bicycle is a function of the time he rides. Study the input and output values shown in the table for this function. Determine whether the function is linear or nonlinear. Explain your decision.

time, min	15	30	45	60	75	90	105	120
distance, mi	2.0	4.5	7.0	10.0	13.0	15.5	18.0	20.0

PRACTICE

For each equation, make a table of values. Then use the table of values to graph the linear function. Refer to the graph and write three ordered pairs not in your table of values that are also solutions of the equation.

1. $y = x - 2$ **2.** $y = -3$ **3.** $y = 4x$ **4.** $y = \dfrac{-2x}{3} + 4$

5. WRITING MATHEMATICS Answer these questions about your work for Exercises 1–4. How did you choose the values to substitute for x to make each table of values? How many ordered pairs did you write in each table before graphing the function? Why did you write that many?

Determine whether each ordered pair is a solution of the equation graphed at the right. Explain why or why not.

6. $(4, 1)$ **7.** $(2, -1)$ **8.** $(-4, 5)$

9. $(12, 7)$ **10.** $(6, 2)$ **11.** $(-11, -8)$

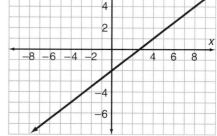

Look at each set of input and output values for the functions below. Then tell whether each function is *linear* or *nonlinear*. Explain your decision.

12. GEOGRAPHY While using a map, Tanika found that these distances on the map represented the corresponding actual distances.

map distance, in.	$\frac{1}{2}$	1	$1\frac{1}{2}$	2	$2\frac{1}{2}$	3
actual distance, mi	50	100	150	200	250	300

13. TAXES A person's tax bracket is the highest rate of federal income tax due on any part of his or her income after deductions. This table shows tax brackets for selected single taxpayers.

taxable income, $	tax rate, %
14,999	15
29,999	28
44,999	28
59,999	31
74,999	31

14. MUSIC Regardless of how many pieces of music Michael is working on, he practices the same amount of time each day, as shown in this table.

pieces of music	practice time, min
1	60
2	60
3	60
4	60
5	60
6	60

EXTEND

Rewrite in the form $y = ax + b$. Make a table of values and graph the equation.

15. $y - x = 5$ **16.** $x + y = 0$ **17.** $y - \dfrac{3x}{4} = 3$ **18.** $2y + x = 4$

TECHNOLOGY Represent each function described below in three forms: a table of at least four pairs of values, an equation, and a graph. Use paper and pencil or a graphing utility to make each graph.

19. CONSTRUCTION To make concrete, for each 3 lb of sand added to cement, 4 lb of gravel are added.

20. CONSUMERISM One mobile phone company charges its customers $24 per month plus $0.24 per minute of air time.

THINK CRITICALLY

Without making a table of values or a graph, tell whether each equation represents a linear function. If it does not, explain why not.

21. $x = 1$ **22.** $y = 1$ **23.** $y = x$ **24.** $y = x^2$

WRITING MATHEMATICS By looking at each of the following, how could you tell whether it represents a linear function?

25. a table   **26.** an equation   **27.** a graph

PROJECT *Connection*   Basketball players must make shots while the opposing team's running and with defense guarding them. How accurately can they shoot without any of these obstacles? For this activity, you will need one member from the girls' basketball team and one member from the boys'.

Use masking tape and a tape measure to mark off distances of 2 ft to 22 ft, in 2-ft increments, from the backboard. The masking tape lines should be parallel to the backboard. Each player will take 5 shots from each line, in ascending order from the backboard. The paired data for each player is the distance in feet from the backboard and the number of shots that go in the basket from that distance.

1. Record the data for each player. Make a color-coded graph of the results.

2. Make a conjecture about the relationship between the distance and the number of successful shots. Is this relationship different for these two players?

3. Repeat the experiment, with two alterations. Have the players alternate turns, and have them pick their distance from the backboard randomly from a deck of index cards. Write the even numbers from 2 to 22 on two sets of index cards—one for each player. Each player will have to use up the entire deck, but in random order. Do you think these alterations will affect the shooting results? Make a conjecture.

4. Graph the results. Does the graph indicate that the random order of distances affected the results? Explain.

Referees are a necessary part of every competitive sports contest. All sports have rules that must be followed. Also, the geometrical nature of the field or court on which some sports are played requires that an impartial judge determine if a player, ball, or puck is fair or foul, in or out of bounds, and so on. These tasks fall to the referee. Sports referees receive special training for the split-second decisions they must make without the benefit of a videotape replay. In many situations, mere inches can have a great effect on the outcome of the game—or an entire season!

In baseball, there are usually more than 200 pitches per game. For almost every one of these pitches, home-plate umpires must decide whether the pitched ball is a ball or a strike. To help them make these calls, major league baseball has officially defined a *rectangular strike zone* as follows.

- The width of the strike zone is the width of home plate, 17 in.
- The height of the strike zone extends vertically from the middle of the batter's knees to the armpits, when the player is in the crouched position in the batter's box, waiting to hit.

A pitch that hits the strike zone is a strike, while a pitch that hits outside the strike zone is a ball. Therefore, the area of the strike zone is critical.

Decision Making

1. Write an equation in which the area of the strike zone A is expressed as a function of the height of the strike zone h.

2. Make a table of at least four values for this function.

3. Use paper and pencil or a graphing utility to graph the function.

4. A batter has a strike zone area of 340 in.² What is the height of his strike zone?

5. Probably the most famous strike zone in baseball history belonged to 3-ft, 7-in. Eddie Gaedel, who had a strike zone height of $1\frac{1}{2}$ in. In his one time at bat in the major leagues, Gaedel saw four balls and no strikes. What was the area of Gaedel's strike zone?

6. Why is the area of the strike zone not a function of the batter's height?

Think Back

Work with a partner. When you use a calculator, you must keep in mind that it is only as efficient as its operator. Estimating the answer in advance may help you notice when you have made a mistake entering numbers or commands.

Without using a calculator, decide whether each calculator display is a reasonable result or an error on the operator's part. Explain.

1. The operator is using the calculator to find the product of 34 times 21.

2. The operator is using the calculator to find the product of 25 times 63.

Explore

Just as good estimation skills help when you use a calculator, good algebra, estimation, mental math, and graphing skills help you to be an effective operator of a graphing utility.

3. Set the range values for the viewing window on your graphing utility to a **standard window** with x- and y-scales of 1 and

   Minimum x-value: −10   Minimum y-value: −10
   Maximum x-value: 10   Minimum y-value: 10

4. Enter the function $y = 3x + 4$. Press the appropriate graphing key. Sketch what appears on the screen.

5. Delete $y = 3x + 4$ from your graphing utility. Enter the function $y = 30x + 450$. Press the appropriate graphing key. Sketch what appears on the screen.

6. If the graph of the function did not appear on the screen, explain what you think happened.

Work with pencil and paper to see if you can figure out why you don't see a graph when you try to graph one function but you do see a graph when you try to graph another function.

7. Set up a table of values using several *x*-values that range from –20 to 20 for $y = 3x + 4$ and $y = 30x + 450$. Then graph both functions on the same pair of coordinate axes. Briefly describe your work.

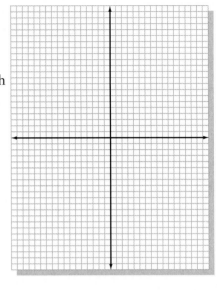

How you set up your axes on the paper determined how much of the graph you could show. Similarly, you can make choices of range values for the viewing window of your graphing utility that allow you to see more of a graph.

Set the range values for the viewing window on your graphing utility as follows. Use an *x*-scale of 2 and a *y*-scale of 100.

Minimum *x*-value: –20	Minimum *y*-value: –200
Maximum *x*-value: 20	Maximum *y*-value: 1100

8. Now try again to graph both functions on your graphing utility. What happened?

Make Connections

9. How are the minimum and maximum *y*-values listed above related to the minimum and maximum *x*-values?

Set the range values and scale for the viewing window that allow you to see the graph of each equation. Use your graphing utility to check.

10. $y = 40x - 500$ **11.** $y = x + 75$

Sometimes a graph displayed in the standard window could be displayed better if the range values and scale were adjusted. Choose a more useful viewing window for the graph of each equation. Use your graphing utility to check.

12. $y = \dfrac{x}{5} + 10$ **13.** $y = 0.05x + 0.3$

Summarize

14. **WRITING MATHEMATICS** Explain to a friend why a graphing utility may not always display the intended graph. Using examples, continue the explanation by describing how to adjust the viewing window, if necessary, to display the intended graph.

THINKING CRITICALLY Suppose you set your graphing utility's range values and graphed a linear function that was clearly displayed. Fill in each blank in Problems 15 and 16 with *always*, *sometimes*, or *never*. Explain your answers.

15. Suppose you were to multiply each of the range values by 10. The graph of the function with the new range values will __?__ appear in the display.

16. Suppose you were to divide each of the range values by 10. The graph of the function with the new range values will __?__ appear in the display.

THINKING CRITICALLY A group of four students was exploring the effects of changing range values on the appearance of the graph of a linear function. Here are the values each student used.

	Student 1	Student 2	Student 3	Student 4
Minimum *x*-value	−20	−20	−20	−20
Maximum *x*-value	20	20	20	20
Minimum *y*-value	−50	−20	−14	−8
Maximum *y*-value	60	25	14	10

17. What do you notice about the values the students used?

18. What will the students be able to determine using this approach?

19. Suggest another step the students could use to further explore the effects of changing range values on the graph. Explain why this step would be useful.

20. **GOING FURTHER** Suppose the maximum *y*-value is set at less than 100. Will the graph of the function $y = x^2 + 100$ be displayed? Why or why not?

4.6 Nonlinear Functions

Explore

1. Look at the equations at the right. How are they the same? How are they different?

$$y = x + 1$$
$$y = x^2 + 1$$

2. **TECHNOLOGY** Use a graphing utility to graph the functions represented by these equations on the same coordinate plane.

3. How are the graphs the same? How are they different?

Build Understanding

In Lesson 4.4 you studied linear functions—functions that can be represented by straight lines. In this lesson you will study some nonlinear functions that have the term x^2 in their equations. In later chapters you will learn more about these functions as well as other types of nonlinear functions.

As with a linear function, you can graph a nonlinear function by making a table of values, plotting the ordered pairs, and connecting the points.

CHECK UNDERSTANDING

Identify each equation in Explore as a linear or a nonlinear function.

EXAMPLE 1

Graph the nonlinear function $y = 2x^2$.

Solution

Make a table of values. Use enough points to be able to draw the graph smoothly.

Plot the ordered pairs that correspond to the values in the table and connect them with a smooth curve.

x	y
−2	8
−1	2
0	0
1	2
2	8

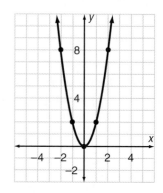

Check your graph by graphing on a graphing utility.

As with a linear equation, you can also tell if an ordered pair is a solution of a nonlinear equation by looking at the graph of the equation.

EXAMPLE 2

Determine whether each ordered pair is a solution of the equation $y = 2x^2$, which was graphed in Example 1.

 a. $(-2, -8)$ **b.** $(2, 8)$ **c.** $(-5, 50)$

Solution

a. $(-2, -8)$ is not a solution of the equation because it does not lie on the graph of the equation.

b. $(2, 8)$ is a solution of the equation because it does lie on the graph of the equation.

c. $(-5, 50)$ is also a solution of the equation because, if you extended the curve drawn in Example 1, you could see that the point does lie on the graph of the equation. ◀

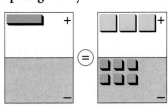

COMMUNICATING ABOUT ALGEBRA

For Example 2a could you have told whether $(-2, -8)$ is a solution without looking at the graph or evaluating the equation? Why or why not?

You can refer to the graph of a nonlinear function to find the output value for a given input value or the input value for a given output value.

EXAMPLE 3

GEOMETRY The area of a circle is a function of its radius r as shown by the equation $y = \pi r^2$. Approximately what is the radius of a circle with an area of 20 square units?

Solution

Find the point on the graph that corresponds to an area of 20 square units. Determine the radius value.

The radius of a circle with an area of 20 square units is approximately 2.5 units.

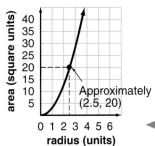

area (square units) vs radius (units). Approximately (2.5, 20) ◀

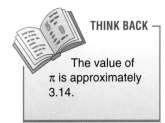

THINK BACK

The value of π is approximately 3.14.

TRY THESE

For each equation, make a table of at least five pairs of values. Use the table to graph the nonlinear function. You may find it necessary to add more ordered pairs to your table in order to draw the curve smoothly. Check your graph by graphing on a graphing utility.

1. $y = 3x^2$ **2.** $y = -x^2$

3. $y = 2x^2 - 4$ **4.** $y = x^2 - x + 1$

5. MODELING Write the equation represented by the Algeblocks. Then make and use a table of values to help you graph the equation.

Determine whether each ordered pair is a
solution of the equation graphed at the
right. Explain why or why not.

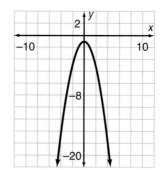

6. $(2, 5)$

7. $(3, -10)$

8. $(-4, -17)$

9. $(0, -1)$

AUTO MECHANICS The graph at the right
shows the distance a car travels in meters
as it starts from rest and continues to
accelerate for 1 min. Refer to the graph
to answer these questions.

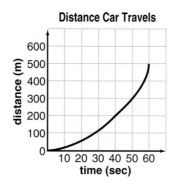

Distance Car Travels

10. About how far does the car travel
in 20 seconds?

11. About how many seconds does it take
for the car to travel 400 m?

PRACTICE

For each equation, make a table of values. Use the table of values to graph the nonlinear
function. Refer to the graph and write three ordered pairs, not in your table of values, that
are also solutions of the equation. Check your graph using a graphing utility.

1. $y = -2x^2$

2. $y = \dfrac{x^2}{2}$

3. $y = 3x^2 + 5$

4. $y = 2x^2 + 3x - 4$

5. WRITING MATHEMATICS Answer these questions about your work for Exercises 1–4.
How did you choose the values to substitute for x to make each table of values? How
many ordered pairs did you write in each table before graphing the function? Why did
you write that many?

LAW ENFORCEMENT The graph at the
right shows the number of traffic accidents
the police department in one town
recorded for the 25-year period from
1970–1995. Refer to the graph to answer
these questions.

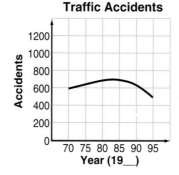

Traffic Accidents

6. About how many accidents occurred
in 1985?

7. What is the first year in which there
were fewer than 600 accidents?

8. During what year was there an increase in accidents? a decrease?

For each equation in Exercises 9–11, make a table of values. Then use the table to graph the equation. Tell whether each equation represents a *linear function*, a *nonlinear function*, or *neither*. Check using a graphing utility.

9. $y = x^3$

10. $y = \pm\sqrt{x}$

11. $y = x^4$

BUSINESS When a potential customer calls a carpet cleaner to get an estimate, the sales representative refers to the graph at the right. Use this graph to solve Exercises 12–15.

12. Mr. Mullins estimates there are 500 square feet of carpeting in his living room and hallway. About how much will it cost him to have his carpet cleaned?

13. Mrs. Santiago was given an estimate of $95. Approximately how many square feet of carpet must she have asked to have cleaned?

14. Why do you think the first part of this graph is a straight line?

15. After 200 square feet, why do you think the remainder of the graph is a curve?

Clean Up Carpet Care, Inc.

EXTEND

SPORTS The height in meters of a ball above its starting point is approximately equal to $rt - 5t^2$, where r is the initial upward velocity in meters per second at which the ball was kicked or thrown and t is the time in seconds after the ball was kicked or thrown. For Exercises 16–20, assume that a football player kicks the ball with an initial upward velocity of 20 m/s.

16. Write an equation using x and y to show the height of the ball as a function of the time after the ball was kicked.

17. TECHNOLOGY Use your graphing utility to graph the equation you wrote in Exercise 16.

18. What is the greatest height above the ground the ball will reach? When will it reach this height?

19. How many seconds after it is kicked will the ball fall back to the ground?

20. How much higher will the ball go if it is kicked with an initial upward velocity of 25 m/s? How much longer will it stay in the air?

THINK CRITICALLY

21. WRITING MATHEMATICS Explain why $y = x^2$ is a curve symmetric to the y-axis and $y = x^3$ is not. Use specific x and y values to illustrate.

For Exercises 22–24, refer to the functions $y = ax^2$ and $y = -ax^2$, when $a \neq 0$.

22. How will the graphs of these functions be the same?

23. How will the graphs of these functions be different?

24. What point will the graphs of both these functions have in common?

MIXED REVIEW

25. REAL ESTATE Mrs. Lake sold five properties during January for $80,000, $97,000, $42,000, $150,000, and p. The mean selling price of the properties Mrs. Lake sold is $89,000. What is the value of p?

Simplify each expression.

26. $7a + 9b - 3ab + 2(a - b)$

27. $4r^2 - 2r + 7r^2 - 3 + 5r$

28. STANDARDIZED TESTS A stereo system is on sale for $90 off the regular price of $750. What is the percent of the discount?

   A. 9%
   B. 12%
   C. 14%
   D. 1.2%

Name the quadrant or axis on which each point is located.

29. $(-3.7, 4)$
   30. $(250, -75)$
   31. $\left(2\frac{1}{2}, 0\right)$
   32. $(-45, -0.03)$

PROJECT *Connection*   Interview your school's athletic director to find out some historical information about won-lost records. Choose five different school teams.

1. Compute the percent of wins for each team for the past five years.

2. Graph the paired data of year and percent of wins for each team. Let the horizontal axis represent years, and let the vertical axis represent winning percents. Use intervals of 10% on the vertical axis. Graph a separate scatter plot for each team.

3. Graph the same data on graphs with a different scale. Use intervals of 5% but keep the same spacing between tick marks that you used in Question 5.

4. Explain how a change in the scale of the vertical axis can be used to make a team look more or less consistent.

4.7 Solve Equations by Graphing

Explore

1. Graph the linear equation $y = 2x - 1$. How many solutions does this equation have?

2. Explain how to use your graph to find the exact value of x for which the value of y is 7.

3. On the same coordinate plane as you used above, graph $y = 7$. Is this a linear equation? How many solutions does $y = 7$ have?

4. The graphs of the two equations intersect at a point. What is the ordered pair for this point? Explain how this point is related to the equation $7 = 2x - 1$.

5. Explain how you could solve the equation $\frac{1}{2}x + 4 = 9$ graphically. Find the solution using your method. Check your solution.

Build Understanding

Recall that an equation such as $2x + 2 = x + 1$ is true when the value of the expression on one side of the equal symbol is the same as the value on the other side. As you saw in Explore, you can think of the expression on each side as a function having the form $y = f(x)$.

CHECK UNDERSTANDING

What value of x makes $2x + 2 = x + 1$ true? How did you find your solution?

EXAMPLE 1

Write each side of the equation $2x - 3 = x + 4$ as a function of the form $y = f(x)$.

Solution

$$2x - 3 \quad = \quad x + 4$$
$$\downarrow \qquad\qquad \downarrow$$
$$y = 2x - 3 \quad \text{and} \quad y = x + 4 \qquad \blacktriangleleft$$

You know that the graph of an equation shows all solutions of the equation. Therefore, if two equations are graphed on the same coordinate plane and those graphs intersect, the point(s) of intersection are the solutions of both equations. You can use this fact to graph and solve a linear equation with the same variable on both sides.

EXAMPLE 2

Use graphing to solve $2x - 3 = x + 4$.

Solution

On the same coordinate plane, graph the functions that represent each side of the equation. From Example 1, the two equations are

$$y = 2x - 3 \quad \text{and} \quad y = x + 4$$

x	y
-3	-9
0	-3
2	1

x	y
-3	1
0	4
2	6

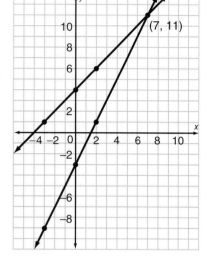

Both graphs intersect at $(7, 11)$, so the solution of $2x - 3 = x + 4$ is $x = 7$.

Check

$$2x - 3 \overset{?}{=} x + 4$$
$$2(7) - 3 \overset{?}{=} 7 + 4$$
$$14 - 3 = 11$$
$$11 = 11 \checkmark$$

You can also use this method to solve some real world problems.

EXAMPLE 3

SPORTS On the second night of a tournament, Justine told her teammates they scored 2 more points from the free throw line than twice the number of points they had scored from the line on the first night. The coach said that if the team had scored 5 more points from the free throw line, they would have tripled the number of points scored from the line on the first night.

a. How many points had the team scored from the free throw line the first night?

b. How many points had the team scored from the free throw line the second night?

Solution

Let x represent the number of points scored from the line on the first night. Write an expression for Justine's description of the points scored from the line the second night and an expression for the coach's description. Then write an equation using the two expressions.

$$\boxed{\begin{array}{c}\text{Justine's}\\\text{Description}\end{array}} = \boxed{\begin{array}{c}\text{Coach's}\\\text{Description}\end{array}}$$

$$2x + 2 = 3x - 5$$

Now write and graph the functions for each side of the equation.

$$y = 2x + 2 \quad \text{and} \quad y = 3x - 5$$

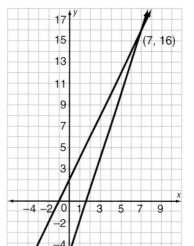

(7, 16)

a. The graphs intersect at $(7, 16)$, so the solution of $2x + 2 = 3x - 5$ is $x = 7$. The team scored 7 points from the free throw line the first night.

b. The ordered pair $(7, 16)$ shows that the value of the expression on each side of the equation is 16 when x is 7. Since either expression describes the number of points on the second night, the team scored 16 points from the free throw line on the second night.

CHECK UNDERSTANDING

Suppose you did not realize that you could find the second night's score from the ordered pair. What other method could you use?

TRY THESE

Write each side of the given equation as a function.

1. $9x - 5 = 22$

2. $5x + 7 = 2x + 30$

3. $3(x - 5) = x + 7$

Graph both sides of each equation. Then use the graphs to solve the equation.

4. $\frac{1}{2}x + 3 = 5$

5. $2(x + 3) = 3x + 9$

6. $4(x - 2) = 4x - 8$

7. MODELING Use graphing to solve the equation represented by these Algeblocks. Check your solution by using Algeblocks to solve the equation.

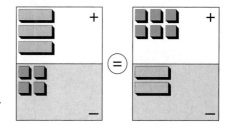

GEOMETRY The perimeter of a particular square increased by 20 units is equal to 7 times the length of a side that is increased by 2.

8. Write an equation that states this.

9. Graph both sides of the equation to find how many units long each side of the square is.

10. WRITING MATHEMATICS Explain how you can use the graph of the function $y = 2x - 5$ to solve the equation $11 = 2x - 5$.

PRACTICE

Write each side of the given equation as a function. Then graph both functions and use the graphs to solve the equations.

1. $-3x + 5 = -10$

2. $4(x - 3) + 1 = 17$

3. $2x - 4 = 2(x - 2)$

4. $0.5x + 3.5 = 2x + 6$

5. **WRITING MATHEMATICS** Solve each equation in Exercises 1–4 algebraically. What advantage does using graphing to solve the equations have over solving them algebraically? What advantage does using algebraic operations have?

AIR TRAFFIC CONTROL One plane left New York headed to Tokyo flying at an average speed of 375 mi/h. Another plane left New York 1 h later following the same route and flying at an average speed of 500 mi/h. If both planes followed the same course, how many hours after it left New York would the second plane catch up to the first plane?

```
 9:45P    ON TIME
 9:30P    ON TIME
 9:40P    ON TIME
 9:50P    ON TIME
10:05P    ON TIME
 9:35P    ON TIME
```

6. Write an equation you could solve to answer the question.

7. **TECHNOLOGY** Use a graphing utility to graph both sides of the equation and solve the problem.

8. How many miles will each plane have flown when they are equidistant from New York?

BUSINESS A video store offers its members two payment options.

- Plan A: Rent each video for $1.99 per night.
- Plan B: Pay an annual fee of $20.00 and rent each video for $0.89 per night. How many videos would a member have to rent in order for Plan B to be more economical?

9. Write an equation you could solve to answer the question.

10. **TECHNOLOGY** Use a graphing utility to graph both sides of the equation and solve the problem.

11. What would be the cost under each plan to rent the number of videos you found in Exercise 10?

EXTEND

Use graphing to solve each equation.

12. $x^2 - 5 = -x + 1$

13. $x^2 + 1 = -x^2 + 1$

Use graphing to tell whether each of the following has a solution.

14. $\dfrac{x}{2} = x - 4$

15. $-4(x + 4) + 5 = 2(-2x + 3)$

16. $3x + 4 = 3x - 2$

17. $x^2 - 2 = -x^2 + 2$

PLUMBING The drain to a 20-gal sink is closed. The hot water faucet is turned on at the rate of 2.5 gal/min. One minute later the cold water faucet is turned on at the rate of 3 gal/min. Use graphing to solve these problems.

18. How many minutes would the faucets have to run for there to be an equal amount of water from each faucet in the sink?

19. Vinje guessed that the water would probably overflow before the amount of hot water equaled the amount of cold water. Decide if she is right and justify your answer.

THINK CRITICALLY

For each of the following conditions, describe the graphs of $y = ax + b$ and $y = cx + d$.

20. There is one x-value for which $ax + b = cx + d$.

21. There are two x-values for which $ax + b = cx + d$.

22. For $a = c$, $ax + b \neq cx + d$.

23. For all real numbers, $ax + b = cx + d$.

For each of the following conditions, describe the graphs of $y = ax^2 + b$ and $y = cx + d$.

24. There is one x-value for which $ax^2 + b = cx + d$.

25. There are two x-values for which $ax^2 + b = cx + d$.

MIXED REVIEW

26. STANDARDIZED TESTS In a survey of 100 people living in the same town, f people reported they had the flu. On the basis of this survey, choose the expression that tells how many people in this town of 8452 could be expected to have the flu.

A. $\dfrac{100f}{8452}$ B. $\dfrac{f}{8452}$ C. $\dfrac{f + 100}{8452}$ D. $\dfrac{8452f}{100}$

Evaluate each expression.

27. $(3 \cdot 4)^2$

28. $(3)(4)^2$

Simplify each expression.

29. $(-c)(-d)(-e)$

30. $\left(\dfrac{-m}{n}\right)\left(\dfrac{o}{-p}\right)$

Evaluate $f(x) = 3x - 5$ at each given value of x.

31. -2

32. 0

33. 3

4.8 Problem Solving File

Qualitative Graphing

In many situations an object's motion is approximately linear; for example, a car traveling along a highway or a jet flying nonstop from one city to another. To give a good description of an object's motion along a straight path, you might include the following.

• How fast the object was moving (at any time)
• If and when the object turned around
• How long the object was in motion
• How far the object is from its starting point (at any time)
• The farthest the object got from its starting point (and when)

You could give all this information in words or with a visual presentation such as a graph. Instead of trying to create an exact picture, you can use a **qualitative graph** to capture many of the important features of an object's trip.

Explore the Problem

1. Work with a small group. Use chalk or tape to lay out a straight path of at least 10 ft (perhaps in your classroom, a school hallway, the gym, or the schoolyard). Mark off the line in feet or half-feet.

2. Each student will "travel" on the line. The student should do a variety of things such as stop, start, slow down, speed up, turn around, walk backward, stand still, and so on.

3. While one student "travels," the other group members should keep track of the number of seconds that have elapsed and the distance the "traveler" is from the starting point. (Do the best you can, but don't worry if you can't measure exactly and continuously.)

4. Use the data collected for each student to sketch a qualitative graph on a coordinate system like the one shown.

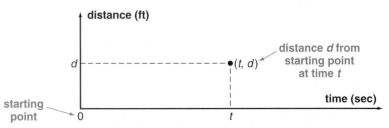

5. Exchange graphs with other groups and try to tell the travel story told by each.

Investigate Further

Below are the graphs of Olivia's and Jorge's trips.

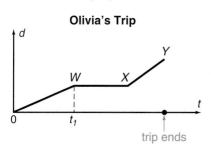

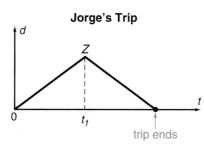

PROBLEM
SOLVING PLAN

- Understand
- Plan
- Solve
- Examine

6. Which student stood still for a period of time? How do you know?

7. Which student moved more rapidly during the first part of the trip (t_1)? How do you know?

8. What did Jorge do at the point marked Z?

9. Did Olivia return to the starting point at the end of the trip? Explain.

10. The graph of Karen's trip is at the right.

    a. During which part of the trip did Karen speed up?

    b. How many times did Karen change direction?

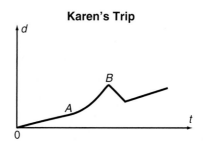

Apply the Strategy

11. Do you think Brian's graph, at the right, is a good representation of his trip? Why or why not?

12. Sketch a graph for the following trip. Rosa stood 3 ft away from the starting point to begin her trip. She walked forward at a slow, steady rate for a few seconds, stood still for a few seconds, turned around sharply, and then walked back very quickly. At the end of her trip, she had walked 5 ft past the starting point. (*Hint*: Consider extending the *d*-axis.)

13. The graph at the right shows Jill's trip and Phil's trip. Compare their trips.

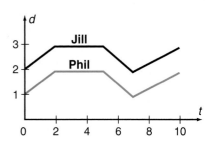

14. Copy the graph for Jill's trip, at the right. Suppose Tod's trip is just like Jill's, except that he starts out 2 seconds after her. Draw Tod's trip on the same coordinate axes as Jill's.

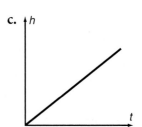

15. The container at the right is being filled with water at a constant rate of flow. In each graph below the time t in seconds that it takes to fill the container is shown along the horizontal axis. The height of the liquid h in centimeters is shown along the vertical axis. Which of the following graphs do you think shows the relationship between height and time? Explain your choice.

a.

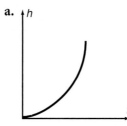

b.

c.

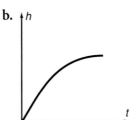

> **PROBLEM SOLVING TIP**
>
> For questions like 15 and 16, acting out the activity can help you understand the problem.

16. **WRITING MATHEMATICS** Sketch several differently shaped bottles or containers. Assume each is to be filled with water flowing at a constant rate. For each, sketch the graph that shows the relationship between height of the water and time. Write a brief explanation for each graph.

REVIEW PROBLEM SOLVING STRATEGIES

MONKEY BUSINESS

1. Ricky is ordering bananas for the monkeys at Jungle World. He knows that four capuchin monkeys will eat 4 lb of bananas every 4 days, three spider monkeys will eat 3 lb of bananas every 3 days, and two howler monkeys will eat 2 lb of bananas every 2 days. How many pounds of bananas should Ricky order for Jungle World's 12 capuchin monkeys, 12 spider monkeys, and 12 howler monkeys for 12 days?

There are several different ways of solving this problem. You can start by figuring out for each type of monkey how much one of them eats in a day. Or, you can organize the data in a table and decide how to "scale up." Whichever method you choose, explain your reasoning and show the computation.

CRYPTO-DIGIT

+ TUB =

2. The following puzzle represents a simple addition problem. Both the addends and the sum are given, but they are disguised as letters. Use the digits 0, 1, 3, 6, 7, 8, and 9. Replace each letter with a digit. The same letter cannot represent different digits, and the same digit cannot be represented by different letters.

```
  S U B
+ T U B
-------
S I N K
```

Answer these questions to help guide your thinking.

a. What number must the letter S stand for? Why?

b. Can the letter T stand for a number that is less than or equal to 7? Why or why not?

c. How can you use what you know about the number that T stands for to help you find the number that U stands for?

d. Now, solve the puzzle.

LINE SEGMENTS

3. How many line segments are in the figure at the right?

a. Work in groups of three students. First, be sure you understand the problem. Then think about whether the line segments are all similar, or if they can be grouped according to some characteristic.

b. Decide on an organized method for counting and recording. Account for all possibilities and do not duplicate any.

c. Review your plan. Is there an easier way or a shortcut you did not see at first?

d. Carry out the work. How many segments did your group find? Compare results with other groups. Discuss and resolve any discrepancies.

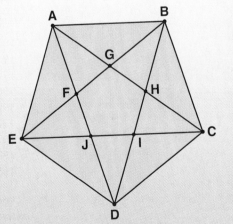

Explore/Working Together

● Work in a group with 5 or 6 students.

1. Measure and record each person's height and foot length.

2. Graph the data (height, foot) on a coordinate plane.

3. Does your graph consist of points or a line? Why?

4. On the basis of your graph, make a statement about the relationship between height and foot length.

Build Understanding

● A graph in which the data are shown as points on a coordinate plane is called a **scatter plot**. You can make a scatter plot by graphing ordered pairs of data.

EXAMPLE 1

SPORTS Refer to the table below to make a scatter plot displaying the number of competitors and the number of nations represented.

SUMMER OLYMPIC GAMES			
Year	Place	Nations Represented	Competitors
1948	London, UK	59	4,099
1952	Helsinki, Finland	69	4,925
1956	Melbourne, Australia	67	3,342
1960	Rome, Italy	83	5,348
1964	Tokyo, Japan	93	5,140
1968	Mexico City, Mexico	112	5,531
1972	Munich, W. Germany	122	7,147
1976	Montreal, Canada	92	6,085
1980	Moscow, USSR	81	5,353
1984	Los Angeles, US	141	7,078
1988	Seoul, S. Korea	160	9,581
1992	Barcelona, Spain	172	10,563

Solution

Summer Olympic Games

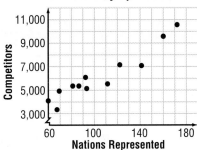

Many times the points in a scatter plot show a general pattern, or **trend**. The pattern formed by the points in the scatter plot in Example 1 approximates a line that rises from left to right.

EXAMPLE 2

Describe the relationship between the number of nations represented in the summer Olympic games and the number of competitors.

Solution

The general rise from left to right indicates that the number of competitors tends to increase as the number of nations increases.

CHECK UNDERSTANDING

Why do you think the points on the scatter plot in Example 1 do not all lie on the same straight line?

COMMUNICATING ABOUT ALGEBRA

Have each member of your group suggest a set of paired data that might show a relationship. Discuss the possible relationship and describe how a scatter plot of the data would look.

TRY THESE

PUBLISHING Use the table below to solve Exercises 1 and 2.

1. Make a scatter plot that shows the relationship of morning newspapers to evening newspapers.

2. Describe the relationship between the number of morning newspapers and the number of evening newspapers.

NUMBER OF DAILY NEWSPAPERS IN U.S.					
Year	Morning	Evening	Year	Morning	Evening
1940	380	1,498	1980	387	1,388
1950	322	1,450	1985	482	1,220
1960	312	1,459	1990	559	1,084
1965	320	1,444	1991	571	1,042
1970	334	1,429	1992	596	995
1975	339	1,436	1993	621	955

PRACTICE

ECONOMICS Use the table at the right to solve Exercises 1 and 2.

1. Make a scatter plot that shows the relationship of production in quadrillion Btu to imports in quadrillion Btu.

2. Describe the relationship between energy production and energy imports.

U.S. ENERGY PRODUCTION AND IMPORTS (in quadrillion Btu)		
Year	Production	Imports
1960	41.49	4.23
1970	62.07	8.39
1975	59.86	14.11
1980	64.76	15.97
1985	64.87	12.10
1990	67.85	18.99
1991	67.34	18.38
1992	66.72	19.45

GEOGRAPHY Use the table at the right to solve Exercises 3 and 4.

3. TECHNOLOGY Use a graphing utility to make a scatter plot for this data.

4. Describe the relationship between the areas of the continents and their populations.

WORLD AREA AND POPULATION, 1992		
Continent	Area (mi² in thousands)	Population (in millions)
Africa	11,687	655
Antarctica	5,100	0
Asia	17,176	3,318
Australia	3,036	18
Europe	4,066	684
North America	9,357	436
South America	6,881	300

5. WRITING MATHEMATICS Compare the tables given for Exercises 1–4 to the scatter plots you made in Exercises 1 and 3. Are the relationships easier to see in the tables or in the scatter plots? Explain.

EXTEND

Use the table at the right to solve Exercises 6–8.

6. Make a scatter plot for this data.

7. Describe the relationship between the number of miles and the number of stations.

8. Suppose you are planning a new mass transit system with 300 mi of rail. About how many stations would you include? Why?

RAPID RAIL SYSTEMS IN THE U.S., 1992		
System	**Route miles**	**Stations**
Baltimore MTA	26.6	12
Chicago TA	191.0	137
Cleveland RTA	38.2	18
Metro Atlanta RTA	67.0	29
Metro Boston RTA	76.7	53
Miami/Dade Co. TA	42.2	21
New York, PATH	28.6	13
New York City TA	492.9	469
New York, Staten Island RT	28.6	22
Philadelphia SEPTA	75.8	76
San Francisco, BART	142.0	34
Washington Metro Area TA	162.1	70

THINK CRITICALLY

Match each pair of *x*- and *y*-values to the appropriate scatter plot below. Explain your thinking.

9. *x*-values: lengths of sides of squares
   y-values: perimeters of squares

10. *x*-values: outside spring temperatures
    y-values: sales of cold drinks

11. *x*-values: students' heights
    y-values: final grades in Science last year

12. *x*-values: outside winter temperatures
    y-values: home heating costs

13. *x*-values: number of hours that have passed today
    y-values: number of hours left today

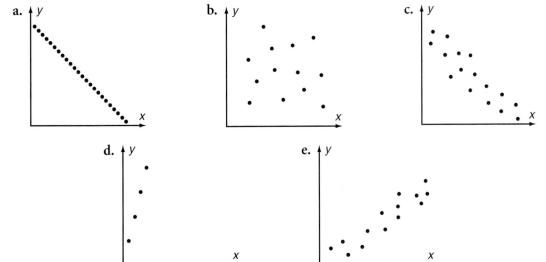

MIXED REVIEW

Use the following matrices for Exercises 14–15.

$$A = \begin{bmatrix} 3 & 5 \\ -4 & 2 \\ 2 & 8 \end{bmatrix} \quad B = \begin{bmatrix} -1 & -4 \\ 4 & 7 \\ 0 & 1 \end{bmatrix}$$

14. Evaluate: $A + B$

15. Evaluate: $B - A$

16. PRINTING A printer received orders for customized stationery in the following quantities: 5,000, 1,000, 15,000, s, and 4,000. The range of these orders is 19,000. What is the value of s?

Evaluate each expression.

17. $|-15|$

18. $-|15|$

19. $-|-15|$

Evaluate $y = x^2 + 2x - 3$ at each given value of x.

20. 6

21. 0

22. -2

Use the distributive property to simplify.

23. $4(x - 3)$ **24.** $-3(a - 7)$ **25.** $4(y + 4y^2)$ **26.** $7(2x - 2)$

27. STANDARDIZED TESTS Choose the letter of the solution to the following equation.

$$6(x + 2) = 4(x + 12)$$

A. 5   **B.** 6   **C.** 18   **D.** 12

28. STANDARDIZED TESTS Choose the letter of the solution to the following equation.

$$4(x + 1) = 6 - 3(1 - 2x)$$

A. $\dfrac{1}{2}$ **B.** $\dfrac{2}{7}$ **C.** $\dfrac{4}{5}$ **D.** 2

Solve for x.

29. $3x - 4 = 5$ **30.** $4(x - 7) = 12$ **31.** $\dfrac{x}{3} - 8 = 16$ **32.** $\dfrac{x}{5} - 16 = -1$

Mathematics is used to determine the level of performance of athletes. Teams, individual players, coaches, and team owners are all judged by analyzing statistics—the data generated by performances on the field of play.

If you've watched a sporting event on television, you've probably seen electronically produced graphics. These charts and graphs are being updated during the event. In addition, charts, graphs, and statistics from previous games are included to add to the interest of the telecast. During a game, the broadcaster is too busy to record and analyze data. Therefore, these tasks are the responsibility of the sports statistician. Some statisticians travel with a broadcasting crew, while others work with one specific team.

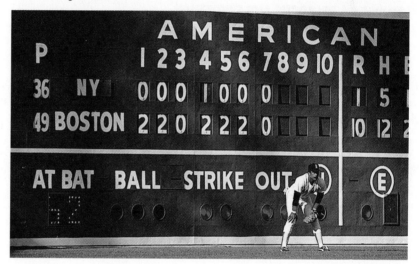

Decision Making

1. Compile a list of as many types of sports statistics as you know or can find out about. Use the library, or consult the coaches at your school, family members, or friends as references. Organize your list by sport. Combine lists with those of your classmates. Then graph the results on a bar graph. Which sport do you think generates the most data? the least data? Explain.

2. Look for sports data in an almanac, encyclopedia, or sports statistics book. Pick three different sets of paired data. Make a scatter plot for each set of data and describe any trends shown by the scatter plots. Make some predictions based on your scatter plots about future sports performances.

• • • CHAPTER REVIEW • • •

VOCABULARY

Choose the word from the list that completes each statement.

1. An equation in the form $y = ax + b$ represents a(n) __?__ function.

2. A graph that does not pass the __?__ test is not the graph of a function.

3. The __?__ of a function includes the y-coordinates of all ordered pairs that are solutions of the equation that represents the function.

4. The equation representing a(n) __?__ function often contains the term x^2.

5. In a function, there is only one output value for each different value in the __?__ of the function.

a. domain

b. linear

c. nonlinear

d. range

e. vertical line

Lesson 4.2   RELATIONS AND FUNCTIONS   pages 171–176

● A **function** is a set of paired data in which there is only one output value for each different input value. Input values make up the **domain** of a function, and output values make up its **range**.

Determine whether each table or set shows a function. For each function, name the values in the domain and in the range.

6.

Input	3	5	7	9	11
Output	2	4	2	4	2

7. $\left\{\left(-5, \frac{1}{5}\right), \left(-4, \frac{1}{4}\right), \left(-3, \frac{1}{3}\right), \left(-4, -\frac{1}{4}\right), \left(-5, -\frac{1}{5}\right)\right\}$

8.

Members	5	10	15	20	25
Nonmembers	25	20	15	10	5

Lessons 4.1 and 4.3   PAIRED DATA AND GRAPHS OF FUNCTIONS   pages 169–170, 177–182

● Paired data can be represented as **ordered pairs** and as points on a graph.

● Use the **vertical line test** to determine whether a graph represents a function.

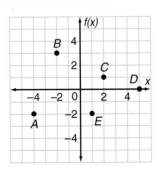

Write an ordered pair to describe each point in the graph at the right.

9. A **10.** B **11.** C **12.** D **13.** E

14. Does the graph show a function? Explain why or why not.

- The graph of a **linear function** is a straight line. To graph a linear function, make a table of input (x) and output (y) values, plot these ordered pairs, and connect them with a line.
- You can also use graphing utilities to display linear functions.

For each equation, make a table of at least four pairs of values. Use the table to graph the linear function. Use the graph to name a solution that is not in your table of values.

15. $y = -x + 3$　　　　**16.** $y = 3x - 2$　　　　**17.** $y = \dfrac{3}{4}x$

- The graphs of **nonlinear functions** are straight lines, but you can graph them the same way as you would a linear function: make a table of values, plot the ordered pairs, and connect the points.

For each equation, make a table of at least four pairs of values. Use the table to graph the linear function. Use the graph to name a solution that is not in your table of values.

18. $y = -2x^2$　　　　**19.** $y = x^2 - 9$　　　　**20.** $y = x^2 - 2x - 3$

- You can solve a linear equation that has the same variable on both sides by graphing. Rewrite each side in the form $y = f(x)$ and graph each function. The point(s) of intersection of the graphs are the solution(s).

Write each side of the given equation as a function. Then graph both functions and use the graphs to solve the equation.

21. $3x - 4 = 8$　　　　**22.** $-2x + 3 = 4x - 9$　　　　**23.** $4(x - 1) = 4x - 4$

- **Qualitative graphs** are useful for representing changes in speed and direction.

24. Sketch a graph for the following trip. Hosni walked for a few minutes. He then jogged for a while, jogged in place while waiting for a red light to turn green, and then jogged some more. Then he walked for a few minutes and returned to his starting place.

- **Scatter plots** are graphs in which paired data are shown as points on a coordinate plane.

25. Make a scatter plot for the data in the table.

26. Describe the relationship, if any, between years and numbers of points.

Olympic Records, Men's Decathlon							
Year	1964	1968	1972	1976	1980	1984	1988	1992
Points	7887	8193	8454	8618	8495	8798	8488	8611

CHAPTER ASSESSMENT

CHAPTER TEST

1. **WRITING MATHEMATICS** Write a paragraph to explain how you could determine whether this set shows a function.

$$\{(-8, 6), (-4, 6), (9, -4), (-8, 6), (-4, 9)\}$$

2. Evaluate $f(x) = -\dfrac{x}{3}$ at $x = 12$.

3. Evaluate $f(x) = (x - 5)^2$ at $x = 3$.

Match each of the ordered pairs to a point on the graph at the right.

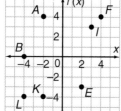

4. $(-2, -4)$

5. $(4, 4)$

6. $(2, -3)$

7. $(-4, 0)$

Make a table of values. Then graph each equation. Tell whether the equation represents a *linear* or *nonlinear* function.

8. $y = x + 4$

9. $y = x^2 - 3$

10. $y = -3x - 1$

11. $y = -\dfrac{3}{4}x + 6$

STANDARDIZED TESTS Refer to the function shown in the table at the right. Tell whether each value is in

Input	Output
7	6
0	2
−1	−9
3	0
−2	−4

   I. the domain
   II. the range

12. 2

   A. I
   B. I and II
   C. II
   D. neither I nor II

13. −3

   A. I
   B. I and II
   C. II
   D. neither I nor II

14. **STANDARIZED TESTS** Which ordered pairs are solutions of the equation graphed at the right?

   I. $(2, 2)$
   II. $(-2, 5)$
   III. $(4, -1)$

   A. II
   B. III
   C. I and II
   D. II and III

15. **WRITING MATHEMATICS** Write a paragraph to explain how you could use graphing to solve the equation $5x + 3 = -2x - 1$.

Refer to the graph at the right. Tell whether each statement is true or false. If it is false, rewrite it to make a true statement.

16. Vicki began walking slower than Mickey.

17. Mickey travelled a greater distance than Vicki.

18. **WRITING MATHEMATICS** Write a paragraph to describe the relationship shown in this scatter plot. Explain how you determined this.

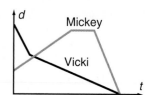

Carl left Dallas traveling at an average speed of 55 mi/h. Karla left Dallas following the same route 1 h later and traveled at an average speed of 65 mi/h. After how many hours will Karla pass Carl?

19. Write an equation you could use to answer the question.

20. Use graphing to solve the problem.

PERFORMANCE ASSESSMENT

USE ALGEBLOCKS Pick a variety of types of equations from this chapter that represent functions. Show how to use Algeblocks to model each equation. Then ask a partner to show the steps for evaluating the function at different values of x to create a table of pairs of values. Check that each step is correct.

ORDER YOUR PAIRS Write sets of ordered pairs or tables of pairs of values that represent linear functions, nonlinear functions, and relations that are not functions. Then ask a partner to identify the type of relation each represents. Discuss any identifications you do not agree with.

GRAPH YOUR INTERESTS Research an area of interest to you—perhaps a sport, a hobby, or a field of science. Make a table of at least eight ordered pairs based on some aspect of your research. Then graph the ordered pairs. Describe any relationship indicated by your graph. Tell whether that relationship is a function.

PICTURING TRIPS Write a brief description of a walking or jogging trip. Ask a partner to draw a qualitative graph to represent the trip. Then ask a third person to describe the trip by looking at the graph. Compare what you wrote to that description. Discuss any differences.

PROJECT ASSESSMENT

PROJECT *Connection* Work with your group to assemble all the material for your Sports Showcase. Decide on one or two central ideas you wish to communicate about the school's teams and be sure each item you use supports these ideas. Talk with your teacher about obtaining permission to use a school bulletin board or display case.

1. In your display, challenge students to try the basketball shot experiment from the Project Connection on page 187. Record the data and plan how you can add it to your display.

2. Include a list of school sports trivia questions.

3. Tie your display to a larger school event, such as a reunion luncheon of former student sports stars or a field day.

4. Use reference books from your school or public library to find data on some of your favorite professional teams. You might use information on attendance, wins or losses, player salaries, player's years on team, or any other interesting statistics. Organize sets of paired data, draw a scatterplot, and discuss the trends shown by the graph.

... • CUMULATIVE REVIEW • ...

Write an ordered pair to describe each point shown on the coordinate plane.

1. W

2. X

3. Y

4. Z

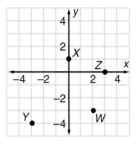

Evaluate each function at the given values of x.

5. $f(x) = 4x - 5$ at $x = 3$, $x = 10$, and $x = -5$.

6. $f(x) = \dfrac{x}{3} + 2$ at $x = 12$, $x = -24$, and $x = 0$.

Solve each equation.

7. $7x - 3 = 25$

8. $\dfrac{2}{3}x = -10$

9. $6x - 2(x + 4) = 3(4x - 3) - 15$

Use the following matrices to answer Questions 10 and 11.

$$A = \begin{bmatrix} -4 & 0 & \frac{3}{4} \\ 2 & -3 & 13 \end{bmatrix} \qquad B = \begin{bmatrix} -7 & -6 & \frac{5}{8} \\ -5 & 9 & -13 \end{bmatrix}$$

10. Find $A + B$.

11. Find $B - A$.

STANDARDIZED TESTS Use Statements I–IV to answer Question 12 regarding the relation $\{(a, 5), (3, 4)(b, 4)\}$.

    I. It cannot be a function since the second value of 4 is repeated.

    II. The values of a and b cannot be equal.

    III. Neither a nor b can be 3.

    IV. Neither a nor b can be 5.

12. Which of the following is true if the relation is to be a function?

    A. I only   **B.** II only

    C. II and IV   **D.** II and III

Graph each function.

13. $f(x) = 2x - 1$

14. $f(x) = \dfrac{-3}{5}x + 3$

15. $f(x) = 2x^2 - 3$

16. WRITING MATHEMATICS Explain why the line $x + y = 7$ cannot pass through Quadrant III.

Use the circle graph to answer Questions 17 and 18.

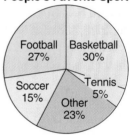

People's Favorite Sport

17. There were 120 responses. How many people chose soccer?

18. What is the probability that the next person asked would choose either football or tennis?

Solve each equation by graphing.

19. $3x + 5 = x - 7$

20. $143 - 6x = 6x - 13$

21. $4 - x^2 = -x - 2$

22. A student recieved a score of 76% on a test. There were 75 questions. How many questions did the student answer correctly?

23. Solve for t in the formula $A = P + Prt$.

24. Simplify the expression.

$$5(2x - 3) + 7x - 4(x + 3) - 10$$

· · · STANDARDIZED TEST · · ·

QUANTITATIVE COMPARISON In each question compare the quantity in Column 1 with the quantity in Column 2. Select the letter of the correct answer from these choices:

A. The quantity in Column 1 is greater.
B. The quantity in Column 2 is greater.
C. The two quantities are equal.
D. The relationship cannot be determined by the information given.

Notes: In some questions, information which refers to one or both columns is centered over both columns. A symbol used in both columns has the same meaning in each column. All variables represent real numbers. Most figures are not drawn to scale.

	Column 1	**Column 2**
1.	The percentage increase from \$4 to \$5	The percentage decrease from \$5 to \$4
2.	$x + 4$	$x + 7$
3.	$|a - b|$	$|b - a|$
4.	xy	$\dfrac{x}{y}$

5. $f(x) = -2x^2 + 5$

	$f(3)$	$f(-3)$
6.	3^4	4^3
7.	$a + 0$	$a \cdot 1$
8.	$-2 - (-5)$	$-5 - (-2)$

9. $\dfrac{x}{-9} = -6$

	0	x

10. $\{(4, -2), (-1, 3), (-6, -5)\}$

The sum of the range values	The sum of the domain values

Column 1   **Column 2**

11.

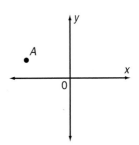

x-coordinate of A	y-coordinate of A
12. $5 - x$	$x - 5$
13. $\dfrac{3}{4} \div \dfrac{2}{3}$	$\dfrac{3}{4} \cdot \dfrac{2}{3}$

14. 2, 3, 6, 6, 6, 9, 10

mean	mode

15. 10 boys and 20 girls in a class

ratio of boys to girls	ratio of girls to total number of students
16. The sum of the coordinates of a point in Quadrant I	The sum of the coordinates of a point in Quadrant III
17. $(50\% \text{ of } 100) \div 4$	$25\% \text{ of } 50$
18. The maximum y-coordinate of $f(x) = -x^2 + 1$	The minimum y-coordinate of $f(x) = x^2 + 1$

19. $3x - 2 = 4x + 5$

$2x$	$x - 1$

20. $x > 0$ and $y < 0$

xy	$x - y$
21. $\left(\dfrac{1}{4}x\right)^2$	$\dfrac{1}{4}x^2$
22. $4 \div 6$	$\dfrac{2}{3}$

5 Linear Inequalities

Take a Look
AHEAD

Make notes about things that look new.

- Explain when the statement "Joan plays the piano and Dan plays the guitar" is true. When is the statement "Joan plays the piano or Dan plays the guitar" true?

- What is a boxplot and when might you use it?

Make notes about things that look familiar.

- What does the prefix in– mean in the term "inequality"? Give some other examples of similar uses of this prefix.

- Find examples of phrases in the chapter that are translated using inequality symbols.

DATA Activity

Product Life Cycle

Advertisers identify four stages in a product's advertising campaign. The stages in this **life cycle model** are *introduction*, *growth*, *maturity*, and *decline*. In the graph on the next page, time *t* is represented along the horizontal axis and sales are represented along the vertical axis. The letters A, B, C, and D correspond to the times when one stage ends and another begins. Think about why the changes shown occur.

SKILL FOCUS

- Use variables.
- Read and write inequalities.
- Interpret a graph.
- Use critical thinking.

GRAPHIC ARTS & Advertising

In this chapter, you will see how:

- **LAYOUT EDITORS** use inequalities to design a printed page with many different elements.
  (Lesson 5.1, page 226)

- **PRODUCTION EDITORS** use algebra to prepare special fold-out pages.
  (Lesson 5.4, page 240)

- **GRAPHIC ARTISTS** follow mathematical guidelines for illustrations of United States paper money.
  (Lesson 5.5, page 246)

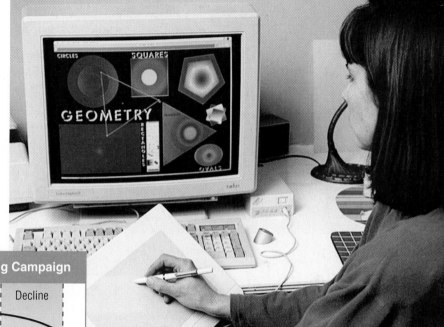

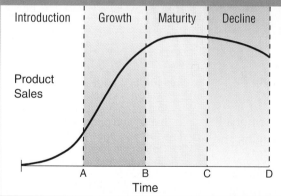

Life Cycle Model of an Advertising Campaign

Use the graph to answer each question. Assume the product being advertised is a snack food.

1. When $t < A$, what is taking place in the advertising campaign? What effect is the campaign having on sales of the product?

2. When $t > B$ and $t < C$, what stage is the campaign in? What is happening to sales during this stage? Why do you think this is so?

3. During which stage do product sales decrease? Represent the time period for this stage using inequalities. What conclusion can you draw about the advertising campaign?

4. **WORKING TOGETHER** A *high learning product* is one for which the consumer needs considerable education to understand the benefits of the new product. When first introduced, CD players, home computers, and microwave ovens were high learning products. How do you think the life cycle graph of such a product would differ from the graph shown? Discuss how each stage differs from the stages of a snack food.

PROJECT

It All "Ads" Up!

Whether you are reading a magazine, listening to the radio, shopping at the mall, or riding along the highway, chances are there will be many product advertisements competing for your attention. Advertising agencies not only design the actual advertisements, they identify market opportunities and suggest strategies for businesses that wish to spread the word about their product or services.

In this project, you will work cooperatively with a group of classmates as a model advertising agency. You will plan an advertising campaign for a product called **S'COOL LUNCH**. Students would purchase these packaged meals at supermarkets and bring them to school to microwave during their lunch period.

PROJECT GOAL

To collect, analyze, interpret, and present information about a product in a mock advertising campaign.

Getting Started

Work in groups of three to five students.

1. Select a name for your agency.

2. The lunches are nutritionally balanced, and packaging and trays are biodegradable. Brainstorm ways to make these lunches appealing to students.

3. Design the front panel of the **S'COOL LUNCH** package. Indicate measurements, graphics, and style of product name.

4. Decide what other information would help you plan the campaign.

Internet Connection

www.swpco.com/
swpco/algebra1.html

PROJECT *Connections*

Lesson 5.1, page 225:
Design and conduct a survey to determine students' lunch habits and preferences.

Lesson 5.2, page 229:
Determine meal contents, cost per meal, and cost per ounce (unit cost).

Lesson 5.6, page 251:
Gather information about advertising rates, design an advertisement, and plan a print campaign based on budget guidelines.

Chapter Assessment, page 259:
Prepare a final presentation based on the four-stage guidelines for a successful advertising campaign.

220

5.1 Inequalities on the Number Line

Explore

An advertising agency is interested in knowing the effectiveness of its campaign for Fiesta Foods, Inc., this year. The change in sales since the campaign began may show the effectiveness of the campaign. The annual sales amount for the year before the new campaign was started is shown on this number line.

Last Year's Sales, dollars

1. What is the dollar amount of last year's sales?

2. Lower annual sales this year than last year may show that the advertising campaign is not very effective. Name an amount less than last year's sales.

3. Name an amount greater than last year's sales.

4. In Question 2, could you have named other lesser amounts? How many others? Where are the points corresponding to these amounts located on the number line above?

5. In Question 3, could you have named other greater amounts? How many others? Where are the points corresponding to these amounts located on the number line above?

6. Can you name an amount that is not less than, not greater than, and not equal to last year's sales?

Build Understanding

The point graphed on the number line in Explore corresponds to 1,000,000. Any other point on that number line corresponds to an amount less than 1,000,000 or more than 1,000,000. This is an example of the *trichotomy property*, also called the *comparison property*.

> **TRICHOTOMY PROPERTY**
>
> For all real numbers *a* and *b*, exactly one of the following is true:
>
> $a = b$, $a < b$, or $a > b$

So far, you have used $<$, $>$, and $=$ to state comparisons of numbers. Comparisons can also be combined to make other symbols.

Inequality Symbol	Read
$\leq$	less than or equal to
$\geq$	greater than or equal to
$\neq$	not equal to
$\not<$	not less than
$\not>$	not greater than

An **inequality** is a statement that two numbers or expressions are not equal. You use inequality symbols to write inequalities. Just like with equations, the solutions of inequalities are values that make the inequality true.

EXAMPLE 1

Determine whether −20 is a solution of $x = 10$, $x < 10$, or $x > 10$.

Solution
Substitute −20 for x in each equation and inequality.

$x = 10$	$-20 \overset{?}{=} 10$	not true
$x < 10$	$-20 \overset{?}{<} 10$	true
$x > 10$	$-20 \overset{?}{>} 10$	not true

So, −20 is a solution of $x < 10$.

Besides −20, there are many other numbers that are solutions of $x < 10$. In fact, the solution of $x < 10$ includes *all* real numbers less than 10.

You can graph the solution to an equation or an inequality on a number line. Of course you graph points by using a *solid dot*. To show that a point is not included you use an *open dot*.

EXAMPLE 2

Graph the solution of $x < 10$ on a number line.

Solution

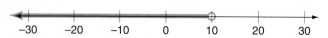

The open dot shows that 10 is not a solution. The arrow pointing to the left shows that the numbers corresponding to all points to the left of 10 are solutions.

EXAMPLE 3

Graph the solution of $c \geq -2$ on a number line.

Solution

A number line from −4 to 4 with a solid dot at −2 and an arrow pointing right.

The solid dot shows that −2 is a solution. The arrow pointing to the right shows that the numbers corresponding to all points to the right of −2 are also solutions. ◄

Inequalities can model real world situations.

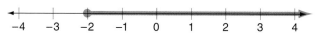

CHECK UNDERSTANDING

How would the graph of $c > -2$ on a number line compare to the graph shown in Example 3?

EXAMPLE 4

METEOROLOGY Throughout the morning the temperature remained at 5°F. In her noon weather report, the meteorologist stated that there had been a sudden change in temperature.

a. Write an inequality to describe the temperature t at noon.

b. Graph the solution to this inequality on a number line.

Solution

a. $t \neq 5$ The temperature could be anything except 5°F.

b. The open dot shows that 5 is not a solution. The arrows to the left and to the right show that all real numbers less than 5 and all real numbers greater than 5 are solutions. ◄

A number line from −1 to 7 with an open dot at 5 and arrows pointing both left and right.

TRY THESE

Determine whether each number is a solution of $c = -3.5$, $c < -3.5$, or $c > -3.5$.

1. 3.5   **2.** −3   **3.** −3.8

Graph the solution of each inequality on a number line.

4. $q > 8$ **5.** $w \neq 0$ **6.** $f \leq 1\frac{1}{2}$ **7.** $s \geq -10$

8. HEALTH CARE In her job as a lab technician, Mrs. Choi earns $7.50 an hour. Her boss just told her she will be getting a raise.
  a. Write an inequality to describe Mrs. Choi's new pay rate r in dollars.
  b. Graph the solution to this inequality on a number line.

9. MANIPULATIVES Use a geoboard. Let P represent the number of pegs on the perimeter of a polygon. Let I represent the number of pegs in the interior of a polygon. Create a polygon with the following characteristics. Make a sketch.

  a. $P \leq 8; I > 5$ **b.** $P > 12; I < 6$

PRACTICE

Determine whether the given number is a solution of the inequality.

1. $h \leq -6; -6$ **2.** $y \neq 0.2; -0.2$ **3.** $k < 100; 100$ **4.** $z \geq 4\frac{1}{2}; 4$

5. WRITING MATHEMATICS Describe at least two ways in which you could show that your answers to Exercises 1–4 are correct.

Graph the solution of each inequality on a number line.

6. $b \geq 1$ **7.** $q \neq -4$ **8.** $w < 2\frac{1}{3}$ **9.** $6\frac{1}{4} \leq c$

10. $x \neq 9.5$ **11.** $\frac{1}{2} < d$ **12.** $15 \geq m$ **13.** $2.2 > r$

14. GOVERNMENT To vote in the United States, a citizen must be 18 years of age or older.

     a. Write an inequality that describes the age a in years of voters in the United States.

     b. Graph the solution to the inequality on a number line.

15. WRITING MATHEMATICS Describe a situation at a service station gas pump where the term "is less than" or "is greater than" might be used.

16. GEOGRAPHY Located on the border of Nepal and China, Mt. Everest has the highest elevation in the world. It rises 29,028 ft above sea level.

     a. Write an inequality that describes the elevations e in feet of all other places in the world.

     b. Graph the solution on a number line.

EXTEND

Tell whether the given number is a solution of the inequality. If it is not, rewrite the inequality using a different inequality symbol so that the number is a solution. (Do not write an equation.)

17. $g > 7.3; 7.3$ **18.** $x \neq -2; -3$ **19.** $v \leq 10; 11$

20. $z \geq -3\frac{1}{4}; -4$ **21.** $3 < m; 5$ **22.** $-4 \geq c; 0$

Write an inequality whose solution is shown by each graph.

23.

24.

25.

26.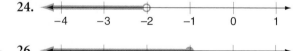

27.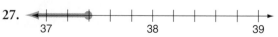

28.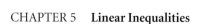

29. LAW ENFORCEMENT Refer to the traffic sign at the right.

   a. Write an inequality that describes the speed s in miles per hour a vehicle could legally be traveling.

   b. Would it be appropriate to graph this inequality on a number line by drawing an open or solid dot and an arrow? Why or why not?

30. SPORTS To win first place in a gymnastic competition, Sonya has to score no less than 8.2 on her final vault. A score of 10 is perfect.

   a. Write an inequality that describes the points p Sonya could receive that would result in a first-place win.

   b. Trevor says he can graph this inequality on a number line by drawing a solid dot and an arrow. Explain why you agree or disagree with Trevor.

THINK CRITICALLY

Tell whether each statement is *always*, *sometimes*, or *never* true for all real numbers a, b, and c. Justify each *sometimes* or *never* answer.

31. $a < a + 1$

32. $b > b - 1$

33. If $a < b$, then $b < a$.

34. If $a \leq b$, then $b \leq a$.

35. If $a < 3$ and $b < 86$, then $a < b$.

36. If $a > 0$ and $b > -4$, then $a > b$.

A layout editor is responsible for making a well-organized and "reader friendly" layout, or design, of pages in books, newspapers, magazines, catalogs, and other printed materials. Many newspaper pages are divided into smaller units, or modules. A module is a rectangular space that might contain a story, an advertisement, a photo, or an illustration. This figure shows one possible modular layout for a newspaper page.

Decision Making

Inequalities such as these can be written to describe how the areas of the modules compare to each other.

$$A > F$$
$$D < C < G$$
$$B + E \geq C$$

1. Work with a partner and write three other inequalities that compare the areas of the modules in this layout.

2. Design a different modular layout of the page above with 7 modules. Write five inequalities that describe the layout you designed.

3. Exchange inequalities, but not layouts, with a partner. Try to draw a modular layout that meets the specifications of your partner's inequalities.

4. Compare results with your partner. Are the inequalities true in both layouts?

5. Suppose the layout shown above is the right-hand page of a newspaper. If you were an advertiser, is there any particular module in which you would prefer to have your advertisement appear? Why?

5.2 Algebra Workshop
Explore Inequalities

Think Back

- Work with a partner. In previous lessons, you worked with simple equations and inequalities like those at the right.

  $x = 8$
  $x < 8$

  1. Graph the solution of the equation and the inequality above on separate number lines.

  2. How do the solutions of the equation and the inequality compare to each other?

Explore

- You can compare an equation and an inequality, such as $x = 8$ and $x < 8$, in another way. Recall that adding the same number to both sides of an equation produces an equivalent equation. So does subtracting the same number from both sides and multiplying or dividing both sides by the same number. Find out if these operations produce inequalities that are true.

  One solution of the inequality $x < 8$ is 7, because $7 < 8$, as shown on this number line.

  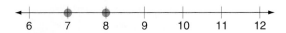

  The number line below shows the result of adding 3 to both sides of the inequality $7 < 8$. Because 10 is to the left of 11, you can see that $10 < 11$.

  $$7 + 3 \overset{?}{<} 8 + 3$$
  $$10 < 11 \text{ true}$$

  3. Selecting a different positive and negative integer, you and your partner should each

     a. add the positive integer to both sides of $7 < 8$
     b. add the negative integer to both sides of $7 < 8$
     c. subtract the positive integer from both sides
     d. subtract the negative integer from both sides
     e. multiply both sides of $7 < 8$ by the positive integer
     f. multiply both sides by the negative integer
     g. divide both sides by the positive integer
     h. divide both sides by the negative integer

4. Use a number line to help you decide whether each inequality is true or not. Record your results in a table like this.

	True or not true?	
New inequality resulting from	**Partner A**	**Partner B**
Adding positive integer		
Adding negative integer		
Subtracting positive integer		

5. What operations resulted in untrue inequalities?

6. Change the inequality symbol to make each untrue inequality true.

7. Substitute a negative solution for x in $x < 8$. Repeat Activities 3–6 using this inequality.

8. Substitute a value for x in $x > -3$ that results in a true inequality. Repeat Activities 3–6.

Make Connections

In the following inequalities, a and b are real numbers, c is a positive real number ($c > 0$), and d is a negative real number ($d < 0$). Based on your findings in Activities 3–8, tell whether each statement is true or false.

9. If $a < b$, then $a - c < b - c$.

10. If $a > b$, then $\dfrac{a}{c} < \dfrac{b}{c}$.

11. If $a < b$, then $a - d > b - d$.

12. If $a > b$, then $ad < bd$.

13. If $a < b$, then $ac > bc$.

Replace ▨ with the inequality symbol that makes each statement true.

14. If $x < 8$, then $x + 10$ ▨ $8 + 10$.

15. If $-x > 2$, then $(-1)(-x)$ ▨ $(-1)(2)$.

16. If $x - 6 \leq -4$, then $x - 6 + 6$ ▨ $-4 + 6$.

17. If $x + 5 < -1$, then $x + 5 - 5$ ▨ $-1 - 5$.

18. If $\frac{3}{4}x \geq -24$, then $\frac{4}{3} \cdot \frac{3}{4}x$ ▨ $\frac{4}{3} \cdot (-24)$.

19. If $-\frac{2}{3}x \geq 18$, then $\left(-\frac{3}{2}\right) \cdot \left(-\frac{2}{3}x\right)$ ▨ $\left(-\frac{3}{2}\right)(18)$.

20. If $-15x < 30$, then $\frac{-15x}{-15}$ ▨ $\frac{30}{-15}$.

THINK BACK

Recall when multiplying factors with the same sign, the product is positive. The product is negative when the factors have different signs.

Summarize

21. WRITING MATHEMATICS Write a paragraph explaining the effect upon an inequality of adding the same real number to each side, subtracting the same real number from each side, multiplying each side by the same real number, or dividing each side by the same real number. If any of these operations results in an untrue inequality, what could you do to make it true?

22. THINKING CRITICALLY What effect upon an inequality does adding zero to each side have? subtracting zero from each side? multiplying each side by zero? Why isn't it appropriate to ask what effect dividing each side by zero has?

23. GOING FURTHER Suppose both sides of $a \leq -4$ and $b \geq -4$ are multiplied by a negative number and the inequality symbols are not reversed. Will the new inequalities be true for *all*, *some*, or *none* of the values for which the original inequality was true? Explain.

PROBLEM SOLVING TIP

For each inequality in Question 23, test different positive and negative numbers to find values for the variable that make the original inequality true.

PROJECT *Connection*   Use the results of your survey to decide on two different meals that you wish to feature in your **S'COOL LUNCH** presentation.

1. Determine the total Calories of the meal and the price at which it will be sold. Find the unit price (cost per ounce) of the meal.

2. Write your price recommendation in the form "A single serving should have a price which is greater than ▨ and less than ▨." Explain your reasoning in setting these price boundaries.

5.3 Solve Inequalities Using One Operation

Explore/Working Together

● Work with a partner. Use Algeblocks. Place the correct inequality symbol on the Sentence Mat.

1. Write the inequality represented by these Algeblocks.

2. Use Algeblocks to solve the inequality you wrote by isolating the variable.

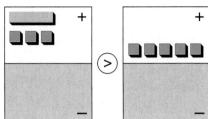

3. Describe the solution of the inequality.

4. How is using Algeblocks to solve an inequality the same as using Algeblocks to solve an equation? How is it different?

Build Understanding

ALGEBLOCKS MODEL

Solve $x - 3 < -2$. Model the inequality. Write the correct inequality sign in the circle.

Add 3 to both sides. Remove zero pairs.

Read the solution: $x < 1$

● You solve an equation by getting the variable alone on one side. You also solve an inequality by getting the variable alone on one side. If an inequality has a number added to or subtracted from a variable, you use the *addition and subtraction properties of inequality* to solve the inequality. These properties are used to write the **equivalent inequalities**.

> **ADDITION AND SUBTRACTION PROPERTIES OF INEQUALITY**
>
> **For all real numbers a, b, and c:**
> If $a > b$, then $a + c > b + c$ and $a - c > b - c$.
> If $a < b$, then $a + c < b + c$ and $a - c < b - c$.

EXAMPLE 1

Solve $x - 3 < -2$. Graph the solution on a number line.

Solution

$$x - 3 < -2 \qquad \text{Write the inequality.}$$

$$x - 3 + 3 < -2 + 3 \qquad \text{Add 3 to both sides.}$$

$$x < 1$$

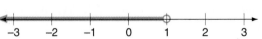

Graph the solution.

Check

Look at the graph and find two solutions of the inequality. Substitute them for the variable in the original inequality and evaluate. Also try a point not on the graph.

Try 0	**Try –2**	**Try 1.5**
$x - 3 < -2$	$x - 3 < -2$	$x - 3 < -2$
$0 - 3 \overset{?}{<} -2$	$-2 - 3 \overset{?}{<} -2$	$1.5 - 3 \overset{?}{<} -2$
$-3 < -2$ ✓	$-5 < -2$ ✓	$-1.5 \not< -2$

The solution is all real numbers less than 1. ◄

You can also apply the addition and subtraction properties of inequality to inequalities with ≥ or ≤.

EXAMPLE 2

Solve $y + 5 \geq 0$. Graph the solution.

Solution

$$y + 5 \geq 0$$

$$y + 5 - 5 \geq 0 - 5 \qquad \text{Subtract 5 from both sides.}$$

$$y \geq -5$$

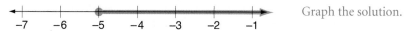

Graph the solution.

To check, substitute numbers such as –8, –4, and 0.
The solution is all real numbers greater than or equal to –5. ◄

In Lesson 5.2, you found that when an inequality is multiplied by a negative number, you must reverse the inequality symbol to obtain an equivalent inequality.

If the variable in an inequality is multiplied or divided by a number, you can use the *multiplication and division properties of inequality* to solve the inequality. Below are the multiplication and division properties of inequality for "is greater than." Notice when c is a negative number the inequality symbol is reversed.

> **MULTIPLICATION AND DIVISION PROPERTIES OF INEQUALITY**
>
> **For all real numbers a, b, and c:**
>
> If $a > b$ and $c > 0$, then $ac > bc$ and $\dfrac{a}{c} > \dfrac{b}{c}$.
>
> If $a > b$ and $c < 0$, then $ac < bc$ and $\dfrac{a}{c} < \dfrac{b}{c}$.

These properties are also true for inequalities with $<$, $\geq$, and $\leq$.

ALGEBRA: WHO, WHERE, WHEN

Chemists use a number line called a *pH scale* to describe properties of a solution. If pH ≥ 0 and pH < 7, then the solution is an acid; the lower the pH number, the stronger the acid. If pH > 7 and pH ≤ 14, the solution is a base. A pH of 7, associated with pure water, is considered neutral. The human stomach with a pH of about 1.6 is more acidic than a lemon at 2.3. In the United States, the natural pH of rainwater ranges from 5.0 to 6.5 depending on where it falls, but the pH of acid rain can be less than 3.0.

CHECK UNDERSTANDING

Write the multiplication and division properties of inequality for "is less than."

In Example 3, if the original inequality had been $-\frac{r}{3} > -1$, would you have reversed the inequality symbol in the solution? Why or why not?

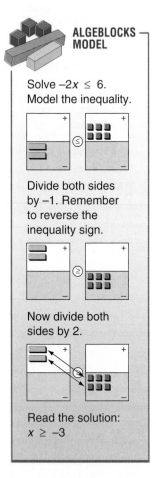

ALGEBLOCKS MODEL

Solve $-2x \leq 6$. Model the inequality.

Divide both sides by -1. Remember to reverse the inequality sign.

Now divide both sides by 2.

Read the solution: $x \geq -3$

EXAMPLE 3

Solve $\frac{r}{3} > -1$. Graph the solution.

Solution

$$\frac{r}{3} > -1$$

$$3\left(\frac{r}{3}\right) > 3(-1) \qquad \text{Multiply both sides by 3.}$$

$$r > -3 \qquad \text{The inequality symbol is } not \text{ reversed.}$$

Substitute numbers such as -6, 0, and 3 to check.
The solution is all real numbers greater than -3. ◀

EXAMPLE 4

Solve $-4s \leq 20$. Graph the solution.

Solution

$$-4s \leq 20$$

$$\frac{-4s}{-4} \geq \frac{20}{-4} \qquad \begin{array}{l}\text{Divide both sides by } -4 \text{ and reverse} \\ \text{the inequality symbol.}\end{array}$$

$$s \geq -5$$

Substitute numbers such as -6, 0, and 5 to check.
The solution is all real numbers greater than or equal to -5. ◀

Some word problems can be translated into inequalities.

EXAMPLE 5

GEOMETRY: THE TRIANGLE INEQUALITY THEOREM
The sum of the lengths of any two sides of a triangle is greater than the length of the third side. In an isosceles triangle, the lengths of two sides are the same. Write and solve an inequality to find the length a of each side in this isosceles triangle. Graph the solution.

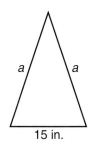

Solution
The sum of the lengths of the two sides is represented by $a + a$ or $2a$.

$$2a > 15$$

$$\frac{2a}{2} > \frac{15}{2} \qquad \text{Divide both sides by 2.}$$

$$a > 7\frac{1}{2}$$

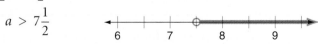

The length of each side a is greater than $7\frac{1}{2}$ in. ◀

Solve and check each inequality. Graph each solution.

1. $a + 2 \leq -5$

2. $\frac{-b}{3} > -2$

3. $c - 0.3 \geq 4.5$

4. $\frac{2}{5}d < 10$

5. $7 \geq m + 3$

6. $-2 < \frac{x}{4}$

7. $w - \frac{5}{6} \geq \frac{1}{6}$

8. $3.1 < t - 1.8$

9. $-z + 3 \geq 0$

10. Auto Mechanics An auto mechanic estimates parts and labor for an auto repair at no more than $300. Parts will cost $59.40. Write and solve an inequality to find the estimate of the cost of labor c.

11. Modeling Write the inequality represented by the Algeblocks at the right. Use Algeblocks to solve. Sketch each step. Describe the solution of the inequality.

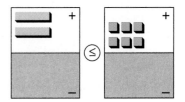

PRACTICE

Solve and check each inequality. Graph each solution.

1. $\frac{3}{4}k \leq 12$

2. $d - \frac{1}{2} < 2$

3. $p - 7 > 9$

4. $-30 < -6n$

5. $\frac{e}{-1.2} < 6$

6. $\frac{-4}{5}b \leq 20$

7. $c + \frac{2}{3} < 1\frac{1}{3}$

8. $z - (-3) > -2.4$

9. $f - (-0.6) \geq 1.2$

10. $\frac{h}{-10} \geq -1$

11. $0 \geq c - 4$

12. $\frac{m}{0.5} \geq -6$

13. Compact Disc Sales After selling 14 copies of a popular compact disc, Mostly Music had at most 11 copies of the disc left. Write and solve an inequality that expresses how many copies of the disc d the store originally had.

Solve and check each inequality. Graph each solution.

14. $0.25 < 5g$

15. $\frac{j}{-8} \leq 3$

16. $-3.2w < 9.6$

17. $\frac{r}{4} \leq -2$

18. $-\frac{1}{3}e > -\frac{5}{6}$

19. $10 \leq q + 1.4$

20. $a - \left(-\frac{1}{3}\right) < 6$

21. $3.6 < \frac{z}{2}$

22. $s - 25 \leq -7$

23. $9 < h + (-4.5)$

24. $-10 > 2.5m$

25. $-3t \geq \frac{1}{2}$

26. $3 + n - 7 \leq 2$

27. $2 - (3 - s) < 4$

28. $3(4d) > -48$

29. Construction The Dysons are planning to build a rectangular patio that is 12 ft long and at least 10 ft wide. Write and solve an inequality to find the area of the patio A in square feet.

30. Writing Mathematics Describe the terms "at least" and "at most." Use a context for your description if you wish.

EXTEND

Solve each inequality. Use integers as replacements for the variable and draw a graph showing at least five solutions.

31. $(h - 3) \leq 5$　　　**32.** $-\frac{1}{2}p > -5$　　　**33.** $-4 \leq g - 7$　　　**34.** $1.2s < 9$

35. TECHNOLOGY Some graphing utilities allow the following entry.

$$y_1 = (x - 4)(x < 3)$$

This yields the graph at the right. Explain this entry and its graph.

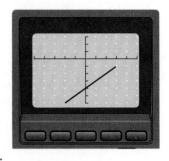

36. CONSUMERISM Ms. Quan must keep a minimum balance of $600 in her checking account to avoid paying a monthly service charge. She has a current balance of $624.32. She is planning to make a car loan payment of $279.38 by check. Write and solve an inequality to find d, the amount she must deposit to maintain at least the minimum balance.

37. ADVERTISING Mark is placing a classified ad in his city's newspaper to advertise his lawn care services. Ads cost $0.75 per word. He wants to spend no more than $25.00 for the ad.
   a. Write an inequality describing the number of words w Mark could have.
   b. Solve the inequality. How many words could Mark have in his ad?

THINK CRITICALLY

In Exercises 38–40, determine whether each statement is *always*, *sometimes*, or *never* true for all real numbers *a, b, c,* and *d.* Justify each *sometimes* or *never* answer.

38. $a - b \leq a + b$　　　　　　　　　　**39.** If $a < b$ and $c < d$, then $a + c < b + d$.

40. If $a > b$ and $c > d$, then $ac > bd$.

41. Why can c be equal to zero in the addition and subtraction properties of inequality but not in the multiplication and division properties of inequality?

42. Write two inequalities equivalent to $x < -3$.

MIXED REVIEW

Find each measure of central tendency for $7.59, $6.95, $8.59, $9.25, $7.75, and $8.25.

43. mean　　　　　　**44.** median　　　　　　**45.** mode　　　　　　**46.** range

Solve each equation.

47. $\frac{e}{0.2} = 8$　　　　　　**48.** $\frac{3}{4}k = -15$　　　　　　**49.** $\frac{1}{3}n = -12$

Tell which quadrant or axis contains each point.

50. $R\left(0, -\frac{2}{3}\right)$　　　　　　**51.** $S(-2.5, 7.4)$　　　　　　**52.** $T\left(2\frac{1}{2}, -1\frac{1}{2}\right)$

53. STANDARDIZED TESTS Which of the following is equivalent to $x + 4 \geq 9$?

　　A. $x \geq 13$　　　　　**B.** $x \leq 5$　　　　　**C.** $x \geq 5$　　　　　**D.** $x \leq -5$

Explore / Working Together

● The manager of a video rental store wants to advertise the store's grand opening. To do this, she plans to mail out fliers containing coupons. She can spend no more than $200 for printing the fliers.

1. Write expressions showing the cost of printing x fliers at Power Print and at Quality Print.

2. Use the expressions you wrote in Question 1 to write two inequalities that the manager can use to find how many fliers she can print at each shop while staying within her budget.

3. How will the methods of solving both inequalities be the same? How will they be different?

QUALITY PRINT

One-time design and setup charge $50.00

Each copy $0.04

POWER PRINT

All copies $0.05 each

Build Understanding

● You have solved equations that require you to use more than one operation. Similarly, you can use the addition, subtraction, multiplication, and division properties of inequality to solve inequalities that require more than one operation.

EXAMPLE 1

Solve $\frac{1}{2}a - 5 > 3$. Graph the solution.

Solution

$$\frac{1}{2}a - 5 > 3$$

$$\frac{1}{2}a - 5 + 5 > 3 + 5 \qquad \text{Add 5 to both sides.}$$

$$\frac{1}{2}a > 8$$

$$2\left(\frac{1}{2}a\right) > 2(8) \qquad \text{Multiply both sides by 2.}$$

$$a > 16$$

Graph the solution.

Check

Try 20	**Try 24**	**Try 14**
$\frac{1}{2}a - 5 > 3$	$\frac{1}{2}a - 5 > 3$	$\frac{1}{2}a - 5 > 3$
$\frac{1}{2}(20) - 5 \overset{?}{>} 3$	$\frac{1}{2}(24) - 5 \overset{?}{>} 3$	$\frac{1}{2}(14) - 5 \overset{?}{>} 3$
$10 - 5 \overset{?}{>} 3$	$12 - 5 \overset{?}{>} 3$	$7 - 5 \overset{?}{>} 3$
$5 > 3 \checkmark$	$7 > 3 \checkmark$	$2 \not> 3$

14 is not a solution.

The solution is all real numbers greater than 16.   ◀

As with solving equations, sometimes you must first combine like terms to solve an inequality. When variable terms are on both sides of an inequality, first use the inequality properties to get them all on the same side.

EXAMPLE 2

Solve $2x - 3 < 7 + 3x$. Graph the solution.

Solution

$$2x - 3 < 7 + 3x$$

$$2x - 3 - 3x < 7 + 3x - 3x \qquad \text{Subtract } 3x \text{ from both sides.}$$

$$-x - 3 < 7 \qquad \text{Combine like terms.}$$

$$-x - 3 + 3 < 7 + 3 \qquad \text{Add 3 to both sides.}$$

$$-x < 10$$

$$(-1)(-x) > (-1)10 \qquad \text{Multiply both sides by } -1 \text{ and}$$

$$x > -10 \qquad \qquad \text{reverse the inequality symbol.}$$

Graph the solution.

To check, try numbers such as −12, −5, and 0.
The solution is all real numbers greater than −10.   ◀

If an inequality contains parentheses, you may need to apply the distributive property as the first step.

COMMUNICATING
ABOUT ALGEBRA

EXAMPLE 3

FOOD SERVICE The manager of Family Fare wants to set sandwich prices so that all members of a family of four can each order a sandwich and a drink for less than $20.00. All drinks are priced at $0.89. Write and solve an inequality to find what prices p the manager should set for the sandwiches.

Could you have begun Example 3 by dividing both sides by 4? What would the result be? What would the next step be? Which solution do you prefer?

Solution
The cost for each person is for 1 sandwich and 1 drink, or $p + 0.89$.
For a family of 4, the total cost is $4(p + 0.89)$.

$$4(p + 0.89) < 20.00$$
$$4p + 3.56 < 20.00 \qquad \text{Apply the distributive property.}$$
$$4p + 3.56 - 3.56 < 20.00 - 3.56 \qquad \text{Subtract 3.56 from both sides.}$$
$$4p < 16.44$$
$$\frac{4p}{4} < \frac{16.44}{4} \qquad \text{Divide both sides by 4.}$$
$$p < 4.11$$

To check, try prices such as $4.25, $4.00, and $3.50.
The manager should set all sandwich prices at less than $4.11. ◄

TRY THESE

Solve and check each inequality. Graph the solution.

1. $7c - 4 \geq 24$ **2.** $2(4 - d) - 5d > 8$ **3.** $\frac{2}{3}(s + 6) < -20$ **4.** $6z + \frac{9}{2} \leq 5z - 7$

5. $8 - 4x \leq 6x - 2$ **6.** $5 > \frac{1}{3}b + 14$ **7.** $\frac{2}{5}x + 3 > \frac{1}{5}x + 1$ **8.** $8y + 9 \leq 4 + 8y$

9. ELECTRICAL CONTRACTING Over the telephone, an electrician told Mrs. Watts that the labor required to repair the electrical problem she described would cost at least $150. He said that this included a base service call charge of $40 plus $25 per hour. Write and solve an inequality to determine how many hours h the electrician estimated for the repair.

10. MODELING Write the inequality represented by the Algeblocks at the right. Use Algeblocks to solve the inequality. Sketch each step. Describe the solution of the inequality.

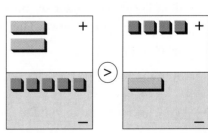

PRACTICE

Solve and check each inequality. Graph the solution.

1. $7 \geq 2 - d$

2. $\dfrac{m}{5} + 3 > 9$

3. $3c - 7 \geq 29$

4. $72 < -3h + 4 - 5h$

5. $106 \geq 19p + 6 + 6p$

6. $\dfrac{2}{3}f - 4 < 12$

7. $\dfrac{5}{8}e - 3 - \dfrac{3}{8}e > -5$

8. $\dfrac{2}{5}(g - 3) \geq -4$

9. $7x + 4 < 39 + 2x$

10. $13 + 5n \geq 25 + 5n$

11. $6(k - 2) > 48$

12. $2x - 5(x + 3) \geq -20$

13. $4a + 6 \geq 7 + 4a$

14. $3(q + 4) - 5(q - 1) < 5$

15. $3y - 7 + 5y < 2 + 8y$

16. $5(7 + r) > 12r$

17. $7(2 - e) \geq 3(e + 8)$

18. $4(p - 3) < 4(p - 4)$

19. $3(3b + 1) - (b - 1) \leq 6(b + 8)$

20. $6(d + 4) - (d - 5) > 5d - 1$

21. WRITING MATHEMATICS Explain why $x > 7$ and $7 < x$ are equivalent statements.

22. TEST SCORES During this semester, Kanika will take eight math tests worth 100 points each. She will earn an A if the average of her scores is at least 90. She will earn a B if the average of her scores is at least 80. So far, she has taken seven tests and received the following scores: 84, 93, 78, 87, 89, 70, and 81.

   a. Write and solve an inequality to determine whether it is possible for Kanika to earn an A. If so, what score p must she receive on the last test?

   b. Rewrite the inequality to determine whether it is possible for Kanika to earn a B. If so, what score p must she receive on the last test?

23. TECHNOLOGY Solve $2x - 3 \leq 3x + 2$ by hand. Enter it into your graphing utility as follows.

   $$y_1 = 2x - 3 \leq 3x + 2$$

   Compare and explain the results.

EXTEND

Identify the error made in solving each inequality that would result in obtaining the incorrect solution shown. Then find the correct solution.

24. $9 - 4a < 25;\ a < -4$

25. $2(h + 4) \geq 12;\ h \geq 4$

26. $2x \leq -12 - 4x;\ x \geq 6$

27. $8t - 2 > 8t + 5;$ all real numbers

Write and solve an inequality for Exercises 28 and 29.

28. Ninety decreased by three times a number n is greater than 48.

29. A number z increased by 27 is less than four times the number decreased by 6.

BUSINESS Mrs. Martinez has been offered a manager's position at an electronics store. If she takes the job, she must decide by which plan she would like to be paid each week.

Plan A: $8.00 per hour and time-and-a-half for overtime (more than 40 hours)
Plan B: $100 plus 6% commission on all sales
Plan C: salary of $360

30. Write and solve an inequality to find out how many overtime hours h Mrs. Martinez would have to work to earn at least as much through Plan A as through Plan C.

31. Write and solve an inequality to find the amount s Mrs. Martinez would have to sell to earn at least as much through Plan B as through Plan C.

32. Suppose Mrs. Martinez is told she will work an average of 45 hours per week. Write and solve an inequality to find the amount s she would have to sell through Plan B to earn more than she would through Plan A. Write and solve an inequality for each.

THINK CRITICALLY

33. **MANIPULATIVES** Use a geoboard and work with a partner. Construct a rectangle of any size. Count the number of pegs in the interior of the rectangle and on the perimeter of the rectangle. Let S represent the sum of those numbers. Write two inequalities of the form

$$a \cdot S + b \geq c \text{ and } d \cdot S - e \leq f$$

where a, b, c, d, e, and f are any integers that form a true inequality. Give these inequalities to your partner. Have your partner construct a rectangle based upon the given information. Compare results.

Write a problem that could be solved using each inequality. Then use the inequality to solve the problem.

34. $2c - 5 \geq 35$

35. $3(m + 1) < 18$

36. If a, b, and c are real numbers and $ax + b < c$, then what is x equal to? Explain.

MIXED REVIEW

37. **EDUCATION** The tuition for one semester at five state universities is $995, $1600, $875, $1250, and t. The mean tuition at these universities is $1144. What is the value of t?

Evaluate each expression.

38. $|-8|$

39. $-|-8|$

40. $-|8|$

41. **STANDARDIZED TESTS** Which coordinate pair is on the graph of the function $y = \frac{1}{2}x - 5$?

   A. $(6, 2)$ **B.** $(6, -2)$ **C.** $(-6, 2)$ **D.** none of these

Solve each inequality.

42. $\frac{1}{3}x - 4 > 8$

43. $y + 3 - 3(y - 1) < 4$

Newspapers, magazines, and book publishers often have production editors. Once the design of a publication has been established, production editors are responsible for editing and placing text, photos, illustrations, and so on to fit the design.

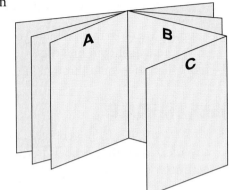

Many times magazines and occasionally books include a fold-out page such as the one shown at the right.

An important thing for a production editor to remember about a fold-out page is that all three parts of the page are not the same width. Part A is the width of normal pages in the magazine or book. Part C must be the narrowest part so that it can be folded in without bending the fold-out. Part B must be narrower than Part A so that, when the outer edges of the magazine or book are trimmed by the cutter, the fold-out is not cut. These differences in width are shown below for a magazine with normal-width pages of x in.

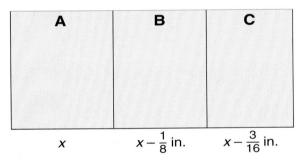

Decision Making

1. Suppose that the width of Parts A, B, and C together must always be less than or equal to $28\frac{3}{4}$ in. Write an inequality that can be used to determine the possible values of x.

2. Solve the inequality you wrote in Problem 1. Use your solution to describe the possible widths of Parts A, B, and C.

3. Describe the dimensions of a fold-out page that could be inserted in your favorite magazine.

5.5 Solve Compound Inequalities

Explore/Working Together

1. A radio station bases its advertising rates on the number of commercial spots an advertiser wants to run each week. The station offers a special rate to those who advertise at least 7 times a week. Write an inequality that describes this number of commercials, c.

2. Advertisers may run up to 20 commercials a week at this rate. Write another inequality that describes this additional information about the number of commercials, c.

3. Can you tell whether a particular number of commercials qualifies for the special rate by looking only at the inequality you wrote in Question 1? Can you tell by looking only at the inequality you wrote in Question 2? Why or why not?

Build Understanding

The number of commercials that qualify for the special rate in Explore can best be described by writing the **compound inequality**

$$c \geq 7 \quad and \quad c \leq 20$$

A compound inequality with the word *and* is a **conjunction**. A conjunction is true only if both of its statements are true.

EXAMPLE 1

Graph $c \geq 7$, $c \leq 20$, and the conjunction $c \geq 7$ *and* $c \leq 20$.

Solution

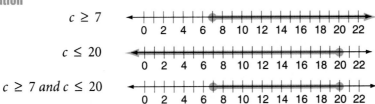

The solution of $c \geq 7$ *and* $c \leq 20$ includes all real numbers that are in the solution of $c \geq 7$ *and* in the solution of $c \leq 20$. Therefore, the solution of $c \geq 7$ *and* $c \leq 20$ is all real numbers greater than or equal to 7 and less than or equal to 20. ◄

CHECK UNDERSTANDING

Are all the solutions shown on the graph in Example 1 appropriate numbers of commercials that an advertiser can run each week at the special rate? Explain.

Geologists classify sand by diameter of the particle. Each sand type listed below is followed by its range of grain sizes s in millimeters.

Very Coarse
$1.0 \le s \le 2.0$
Coarse
$0.5 \le s < 1.0$
Medium
$0.25 \le s < 0.5$
Fine
$0.125 \le s < 0.25$
Very Fine
$0.0625 \le s < 0.125$

Note that each type contains particles ranging from a smallest size up to twice that size. This kind of system takes into account differences in relation to size; for example, 0.1 mm makes a big difference in a 0.5-mm grain but is negligible in a 5-mm pebble.

COMMUNICATING ABOUT ALGEBRA

Could you rewrite the compound inequality in Example 3 more compactly by using only numbers and inequality signs? Explain.

Since $7 \le c$ is the same as $c \ge 7$, a more compact way of writing $c \ge 7$ and $c \le 20$ is $7 \le c \le 20$. Read $7 \le c \le 20$ as "c is between 7 and 20, inclusive." Read $7 < c < 20$ as "c is between 7 and 20."

EXAMPLE 2

Write the compound inequality $a > -3$ and $a \le 5$ in a more compact form. Then graph the solution.

Solution

$-3 < a \le 5$
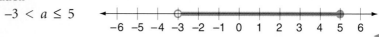

So far, the compound inequalities in this lesson have included the word *and*. Compound inequalities may also be written using the word *or*. A compound inequality with the word *or* is a **disjunction**. A disjunction is true if one or the other or both statements are true.

EXAMPLE 3

Graph the solution of $w > 4$ *or* $w < -4$.

Solution

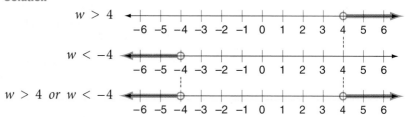

The solution includes all real numbers in the solution of $w > 4$ *or* in the solution of $w < -4$. Therefore, the solution of $w > 4$ *or* $w < -4$ includes all real numbers greater than 4 or less than -4. ◄

An efficient method for solving some compound inequalities is to work with both simple inequalities at the same time.

EXAMPLE 4

Solve $-8 < 2x - 6 < 4$. Graph the solution.

Solution

Use the properties of inequalities to isolate the variable x between the two inequality signs.

$$-8 < 2x - 6 < 4$$
$$-2 < 2x \quad\quad < 10 \quad\quad \text{Add 6 to each side of each inequality.}$$
$$-1 < x \quad\quad\;\; < 5 \quad\quad \text{Divide each side of each inequality by 2.}$$

Graph the solution.

The solution is all real numbers between -1 and 5. ◄

In Example 4 you could also have found the solution by working separately on each of the simple inequalities. However, in the next example, you must solve each part separately.

EXAMPLE 5

Solve $4 + 3x \le -11$ *or* $4x + 6 > 18$. Graph the solution.

Solution

Solve each simple inequality.

$$4 + 3x \le -11 \qquad\qquad or\ 4x + 6 > 18$$
$$3x \le -15 \quad \text{Subtract 4.} \qquad\qquad 4x > 12 \quad \text{Subtract 6.}$$
$$x \le -5 \quad \text{Divide by 3.} \quad or \qquad x > 3 \quad \text{Divide by 4.}$$

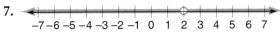

Graph the solution.

The solution is all real numbers that are either less than or equal to –5 or greater than 3. ◄

TRY THESE

Graph each compound inequality.

1. $d > -3$ and $d \le 2$ **2.** $x < -3$ or $x \ge 0$ **3.** $1.5 \le z \le 4$

4. $p \ge -2$ and $p < -3$ **5.** $-1 \le k \le 6$ **6.** $0 < 2z - 4 < 18$

Write the compound inequality for each graph.

7.
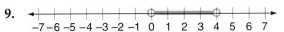

8.
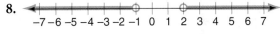

9.
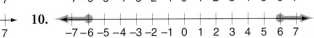

10.

Solve each compound inequality.

11. $4 < x + 6 < 10$ **12.** $7 \le 2x - 3 \le 9$

13. $6 + 2x < -12$ or $3x + 8 > 20$ **14.** $4 - 2x < 6$ or $3x + 7 < -2$

15. Graph the solution of $s \le 0$ *and* $s > 6$.
  a. What do you notice about the graph of this conjunction?
  b. Rewrite the compound inequality as a disjunction.
  c. For what values is the disjunction true?

16. SPORTS A regulation basketball weighs at least 20 oz but no more than 22 oz.

  a. Write a compound inequality that describes the weight w of a basketball in ounces.

  b. Graph the inequality on a number line and describe the solution.

Determine whether the given number is a solution of each compound inequality.

1. $2 < r < 9$; 9

2. $h \geq 5$ or $h < -1$; 5

3. $-3 \leq b \leq 0$; -2

4. $d > -1\frac{1}{2}$ and $d < 2$; -1

5. $k > 2$ or $k \leq -3$; 0

6. $f > 6$ and $f < 4$; 7

Write the compound inequality for each graph.

7.

8.

9.

10.

Graph each compound inequality.

11. $2.3 < v < 3.2$

12. $c < -2$ or $c \geq 6$

13. $t < 5$ and $t > \frac{1}{2}$

14. $q \geq 3$ and $q \leq 3$

15. $a > -3$ or $a < 4$

16. $-2 < x < -6$

Solve each compound inequality. Graph the solution.

17. $3a + 2 < -13$ or $3a + 5 \geq 2$

18. $u + 3 > 2$ and $u + 3 \leq 7$

19. $3j + 5 < 20$ or $2j - 1 > 13$

20. $-3 \leq 2c + 9 \leq 7$

21. $5 < \frac{1}{2}b - 7 < 9$

22. $h - 3 < 6$ and $h - 3 > 12$

23. NUMBER RELATIONSHIPS Seven times the sum of an integer and 12 is at least 50 and at most 68. Write and solve the compound inequality that expresses this relationship.

Solve each compound inequality.

24. $-8 < k + 3 < 5$

25. $2 - x > -3$ or $3x + 1 > 22$

26. $-15 \leq -2b + 5 \leq 3$

27. $3 - x > -5$ or $2x + 2 > 28$

28. $2 - 3x > -6$ or $-x + 8 < 16$

29. $2 - x < -3$ and $4 - 3x > -17$

30. WRITING MATHEMATICS Write a paragraph explaining how you can identify the compound inequality solution that is represented on a number line graph. Use examples to illustrate if you wish.

31. HEATING/COOLING The Robinsons turn their furnace on whenever the outside temperature falls below 55°F. They turn their air conditioning on whenever the outside temperature rises above 80°F.

   a. Write a compound inequality that describes the outside temperatures t for which the heating or cooling system is turned on.

   b. Graph the inequality on a number line and describe the solution.

Solve each compound inequality. Graph the solution.

32. $-5 \le \dfrac{4 - 3m}{2} < 1$

33. $0 < 2 - \dfrac{3}{4}d \le \dfrac{1}{2}$

34. $-2 \le \dfrac{3 + 2y}{-4} < 2$

35. $3 - x > -5$ or $2x - 2 > 28$ **36.** $-20 < 5(t + 3) < 5$ **37.** $-20 - 2b < -4b < -4$

38. ENTERTAINMENT Raisa plans to spend at least $15 but no more than $50 on CDs. CDs cost $9 each at CD City.
   a. Write a compound inequality to find the number of CDs c Raisa could buy.
   b. Graph the compound inequality to find how many CDs she could buy.

39. HEALTH One set of guidelines recommends that adults 35 years of age and over who are 6 ft tall should weigh no less than 155 lb and no more than 199 lb, depending upon their gender and frame size. A 6-ft adult was told to lose 30 lb to have a healthy weight.
   a. Write a compound inequality that describes the current weight w of this adult.
   b. Graph the compound inequality and describe the solution.

THINK CRITICALLY

GEOMETRY Recall the following geometric figures.

  point   ray   line segment   line

40. Write and graph an inequality to represent each geometric figure named.

   a. a ray   **b.** two rays   **c.** a line segment   **d.** a point   **e.** a line

41. Solve and graph.
   a. $2x > 5$ and $x + 7 > 6$ and $3x + 2 < 20$
   b. $6 - x < 9$ or $-3x < 18$ or $-2 > x + 6$

42. If $x > 2$ and $x < 10$, what is the solution of the disjunction $x < 2$ or $x > 10$?

43. If $x > 2$ and $x < 10$, what is the solution of the disjunction $x > 2$ or $x < 10$?

MIXED REVIEW

44. ENTERTAINMENT The length of the selections on a CD are 4:25, s, 5:10, 3:75, and 4:50. The range of the length of the selections is 1:25. What is the value of s?

Solve each equation.

45. $5b + 8 = -2b - 6$

46. $3(4m - 2) = 0.5(8m + 4)$

Make a table of values and graph each function.

47. $y = -2z - 3$

48. $y = \dfrac{x}{3} + 4$

49. STANDARDIZED TESTS Which is a solution of the compound inequality $y > 6$ or $y < -2$?

   A. 2   **B.** 0   **C.** -1 **D.** none of these

Career
Graphic Artist

A **graphic artist** uses visual forms to communicate ideas and emotions. To design an advertisement that will capture the interest—and money—of consumers, a graphic artist uses information about a product or service along with his or her imagination. The artist must also follow regulations about what can appear in advertisements.

For example, the United States Secret Service imposes these restrictions on illustrations of United States paper money.

- Printed illustrations of paper money may only be used for numismatic, educational, historic, and newsworthy purposes. They may not be used for decorating or for advertising.

- Paper money illustrations must be in black and white and must be less than $\frac{3}{4}$ times or more than $1\frac{1}{2}$ times the size of the genuine bill.

Decision Making

1. Printed illustrations of coins may be of any size and may be used for any purpose. Why do you think the government allows this but does not allow unrestricted use of illustrations of paper money?

Measure and record the length and width of a dollar bill in inches or centimeters.

2. Write a compound inequality to describe the allowable length y of an illustration of a dollar bill in printed material.

3. Write a compound inequality to describe the allowable width w of an illustration of a dollar bill in printed material.

4. Write a compound inequality to describe the length y of an illustration of a dollar bill that is *not* allowed in printed material.

5. Write a compound inequality to describe the width w of an illustration of a dollar bill that is *not* allowed in printed material.

6. Refer to the compound inequalities you wrote. Draw a template that a graphic artist could use as a guide to see the allowable sizes of a dollar bill.

Explore/Working Together

Manufacturers advertise a product on different radio and TV stations and during different programs depending upon the age group that will most likely use the product. Twenty people who recently purchased a personal computer were surveyed. Their ages are

32	21	40	52	22	60	25	34	35	26
22	30	58	44	55	25	37	54	23	21

1. Find the median age of the computer buyer.

2. What information about the age of personal computer buyers does the median give you?

3. Is the median age of buyers enough information for a manufacturer to select stations or programs on which to advertise? Explain.

> **SPOTLIGHT ON LEARNING**
>
> **WHAT?** In this lesson you will learn
> - to display data in boxplots.
> - to use inequalities to describe data displayed in boxplots.
>
> **WHY?** Using inequalities based on boxplots helps you solve problems about business, aviation, and climate.

Build Understanding

A better representation of a set of data than the median is given by **quartiles**, three numbers that divide the set of data into four equal parts, or quarters. The **first quartile**, Q_1, is the median of the lower half; the **second quartile**, Q_2, is the median of the whole set; and the **third quartile**, Q_3, is the median of the upper half.

EXAMPLE 1

EDUCATION The test scores received by the students in one math class are shown at the right. Find the quartiles for this set of data.

B. Ashley	94	N. Lee	79
J. Basso	71	G. McNamara	68
R. Cappello	85	K. Mubarak	77
K. Chan	99	M. Norris	82
L. Drake	23	S. Osmunsen	85
R. Hightower	67	J. Ross	89
B. Isaacs	65	D. Wolfe	87
M. Jacobiak	98		

> **THINK BACK**
>
> The **median** of a set of data is the middle number when the data are arranged in numerical order. When the data set has an even number of items, the median is the average of the two middle numbers of the ordered set.

CHECK —
UNDERSTANDING

Find the quartiles of
the set of data given
in Explore. What
additional information
do the quartiles give
about the age of the
personal computer
buyers surveyed?

Solution

Arrange the data in numerical order. Find the median of the whole set
of data. Then find the median of the lower half and the median of the
upper half of the set of data. Label the quartiles Q_1, Q_2, and Q_3.

23 65 67 68 71 77 79 82 85 85 87 89 94 98 99
$\qquad\qquad\quad\uparrow\qquad\qquad\qquad\quad\uparrow\qquad\qquad\qquad\uparrow$
$\qquad\qquad\quad Q_1\qquad\qquad\qquad\quad Q_2\qquad\qquad\qquad Q_3$
median of lower half   median   median of upper half ◄

A **boxplot**, or box-and-whisker plot, is a graph that uses the three
quartiles of a set of data to provide a visual display of the data. An
asterisk shows any **outliers**, values far from the data set. **Whiskers**
are lines that show the range of the data.

EXAMPLE 2

Make a boxplot of the math test scores from Example 1.

Solution

Draw a number line that includes the lowest and highest values. Then
draw a box with ends that line up with Q_1 and Q_3. Draw a line through
the box to show Q_2.

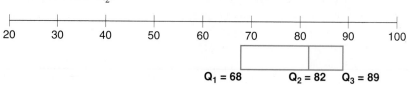

Use an asterisk to show any outliers. In this example, 23 is the only outlier.

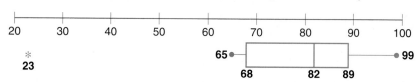

Draw whiskers from the box to show the range of the data. The range
is $99 - 6 = 34$. Notice that the length of the box represents the
middle half of the data. A completed boxplot does not need to show
the number line. ◄

Inequalities can be used to describe data shown in boxplots.

EXAMPLE 3

Refer to the boxplot in Example 2. Write an inequality that describes
the middle half of the math test scores s.

Solution

Since the ends of the box are at 68 and 89, the inequality
$68 \le s \le 89$ describes all the scores in the middle half. ◄

COMMUNICATING —
ABOUT ALGEBRA

When the data set
has an odd number
of items, will the
quartiles always be
items from the set?
What about when
there are an even
number of data?
Discuss any other
differences that relate
to finding quartiles for
these sets.

CHECK —
UNDERSTANDING

Since the box in the
boxplot in Example 2
represents the
middle *half* of the
test scores, why is it
not *half* the length of
the entire boxplot?

BUSINESS Use the data at the right for Exercises 1–3.

Commissions Earned by Salespeople, July 1995	
Salesperson	**Commission, $**
Frank	550
Louise	675
Hani	400
Bonita	575
Adrienne	480
Brett	700
Hai	650
José	950
Helena	500
Lori	525

1. Find the quartiles.

2. Make a boxplot. Identify any outliers.

3. To assure prospective salespeople about the amount of commission they can earn, the sales manager mentions the commissions earned by the 25% of the sales force with the lowest sales. Write an inequality that describes the commissions c that the sales manager would mention.

PRACTICE

AVIATION Use the data below for Exercises 1–3.

20 Busiest Airports, 1993			
Airport	**Passengers, millions**	**Airport**	**Passengers, millions**
Charles de Gaulle, Paris	26	Los Angeles International	48
Chicago O'Hare International	65	Miami International	29
Dallas/Ft. Worth International	50	Minneapolis-St. Paul International	23
Detroit Metropolitan Wayne County	24	Newark International	26
Frankfurt, Germany	33	Orly, Paris	25
Hartsfield Atlanta International	48	Osaka International	23
Heathrow Airport, London	48	San Francisco International	33
Hong Kong International	25	Sky Harbor International, Phoenix	24
J. F. Kennedy International, NY	27	Stapleton International, Denver	33
Logan International, Boston	24	Tokyo-Haneda International	42

1. Find the quartiles.

2. Make a boxplot. Identify any outliers. What is the range?

3. Write an inequality describing the millions of passengers p at the top quarter of the busiest airports.

HEALTH Use the data at the right for Exercises 4–6.

4. Find the quartiles.

5. Identify any outliers. Then make a boxplot. What is the range?

6. WRITING MATHEMATICS Based on the boxplot from Exercise 5, write a paragraph comparing and contrasting life expectancy throughout the world.

Life Expectancy, 1990–1995			
World Region	**Years**	**World Region**	**Years**
Africa		Asia	
Eastern	49	Eastern	72
Middle	51	Southeastern	63
Northern	61	Southern	59
Southern	63	Western	66
Western	51	Europe	
Latin America		Eastern	71
Caribbean	60	Northern	76
Central America	69	Southern	76
South America	67	Western	76
North America	76	Oceania	73
		Former USSR	70

SPORTS The boxplots below summarize the bowling scores of two teams. Use the boxplots for Exercises 7–9.

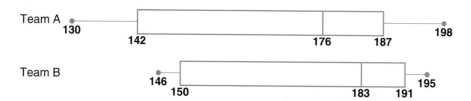

Team A
130   142   176   187   198

Team B
146   150   183   191   195

7. Which team had the highest score? the lowest score? What were the scores?

8. Which team had the higher median? What was it?

9. For which team were the scores in the middle 50% distributed more evenly? Which team did better? Explain.

⌐ EXTEND

CIVICS Use the data below for Exercises 10–14.

Electoral Votes, 1992							
State	Electoral Votes	State	Electoral Votes	State	Electoral Votes	State	Electoral Votes
Alabama	9	Idaho	4	Missouri	11	Pennsylvania	23
Alaska	3	Illinois	22	Montana	3	Rhode Island	4
Arizona	8	Indiana	12	Nebraska	5	South Carolina	8
Arkansas	6	Iowa	7	Nevada	4	South Dakota	3
California	54	Kansas	6	New Hampshire	4	Tennessee	11
Colorado	8	Kentucky	8	New Jersey	15	Texas	32
Connecticut	8	Louisiana	9	New Mexico	5	Utah	5
Delaware	3	Maine	4	New York	33	Vermont	3
District of Columbia	3	Maryland	10	North Carolina	14	Virginia	13
		Massachusetts	12	North Dakota	3	Washington	11
Florida	25	Michigan	18	Ohio	21	West Virginia	5
Georgia	13	Minnesota	10	Oklahoma	8	Wisconsin	11
Hawaii	4	Mississippi	7	Oregon	7	Wyoming	3

10. **TECHNOLOGY** Use a graphing utility to make a boxplot of this data.

11. Find the quartiles.

12. What variations in the boxplot might occur, depending on the utility used to graph the data?

13. Write an inequality that describes each of the following parts of the data. Let v represent electoral votes.

    a. 75%   **b.** 50%   **c.** 100%

14. Suppose you are working on a national political campaign. Based on the boxplot you graphed, in what portion of the states would you concentrate your campaign efforts? Why?

THINK CRITICALLY

CLIMATE Each boxplot shows the average monthly temperature for one year for a different city. Match each plot with the appropriate city. Explain your thinking.

15.

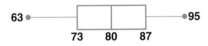

64
57 • 73
61.5 69.5

a. Minneapolis, MN
b. Phoenix, AZ
c. St. Louis, MO
d. San Francisco, CA

16.

20 • 83
32.5   58   75

17.

38 • 89
48   68   82

18.

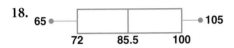

65 • 105
72   85.5   100

19. Create a data set of 20 numbers that could be represented by this boxplot. Explain your reasoning.

63 • 95
73   80   87

PROJECT *Connection*   Work with the other members of your group to write and design a print advertisement for **S'COOL LUNCH**.

1. Contact the advertising departments of at least three newspapers or magazines. Explain that you are working on a school project and ask for the rate schedule for placing ads. Record all the various options for size, color, weekday or weekend, and so on. Prepare a chart showing how the costs compare.

2. Make a decision on the characteristics of your ad. Lay out the plan on graph paper. Indicate where titles, text, and pictures or photos will appear. Include all measurements and colors.

3. Assume your client has budgeted $40,000 for this advertising. This total must include your fee, which is 12.5% of the cost for placing an ad. Decide where you will place the ads and how long they will run. How did you make your decision? Compute the amount of your fee and then determine the total cost to the client. Write a detailed cost summary.

Use Inequalities

Inequalities can be used to model real world situations and give information concerning the range of possible solutions to a problem. Read the problem carefully so that you can use a variable to write the correct inequality based on the given information.

Problem

Between them, Fran and Stan have 100 celebrity autographs. If Stan has at least two-thirds as many autographs as Fran, at least how many autographs does Stan have? What is the greatest number Fran can have?

Explore the Problem

1. Let a represent the number of autographs that Fran has. Using the information that there are 100 autographs in all, write an expression for the number that Stan has.

2. The problem also tells you how Stan's number compares to Fran's number. Which words can you translate using an inequality symbol? Write the inequality.

3. Solve the inequality. How must you restrict the range of values for a so that they make sense in this problem?

4. How many autographs does each person have? As part of your check, see whether the conditions of the problem would be satisfied if Fran's number were the next integer. By doing this, you make sure you have found the greatest number she can have.

5. Explain why it was preferable to begin the solution by letting a represent Fran's number rather than Stan's number.

6. **WRITING MATHEMATICS** Create a problem similar to the one you just solved. Use a situation where the answer does not have to be an integer. Have a classmate solve and check.

PROBLEM SOLVING TIP

Think about the equation you would write if the Problem stated that Stan had exactly two-thirds as many autographs as Fran. Then decide how to use an inequality symbol to express what the problem actually states.

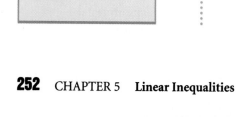

Investigate Further

7. HEALTH Recall the formulas that relate Fahrenheit and Celsius temperatures.

$$F = \frac{9}{5}C + 32 \qquad\qquad C = \frac{5}{9}(F - 32)$$

Medical professionals generally use the term "fever" when a person's oral temperature exceeds 98.6°F. If you are using a Celsius thermometer, what readings would indicate a fever?

PROBLEM SOLVING PLAN

• Understand
• Plan
• Solve
• Examine

   a. Write an inequality that describes the range of Fahrenheit temperatures that would indicate a fever.

   b. If you want to find a range of corresponding Celsius temperatures, what expression can you substitute for F? Write the inequality.

   c. Solve the inequality you wrote above. What Celsius temperatures would indicate a fever?

8. BIOLOGY In an experiment to monitor breathing rate, you are told to keep the water temperature of a bowl containing goldfish between 15°C and 24°C. What range of Fahrenheit temperatures will be safe for the fish?

   a. Write two inequalities to describe the range of Celsius temperatures that would be dangerous for the fish.

   b. Substitute an expression in each inequality so that you can find a corresponding range of Fahrenheit temperatures.

   c. Solve the inequalities. Which Fahrenheit temperatures are unsafe for the fish? Which are safe?

Apply the Strategy

9. TRAVEL Two cars start from the same point at the same time and go in opposite directions. One car averages 87 km/h and the other averages 78 km/h. How long must they travel to be at least 858 km apart?

10. CARPENTRY The length of a rectangular table is 25 cm more than twice the width. The perimeter of the table is no more than 335 cm. Find the maximum dimensions of the table if each dimension is an integer. Explain your answer.

ALGEBRA: WHO, WHERE, WHEN

The amount of variation permitted in the size of a manufactured item is called *tolerance*. Since manufacturing variables make it impossible to produce an item of the exact dimensions noted by a designer on a drawing, the designer must be satisfied with items between a maximum and minimum size. For example, the diameter *d* of a rod that is specified as 3.00 mm may be acceptable if it falls within the range $2.95 \le d \le 3.05$ or $d = 3.00 \pm 0.05$. This method of stating tolerance is called *bilateral tolerance* because variation from a basic size is given in plus and minus values.

11. **POLITICS** A box contains 360 campaign buttons. Some buttons are red, and the rest are blue. If there are no more than $1\frac{1}{2}$ times as many red buttons as blue ones in the box, what is the maximum number of red buttons in the box? What is the minimum number of blue buttons there could be?

12. **CHEMISTRY** The directions in a laboratory manual state that the temperature of a certain solution should be maintained in the range $86°F \le t \le 104°F$. What is the corresponding range of Celsius temperatures?

13. **CHEMISTRY** The melting point of a substance is the temperature at which it changes from a solid to a liquid state. The boiling point is the temperature at which it changes from a liquid to a gaseous state. The melting point of diamond is 3700°C, and the boiling point is 4200°C. For what range of Fahrenheit temperatures *t* will diamond be in the liquid state?

14. **GAS MILEAGE** A certain car holds 20 gal of gas and averages 22 mi/gal. If the car runs out of gas after having traveled at least 330 mi one day, what are the possible amounts *a* of gas that were in the car's tank at the start of the day?

15. **HEALTH CLUB MEMBERSHIP** Angela now pays $20 per week to belong to a health club. A new club is available for $17.50 per week, but it requires a $15 enrollment fee. For what lengths of membership *m* in the new club would Angela save money?

16. **BAGEL SALES** A bagel store owner reads an industry study in *Bagel Times* magazine. The study reports that if the price of a large sesame bagel is set at *p* cents, where $40 \le p \le 200$, then a store in a busy mall will sell $1000 - 5p$ bagels daily.

a. If the owner wants to sell at least 500 bagels a day at the minimum price of 40 cents, what are the possible prices for bagels?

b. If the owner prices a bagel at the midpoint of the range of prices you determined above, what are the daily sales predicted by the model?

REVIEW PROBLEM SOLVING STRATEGIES

A TALL STORY

1. There are 200 students in the schoolyard, arranged in 10 rows of 20 students each. From each of the 20 columns formed, the shortest student is asked to come forward, and the tallest of these 20 short students is given the letter A. These students return to their original places. Next, the tallest student in each row is asked to come forward and from these 10 tall students, the shortest is given the letter B. Assuming that A and B are not the same student, which of the two is taller? How do you know?

   a. What strategies will you use to solve this problem?

   b. If A and B were originally in the same column, who is taller? Why?

   c. If A and B were in the same row, who is taller? Why?

   d. If A and B are in different rows and columns, how can you compare their heights? Who is taller? Why?

   e. Suppose there were 1000 students arranged in 20 rows of 50 students. If the same procedure is used to determine A and B, do you have to solve the problem again to decide who is taller, A or B? Explain.

7 Makes Magic

2. Work with a partner.

   Make copies of the number strip below. You will need at least seven of them, but it will be helpful to have extras on hand.

   | 1 | 2 | 3 | 4 | 5 | 6 | 7 |

   The problem is to find a way to cut seven of these strips into the fewest number of pieces so that the pieces may be arranged to form a magic square, with the seven rows, seven columns, and two diagonals adding up to the same number. Pieces cannot be turned upside down or on their sides.

   a. In a seven-by-seven square, using each of the numbers 1, 2, 3, 4, 5, 6, and 7 exactly seven times, what arrangement of the numbers will guarantee that your square will be magic?

   b. What do you think the magic sum for this square will be? Verify your prediction when you find a square that works.

COMPUTER REPAIR PUZZLER

3. A technician inspected 58 computers. Suppose that 36 needed new disk drives and 31 needed new screens.

   a. What is the least number of computers that could have needed both parts?

   b. What is the greatest number of computers that could have needed both parts?

   c. What is the greatest number of computers that could have needed neither part?

• • • CHAPTER REVIEW • • •

VOCABULARY

Match the letter of the word in the right column with the description at the left.

1. the median of the upper half of a data set
2. For all real numbers a, b, and c, if $a > b$, then $a + c > b + c$.
3. For all real numbers a and b, either $a = b$, $a < b$, or $a > b$.
4. an inequality with *and* or *or*
5. a statement that two numbers or expressions are not equal

a. trichotomy property

b. inequality

c. compound inequality

d. addition property of inequality

e. third quartile

Lesson 5.1 INEQUALITIES ON THE NUMBER LINE   pages 221–226

- For any real numbers a and b, either $a = b$, $a < b$, or $a > b$.
- The graph of an inequality is the set of all real numbers that make the inequality true.

Graph the solution of each inequality on a number line.

6. $n \geq 3$ **7.** $c \neq -1$ **8.** $y < 0$ **9.** $-2.5 > w$

Lesson 5.2 EXPLORE INEQUALITIES   pages 227–229

- Adding or subtracting a positive real number from both sides of an inequality does not change the direction of the inequality symbol.
- Multiplying or dividing both sides of an inequality by a positive real number does not change the direction of the inequality symbol.
- Multiplying or dividing both sides of an inequality by a negative real number reverses the direction of the inequality symbol.

Write the inequality symbol that makes each statement true.

10. If $x < 9$, then $x - 4$ ▨ $9 - 4$.

11. If $-3x \geq 21$, then $\dfrac{-3x}{-3}$ ▨ $\dfrac{21}{-3}$.

12. If $-x > 1$, then x ▨ -1.

13. If $5x - 3 > 22$, then x ▨ 5.

Lesson 5.3 SOLVE INEQUALITIES USING ONE OPERATION   pages 230–234

- To solve an inequality, use the addition, subtraction, multiplication, and division properties of inequality.

Solve and check each inequality. Graph each solution.

14. $a - (-3) > 8$ **15.** $\dfrac{5}{6}x < 5$ **16.** $-8 \leq \dfrac{4}{5}y$ **17.** $-4.2 > -1.4w$

- To solve inequalities with two or more operations, combine like terms where necessary. Then use the inequality properties to get the variable terms on one side of the inequality symbol.

- Apply the distributive property if an inequality contains parentheses.

Solve and check each inequality.

18. $\frac{x}{4} + 2 > 7$

19. $3 - 2b < 15$

20. $4x + 2 \leq 47 + x$

21. $3(r + 2) > 4(r + 1)$

22. $\frac{x}{2} + 3 > 9 + \frac{x}{3}$

23. $3x + 4 < 4x - 8$

24. $2(x + 3) > 5x + 5$

25. $2(y + 2) \geq 4(y + 3)$

- A compound inequality written with *and* is a **conjunction**. The solution of a conjunction includes all real numbers that are in the solution of *both parts* of the inequality.

- A compound inequality written with *or* is a **disjunction**. The solution of a disjunction includes all real numbers that are in the solution of *either part* of the inequality.

Solve each compound inequality. Graph the solution.

26. $x + 8 > 6$ and $x + 2 < 10$

27. $2x - 6 > 8$ or $3x - 4 < 5$

28. $4 + y \leq 1$ or $37 \leq 12 + 5y$

29. $x - 4 > 1$ and $x + 4 < -8$

30. $4x - 7 \geq 13$ or $2x + 8 \leq 6$

31. $13 + 4x \leq 1$ or $28 \leq 14 + 7x$

- A **boxplot**, or box-and-whisker plot, is a graph that provides a visual display of data. The boxplot shows the median, the upper quartile, and the lower quartile of the data, as well as the range. The length of the box represents the middle half of the data.

- Inequalities can be used to describe data shown in boxplots.

32. Make a boxplot for the following test scores.

    61, 62, 65, 68, 72, 75, 77, 77, 82, 83, 85, 85, 86, 87, 92, 95, 98

33. Write an inequality describing the middle half of the data.

- You can use inequalities to model real world situations involving a range of possible solutions to a problem.

Use an inequality to solve.

34. The length of a rectangle is 47 m. What width will produce a perimeter that is no less than 108 m?

35. The length of a rectangular garden plot is 15 ft more than three times its width. The perimeter of the garden is no more than 94 ft. Find the maximum dimensions of the garden if each dimension is an integer.

CHAPTER ASSESSMENT

CHAPTER TEST

Graph each inequality.

1. $y \leq 2$ 2. $x \neq 0$ 3. $x > -3$

Solve each inequality. Graph each solution.

4. $x - 4 < 7$ 5. $x + 7 \geq -1$ 6. $1.3n \geq 5.2$

7. $-4n < -8$ 8. $\frac{1}{3}y \leq 3$ 9. $-\frac{3}{4}x < 6$

Solve each inequality.

10. $4 - n \leq -2$ 11. $2y + 3 \leq 9$

12. $9 - 4x < 7$ 13. $4(x + 2) < 6(x - 1)$

14. $0.37x + 0.17 > 3.5$ 15. $4x - 2 < 13 - x$

16. $2x - 2 > 3x + 6$ 17. $\frac{2}{3}x - 4 > 2$

18. $x + 5 \geq 6(x - 4) + 9$

19. $2x + 17 \leq 5(x - 4) - 8$

20. WRITING MATHEMATICS Write a paragraph explaining how to solve the inequality $-3x + 4 > 13$.

Solve each compound inequality. Graph each solution.

21. $x + 5 > 4$ and $x + 7 < 12$

22. $3x - 1 > 5$ or $2x - 7 < -11$

23. $3 - y \geq 4$ or $22 \leq 3y + 4$

24. $2x + 3 > 7$ and $x + 9 < 16$

25. $5y - 2 < 3$ or $-3 < 3(y - 4)$

26. $0.7a - 1.3 \leq 3.6$ and $2.3a + 1.4 \geq -7.8$

Write an inequality to express each statement.

27. A number y is at least 4.

28. A number n is 24 at most.

29. 3 more than 2 times a number x is no more than 13.

30. At Mel's Music the cost c of CDs is at least $11.95 but never more than $17.95.

Write and solve an inequality for each situation.

31. Lisa and Keith work at a novelty shop. Lisa worked 5 more hours than Keith. Together they worked at least 37 h. What is the least number of hours each person worked?

32. The regular-size box of BestGrain cereal weighs 4 oz less than the economy-size box. The giant-size box weighs 6 oz more than the economy size. The three boxes together weigh at most 38 oz. What is the most that the regular box can weigh?

33. The sum of two even integers is less than 50. What is the greatest possible value for each of the integers?

Use the boxplot of test scores to answer questions 34–38.

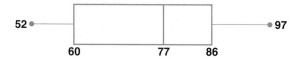

34. What is the range of the data?

35. What is the median of the data?

36. Between what two values is the middle half of the data?

37. At what data point is the first quartile?

38. At what data point is the third quartile?

39. What percentage of the scores are above 86?

40. What percentage of the scores are below 60?

PERFORMANCE ASSESSMENT

USE ALGEBLOCKS Select at least five different inequalities from this chapter. Model each inequality on a Sentence Mat with Algeblocks. Ask a partner to record the algebraic steps necessary for solving the inequalities. Together determine a way you can use Algeblocks to check the solutions.

USE NUMBER LINES Start with the true inequality 1 < 3.

a. Explain how the diagram below illustrates the addition property of inequality.

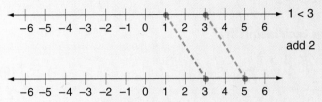

b. Draw a similar diagram to illustrate the subtraction property of inequality. Start with 1 < 3 and subtract 3.

INEQUALITIES IN ADVERTISING *The Times* and *The Chronicle* are two daily newspapers in which you wish to place your advertisement. *The Times* charges a flat fee of $585 plus $90 per day. *The Chronicle* charges $155 per day. Let *d* represent the number of days you plan to advertise. Set up two inequalities with these relationships.

$$\text{The Chronicle} > \text{The Times}$$
$$\text{(cost)} \qquad \text{(cost)}$$

$$\text{The Chronicle} < \text{The Times}$$
$$\text{(cost)} \qquad \text{(cost)}$$

Solve these inequalities for *d* and interpret the solution. In which newspaper does it cost less to advertise? When might it be smarter to advertise in the more expensive paper?

MAGAZINE ANALYSIS Work in groups of three. Gather data about magazines in one category such as news, fashion, entertainment, sports, or some other area of interest. Make a list of characteristics to focus on—for example, number of pages, number of full-page advertisements, dimensions, price, average length of feature articles, and so on. Write six inequalities that give information based on your data. Exchange work with another group and examine their inequalities. Work with your group to write a short summary of the information you are able to learn from the inequalities.

PROJECT ASSESSMENT

PROJECT *Connection* In his book *Advertising: Planning, Implementation, and Control* (Cincinnati, Ohio: South-Western Publishing, 1993), author David Nylen discusses the four stages of a successful advertising campaign. In the *analysis* stage a set of data is gathered, analyzed, and interpreted. In the *direction* stage objectives and target audience are identified and defined. During the *advertising program* advertisements are created, schedules planned, and budgets produced. Finally, the *evaluation* stage provides ways to monitor and evaluate advertising effectiveness.

Your advertising agency should be ready to make a sales presentation to the **S'COOL LUNCH** Company. Prepare a report and a display board to show that you have developed a well thought out campaign. Include each of the four stages of the advertising process.

· · · CUMULATIVE REVIEW · · ·

Graph each inequality.

1. $x > -2$

2. $n \le 5$

Solve each inequality.

3. $y + 5 > 8$

4. $2x \le -12$

5. $4(3 - c) \le -4$

Use the graph below to answer Questions 6–8.

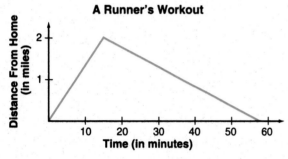

A Runner's Workout

6. How many miles did the runner run?

7. About how much time did it take the runner to reach the farthest point?

8. STANDARDIZED TESTS Which of the following best describes the runner's workout on this day?

   A. The runner started off slowly, then sprinted home.

   B. The runner started off running fast, then walked home.

   C. The runner took an equal amount of time going and returning.

   D. This cannot be determined from this graph.

Let $A = 5$, $B = 4$, and $C = 6$. Evaluate each expression in BASIC.

9. $C * \dfrac{B}{A}$

10. $A * C - B$

11. Make a line graph for the following high temperatures during one week.

   Sun, 73; Mon, 78; Tue, 74; Wed, 66; Thur, 71; Fri, 77; Sat, 82

Simplify each expression.

12. $4(x - 12) - 3(5x + 3) + 30$

13. $-3c + \dfrac{2}{3}(6c - 5) - \dfrac{1}{4}(12c - 2)$

Solve each compound inequality. Graph each solution.

14. $3x - 7 < 5$ and $5x + 11 > 1$

15. $7 - 2n \le 13$ or $4(n + 6) \le 8$

Evaluate each expression.

16. $-9 - 8$

17. $-9 \cdot (-8)$

18. $-38 - (-145)$

19. $(-4)(-5) - 6$

Write and solve an inequality or an equation for each situation.

20. A car rental company charges $24.95 per day plus $0.10 per mile. A business traveler is allotted $50.00 each day for travel expenses. What is the greatest number of miles the businessperson can travel without going over the allotted $50.00?

21. Two less than six times a number is –32. Find the number.

22. The sum of two consecutive odd integers has to exceed 72. What are the least possible values for the integers?

23. A pair of jeans was on sale for $27.00. This represented a 20% savings. How much were the jeans originally?

STANDARDIZED TEST

STANDARD FIVE-CHOICE Select the best choice for each question.

1. The number line that shows the solution to $-4x > -12$ is

   A.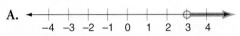

   B.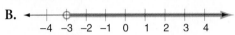

   C.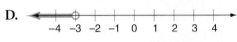

   D.

   E.

2. Which of the following illustrates the associative property of multiplication?

   A. $(3 + 4) + 5 = 3 + (4 + 5)$
   B. $7 \cdot 4 = 4 \cdot 7$
   C. $3(x + 2) = 3 \cdot x + 3 \cdot 2$
   D. $(-3 \cdot 4) \cdot 5 = 5 \cdot (-3 \cdot 4)$
   E. $(7 \cdot 2) \cdot 4 = 7 \cdot (2 \cdot 4)$

3. To solve for y in the equation $2y - 5 = 7$, you have to

   A. subtract 5, then multiply by 2
   B. add 5, then divide by 2
   C. divide by 2, then add 5
   D. subtract 2, then multiply by –5
   E. add 5, then multiply by 2

4. The median of a set of whole-number data values is 7.5. Consider Statements I–IV.

   I. The number of data values is odd.
   II. The number of data values is even.
   III. The mean is 7.5.
   IV. The mode cannot be 7.5.

   Which statement is true?

   A. I only   **B.** II only   **C.** I and III
   D. II and IV   **E.** II, III, and IV

5. Given that $x > y$, which of the following is false?

   A. $x + 4 > y + 4$
   B. $x - 15 > y - 15$
   C. $6x > 6y$
   D. $\frac{1}{3}x < \frac{1}{3}y$
   E. $\frac{y}{3} < \frac{x}{3}$

6. $(-2)^4 - 2^4 =$

   A. 0   **B.** 32   **C.** –32
   D. 16   **E.** –16

7. Where is the point $(-3, 0)$ located?

   A. in Quadrant I
   B. in Quadrant II
   C. in Quadrant III
   D. on the x-axis
   E. on the y-axis

8. –3 is a solution to which of the following?

   A. $x < 1$
   B. $x \geq -2$
   C. $3x < -9$
   D. $-4x \leq 0$
   E. two of the above

9. *Multiplicative inverse* is another name for

   A. opposite
   B. 0
   C. reciprocal
   D. 1
   E. division

10. In a random selection of people at a ballgame, 5 out of 8 were male. What is the probability that the next person selected will be female?

    A. 3%   **B.** 37.5%   **C.** 50%
    D. 62.5%   **E.** 80%

6 Linear Functions and Graphs

Take a Look
AHEAD

Make notes about things that look new.
- In what different forms can the equation of a line be given? What information do you need to write each of these forms?
- Can you measure the "steepness" of a line? What word describes the measure of a line's steepness?

Make notes about things that look familiar.
- What information can you find from the graph of a line?
- How are the mathematical meanings of the terms "slope," "intercept," "parallel," and "variation" similar to their meanings in everyday language?

DATA*Activity*

Clean Up the Environment

For many people, cleaning up the environment is a full-time job. There are thousands of public and private companies engaged in a wide variety of environmental activities including laboratory testing services, collection and disposal of solid waste, water purification, air pollution control, resource recovery, instrument manufacturing, and development of alternate energy sources. The table on the next page shows how the environmental industry has grown.

SKILL FOCUS

- ▶ Add, subtract, multiply, and divide real numbers.
- ▶ Use data to determine average change and make predictions.
- ▶ Find the percent of a number.
- ▶ Write numbers in scientific notation.
- ▶ Determine scales for graphing data.

SAVE THE PLANET

In this chapter, you will see how:

- **UTILITY COMPANY SERVICE REPRESENTATIVES** use tables and graphs to provide billing plans that fit a family's budget. (Lesson 6.2, page 275)

- **APPLIANCE SALESPEOPLE** use a special formula to choose the air conditioner best suited to a customer's needs. (Lesson 6.3, page 284)

- **RECYCLING PLANT MANAGERS** use equations to help plan for the future. (Lesson 6.4, page 292)

Environmental Industry Revenues and Employment 1988–1992		
Year	Revenue (in billions)	Employment (in thousands)
1988	$100.3	857
1989	$114.6	967
1990	$126.0	1047
1991	$128.7	1052
1992	$133.7	1073

Use the table to answer the following questions.

1. What was the average yearly increase in revenue from 1988 to 1992? Use your result to predict total revenue for the year 1998. Do you think this prediction will be too high or too low? Why?

2. What was the average yearly increase in employment for the period? Use your result to predict employment for the year 2000. Do you think this prediction will be too high or too low? Why?

3. In 1990, hazardous waste management accounted for $13.3 billion of revenue. What percent of total revenues for that year does this category represent? Round to the nearest tenth of a percent.

4. Express the revenue and employment totals for 1992 in scientific notation.

5. **WORKING TOGETHER** What difficulty would you encounter if you tried to display the revenue and employment data using a double-line graph?

Earth Day Magazine

Many holidays celebrated during the year remind you of the *past*—they commemorate a historical event. Earth Day, observed in April, is about both the present and the future. It is a day dedicated to raising people's environmental consciousness, and your class can help.

PROJECT GOAL

To combine research, experimentation, technology, algebra, and art skills to produce an informative class publication about saving the environment.

Getting Started

Your entire class will work together on the Earth Day magazine. All of you must decide how to delegate assignments fairly. Ideally, each student or group should concentrate on one aspect according to their talents and interest. Here are some ideas.

1. Write to the U.S. Environmental Protection Agency (EPA) and your local power company requesting information on environmental problems and conservation.

2. Research environmental topics such as air and water pollution, endangered species, rain forests, fossil fuel shortages and alternate energy sources, the ozone layer, solid waste disposal and recycling, and overpopulation.

3. Write articles for the magazine based on your research. Create graphs and tables to explore trends.

4. Begin a column of "Helpful Environmental Hints" that you will add to throughout the project.

PROJECT Connections

Lesson 6.4, page 291:
Explore water usage for different activities in the home to discover ways to conserve water.

Lesson 6.6, page 306:
Survey students on environmental issues relating to automobile use.

Lesson 6.7, page 311:
Learn how much electrical energy different appliances use, what it costs to run them, and how to conserve energy.

Chapter Assessment, page 319:
Combine project results in an Earth Day magazine.

Internet Connection

www.swpco.com/
swpco/algebra1.html

Think Back

- Recall that the graph of a linear equation in two variables is the solution of the equation and is a straight line.

  Work with a partner. Use any method you choose, including a table of values or a graphing utility.

1. For which of the following equations will the graph pass through the point (2, 3)? How did you decide?

   a. $y = x - 1$ **b.** $x + y = 5$

   c. $y - x = 1$ **d.** $2x + 3y = 13$

   e. $2y - x = 1$

2. Which of the following points lie on the graph of the equation $3y + 2x = 5$? How did you decide?

   a. $(1, 1)$ **b.** $(-3.5, 4)$

   c. $\left(2, -\dfrac{1}{2}\right)$ **d.** $(4, -3)$

   e. $\left(\dfrac{3}{5}, 0\right)$

Explore

3. Work in a group. Each student should draw x- and y-axes in exactly the same position on separate sheets of graph paper. Each student should set up a table of values and graph at least one of the following equations.

   a. $y - 3x = -1$ **b.** $y = 3x - 1$ **c.** $-3x + y = -1$

   d. $1 - 3x + y = 0$ **e.** $y + 1 = 3x$

4. Take any two graphs from Question 3 and hold them up to the light. Adjust the graphs so that the coordinate axes line up with each other. Describe what you see.

5. Position the remaining graphs, one at a time, over the first graph. What do you notice?

6. How could you show algebraically what you discovered in Questions 4 and 5?

THINK BACK

Recall that two lines can either intersect (cross each other) or be parallel (not cross each other even if the line is extended forever).

Algebra Workshop

Your graphing utility allows you to see the graphs of several equations on the same viewing screen. However, since the graphs are not labeled, it will be helpful to graph one equation at a time. Then sketch and label the graph shown on the viewing screen on graph paper.

7. Enter and graph these linear functions one at a time on your graphing utility. Then sketch and label each graph on graph paper using the same axes. Describe what you see.

    a. $y = 2x + 5$ **b.** $y = 2x + 1$ **c.** $y = 2x - 1$

8. Predict the location of the graph of $y = 2x + 4$. Sketch the graph on the same axes you used for the graphs in Question 7. How did you decide where to sketch the graph? Use your graphing utility to check your graph.

9. Set your graphing utility so integers are displayed when using the TRACE feature. For each line in Questions 7 and 8, trace through several integer values of x. As the value of the x-coordinate increases by 1 unit, by how much does the y-coordinate increase?

10. Predict whether the graph of $y = 5x + 1$ will be parallel to the other graphs in Questions 7 and 8. Justify your prediction.

Make Connections

11. WRITING MATHEMATICS From the work you did in Questions 3–10, create a definition of the term *parallel* as it applies to graphs of linear equations. Then explain how you can look at two equations and decide, before graphing, that their graphs are parallel.

12. Graph $y = 3x + 2$ and $y = 3x - 1$. Does what you see support the definition of *parallel* you wrote for Question 11? Explain.

13. Graph $y = 4x - 1$ and $y = 2x - 1$. Does what you see support the definition of *parallel* you wrote for Question 11? Explain.

14. For which of the following graphs does y decrease as x increases?

    a.   **b.**   **c.**

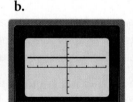 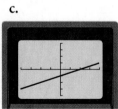

15. Predict whether the graphs of $y = 4x$ and $y = -4x$ will be parallel or will intersect. Explain your reasoning. Then use your graphing utility to test your prediction.

16. **a.** Make a table of values and graph the equation $y = 3x + 1$ on graph paper.

   b. Make a third column in the table of values by adding 4 to each y-value as shown at the right. Call the third column $y + 4$.

   c. Describe the location of each point $(x, y + 4)$ in relation to each point (x, y).

   d. Describe the relationship between the graphs of the equations described in parts 16a and 16b.

x	y	$y + 4$

17. **MODELING** A mover sometimes places a board over a staircase to roll heavy objects upstairs. How can you use the stair measurements shown at the right to determine which set of stairs will require more effort?

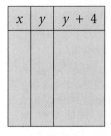

5 in.

1 ft

Staircase A

18. **CONSTRUCTION** An architect is considering different roof designs. The pitch (slant) of each roof is given by the following equations.

   a. $y = 2x$ **b.** $y = 0.8x$ **c.** $y = 1.5x$

   Which roof is probably best for an area in which there is a lot of snow? Explain your reasoning.

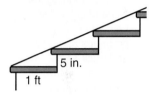

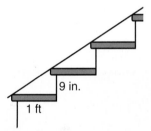

9 in.

1 ft

Staircase B

Summarize

19. **WRITING MATHEMATICS** Explain how you can determine whether the graphs of two linear equations are parallel by examining tables of values for each equation.

20. **THINKING CRITICALLY** For which graph does y increase faster than x? For which graph does y increase at the same rate as x? For which graph does y decrease as x increases?

a.
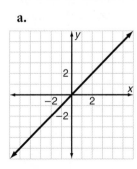

b.
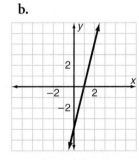

c.
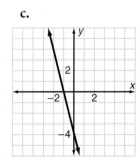

21. **GOING FURTHER** For which equation does the value of y increase faster than the value of x? Explain your reasoning.

   $y = 10x - 1$ or $y = x + 15$

6.2 The Slope of a Line

Explore/Working Together

Top of the Mountain Ski Resort is almost ready to open for business. First the slopes must be ranked according to steepness so that skiers can choose the one appropriate to their skill level.

1. Work with your group. Rank the six ski slopes in order of steepness from least to greatest.

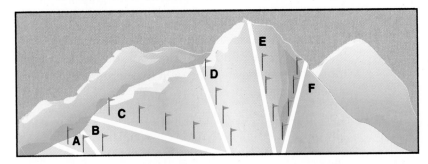

2. How could you use the coordinate plane to measure the steepness of one of the slopes?

3. Work independently. Use your method to compare the steepness of ski slopes A and E. Discuss your results with other members of your group. When your results agree, use your method to compare the steepness of the other slopes on the mountain. How do your results compare to your original rankings?

Build Understanding

As you can see in Explore, the term *slope* has everyday meanings as in "ski slope" or "slope of a roof." Slope relates to steepness. In mathematics, slope is expressed as a number that represents the steepness of a line.

For a given distance along a line, the **slope** of the line is the ratio of the number of units the line rises or falls vertically (the **rise**) to the number of units the line moves horizontally from left to right (the **run**). On a coordinate plane, the slope of the line is the *ratio* of the change in the *y*-coordinates of the two points to the change in the *x*-coordinates of the two points.

For a line containing two points (x_1, y_1) and (x_2, y_2),

$$\text{slope} = \frac{\text{rise}}{\text{run}} = \frac{\text{change in } y}{\text{change in } x} = \frac{y_2 - y_1}{x_2 - x_1}$$

ALGEBRA: WHO, WHERE, WHEN

In some branches of mathematics, the slope ratio is denoted by $\frac{\Delta y}{\Delta x}$. The Greek letter Δ is read "delta". It corresponds to the letter D in the Roman alphabet and stands for *difference*.

Either point on the line can be labeled (x_1, y_1) or (x_2, y_2). However, when you subtract values in the ratio for slope, you must use the same order for both numerator and denominator.

EXAMPLE 1

Determine the slope of a line that passes through the points $(2, 2)$ and $(8, 4)$.

Solution

Let (x_1, y_1) be $(2, 2)$. Let (x_2, y_2) be $(8, 4)$.

$$\text{slope} = \frac{y_2 - y_1}{x_2 - x_1} = \frac{4 - 2}{8 - 2} = \frac{2}{6} = \frac{1}{3}$$

The slope of the line that passes through the points $(-3, 1)$ and $(3, 3)$ is $\frac{1}{3}$. ◄

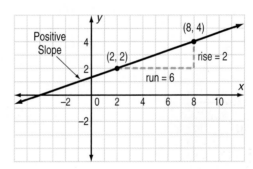

In Example 1, the line slants upward from left to right. It has **positive slope.** For every 3-unit change to the right, it slants *up* one unit.

EXAMPLE 2

Determine the slope of a line that passes through the points $(-2, 7)$ and $(2, -1)$.

Solution

Let (x_1, y_1) be $(-2, 7)$. Let (x_2, y_2) be $(2, -1)$.

$$\text{slope} = \frac{y_2 - y_1}{x_2 - x_1}$$

$$= \frac{-1 - 7}{2 - (-2)} = \frac{-8}{4} = -2$$

The slope of the line that passes through the points $(-2, 7)$ and $(2, -1)$ is -2. ◄

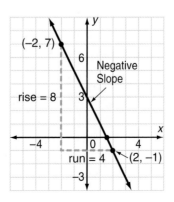

CHECK UNDERSTANDING

Suppose you labeled in reverse the points in Examples 1 and 2 that you chose as (x_1, y_1) and (x_2, y_2). How does that change your result?

The line in Example 2 slants downward from left to right. It has **negative slope.** For every 1-unit change to the right, it slants *down* 2 units.

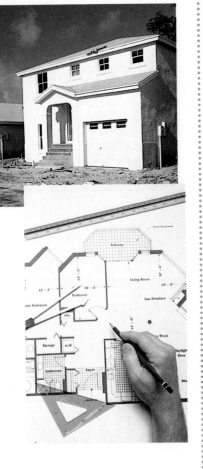

You can compare the steepness of two lines by comparing the absolute value of their slopes. The slope with the greater absolute value is steeper.

EXAMPLE 3

Which is steeper, the line in Example 2 or the line that passes through the points $(-1, 5.9)$ and $(0, 8)$?

Solution

$$\frac{y_2 - y_1}{x_2 - x_1} = \frac{8 - 5.9}{0 - (-1)} = \frac{2.1}{1} = 2.1$$

The slope of the line in Example 2 is -2. Compare $|-2|$ and $|2.1|$.

$$|-2| = 2 \qquad |2.1| = 2.1$$

Since $2 < 2.1$, the steeper line passes through $(0, 8)$ and $(-1, 5.9)$. ◄

The slopes of horizontal and vertical lines present special cases.

EXAMPLE 4

a. Find the slope of the line that passes through the points $(2, 4)$ and $(5, 4)$.

b. Find the slope of the line that passes through the points $(1, 6)$ and $(1, -3)$.

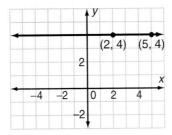 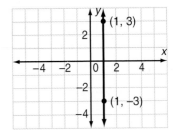

Solution

a. $\dfrac{y_2 - y_1}{x_2 - x_1} = \dfrac{4 - 4}{5 - 2} = \dfrac{0}{3} = 0$

The line is horizontal and has slope equal to zero.

b. $\dfrac{y_2 - y_1}{x_2 - x_1} = \dfrac{-3 - 3}{1 - 1} = \dfrac{-6}{0}$

The line is vertical and has an undefined slope, since division by zero is undefined. ◄

> **The slope of any horizontal line is zero.**
> **The slope of any vertical line is undefined.**

Slope can be applied in real world problems to describe a constant or an average rate of change over time. The rate can be represented as the ratio of the change in the y-quantity to the change in the x-quantity.

EXAMPLE 5

CAR SPEED CHANGES Some race cars use parachutes to help slow their cars down at the end of a race. One car traveled 345 ft during the first second following the opening of its parachute. After 8 seconds, the car had traveled a total of 1640 ft. What was the average rate of change in the distance covered during that time?

Solution

The points (1, 345) and (8, 1640) represent the two times and distances. Let (x_1, y_1) and (x_2, y_2) represent these points, respectively. Use the slope formula to find the average rate of change.

$$\text{rate of change} = \frac{y_2 - y_1}{x_2 - x_1} \quad \begin{array}{l}\text{change in distance}\\ \text{change in time}\end{array}$$

$$= \frac{1640 - 345}{8 - 1} \quad \text{Substitute known values.}$$

$$= \frac{1295}{7} = 185 \quad \text{Simplify.}$$

The *average rate of change* in the distance covered by the car after its parachute opens is 185 feet per second. However, since the car is slowing down, the change in distance covered during any 1-second period is likely to be different from the average. ◄

COMMUNICATING ABOUT ALGEBRA

Speed of cars is usually given in miles per hour. To see the change in speed of the car in Example 5, convert feet per second to miles per hour. Follow these steps on a calculator.

1. 345 ÷ 5280
   The answer is 0.0653409091.

2. Multiply the answer by 3600. The result is 235.2272727.

A car traveling 345 ft/s is traveling approximately 235 mi/h. Discuss the reason for using each step.

TRY THESE

Match each line in the graph with its slope.

1. slope 4

2. slope 0.5

3. slope 2

4. slope −1

5. slope −5

6. slope −3.5

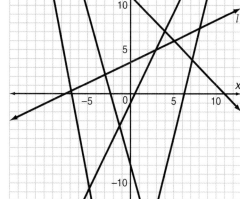

Find the slope of the line passing through the given points.

7. (6, 3), (0, 0)

8. (2, 1), (5, 4)

9. (2, 6), (−8, 6)

10. (1, −5), (4, 10)

11. (4, 7), (4, −1)

12. (−2, −1), (3, −6)

Each line passes through the given points. For each pair, determine the steeper line.

13. line *a*: (3, 1), (6, 8)
line *b*: (−5, 2) (0, 7)

14. line *c*: (−1, 9), (2, 2)
line *d*: (3, 8), (7, 1)

15. line *e*: (4, 12), (12, 4)
line *f*: (−3, −2), (5, −8)

16. CAR AND TRUCK SALES Find the average rate of change for the total number of cars and light trucks (vehicles) sold in the United States during the years given. Include the unit of measure in your answer.

1983: 12,400,000   1993: 14,100,000

17. WRITING MATHEMATICS On a coordinate plane, label the horizontal axis from 0 to 10 seconds. Label the vertical axis from 0 to 2000 ft. Beginning at the point (0, 0), graph two lines representing takeoffs of two airplanes. Using slope, compare the takeoffs of the planes.

PRACTICE

Find the slope of the line passing through the given points.

1. (5, 4), (1, 2)

2. (1, 2), (4, 5)

3. (5, 1), (2, 9)

4. (0, 0), (−7, −3)

5. (1, −2), (2, 1)

6. (1.6, −3), (1.6, 6)

7. (−7, 14), (−1, −4)

8. (4, 2), (−2, 2)

9. (−3, 10), (−3, 13)

Each line passes through the given points. For each pair, name the steeper line.

10. line *a*: (6, 2), (12, 16)   line *b*: (−3, 0), (1, 6)

11. line *c*: (−1, 7), (3, 3)   line *d*: (0, −5), (8, −1)

12. line *k*: (1, 2), (5, 7)   line *l*: (−4, 1), (0, 6)

13. line *p*: (−2, 8), (1, 0)   line *q*: (2, 8), (6, 0)

Choose two points on each line and use them to determine the slope of the line.

14.

15.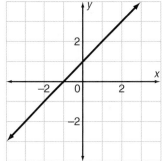

16.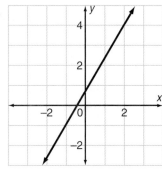

Find the average rate of change for the given data. Include the unit of measure.

17. homes with cable TV

        1970: 4,500,000   1988: 45,000,000

18. fishing licenses issued

        1980: 26,100,000   1988: 30,400,000

19. daily personal trash (Round to the nearest hundredth.)

        1960: 2.9 lb   1988: 3.5 lb

20. U.S. auto registrations (Round to the nearest whole number.)

        1960: 61,700,000   1987: 139,000,000

Visualize the line containing each pair of points. Determine whether the slope of the line is _positive_, _negative_, _zero_, or _undefined_.

21. $(7, 8), (-2, 5)$ **22.** $(-4, -1), (0, -5)$

23. $(12, 4), (-2, -6)$ **24.** $(-9, 15), (2, -3)$

25. $(-6, -3), (13, -3)$ **26.** $(-14, -1), (-5, 1)$

On a set of coordinate axes, plot each point. Then use the slope to graph the line.

27. $(0, 0)$; slope $= 1$ **28.** $(3, 5)$; slope $= -\dfrac{1}{2}$

29. $(2, 1)$; slope $= 2$ **30.** $(0, 3)$; slope $= -1$

ENGINEERING The highest mountain on the Atlantic Coast of the United States may one day be on Staten Island, New York. The mountain is a landfill and is being constructed from garbage.

31. The diagram at the right shows that when the mountain is completed, it will be 435 ft high with a slope of $\dfrac{1}{3}$. What is the distance x from the base of the mountain to the point directly below the summit?

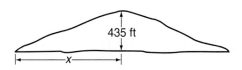

32. The landfill mountain was originally intended to have a slope of $\dfrac{2}{5}$. However, the slope was reduced to $\dfrac{1}{3}$ because of fear of landslides. If the distance x that you determined in Exercise 31 had remained constant (as was planned), what would have been the height of the mountain?

33. AZTEC PYRAMIDS The largest pyramid in terms of volume is Quetzalcoatl outside of Mexico City, Mexico. It has a square base and a height of 177 ft. The slope of each face is approximately 0.25. What is the length of each side of the base?

34. WRITING MATHEMATICS The slope of a certain bicycle wheel spoke is 2 when the bicycle is not in motion. What will the slope of the same spoke be when the wheel has made a one-half turn? a complete turn? Use diagrams to explain your reasoning.

EXTEND

AIRLINE FLIGHTS Use the graph for Exercises 35–38.

Airline Flights
(in millions of takeoffs and landings at U.S. airports)

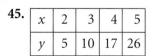

35. Were there more flights each year than during the previous year? Explain your answer in terms of the slope of each section of the graph.

36. During which time interval was the increase in the number of flights the greatest? Use slope to explain.

37. What was the approximate average rate of change per year in the number of flights from 1982–1992?

38. Was the rate of change greater during the first or second half of the period covered by the graph? Explain your answer using slope.

39. **HIGHWAY CONSTRUCTION** A steep section of highway has a slope of 0.07, which is called a 7% grade on highway signs. Over a horizontal distance of 0.75 mi, the road rises to an elevation of 2500 ft. What is the elevation at the beginning of the section of roadway? (*Hint:* Draw a diagram and convert distances to the same units.)

THINK CRITICALLY

40. **MISSING AREA** The square below is divided into two congruent trapezoids and two congruent triangles. The rectangle is formed by rearranging the four figures.

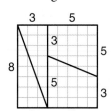

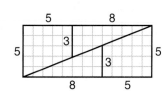

a. What is the area of the square? the rectangle?

b. Moving the trapezoids and triangles does not change their area. Which figure is incorrectly drawn? Use slope to explain.

MIXED REVIEW

Evaluate each expression when $a = -3$, $b = -5$, and $c = 4$.

41. $a^2 - 5ab$

42. $\dfrac{ab - c^2}{7}$

43. $-\dfrac{b}{5}(a^2 c)$

Decide from each table of values whether the graph of the function will be *linear* or *nonlinear*.

44.

x	0	1	2	3
y	3	5	7	9

45.

x	2	3	4	5
y	5	10	17	26

46.

x	2	4	6	8
y	1	7	17	31

47. **STANDARDIZED TESTS** A shipping company limits the loaded weight of a particular size carton to 23 lb. If the empty carton weighs 3.5 lb, what is the greatest number of 2.5-lb books that can be shipped in the carton?

A. 6 **B.** 7 **C.** 8 **D.** 9

Saving energy is a major concern for utility companies. They hire customer service representatives to work closely with customers, helping them to save money and avoid seasonal fluctuations in their utility bills.

One service that a utility company representative may be able to offer is balanced billing. Under this plan, a family's average monthly expense for the previous year is determined and that amount is billed monthly throughout the next year. Any needed adjustments or corrections are made at the end of the year.

Decision Making

The table below shows a family's combined monthly electric and natural gas bills for one year.

Month	Billing	Month	Billing	Month	Billing
January	$321	May	$103	September	$197
February	$356	June	$157	October	$177
March	$277	July	$288	November	$244
April	$154	August	$387	December	$301

1. Do the data appear to represent a linear function? How did you decide?

2. What do you think might account for the fluctuations in the monthly expenses?

3. If a customer service representative offered this family the balanced billing option, what would their monthly bill be for that year, based on billings for the year showing in the table? Round to the nearest dollar.

4. If the family chose balanced billing, what would the graph of their monthly expenses look like?

5. What would be the slope of the graph of their monthly expenses? What would that slope indicate?

6.3 Slope-Intercept Form; Parallel and Perpendicular Lines

Explore

You have already learned about slope, one characteristic of a graph of a line. Another characteristic of a graph of a line is the y-intercept. The y-intercept is the y-coordinate of the point where a graph crosses the y-axis. At this point, the value of the x-coordinate is zero. In the graph at the right, the y-intercept is 1.

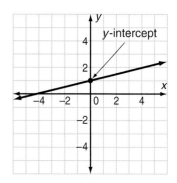

Equation	In "$y =$" Form	Slope	y-Intercept
$y = 3x + 4$	$y = 3x + 4$		
$2y = 8x - 10$			
$x + y = 6$			
$y = x$			

1. For each equation, follow these steps to complete the table.

   a. If it is not already done, solve for y in terms of x.

   b. Graph the equation using your graphing utility.

   c. Set your graphing utility so integers are displayed when using the TRACE feature. Find and record two points along the graph.

   d. Use the two points to determine the slope of the graph.

   e. Find the y-intercept from the graph.

2. Describe the patterns you see in the chart.

Build Understanding

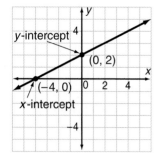

The y-coordinate of the point where a graph crosses the y-axis is called the **y-intercept**. The x-coordinate of the y-intercept is 0. Likewise, the x-coordinate of the point where a graph crosses the x-axis is called the **x-intercept**. The y-coordinate of the x-intercept is 0. On the graph shown at the left, the x-intercept is the point $(-4, 0)$ and the y-intercept is the point $(0, 2)$.

You can use the x- and y-intercepts to graph any linear equation easily.

EXAMPLE 1

Use the x- and y-intercepts to graph $x - y = -6$.

Solution

First determine the x- and y-intercepts.

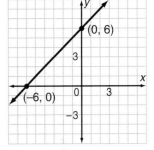

To find the y-intercept, let $x = 0$.

$$x - y = -6$$
$$-y = -6$$
$$y = 6$$

The y-intercept is the point $(0, 6)$.

To find the x-intercept, let $y = 0$.

$$x - y = -6$$
$$x = -6$$

The x-intercept is the point $(-6, 0)$.

To graph the equation, plot the points $(0, 6)$ and $(-6, 0)$ and connect them. ◄

You have already found the slope of a line using two points on a graph. As you saw in Explore, you can determine slope by examining an equation that has been solved for y in terms of x. When an equation is in the form $y = mx + b$, both m and b are real numbers.

$$y = mx + b$$

slope y-intercept

SLOPE-INTERCEPT FORM

A linear equation of the form $y = mx + b$ is in *slope-intercept form*, where m is the slope and b is the y-intercept of the graph.

The slope-intercept form allows you to visualize or draw the graph of the linear equation quickly.

EXAMPLE 2

Find the slope and y-intercept of $3x = 4y + 8$. Then graph the line.

Solution

Solve the equation for y in terms of x.

$$4y + 8 = 3x \quad \text{Write the equation with the } y\text{-term on the left.}$$
$$4y = 3x - 8 \quad \text{Subtract 8 from both sides.}$$
$$y = \frac{3}{4}x - 2 \quad \text{Divide both sides by 4.}$$

The equation is now in slope-intercept form. The slope m of the line is $\frac{3}{4}$. The y-intercept b is -2.

Use the values for m and b to draw the graph. Plot the y-intercept at $(0, 2)$. Then use the slope to locate another point on the graph.

The slope is $\frac{3}{4}$. Move 3 units up and 4 units right, since the slope is positive. Plot the point and draw a line through the two points.

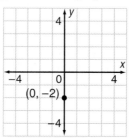

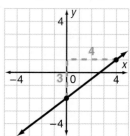

If you know the slope of a line and its y-intercept, you can write an equation for the line and draw its graph.

EXAMPLE 3

Write an equation for each line with the given slope and y-intercept. Graph the lines on the same axes.

line a: slope 2.5   line b: slope 2.5
     y-intercept -6 y-intercept -2

Solution

Use the slope-intercept form for each line.

$y = mx + b$ $y = mx + b$
$y = 2.5x + (-6) \quad m = 2.5,$ $y = 2.5x + (-2) \quad m = 2.5,$
$y = 2.5x - 6 \quad\quad b = -6$ $y = 2.5x - 2 \quad\quad b = -2$

Plot the y-intercept and then use the slope to graph each line. The slope 2.5 equals $\frac{5}{2}$ so the rise is 5 and the run is 2.

In Example 3, lines *a* and *b* will never intersect because $2.5x - 2$ will never equal $2.5x - 6$. Lines in the same coordinate plane that do not intersect are called **parallel lines**. Lines *a* and *b* are parallel. They have the same slope but different *y*-intercepts.

> **Two nonvertical lines are *parallel* if and only if they have the same slope.**

All horizontal lines are parallel with slope 0. All vertical lines are parallel, even though the slope of a vertical line is undefined.

Two lines in the same plane may also intersect. Lines that intersect to form right angles are called **perpendicular lines**. You can determine whether two lines are perpendicular by examining their slopes.

EXAMPLE 4

Write an equation for each of the two perpendicular lines with the given slopes and *y*-intercepts. Graph the lines on the same axes.

line *c*: slope 3   line *d*: slope $-\dfrac{1}{3}$

        y-intercept −4   *y*-intercept 3

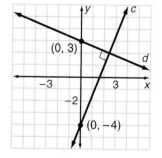

Solution

$$y = mx + b$$
$$y = 3x + (-4) \quad \begin{matrix} m = 3, \\ y = -4 \end{matrix}$$

$$y = mx + b$$
$$y = -\dfrac{1}{3}x + 3 \quad \begin{matrix} m = -\dfrac{1}{3}, \\ y = 3 \end{matrix}$$

Plot the *y*-intercept and use the slope to graph each line. ◄

In Example 4, the product of the slopes of lines *c* and *d* is $3\left(-\dfrac{1}{3}\right) = 1$. Notice that $-\dfrac{1}{3}$ is the negative reciprocal of 3.

> **Two lines are *perpendicular* if the slopes of the lines are negative reciprocals of each other. That is, the product of their slopes is −1.**

So, lines *c* and *d* are perpendicular.

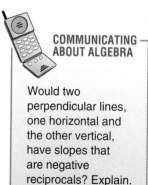

COMMUNICATING ABOUT ALGEBRA

Would two perpendicular lines, one horizontal and the other vertical, have slopes that are negative reciprocals? Explain.

You can use the slope of a line and its y-intercept to model in real world problems about population growth.

EXAMPLE 5

POPULATION GROWTH From 1950 to 1990 the average change in the population of North America was approximately 2.8 million people per year. In 1990 this population was about 278 million. Write and graph the equation that models the population growth P.

Solution
Let $x = 0$ represent the year 1950 so that $x = 40$ represents 1990. Then the total population in 1950 is the P-intercept. Find b.

$$P = mx + b$$
$$278 = 2.8(40) + b \qquad \text{Substitute known values for 1990.}$$
$$278 = 112 + b \qquad \text{Simplify.}$$
$$166 = b \qquad \text{Solve for } b.$$

So, the linear equation $P = 2.8x + 166$ models the population growth over the period 1950 to 1990. The graph is shown at the left. ◄

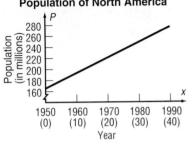

Population of North America

TRY THESE

Identify the x-intercept and y-intercept of each line.

1. 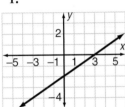   **2.** 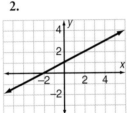   **3.** 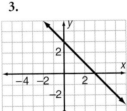   **4.**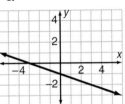

For the given intercepts, graph the line.

5. x-intercept: -2
    y-intercept: -4

6. x-intercept: -1
    y-intercept: 3

7. x-intercept: 5
    y-intercept: -5

8. x-intercept: 2
    y-intercept: 3

Find the slope and y-intercept of each. Then graph the line.

9. $y - 4x = 8$ **10.** $2y + 3x = 6$ **11.** $5x - 2y = 10$ **12.** $8 - x = 2y$

Determine the negative reciprocal of each number.

13. -20 **14.** $\dfrac{3}{5}$ **15.** 1.25 **16.** $-\dfrac{1}{6}$

Determine whether the graphs will be *parallel*, *perpendicular*, or *neither*.

17. $y = \dfrac{2}{3}x - 6$
    $2x - 4 = 3y$

18. $y = \dfrac{1}{2}x + 4$
    $4x - 3y = 12$

19. $2y = 10x - \dfrac{2}{5}$
    $\dfrac{1}{5}x + y = 3$

Write an equation for each line with the given slope and *y*-intercept.

20. slope 4, *y*-intercept −3

21. slope $\frac{1}{3}$, *y*-intercept 1

22. slope −2, *y*-intercept $-\frac{1}{2}$

23. slope $\frac{4}{5}$, *y*-intercept 0

24. RECYCLING In 1960 approximately 8 million tons of solid trash were recycled. By the year 2000, this amount is expected to grow to 55 million tons. The equation $T = 1.175x + 8$ gives an estimate of the growth of recycling over the time period. *T* represents the total number of tons in millions and *x* the year, where 1960 is represented by 0 and 2000 is represented by 40. In the equation, what is the slope? What is the *T*-intercept? What does each value represent?

25. WRITING MATHEMATICS The graph of a line passes through the points (0, 6) and (8, −10). Explain how you would find the *y*-coordinate of the point (20, *y*) on the same line. (*Hint:* Since all three points are on the same line, the slope from (0, 6) to (8, −10) is the same as the slope from (0, 6) to (20, *y*).)

PRACTICE

For the given intercepts, graph the line.

1. *x*-intercept: 4
y-intercept: 3

2. *x*-intercept: −3
y-intercept: −3

3. *x*-intercept: 4
y-intercept: 6

4. *x*-intercept: 8
y-intercept: −5

Find the slope and *y*-intercept of each. Then graph the line.

5. $y - 7 = -x$

6. $5y = -2x + 10$

7. $x + 2y = 3$

8. $5x - y = 15$

Match each equation with its graph.

9. $y + 1 = x$

10. $y - x = 1$

11. $x + y = 1$

12. $y = -x - 1$

a.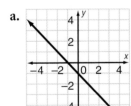

b.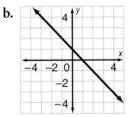

c.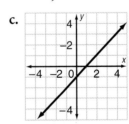

d.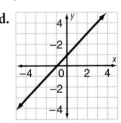

Write an equation for each line with the given slope and *y*-intercept.

13. slope −2.5, *y*-intercept 0

14. slope 3.1, *y*-intercept −0.5

15. slope $-\frac{7}{8}$, *y*-intercept $\frac{3}{5}$

16. slope 9.1, *y*-intercept −1.7

Determine the slope of a line parallel to the line of the given equation.

17. $y = 3x - 1$

18. $7y = 5x - 3$

19. $x - 3y = 9$

20. $4 - 2x = 6y$

Determine the slope of a line perpendicular to the line of the given equation.

21. $y = x$ 　　　　 **22.** $y = -\dfrac{1}{2}x + 1$ 　　 **23.** $2y - x = 5$ 　　　 **24.** $3x + 2y = 4$

25. TENNIS SHOE SALES During the period 1983 through 1988, the total value of tennis shoe sales in the United States rose from \$1.8 billion to \$4.3 billion. The equation $S = 0.5x + 1.8$ approximates the yearly change in sales. S represents total sales in a given year and x represents the year where $1983 = 0$ and $1988 = 5$.

　a. What does the slope of the graph indicate? What does the S-intercept indicate?

　b. Suppose that in 1986, the actual sales of tennis shoes in the United States had totaled \$2.93 billion. By how much would the estimate given by the equation $S = 0.5x + 1.8$ have differed from the actual amount?

26. WRITING MATHEMATICS Describe the graphs of equations containing each of the following pairs of points.

　a. $(0, -3)$ and $\left(0, 4\,\dfrac{1}{2}\right)$ 　　　　　　**b.** $(-2.5, 0)$ and $(2.5, 0)$

EXTEND

Write the slope-intercept equation of the line that will be parallel to the given line and pass through the given point.

27. $y = 3x - 1; (0, 0)$ 　　　 **28.** $x - 3y = 9; (6, 3)$ 　　　 **29.** $2y = -4x + 7; (1, -5)$

Write the slope-intercept equation of the line that will be perpendicular to the given line and pass through the given point.

30. $y = x; (1, 3)$ 　　　　 **31.** $2y = 3x + 5; (-3, 0)$ 　　　 **32.** $\dfrac{1}{2}x = 1 - y; (1, -5)$

33. GEOMETRY In a parallelogram both pairs of opposite sides are parallel. Yoko wrote these four equations to graph the sides of a parallelogram.

$$y = 3x + 2 \qquad y = 2x - 2 \qquad y = -\frac{1}{3}x + 1 \qquad y = -\frac{1}{2}x - 2$$

　a. Without graphing, how could you determine whether these four equations can be the sides of a parallelogram?

　b. Can Yoko's equations form the sides of a parallelogram? Explain.

34. ACCIDENTS Death rates from all types of accidents have dropped dramatically in the United States. The equation $A = -0.5x + 82.9$ models the change in accidental death rates A per hundred thousand people for the years 1900 through 1992, where 1900 is represented by $x = 0$. What information can you draw from the equation?

SUBMARINE MANEUVERS The graph describes the rise of a submarine.

35. What does the y-intercept represent?

36. If the graph were extended, what would the x-intercept represent?

37. As x increases, y decreases. Will graphs of such relationships always have negative slope? Explain your reasoning.

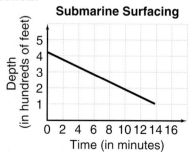

Submarine Surfacing

Depth (in hundreds of feet) vs. Time (in minutes)

THINK CRITICALLY

On a coordinate plane, copy line *n* from the graph at the right. On the same coordinate plane, draw line *n'* in the position line *n* would occupy after it is rotated counterclockwise 90° about the origin. Use your graph to answer Exercises 40–43.

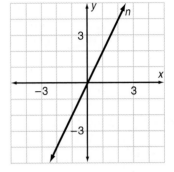

38. What is the slope of line *n*?

39. What are the coordinates of the point on line *n'* that corresponds to point (2, 4) on line *n*?

40. What is the slope of line *n'*?

41. What is the relationship between the slopes of lines *n* and *n'*? What is the product of the slopes?

42. Describe each pair of lines.

    a. Two lines have the same value for *m*, but each has a different value for *b* in the equation $y = mx + b$.

    b. Two lines have the same value for *b*, but each has a different value for *m*.

43. Write an equation in slope-intercept form and graph it on a graphing utility. Describe the steps you could use to create a rectangle by writing three more equations.

MIXED REVIEW

Several students were asked to estimate the ratio of the amount of free time they spend reading versus watching television. Their ratios were as follows: Jaclyn, 1:3; Sara, 1:1; Connie, 5:3; Dennis, 1:2; Edgar, 7:4, Samira, 2:3; Andie, 3:4.

44. During one day of free time, which student spends the greatest amount of time reading?

45. Approximately what percent of the students spend more time watching TV than reading?

46. If you telephoned Samira during her free time, what is the probability that you would find Samira reading?

47. In 6 hours of free time, approximately how much time does Jaclyn estimate that she watches TV?

Name the property illustrated.

48. $-4(5 + 6) = -4(5) + -4(6)$

49. $-18 + (7 + 45) = (-18 + 7) + 45$

50. $6(-12) = -12(6)$

51. $8 + (-8) = 0$

52. STANDARDIZED TESTS   Which is the correct solution for the following equation?

$$\frac{1}{2}x - 7 = 4$$

   A. 21   **B.** 14   **C.** 7   **D.** 42

People use various appliances such as washers, dryers, refrigerators, and air conditioners in their homes. An appliance salesperson helps consumers make decisions about purchasing and using their appliances in an energy-efficient way. A good salesperson must know how the appliances can conserve energy.

Sold according to their BTU (British Thermal Units) rating, air conditioners not only cool but also remove moisture from the air. The air conditioner with the right BTU rating will remain on just long enough to both cool and dehumidify a room. If an air conditioner has too low a rating, it may not cool sufficiently. If the rating is too high, the air conditioner may shut off before removing the humidity.

Many salespeople use the formula known as "W.H.I.L.E. divided by 60" to compute the appropriate BTU rating for a given room. In this formula, W represents the width of the room; H its height; I its level of insulation (either 10 for well insulated or 18 for poorly insulated); L the length of the room; and E the exposure (16 for north, 17 for east, 18 for south, and 20 for west).

Decision Making

1. Write the formula above algebraically.

2. To the nearest 500, what BTU rating is appropriate for a poorly insulated room facing south that is 18 ft long, 20 ft wide, and 9 ft high?

Room requirements cannot always exactly match air conditioner BTU ratings. An air conditioner with a BTU rating within 5% of the room requirement is recommended.

3. For a BTU rating of R, what equation describes L_1, the lower limit for the recommended BTU range?

4. What equation describes L_2, the upper limit for the recommended BTU range?

5. Are the graphs of the two equations parallel? Explain your answer.

6. Work with a small group. Find the measurements and other information you need to compute the BTU rating of your classroom or another room of your choosing.

6.4 The Point-Slope Form of a Linear Equation

Explore/Working Together

● First, answer each of the questions below on your own. Use any of the following techniques: visualization, a coordinate plane, a table of values, or a graphing utility.

After you have finished, compare and discuss your answers with those of other members of your group. If there are answers where you cannot agree, compare the techniques you used to see whether another technique would be helpful.

1. How many lines can be drawn that have slope 8 and *y*-intercept 3?

2. How many lines have a slope of 7?

3. How many lines can be drawn with *y*-intercept 5?

4. How many lines can be drawn through the point $(6, -7)$?

5. How many lines can be drawn through the points $(1, 2)$ and $(4, 17)$?

6. How many lines can be drawn through the points $(1, 1)$, $(3, 3)$, and $(4, 17)$?

7. How many lines can be drawn perpendicular to the line $y = 2x + 5$?

8. How many lines can be drawn through the point $(1, -2)$ that are perpendicular to the line $y = x + 1$?

9. How many lines are both parallel and perpendicular to the line $y = 2x - 9$?

10. How many lines can be drawn with slope 6 through the point $(-4, 1)$?

11. Reread Questions 1, 5, and 10. For which of the lines described could you write an equation in the form $y = mx + b$ immediately from the information given?

SPOTLIGHT ON LEARNING

WHAT? In this lesson you will learn
- to write a linear equation and to graph, given the slope and any one point, or given two points.
- to write a linear equation in point-slope form.

WHY? Writing linear equations in point-slope form can help you solve problems about population density, newspaper sales, and weather.

Build Understanding

● The equation of a line in slope-intercept form is $y = mx + b$. If you know the slope and any point, then you know three of the four variables in the equation. So you can find the missing variable, the *y*-intercept *b*.

Therefore, given the slope and a point the line passes through, you can determine the equation of the line.

EXAMPLE 1

Write an equation of the line in slope-intercept form with slope 3 that passes through the point (–2, 4). Graph the line.

Solution

Examine the table. You want to write the equation in the form $y = mx + b$. First, solve for b.

Variable	y	m	x	b
Value	4	3	–2	unknown

$$y = mx + b \qquad \text{Use the slope-intercept form.}$$
$$4 = 3(-2) + b \qquad \text{Substitute 4 for } y, -2 \text{ for } x, \text{ and 3 for } m.$$
$$4 = -6 + b \qquad \text{Simplify.}$$
$$10 = b \qquad \text{Solve for } b.$$

Once you know the value of b, you can write an equation for the line.

$$y = 3x + 10$$

The graph is shown at the right. The graph passes through the given point (–2, 4), and through (0, 10), the y-intercept, and has slope 3.

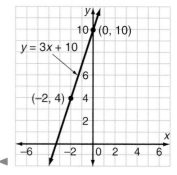

You can also write an equation in slope-intercept form if you know two points on the graph.

EXAMPLE 2

Write an equation in slope-intercept form of the line that passes through the points (4, –2) and (8, 1). Graph the line.

Solution

You can draw the graph using the two points. To write the equation, use the coordinates of the two points to find the slope.

$$m = \frac{y_2 - y_1}{x_2 - x_1} = \frac{1 - (-2)}{8 - 4} = \frac{3}{4}$$

Substitute for m, x, and y to find b.

$$y = mx + b \qquad \text{Slope-intercept form.}$$
$$-2 = \frac{3}{4}(4) + b \qquad \text{Substitute 4 for } x, -2 \text{ for } y, \text{ and } \frac{3}{4} \text{ for } m.$$
$$-2 = 3 + b \qquad \text{Simplify.}$$
$$-5 = b \qquad \text{Solve for } b.$$

Write the equation.

$$y = \frac{3}{4}x - 5$$

Notice on the graph that the slope is $\frac{3}{4}$.

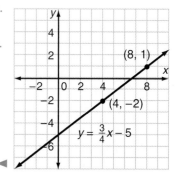

Using one point and the slope, you can write a linear equation to model a real world situation.

EXAMPLE 3

GREENHOUSE EFFECT In 1896 the Swedish chemist Svante Arrhenius coined the term "greenhouse effect" to describe the increase in heat-trapping gases in the atmosphere.

One such gas, carbon dioxide (CO_2), is closely monitored today. From 1970 through 1990, the concentration of CO_2 increased by approximately 1.7 parts per million (ppm) per year. By 1990 the concentration was 355 ppm. Write an equation in slope-intercept form that relates the concentration of carbon dioxide y to the year x.

Solution

Assume a constant increase in CO_2 levels. Since the concentration changed 1.7 parts per million per year, the average rate of change is 1.7. Therefore the slope m is 1.7.

Let the year 1970 be represented by $x = 0$. Then 20 represents the year 1990. Since 55 represents the concentration of CO_2 during year 20, a point on the graph is (20, 355).

$y = mx + b$	Use the slope-intercept form.
$355 = 1.7(20) + b$	Substitute 20 for x, 355 for y, and 1.7 for m.
$355 = 34 + b$	Simplify.
$321 = b$	Solve for b.
$y = 1.7x + 321$	Write the equation in slope-intercept form. ◄

You can use the definition of slope $m = \dfrac{y_2 - y_1}{x_2 - x_1}$ to write the *point-slope form of a linear equation.* Let (x_1, y_1) be a known point on the line. Let (x_2, y_2) represent any other point (x, y) on the line.

> **POINT-SLOPE FORM OF A LINEAR EQUATION**
>
> The *point-slope form* of the equation of a line is
>
> $$(y - y_1) = m(x - x_1),$$
>
> where (x_1, y_1) is a known point on the line and (x, y) is any other point on the line.

If you know the coordinates of two points or the slope and the coordinate of one point, you can write an equation of a line in point-slope form.

EXAMPLE 4

Write the point-slope form of the equation of a line that passes through the points $(4, -2)$ and $(-2, 10)$, then convert it to slope-intercept form.

Solution

First, find the slope.

$$m = \frac{y_2 - y_1}{x_2 - x_1} = \frac{10 - (-2)}{-2 - 4} = \frac{12}{-6} = -2$$

Use the slope and either of the given points.

$y - y_1 = m(x - x_1)$	Use the point-slope form.
$y - (-2) = -2(x - 4)$	Substitute 4 for x, -2 for y.
$y + 2 = -2(x - 4)$	Simplify.

To convert the equation to slope-intercept form, solve for y.

$y + 2 = -2(x - 4)$	
$y + 2 = -2x + 8$	Use the distributive property.
$y + 2 - 2 = -2x + 8 - 2$	Subtract 2 from both sides.
$y = -2x + 6$	Write in slope-intercept form. ◀

Linear equations in point-slope form can model real world situations.

EXAMPLE 5

ENERGY SOURCES The percent of energy used by industries worldwide that comes from oil has been decreasing. In 1973, industries got 47% of their energy from oil. By 1990, the percentage had decreased to 38%. Write a linear equation that relates the percent of energy from oil y to the year x. Let the year 1970 be represented by $x = 0$.

Solution

Use the information above to write two points, $(3, 47)$ and $(20, 38)$.

Find the slope of the line.

$$m = \frac{38 - 47}{20 - 3} = -\frac{9}{17}$$

Write an equation in point-slope form, using either of the points.

$$y - 47 = -\frac{9}{17}(x - 3)$$

The equation can also be written in slope-intercept form.

$y - 47 = -\dfrac{9}{17}x + 1\dfrac{10}{17}$	Use the distributive property.
$y = -\dfrac{9}{17}x - 48\dfrac{10}{17}$	Solve for y. ◀

The negative slope reflects the fact that the percent of energy obtained from oil has been decreasing.

Write an equation in slope-intercept form of the line that has the given slope and passes through the given point.

1. $m = 0.5$, $(9, 8)$ **2.** $m = 2$, $(-4, 1)$ **3.** $m = -2$, $(1, 5)$

4. $m = \frac{1}{3}$, $(3, 3)$ **5.** $m = 0$, $(-2, -6)$ **6.** $m = 1.5$, $(4, 2)$

Write the slope-intercept form of the equation of the line that passes through the given points.

7. $(3, 1), (5, 4)$ **8.** $(3, 0), (-3, -6)$ **9.** $(-2, -1), (-4, -4)$ **10.** $(1, 4), (4, -2)$

Use the given slope and the given point to write the equation in point-slope form. Then rewrite it in slope-intercept form.

11. $m = -3$, $(1, 4)$ **12.** $m = 4$, $(-2, -2)$ **13.** $m = \frac{1}{2}$, $(-2, 1)$ **14.** $m = \frac{2}{3}$, $(-2, -3)$

15. FARM WORKERS From 1920 through 1960, the percent of Americans working on farms decreased by about 0.5% per year. In 1940, approximately 17% of the work force was on farms. Write an equation that relates the percent of Americans working on farms y to the year x. Let $x = 0$ represent 1900.

16. WRITING MATHEMATICS Apply the equation you wrote for Exercise 15 to the year 1990. How can you explain the results?

Solve each equation for b.

1. $-0.5 = -2(-1) + b$ **2.** $7 = 3(-2) + b$ **3.** $4.5 = 2(-3) + b$ **4.** $0 = -7(-1) + b$

Write an equation of the line that has the given slope and passes through the given point. Use slope-intercept form.

5. $m = -4$, $(0, 2)$ **6.** $m = \frac{1}{2}$, $(1, -4)$ **7.** $m = -1.5$, $(-5, 5)$

8. $m = \frac{5}{2}$, $(-4, 2)$ **9.** $m = 6$, $(3, -0.5)$ **10.** $m = -0.1$, $(2, 3)$

For each graph, write an equation in point-slope form. Then rewrite it in slope-intercept form.

11.

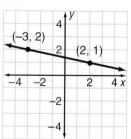

12.

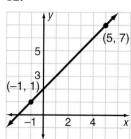

13.

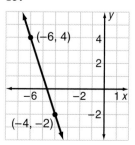

14.

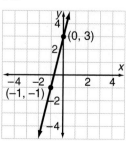

15. **POPULATION DENSITY** The population density of a country is the ratio of the total population to the total area of the country. The population density of the United States has increased at a fairly constant rate during the twentieth century. In 1930, the density was approximately 40 people per square mile. By 1990, the density had increased to 70. Using $x = 0$ as 1900, write an equation in slope-intercept form that relates the population density of the United States y to the year x.

16. **NEWSPAPERS** From 1983 through 1991, the number of evening newspapers published in the United States decreased at an average rate of approximately 30 per year. In 1988 there were 1141 evening newspapers. Let 0 represent 1980. Write an equation in slope-intercept form that relates the number of evening newspapers y to the year x.

17. **WIND CHILL** The combination of cold air and wind makes the temperature feel colder than it actually is. The result of this combination is called the "wind chill." At a constant wind speed of 20 mi/h, for example, a temperature of 25°F feels like –3°F. A temperature of 5°F feels like –31° F. Use the given information to write a linear equation in slope-intercept form that relates wind chill y to the actual temperature x for a constant wind speed of 20 mi/h.

18. **COLD WEATHER** The label on a ski parka you have just bought claims that it will protect you to a temperature of –20°F. If the wind speed is 20 mi/h and the wind chill is –46°F, can you feel safe wearing the parka?

19. **WRITING MATHEMATICS** Make up a wind-chill problem of your own that can be solved using the equation you wrote for Exercise 17.

EXTEND

Write an equation in slope-intercept form for the line that passes through the given points.

20. $(-2, 8)$, $(7, 8)$ 21. $(-6, 1)$, $(6, -2)$ 22. $(3, -4)$, $(-9, 2)$

23. $(2, 0)$, $(1, -8)$ 24. $(2.5, -1)$, $(0.5, 1)$ 25. $(5, 0)$, $(3, -2.5)$

26. $(-2, -5.5)$, $(1.5, 1.5)$ 27. $(6, -1)$, $(4, -2.5)$ 28. $(-4.2, -4)$, $(-2.2, 6)$

29. Write an equation in point-slope form for the line that has slope $-\dfrac{7}{2}$ and passes through the point $(-1, 2)$.

30. Write an equation in slope-intercept form that is equivalent to the one you wrote in Exercise 29.

31. **JEWELRY SALES** A jewelry store began selling a style of earring in January 1993. During December 1994 the store sold 185 pairs. During the entire year, sales increased at an average rate of 10 pairs per month.

    a. Write an equation that relates sales y to the number of months the store sells the earrings x. Let 0 represent January 1994.

    b. Assume that sales continued to increase at the same rate. How many pairs were sold in April 1995?

THINK CRITICALLY

32. The equation of a certain line can be written in point-slope form in either of the following ways: $y - 1 = x - 2$ or $y + 5 = x + 4$. From the equations, name two points on the graph of the equation.

33. How would you show algebraically that the two equations in Exercise 32 are equivalent? Solve the equation.

34. Write equations in point-slope form for the lines that are perpendicular to the one in Exercise 32 and pass through the points you found in that problem.

35. Use the equations from Exercise 34. The x-coordinate of a point on the graph is $x = a$. Write the y-coordinate of the point in terms of a.

MIXED REVIEW

Evaluate each expression when $w = 21$, $x = 12$, and $y = 8$. Use a calculator and round each answer to the nearest hundredth.

36. $\sqrt{w} - \sqrt{y}$ **37.** $\sqrt{w - y}$ **38.** $\sqrt{x} + \sqrt{y}$ **39.** $\sqrt{x + y}$

Write the value of each expression.

40. $|4|$ **41.** $\left|-\dfrac{1}{3}\right|$ **42.** $|0|$ **43.** $|-9|$ **44.** $-|-2.4|$

Add or subtract.

45. $-\dfrac{3}{8} + 2\dfrac{4}{5}$ **46.** $-3\dfrac{1}{4} - \left(-8\dfrac{2}{5}\right)$ **47.** $6\dfrac{8}{10} - \left(-2\dfrac{1}{6}\right)$

48. STANDARDIZED TESTS Which is the correct solution of the inequality?

$$x + 18 \geq 3x + 6$$

    A. $x \leq -6$ **B.** $x \leq 12$ **C.** $x \geq -6$ **D.** $x \leq 6$

PROJECT *Connection*   Many people believe environmental problems are beyond their control. Actually, if you *use* natural resources you can *save* natural resources. For this activity, you will need an empty gallon container and a watch or timer.

1. Run a sink faucet and record the time it takes to fill the container. Determine the number of gallons per minute used.

2. Record the time it takes to brush your teeth.

3. Estimate how much water you waste if you leave the faucet on while you brush your teeth.

4. Based on this estimate, determine how much water all the students in your school would waste in a year if they brushed their teeth two times a day and left the water running. What would be the total waste for your whole city or state?

5. Use a similar method to estimate water waste for activities such as dishwashing or shaving.

6. Write a summary of your results and recommendations for your magazine.

Career
Recycling Plant Manager

Landfills are closing throughout the country as they reach capacity. In response, Waste-to-Energy resource recovery plants are becoming more numerous. These plants convert trash into fuel. Operators of these plants must know how much trash they can expect on an average day.

The table shows how the amount of trash produced by an average American grew over a 20-year period.

Year	Pounds, per person per day
1970	3.27
1980	3.65
1985	3.77
1986	3.88
1987	4.01
1988	4.12
1989	4.20
1990	4.30

Decision Making

1. Choose any two rows from the table. Write an equation that estimates the growth in trash y over the years x shown. Let 1960 be represented by $x = 0$.

2. Approximately how many pounds of trash will the average American produce in the year 2003?

3. In what year will the average American produce approximately twice as much trash as was produced in 1970?

4. How much trash will the average American generate this year?

5. How many tons of trash will the students in your school produce this year?

6.5 The Standard Form of a Linear Equation

Explore

- You can use the statistical features of a graphing utility to draw a graph given data.

  On their third test, students scored the following grades: 87, 85, 84, 79, 81, 79, 81, 65, 78, 81, 84, 87, 78, and 81. You can think of each score as an ordered pair of data in which the first value, or x-value, is always 3.

  1. Enter the 14 data points. If your graphing utility lists data in a table, list x-values in the first column and y-values in the second.

  2. Set the range for the viewing window for x from 0 to 5 and for y from 60 to 100.

  3. Draw a scatter plot of the data.

  4. How many points are plotted on the graph? Can you explain why?

  5. Use any two of the data points to determine the slope of the line.

  6. Draw a line through the points.

  7. Would a student's score of 67 be represented on this line? Explain.

  8. Would a score of 87 on the fourth test be on this line? Explain.

  9. Why couldn't this graph be drawn by entering the equation in slope-intercept form?

Build Understanding

- You have studied and applied two forms of linear equations.

  slope intercept: $y = mx + b$ where m is the slope and b is the y-intercept

  point-slope: $y - y_1 = m(x - x_1)$, where (x_1, y_1) are the coordinates of a given point on the line and m is the slope.

  In addition to these two forms of linear equations, a line may also be represented by a linear equation in *standard form*.

  > **STANDARD FORM**
  >
  > The *standard form* of a linear equation is $Ax + By = C$ where A, B, and C are integers and A and B are not both zero.

Since all three forms of a linear equation can represent the same graph, it is possible to convert any one form into another.

EXAMPLE 1

Write $y = \frac{1}{3}x - 2$ in standard form.

Solution

Converting from slope-intercept form to standard form involves placing both variables on the left side of the equation.

$$y = \frac{1}{3}x - 2 \qquad \text{Slope-intercept form.}$$

$$3y = x - 6 \qquad \text{Multiply both sides by 3.}$$

$$-x + 3y = -6 \qquad \text{Subtract } x \text{ from both sides.}$$

The equation is in standard form with $A = -1$, $B = 3$, and $C = -6$. ◄

You can use the standard form to write the equation of a line with undefined slope or a line with a slope of 0.

EXAMPLE 2

Write an equation for the line that passes through

a. the points $(3, 4)$ and $(3, -2)$ **b.** the point $(2, 4)$ and is parallel to the x-axis

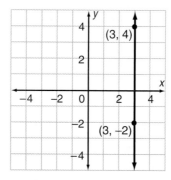

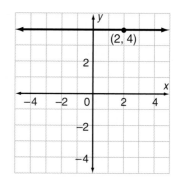

Solution

a. Use the given points to find the slope: $m = \frac{-2 - 4}{3 - 3} = \frac{-6}{0}$.

The slope is undefined. The graph is a vertical line. Examine the graph. You can see that for every y-value the x-value is 3. The equation of the vertical line is $x = 3$. The equation is of the form $Ax + By = C$ where A is 1, B is 0, and C is 3.

b. The slope of a line parallel to the x-axis is 0. The graph of $y = 4$ is a horizontal line 4 units above the x-axis. The equation $y = 4$ is in standard form where $A = 0$, $B = 1$, and $C = 4$. ◄

294 CHAPTER 6 **Linear Functions and Graphs**

The standard form of a linear equation is often the best choice for representing problem conditions.

COMMUNICATING ABOUT ALGEBRA

Discuss why the term By is not represented in the equation $x = 3$ for a vertical line.

EXAMPLE 3

BASKETBALL SCORES In a recent basketball game, a team scored 60 points without making any free throws. The 60 points were a combination of 2-point goals and 3-point goals.

a. Write an equation that represents the different combinations of 2- and 3-point goals that could have been scored.

b. Graph the equation and find several combinations of 2- and 3-point goals that could have been scored.

Solution

a. Use the standard form $Ax + By = C$. Assign variables and constants as follows.

> A: points for each 2-point goal scored (= 2)
>
> x: number of 2-point goals scored
>
> B: points for each 3-point goal scored (= 3)
>
> y: number of 3-point goals scored
>
> C: total points scored (= 60)

Write the equation.

$$2x + 3y = 60$$

b. Use the standard form to find the x- and y-intercepts.

$$2x + 3y = 60$$
$$2x = 60 \quad \text{Let } y = 0.$$
$$x = 30$$

The x-intercept is 30.

$$2x + 3y = 60$$
$$3y = 60 \quad \text{Let } x = 0.$$
$$y = 20$$

The y-intercept is 20.

From the graph at the right, you can read possible combinations.

Point Combinations

2-point goals, x	0	4	9	15	18	24	30
3-point goals, y	20	18	14	10	8	4	0

Note that every number in the table is an integer. In this problem, only integral values for x and y make sense. ◄

The form of the equation you decide to use depends on the purpose.

FORM	USES
Slope-intercept	Graphing using the slope and *y*-intercept Graphing with a graphing utility Writing equation from slope and *y*-intercept
Point-slope	Writing equation either when one point and slope or two points are known
Standard Form	Writing equation of a line with undefined slope (vertical line); solving certain applied problems

TRY THESE

Write each equation in standard form.

1. $y = 8x + 5$

2. $2y = 5x - 1$

3. $3x = 2y$

4. $y = \dfrac{1}{2}x + 4$

5. $y - 9 = 0$

6. $y - 2 = 3(x - 4)$

7. $x + 7 = 0$

8. $y = -\dfrac{2}{3}x + \dfrac{1}{2}$

9. $3y = \dfrac{1}{2}x - 2$

Write an equation in standard form for the line that passes through the given points.

10. $(-1, 3)$, $(3, 5)$

11. $(2, 9)$, $(4, 7)$

12. $(-1, 1)$, $(-4, -2)$

13. $(-3, -1)$, $(-3, 7)$

14. $(6, -2)$, $(-2, -2)$

15. $(-3, -3)$, $(3, 3)$

Write an equation for each graph.

16.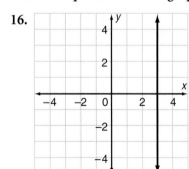

17.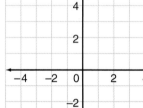

18.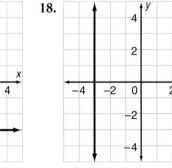

19. RECORDING TIMES Sam wants to transfer his music onto higher-quality tapes. He plans to use both 60- and 90-min tapes and has to transfer 1260 min of music.

   a. Write an equation in standard form that represents the different combinations of 60- and 90-min tapes Sam could use.

   b. Graph your equation from 19a. Copy the table at the right. Use the graph to complete the table.

60-minute tapes, *x*	0	6			
90-minute tapes, *y*			6	4	0

20. WRITING MATHEMATICS Write any linear equation in slope-intercept form. Then convert the equation into standard form. Use whichever form of the equation you feel makes it easier to sketch a graph. Write an explanation of why you chose the form you did.

PRACTICE

Use the given information to write an equation in standard form.

1. $m = -1$, $(-2, 2)$ **2.** $m = \frac{1}{2}$, $(4, 1)$ **3.** $(3, 3)$, $(-1, -3)$

4. $(0, 2)$, $(5, 7)$ **5.** $m = -2$, $(8, 2)$ **6.** $m = -\frac{1}{2}$, $(-2, 1)$

7. $(0, -3)$, $(3, -3)$ **8.** $m = 0$, $(4, 2)$ **9.** $(-1, -1)$, $(-2, -3)$

10. $(4, -5)$, $(4, 7)$ **11.** $(-7, 0)$, $(3, 0)$ **12.** $(0, 1)$, $(0, 11)$

13. AMUSEMENT PARK The theme park that you and your friends are visiting has two kinds of rides: those costing $1 and those costing $2.

     a. Write an equation that represents the different combinations of $1-rides x and $2-rides y available to you if you have $18 to spend.

     b. Graph the equation in 13a. Copy the table. Use your graph to complete the table.

x	0	2	4	6		
y			7		3	0

14. WRITING MATHEMATICS Use the price list at the right. Select two items and write an equation for the combinations of these items you could buy if you had $24 to spend. Explain each step you use. Define each term in the equation.

Price List	
Avocado $1.00	Pineapple $3.00
Mango $2.00	Watermelon $4.00

EXTEND

Write each equation in standard form.

15. $y = -0.2x + 1.5$ **16.** $y = 1.4x - 2.8$ **17.** $y = 3.6x + 8$

18. $y = 0.25x - 1.25$ **19.** $y = -1.35x + 0.15$ **20.** $y = -2.44x - 1$

21. $3y = 4.5x - 1.75$ **22.** $2.5y = -1.5x + 2$ **23.** $0.2y = 3x - 15$

24. $x = 2.5y - 3$ **25.** $2x = 15 - 3.2y$ **26.** $-x = 2.4 + 3y$

27. BIRTH RATE The table gives birth rates in the United States for the years 1977 through 1992. Rates are the number of births per 1000 population.

Year	Birth Rate	Year	Birth Rate	Year	Birth Rate	Year	Birth Rate
1977	15.4	1981	15.8	1985	15.8	1989	16.2
1978	15.3	1982	15.9	1986	15.5	1990	16.7
1979	15.9	1983	15.5	1987	15.7	1991	16.2
1980	15.9	1984	15.5	1988	15.9	1992	15.9

If you use a line to show the relationship between birth rates y and the years x in the table, what is the approximate slope of the line? What would be its approximate y-intercept? Explain how you determined these values. Write an equation in standard form for the line you chose to represent the relationship in the table.

MINIMUM WAGE The graph at the right reflects the changes in the Federal minimum wage for the years 1957 through 1993. Use the graph to answers Exercises 28–32.

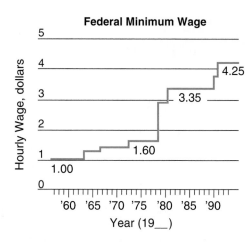

Federal Minimum Wage

28. The Federal Government first set the minimum wage at $0.25 in 1938. To the nearest cent, what was the average annual change in the minimum wage from that year until 1960?

29. To the nearest whole number, how many times greater was the average annual change in minimum wage from 1960 to 1993 than during the years 1938–1960?

30. What equation represents the minimum wage w for the years 1972–1978?

31. What equation represents the minimum wage lasting for the greatest number of years?

32. Between $1.00 and $4.25, how many other amounts were set as the minimum wage from 1957 through 1991? How did you decide?

THINK CRITICALLY

Write one or more equations to describe the graphs in each question.

33. horizontal and vertical lines that intersect at the point $(2, -5)$

34. a horizontal line that passes through the point $(0, -3)$

35. a vertical line that is as far from the y-axis as the line $x = -1$

36. a line on which every x-value is 4

37. a line on which every y-value is -8

38. horizontal and vertical lines that intersect at the origin

MIXED REVIEW

39. **STANDARDIZED TESTS** Identify the median for the data set: 17, 19, 21, 24, 16, 29, 12, 28

    A. 20.75   **B.** 20   **C.** 19   **D.** 21

Round each answer to the nearest tenth.

40. What number is 35% of 125?

41. What percent of 120 is 250?

42. 221 is 24 percent of what number?

43. What number is 12% of 18?

Solve.

44. $-2x - 18 = -44$ 45. $45 - 3x = -75$ 46. $8x - 3(2x + 1) = -9$

47. In New York City, the average price of a slice of pizza increased from $0.10 in 1950 to $1.35 in 1995. What was the percent increase over that time?

6.6 Explore Statistics: Scatter Plots; Lines of Best Fit

Explore

● A mathematics teacher asked his students how much television they watched the night before an exam. The data in the table below show the number of hours the students watched TV and their grades on the exam.

Student	A	B	C	D	E	F	G	H	I	J	K	L
Hours watched	1	3	0	1	2	2	3	4	2	2	1	3
Test score	88	71	87	80	78	76	70	68	84	90	93	74

1. Examine the data in the table. How would you describe the relationship between hours watched and test scores?

2. If you graphed the data as a set of ordered pairs, which would be the independent variable? Which would be the dependent variable?

3. Plot the points on a coordinate plane. In general, as the number of hours of TV watching is increased, what happened to test scores?

4. Find the average value of each variable to the nearest tenth and mark that point on the plane. Label the x average $\bar{x}$ and the y average $\bar{y}$.

5. Place a thin, straight object on the grid (an uncooked spaghetti strand works well) so that it passes through the average point $(\bar{x}, \bar{y})$. Rotate it about the average point until it seems to best fit the data.

6. How many data points are on the line? How many are above? How many are below?

7. What do you think this line represents?

Build Understanding

● In Explore, you estimated the **line of best fit.** That is, you found a line that approximated a trend for the data in a scatter plot. The data in Explore show a negative relationship between test scores and the number of hours of TV watched the night before a test.

A relationship can be positive or negative and strong or weak. Some scatter plots show no relationship at all.

The **coefficient of correlation** is a statistical measure of how closely data fit a line. Most graphing utilities can calculate the coefficient of correlation r for you. The coefficient of correlation ranges from -1 to 1.

COEFFICIENT CORRELATION

−1   The data show a perfect *negative correlation*. All data points are on the line of best fit which has a *negative slope*. An increase in x gives a decrease in y in a linear relation.

0   The data show *no correlation* between x and y. The data are randomly scattered. No trend appears in the data.

1   The data show a perfect *positive correlation*. All data points are on the line of best fit which has a *positive slope*. An increase in x gives an increase in y in a linear relation.

The coefficient of correlation r is usually a number other than −1, 0, or 1. The closer the correlation coefficient is to −1, or 1, the stronger the relationship is between x and y. The closer the correlation coefficient is to 0, the weaker the relationship is between x and y.

EXAMPLE 1

Match each scatter plot with one of the following values as the best estimate of the coefficient of correlation: −0.2, 0.8, −0.9, and 0.3.

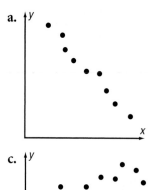
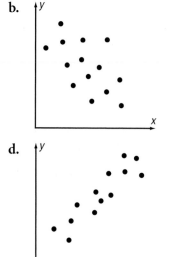

Solution

Figures a and b show a negative correlation. As x increases, y decreases. The data points in b are more scattered than those in a. Therefore, the coefficient of correlation for a is −0.9 and the coefficient of correlation for b is −0.2

Figures c and d show a positive correlation. As x increases, y increases. The data points in d are more clustered than those in c. Therefore, the coefficient of correlation for c is 0.3 and the coefficient of correlation for d is 0.8 ◀

A graphical approach can help you approximate the line of best fit. You can then write an equation for this line. The equation can help you make predictions based on the data.

COMMUNICATING ABOUT ALGEBRA

EXAMPLE 2

BOTTLED WATER SALES The table at the right compares the daily high temperatures to the daily sales of bottled water in Huntsville during one summer.

Daily High Temperature, x	Bottled Water Sales, y
66°F	141
70°F	149
74°F	159
78°F	165
82°F	175
86°F	180
90°F	193
94°F	195
98°F	210

a. Approximate the line of best fit for the data in the table.

b. Predict how many bottles of water the market can expect to sell on a day when the temperature reaches 101°F.

As the number of data points increases, do you think it becomes easier or more difficult to see a correlation?

Solution

a. Represent the data in the table as ordered pairs. Then plot the points.

The graph shows a positive correlation. Water sales increase as the temperature increases. Find the mean of the x- and y-coordinates to the nearest integer. Call the point $(\bar{x}, \bar{y})$.

$$\bar{x} = 82, \bar{y} = 174$$

The line of best fit will pass through (82, 174). Pivot a straightedge about this point and draw the line that appears to best fit the data. Approximate the coordinates of another point on the line, such as (97, 205).

Use the points to find the slope.

$$m = \frac{205 - 174}{97 - 82} = 2.07 \quad \text{to nearest hundredth}$$

Find the y-intercept.

$$y = mx + b \qquad \text{Write the slope-intercept form.}$$
$$174 = 2.07(82) + b \qquad \text{Substitute known values.}$$
$$b = 4.26 \qquad \text{Solve for } b.$$

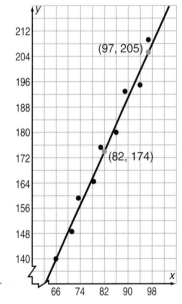

Therefore, an approximate equation of the line of best fit is

$$y = 2.07x + 4.26$$

b. Substitute 101 for x in your equation. The resulting value of y, 213.33, tells you that, based on current data, the market can expect to sell 213 bottles of water. ◀

With a graphing utility, you can determine the actual line of best fit.

EXAMPLE 3

GOVERNMENT The table gives the total number of African-Americans serving as state governors, state legislators, and U.S. Representatives y in selected years x.

Year, x	1970	1975	1980	1985	1990	1993
Total Number Serving, y	182	299	326	407	447	571

a. Find the equation for the line of best fit and the coefficient of correlation.

b. Use the equation to predict the number of African-Americans who will be serving in these positions in the year 2010.

Solution

a. Enter the data in your graphing utility. Let $x = 0$ represent 1970. Use the statistics features of your graphing utility to do the linear regression.

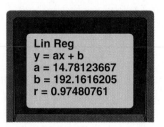

Lin Reg
y = ax + b
a = 14.78123667
b = 192.1616205
r = 0.97480761

The variables on the screen represent the following:

a: slope of best fit
b: y-intercept of line of best fit
r: coefficient of correlation

With each variable rounded to the nearest tenth, the equation for the line of best fit for this data is

$$y = 14.8x + 192.2$$

The high coefficient of correlation of 0.974 tells you that the data points lie very close to the graph of the equation.

Note that once the data are entered, the graphing utility will display the graph of the equation along with the scatter plot.

b. Substitute the year 2010 into the equation. Since $x = 0$ represents 1970, $x = 40$ represents 2010.

$$y = 14.8(40) + 192.2$$
$$y = 784.2$$

Based on current data, the number of African-Americans serving as governors, state legislators, and U.S. Representatives will be 784 in 2010. ◀

PROBLEM SOLVING TIP

When computing a linear regression, some graphing utilities use the form $y = a + bx$, where b is the slope and a is the y-intercept. Be sure to check your calculator.

State whether you believe each of the following sets of variables would show *positive*, *negative*, or *zero correlation*. Explain your reasoning.

1. number of cars registered and carbon monoxide emissions

2. weight of a car and its gas mileage

3. a person's height and weight

4. an adult hand size and his or her salary

5. temperature and snow ski sales

6. **PERSONAL INCOME** The table and scatter plot show the change in personal income of individuals from 1975 to 1990.

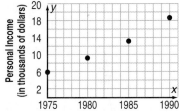

Year	1975	1980	1985	1990
Personal Income	$6,081	$9,916	$13,890	$18,477

The *x*-axis shows years, with $x = 0$ representing 1975, $x = 5$ representing 1980, and so on. The *y*-axis shows income in thousands of dollars. Write an equation of the line that best approximates the data. Then use the equation to predict personal income in 1998.

7. **PERSONAL INCOME** Use the equation from Exercise 6. According to current data, in what year will the personal income of the average American pass $30,000?

8. **WRITING MATHEMATICS** When working with years, smaller numbers are often substituted for larger ones. The data in the table represent the number of students *y* in each graduating class at Eastview High School from 1987–1995. If you had to graph the data, what values would you choose to represent the years? Explain why you chose those numbers.

Year	1987	1988	1989	1990	1991	1992	1993	1994	1995
Number of Students	456	312	325	290	501	480	362	400	437

PRACTICE

Determine which scatter plots reflect a linear relationship. For those that do, determine whether the correlation is *positive*, *negative*, or *zero*.

1.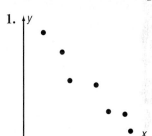

2.

3.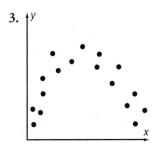

Determine which scatter plots reflect a linear relationship. For those that do, determine whether the correlation is *positive*, *negative*, or *zero*.

4.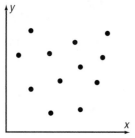

5.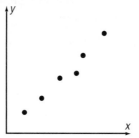

6.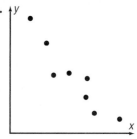

7. Two points on the line of best fit for a set of data are (2.1, 14) and (6.1, 33). Find the equation of the line of best fit.

8. A line of best fit for a set of data has slope 4.1 and passes through the point (10, 8). Find the equation for the line of best fit.

9. Draw a scatter plot for the following set of data.

x	1.0	2.0	2.5	3.1	4.2	5.0	7.0	8.3
y	1.1	3.9	5.4	7.1	10.6	13.2	19.1	22.1

One of the following represents the coefficient of correlation r for the data above. Choose the one that does and justify your response.

a. 0.99   **b.** −0.98

c. 0.62   **d.** 0.08

10. WRITING MATHEMATICS The table shows the average number of miles per gallon of gasoline of motor vehicles for each year given. Use the data to create and answer at least three questions about automotive fuel efficiency.

Year	1982	1983	1984	1985	1986	1987	1988	1989	1990
Average MPG	16.65	17.17	17.83	18.20	18.27	19.20	19.87	20.31	20.92

EXTEND

RECYCLING CAMPAIGN A local radio station encourages listeners to drop off soft drink cans at the studio. They also monitor the number of cans found in trash bins. The weekly data is shown below. Use this data for Exercises 11–17.

Cans Redeemed, x	324	309	310	288	523	509	210	251
Cans Found in Trash, y	275	300	281	312	70	100	400	340

11. Are the data positively or negatively correlated? How did you decide?

12. Use a graphing utility to find the line of best fit for the data.

13. Use your graphing utility to find the coefficient of correlation. Interpret the coefficient.

14. What does the slope of −1 tell you about the relationship between cans redeemed and cans thrown away?

15. In the drive's ninth week, 298 cans were collected. Predict how many cans they will find in the trash.

16. According to the table, 400 cans were found in the trash in the seventh week. By how much does this differ from the estimate given by the equation for the line of best fit?

17. The station would like to know whether the drive is having a positive effect on its collections. The manager plans to extend the drive. Number the weeks 1–8 and determine whether the weeks and weekly redemption totals are positively correlated.

18. NEWSPAPERS On page 290 you solved Exercise 16 about evening newspapers in the United States. Use the data in the following table to find a line of best fit. Then compare the equation you find with the one you found for the earlier exercise. Use each equation to predict the number of evening newspapers that will be published in the year 2015. By how much do your predictions differ?

Year	1983	1984	1985	1986	1987	1988	1989	1990	1991
Number of Evening Papers	1284	1257	1220	1188	1166	1141	1125	1084	1042

THINK CRITICALLY

PERCENT ERROR For each data point in a scatter plot there is an actual value that is given by the data and an estimated value that is calculated from the equation for the line of best fit. To calculate the percent error of the estimated value, find the absolute value of the difference (that is, the *residual*) between the actual and estimated values, divide this by the actual value, and multiply the result by 100. For Exercises 19 and 20, use the data and equation from Example 2 of this lesson.

19. The actual average number of bottles of water sold when the temperature is 70°F is 149. What is the estimated number from the line of best fit? By what percent does this estimate differ from the actual number?

20. What is the percent difference when the temperature is 98°F? What does this tell you about the relationship of the data point to the line of best fit, as compared to the data point for 70°F?

MIXED REVIEW

Perform the indicated operation.

21. $\begin{bmatrix} 1 & 0 & 1 \\ 4 & 3 & -1 \\ 7 & -5 & -3 \end{bmatrix} + \begin{bmatrix} 2 & -2 & 3 \\ -4 & -3 & 5 \\ -3 & 9 & -1 \end{bmatrix}$

22. $\begin{bmatrix} 8 & -3 & 5 \\ 2 & -1 & 4 \\ -5 & 9 & -6 \end{bmatrix} - \begin{bmatrix} 6 & -4 & 1 \\ 3 & -2 & 6 \\ 2 & -1 & 3 \end{bmatrix}$

Evaluate when $x = 3$, $y = -2$, and $z = 12$.

23. $x^2 - 2xy + y^2$

24. $z^2 + 5x - xyz + xz$

25. $x^4 - x^2y + 2xy^2 + yz$

Solve.

26. $4 + x > 12$

27. $4x - 4 < x + 5$

28. $2.2x \le 3x - 4$

29. STANDARDIZED TESTS For which pair of points is the slope of the line containing the points *negative*?

    A. $(-5, 0), (2, 4)$ **B.** $(-3, 2), (7, 2)$ **C.** $(-1, 2), (3, -4)$ **D.** $(-7, -2), (-3, 6)$

Write the slope and *y*-intercept for each equation.

30. $y = 0.25x - 25$

31. $y = 4.2x + 0.3$

32. $y = x$

33. $y = -x - 1$

34. The equation for the path of an airport runway is $y = 1.2x + 7$. Another runway is parallel to it, with *y*-intercept 12 units greater. A third runway is perpendicular to the first and passes through the point $(-2, 4.6)$. Write the equations of the other two runways.

PROJECT *Connection* Many students are eager to get their driver's license. However, many environmental problems are caused by automobiles. Work with your group to brainstorm questions you could use to survey students about cars and the environment. Some sample questions are listed.

- Should highways have carpool lanes?
- Should the legal driving age be raised?
- Should annual vehicle mileage be restricted or taxed?
- Should tolls be raised to reduce traffic?
- Are you willing to help save the earth by not driving?

1. Think of other questions, then create a survey form. Have students include their age, grade, sex, and whether or not they have a driver's license.

2. Distribute your survey form to at least 50 randomly selected students. Why is random selection important? Why might answers differ between freshmen and seniors?

3. Collect, tally, and analyze the data. Decide how to use tables and graphs to communicate your findings. Write a summary and interpretation for your magazine.

6.7 Direct Variation

Explore

1. Enter the equations $y = 2.4x$ and $y = 2.4x + 2$ into your graphing utility.

2. Copy and complete the two tables of values for the functions. You may wish to use the TRACE feature to locate the ordered pairs. If so, set your graphing utility so integers are displayed for x.

$y = 2.4x$		
x	y	$\frac{y}{x}$
1		
3		
5		
7		

$y = 2.4x + 2$		
x	y	$\frac{y}{x}$
1		
3		
5		
7		

3. Describe the difference between the third columns in the tables.

4. Delete the constant 2 from the equation $y = 2.4x + 2$ and replace it with any other nonzero number. Graph the new equation.

5. Complete another table like the first two for your new equation. Describe the result. Then compare your work with that of other students.

6. Under what condition do you think the quotient of y and x will be a constant?

SPOTLIGHT ON LEARNING

WHAT? In this lesson you will learn
- to apply linear relationship concepts to direct variation problems.

WHY? Understanding direct variation can help you solve problems involving building construction, typing speed, physics, and chemistry.

Build Understanding

The functions you examined in Explore were all linear and the graphs were all straight lines. However, the quotient $\frac{y}{x}$ was constant only for the equation $y = 2.4x$. Because of the constant quotient, $y = 2.4x$ is an example of a **direct variation**.

> **DIRECT VARIATION**
>
> When y *varies directly as* x and k is the constant of variation, an equation can be written in the form
> $$y = kx \quad \text{where } k \neq 0$$

If you know one pair of values for x and y in a direct variation, you can determine the constant of variation k. Then you can write an equation for the direct variation and use it to calculate other values.

EXAMPLE 1

Assume y varies directly as x. When $x = 9$, $y = 5.4$.
Find y when $x = 24$.

Solution

Use the equation $y = kx$ to determine the value of k, the constant of variation.

$y = kx$	Write the equation.
$5.4 = 9k$	Substitute known values.
$0.6 = k$	Divide both sides by 9.
$y = 0.6x$	Write the equation using 0.6 for k.
$y = 0.6(24)$	Substitute 24 for x.
$y = 14.4$	

When $x = 24$, $y = 14.4$. ◄

You can use a table of values to determine whether two variables vary directly.

EXAMPLE 2

Examine the following tables of values. If a table shows direct variation, determine the constant of variation and write the equation of direct variation.

a.

x	y
3.5	6.65
4.1	7.79
6.9	13.11
8.1	15.39

b.

x	y
2.7	8.37
4.8	16.80
5.6	19.04
6.7	22.11

Solution

a. Determine the ratio $\frac{y}{x}$ for each row of the first table.

$$\frac{6.65}{3.5} = 1.9 \qquad \frac{7.79}{4.1} = 1.9$$

$$\frac{13.11}{6.9} = 1.9 \qquad \frac{15.39}{8.1} = 1.9$$

Since each ratio is 1.9, the values show direct variation. The constant of variation is 1.9 and the equation of direct variation is $y = 1.9x$.

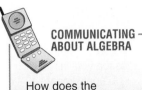

COMMUNICATING ABOUT ALGEBRA

How does the equation of a direct variation compare to the slope-intercept form of a linear equation?

b. Determine the ratio $\dfrac{y}{x}$ for each line of the second table.

$$\frac{8.37}{2.7} = 3.1 \qquad \frac{16.8}{4.8} = 3.5$$

$$\frac{19.04}{5.6} = 3.4 \qquad \frac{22.11}{6.7} = 3.3$$

Since the ratios are not equal, y does not vary directly as x. ◄

CHECK UNDERSTANDING

In Example 2b, how many ratios are needed to determine that the variables do not vary directly?

Problems involving direct variation can be solved using the equation $y = kx$.

EXAMPLE 3

MAP SCALES On a map with scale 1 cm : 50 km two cities are shown to be 3.75 cm apart. What is the actual distance between the two cities?

Solution
In this situation the constant of variation k is 50. Let x represent 3.75.

$y = kx$ — Write the equation.
$y = 50(3.75)$ — Substitute 3.75 for x and 50 for k.
$y = 187.5$

The cities are 187.5 km apart. ◄

THINK BACK

In Example 3, the scale 1 cm: 50 km means the ratio of 1 centimeter to every 50 kilometers.

TRY THESE

In each case, y varies directly as x.

1. When $x = 12$, $y = 15$. Find y when $x = 40$.

2. When $x = 25$, $y = 10$. Find y when $x = 18$.

3. When $x = 50$, $y = 14$. Find y when $x = 85$.

For each set of data, decide whether y varies directly as x. If it does, find the constant of variation and write the equation that describes the relationship.

4.

x	y
4	7
6	9
10	15
18	27

5.

x	y
84	67.2
60	48.0
48	38.4
20	16.0

6.

x	y
0.3	1.2
0.7	2.8
12	48.0
21	84.0

Determine whether each equation represents direct variation. Write *yes* or *no*.

7. $x - y = 3$

8. $3x = y$

9. $2y = 9x$

10. $y = 2x - 1$

11. $y = \dfrac{x}{3}$

12. $y - x = 1$

13. $xy = 8$

14. $x + 2 = y$

15. TYPING SPEED On a typing test, Jack typed 51 words correctly in 3 min. The number of correctly typed words varies directly as the length of time. How many words can Jack type correctly in 10 min?

16. WRITING MATHEMATICS Use the following information to write an equation of direct variation. Ben is calculating the distance he can drive at a steady rate of 55 mi/h. Use your equation to describe the relationship between distance and time.

PRACTICE

In each case, y varies directly as x. Find the constant of variation and use it to write the equation of direct variation.

1. $x = 8, y = 14$ **2.** $x = -3, y = 9$ **3.** $x = 5, y = 12$

4. $x = -8, y = -6$ **5.** $x = 1.5, y = 6$ **6.** $x = 2.5, y = 7$

7. $x = 9, y = -1.8$ **8.** $x = -0.5, y = 3$ **9.** $x = 4, y = -3$

In each case, y varies directly as x.

10. If $y = 1$ when $x = 1.5$, find x when $y = 8$.

11. If $y = -5$ when $x = -8$, find y when $x = 18$.

12. If $y = 12$ when $x = 16$, find y when $x = 48$.

13. If $x = -10$ when $y = 4$, find x when $y = 14$.

14. If $x = -4.5$ when $y = 13.5$, find x when $y = 24$.

15. BUILDING CONSTRUCTION To help consumers compare various insulating materials used in building construction, each material is rated with an R-value. The R-value gives the material's rate of resistance to heat flow for 1 cm of thickness. So, the total R-value of a material varies directly with its thickness. A wall made of fiberglass that is 2.5 cm thick has an R-value of 3.05.

    a. What is the per centimeter R-value for fiberglass?

    b. What is the total R-value of a fiberglass wall with a thickness of 3 cm?

    c. A 6-cm space filled with loose foam insulation has an R-value of 11.34. What is the R-value of a 4.0-cm space filled with loose foam?

16. GEOMETRY The circumference of a circle varies directly as the length of its diameter. In the equation for the circumference of a circle, what is the value of the constant of variation?

17. COMPARISON SHOPPING A 5-lb bag of sugar costs $1.79. A shopper finds that an 8-lb bag is priced at $2.76. Do the weight and price of sugar vary directly?

18. PHYSICS Hooke's law states that the distance an object is stretched varied directly with the force exerted on the object. Suppose the force exerted by a 7-kg mass stretches a certain spring 5 cm. To the nearest 0.1 cm, how far will the spring be stretched by the force of an 11-kg mass?

19. WRITING MATHEMATICS Kim thinks she can use the proportion $\dfrac{50 \text{ km}}{1 \text{ cm}} = \dfrac{d}{3.75 \text{ cm}}$

to solve the problem in Example 3. She says that direct variation and proportion are similar to each other. Is what Kim says true? Justify your response.

EXTEND

20. CHEMISTRY A law discovered by Jacques Charles, a French scientist of the late 18th and early 19th centuries, states that for a fixed amount of a gas, the volume varies directly as the temperature. If a gas has a volume of 550 mL at 40°C, what is its volume at 25°C?

21. GRAVITY The gravitational pull on an object is not the same on all planets. A 175-lb person would weigh only 66.5 lb on Mars. Assume that weight on Earth varies directly as weight on Mars. What is the constant of variation?

22. GRAVITY The constant of variation for the pull of gravity on Jupiter compared to Earth is 2.54. If a probe weighs 381 lb on Jupiter, how much does it weigh on Earth?

23. GEOMETRY Examine the diagrams of two "staircases," as shown. The height of staircase A is 3 units. The height of staircase B is 6 units. For these staircases, are either of the relationships below examples of direct variation? Explain.
  a. height and surface area
  b. height and volume

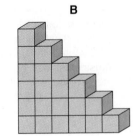

THINK CRITICALLY

24. Will the graph of a direct variation always pass through the point (0, 0)? Explain.

25. Why will the graph of a direct variation never have an undefined slope?

26. Can the graph of a direct variation ever have a slope of 0? Explain.

PROJECT *Connection*   Electrical use is measured in *watts*, and every appliance is rated by the number of watts it uses in one hour. Utility companies bill customers by the *kilowatt-hour*. A kilowatt is 1000 watts, and a kilowatt-hour means 1000 watts were used for one hour.

1. Find out the cost of one kilowatt-hour of electrical use in your area.

2. Prepare a list of the rating in watts of at least ten appliances in your home or school. This information may be on the back of the appliance or in the product booklet. You may also use typical appliance ratings found in utility company literature.

3. Create graphs showing the cost of running each appliance as a function of the number of hours used. Represent each function as an equation. Interpret the graphs using terms from this chapter.

4. Estimate the annual cost of power for the appliances used in your home. (Begin by estimating average weekly usage.)

5. Add suggestions for saving electrical energy to your list of helpful hints.

6. Use your work above as the basis for an article in your magazine.

Misleading Graphs

One advantage of a statistical graph (such as a line graph) over a table of numbers is that it can provide information at a glance. But sometimes this can be a disadvantage. There are ways to change the appearance of a graph to create different impressions.

Problem

The table below shows the number of paid admissions each month for the past year at the Old West theme park.

Month	Admissions	Month	Admissions
January	503,238	July	521,654
February	506,790	August	523,809
March	509,515	September	525,248
April	512,376	October	527,391
May	516,633	November	532,457
June	519,312	December	536,112

Make a line graph for the data that shows how admissions have changed during the year.

Explore the Problem

Work with a partner. You will each make a line graph showing months on the horizontal axis and admissions on the vertical axis. Both graphs must be the same size.

1. One partner will divide the vertical axis into intervals of 50,000 to show a range from 0 to 550,000.

2. The other partner will divide the vertical axis into intervals of 5,000 to show a range from 500,000 to 550,000.

3. Compare the two graphs. What impression does each graph give?

4. If you were a manager planning to include a graph in a report for investors, which graph would you use? What purpose might someone who chose to use the other graph have?

5. On which graph do the line segments appear to have a steep slope? Are the slopes of equivalent segments on the two graphs actually different? Illustrate your answer by computing the slope of the segment for April to May on each graph. (Use 30 for the change in x.)

6. Explain why the two admissions graphs, although based on the same data, appear so different.

7. WRITING MATHEMATICS Write a short paragraph explaining what you should look for if you want to be sure of drawing correct conclusions about data presented in a line graph.

> ┌─ PROBLEM ─────
> SOLVING PLAN
>
> • Understand
> • Plan
> • Solve
> • Examine

Investigate Further

● Other types of statistical graphs can also be misleading. Consider the bar graph below, used in a brochure created for prospective students at a trade school.

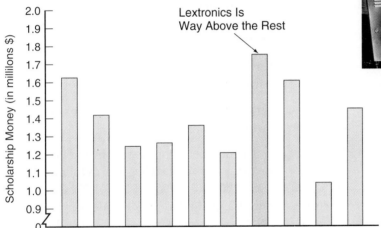

Scholarship Money at Ten Trade Schools

Lextronics Is
Way Above the Rest

8. Approximately how many times as high is the tallest bar than the shortest bar?

9. Make a list of the scholarship money available at each of the ten schools. Use the data in the graph and round each number to the nearest $100,000. For example, the first number would be $1,600,000.

10. Based on your numbers, how does the greatest figure compare to the least?

11. What caused the distortion in the graph?

12. Work with a partner. Decide how to redraw the graph so that it is about the same overall size but gives a more honest picture of the situation. Then carry out your ideas.

Apply the Strategy

13. QUIZ SCORES Use the table at the right showing Franco's weekly quiz scores.

a. Make a line graph so that Franco's quiz scores appear to have varied greatly from week to week.

b. Make a line graph so that Franco's quiz scores appear to have varied only a little from week to week.

Franco's History Quiz Scores			
Week	Score	Week	Score
1	76	7	84
2	79	8	85
3	81	9	89
4	86	10	82
5	83	11	84
6	80	12	78

14. MUSIC BUSINESS A songwriter earned the following yearly totals in royalties over a four-year period: 1993, $31,000; 1994, $33,000; 1995, $36,000; 1996, $40,000.

a. Draw a vertical bar graph that shows the actual trend in the writer's earnings.

b. Draw a vertical bar graph so that the writer's earnings appear to be growing substantially.

15. PIZZA SALES The number of pizzas sold by the Pizza Tower chain for each of the three years since it began is as follows: 1994, (100,000); 1995, (200,000); 1996, (300,000). To show the sales increase visually, the chain's president made the pictograph below.

Pizza Tower Sales

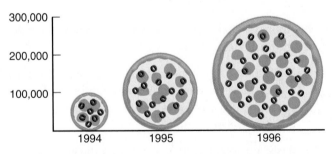

300,000
200,000
100,000
1994   1995   1996

a. Is the president justified in claiming the graph presents the data accurately? How might the graph be misleading?

b. How could the president redraw the graph to avoid misleading anyone?

16. WRITING MATHEMATICS Find examples of misleading graphs in newspapers or old magazines. Paste each graph on a piece of paper and write an explanation of what is wrong with the graph.

REVIEW PROBLEM SOLVING STRATEGIES

FOUR 4's MAKE 30

1. **a.** Work in groups of four students. Using four 4s, write expressions to represent as many whole numbers as you can, from 1 to 18. You may use any operation $(+, -, \times, \div)$, parentheses, exponents, square roots, and factorials $(4! = 4 \cdot 3 \cdot 2 \cdot 1 = 24)$. For example,

$$1 = \frac{4 + 4}{4 + 4} \text{ or } \frac{4 \cdot 4}{4 \cdot 4} \text{ or } \frac{44}{44}$$

(You may use decimals less than 1 without writing the leading zero.) If you find more than one way for some numbers, include them.

After your group has worked on the problem for a while, compare results with other groups. Discuss strategies you used to expand your lists.

b. Try writing expressions for whole numbers from 19 to 30.

2. Nina's home is 17 km from her office. She allows herself 1 h to travel to work and, as part of her fitness program, combines walking with riding the bus. Her goal is to spend as much time as she can walking along the bus route, then catch the bus to complete the trip. Buses average 30 km/h and come along every 6 min; a bus will stop anywhere along its route to pick up a passenger. Nina walks at the rate of 5 km/h.

a. Can Nina walk 5 km before catching the bus? What clue does this give you about the distance she can walk?

b. Is the distance she can walk more or less than 3 km? Explain.

c. Can Nina expect to get on a bus the moment she stops walking? How should she plan for this?

d. Where along the bus route should Nina get on? Explain the reasoning you used in reaching your decision.

Who Did It?

3. A crime was committed, and only one of the four suspects being questioned by the poice is guilty. Here are the statements they made.

Adam: "Barbara did it."

Barbara: "Dan did it."

Carol: "I didn't do it."

Dan: "Barbara lied when she said I did it."

a. If only one statement is true, who is guilty? Explain.

b. If only one statement is false, who is guilty? Explain.

• • • CHAPTER REVIEW • • •

VOCABULARY

Choose the word from the list that completes each statement.

1. The line of best fit for a data set is often called a ___?___ line.

2. Two variables that have a constant quotient are said to be in ___?___.

3. An equation in ___?___ can be used to write the equation of a line with undefined slope.

a. standard form

b. regression

c. direct variation

Lesson 6.1 RELATIONSHIPS OF LINES   pages 265–267

● Two lines are parallel when they both have the same change in y per unit change in x.

4. The graph of which equation below is parallel to the graph of $y = 4x + 3$?

    a. $y = x + 3$ **b.** $y = 3x + 3$ **c.** $y = 4x + 2$

Lesson 6.2 THE SLOPE OF A LINE   pages 268–275

● The slope of a line containing points (x_1, y_1) and (x_2, y_2) is $\dfrac{y_2 - y_1}{x_2 - x_1}$.

● A line slanting down from left to right has *negative* slope. A line slanting up from left to right has *positive* slope.

● The slope of a horizontal line is zero. The slope of a vertical line is undefined.

Find the slope of the line passing through the given points.

5. $(8, 4)$ and $(3, 2)$ 6. $(2, 3)$ and $(5, 9)$ 7. $(3, 8)$ and $(4, 5)$

Lesson 6.3 SLOPE-INTERCEPT FORM; PARALLEL AND PERPENDICULAR LINES   pages 276–284

● An equation of the form $y = mx + b$ is in **slope-intercept form**. The slope of the line is m. The y-intercept is b.

● Two nonvertical lines are **parallel** if and only if they have the same slope. Two lines are **perpendicular** if the product of their slopes is –1.

Write an equation for the line.

8. slope, $\dfrac{3}{4}$; y-intercept, 8   9. slope, –2; y-intercept, 3   10. slope, $-\dfrac{1}{3}$; y-intercept, –2

Indicate whether the graphs of each pair of equations will be *parallel*, *perpendicular*, or *neither*.

11. $\begin{cases} y = 3x + 2 \\ y + \dfrac{1}{3}x = 5 \end{cases}$ 12. $\begin{cases} y - 3x = 4 \\ y = 3x + 2 \end{cases}$ 13. $\begin{cases} 5y = x - 7 \\ y = 3x + 2 \end{cases}$

Lesson 6.4 THE POINT-SLOPE FORM OF A LINEAR EQUATION pages 285–292

- The **point-slope form** of a linear equation of a line is $(y - y_1) = m(x - x_1)$, where (x_1, y_1) is a known point on the line and (x, y) is any other point on the line.

Use the given information to write an equation in point-slope form and in slope-intercept form.

14. $(6, 2)$; slope, -9

15. $(3, 3)$, $(4, 9)$

16. $(9, 2)$; slope, $-\dfrac{1}{2}$

Lesson 6.5 THE STANDARD FORM OF A LINEAR EQUATION pages 293–298

- The **standard form** of a linear equation is $Ax + By = C$, where A, B, and C are integers and A and B are not both zero.

Use the given information to write an equation in standard form.

17. $m = 2$; $(2, 5)$

18. $(4, 3)$, $(2, 7)$

19. $(5, 4)$, $(-3, 1)$

20. $m = -\dfrac{3}{2}$; $(-2, 7)$

Lesson 6.6 EXPLORE STATISTICS: SCATTER PLOTS; LINES OF BEST FIT pages 299–306

- The **coefficient of correlation** is a measure of how closely data fit a line. The range for the coefficient of correlation between the x- and y-values is from -1 (perfect negative correlation) to 1 (perfect positive correlation). A coefficient of correlation of 0 indicates that there is no correlation between the x- and y-values.

Determine whether the x- and y-values have a *positive*, *negative*, or *zero* correlation.

21.

22.
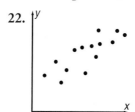

23.
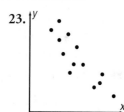

Lesson 6.7 DIRECT VARIATION pages 307–311

- When y varies directly as x varies, $y = kx$ $(k \neq 0)$. In this equation, called an **equation of direct variation**, k is the **constant of variation**.

Find the constant of variation and use it to write the equation of direct variation.

24. $x = 6$, $y = 4$

25. $x = 2.5$, $y = 8$

26. $x = 4$, $y = -7$

Lesson 6.8 MISLEADING GRAPHS pages 312–315

- The appearance of a graph can be misleading, depending upon the scale used.

27. Use the information at the right to create two graphs. In one, make it appear that sales rose steeply from month to month. In the second, make it appear that sales varied little from month to month.

Gloria's Gloves Sales						
Month	1	2	3	4	5	6
Sales, in 100s	72	65	81	84	75	79

CHAPTER ASSESSMENT

CHAPTER TEST

Write the slope-intercept form of an equation for a line containing the given point and having the given slope.

1. $(2, 3)$; $m = -2$ **2.** $(-2, 2)$; $m = \dfrac{1}{2}$

3. $(0, 0)$; $m = \dfrac{2}{3}$ **4.** $(-3, -3)$; $m = 3$

5. $(2, -1)$; $m = 2$ **6.** $(2, 10)$; $m = 3$

Write the slope-intercept form of the equation of the line that passes through the given points. Then rewrite the equation in standard form.

7. $(-5, 0)$ and $(0, 5)$ **8.** $(1, -3)$ and $(-5, 3)$

9. $(-4, 2)$ and $(2, 4)$ **10.** $(-2, 4)$ and $(0, 7)$

11. $(-6, 4)$ and $(6, -2)$ **12.** $(4, 6)$ and $(1, -3)$

13. $(-6, -2)$ and $(6, 2)$ **14.** $(2, -2)$ and $(-2, 4)$

15. STANDARDIZED TESTS Which point lies on the line $3y + 2x = 5$?

    A. $(1, 1)$ **B.** $(-1, 4)$

    C. $(2, -1.2)$ **D.** $(4, -3)$

16. WRITING MATHEMATICS Write a paragraph explaining how to distinguish equations of parallel lines from equations of perpendicular lines.

Determine whether the graphs of each pair of equations are *parallel*, *perpendicular*, or *neither*.

17. $2x + y = 6$
$\ 2x + y = 4$

18. $2x + y = 6$
$\ y = \dfrac{1}{2}x - 4$

19. $y = 3x - 6$
$\ y = -3x + 8$

20. $y = 2x + 4$
$\ 3y = 6x + 6$

21. STANDARDIZED TESTS Which of the following equations has a graph that does not go through the point $(2, 3)$?

    A. $y = x - 1$ **B.** $x + y = 5$
    C. $y - x = 1$ **D.** $2x + 3y = 13$

Write the slope-intercept equation of a line perpendicular to the given line and passing through the given point.

22. $y = 2x + 4$; $(0, 1)$ **23.** $y = \dfrac{2}{3}x$; $(0, 0)$

24. $3y = x + 6$; $(3, 3)$ **25.** $y = 2x + 5$; $(3, 1)$

Solve the problem.

26. Shana can buy juice glasses for \$2 apiece and water glasses for \$4 each. She plans to spend exactly \$60.

    a. Write an equation to represent the different combinations of glasses Shana can buy.

    b. Graph your equation.

    c. Use your graph to complete the table below.

x	0	2	4	6	12	16	20	30
y	15	14						

27. Use the table below to find the constant of variation and write an equation for the variation.

x	3.24	4.23	4.77	6.66	12.33
y	0.36	0.47	0.53	0.74	1.37

28. WRITING MATHEMATICS Write a paragraph explaining how the way in which data is presented in a graph can lead to misleading conclusions.

PERFORMANCE ASSESSMENT

SLOPE MODELS The "staircase" design below is constructed from unit squares.

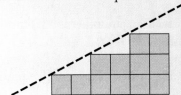

a. What is the slope of the staircase shown? Explain your reasoning.

b. Use unit squares cut from paper, or shade squares on graph paper. Show at least three staircase designs with different slopes.

DRAWING DIRECTIONS Use graph paper. Label point A in the lower left corner.

a. Draw segment AB with a very steep positive slope.

b. Draw line segment BC with a gradual positive slope.

c. Draw line segments CD then DE with different negative slopes.

d. Draw line segment EF with a slope of 0.

e. Draw any other line segment FG.

f. Which segments of your graph shows an increase? Which show no change? Which show a decrease?

STUDENT SURVEY How does the time students in your school spend exercising compare to the time they spend watching television and listening to the radio or CDs? Work with a small group to conduct a survey of how these two variables are related.

a. prepare a log sheet students can use to record their hours at each activity for a week.

b. Randomly select a group of 30 students to survey.

c. Collect the data and display it in a scatter plot.

d. Use a graphing utility to find the line of best fit and the coefficient of correlation.

e. Discuss whether the relationship is linear and how good your linear equation is as a prediction.

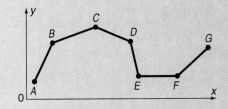

A GOOD PLOT Use the data in the table at the right.

a. Draw a scatter plot for the data. Tell whether the data is positively or negatively correlated.

b. Determine the average values of x and y.

x	1	2	3	4	5	6	7	8	9	10
y	100	95	80	80	75	70	75	55	45	30

c. Use a straight edge to approximate a location for the line of best fit and draw it in.

d. Estimate the coordinates of a point on the line you drew. Then write the equation of the line in slope-intercept form. Give the slope and y-intercept.

PROJECT ASSESSMENT

 Work together as a class and in groups to prepare your Earth Day magazine. Here are some suggestions for how to proceed.

1. Assemble articles, data, artwork, and conclusions from the project activities. Edit the material and create additional graphs and tables if necessary.

2. Decide how to divide the magazine into appropriate sections. Then determine the layout of each page.

3. Make a list of all the tasks that must be done (such as keyboarding the articles, preparing the artwork, and proofreading) Ask students to sign up so all work is covered.

4. Use the magazine to announce a recycling logo contest in your school. Students should submit black and white logos with explanations of the symbolism in the logo.

5. Arrange to distribute the magazine.

CUMULATIVE REVIEW

Find the slope of each of the lines described.

1. passing through $(1, 3)$ and $(6, 5)$

2. passing through $(-2, 7)$ and $(1, -5)$

3. $y = \frac{3}{4}x - 2$

4. perpendicluar to the line $y = -3x - 2$

5. a vertical line

Use the box-and-whisker plot below to answer Questions 6–9.

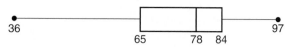

6. What is the range of the data?

7. What is the median test score?

8. What is $Q_3 - Q_1$?

9. **STANDARDIZED TESTS** Between which scores are the greatest number of test scores located?

    A. 36 and 65
    B. 65 and 78
    C. 78 and 84
    D. 84 and 97
    E. they all have the same number of test scores

Perform the matrix operations.

10. $\begin{bmatrix} 5 & -3 \\ -2 & 8 \end{bmatrix} - \begin{bmatrix} -1 & -9 \\ 6 & 3 \end{bmatrix}$

11. $\begin{bmatrix} 5.8 & 1.6 \\ -7.1 & -1.6 \\ 4 & -9.5 \end{bmatrix} + \begin{bmatrix} -4.6 & -1.6 \\ -0.7 & -1.6 \\ 2.7 & 4.8 \end{bmatrix}$

Solve each equation.

12. $\frac{x}{7} = -8$

13. $8(2n - 3) = -3(4 - 5n)$

14. Eleven less than a number is 13. Find the number.

Use a graphing utility to solve each equation.

15. $2x - 4 = x + 3$

16. $60 - 3x = 3x + 24$

17. $x^2 - 4 = x + 2$

Evaluate the function at the given values of x.

18. $f(x) = 2(x - 6)$ at $x = -2$, $x = 5$, and $x = \frac{1}{2}$.

19. $f(x) = \frac{x}{5} - 4$ at $x = 1$, $x = 20$, and $x = -5$.

Write the equation of each line in slope-intercept form.

20. passing through $(-2, -3)$ and $(4, 5)$

21. passing through $(3, -1)$ and parallel to the x-axis

22. **TECHNOLOGY** Which form for the equation of a line is most useful when using a graphing calculator? Explain your choice.

23. Solve for b in the equation $2a - 3b = 7$.

24. When $x = 18$, $y = 27$. If y varies directly with x, find y when $x = 60$.

25. The number of items a machine can produce varies directly with the amount of time the machine is in operation. After 6 minutes, the machine has produced 550 items. How long will it take for the machine to produce 2475 items?

26. **WRITING MATHEMATICS** Explain why the points $(3, 5)$ and $(5, 3)$ do not name the same point.

· · · STANDARDIZED TEST · · ·

STUDENT PRODUCED ANSWERS Solve each question and on the answer grid write your answer at the top and fill in the ovals.

Notes: Mixed numbers such as $1\frac{1}{2}$ must be gridded as 1.5 or 3/2. Grid only one answer per question. If your answer is a decimal, enter the most accurate value the grid will accommodate.

1. Find the slope of the line $3x - 4y = 12$.

2. If $f(x) = 2x^2 - 3x - 12$, find $f(-3)$.

3. On one particular school day, 8% of all the students were absent. This was 24 students. How many students attend the school?

4. Use the order of operations to evaluate $72 \div 2 - 4 \cdot 3^2$.

5. The four corners of a rectangle are located at $(8, 2)$, $(8, -5)$, $(-3, -5)$ and $(-3, 2)$. Disregarding the units, find the sum of the perimeter and the area of the rectangle.

6. If $3x - 8(kx + 2) = 19x - 16$, then what is the value of k?

A baseball player's results for his last 80 at-bats are shown in the table below. Use these data to answer Questions 7 and 8.

Result	Frequency
Single	12
Double	3
Triple	1
Home Run	2
Out	62

7. In his next at-bat, what is the probability that the player will get a single? Enter the answer as a decimal.

8. In a circle graph, what percent of the circle would represent the player's outs?

9. On the first nine holes, a golfer scored 3, 5, 4, 4, 3, 4, 5, 4, and 5. Find the mean number of strokes per hole.

10. Find the least integer that satisfies $4x + 2.5 < 6x + 8.2$.

11. Deborah had $1500 to invest. She deposited some of the money into an account that pays 6% annual interest and the rest into an account that pays 8% annual interest. If Deborah earns $111.60 total interest, how much did she invest at 8%?

12. If the points $(t, 6)$ and $(2, t)$ are both on the line $4x + 2y = b$, determine the value of t.

13. Consider the points on the graph of the equation $y = x^2$. For any $y > 0$, what is the quotient of the x-coordinates?

14. When you graph the solutions to $7 - 6x > 9$, what is the first integer you will show?

15. Find the x- and y-intercepts of the graph of the line $5x + 2y = 15$. Enter the product of the nonzero coordinates.

16. A student had test scores of 78, 85, 82, and 81 on four tests. To earn a B, the student has to have an average of at least 80. What is the least score the student can get on the fifth test and still recieve a B?

17. A person has 24 coins, all dimes and quarters. There were three times as many quarters as dimes. How much money did the person have? Enter the amount, omitting the dollar sign.

18. Find the slope of any line perpendicular to the graph of the line $5x - 7y = 35$. Enter the answer as a fraction.

19. Find the value of x in the matrix.
$$\begin{bmatrix} y & 5 \\ -2 & 3z \end{bmatrix} - \begin{bmatrix} -2 & y + z \\ 4 & 5 \end{bmatrix} = \begin{bmatrix} 5 & x \\ -6 & 10 \end{bmatrix}$$

Take a Look
AHEAD

Make notes about things that look new.
- Do you think the methods for solving systems of equations are the same or different than for single equations? Why?
- What are names of the solving methods in this chapter? Does the name tell you anything about the method?

Make notes about things that look familiar.
- Do you recognize the types of equations that are graphed in this chapter? Describe what you see.
- Find some different ways that addition is used in this chapter.

DATA*Activity*

City Streets

Municipal planners usually try to lay out streets and avenues using a systematic method so that people can easily find their way around. For example, in the New York City borough of Manhattan, avenues such as Fifth or Park run north and south, while streets such as 14th or 42nd run east and west. Moreover, there is a helpful little guide that helps people locate the intersecting street nearest to a given building number. A portion of this guide is shown on the next page.

SKILL FOCUS

▶ Add, subtract, multiply, and divide real numbers.

▶ Interpret maps and street directions.

▶ Solve multiple-step problems.

▶ Work backwards.

CITIES *and* MUNICIPALITIES

In this chapter, you will see how:

- **SANITATION MANAGERS** use graphs and equations to analyze costs of waste disposal programs. (Lesson 7.2, page 334)

- **TRAFFIC CONTROLLERS** use formulas to help regulate the flow of motor vehicles. (Lesson 7.3, page 341)

- **CIVIC LEADERS** use algebra to help plan fund-raising events. (Lesson 7.6, page 359)

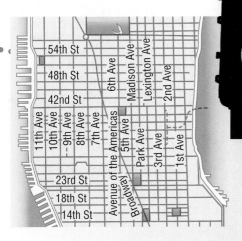

KEY TO MANHATTAN STREET NUMBERS

Cancel last figure of address. Divide remainder by two and add the key number below. Result is approximately nearest street.	Broadway–1 to 754, Below 8th St.
	754 to 858.............DEDUCT 29
	858 to 958.............DEDUCT 25
	Above 958............DEDUCT 31
Ave A,B,C,D.....................3	Columbus Ave......................59
1st Ave...............................3	Lexington Ave.....................22
2nd Ave..............................3	Madison Ave.......................27
3rd Ave..............................10	Park Ave............................34
4th Ave................................8	West End Ave......................59
5th Ave–1 to 200................13	
201 to 400......................16	EXCEPTIONS: Cancel last figure of address and add or deduct key.
401 to 600......................18	
601 to 775......................20	
Ave. of Americas........DEDUCT 12	5th Ave–775 to 1286.......DEDUCT 18
7th Ave–1 to 1800..............12	Riverside Drive–1 to 567.....Add 73
Above 1800.....................20	Above 567.................Add 78
8th Ave................................9	Central Park West.........Add 60
9th, 10th Ave......................13	Going east and west, street numbers begin at Fifth Avenue with the number 1 and increase by 100 per Avenue.
11th Ave............................15	
Amsterdam Ave..................59	

Use the map and guide to answer each question.

1. Janet is going to a restaurant at 546 Fifth Ave. What is the nearest street?

2. Lucas is walking to Grand Central Station at 42nd Street and Lexington Avenue. His current location is 800 Lexington Avenue. In which direction should he walk and how many blocks?

3. Carmen has a job interview at 1648 Broadway, so she took the Broadway bus to 113th Street. Do you think Carmen found the right address there? Explain.

4. **WORKING TOGETHER** Carnegie Hall, a location famous for concerts and other performances, is at 57th Street and Seventh Avenue. Work with a partner to estimate the address of Carnegie Hall.

323

City Planning

Mayors, council members, managers, and the local chamber of commerce work very hard to make their city an attractive place for people to live and work. By "marketing" their city—promoting its best features, such as climate, clean air, or low taxes—these professionals try to attract businesses and individuals to the area. More residents and commerce in a city translates into a better economy and more city services.

PROJECT GOAL

To analyze features of cities using statistics and algebra and to prepare a promotional brochure for an imaginary city.

Getting Started

Work in groups of four or five students. Use these ideas to get started.

1. Brainstorm a list of features that make a city attractive to individuals, families, or businesses.

2. To learn more about cities, research the "most livable" cities in the United States. Find out what each of these cities offers to businesses, individuals, and families.

PROJECT *Connections*

Lesson 7.2, page 333:
Use graphing to compare revenues generated by different sales tax rates.

Lesson 7.3, page 340:
Analyze housing costs in different United States cities.

Lesson 7.5, page 352:
Create visual displays for several types of population statistics.

Chapter Assessment, page 367:
Prepare a promotional brochure for a city.

3. Discuss all the ideas and agree on a procedure for selecting a final list of ten to fifteen positive features for a city. Some features are more important to businesses while others appeal to families or individuals, so try to include a balance.

4. Select guidelines that are consistent with one another.

5. Design a city with the features you selected. Decide on a name, location, area, and population of your city. Throughout this project, you will decide about other aspects of your city.

Internet Connection

www.swpco.com/ swpco/algebra1.html

7.1 Algebra Workshop
Explore Systems of Linear Equations

Think Back

Work with a partner. Use the graph to answer each question.

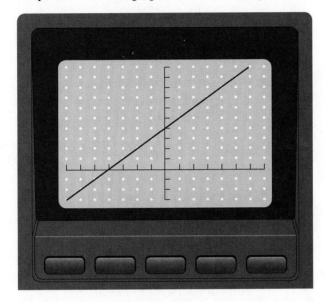

SPOTLIGHT ON LEARNING

WHAT? In this lesson you will learn
- to graph pairs of linear equations using a graphing utility.
- to interpret the point of intersection of two graphs.
- to solve systems of linear equations using a graphing utility.

WHY? Graphing systems of linear equations can help you pinpoint the location of a forest fire.

1. Approximate the y-coordinate of the point on the line with the given x-coordinate.

    a. -4 **b.** 6 **c.** -2

2. Approximate the x-coordinate of the point on the line with the given y-coordinate.

    a. -3 **b.** 1 **c.** 8

3. What is the relationship between the x- and y-coordinates of a point on a line and the equation of the line?

4. What is an equation of the line in the graph?

Explore

5. Use a graphing utility to graph the line $y = 2x - 0.8$.

6. Set your graphing utility so that integers are displayed when using the TRACE feature. Make a table of about 8 ordered pairs (x, y) that are on the line $y = 2x - 0.8$.

7. Delete the graph of the equation $y = 2x - 0.8$ and graph only the equation $y = -x + 5.2$.

Algebra Workshop

8. Trace along the line and make a table of about 8 ordered pairs (x, y) that are on the line $y = -x + 5.2$.

9. Examine your tables of ordered pairs. Write the coordinates of a point that is in both tables. If you cannot find such a point, write the coordinates of a point close to points on both tables.

10. Without graphing, how do you know that the graphs of $y = 2x - 0.8$ and $y = -x + 5.2$ will intersect?

11. Where do you think the lines will intersect? Check by graphing both lines on the same set of axes. Use the TRACE feature to find the point of intersection.

12. Clear the graph screen. On the same set of axes, graph the equations $y = -0.4x + 0.6$ and $y = -1.2x - 3$. Use the TRACE feature and, if necessary, the ZOOM feature to find the point of intersection.

13. Clear the graph screen. On the same set of axes, graph the equations $y = 1.7x + 3.3$ and $y = 1.7x - 2.8$. (If you used the ZOOM feature in Question 12, you may need to reset the range.) What happens when you try to find the point of intersection? Explain.

14. Write a pair of equations and have your partner find the point of intersection of the graphs of the equations. How can you be sure the pair of equations you write will intersect?

Make Connections

15. Check to see that the ordered pair $(2, 3)$ satisfies each of the equations $y = 2x - 1$ and $y = 3x - 3$. Explain how you made the check.

16. Without graphing the equations $y = 2x - 1$ and $y = 3x - 3$, how do you know that the point $(2, 3)$ is on each graph?

17. Complete the statement: If the coordinates of a point satisfy two linear equations, then that point is the __?__ of the graphs of the equations.

18. The graphs of the equations $y = 6x - 5$ and $y = 2x - 1$ intersect at the point $(1, 1)$. How do you know that the point $(1, 1)$ is on each graph?

19. How do you know that $x = 1$ and $y = 1$ are a solution of $y = 6x - 5$ and $y = 2x - 1$?

20. Complete the statement: If two lines intersect, then the coordinates of the point of intersection __?__.

Summarize

21. **WRITING MATHEMATICS** You are given two linear equations and you want to find an ordered pair (x, y) that is a solution of both equations. Describe a method you could use to find the ordered pair using a graphing utility. How could you check the ordered pair?

22. **MODELING** Use a graphing utility to show that $x = 2$ and $y = 1$ are a solution of $y = 2x - 3$ and $y = 3x - 5$.

23. Find an ordered pair (x, y) that is a solution of $y = 4x + 7$ and $y = -2x + 1$.

24. **CRITICAL THINKING** Suppose that an infinite number of ordered pairs satisfy two given linear equations. Describe the graphs of the equations.

25. **CRITICAL THINKING** Suppose that no ordered pair satisfies two given equations. Describe the graphs of the equations.

26. **CRITICAL THINKING** A forest ranger spots a fire and radios his position and the compass direction of the fire to headquarters. Another ranger spots the same fire and radios her position and the compass direction of the fire to headquarters. Explain how headquarters can pinpoint the location of the fire.

27. **GOING FURTHER** The graph of the equation $y = x^2 + x + 2$ is a curve called a *parabola*. Use a graphing utility to find where the parabola and the graph of the equation $y = x + 3$ intersect.

Explore

The table at the right gives Census Bureau figures for the populations of Baltimore and San Diego in 1970 and 1990.

Population, in millions		
	1970	1990
Baltimore	0.906	0.736
San Diego	0.697	1.111

1. Draw a set of coordinate axes showing the years from 1970 to 1990 on the horizontal axis and population on the vertical axis. Then graph the data in the table.

2. Draw a line segment showing the change in population for each city from 1970 to 1990. What assumption are you making when you show the change as a line?

3. What is the significance of the point where the line segments intersect?

4. When were the populations of Baltimore and San Diego the same? What were the populations at that time?

5. Suppose you drew line segments showing the population changes of two other cities from 1970 to 1990. Would the segments intersect? Explain.

CHECK UNDERSTANDING

What actual numbers of people do the decimals 0.906 and 0.736 in the table represent? What increments could you use along the y-axis for population?

Build Understanding

In Explore, you worked with two equations that had the same variables. Until now, you have looked at individual equations mostly as separate items unrelated to other equations. Look at these two equations.

$$x + y = 7 \qquad x - y = 3$$

Think of them as a closely related pair. Two or more linear equations that are considered *together*, such as these two equations, form a **system of linear equations** or simply a **linear system**.

From your earlier work, you know that there are an infinite number of ordered pairs (x, y) that are solutions of the equation $x + y = 7$ and also for the equation $x - y = 3$. Here are some of them.

$$x + y = 7 \rightarrow (1, 6), (2, 5), (3, 4), (4, 3), \mathbf{(5, 2)}, (6, 1), (7, 0)$$
$$x - y = 3 \rightarrow (8, 5), (7, 4), (6, 3), \mathbf{(5, 2)}, (4, 1), (3, 0), (2, -1)$$

Notice that $(5, 2)$ is a solution of *both* equations. So, the ordered pair $(5, 2)$ is a solution of this system.

An ordered pair that is a solution of all the equations in a system is a *solution of the system.*

COMMUNICATING ABOUT ALGEBRA

Suppose you find that an ordered pair does not satisfy the first equation in a system. Can you conclude that the ordered pair is not a solution of the system? Why or why not?

EXAMPLE 1

Determine whether $(-2, 2)$ is a solution of the system.

$$\begin{cases} y = -3x - 4 & \textbf{Equation 1} \\ y = -2x - 2 & \textbf{Equation 2} \end{cases}$$

Solution

To determine whether $(-2, 2)$ is a solution, substitute -2 for x and 2 for y in both equations in the system.

Equation 1	**Equation 2**
$2 \stackrel{?}{=} -3(-2) - 4$	$2 \stackrel{?}{=} -2(-2) - 2$
$2 \stackrel{?}{=} 6 - 4$	$2 \stackrel{?}{=} 4 - 2$
$2 = 2 \checkmark$	$2 = 2 \checkmark$

So, $(-2, 2)$ is a solution of the system. ◄

Look at the graphs of both the equations in Example 1. Notice that the graphs intersect at $(-2, 2)$, which is a solution of the system. This suggests that one way to solve a linear system is to graph the equations and find the point of intersection. You can graph a system using paper and pencil or a graphing utility.

CHECK UNDERSTANDING

When a linear equation is written in slope-intercept form, how can you find the slope of the line? the *y*-intercept? How can you draw the graph?

EXAMPLE 2

Solve the system by graphing.

$$\begin{cases} 2x + y = 3 \\ x - y = 3 \end{cases}$$

Solution

Write the equations in slope-intercept form, $y = mx + b$, so that they can be graphed easily.

$$\begin{cases} y = -2x + 3 \\ y = x - 3 \end{cases}$$

The graphs appear to intersect at the point $(2, -1)$.

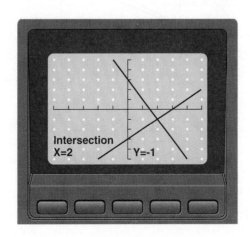

Intersection
X=2   Y=-1

Substitute 2 for x and -1 for y into each equation.

Equation 1	**Equation 2**
$2x + y = 3$	$x - y = 3$
$2(2) + (-1) \stackrel{?}{=} 3$	$2 - (-1) \stackrel{?}{=} 3$
$4 - 1 \stackrel{?}{=} 3$	$2 + 1 \stackrel{?}{=} 3$
$3 = 3 \checkmark$	$3 = 3 \checkmark$

The solution is $(2, -1)$. ◄

You can use systems of equations to model real world situations.

EXAMPLE 3

ECONOMICS When she graduated from college, Marci was offered two jobs. One paid a starting salary of $30,000 annually, plus guaranteed increases of $2,000 a year. The second paid a starting salary of $25,000, plus guaranteed increases of $2,500 per year. Considering only this information, which job should she take?

Solution

Let x represent the number of years since Marci began work. Let y represent her annual salary. Make a verbal model. Then write two equations.

$$\boxed{\text{Annual salary}} = \boxed{\text{Yearly increase}} + \boxed{\text{Starting salary}}$$

$$\begin{cases} y = 2{,}000x + 30{,}000 & \textbf{Salary 1} \\ y = 2{,}500x + 25{,}000 & \textbf{Salary 2} \end{cases}$$

Graph the system. If you use a graphing utility, set an appropriate range for the viewing window, such as:

Xmin	0	Ymin	0
Xmax	15	Ymax	60000
Xscl	1	Yscl	5000

You may need to use the ZOOM and TRACE features.

The graph shows that although Salary 2 is lower than Salary 1 at first, it increases at a greater rate than Salary 1. The point of intersection $(10, 50{,}000)$ tells you that both salaries will be $50,000 in 10 years (that is, when $x = 10$). Thereafter, Salary 2 is greater than Salary 1.

PROBLEM SOLVING TIP

In Example 3, you can use patterns to write the equations that express Marci's salaries.

The graph shows:
- $y = 2{,}500x + 25{,}000$
- $y = 2{,}000x + 30{,}000$
- $(10, 50{,}000)$

If Marci intends to stay with the company and is looking for higher salaries later in life, she should consider taking the lower starting salary. If she is uncertain that the job will be permanent, she should take the higher starting salary. Over the first 10 years Marci will have collected $27,500 more by having taken the higher starting salary. ◄

TRY THESE

Determine whether the ordered pair is a solution of the system of equations.

1. $(3, 5)$ $\begin{cases} y = 4x - 7 \\ y = -x + 8 \end{cases}$

2. $(2, -3)$ $\begin{cases} 2x + y = 1 \\ 3x - y = 3 \end{cases}$

3. $(0, 2)$ $\begin{cases} y = -x + 2 \\ 3y = -3x + 6 \end{cases}$

4. $(-3, 2)$ $\begin{cases} y = x + 5 \\ y = x + 3 \end{cases}$

Solve the system by graphing.

5. $\begin{cases} x + y = 10 \\ x - y = 4 \end{cases}$

6. $\begin{cases} y = x - 4 \\ y = 3x - 6 \end{cases}$

7. $\begin{cases} 4x - y = -3 \\ -3x + 2y = -4 \end{cases}$

HOUSE REPAIRS Thrifty Rentals rents a sander for a flat fee of $8, plus $3 per hour. Premier Rentals rents the same sander for a flat fee of $5, plus $4 per hour.

8. Write a linear system you can solve to find the length of rental time for which both firms charge the same amount.

9. Solve the system by graphing.

10. Suppose your neighbor hires you to sand her wood deck. Which sander would you rent? Why?

11. WRITING MATHEMATICS Is it always easier to use a graphing utility to solve a system? When might you prefer to use pencil and paper? Write a list of advantages and disadvantages for each method.

PRACTICE

Determine whether the ordered pair is a solution of the system of equations.

1. $(1, 2)$ $\begin{cases} y = 3x - 1 \\ y = -x + 3 \end{cases}$

2. $(0, 4)$ $\begin{cases} 5x - y = 1 \\ 3x + 1 = y \end{cases}$

3. $(-2, 3)$ $\begin{cases} y = 3x + 3 \\ y = -x + 1 \end{cases}$

4. $(0.5, 1)$ $\begin{cases} 2x + 3y = 4 \\ -8x - 2y = -6 \end{cases}$

FABRIC Sue bought 3 yd of cotton fabric and 2 yd of rayon fabric at Fabrics Inc. The bill, not including tax, was $28. Her friend bought 2 yd of cotton and 1 yd of rayon for $17. Let x represent the unit price of cotton fabric and let y represent the unit price of rayon fabric.

5. Write a linear system you can solve to find the unit prices of cotton and rayon fabric.

6. Solve the system by graphing.

7. What was the unit price of cotton fabric? of rayon fabric?

Solve the system by graphing.

8. $\begin{cases} y = 2x - 4 \\ y = 5x - 13 \end{cases}$

9. $\begin{cases} x + y = 8 \\ x - y = -2 \end{cases}$

10. $\begin{cases} x - 2y = 4 \\ 3x + y = 5 \end{cases}$

11. $\begin{cases} 5x + 3y = 3 \\ 2x - y = 10 \end{cases}$

12. $\begin{cases} 3x - y = 0 \\ x + 2y = -21 \end{cases}$

13. $\begin{cases} 5x + 4y = 12 \\ 3x - 4y = 4 \end{cases}$

14. $\begin{cases} 5x + 3y = 1 \\ 3x + y = -1 \end{cases}$

15. $\begin{cases} 5x + 6y = 24 \\ x - 3y = 9 \end{cases}$

16. $\begin{cases} -3x + 2y = 8 \\ x + 2y = -8 \end{cases}$

CAR RENTAL Prestige Car Rentals charges $44 per day plus 6¢ per mile to rent a mid-sized vehicle. Getaway Auto charges $35 per day plus 9¢ per mile for the same car. Let x represent the number of miles and let y represent the total rental charge.

17. Write a system of linear equations representing the prices for one day charged by both companies. Write the equations as functions of the number of miles driven.

18. Graph to find the number of miles for which both companies charge the same price.

19. Suppose you need to rent a car for a day. Which company would you rent from? Give reasons for your answer.

20. **WRITING MATHEMATICS** Write a paragraph to a friend who was absent, explaining how to solve a system of linear equations graphically.

SAVINGS Nilda has $250 in her savings account. She plans to save $15 per week from her salary. Iona has only $200 in her account but can save $20 a week from her paycheck. Let x represent the number of weeks and y represent the total amount in the savings account.

21. Write a system of linear equations representing the amount each person will have in her savings account at the end of any week if no money is taken out.

22. Solve the system by graphing.

23. How many weeks will it take before the amount in each savings account will be the same? How much money will be in each account?

EXTEND

Somewhere on the graph of the equation $y = x - 1$ is a point whose x- and y-coordinates have a sum of 7.

24. Write a system of equations you could solve to find the coordinates of the point.

25. Solve the system.

GEOMETRY The graphs of the equations $y = 3$, $y = 3x + 3$, and $y = -x + 11$ contain the sides of a triangle.

26. Graph the triangle and name the coordinates of its vertices.

27. Find the area of the triangle.

28. GEOMETRY Find the vertices of the triangle formed by the graphs of the equations $x = 0$, $x + 2y = 8$, and $2y = 3x$.

29. Where on the graph of $y = 3x - 2$ is the y-coordinate 4 less than the x-coordinate?

THINK CRITICALLY

30. Give an example of a system of two linear equations that has $(2, -4)$ as its only solution.

31. The solution of a system of two distinct linear equations is $(-3, -5)$. Are there any other solutions? Explain. How many different systems have $(-3, -5)$ as a solution?

32. The points $(4, 9)$, $(8, 9)$, $(4, 1)$, and $(8, 1)$ form a rectangle. Find the point where the diagonals of the rectangle intersect.

33. Two equations with different slopes each have y-intercept b. Write the ordered pair that is the solution of the system made up of those equations.

PROJECT *Connection*   Taxes are an important consideration for businesses or individuals, and rates vary widely. The sales tax in a city consists of the state tax plus the city tax. Thus, in a state with a 5% rate, one city may add another 0.75% for a total of 5.75%, while another city may add 3.25% for a total of 8.25%. Higher sales taxes may be justified if revenues are used for city improvements such as bus systems, civic centers, or social welfare programs.

1. Decide on a sales tax rate for your city.

2. Draw a graph showing money spent by city residents and the resulting sales tax revenue generated for rates of 5.75%, 8.25%, and your city's rate. Where do all the lines intersect?

3. Write a paragraph that promotes your city's sales tax rate as a well-planned strategy for a successful community.

Each year, Americans produce enough trash to fill a line of 10-ton disposal trucks reaching half way to the moon. What to do with all the trash is a problem that faces every city sanitation manager. Not only is the amount of trash growing annually, but budget resources are increasingly tight. Sanitation departments must search for alternatives to traditional disposal methods, such as landfills. A large part of sanitation department budgets are devoted to landfill expenses. The projected revenues (income) and expenses for landfill for one city are shown at the left.

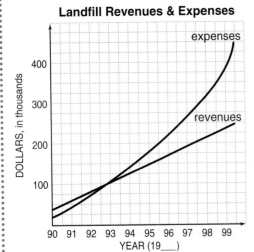

Landfill Revenues & Expenses

Decision Making

1. Which graph shows a constant rate of increase over time? Which shows a changing rate of increase? How can you tell?

2. When do landfill expenses exceed revenues for the first time? How can you tell?

The manager is investigating two recycling programs. In one, each household would separate its own recyclables for collection. In the other, separation would be done by the city at a central facility. Because the first plan requires special trucks and additional collection bins at each residence, its start-up costs are higher than those of the second plan.

Separation Plan	Start-up Costs, X $100,000	Yearly Cost, X $100,000
Individual residences	2.16	1.48
Central facility	0.71	1.95

3. Write an equation for each plan, describing its total cost as a function of the number of years n the plan has been in operation.

4. Graph the cost equation for each plan.

5. During which year will the cost of the central facility plan exceed that of the individual residence plan?

6. Which plan should the city choose? Defend your answer.

7.3 Solve Linear Systems by Substitution

Explore/Working Together

- Work with a partner. Use pencil and paper or a graphing utility.

  1. Graph the system of linear equations.

  $$\begin{cases} y = x \\ y = -x + 2 \end{cases}$$

  2. Name the point of intersection of the graphs.

  3. Graph $7x + 7y = 48$ on the same set of axes. What happens when you try to locate the point of intersection of the new line with the graph of $y = x$?

  4. Why are some systems of linear equations easy to solve graphically while others are not?

Build Understanding

- Graphing works well when the solution to a system of equations consists of integers or fractions that are easy to determine from a graph. Often, however, the coordinates of the point of intersection are hard to determine and you can only approximate the solution to a system. To find an exact solution, you need to use a different method.

One method you can use is the **substitution method**. To solve a system by substitution, you need to solve one equation for one variable in terms of the other and then substitute for that variable in the other equation. The result will be an equation with only one variable. Solve for that variable.

Substitute the value of that variable in either original equation and then solve for the variable you do not know. Be sure to check your solution in both original equations.

Solve the system.
$$\begin{cases} y = x - 4 \\ y = 2 - x \end{cases}$$
Model $y = x - 4$.

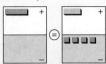

For y on left side, substitute values from right side. Model $y = 2 - x$ on the right side.

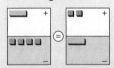

Complete the modeling to solve for x. You should read the solution $2x = 6$; $x = 3$. Then substitute for x in the first equation and solve for y: $y = -1$.

CHECK UNDERSTANDING

In Example 1, why did -1 from the "substitution" step become $+1$ in the "use the distributive property" step?

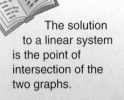
THINK BACK

The solution to a linear system is the point of intersection of the two graphs.

EXAMPLE 1

Solve the system using the substitution method.

$$\begin{cases} y = x - 1 & \textbf{Equation 1} \\ 2x - y = 4 & \textbf{Equation 2} \end{cases}$$

Solution

Equation 1 is already solved for y. So begin by *substituting* the expression $x - 1$ for y in Equation 2.

$$\begin{cases} 2x - y = 4 & \text{Equation 2} \\ 2x - (x - 1) = 4 & \text{Substitute } (x - 1) \text{ for } y. \end{cases}$$

Each of the original equations had two variables. Substitution gives an equation with only one variable, x. Solve for x.

$$\begin{aligned} 2x - x + 1 &= 4 & \text{Use the distributive property.} \\ x + 1 &= 4 & \text{Combine like terms.} \\ x &= 3 & \text{Solve for } x. \end{aligned}$$

Now substitute 3 for x in either of the original equations. In this case Equation 1 looks easier to use.

$$\begin{aligned} y &= x - 1 & \text{Equation 1} \\ y &= 3 - 1 & \text{Substitute 3 for } x. \\ y &= 2 & \text{Solve for } y. \end{aligned}$$

The solution to the system of equations is $(3, 2)$.

Check

Check the solution in both original equations.

Equation 1	**Equation 2**
$y = x - 1$	$2x - y = 4$
$2 \stackrel{?}{=} 3 - 1$	$2(3) - 2 \stackrel{?}{=} 4$
$2 = 2 ✓$	$6 - 2 \stackrel{?}{=} 4$
	$4 = 4 ✓$

You can also check the solution graphically. ◀

In Example 1, one equation was already solved in terms of the other variable. More often, however, you will have to solve one equation for one variable before you can substitute. Since you will obtain the same solution no matter which equation you choose to solve for the variable, you should choose the equation with the variable easiest to isolate.

EXAMPLE 2

Solve the system by substitution.

$$\begin{cases} 4x + 3y = 4 & \textbf{Equation 1} \\ 3x + y = -2 & \textbf{Equation 2} \end{cases}$$

Solution

The variable easiest to isolate is y in Equation 2.

$$3x + y = -2 \qquad \text{Equation 2}$$
$$y = -3x - 2 \qquad \text{Solve for } y.$$
$$4x + 3(-3x - 2) = 4 \qquad \text{Substitute } (-3x - 2) \text{ for } y \text{ in Equation 1.}$$
$$4x - 9x - 6 = 4 \qquad \text{Use the distributive property.}$$
$$-5x - 6 = 4 \qquad \text{Combine like terms.}$$
$$-5x - 6 + 6 = 4 + 6 \qquad \text{Add 6 to both sides.}$$
$$-5x = 10 \qquad \text{Simplify.}$$
$$x = -2 \qquad \text{Divide both sides by } -5.$$

Having found x, you can now substitute in Equation 2.

$$3(-2) + y = -2 \qquad \text{Substitute } -2 \text{ for } x.$$
$$-6 + y = -2 \qquad \text{Simplify.}$$
$$y = 4 \qquad \text{Solve for } y.$$

The solution to the system of equations is $(-2, 4)$. Check the solution in both original equations. A graphical solution is shown at the right. ◄

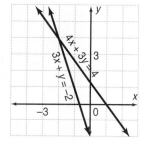

You can use substitution to solve real world problems.

EXAMPLE 3

MARKETING Josh cannot find a publisher for a book he wrote, so he decides to publish the book himself. He calculates that his fixed costs will be $2665 and that the unit cost of each book will be an additional $1.70. He intends to sell the book for $4.95. How many copies must he sell to break even? ("Break even" means that income equals expenses.)

Solution

Let x represent the number of copies Josh must sell to break even. Let y represent the break-even amount. Write two equations, one representing expenses and one representing income.

$$y = 1.7x + 2665 \qquad \textbf{Expenses equation}$$

$$y = 4.95x \qquad \textbf{Income equation}$$

$$(4.95x) = 1.7x + 2665 \qquad \text{Substitute } 4.95x \text{ for } y \text{ in the expenses equation.}$$

$$3.25x = 2665 \qquad \text{Subtract } 1.7x \text{ from both sides.}$$

$$x = 820 \qquad \text{Divide both sides by } 3.25.$$

Josh must sell 820 copies to break even. ◄

Use the substitution method to solve the system.

1. $\begin{cases} x = 5y \\ x - 3y = 6 \end{cases}$

2. $\begin{cases} x = -2y - 2 \\ 3x - 2y = 10 \end{cases}$

3. $\begin{cases} x - y = 1 \\ y = -x + 5 \end{cases}$

4. $\begin{cases} y + 6 = 3x \\ 9x - 2y = 3 \end{cases}$

5. $\begin{cases} 3x - 2y = -3 \\ 3x + y = 3 \end{cases}$

6. $\begin{cases} 3x - 4y = 8 \\ 4x + y = 17 \end{cases}$

ECONOMICS Ricardo invested part of $8000 in a money market account and the remaining part in a mutual fund. At the end of one year, the money market had yielded 6% on the amount invested, and the mutual fund had yielded 8%. The total yield on the two accounts was $530.

7. How much did Ricardo invest at each rate? Let x represent the amount invested at 6% and let y represent the amount invested at 8%. Write a system of linear equations you could solve to find the amounts.

8. Solve the system.

9. WRITING MATHEMATICS You are about to use the substitution method to solve Exercise 5. Describe how to decide which variable to solve for.

PRACTICE

Use the substitution method to solve the system.

1. $\begin{cases} x - 2y = 16 \\ 4x + y = 1 \end{cases}$

2. $\begin{cases} x + 2y = 6 \\ 4x + 3y = 4 \end{cases}$

3. $\begin{cases} y = -4x + 5 \\ 2x - 3y = 13 \end{cases}$

4. $\begin{cases} 5x - y = -23 \\ 3x - y = -15 \end{cases}$

5. $\begin{cases} -2x + y = 2 \\ 2x + 3y = 6 \end{cases}$

6. $\begin{cases} x + 5y = 11 \\ 4x - y = 2 \end{cases}$

7. $\begin{cases} 2x + 3y = 11 \\ 3x + 3y = 18 \end{cases}$

8. $\begin{cases} 5x + 3y = 4 \\ 4x - 2y = 1 \end{cases}$

9. $\begin{cases} 5x + 7y = -3 \\ 2x + 14y = 2 \end{cases}$

10. $\begin{cases} 3x + 6y = 6 \\ 2x = 2y \end{cases}$

11. $\begin{cases} x = 10y \\ \dfrac{1}{2}x = 3y + 2 \end{cases}$

12. $\begin{cases} x - 7 = y \\ \dfrac{1}{4}x - 1 = y \end{cases}$

GEOMETRY The measure of one of the two acute angles in a right triangle is five times the measure of the other acute angle.

13. Let x represent the measure of one acute angle and let y represent the measure of the other. Write a system of equations you could solve to find the measures of the angles.

14. Solve the system.

PLUMBING COSTS Reliable Plumbing charges $50 for a service call, plus an additional $40 an hour for labor. Paula's Plumbing charges $30 for a service call, plus an additional $45 an hour for labor.

15. Let x represent the number of hours of labor and let y represent the total charge. Write a system of equations you could solve to find the length of a service call for which both companies charge the same amount.

16. Solve the system.

17. Which company would you use? Why?

18. WRITING MATHEMATICS Write a paragraph describing a linear system that you would prefer to solve by graphing. Describe another linear system that you would prefer to solve by substitution. Provide reasons for your choice.

EXTEND

Solve the system by simplifying and then using the substitution method.

19. $\begin{cases} 2(x + 1) + 3(y - 3) - 4y + 4 = -5 \\ 4(2x + y) - 3(x + y) - x = 5 \end{cases}$

20. $\begin{cases} 4(2x + 2y - 1) - 3(x + 2y + 4) - 2\left(x + \dfrac{1}{2}y\right) = 0 \\ 2(x + y + 1) + 3(x - y - 2) + 2(3 - 2x) = -2 \end{cases}$

21. Write a system of linear equations that has the solution $(3, -5)$. Then show how to solve the system by substitution.

22. CIRCUS TICKETS In one day, the Easy Ticket Agency sold 395 tickets to the Barcelona Circus, some at $28 and some at $22. The total amount the agency collected that day was $10,130. How many tickets were sold at each price?

23. NUMBER THEORY The sum of the digits of a two-digit number is 8. The number is seven times the ones digit. Find the number. (*Hint*: Let x represent the tens digit and let y represent the ones digit. Then you can represent the number as $10x + y$.)

24. CHEMISTRY Solution A is 10% acid. Solution B is 35% acid. A chemist wants to mix portions of each solution to obtain 200 cm³ of liquid that is 25% acid. How much of each solution should the chemist use?

SPORTS TRIP Roy can row his boat 4 mi upstream in 2 h and can make the return trip in 1 h.

25. Find the speed of the current c and the rate of the boat r in still water. (*Hint*: Use $d = rt$. Represent the rate with the current and against the current.)

26. Roy's friend's boat can go 10 mi downstream in 50 min and 12 mi upstream in $1\dfrac{1}{2}$ h. Find the speed of the current and the rate of the boat in still water.

Use the substitution method to solve for x, y, and z.

27. $\begin{cases} 2x - 3y - z = -1 \\ 5y - 3z = 7 \\ 5z = 5 \end{cases}$

28. $\begin{cases} 4x - y + 3z = 4 \\ -3y + 2z = 1 \\ -2z = -10 \end{cases}$

29. Carmen is solving the system $2x - y = 4$ and $y = 4x - 10$ by substitution. She obtained the equation $-2x - 10 = 4$. What did she do wrong?

30. The solution to the system $Ax + 3y = 7$ and $2x - By = 13$ is $(-1, 3)$. Find A and B.

THINK CRITICALLY

Use the substitution method to solve. (*Hint*: Let m represent $\dfrac{1}{x}$ and let n represent $\dfrac{1}{y}$.)

31. $\begin{cases} \dfrac{1}{x} + \dfrac{1}{y} = \dfrac{3}{4} \\ \dfrac{3}{x} - \dfrac{1}{y} = \dfrac{1}{4} \end{cases}$

32. $\begin{cases} \dfrac{2}{x} - \dfrac{2}{y} = \dfrac{1}{2} \\ \dfrac{1}{x} + \dfrac{5}{y} = \dfrac{3}{4} \end{cases}$

33. The line $Ax + By = 9$ passes through the points $(2, 1)$ and $(-3, 3)$. Find A and B.

34. The line $Ax + By + 3 = 0$ passes through the points $(-1, 2)$ and $(3, 0)$. Find A and B.

35. What happens when you try to solve the system $6x - 3y = -8$, $y = 2x - 3$? Why?

PROJECT *Connection* The cost of housing is an important factor in the decision to relocate to a new city. Most real estate companies compile an index comparing average cost of a certain size house or apartment in different U.S. cities. The sample index below shows the average price of a 2200-ft^2 house with 4 bedrooms, $2\frac{1}{2}$ baths, a family room, and a two-car garage.

1. Contact a real estate agent to determine the average price of a 4-bedroom house in your area. Also ask if the company has a current index for U.S. cities.

2. To best market your ideal city, where do you think it should place in the rankings? Why? Are there any disadvantages to having one of the lowest prices? Decide on an average price that corresponds to the ranking you want.

City	Cost
Dallas	$127,966
Houston	$133,731
Las Vegas	$134,625
Phoenix	$137,350
Denver	$139,753
Salt Lake City	$144,967
Albuquerque	$148,250

3. Create a graphic display to include all the information on your city's housing costs in relation to other places.

Career
Traffic Controller

Traffic jams, streets snarled with cars, clogged intersections—city driving is rarely the time-saver it was meant to be. Surveys show that in cities such as Washington, D.C., the average speed of cars is actually *slower* than the rate of horse-drawn vehicles a century ago.

To combat modern traffic problems, city public works departments employ *traffic controllers*. Among their many responsibilities are the maintenance and timing of traffic lights. The length of time a light remains green can spell the difference between smoothly flowing traffic and a traffic jam. To determine the green-light time interval, traffic controllers sometimes use **Greenshield's formula**

$$t = 2.1n + 3.7$$

where t represents the time in seconds a light should remain green to maintain traffic flow and n represents the average number of cars per lane that must move through the light.

Decision Making

At a busy intersection in Clifton, a traffic controller collected the data shown in the table at the left on the buildup of cars while the light was red.

Red-light time, seconds	Number of cars, per lane
3.0	0
7.3	2
11.5	4
15.9	6

1. Let x represent the number of cars per lane and let y represent the red-light time. Use a graphing utility to find the line of best fit for the data.

2. Find the correlation coefficient. Is the equation linear?

3. If the light remained red for 35 seconds, how many cars would you expect to find in each lane?

4. Rewrite Greenshield's formula using x for n and y for t. Use substitution to find the point of intersection of the graph of the equation you wrote in Question 1 and the graph of Greenshield's formula. Round to the nearest hundredth.

5. How long should the light remain green at the intersection? How long should the light remain red?

6. About how many cars in each lane will move through the green light during each cycle?

Explore/Working Together

● Work with a partner. Use Algeblocks.

1. Model each of the following equations on a sentence mat.
   a. $x - 3 = 2$ **b.** $x + 1 = 6$

2. Solve the equations. What are your results?

3. Model the equation in Question 1a again. Without removing the blocks, add blocks to model the equation in Question 1b on the *same* mat. What single equation is represented by the combination of blocks from both equations?

4. Solve the new equation from Question 3.

5. Compare the solutions of the equations you added with the solution of the "sum" equation in Question 3. How did adding the equations affect the solutions?

6. Write two equations with the same solution. Model the equations on the same sentence mat and solve the "sum" equation. What is true about the solution?

Build Understanding

● As you saw in Explore, you can add two equations left side to left side and right side to right side. You can add or subtract equations to eliminate a variable. This is called the **elimination method** for solving systems of linear equations. When the coefficients of the variable you want to eliminate are *opposites*, add the equations.

EXAMPLE 1

Solve the system using the elimination method.

$$\begin{cases} 2x + 5y = -4 & \textbf{Equation 1} \\ 4x - 5y = 22 & \textbf{Equation 2} \end{cases}$$

Solution

The coefficients of the y-terms are opposites. By adding the equations, you can eliminate the y-term. This leaves an equation with only one variable, which you can solve for x.

$$\begin{array}{rl} 2x + 5y = -4 & \\ 4x - 5y = 22 & \\ \hline 6x = 18 & \text{Add the equations.} \\ x = 3 & \text{Solve for } x. \end{array}$$

To find y, substitute 3 for x in either of the original equations.

$$2(3) + 5y = -4 \quad \text{Substitute for } x \text{ in Equation 1.}$$
$$6 + 5y = -4 \quad \text{Simplify.}$$
$$5y = -10 \quad \text{Add } -6 \text{ to both sides.}$$
$$y = -2 \quad \text{Solve for } y.$$

Check the solution, $(3, -2)$, in both original equations. ◄

If the coefficients of the variable you want to eliminate are *identical*, subtract the equations. If the equations are not in standard form $Ax + By = C$, begin by writing them in standard form.

CHECK UNDERSTANDING

Why should equations be written in standard form before they are added or subtracted?

EXAMPLE 2

Solve the system using the elimination method.

$$\begin{cases} 9x - 11 = 2y & \textbf{Equation 1} \\ 5x - 2y = 15 & \textbf{Equation 2} \end{cases}$$

Solution

$$\begin{cases} 9x - 2y = 11 \\ 5x - 2y = 15 \end{cases} \quad \text{Write Equation 1 in standard form so like terms are arranged in columns.}$$

The coefficients of the y-terms are *identical*. Subtract the equations to eliminate the y-term.

$$\begin{array}{r} 9x - 2y = 11 \\ 5x - 2y = 15 \\ \hline 4x = -4 \quad \text{Subtract the equations.} \\ x = -1 \quad \text{Solve for } x. \end{array}$$

$$9x - 2y = 11 \quad \text{Use Equation 1.}$$
$$9(-1) - 2y = 11 \quad \text{Substitute } -1 \text{ for } x.$$
$$-9 - 2y = 11$$
$$-2y = 20$$
$$y = -10 \quad \text{Solve for } y.$$

Check the solution, $(-1, -10)$, in both original equations. ◄

THINK BACK

Remember that subtracting one number from another is the same as adding the opposite of the number.

COMMUNICATING ABOUT ALGEBRA

Find an illustration of a system of linear equations in the picture of the airport parking facility. Describe why you think your choice is a linear system.

7.4   **Solve Linear Systems by the Elimination Method**   **343**

You may find that neither variable has coefficients with equal absolute values, so you cannot immediately add or subtract to eliminate a variable. However, you may be able to obtain coefficients that you can add or subtract by multiplying both sides of one of the equations by a constant.

EXAMPLE 3

Solve the system using the elimination method.

$$\begin{cases} 3x + 8y = -1 & \textbf{Equation 1} \\ 5x + 2y = -13 & \textbf{Equation 2} \end{cases}$$

Solution

Neither variable has coefficients with equal absolute values. But you can obtain y-terms with coefficients of 8 by multiplying both sides of Equation 2 by 4. Then subtract to eliminate the y-term.

$$
\begin{array}{llll}
3x + 8y = -1 & \rightarrow & 3x + 8y = -1 & \\
5x + 2y = -13 & \rightarrow & \underline{20x + 8y = -52} & \text{Multiply \textit{both sides} by 4.} \\
& & -17x \quad\;\; = 51 & \text{Subtract the equations.} \\
& & x \quad\quad\;\; = -3 & \text{Solve for } x.
\end{array}
$$

$$
\begin{array}{ll}
3(-3) + 8y = -1 & \text{Substitute } -3 \text{ for } x \text{ in Equation 1.} \\
8y = 8 & \text{Simplify.} \\
y = 1 & \text{Solve for } y.
\end{array}
$$

Check the solution, $(-3, 1)$, in both original equations. ◄

You may have to multiply *each* equation by a constant in order to obtain coefficients with equal absolute values.

EXAMPLE 4

Solve the system using elimination.

$$\begin{cases} 3x + 2y = 7 & \textbf{Equation 1} \\ 4x + 7y = -8 & \textbf{Equation 2} \end{cases}$$

Solution

You can obtain x-terms with coefficients of 12 if you multiply both sides of Equation 1 by 4 and both sides of Equation 2 by 3.

$$
\begin{array}{llll}
3x + 2y = 7 & \rightarrow & 12x + 8y = 28 & \text{Multiply both sides by 4.} \\
4x + 7y = -8 & \rightarrow & \underline{12x + 21y = -24} & \text{Multiply both sides by 3.} \\
& & -13y = 52 & \text{Subtract the equations.} \\
& & y = -4 & \text{Solve for } y.
\end{array}
$$

$$
\begin{array}{ll}
3x + 2(-4) = 7 & \text{Substitute } -4 \text{ for } y \text{ in Equation 1.} \\
3x = 15 & \text{Simplify.} \\
x = 5 & \text{Solve for } x.
\end{array}
$$

Check the solution, $(5, -4)$, in both original equations. ◄

Elimination may be the best method for solving a real world problem.

EXAMPLE 5

GROCERY SHOPPING At Monpaz Natural Foods, 3 lb of walnuts and 6 lb of almonds sell for $51. Five lb of walnuts and 4 lb of almonds sell for $55. Find the price of 1 lb of walnuts and 1 lb of almonds.

Solution

Let w represent the price of 1 lb of walnuts and let a represent the price of 1 lb of almonds.

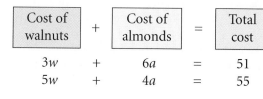

Cost of walnuts	$+$	Cost of almonds	$=$	Total cost

verbal model

$$3w \ + \ 6a \ = \ 51$$
$$5w \ + \ 4a \ = \ 55$$

3 lb walnuts and 6 lb almonds sell for $51

5 lb walnuts and 4 lb almonds sell for $55

$3w + 6a = 51 \rightarrow 15w + 30a = 255$ Multiply both sides by 5.
$5w + 4a = 55 \rightarrow 15w + 12a = 165$ Multiply both sides by 3.

$$18a = 90$$ Subtract the equations.
$$a = 5$$ Solve for a.

$3w + 6(5) = 51$ Substitute 5 for a in the first equation.
$w = 7$ Solve for w.

Walnuts sell for $7 per pound. Almonds sell for $5 per pound. ◄

PROBLEM SOLVING TIP

Make a verbal model.

COMMUNICATING ABOUT ALGEBRA

In Example 5, the variable w was eliminated. Explain to another student how you could eliminate the variable a instead of w.

TRY THESE

Describe two ways you can change the system so that you can solve by elimination.

1. $\begin{cases} 4x - 5y = 2 \\ 7x + 3y = 8 \end{cases}$

2. $\begin{cases} -2x + 8y = 3 \\ 3x - 9y = 1 \end{cases}$

Solve the system using the elimination method.

3. $\begin{cases} x - y = 5 \\ 2x - y = 13 \end{cases}$

4. $\begin{cases} 3x + 2y = 41 \\ 5x - 2y = 15 \end{cases}$

5. $\begin{cases} 4y = 10 - 5x \\ 6x + 22 = 2y \end{cases}$

6. $\begin{cases} 2x - 3y = 8 \\ 3x + 2y = -1 \end{cases}$

7. $\begin{cases} -8x + 4y = 0 \\ 3x - 2y = 6 \end{cases}$

8. $\begin{cases} 2y = 1 - 3x \\ 3y = -(1 + 4x) \end{cases}$

9. **BRIDGE REPAIR** Sixteen workers are employed on a bridge-repair project, some at $200 per day and some at $165 per day. The daily payroll is $2745. Find the number of workers employed at each wage.

10. **WRITING MATHEMATICS** Write a word problem that you can solve using a linear system and that has the solution (9, 36).

PRACTICE

Describe two ways you can change the system so that you can solve by elimination.

1. $\begin{cases} 5x - 6y = 9 \\ 3x + 7y = 1 \end{cases}$

2. $\begin{cases} -8x + 3y = 2 \\ 9x - 7y = 5 \end{cases}$

Solve the system using the elimination method.

3. $\begin{cases} 2x + 2y = -2 \\ 5x - 2y = 9 \end{cases}$

4. $\begin{cases} 2x + 2y = 8 \\ 2x - y = 5 \end{cases}$

5. $\begin{cases} 3x + 3y = 9 \\ 4x - 3y = -16 \end{cases}$

6. $\begin{cases} 8x - 3y = 17 \\ -7x + 6y = 2 \end{cases}$

7. $\begin{cases} 7x - 10y = -1 \\ 3x + 2y = -13 \end{cases}$

8. $\begin{cases} 4x - 3y = 15 \\ 8x + 2y = -10 \end{cases}$

9. $\begin{cases} 2x + 8y = -1 \\ -10x + 4y = 16 \end{cases}$

10. $\begin{cases} 5x + 3y + 9 = 0 \\ 3x - 4y + 17 = 0 \end{cases}$

11. $\begin{cases} 6x - 5 = 2x - 7y \\ 2x = 5y - 6 \end{cases}$

12. $\begin{cases} 4x - 3y = 9 \\ -3x + 5y = 7 \end{cases}$

13. $\begin{cases} 3x + 4y = -1 \\ 7x + 9y = 0 \end{cases}$

14. $\begin{cases} 4x - 2y = -19 \\ -6x - 3y = 1.5 \end{cases}$

15. **RETAILING** Mercury running shoes sell for $79 a pair. Whirlwinds sell for $65 a pair. In a single day, Sports Stop brought in $6289 on sales of 89 pairs of Mercurys and Whirlwinds. How many pairs of each shoe did the store sell?

16. **CONSUMERISM** Seven lb of cheddar cheese and 8 lb of Swiss cheese sell for $50. Five lb of cheddar and 4 lb of Swiss sell for $31. Find the price per pound of cheddar cheese and the price per pound of Swiss cheese.

17. **WRITING MATHEMATICS** The solution to an equation is $x = 5$. Suppose that you multiply both sides of the equation by 3 to produce a new equation. Is the solution to the new equation $x = 5$, $x = 15$, or a value that depends on the equation you began with? How is this fact related to the elimination method? Write a few sentences explaining your answer.

EXTEND

Solve the system using the elimination method.

18. $\begin{cases} 3(x - y) - \dfrac{1}{3}(x + y) = 30 \\ x + y + \dfrac{5}{3}(x - y) = 22 \end{cases}$

19. $\begin{cases} \dfrac{x}{2} - \dfrac{y}{5} = 1 \\ y - \dfrac{x}{3} = 8 \end{cases}$

20. $\begin{cases} 7x + 3(y - 3) = 5(x + y) \\ 7(x - 1) - 6y = 5(x - y) \end{cases}$

21. $\begin{cases} 3x - 2(y + 3) = 2 \\ 3x + 7y = 1 \end{cases}$

22. **SURVEYING** The perimeter of a parking lot is 200 yd. The length of the lot is 16 yd greater than the width. Find the dimensions of the lot.

23. **BOTANY** In January 1970, a 124-ft Douglas fir tree and a 70-ft coast redwood were growing side by side in northern California. The fir grew at the rate of 3.2 ft per year, the redwood at the rate of 6.5 ft per year. In what year did the redwood surpass the fir in height?

24. CHEMISTRY Find the amount of a 20% salt solution and the amount of a 30% salt solution that must be combined to produce 500 mL of a 24% salt solution.

TRANSPORTATION ENGINEERING After analyzing a bus route, a transportation engineer wrote the equations $N = 1200 - 10t$ and $t = 18 + 0.001N$, where N represents the number of seats needed per hour for passengers and t represents the time in minutes to complete the route. The solution to this system of linear equations represents the "equilibrium point," the point where the demand for bus service by riders equals the number of available seats.

25. Which equation shows that N decreases as t increases?

26. What circumstances would cause a decrease in N as t increases?

27. Solve the system. Explain what your solution means.

28. Suppose street conditions improve so the travel time equation is $t = 15 + 0.001N$. Without solving, describe how N will be affected.

THINK CRITICALLY

Solve the system.

29. $\begin{cases} 5x + 4y = 9a + b \\ 7x - 6y = a + 13b \end{cases}$

30. $\begin{cases} 4ax - 3cy = 3b + a \\ ax - cy = b \end{cases}$

31. The sides of a triangle are formed by the x-axis, the line $y = 3x$, and the line $y = -\frac{1}{2}x + 7$. Find the area of the triangle.

32. When you graph a system of three linear equations, what outcomes are possible? Draw a sketch illustrating each possible outcome. How many solutions are possible?

33. Find the vertices of the triangle formed by the graphs of the equations $x - 2y = -6$, $x + 6y = -6$, and $3x + 2y = 14$.

34. Keisha solved the system as shown. What did she do incorrectly? What is the solution?

$$
\begin{array}{ll}
2x - 3y = 13 & \rightarrow \quad 2x - 3y = 13 \\
x + 4y = 1 & \rightarrow \quad \underline{2x + 8y = 2} \\
& 5y = 11
\end{array}
$$

MIXED REVIEW

Use the distributive property to simplify each expression.

35. $5(a - 7)$ **36.** $-2(x - 6)$ **37.** $2(b + 3b^2)$ **38.** $4(2m - 1)$

39. Distance varies directly as time. The Space Shuttle travels 93,500 mi in 5 h. How far does it travel in 4 h?

40. STANDARDIZED TESTS Which is the solution of the inequality $-x - 5 < 4$?

     A. $x > 1$ **B.** $x < -1$ **C.** $x > 9$ **D.** $x > -9$

41. Solve the linear system $3x - 2y = -4$ and $7x - 3y = 9$ using the elimination method.

7.5 Special Systems of Equations

Explore/Working Together

The map shows a section of midtown Manhattan. Several Manhattan streets are known by two names. Three tourists each hail a cab at the corner of 7th Avenue and 47th Street and announce their destinations to the drivers. Describe the shortest route to each destination, or explain why the driver cannot follow the tourist's instructions.

1. Tourist A: "Take me to the intersection of 52nd Street and 7th Avenue."

2. Tourist B: "Take me to the intersection of 59th Street and Central Park South."

3. Tourist C: "Take me to the intersection of 5th Avenue and the Avenue of the Americas."

4. What are other possibilities for how the graph of two lines might appear? Give examples from the map.

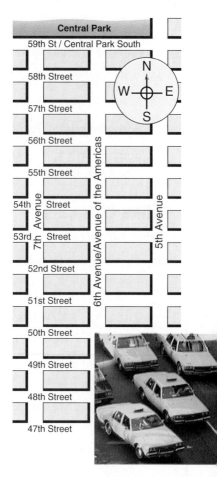

Build Understanding

A system of equations with *at least one* solution is a **consistent** system. If the system has *exactly one* solution, the system is **independent**. Algebraically, this unique solution represents values of the variables that satisfy all the equations in a system. Graphically, this solution is represented by the point where the graphs of the equations intersect. Most of the linear systems that you have solved so far have had unique solutions and so have been both consistent and independent.

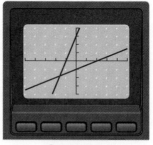

Consistent
Independent

Two other types of systems are possible.

1. The graphs of the equations do not intersect. Such a system does not have a solution.

2. The graphs of the equations are the same line. Every point on the line is a solution of the system.

EXAMPLE 1

Use a graphing utility to determine the number of solutions.

$$\begin{cases} y = 0.5x + 4 \\ y = 0.5x - 2 \end{cases}$$

Solution

Graph the equations. Notice that the lines have the same slope but different y-intercepts. Lines with the same slope but different y-intercepts are parallel and cannot intersect. Therefore, the system has no solution.

A system with no solution is called an **inconsistent system**.

If you try to solve this system using elimination, the result is a false equation.

$$\begin{array}{l} y = 0.5x + 4 \\ y = 0.5x - 2 \\ \hline 0 = 6 \end{array}$$ Subtract the second equation from the first.

The false equation tells you that the system is inconsistent.

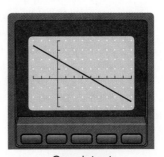
Inconsistent

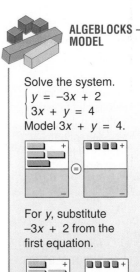
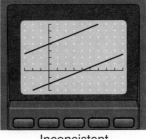
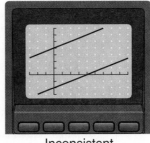
You have seen that a linear system can have one solution or no solutions. The following example illustrates the third type of system.

EXAMPLE 2

Use a graphing utility to determine the number of solutions.

$$\begin{cases} x + 2y = 4 \\ 2x + 4y = 8 \end{cases}$$

Solution

$$\begin{cases} y = -0.5x + 2 \\ y = -0.5x + 2 \end{cases}$$ Write each equation in slope-intercept form.

Graph the equations. The system consists of two equations, one a multiple of the other. Any point on the graph of $x + 2y = 4$ is also on the graph of $2x + 4y = 8$. Therefore, any ordered pair that satisfies one equation will also satisfy the other equation. There are an infinite number of solutions to this linear system.

A system of equations with the same graph is called a **dependent system**. Because such a system has at least one solution it is also *consistent*.

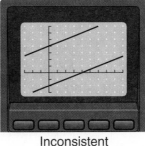
Consistent
Dependent

CHECK UNDERSTANDING

Why do you want to rewrite the equations in slope-intercept form?

CHECK UNDERSTANDING

In what way are the graphs of an inconsistent system similar to the graphs of a dependent system? How do they differ?

COMMUNICATING ABOUT ALGEBRA

You have seen that a linear system can have zero, one, or an infinite number of solutions. Can it have two solutions? Can it have more than two but less than an infinite number? Explain your answer.

If you try to solve this system using elimination, the result is the equation $0 = 0$.

$$\begin{array}{lll} x + 2y = 4 & \rightarrow & 2x + 4y = 8 \\ 2x + 4y = 8 & \rightarrow & \underline{2x + 4y = 8} \\ & & 0 = 0 \end{array}$$

Multiply both sides by 2.
Subtract.

A true equation, like $0 = 0$ or $-3 = -3$, tells you that a linear system is dependent.

EXAMPLE 3

BUSINESS The sales department at Synergy Computers collected the following data about the growth of Synergy and of the firm's chief competitor, Compco.

Number of Units Sold, in hundred thousands				
	1991	1992	1993	1994
Synergy	2.3	2.7	3.1	3.5
Compco	2.9	3.3	3.7	4.1

If the companies maintain their current rates of growth, when will Synergy overtake Compco in sales?

Solution

Let x represent the year number (year $1991 = 0$). Let y represent the number of units sold in hundred thousands. You can use linear regression to determine these two equations. You can also use other methods you learned in Chapter 6 to write an equation of a line, given two points.

$$\begin{cases} y = 0.4x + 2.3 & \textbf{Synergy sales equation} \\ y = 0.4x + 2.9 & \textbf{Compco sales equation} \end{cases}$$

The graphs of the equations are shown at the right. The slopes are equal, so the lines are parallel. The parallel lines indicate that the two companies are growing at the same rate. If the growth rates do not change, Synergy will never overtake Compco in sales. ◀

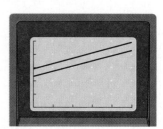

TRY THESE

If the system has a unique solution, find it. If it does not, state whether the system is *inconsistent* or *dependent*.

1. $\begin{cases} x + y = 13 \\ x - y = 5 \end{cases}$

2. $\begin{cases} -2y + 4 = -x \\ 3x + 12 = 6y \end{cases}$

3. $\begin{cases} y = 2x \\ 2y = 4x + 2 \end{cases}$

State whether the system shown is *independent*, *inconsistent*, or *dependent*.

4.

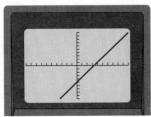

5.

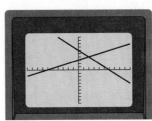

6.

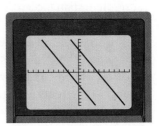

FARMING Consolidated Grains pays farmers $5.60 per bushel for soybeans, minus a $35 processing fee. The Soybean Co-op charges a $25 fee and pays $5.60 per bushel.

7. Write an equation expressing the amount y paid by each buyer as a function of x, the number of bushels purchased.

8. How many bushels of soybeans must a farmer sell before the payment from Consolidated exceeds the payment from the Co-op? Explain your answer.

9. WRITING MATHEMATICS Make a chart summarizing the possible types of graphs of a system of two linear equations. Include a verbal description and a drawing of each graph, tell the number of solutions, and characterize the systems by using the terms *dependent*, *independent*, *consistent*, and *inconsistent*.

PRACTICE

If the system has a unique solution, find it. If it does not, state whether the system is *inconsistent* or *dependent*.

1. $\begin{cases} y = \dfrac{2}{3}x + 5 \\ y = \dfrac{2}{3}x - 2 \end{cases}$

2. $\begin{cases} 4x - y = 3 \\ 2x - y = -1 \end{cases}$

3. $\begin{cases} x - 2y = 4 \\ 3y - x = y - 8 \end{cases}$

4. $\begin{cases} y = -x + 4 \\ 3y = -3x + 12 \end{cases}$

5. $\begin{cases} 2(6x + 10y) + 8 = 0 \\ -2 = 3x + 5y \end{cases}$

6. $\begin{cases} y = -4x + 2 \\ x + 2y = 11 \end{cases}$

State whether the system shown is *independent*, *inconsistent*, or *dependent*.

7.

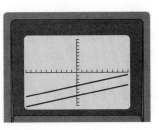

8.

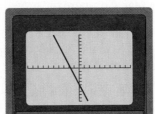

9.

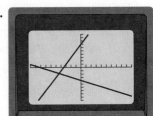

10. **NAVIGATION** The map shows a section of the Indian Ocean. The system of equations $2x + 3y = 96$ and $x + 2y = 42$ gives the courses of two freighters headed in a southeasterly direction where x represents latitude and y represents longitude. Will the freighters meet? If so, where? If not, why not?

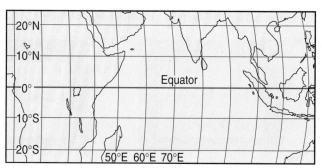

11. **WRITING MATHEMATICS** Write a real world problem that involves a system of inconsistent linear equations. Interpret what the inconsistency means in the situation.

EXTEND

12. The equations $12x + 10y = 20$ and $ax + 2y = b$ form a dependent system. Find a and b.

13. The equations $y = 3x - 5$ and $5y = ax + b$ form an inconsistent system. Find a. Find a value that b cannot have. Why?

14. Write a system of equations based on the tables of values shown at the right. Then solve the system.

15. The equations $5x - 4y = 7$ and $ax + 4y = -1$ form an independent system with the solution $(3, 2)$. Find a.

x	y	x	y
1	−4	1	−9
2	−6	2	−12
3	−8	3	−15
4	−10	4	−18

THINK CRITICALLY

Describe each linear system in terms of the slopes and y-intercepts of the graphs of the equations in the system.

16. independent

17. inconsistent

18. dependent

19. **INDUSTRY** Sawmills A, B, and C can turn out 11,900 board feet of lumber per day. Mills A and B together produce 7,700 board feet daily, while mills B and C together produce 8,300 board feet daily. Find the production capacity of each mill.

PROJECT *Connection*   Population statistics are usually included in a description of a city. To create statistics for your imaginary city, answer the questions below for the city in which you live.

1. What is the current population? Has the population been increasing or decreasing and, if so, by how much each year? Write and graph a linear equation to express the population as a function of the year. Compare the growth rate to other cities.

2. What is the age structure of the population? That is, how many people are under 20, under 50, over 70, and so on? Make a visual display of this data. Why is this information important to individuals and businesses?

3. Which different ethnic groups are represented in the population? Describe how this diversity enriches the culture of your city in terms of foods, restaurants, festivals, the arts, and shopping.

7.6 Explore Determinants of Systems

Explore/Working Together

- Work with a partner. Use this linear system.

$$\begin{cases} 4x + 6y = -1 \\ 3x + 5y = -1 \end{cases}$$

1. Solve the system by graphing, substitution, and elimination.

2. Was any one of the methods easier to use than the others? Was any one harder? Give reasons for your answers.

3. What are the advantages and disadvantages of each method?

Build Understanding

- In Chapter 1, you learned that a *matrix* is a rectangular array of values. The values are called *elements*. The numbers of rows and columns in a matrix specify its *dimensions*. The matrix shown here has elements 4, −1, 0, 3, 6, and 2, and its dimensions are 2×3. A matrix of n rows and m columns has dimensions $n \times m$.

$$\begin{bmatrix} 4 & -1 & 0 \\ 3 & 6 & 2 \end{bmatrix}$$
Matrix

Recall that a matrix with the same number of rows and columns is a square matrix. A square matrix has dimensions $n \times n$. Associated with the 2×2 square matrix $A = \begin{bmatrix} a & b \\ c & d \end{bmatrix}$ is a number called the **determinant** of A. The determinant of a matrix is symbolized by using vertical bars in place of the matrix brackets. Thus,

$$\text{determinant } A = \begin{vmatrix} a & b \\ c & d \end{vmatrix}$$

The value of a 2×2 determinant is the difference of the products of the diagonal entries.

$$\text{determinant } A = \begin{vmatrix} a & b \\ c & d \end{vmatrix}$$

$$= ad - bc$$

EXAMPLE 1

Evaluate the determinant: $\begin{vmatrix} 3 & -5 \\ 2 & 4 \end{vmatrix}$

Solution

$$ad - bc = 3(4) - (-5)(2)$$
$$= 12 - (-10) = 22$$

The value of the determinant is 22. ◄

SPOTLIGHT ON LEARNING

WHAT? In this lesson you will learn

- to use a determinant to solve linear systems.

WHY? Determinants of matrices help you solve problems about quality control, employment, baseball, and fund-raising for charity.

ALGEBRA: WHO, WHERE, WHEN

A method of detached coefficients similar to determinants has been traced to twelfth-century Chinese mathematics. The earliest Chinese abacus was a flat piece of wood divided into squares. As the Chinese studied patterns involved in solving systems of equations, they discovered that they could arrange numerical coefficients in a square array similar to those on their counting board.

One of the important applications of determinants is in solving systems of equations. The method used is called **Cramer's rule**, after the Swiss mathematician Gabriel Cramer (1704–1752), and is used in Example 2. Notice that to use Cramer's rule, you must first write both equations in standard form, $Ax + By = C$.

EXAMPLE 2

Use determinants to solve the system.

$$\begin{cases} 3x - 7y = -6 \\ x + 2y = 11 \end{cases}$$

Solution

Write the coefficients of x and y in a determinant D.

$$D = \begin{vmatrix} 3 & -7 \\ 1 & 2 \end{vmatrix} \quad \begin{array}{l} \text{—} y\text{-coefficients} \\ \text{—} x\text{-coefficients} \end{array}$$

Write another determinant. Use D and replace the x-column with the constants from the equations. Label it D_x.

$$D_x = \begin{vmatrix} -6 & -7 \\ 11 & 2 \end{vmatrix} \begin{array}{l} \text{Replace} \\ x\text{-coefficients} \\ \text{with constants.} \end{array}$$

Write a third determinant. Use D and replace the y-column with the constants from the equations. Label it D_y.

$$D_y = \begin{vmatrix} 3 & -6 \\ 1 & 11 \end{vmatrix} \begin{array}{l} \text{Replace} \\ y\text{-coefficients} \\ \text{with constants.} \end{array}$$

If $D \neq 0$, the solution of the system is (x, y), where $x = \dfrac{D_x}{D}$ and $y = \dfrac{D_y}{D}$. Solve for x and y.

$$x = \frac{D_x}{D} = \frac{\begin{vmatrix} -6 & -7 \\ 11 & 2 \end{vmatrix}}{\begin{vmatrix} 3 & -7 \\ 1 & 2 \end{vmatrix}} = \frac{-6(2) - (-7)(11)}{3(2) - (-7)(1)} = \frac{65}{13} = 5$$

$$y = \frac{D_y}{D} = \frac{\begin{vmatrix} 3 & -6 \\ 1 & 11 \end{vmatrix}}{\begin{vmatrix} 3 & -7 \\ 1 & 2 \end{vmatrix}} = \frac{3(11) - 1(-6)}{3(2) - (-7)(1)} = \frac{39}{13} = 3$$

The solution is $(5, 3)$. You should check the solution in both original equations. ◀

Together with graphing, substitution, and elimination, determinants give you four methods for solving a system of linear equations. When you solve a real world problem, choose the solution method that seems most efficient to use.

EXAMPLE 3

QUALITY CONTROL Workers at Kitchen Industries can assemble an
electronic juicer in 3 minutes and a food processor in 5 minutes.
Quality control engineers inspect each juicer for 2 minutes and each
food processor for 6 minutes. The assembly line operates 7 hours a
day, and the quality control department operates 6 hours a day.
How many juicers and food processors should the company
produce daily so that every assembled product is inspected for
quality on the day it is made?

Solution

The assembly line is in operation $7 \cdot 60 = 420$ minutes per day.
Quality control works $6 \cdot 60 = 360$ minutes per day.

Let x represent the number of juicers produced.
Let y represent the number of food processors produced.

$$\boxed{\begin{array}{c}\text{Time to}\\\text{produce juicers}\end{array}} + \boxed{\begin{array}{c}\text{Time to produce}\\\text{food processors}\end{array}} = \boxed{\text{Total time}}$$

$$\begin{cases} 3x + 5y = 420 & \textbf{Assembly equation} \\ 2x + 6y = 360 & \textbf{Quality control equation} \end{cases}$$

Use determinants to solve the system.

$$x = \frac{D_x}{D} = \frac{\begin{vmatrix} 420 & 5 \\ 360 & 6 \end{vmatrix}}{\begin{vmatrix} 3 & 5 \\ 2 & 6 \end{vmatrix}} = \frac{420(6) - 5(360)}{3(6) - 2(5)} = \frac{720}{8} = 90$$

$$y = \frac{D_y}{D} = \frac{\begin{vmatrix} 3 & 420 \\ 2 & 360 \end{vmatrix}}{\begin{vmatrix} 3 & 5 \\ 2 & 6 \end{vmatrix}} = \frac{3(360) - 2(420)}{3(6) - 2(5)} = \frac{240}{8} = 30$$

The company should produce 90 juicers and 30 food processors daily. ◀

TRY THESE

Evaluate each determinant.

1. $\begin{vmatrix} 1 & 2 \\ 3 & 4 \end{vmatrix}$

2. $\begin{vmatrix} 0 & 1 \\ 0 & 1 \end{vmatrix}$

3. $\begin{vmatrix} 2 & 5 \\ 4 & -1 \end{vmatrix}$

4. $\begin{vmatrix} -6 & 2 \\ 7 & -3 \end{vmatrix}$

5. $\begin{vmatrix} 5 & -2 \\ -4 & 8 \end{vmatrix}$

6. $\begin{vmatrix} 3 & -3 \\ -3 & 3 \end{vmatrix}$

Use determinants to solve the system.

7. $\begin{cases} 5x + 6y = 1 \\ 2x + y = 6 \end{cases}$

8. $\begin{cases} -4x + 3y = 6 \\ 3x + 4y = 8 \end{cases}$

9. $\begin{cases} 2y = x - 4 \\ 2x = 1 - 3y \end{cases}$

10. **RETAIL** At Music City's Anniversary Sale, all CDs are on sale for one price and all cassettes are on sale for a different price. Juanita bought five CDs and six cassettes for $76. Joel bought seven CDs and four cassettes for $80. Write a system of linear equations and use determinants to find the sale prices of CDs and cassettes.

11. **WRITING MATHEMATICS** Write a paragraph describing how to solve Exercise 9 using determinants. Discuss advantages and disadvantages of this method.

PRACTICE

Evaluate each determinant.

1. $\begin{vmatrix} 0 & 1 \\ 3 & 2 \end{vmatrix}$

2. $\begin{vmatrix} 5 & 2 \\ 1 & 4 \end{vmatrix}$

3. $\begin{vmatrix} 4 & 4 \\ 4 & 4 \end{vmatrix}$

4. $\begin{vmatrix} 2 & 2 \\ 5 & 5 \end{vmatrix}$

5. $\begin{vmatrix} -1 & 3 \\ 4 & -8 \end{vmatrix}$

6. $\begin{vmatrix} 6 & 4 \\ 3 & 2 \end{vmatrix}$

7. $\begin{vmatrix} -6 & 8 \\ 5 & -7 \end{vmatrix}$

8. $\begin{vmatrix} -10 & 9 \\ -8 & 6 \end{vmatrix}$

9. $\begin{vmatrix} 12 & 11 \\ 9 & -6 \end{vmatrix}$

10. $\begin{vmatrix} x & y \\ 2 & 3 \end{vmatrix}$

11. $\begin{vmatrix} 2k & 4 \\ 3k & 2 \end{vmatrix}$

12. $\begin{vmatrix} 3m & 2m \\ 7 & 5 \end{vmatrix}$

13. **WRITING MATHEMATICS** Write a note to a friend explaining how to evaluate a 2×2 determinant.

Use determinants to solve the system.

14. $\begin{cases} 4x - y = 9 \\ x - 3y = 16 \end{cases}$

15. $\begin{cases} 3x - 5y = -23 \\ 5x + 4y = 11 \end{cases}$

16. $\begin{cases} x + 4y = 13 \\ 5x - 7y = -16 \end{cases}$

17. $\begin{cases} 3x + y = -1 \\ 2x - 3y = -8 \end{cases}$

18. $\begin{cases} 2y + 3 = 9x \\ 3x - y = 6 \end{cases}$

19. $\begin{cases} x - 7 = 3y \\ 2(3y - 7) = 5x \end{cases}$

20. $\begin{cases} 3(x + y) = 2 + 10y \\ 2(3x - 2) = 13y \end{cases}$

21. $\begin{cases} 9x = 6y \\ 3x - 4y = -18 \end{cases}$

22. $\begin{cases} 4x + 6y = 16 \\ x = 2y + 1.2 \end{cases}$

23. $\begin{cases} 3x + 2y = 24 \\ 15x - 2y = 48 \end{cases}$

24. $\begin{cases} x - 6y = 3 \\ x + 2y = 5 \end{cases}$

25. $\begin{cases} 5x - 2y = 1 \\ 4x + 5y = 47 \end{cases}$

26. **EMPLOYMENT** At the end of each day, a building contractor paid each employee in cash, using just two denominations of bills. One day Julius received nine bills of the larger denomination and eight of the smaller for a total of $220. Keith received seven of the larger bills and six of the smaller for a total of $170. Write a system of equations and use determinants to find the denominations of bills used by the contractor.

27. **CLOTHING SALE** Frankie sold 40 pairs of slacks during a sale. Wool blend slacks were priced at $30. Chinos sold at $22 a pair. How many of each kind did Frankie sell?

28. BASEBALL There were 30,000 paid spectators at a baseball game. Receipts totaled $530,000. Tickets for bleacher seats sold for $15 and tickets for the other seats sold for $25. Use determinants to find how many tickets of each kind were sold.

29. NUTRITION Protein burned in the body yields a constant number of calories per gram. This is also true for carbohydrates. Use the information in the table to write a system of equations. Solve the system using determinants to find

Food	Protein, g	Carbohydrates, g	Calories from Protein and Carbohydrates
Milk, 1 cup	8	11	76
Navy Beans, 1 cup	15	40	220

the number of calories yielded by each gram of protein and by each gram of carbohydrate.

30. WRITING MATHEMATICS Write a word problem that you can solve using determinants. Then solve the problem.

EXTEND

Evaluate.

31. $2\begin{vmatrix} 0 & 4 \\ 6 & 2 \end{vmatrix} - 3\begin{vmatrix} 3 & 5 \\ -4 & -5 \end{vmatrix}$

32. $-5\begin{vmatrix} -1 & -2 \\ -3 & -4 \end{vmatrix} + 4\begin{vmatrix} 6 & 2 \\ -3 & -2 \end{vmatrix}$

33. MONEY MATTERS André gave Franco $3. They both then had equal amounts of money. Franco then gave $5 to André, and André then had twice as much money as Franco did. Find how much money each had originally.

Solve for x.

34. $\begin{vmatrix} 6 & 4 \\ x & x \end{vmatrix} = \begin{vmatrix} 10 & 6 \\ 2 & 2 \end{vmatrix}$

35. $-\begin{vmatrix} 5 & x+1 \\ 2 & x-1 \end{vmatrix} = \begin{vmatrix} 4 & x-1 \\ 3 & x+1 \end{vmatrix}$

TECHNOLOGY In yet another solving method, you can use matrices and appropriate software or calculator technology. Consider the system: $\begin{cases} 5x + 2y = -7 \\ 3x + 4y = 25 \end{cases}$

Let matrix *A* be the **coefficient matrix**. Let matrix *B* be the **constant matrix**.

$A = \begin{bmatrix} 5 & 2 \\ -3 & 4 \end{bmatrix}$ $B = \begin{bmatrix} -7 \\ 25 \end{bmatrix}$

The inverse of the coefficient matrix A^{-1} multiplied by the constant matrix *B* equals the solution matrix.

$\begin{bmatrix} x \\ y \end{bmatrix} = A^{-1}B$

The solution is $(-3, 4)$

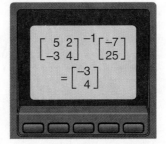

Use technology to solve each system.

36. $\begin{cases} 7x + 2y = 4 \\ 5x + 3y = -5 \end{cases}$

37. $\begin{cases} 3x - 2y = -2 \\ -6x + 5y = 14 \end{cases}$

38. $\begin{cases} -7x - 12y = 13 \\ 4x + 9y = -1 \end{cases}$

THINK CRITICALLY

Evaluate each determinant if $\begin{vmatrix} a & b \\ c & d \end{vmatrix} = 4$. Explain your reasoning.

39. $\begin{vmatrix} c & d \\ a & b \end{vmatrix}$

40. $\begin{vmatrix} b & a \\ d & c \end{vmatrix}$

41. $\begin{vmatrix} d & c \\ b & a \end{vmatrix}$

42. Use determinants to solve the system $3x - 5y = 2$ and $9x - 15y = 6$. What happens? Why?

MIXED REVIEW

43. In a survey of 200 potential voters, 78 expressed a preference for candidate Bryant, 86 chose candidate Sanchez, and 36 chose candidate Grant. Based on the results of the survey, what is the probability that a potential voter, chosen at random, will favor Bryant?

44. In one year, Sarah earned $112 in interest on her savings. If the bank paid 4% interest compounded annually, how much did she have in her account?

Plot each point on a coordinate plane.

45. $A\,(-3, 5)$

46. $B\,(4, 2)$

47. $C\,(2, -1)$

48. $D\,(-2, -5)$

49. STANDARDIZED TESTS Which is the equation of the line passing through the point $(2, 3)$ that is parallel to the line $y = 4x - 6$?

A. $y = 4x + 5$ **B.** $y = 4x - 5$ **C.** $y = \frac{1}{4}x - 5$ **D.** $y = -4x - 5$

50. Use determinants to solve the linear system.

$$\begin{cases} 3x + 7y = 1 \\ 5x + 9y = 7 \end{cases}$$

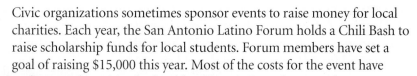

Civic organizations sometimes sponsor events to raise money for local charities. Each year, the San Antonio Latino Forum holds a Chili Bash to raise scholarship funds for local students. Forum members have set a goal of raising $15,000 this year. Most of the costs for the event have been underwritten by local businesses. Costs that must be met through ticket sales will be $2400 for advertising and $1280 in miscellaneous costs, plus $2.25 for each meal served at the event. Tickets sell for $8.

The week before the event, the president of the forum asks the organizing committee to calculate the number of tickets that must be sold in order to break even and to meet the organization's fund-raising goal.

Decision Making

1. Let x represent the number of people attending the event. Write an equation for the revenues from ticket sales.

2. Write an equation for the cost of the event.

3. Solve the system of equations you have written. Explain what the solution means.

4. How many tickets must be sold for the organization to meet its $15,000 fund-raising goal? Explain how you got your answer.

5. How might the amount that is raised be affected if the organization raises the price of a ticket? How might the amount be affected if the ticket price is lowered? Explain your answers.

7.7 Problem Solving File

Use Variables

Choose a Method: One Equation or a System

Sometimes you can solve a problem either with a one-variable equation or with a system of equations with two or more variables. As you become a more experienced problem solver, you will choose the most efficient and effective method for solving a particular problem.

Problem

A worker collected a total of 35 dimes and quarters from a parking meter. The coins were worth $5.15. How many of each coin were there?

Explore the Problem

<table>
<tr><td colspan="4">

SPOTLIGHT ON LEARNING

WHAT? In this lesson you will learn
• to solve problems using either an equation in one variable or a system of equations in two variables.

WHY? Choosing whether to use one equation or a system of equations helps you solve problems about jewelry store prices, fast food, and money.

</td></tr>
</table>

● First, try using one variable and one equation.

1. Let x represent the number of dimes. In terms of x, how can you represent the number of quarters?

2. What is the total value in cents of the dimes?

3. What is the total value in cents of the quarters? You can complete a table like the one shown if you wish.

	Number of coins	Value of coins	Total value
Dimes			
Quarters			

4. What is the value in cents of the whole collection of coins? Write an equation relating the value of the dimes and the value of the quarters to the total value.

5. Solve the equation. How many of each coin are there?

Now, solve the same problem using two variables and two equations.

6. Let d represent the number of dimes and let q represent the number of quarters. Write an equation for the total number of coins.

7. What is the value in cents of the dimes? What is the value in cents of the quarters? Use these expressions to write an equation for the total value of the coins.

8. Solve the system of two equations. What method did you use?

9. **WRITING MATHEMATICS** When you are given a problem, how will you decide whether to use one or two variables? In your opinion, what are the advantages and disadvantages of each approach?

10. What other method could you use to solve this problem?

Investigate Further

● Consider the following problem.

The sum of two numbers is 50. Three times the larger number decreased by twice the smaller is 60. What are the two numbers?

PROBLEM
SOLVING PLAN

• Understand
• Plan
• Solve
• Examine

11. In the following system of equations, what do x and y represent?

$$\begin{cases} x + y = 50 \\ 3x - 2y = 60 \end{cases}$$

12. Solve the system. Which method did you use?

13. Suppose you wanted to solve this problem using one variable. How would you represent each number in terms of the same variable?

14. Write one equation to express the information you have not yet used in the problem.

15. Solve the equation. Compare this solution with your results in Question 12.

16. Solve the following problem. First use two variables, then one.

The sum of two numbers is 87. The greater number is 3 more than twice the smaller. Find the numbers.

17. JEWELRY STORE A boutique owner bought 6 bracelets and 8 necklaces for $140. A week later, at the same prices, she bought 9 bracelets and 6 necklaces for $132. Find the price of each item.

a. Would you use one variable to solve this problem? Explain.

b. Solve the problem.

18. WRITING MATHEMATICS
Copy the following problem and explain the different ways in which it could be solved.

Shoji has $7000 invested in savings bank Certificates of Deposit (CDs). One CD pays 4% simple interest and the other pays 5.5%. If his annual income from the two CDs is $337, how much is invested at each rate?

Apply the Strategy

Solve each problem using the method you choose.

19. The dimes and quarters in Tim's piggy bank are worth $11.60. He has 32 more dimes than quarters. How many coins are in the bank?

20. Lisa has a collection of 16 coins consisting of nickels and quarters. If her collection is worth $2.20, how many of each coin does she have?

21. The sum of two numbers is 104. The greater number is one less than twice the smaller number. Find the numbers.

22. If five times the smaller of two numbers is subtracted from twice the greater number, the result is 16. If the greater number is increased by three times the smaller number, the result is 63. Find the numbers.

23. One number is 15 more than another. The sum of twice the greater number and three times the smaller number is 100. Find the product of the numbers.

24. RETAILING A store owner has some almonds worth $6.00 per kilogram and some cashews worth $9.00 per kilogram. She wishes to make a mixture of 80 kg that she can sell for $6.60 per kilogram. How many kilograms of each variety should she use?

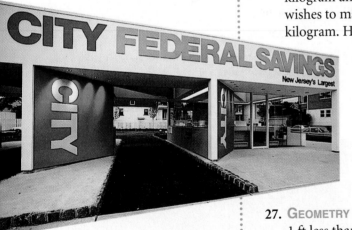

25. PERSONAL FINANCE One year Roberto and his wife Melinda together earned $57,000. If Roberto earned $4,000 less than Melinda that year, how much did he earn?

26. INVESTING Mr. Soong invested $14,000, part at 5% and part at 8% simple interest. His total annual income from both investments was $1,000. Find the amount invested at each rate.

27. GEOMETRY A rectangle has a perimeter of 38 ft. The length is 1 ft less than three times the width. Find the dimensions.

28. FAST FOOD If you buy 7 hamburgers and 3 slices of pizza, you get $1 change from a $20 bill. If you buy 8 hamburgers and 2 slices of pizza, you get 50¢ change from a $20 bill. How much change would you get from $20 if you bought 4 hamburgers and 4 slices of pizza?

29. MONEY Andrea has $446 in $10, $5, and $1 bills. There are 94 bills in all and 10 more $5 bills than $10 bills. How many of each bill does Andrea have?

REVIEW PROBLEM SOLVING STRATEGIES

PINE TREE PUZZLE

A forest ranger told some campers, "That pine tree is twice as old as I was when the tree was as old as I am now. Right now, our two ages add up to 49 years."

How old is the pine tree and how old is the ranger?

a. First, try solving the problem with a guess-and-check strategy. Suppose you guess that the tree is 30 years old. How old must the ranger be? Why?

b. Using your results from Question a, find how many years ago the tree would have been as old as the ranger is now. How old would the ranger have been then? If the tree is 30 years old, is it twice as old as the ranger was then?

c. Do you think the tree is more or less than 30 years old? Keep trying until you find ages that work.

d. Next, solve the problem again, this time using algebra. Will you write one or two equations? Show how you set up and solve the problem.

e. Which solution method do you prefer? Explain.

RED AND BLUE I

Three wrapped packages contain socks. One package has all blue socks, one has all red socks, and one has a mixture of blue and red socks. All three packages are mislabeled. Show how you could correct all the labels by opening only one package and seeing only one sock from that package.

a. Suppose you opened the package labeled Blue and saw a blue sock. What would that tell you?

b. Suppose you opened the package labeled Blue and saw a red sock. What would that tell you?

c. Suppose you opened the package labeled Blue and Red and saw a blue sock. What would this tell you? Why?

d. From which package should you make your selection? Why?

RED and BLUE II

Here is a row of ten disks. Half of the disks are red and half are blue. At the end of the row of disks there are two empty spots.

Move two adjacent disks (without changing their position with respect to one another). Continue, moving disks two at a time, until all the red disks are together and all the blue disks are together in one continuous row. When you are finished, there must be two empty spots at one end of the row.

· · · CHAPTER REVIEW · · ·

VOCABULARY

Choose the word from the list that correctly completes each statement.

1. A system of equations that has at least one solution is called a(n) __?__ system.

2. A system of equations that has exactly one solution is called a(n) __?__ system.

3. A system of equations that has no solution is called a(n) __?__ system.

4. A system of equations that has an infinite number of solutions is called a(n) __?__ system.

a. independent

b. dependent

c. consistent

d. inconsistent

Lesson 7.1   EXPLORE SYSTEMS OF LINEAR EQUATIONS   pages 325–327

- A point whose coordinates satisfy two linear equations is the intersection of the graphs of those equations.

In Exercises 5 and 6, determine if $(4, -2)$ is the solution to each system.

Use a graphing utility to find an ordered pair that satisfies both equations.

5. $\begin{cases} y = -2x + 6 \\ y = -x + 2 \end{cases}$

6. $\begin{cases} 7x - y = -8 \\ y = -4x - 3 \end{cases}$

7. $\begin{cases} -x + y = 2 \\ y = 3x - 8 \end{cases}$

8. $\begin{cases} -2x + y = 5 \\ -3x + y = 1 \end{cases}$

Lesson 7. 2   USE GRAPHS TO SOLVE SYSTEMS OF LINEAR EQUATIONS   pages 328–334

- Two or more linear equations with the same variables form a **system of linear equations**.
- You can solve a system of linear equations by graphing the equations and finding the point of intersection.

Solve the system by graphing.

9. $\begin{cases} 2x + y = 2 \\ y = x - 1 \end{cases}$

10. $\begin{cases} y = 2x + 7 \\ -x + y = 1 \end{cases}$

11. $\begin{cases} y = -2x \\ 4x + y = 2 \end{cases}$

12. $\begin{cases} -3x + y = 10 \\ -x + y = -2 \end{cases}$

Lesson 7.3   SOLVE LINEAR SYSTEMS BY SUBSTITUTION   pages 335–341

- To solve a linear system by substitution, solve either equation for one variable in terms of the other. Substitute the new expression in the other equation and solve. Substitute that solution in an original equation and solve. Check the solution in both original equations.

Use the substitution method to solve each system.

13. $\begin{cases} x + 2y = -4 \\ 3x - 2y = 12 \end{cases}$

14. $\begin{cases} x - 2y = 4 \\ x + 4y = 7 \end{cases}$

15. $\begin{cases} x - y = 5 \\ x + 3y = -3 \end{cases}$

16. $\begin{cases} y - 2x = -3 \\ y + 5 = 3x \end{cases}$

17. Victor is half as old as Maria. The sum of their ages is 54. How old is Victor?

Lesson 7.4 SOLVE LINEAR SYSTEMS BY THE ELIMINATION METHOD   pages 342–347

- To solve a linear system using elimination, write both equations in standard form. Add or subtract the equations to eliminate a variable. Solve the resulting equation for the remaining variable. Then substitute that value in either original equation and solve for the other variable. Check the solution in both original equations. You may need to multiply one or both equations by a constant or constants before adding or subtracting the equations.

Use the elimination method to solve each system.

18. $\begin{cases} 3x + 2y = 18 \\ x - 2y = -6 \end{cases}$ **19.** $\begin{cases} x + y = 3 \\ 2x + 3y = 1 \end{cases}$ **20.** $\begin{cases} 8x + 3y = -27 \\ 6x - 9y = -9 \end{cases}$ **21.** $\begin{cases} x - y = -3 \\ 2x - 4y = 22 \end{cases}$

22. Between them, Chen and Lisa have a total of $90 in cash. Chen gives Lisa $8 for a book. After that, each friend has exactly the same amount of money. How much did each have at first?

Lesson 7.5 SPECIAL SYSTEMS OF EQUATIONS   pages 348–352

- A system of equations whose graphs intersect in a point has a unique solution and is consistent and independent. A system whose graphs are parallel lines has no solution and is inconsistent. When the equations have the same graph, the system has an infinite number of solutions and is dependent.

If the system has a unique solution, find it. If it does not, state whether the system is *inconsistent* **or** *dependent.*

23. $\begin{cases} x + y = 3 \\ x - y = 7 \end{cases}$ **24.** $\begin{cases} -3x + y = -6 \\ x + 1 = \dfrac{1}{3}y \end{cases}$ **25.** $\begin{cases} x + 3y = -4 \\ 2(3x + 9y) = -24 \end{cases}$

Lesson 7.6 EXPLORE DETERMINANTS OF SYSTEMS   pages 353–359

- Associated with each $n \times n$ matrix is a number called the **determinant** of the matrix. The value of a 2×2 determinant is the difference of the products of the diagonal entries. You can use Cramer's rule to solve a system of linear equations by determinants.

Use determinants to solve the system.

26. $\begin{cases} 3x + y = -4 \\ 2x + 4y = -6 \end{cases}$ **27.** $\begin{cases} 2x + 5y = 9 \\ 3x - 4y = 2 \end{cases}$ **28.** $\begin{cases} 4x = 3y \\ 4y - 5x = 1 \end{cases}$

Lesson 7.7 CHOOSE A METHOD: ONE EQUATION OR A SYSTEM   pages 360–363

- Some problems may be solved either by an equation in one variable or by a system of linear equations in two or more variables.

29. A rectangle has a perimeter of 180 in. The length is four times the width. Find the dimensions.

30. When four times the smaller of two numbers is added to three times the greater number, the result is 130. If the greater number is decreased by twice the smaller number, the result is −10. Find the numbers.

CHAPTER ASSESSMENT

CHAPTER TEST

Use the substitution method to solve the system.

1. $\begin{cases} x + 2y = 4 \\ 2x + 3y = 7 \end{cases}$ 2. $\begin{cases} b + 4c = -5 \\ 2b - c = 8 \end{cases}$

3. $\begin{cases} x - 2y = 0.5 \\ 2x - y = -0.5 \end{cases}$ 4. $\begin{cases} 3x + y = 1 \\ 6x + y = 3 \end{cases}$

5. Alana is three times as old as Leon. Their ages differ by 12 years. How old is Alana?

6. **WRITING MATHEMATICS** Write a paragraph explaining the difference between a linear system that is consistent but dependent and one that is inconsistent.

7. A travel agency offers two New York trip plans. The Regular plan includes hotel accommodations for three nights and two pairs of theater tickets for $645. The Deluxe plan includes accommodations for five nights and four pairs of theater tickets for $1135. What prices are being used for nightly accommodations and for pairs of tickets?

8. **STANDARDIZED TESTS** Which of the following descriptions fits the system?

$$\begin{cases} 10x - 6y = 42 \\ 15x - 63 = 9y \end{cases}$$

   A. inconsistent and independent
   B. consistent and independent
   C. consistent and dependent
   D. inconsistent and dependent

Use the elimination method to solve the system.

9. $\begin{cases} 2x - 3y = 1 \\ 2x + 3y = 7 \end{cases}$ 10. $\begin{cases} x + 2y = -6 \\ 3x - 4y = 32 \end{cases}$

11. $\begin{cases} x + y = 0.5 \\ 2x + 4y = -5 \end{cases}$ 12. $\begin{cases} -4m + 3n = -25 \\ -3m - 2n = -6 \end{cases}$

13. $\begin{cases} s = -3r \\ 3r + 4s = -9 \end{cases}$ 14. $\begin{cases} 3x = -2y \\ -2x + 3y = 13 \end{cases}$

15. The perimeter of a rectangle is 158 cm. The length is 5 cm greater than the width. Find the dimensions of the rectangle.

16. At the spring play, a total of 240 tickets were sold. Adult tickets were $3 and children's tickets were $1.50. A total of $510 was taken in. How many of each kind of ticket were sold?

17. **STANDARDIZED TESTS** Which is the solution of the system? $\begin{cases} 0.5x = 0.25y \\ 0.25x + 0.5y = 2.5 \end{cases}$

   A. $(1, 3)$ B. $(2, 4)$
   C. $(-0.5, -0.25)$ D. no solution

Solve the system by graphing.

18. $\begin{cases} x + y = 6 \\ x - y = 2 \end{cases}$ 19. $\begin{cases} 2x + 4y = -10 \\ 3x - y = 13 \end{cases}$

Evaluate the determinant.

20. $\begin{vmatrix} 3 & -4 \\ 6 & 2 \end{vmatrix}$ 21. $\begin{vmatrix} 7 & 2 \\ -2 & 3 \end{vmatrix}$

Use determinants to solve the system.

22. $\begin{cases} 6x + 5y = -1 \\ x + 2y = -6 \end{cases}$ 23. $\begin{cases} 3x + y = -15 \\ 2x - 12 = 3y \end{cases}$

24. $\begin{cases} 3x - 4y = -1 \\ 8x - 5y = \frac{1}{6} \end{cases}$ 25. $\begin{cases} 0.3x + 0.2y = 0.7 \\ x - 0.5y = 0 \end{cases}$

26. **WRITING MATHEMATICS** Write a paragraph explaining how to solve a linear system using Cramer's rule.

27. Linda invested $5000, part at 8% and part at 6%. The total interest earned for a year was $360. How much did she invest at each rate?

28. The difference between two numbers is 22. Twice the smaller number plus three times the greater is 246. What are the numbers?

PERFORMANCE ASSESSMENT

USE ALGEBLOCKS Model each equation in the linear system below on a separate Sentence Mat. Sketch your models.

$$\begin{cases} 2x + 2y = 12 & \textbf{Equation 1} \\ y = 3x - 2 & \textbf{Equation 2} \end{cases}$$

Solve the system using substitution. Model and sketch each step.

EQUATION–A–ROUND Work in groups of four or five students. Each student writes on an index card a system of linear equations in two variables in standard form. Student 1 and Student 2 show their equations to the group. Each group member determines the solution of each system, if there is one. If there is no solution, other group members should explain why this is true. The group should compare work, agree on a correct answer, and record the equations and answers. Continue until all possible pairs of student equations have been explored.

EQUATION ART Work individually or in pairs. Parallel, perpendicular, and intersecting lines often appear in art and architecture. Find examples of linear designs in old magazines. Lay a transparent coordinate grid over each design. Try to determine the equations of a pair of lines in the design and their point of intersection, if there is one. Make a collage of your examples. Cut them out.

SPECIAL SYSTEMS Graph Systems A and B below on the same coordinate plane. Use a different color for the graph of each system.

A. $\begin{cases} 3x + 5y = 9 \\ -2x + 3y = 13 \end{cases}$ **B.** $\begin{cases} x = -2 \\ y = 3 \end{cases}$

Interpret your graphical representation. Explain how the systems are related. Then justify your conclusion algebraically.

PROJECT ASSESSMENT

 Prepare a trifold brochure that can be used to market your city. With the other members of your group, discuss how you can best convince businesses and individuals to relocate to your city.

1. Decide how you will share the work of writing and illustrating the items for your brochure. Include results from previous project activities and create additional material to make the presentation effective.

2. Create a catchy slogan for your city. Try to organize the brochure so that each item can be linked with the idea the slogan conveys.

3. After all the brochures in your class are completed and reviewed, take a class vote on the best city. Determine a first- and second-place winner.

CUMULATIVE REVIEW

Solve each system of equations.

1. $\begin{cases} 4x + y = 11 \\ 5x - 3y = 1 \end{cases}$ 2. $\begin{cases} 3x - 2y = -4 \\ x - 5y = 3 \end{cases}$

Evaluate each determinant.

3. $\begin{vmatrix} 4 & 5 \\ 2 & 3 \end{vmatrix}$ 4. $\begin{vmatrix} -2 & 7 \\ -3 & 9 \end{vmatrix}$

5. STANDARDIZED TESTS Which of the following values for the correlation coefficient seems to best represent the relationship between the two variables graphed on the scatter plot below?

   A. 1
   B. 0.5
   C. 0
   D. −0.5
   E. −1

Solve each inequality. Then graph the solution.

6. $n - 8 > -6$ 7. $6c \le -12$

8. $15x - 4(3x - 1) \ge 5x + 6$

Evaluate each expression.

9. $\dfrac{6(3 + 2) - 3^2}{2^3 - 1}$ 10. $\dfrac{|-4 - (-6)|}{|-4| - |-6|}$

11. $3 + 4 \cdot 5 - 6 \div 3 \cdot 4 + 3^2 - 10$

Graph each system of equations. Then find the solution.

12. $\begin{cases} 2x + 3y = 6 \\ x - y = 3 \end{cases}$ 13. $\begin{cases} 2x - y = -7 \\ x + 2y = 4 \end{cases}$

14. Kathryn put her nickels and dimes in a piggy bank. One day she emptied it and counted 25 coins, which totaled $2.05. How many dimes did Kathryn have?

15. The sum of two numbers is 2. Their difference is 10. Find the numbers.

Use the graph for Questions 16–18.

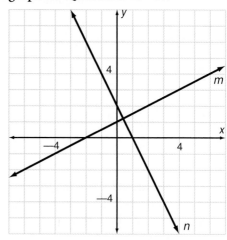

16. Write the equation of line n in slope-intercept form.

17. Write the equation of line m in point-slope form.

18. WRITING MATHEMATICS How are lines m and n related? How do you know?

State the domain and range of each function.

19. $\{(1, 5), (2, 6), (3, 7), (4, 8)\}$

20. $f(x) = 3x - 1$ 21. $f(x) = x^2 - 4$

The pictograph shows how many aluminum cans were collected by certain students. Use this data to answer Questions 22–24.

22. How many aluminum cans were collected by Benita?

23. Francisco collected 90 aluminum cans. Francisco's name is to be added to the pictograph. How many symbols would need to be shown to represent the number of cans that Francisco collected?

24. How many aluminum cans were collected by the class (including Francisco)?

STANDARDIZED TEST

STANDARD FIVE-CHOICE Select the best choice for each question.

1. On average, Mike misses 4 out of 25 free throws during basketball games. What is the probability that he will make his next free throw?

   A. 0.16
   B. 0.5
   C. 0.21
   D. 0.04
   E. 0.84

2. Which of the following is *not* in the range of the function $f(x) = |x| - 3$?

   A. 3
   B. 1
   C. −2
   D. −4
   E. two of the above

3. Find the value of x so that
   $15 - (3x - 5) = 2(5 - 2x) - 4$.

   A. −14
   B. 14
   C. 2
   D. −2
   E. no solution

4. Determine the value of $A - B$.

   $$A = \begin{vmatrix} 5 & 7 \\ 3 & -2 \end{vmatrix} \qquad B = \begin{vmatrix} -4 & 2 \\ -5 & 6 \end{vmatrix}$$

   A. −23
   B. −17
   C. −45
   D. 23
   E. −3

5. The line $4x - 7y = 28$ crosses the x-axis at which point?

   A. $(7, 0)$ B. $(0, 7)$ C. $(0, -4)$
   D. $(-4, 0)$ E. $(-7, 0)$

6. A student made this frequency table of the responses to the question "How many brothers and sisters do you have?" What is the median number of brothers and sisters of the respondents?

Number	Frequency
0	4
1	5
2	4
3	3
4	1
7	1

   A. 3.5   B. 1   C. 1.5
   D. 1.8333 . . . E. 2

7. The sum of three consecutive odd integers cannot exceed 36. What is the greatest possible value of the largest integer?

   A. 9
   B. 10
   C. 13
   D. 14
   E. cannot be named—it is infinite

8. The solution to the following system of equations is
   $$\begin{cases} 3x - 5y = 4 \\ 7y + 2x = -18 \end{cases}$$

   A. $(3, 1)$ B. $(0, -9)$ C. $(8, 4)$
   D. $(-2, -2)$ E. $(-9, 0)$

9. All of the following are solutions of $-2x + 1 > -7$ except

   A. 5   B. 2   C. 0
   D. −6   E. two of the above

10. There were 28,575 people at the stadium. This was about 60% of capacity. What is the capacity of the stadium?

    A. 35,000   B. 47,000   C. 50,000
    D. 60,000   E. 18,000

Take a Look AHEAD

Make notes about things that look familiar.

- How do you think solving a system of inequalities will be similar to solving a system of equations? How might it be different?
- Review the meaning of the inequality symbols. Make a chart of different word phrases associated with each symbol.

Make notes about things that look new.

- Find some graphs that are different from graphs in previous chapters. Explain what is different about them.
- Use the word constraint in a sentence. Try to give a definition or some synonyms.

DATA Activity

ATM Users

Conveniently located automated teller machines (ATMs) allow customers to bank anytime, day or night. However, there are disadvantages to this banking method. Customers must be careful not to lose their ATM card or to reveal their personal identification number. Computer breakdowns or electrical outages can cause widespread problems. As a result, some people are still reluctant to use ATMs. A bank conducted a survey to gather information about the age and income of ATM users.

SKILL FOCUS

- ▶ Calculate experimental probabilities and use values to make predictions.
- ▶ Write numbers as decimals, fractions, and percents.
- ▶ Draw conclusions from data.
- ▶ Make a circle graph.

HELP WANTED

Be a $mart $hopper

In this chapter, you will see how:

- **VIDEO SHOP OWNERS** use inequalities to determine store income. (Lesson 8.2, page 385)

- **RETAIL STORE BUYERS** use probability to make purchasing decisions. (Lesson 8.6, page 403)

ATM USAGE Distribution by Age	
Age (years)	Percent Using ATMs
Under 25	7
25 to 34	32
35 to 44	27
45 to 54	15
55 to 64	11
65 to 74	6
75 and over	2

The tables above and below show the results of the survey conducted by the bank. Use the tables to answer each question.

ATM USAGE Distribution by Income	
Household Income	Percent Using ATMs
Under $15,000	14
$15,000 to $24,999	16
$25,000 to $34,999	18
$35,000 to $49,999	23
$50,000 to $74,999	18
$75,000 and over	11

1. Based on this survey, how many in a group of 200 people aged 45 to 54 might be expected to be ATM users?

2. Based on this survey, what is the experimental probability that a person selected at random with a household income of $20,000 per year will be an ATM user? Write your answer as a fraction in lowest terms.

3. How would you describe the person most likely to be an ATM user?

4. **WORKING TOGETHER** For each table, display the information in an accurately drawn circle graph.

PROJECT

Consumer Report for Teens

Have you ever heard the expression "Read the fine print?" What do you think it means? In a world that offers choices for everything from breakfast cereals to credit cards, you must read labels, advertisements, and contracts carefully before you make a decision. In this project, you will investigate consumer issues and report on where the "real deals" can be found.

PROJECT GOAL

Identify and explore consumer issues of interest to young people.

Getting Started

Work with a group.

1. Begin by brainstorming a list of things you would like to investigate and report on. For example, you may want to compare computer network plans, discount stores, restaurant meals, or air travel costs. Skim the Project Connections so that you do not duplicate the issues in them.

2. Select a final list of three to five issues to investigate.

3. Discuss the format to use for your consumer report. Consider possibilities such as a panel discussion, taped "radio" segments, a video program, or a frequent newsletter for students. Your presentation can be entertaining as well as informative.

4. Begin a *Tips for Teens* list that includes warnings about stores with deceptive practices, notifications of special sales and free concerts, or information about getting the most out of economy sizes of products.

PROJECT *Connections*

Lesson 8.2, page 384:
Collect data about checking accounts at different banks and compare costs.
Lesson 8.4, page 393:
Evaluate negative option purchase plans and analyze advantages and disadvantages.
Lesson 8.6, page 402:
Explore the added costs involved with installment buying and determine annual percentage rate of interest.
Chapter Assessment, page 409:
Prepare and extend the Teen Consumer Report.

Internet Connection

www.swpco.com/
swpco/algebra1.html

8.1 Graph Linear Inequalities in Two Variables

Explore/Working Together

Work with a partner. Assume that marigolds are $1.00 each and geraniums are $2.00 each.

Ken is planting a flower box garden. He can spend up to $10. Let x represent the number of marigolds he can buy, and let y represent the number of geraniums he can buy.

1. **a.** If Ken buys only marigolds, what is the greatest number of plants he can buy? the smallest number? Write an ordered pair (x, y) for each of your solutions.

   b. If Ken buys only geraniums, what is the greatest number he can buy? What is the smallest number he can buy? Write an ordered pair for each of your solutions.

   c. If Ken buys 3 geraniums, what is the greatest number of marigolds he can buy? What is the smallest number he can buy? Write an ordered pair for each of your solutions.

2. Graph all of the points you found for Questions 1a, 1b, and 1c. Find all combinations of marigolds and geraniums Ken can consider. Graph those points as well.

3. What pattern do you notice on your graph?

4. Suppose Ken spends exactly $10 on marigolds and geraniums. Write an equation that represents the number of marigolds and geraniums Ken can buy.

5. Graph the equation you wrote for Question 4 on the same coordinate plane as the points for the ordered pairs.

SPOTLIGHT ON LEARNING

WHAT? In this lesson you will learn
- to determine whether ordered pairs are solutions of linear inequalities.
- to graph linear inequalities.

WHY? Graphing linear equalities in two variables can help you to solve problems in geometry, business, sports, health care, and technology.

Build Understanding

If you replace the equal symbol in any linear equation of two variables with $<, >, \leq, \geq$, or $\neq$, you create a **linear inequality** in two variables.

The points you graphed in Explore are solutions to the linear inequality $x + 2y \leq 10$. In this case, the solutions are found on or below the line $x + 2y = 10$. There are no ordered pairs above the line that are solutions to this linear inequality.

To find whether an ordered pair is a solution to a linear inequality, substitute the values for x and y into the inequality and evaluate.

EXAMPLE 1

Determine whether each ordered pair is a solution of $y < 3x - 2$.

a. $(3, 6)$ **b.** $(-4, 2)$ **c.** $(4, 10)$

Solution

a. $y < 3x - 2$
$6 \overset{?}{<} 3(3) - 2$
$6 \overset{?}{<} 9 - 2$
$6 < 7$ **true**

b. $y < 3x - 2$
$2 \overset{?}{<} 3(-4) - 2$
$2 \overset{?}{<} -12 - 2$
$2 < -14$ **not true**

c. $y < 3x - 2$
$10 \overset{?}{<} 3(4) - 2$
$10 \overset{?}{<} 12 - 2$
$10 < 10$ **not true**

So, $(3, 6)$ is a solution of $y < 3x - 2$, but $(-4, 2)$ and $(4, 10)$ are not. ◀

The graph of a linear equation divides the coordinate plane into two regions or **half-planes**. The line itself is the **boundary** of each of these two half-planes.

The graph of a linear inequality in two variables includes all points in a half-plane. Shaded half-planes show where solutions can be found. When the inequality symbol is $\leq$ or $\geq$, draw the boundary as a solid line. When the inequality symbol is $<$ or $>$, draw the boundary as a dashed line.

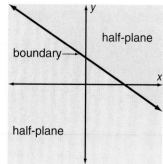

EXAMPLE 2

Graph: $y < 2x + 1$

Solution

Graph the corresponding linear equation $y = 2x + 1$. Because the inequality symbol is $<$, the boundary line is not a part of the solution. Show the boundary as a dashed line.

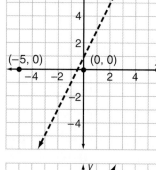

Test points on both sides of the boundary line to determine which half-plane to shade. Choose points that will be easy to substitute, such as $(0, 0)$ and $(-5, 0)$.

$(0, 0)$	$(-5, 0)$
$y < 2x + 1$	$y < 2x + 1$
$0 \overset{?}{<} 2(0) + 1$	$0 \overset{?}{<} 2(-5) + 1$
$0 < 1$ **true**	$0 < -9$ **not true**

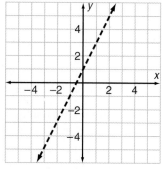

The point $(0, 0)$ is a solution of the inequality, so all other points in the half-plane containing $(0, 0)$ are also solutions of $y < 2x + 1$. Shade that half-plane. ◀

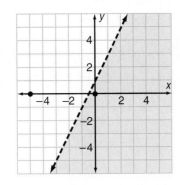

A half-plane that does not include the boundary line, such as the one in Example 2, is an **open half-plane**. A **closed half-plane** is one that includes the boundary line, as in Example 3 below.

 **CHECK UNDERSTANDING**

Look at the graph you drew in Explore. What part of the coordinate plane contains the points representing combinations of marigolds and geraniums that cost more than a total of $10?

EXAMPLE 3

GEOMETRY A certain rectangle has a perimeter of at least 30 cm.

a. Write a linear inequality that represents this situation. Then graph the solution of the inequality.

b. Refer to the graph and name three possible combinations of length and width of this rectangle.

Solution

a. Let x equal the length of the rectangle in centimeters and y equal the width of the rectangle in centimeters.

$$2x + 2y \geq 30$$

Graph the corresponding linear equation. Because the inequality symbol is $\geq$, the boundary is part of the solution. Show the boundary as a solid line. Test $(0, 0)$ and $(10, 10)$ to see whether either point is a solution of the inequality.

(0, 0)
$$2x + 2y \geq 30$$
$$2(0) + 2(0) \overset{?}{\geq} 30$$
$$0 \not\geq 30$$

(10, 10)
$$2x + 2y \geq 30$$
$$2(10) + 2(10) \overset{?}{\geq} 30$$
$$40 \geq 30$$

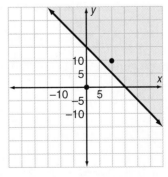

Therefore, shade the half-plane containing the point $(10, 10)$.

THINK BACK

If a linear equation is not in slope-intercept form, $y = mx + b$, rewrite it in that form before graphing it on a graphing utility. In Example 3, the slope-intercept form of the corresponding linear equation is $y = -x + 15$.

b. Both the length and the width of a rectangle must be positive numbers. Therefore, only the ordered pairs in the part of the shaded region that is within Quadrant I represent possible combinations of length and width. Some possible dimensions of the rectangle are 1 cm × 20 cm, 4 cm × 16 cm, or 16.2 cm × 20.5 cm. ◀

Determine whether the ordered pair is a solution of the inequality.

1. $y > -4x$; $(-4, 16)$

2. $3x - 2y \geq 12$; $(2, -3)$

3. $6x + 3 < 5y$; $(0, 0)$

Graph each inequality.

4. $y \leq 0.5x - 3$

5. $3x + 4y \geq 8$

6. TICKET GIVEAWAY The Spanish Club at Riverdale High School has received 40 free tickets to the International Festival to be held in Riverdale. These tickets will be given to the first students—club members or nonmembers—who request them.

    a. Write and graph an inequality that represents the number of club members and the number of nonmembers who could receive the tickets.

    b. Based on your graph, name three possible combinations of members and nonmembers who could receive tickets.

7. MODELING Write the inequality whose solution is represented by the graph below.

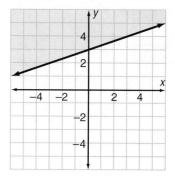

Determine whether the ordered pair is a solution of the inequality.

1. $y > 5x - 10$; $(3, 5)$

2. $2y \leq x$; $(0, 1)$

3. $2x + 4y \geq 7$; $(-1, 3)$

4. $3x - y \leq -7$; $(0, 0)$

Graph each linear inequality.

5. $y \leq x - 2$

6. $y > \dfrac{4}{5} x - 4$

7. $3x + 5y < 20$

8. $2x - y \geq 3$

9. $4x - 3y > 0$

10. $5x + 10y \leq -30$

Write the inequality whose solution is represented by each graph.

11.

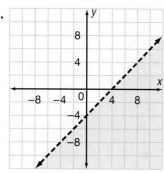

12.

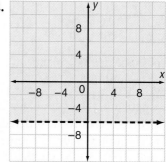

13.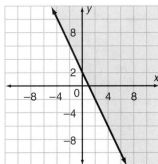

14.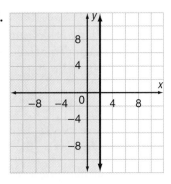

15. WRITING MATHEMATICS How is graphing a linear inequality in two variables like graphing an inequality in one variable? How is it different?

16. SPORTS The Spartans scored 20 points, but they still lost the football game to the Eagles. The Eagles scored field goals (3 points each), as well as touchdowns that were each followed by the extra conversion point (7 points in all).

  a. Write and graph an inequality that represents the combination of field goals and touchdowns the Eagles could have scored.

  b. Based on your graph, name three possible combinations of field goals and touchdowns the Eagles could have scored.

EXTEND

Equivalent inequalities are inequalities that have the same solution set. Determine whether each pair of inequalities is equivalent.

17. $y > x$
$y < -x$

18. $y \geq 3x + 1$
$3y \geq 9x + 3$

19. $2y > x - 3$
$x - 3 < 2y$

20. $3y > 2x + 6$
$4y \geq 6x + 8$

21. WRITING MATHEMATICS Describe how to decide whether the inequalities in Exercise 19 are equivalent.

22. HEALTH CARE A doctor's office schedules 30 min for each new patient and 15 min for each returning patient up to a total of 7 h each day.

  a. Write and graph an inequality that represents the combination of new and returning patients that could be scheduled for one day.

  b. Based on your graph, name three possible combinations of new and returning patients that could be scheduled.

23. **BUSINESS/TECHNOLOGY** A shoe store makes an average profit of $20 on each pair of regularly priced shoes it sells. The store has an average profit of $5 on each pair of sale priced shoes it sells. The store's goal is to make a daily profit of at least $1000.

a. Write an inequality that represents the combinations of regularly priced and sale priced shoes that must be sold for the store to achieve its goal. Use a graphing utility to graph this inequality.

b. Based on your graph, name three possible combinations of the two kinds of shoes the store might sell to achieve its goal.

THINK CRITICALLY

Describe each graph. Be sure to include whether the line is dashed or solid and which half-plane is shaded.

24. $y < mx + b$

25. $y > mx + b$

26. $y \leq mx + b$

27. $y \geq mx + b$

28. Write a linear inequality whose graph is all of Quadrants I and II, as well as the x-axis.

29. Write a linear inequality whose graph is all of Quadrants I and IV.

30. **WRITING MATHEMATICS** Graph the inequality you wrote for Exercise 29 on both a coordinate plane and a number line. How do the two graphs relate to each other?

31. **WRITING MATHEMATICS** Write a problem that could be solved by graphing $2x + 0.5y < 10$. Then graph the inequality and name three possible sets of ordered pairs that are in the solution of the inequality.

MIXED REVIEW

32. **STANDARDIZED TESTS** What is the value of the underlined variable in the formula for the given value of A?

$$A = \pi \underline{r}^2 \text{ when } A = 19.625$$

   A. 6.247   **B.** 2.5   **C.** 7.85   **D.** 39

Graph each compound inequality.

33. $q < -1$ or $q \geq 1$

34. $-2.3 \leq a \leq 6$

Solve the system using elimination.

35. $\begin{cases} 3x + 6y = 3 \\ 2x - 5y = -16 \end{cases}$

36. $\begin{cases} 8x - 4y = 2 \\ 4x + 8y = 26 \end{cases}$

Graph each linear inequality.

37. $2x + y \geq -4$

38. $y < 3x - 2$

Explore/Working Together

- Work with a partner. Let x represent the number of lawn seats and y represent the number of reserved seats.

  1. Tickets for the Dark Nights concert are $15 for lawn seats and $25 for reserved seats. Each person waiting in line for tickets will buy at least one ticket. Some will buy both types of tickets. Write an inequality that represents the amount of money each person will spend.

  2. Make a list of five ordered pairs that are solutions of the inequality you wrote.

  3. Cross out any ordered pairs you wrote that could not represent a combination of lawn seat and reserved seat tickets.

  4. Because the concert is expected to be a sellout, each person is limited to buying 6 tickets. Write an inequality that represents the number of tickets each person will buy.

  5. Cross out any ordered pairs that remain on your list that are not solutions of the inequality you wrote in Question 4.

  6. For an ordered pair to remain on your list, what must be true about it?

Build Understanding

- Like linear equations, two or more linear inequalities considered together form a **system of linear inequalities**. If you graph both inequalities in the system on the same coordinate plane, the region in which their graphs overlap contains all the solutions of the system. An ordered pair that is a solution of all the inequalities in a system is a solution of the system. The two inequalities you wrote in Explore form a system of linear inequalities.

What parts of the graph shown in Example 1 contain the points that represent the combinations of lawn and reserved seats that each person in Explore could buy?

EXAMPLE 1

Graph: $\begin{cases} 15x + 25y \geq 15 \\ x + y \leq 6 \end{cases}$

Solution

First, graph the linear inequality $15x + 25y \geq 15$.

Then graph $x + y \leq 6$ on the same coordinate plane.

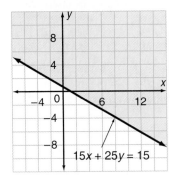

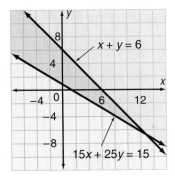

The region with the double shading and the boundary lines of this region contain the points that are solutions of the system.

You can solve real world problems using a system of linear inequalities.

EXAMPLE 2

CONSUMERISM Mrs. Fuentes wants to buy at least 10 books. Each paperback book costs an average of $10, and each hardcover book costs an average of $20. Mrs. Fuentes is planning to spend less than $250 on books.

a. Write a system of linear inequalities that represents this situation. Then graph the solution of the system.

b. Based on your graph, name three possible combinations of paperback and hardcover books Mrs. Fuentes could buy.

Solution

a. Let x represent the number of paperback books and y represent the number of hardcover books. The number of books Mrs. Fuentes wants to buy is $x + y \geq 10$. The amount she plans to spend is $10x + 20y < 250$. So, the system of inequalities that represents the situation is

$$\begin{cases} x + y \geq 10 \\ 10x + 20y < 250 \end{cases}$$

Graph $x + y = 10$ with a solid line.
Test points on either side of the line,
such as $(0, 0)$ and $(10, 10)$, to see
which half-plane to shade. Shade
above the line, because the point
$(10, 10)$ is above the line and makes
$x + y \geq 10$ true.

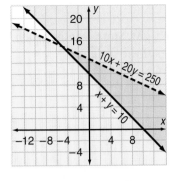

Then graph $10x + 20y = 250$
with a dashed line. Testing points
above and below the line shows
that the half-plane below the line should be shaded. The solution
is the double-shaded region.

b. Select any three points in the double-shaded region. The points
representing 12 paperback books and 6 hardcover books, 16
paperback books and 0 hardcover books, and 0 paperback books
and 10 hardcover books are all within the double-shaded region. ◄

TRY THESE

Graph each system of linear inequalities.

1. $\begin{cases} y > 3x - 1 \\ y \leq -\dfrac{2}{5}x + 3 \end{cases}$

2. $\begin{cases} 2x + y < 4 \\ 6x - 3y > 18 \end{cases}$

3. $\begin{cases} y < x \\ y \geq -x + 1 \end{cases}$

4. $\begin{cases} 2x - 3y \leq 9 \\ 2y + x < 6 \end{cases}$

5. $\begin{cases} x - y \leq 4 \\ x + y < 3 \end{cases}$

6. $\begin{cases} y \leq -2x + 3 \\ y > 4x - 1 \end{cases}$

7. $\begin{cases} x + y < -1 \\ 3x - y > 4 \end{cases}$

8. $\begin{cases} y > 2x \\ y - x \leq 5 \end{cases}$

9. $\begin{cases} x + 2y \leq 4 \\ x - 2y > 6 \end{cases}$

10. MODELING Write the system of linear inequalities whose solution is
represented by the graph at the right.

11. Determine whether each ordered pair is a solution of the system
represented by the graph at the right.

    a. $(0, 0)$ **b.** $(-4, -6)$ **c.** $(-4, 6)$ **d.** $(4, 10)$

RECREATION While traveling,
Mr. Komuro will not drive more than
8 h each day. However, he likes to
cover at least 400 km. He averages
60 km/h on the expressways and 40 km/h on the highways.

12. Write and graph a system of linear inequalities that represents
the hours and kilometers Mr. Komuro could drive each day.

13. Based on your graph, name three possible combinations of
roadways Mr. Komuro could drive each day.

PRACTICE

Refer to the graph at the right. State whether each ordered pair is a solution of the system.

$$\begin{cases} y \geq x - 4 \\ 2x + 3y < 9 \end{cases}$$

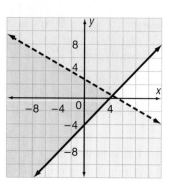

1. $(5, -2)$ 2. $(-1, 2)$

3. $(10, 3)$ 4. $(-10, -6)$

Graph each system of inequalities.

5. $\begin{cases} y > -3 \\ y \leq 2x + 4 \end{cases}$ 6. $\begin{cases} y < 2x + 4 \\ y > 2x - 3 \end{cases}$ 7. $\begin{cases} x + 4y \geq -8 \\ x - 2y \geq -6 \end{cases}$ 8. $\begin{cases} 2x + 5y > -10 \\ 3x - 4y \geq 12 \end{cases}$

9. $\begin{cases} y \geq 1 - x \\ y \leq x - 1 \end{cases}$ 10. $\begin{cases} 5x + 2y \geq 12 \\ 2x + 3y \leq 10 \end{cases}$ 11. $\begin{cases} x + y \leq 9 \\ x - y \geq 3 \end{cases}$ 12. $\begin{cases} x - y \leq -6 \\ 3x - y > -2 \end{cases}$

13. $\begin{cases} y > 4x - 1 \\ y < -2x + 3 \end{cases}$ 14. $\begin{cases} y < -3x + 2 \\ y \geq \dfrac{x}{3} + 5 \end{cases}$ 15. $\begin{cases} y < 3x + 3 \\ y > -3x + 3 \end{cases}$ 16. $\begin{cases} x - 6y \leq -5 \\ 2x - 3y > 0 \end{cases}$

Write the system of inequalities whose solution is represented by each graph.

17.
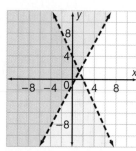

18.
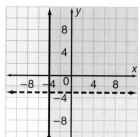

19.
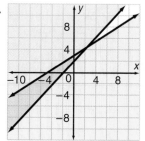

20. **WRITING MATHEMATICS** Explain how you determined the systems of inequalities for the graphs shown in Exercises 17–19.

21. **BUSINESS** During the spring Mr. Wilson assembles lawn furniture. It takes him $\dfrac{3}{4}$ h to assemble a chair and 1 h to assemble a table. He earns $50 for each chair and $80 for each table. He works no more than 50 h each week, but he likes to assemble enough lawn furniture to generate more than $3000 income.

   a. Write and graph a system of linear inequalities that represents the combination of chairs and tables Mr. Wilson assembles in a week.

   b. Based on your graph, name three possible combinations of chairs and tables Mr. Wilson could assemble in a week.

EXTEND

State whether the indicated part of the coordinate plane at the right includes a solution of $y < x + 5$, of $2x + y \leq 3$, of the system of both inequalities, or of neither inequality.

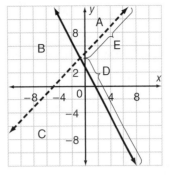

22. A

23. B

24. C

25. D

26. E

Write a system of linear inequalities whose solution represents each of these parts of the coordinate plane.

27. Quadrant I

28. Quadrant III and the axes that border it

29. Quadrant IV and the x-axis that borders it

Graph each system of linear inequalities.

30. $\begin{cases} y < x \\ y \geq -x - 3 \\ 3x + 5y < 15 \end{cases}$

31. $\begin{cases} 2y - x \geq 0 \\ x + 5y < 15 \\ y > x + 1 \end{cases}$

32. $\begin{cases} x > -1 \\ y \leq x + 2 \\ y > 2x \end{cases}$

33. $\begin{cases} x > 0 \\ x + y \leq 5 \\ x - y > 3 \end{cases}$

34. WEATHER The average monthly temperature is the average of the average monthly high temperature and the average monthly low temperature. During July, the average temperature in Buenos Aires, Argentina was less than 60°F. The average low was greater than 40°F.

a. Write and graph a system of three linear inequalities that represents this situation. (*Hint:* Let x represent the average low temperature and y represent the average high temperature. Then write one inequality to compare the average low and the average high.)

b. Based on your graph, name three possible combinations of average monthly high and low temperatures for that July in Buenos Aires.

35. FINANCE Mr. and Mrs. Patel plan to invest a maximum of $10,000. They will invest part of this amount in a mutual fund that pays an average of 8% annual interest and part in certificates of deposit that pay 5% annual interest. They would like to earn at least $500 interest the first year. Because the mutual fund has been paying a higher interest, they will invest more in the mutual fund than in the certificates of deposit.

    a. Write and graph a system of three linear inequalities that represents this situation.

    b. Based on your graph, name three possible combinations of amounts the Patels could invest in the mutual fund and in certificates of deposit.

THINK CRITICALLY

Write and graph a system of two inequalities whose solution is described by each of the following. If it is not possible to do so, explain why.

36. all points in a region bounded by two lines that do not intersect

37. no solution

38. all points in one line

39. all points in the coordinate plane

PROJECT *Connection* Many students have part-time jobs to earn their own money. A good way to keep track of that money is by opening a checking account. Different banks may offer the same basic accounts, but service charges, interest rates, and minimum balance requirements can vary widely.

    1. The different types of checking accounts usually offered by banks are cost-per-check accounts, minimum balance accounts, free checking accounts, and negotiable order of withdrawal (NOW) accounts. Make a table of information about each type of account at several local banks, including costs per check, monthly service charges, minimum balances, and possible interest rates.

    2. Express the total monthly cost of each account as a function of number of checks used. Graph the functions. Are the graphs linear? Explain.

    3. For interest-bearing accounts, write a verbal model and then a formula for determining the total amount earned or charged monthly.

    4. Use your findings to name the "Top Two Banks for Teens."

Career
Video Shop Owner

Shop owners are faced with decisions about inventory and pricing. For a video shop owner, considerations include determining the number of copies of new films needed, deciding how to price the newest films relative to older ones, determining how to dispose of videos that are no longer rented, and whether or not to begin a club.

Decision Making

1. Some video shops charge a set rate per day for rentals. Videos 'n More charges $3.00 per day or less, depending on whether the film is a new release, an older film, or a public service video. If a customer rents x videos for one day, write an inequality that represents the store's income, y.

2. Videos 'n More is considering offering the option of a club with a membership fee of $20.00 per year. Club members would be charged $2.00 or less per video, depending on the type of video. Suppose a club member rents x videos for one day. Write an inequality that represents the store's income, y. Write another inequality that describes at least how much annual income the store will receive from a club member.

3. Graph the three inequalities on the same coordinate plane. Create an ad the store owner could use if the store offered the club option. (*Hint*: Determine the fewest number of rentals for which the two plans produce the same income.)

4. If a customer rented at most two videos per month, which plan would be best?

5. If you were the owner of Videos 'n More, would you offer the club option to your customers? Explain your reasoning.

8.3 Algebra Workshop

Use Graphing Utilities to Solve Systems of Inequalities

Think Back

● Work with a partner. Use a graphing utility.

1. Graph: $\begin{cases} 3x - 2y = 7 \\ 4x + 6y = 5 \end{cases}$

2. Refer to your graph to find the solution of the system of equations. How did you determine the solution?

You can also use a graphing utility to solve systems of inequalities.

Explore

● 3. Use a graphing utility to graph these systems of linear inequalities, which correspond to the system of linear equations you graphed in Explore. (If your graphing utility does not shade for inequalities, copy the graph of the boundary lines onto paper and shade the half-planes.)

   a. $\begin{cases} 3x - 2y \le 7 \\ 4x + 6y \le 5 \end{cases}$ **b.** $\begin{cases} 3x - 2y < 7 \\ 4x + 6y < 5 \end{cases}$ **c.** $\begin{cases} 3x - 2y > 7 \\ 4x + 6y \ge 5 \end{cases}$

4. At what point do the boundary lines intersect? Does this point represent a solution of each system of inequalities?

5. Do any other parts of these graphs represent solutions of the systems? If so, describe them.

6. Use your graphing utility to graph this system of linear equations and the corresponding system of inequalities.

   a. $\begin{cases} y = 4 \\ -x + 3y = 6 \\ 2x + y = -12 \end{cases}$ **b.** $\begin{cases} y \le 4 \\ -x + 3y \ge 6 \\ 2x + y \ge -12 \end{cases}$

Refer to the graph of the system in Question 6a.

7. Name three points at which the lines intersect.

8. Do these points represent solutions of the system? Explain.

Refer to the graph of the system in Question 6b.

9. Name three points at which the lines intersect.

10. Does any part of this graph represent solutions of the system? If so, which part?

Making Connections

Refer to the graphs you have drawn in this workshop to answer these questions.

ALGEBRA: WHO, WHERE, WHEN

In calculus, inequalities are important in defining and determining the limit (maximum or minimum value) of any function.

11. How are the graphs of corresponding systems of linear equations and inequalities the same? How are they different?

12. How are the representations of the solutions of systems of linear equations and systems of linear inequalities the same? How are they different?

13. How does the number of solutions for a system of linear equations compare with the number of solutions for a system of linear inequalities?

Summarize

14. **MODELING** Write the system of inequalities represented by the graph at the right.

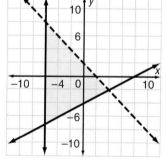

15. **WRITING MATHEMATICS** Consider the graph of the system of linear equations that corresponds to this system of linear inequalities.

$$\begin{cases} y \le 4x + 1 \\ y < -\dfrac{1}{2}\,x \end{cases}$$

Does the intersection point of the graph represent a solution to the system of inequalities? Explain your reasoning.

THINKING CRITICALLY Match each graph below to the system of inequalities it represents.

16.
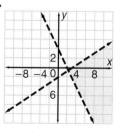

17.
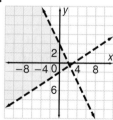

18.
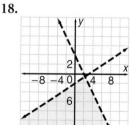

a. $\begin{cases} 2x + y < 4 \\ 2x - 3y > 6 \end{cases}$

b. $\begin{cases} 2x + y < 4 \\ 2x - 3y < 6 \end{cases}$

c. $\begin{cases} 2x + y > 4 \\ 2x - 3y > 6 \end{cases}$

19. **GOING FURTHER** Write a system of three inequalities that forms a triangle with its base parallel to the x- or y-axis. Find the area of your triangle.

ALGEBRA: WHO, WHERE, WHEN

When you make a telephone call, the circuiting system searches available paths for the fastest route using the *simplex method*, an application of linear programming developed in 1947 by mathematician George Danzig. The computer visualizes the network of possible connections as edges of a geometric solid, checking each corner as a possible connection for the call.

Explore/Working Together

Work with a partner.

1. Graph: $\begin{cases} 4x - 2y > -10 \\ x + y < 5 \\ 2x - 4y < 16 \end{cases}$

2. Select ten different points that represent solutions of this system of inequalities. Which point has the greatest sum of the coordinates? Which has the least sum?

3. In which region(s) of the graph did you find the points having greater coordinate sums? having lesser sums?

4. Write the ordered pairs for the vertices of the triangle formed by the boundary lines of the system. How do the sums of the coordinates of these points compare with the greatest and least sums you found in Question 2?

5. Examine the graph of a system of inequalities at the right. What do you think the greatest sum of the coordinates of a point in the shaded region is? What do you think the least sum is? Explain your thinking.

6. Test points within the shaded region. Was your thinking in Question 5 correct?

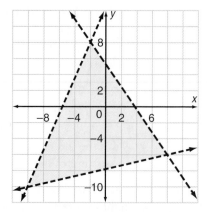

Build Understanding

Linear programming is a method used in business and government to help manage resources and time. **Constraints** are conditions that limit business activity. In linear programming such constraints are represented by inequalities. The intersection of the graphs of a system of constraints includes all possible solutions and is known as the **feasible region**. The vertices of the feasible region represent minimum and maximum values, as you found in Explore.

EXAMPLE 1

FUND RAISING The Student Council sells hot dogs and beverages at a concession stand during sports events. The following constraints must be considered.

- At each game council members always sell at least half as many hot dogs as beverages.

- At each game they always sell at least 150 beverages.

- There is space to refrigerate only 300 hot dogs for each game.

a. Should council members expect to sell 250 hot dogs and 400 beverages?

b. Should they expect to sell 200 hot dogs and 500 beverages?

c. What is the least number of hot dogs council members should expect to sell at a game?

d. What is the greatest number of beverages they should expect to sell at a game?

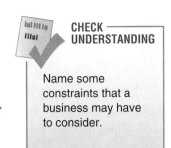
CHECK UNDERSTANDING

Name some constraints that a business may have to consider.

THINK BACK

Recall the inequality symbols:
$\leq$: "is at most"
$<$: "is less than"
$\geq$: "is at least"
$>$: "is greater than"
$\neq$: "is not equal to"

Solution

Let x represent the number of hot dogs and y represent the number of beverages. Write an inequality to represent each constraint.

- at least half as many hot dogs as beverages $x \geq \dfrac{1}{2}y$ or $y \leq 2x$

- at least 150 beverages $y \geq 150$

- space for only 300 hot dogs $x \leq 300$

Graph the system of inequalities. Write ordered pairs for the vertices of the shaded region. The triangular shaded region is the feasible region and represents all solutions that satisfy the constraints.

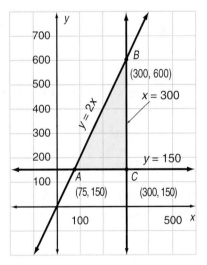

a. The ordered pair (250, 400) represents 250 hot dogs and 400 beverages. The point representing this ordered pair is within the feasible region, so this is a possible expectation.

b. The ordered pair (200, 500) represents 200 hot dogs and 500 beverages. The point representing this ordered pair is *not* within the feasible region, so this is not a possible expectation.

c. The ordered pair for vertex A (75, 150) represents the least number of hot dogs and beverages. So the least number of hot dogs the Student Council should expect to sell is 75.

d. The ordered pair for vertex B (300, 600) represents the greatest number of hot dogs and beverages. So the greatest number of beverages the Student Council should expect to sell is 600. ◄

Solutions to linear programming problems are often restricted to the first quadrant of the coordinate plane. The feasible region for Example 1 is completely within the first quadrant. When this is not the case, include $x \geq 0$ and $y \geq 0$ as constraints, if necessary.

Business people use linear programming to determine how to make the most (maximum) profit with the least (minimum) cost. You can write an equation representing a quantity such as profit or cost. This equation is called the **objective function**. By substituting values within the constraints, you can find the values that **maximize** or **minimize** the objective function. The maximum or minimum values of the objective function will occur at or near a vertex of the feasible region.

CHECK
UNDERSTANDING

Describe the set of numbers that lie within a region if the inequalities $x \geq 0$ and $y \geq 0$ are included as boundaries for that region.

EXAMPLE 2

PROFIT   At its concession stand, the Student Council makes a profit of $0.65 on each hot dog sold and $0.35 on each beverage sold. The objective function for profit is $P = 0.65x + 0.35y$. What is the maximum profit the Student Council should expect to make?

Solution

Substitute the coordinates for the vertices of the feasible region in the objective function.

Vertex	$P = 0.65x + 0.35y$	Profit, P
(75, 150)	$0.65(75) + 0.35(150)$	$101.25
(300, 150)	$0.65(300) + 0.35(150)$	$247.50
(300, 600)	$0.65(300) + 0.35(600)$	$405.00

The maximum profit is $405.00. To do this, the Council must sell 300 hot dogs and 600 beverages, the coordinates at point B.

TRY THESE

Graph the given constraints to determine the feasible region. Determine whether each point is within the feasible region.

1. Constraints: $y \leq x - 3$, $x \leq 8$, $y \geq 0$; point $(1, 4)$

2. Constraints: $y \leq 2x + 6$, $y \geq -3 - 2x$, $x \leq 0$, $y \geq 0$; point $(-1, 2)$

For the given vertices of each feasible region in Exercises 3–7, find the minimum and maximum value of the objective function and name the coordinates at which each occurs.

Objective Function	Vertices
3. $P = x + 3y$	$(0, 0), (3, 0), (0, 7), (4, 2)$
4. $P = 2x + 5y$	$(2, 3), (3, 6), (1, 6), (0, 5)$
5. $P = 4x + 2y$	$(0, 1), (9, 1), (2, 7), (6, 3)$
6. $C = 5x + y$	$(10, 10), (10, 20), (20, 20), (30, 10)$
7. $C = 3x + y$	$(0, 0), (0, 8), (4, 3), (5, 0)$

8. Graph the given constraints to determine the feasible region. Then find the minimum cost C for a business whose cost is given by the objective function $C = x + 2y$. Name the point at which it occurs.

$$x + y \le 6 \quad 2x - y \le 4 \quad x \ge 2 \quad y \ge 3$$

9. Graph the given constraints to determine the feasible region. Then find the maximum profit P given by the objective function $P = 2x + y$. Name the point at which it occurs.

$$y \le x + 6 \quad y \ge 3x - 1 \quad x \ge 0 \quad y \ge 0$$

10. WRITING MATHEMATICS Why should the feasible region for many linear programming problems be restricted to Quadrant I?

PRACTICE

For Exercises 1–3, graph the given constraints to determine the feasible region. Then determine whether each point is within the feasible region.

1. Constraints: $y \le -2x + 7$, $y \ge -x + 2$, $x \ge 0$, $y \ge 0$; point $(1, 4)$

2. Constraints: $3y - x \ge 0$, $y \le -2x + 9$, $x \ge 0$, $y \ge 0$; point $(5, 4)$

3. Constraints: $2x - y \le 2$, $2y - x \le 8$, $y \ge 2$; point $(1, 3)$

For the given vertices of each feasible region in Exercises 4–8, find the minimum and maximum value of the objective function and name the coordinates at which each occurs.

Objective Function	Vertices
4. $P = x + 4y$	$(0, -1), (4, 0), (3, 2), (0, 1)$
5. $P = 2x - 3y$	$(3, 1), (2, 4), (0, 5), (0, 0)$
6. $P = -x + 6y$	$(-2, 0), (3, 0), (4, 3), (0, 1)$
7. $R = 3x - 2y$	$(0, 0), (-2, -3), (-3, 3), (0, 1)$
8. $C = 4x + 5y$	$(0, 2), (2, 3), (4, 0), (0, 6)$

9. Graph the given constraints to determine the feasible region. Then find the maximum value of the objective function $P = x + 4y$. Name the point at which it occurs.

$$y + x \ge 0 \quad x + 2y \le 6 \quad x \le 0 \quad y \le 3$$

10. Graph the given constraints to determine the feasible region. Then find the minimum value of the objective function $C = 4x + 3y$. Name the point at which it occurs.

$$2y \le 22 - x \quad 3x \le 30 - 2y \quad x \le 6 \quad y \ge 3$$

11. Graph the given constraints to determine the feasible region. Then find the maximum value of the objective function $P = 4x + y$. Name the point at which it occurs.

$$x \ge 4 \quad 4y - x \ge 4 \quad y \ge -3x + 53 \quad y - x \le 5$$

12. Graph the given constraints to determine the feasible region. Then find the minimum value of the objective function $C = x + 5y$. Name the point at which it occurs.

$$3y + 2x \ge 3 \quad x \ge 0 \quad x \le 6 \quad y + x \le 16 \quad 3y - 4x \le 9$$

DECISION MAKING A shop owner can use either of two objective functions for determining her costs.

$$C = 4x + 4y \quad \text{or} \quad C = 3x + 2y$$

She has determined the constraints as follows.

$$y + 2x \leq 12 \quad x + y \leq 8 \quad x \geq 0 \quad y \geq 2$$

13. Graph the given constraints to determine the region. Name the vertices of the feasible region.

14. Use the vertices to determine which objective function she should use to minimize her costs.

FARMING Manuel needs to plant 210 acres on his farm. From crop x he will earn $400 an acre and from crop y he will earn $350 an acre. He must plant at least 40 acres of crop x and 50 acres of crop y. Soil conditions do not permit him to plant more than 80 acres of crop x. State regulations will allow the acres for crop y to be no more than twice the number of acres for crop x.

$$x \geq 40 \quad y \geq 50 \quad x \leq 80 \quad y \leq 2x \quad x + y \leq 210$$

15. Graph the given constraints to determine the feasible region. Name the vertices of the feasible region.

16. Use the objective function $P = 400x + 350y$ to find how many acres he should plant to maximize his earnings.

EXTEND

For the given objective function and constraints in Exercises 17–20, state whether there is a minimum value, a maximum value, or both. If a minimum or maximum value does not occur, explain why.

17. $T = 6x + 3y$
    Constraints: $3x + y \geq 8$, $2y - 4 \leq 8x$, $x \geq 0$, $y \geq 0$

18. $W = 3x - 4y$
    Constraints: $2x \leq y + 5$, $0 \leq y \leq 3$, $x \geq 0$

19. $R = x + 3.5y$
    Constraints: $y \leq 2x + 1$, $y \geq -x + 1$, $x \geq 0$, $y \geq 0$

20. For the given objective function and constraints, find the minimum and name the point at which it occurs.

    $C = x + y$
    Constraints: $3x + 2y \geq 12$, $x + 3y \geq 11$, $x \geq 0$, $y \geq 0$

21. Substitute the minimum value M you determined in Exercise 20 and graph the equations $x + y = M - 1$ and $x + y = M - 2$. What do you notice?

THINK CRITICALLY

JEWELRY MAKING Sherry is starting a business making friendship bracelets and necklaces. She must consider the following constraints.

- Materials cost $0.50 for a bracelet and $1.25 for a necklace. She has no more than $10.00 with which to begin.

- She believes she can make at least 2 necklaces per week.

- She believes she can make at least 5 bracelets per week.

- Also, Sherry expects to be able to make a profit of $1.00 on each bracelet and $2.00 on each necklace.

22. Write inequalities to represent the constraints. Then graph the system of inequalities.

23. How is the objective function $P = x + 2y$ related to the data in the Exercise?

24. Use the objective function given in Exercise 23 to find the maximum profit Sherry should expect to make. How many necklaces and how many bracelets would she have to make to maximize her profit?

MIXED REVIEW

25. STANDARDIZED TESTS Which of the following expressions has a value of -6?

   A. $(-3)(-5) - (1)(-9)$ **B.** $(-3)(-5) - (1)(-9)$

   C. $(-3)(5) - (-1)(9)$ **D.** $(3)(5) + (-1)(9)$

Evaluate the function for $x = -2$.

26. $y = x^2 - 4x + 7$ 27. $y = x^4 - 3x^3 + x - 4$

28. $y = 3x^2 - 7x + 12$ 29. $y = 2x^5 - 4x^3 + 12x - 6$

PROJECT *Connection*

An ad for a book, CD, or video club may read "Receive six selections for $2.99." By accepting the merchandise, you may also be agreeing to purchase additional selections under the club's *negative option plan*. These plans work as follows: If you want the selection, do nothing. If you do not want it, you must direct the seller not to send it by returning a "negative option" form that comes with the announcement of the selection.

1. Find ads for several different clubs. Make a table summarizing features such as what you receive as a joining bonus, how many selections you must buy over what period of time, whether future selections are offered at a discount from list price, how and when you can cancel your membership, if there are shipping charges, and how you must notify the seller that you do not want a selection.

2. Compare the total cost for the amount of merchandise you receive from the plan with the cost if the same items were purchased at local stores. Draw graphs showing the total costs of different numbers of items. Do the clubs really offer good deals? What are the advantages and risks of belonging?

Use Linear Programming

All businesses want to maximize profits and minimize costs. Linear programming is a model for determining maximum profit or minimum costs under a set of constraints. Recall that the value to be maximized or minimized is the objective function.

Problem

The Spirit Club is selling both long- and short-sleeved T-shirts. The club pays $4 per short-sleeved shirt and $5 per long-sleeved shirt. Each shirt is then silk-screened. From experience, members know the following.

• They always sell at least as many long- as short-sleeved shirts.
• They can silk-screen at least 30 shirts per week but no more than 50.
• Their supplier can provide them with no more than 25 long-sleeved shirts per week.

How many of each type of shirt should the Spirit Club produce weekly to minimize their costs?

CHECK UNDERSTANDING

For this problem, are the constraints $x \geq 0$ and $y \geq 0$ needed? If so, how do they appear in the graph?

Explore the Problem

First, define the variables. Let x represent the number of short-sleeved shirts produced, and let y represent the number of long-sleeved shirts produced.

1. Express the constraints in terms of x and y.

Next, write an objective function for cost.

2. If the cost of producing one short-sleeved shirt is $4, write an expression for the cost of producing x short-sleeved shirts.

3. At a cost of $5 per long-sleeved shirt, write an expression for the cost of producing y long-sleeved shirts.

PROBLEM SOLVING TIP

Set your graphing utility so that integers are displayed when using the TRACE feature.

4. Since the only costs being considered are for the purchase of shirts, what equation represents the club's total costs C?

5. Graph the constraints on a graphing utility. Use the TRACE feature to find the vertices of the feasible region.

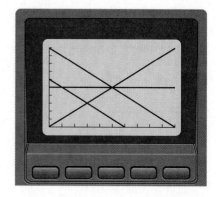

6. Copy and complete the following table to evaluate the objective function $C = 4x + 5y$ for the (x, y) value at each vertex.

Vertex	$C = 4x + 5y$	Cost
(15, 15)	4(15) + 5(15)	_____
(5, 25)	4(5) + 5(25)	_____
(25, 25)	4(25) + 5(25)	_____

PROBLEM
SOLVING PLAN

• Understand
• Plan
• Solve
• Examine

7. At which vertex is the cost function a minimum?

8. What combination of shirts should the club produce to minimize its costs? What will be the minimum cost?

CHECK
UNDERSTANDING

How does a business determine its profit?

Investigate Further

Given the constraints in the problem, assume that the club can sell short-sleeved shirts for $6 and long-sleeved shirts for $8.

9. What is the club's profit on one short-sleeved shirt? on one long-sleeved shirt?

10. Write an expression for the profit on x short-sleeved shirts. Write an expression for the profit on y long-sleeved shirts.

11. Write the objective function for the club's profit from both short- and long-sleeved shirts.

12. Copy and complete the following table.

Vertex	$P = 2x + 3y$	Profit
(15, 15)	2(15) + 3(15)	_____
(5, 25)	2(5) + 3(25)	_____
(25, 25)	2(25) + 3(25)	_____

13. At which vertex will the maximum profit be earned? What will the profit be for that number of shirts?

14. Consider this statement, in reference to the original problem and Questions 1–13: "If club members want to maximize their profits, they will have to spend more than the minimum. If they want to minimize their costs, they will earn less than the maximum profit." How does this statement apply to the problem?

Apply the Strategy

15. SKI MANUFACTURE A manufacturer makes two types of snow skis, slalom and cross-country. It takes 6 h to make a pair of cross-country skis and 4 h to make a pair of slalom skis. One hour is required to finish each type. The maximum number of hours of labor available is 96 for manufacture and 20 for finishing. The profit is $45 for each pair of cross-country skis and $30 for each pair of slalom skis.

  a. Express the constraints in terms of x and y.

  b. Graph the constraints. Find the vertices of the feasible region.

  c. Write an objective function for profit P and evaluate it for the (x, y) value at each vertex.

  d. To maximize profit, how many pairs of each type of ski should the company produce? What will the maximum profit be?

16. SCHOOL FIELD TRIP A school is planning a field trip and will rent buses and vans for the trip. Each bus carries up to 60 students, requires 4 chaperons, and costs $1000 to charter. A van carries up to 10 students, requires 2 chaperons, and costs $100 for gas. At least 300 students plan to attend, and 36 parents will chaperon.

  a. Express the constraints in terms of x and y.

  b. Graph the constraints. Find the vertices of the feasible region.

  c. Write an objective function for cost C and evaluate it for the (x, y) value at each vertex.

  d. To minimize cost, how many of each type of vehicle should the school rent? What is the minimum cost?

17. EXERCISE EQUIPMENT A sports equipment company makes stationary bicycles and rowing machines. On a given day, the company can make no more than 110 pieces. The cost to produce a bicycle is $75; a rower, $125. The company can spend no more than $10,000 per day on production. A bicycle sells for $125, and a rower sells for $200. To maximize profit, how many of each type should the company produce per day? What is the maximum profit?

18. WRITING MATHEMATICS Describe two services or products you could provide. Determine constraints that might apply to your business. Base the constraints on performance or production time, operating costs, and profits. Determine a relationship between the constraints that makes sense.

REVIEW PROBLEM SOLVING STRATEGIES

YIP, WOOF, Grr!

A Strange and Moving Story

1. The diagram shows the room arrangement at the Hotel Strange. Room 102 is currently empty. Luggage for five different families is in the other five rooms.

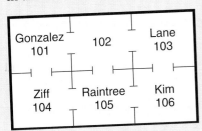

| Gonzalez 101 | 102 | Lane 103 |
| Ziff 104 | Raintree 105 | Kim 106 |

Mr. Bumbler, the desk clerk, realizes that the Lanes and the Kims were given each other's rooms, so their luggage must be switched. To prevent further mixups, Mr. Bumbler insists that luggage for two families cannot be in the same room at the same time. Because of the odd arrangement of rooms, Mr. Bumbler knows that some of the other families will also have to be moved. He is not concerned about the other families returning to their original rooms, as long as the Lanes and the Kims are switched. Finally, Mr. Bumbler wants to complete the switch in as few moves as possible.

a. Work with a partner. Decide on a strategy for solving the problem. Also decide on a method for recording the moves.

b. Compare your solution with those of others. Did anyone make the switch in fewer moves?

2. There are 100 dogs at Pampered Paws Kennel. Each day, the dogs eat 100 lb of food. Large dogs eat 3 lb a day, medium dogs eat 2 lb, and small dogs eat $\frac{1}{2}$ lb. How many dogs of each size are at the kennel if there are five times as many medium dogs as large ones?

a. How many unknowns are there in the problem? How will you represent them algebraically?

b. How many facts are given in the problem? Is this enough to solve the problem? Why?

c. How many of each size dog are there?

d. Could you have solved this problem using another strategy? If so, do you think one method is more efficient?

A Burning Question

3. After his electricity went out, Mr. Choi took out two candles. Both were the same length, but the label on candle A said that it would burn for 4 h while the label on candle B said it would burn for 5 h. Mr. Choi lit both candles at the same time and let them burn until the electricity was restored. He discovered that what remained of one candle was exactly four times the length of what was left of the other. How long were the candles burning? You may use the questions below to guide your thinking, or solve the problem on your own and explain your method.

a. Let t represent the number of hours that the candles were burning, and let x represent the original length of each candle. How much of its original length does candle A burn in t hours? How long is candle A after t hours?

b. How much of its original length does candle B burn in t hours? How long is candle B after t hours?

c. After t hours, how is the remaining length of candle B related to the remaining length of candle A?

d. How long were the candles burning? How much of the original length of each candle remained?

Explore

Imagine playing a game with one standard die and a spinner like the one shown. You or your partner chooses the die or the spinner. The other person then takes the other objects. A game consists of one throw of the die and one spin of the spinner. Whichever player shows the greater number wins the game.

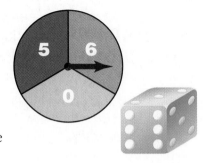

1. If you had first choice, which would you choose to have the greater chance of winning?

2. What would you need to know in order to have enough information to choose?

3. How many games would you want to play before deciding which, if either, player has an advantage?

4. Use a die and a spinner like the one shown. Play the game until you are convinced that one is better. Which one is it?

Build Understanding

The game described in Explore is a **probability experiment**. To find out without experimenting which device is more favorable involves **theoretical probability**.

Recall that the **probability of an event**, $P(E)$, is the ratio

$$P(E) = \frac{\text{number of favorable outcomes}}{\text{total number of outcomes in the sample space}}$$

In the Explore game, each combination of numbers that results from a throw of the die and a spin of the spinner is a possible **outcome** of the experiment. Each of the outcomes is equally likely. Any one of the possible outcomes of the experiment constitutes an **event**. Together, all possible outcomes of a probability experiment make up the **sample space** for the experiment.

EXAMPLE 1

What is the sample space for the game in Explore in which the die is rolled once and the spinner is spun once?

Solution

Two methods for determining the sample space are shown below.

a. Use a tree diagram.

Spinner	Die	Outcomes	Spinner	Die	Outcomes	Spinner	Die	Outcomes
0	1 →	0, 1	5	1 →	5, 1	6	1 →	6, 1
	2 →	0, 2		2 →	5, 2		2 →	6, 2
	3 →	0, 3		3 →	5, 3		3 →	6, 3
	4 →	0, 4		4 →	5, 4		4 →	6, 4
	5 →	0, 5		5 →	5, 5		5 →	6, 5
	6 →	0, 6		6 →	5, 6		6 →	6, 6

b. Use a table of ordered pairs.

		Die					
		1	**2**	**3**	**4**	**5**	**6**
	0	(0, 1)	(0, 2)	(0, 3)	(0, 4)	(0, 5)	(0, 6)
Spinner	**5**	(5, 1)	(5, 2)	(5, 3)	(5, 4)	(5, 5)	(5, 6)
	6	(6, 1)	(6, 2)	(6, 3)	(6, 4)	(6, 5)	(6, 6)

Both methods of display show that the sample space for the experiment or game in Explore consists of 18 possible outcomes. ◄

You can also use the **fundamental counting principle** to count outcomes in a sample space.

> ┌─ FUNDAMENTAL COUNTING PRINCIPLE ─
>
> **If one activity can be done in m ways and, for each of the m ways, a second activity can be done in n ways, then both can be done in mn ways.**
>
> **total number of possible outcomes = $m \cdot n$**

The fundamental counting principle can be applied to more than two activities.

EXAMPLE 2

Use the fundamental counting principle to determine the number of possible outcomes for the game in Explore.

Solution

The game consists of two activities, rolling the die and spinning the spinner. For one roll of the die there are 6 possible outcomes. For one spin of the spinner there are 3 possible outcomes. Using the fundamental counting principle, there are $3 \cdot 6$ or 18 possible outcomes. ◄

ALGEBRA: WHO, WHERE, WHEN

The modern theory of probability began in a series of letters written in 1654 between the mathematician Blaise Pascal and another mathematician, Pierre Fermat. The discussion centered on a problem: A player is given eight throws of a die to throw a one. The game is interrupted after only three throws. How should the player be compensated?

ALGEBRA: WHO, WHERE, WHEN

The earliest book dealing with the theory of probability was the *Art of Conjecturing* (1713) by Jacques Bernoulli.

You can use theoretical probability to find out which you would choose in order to have the greater chance of winning, the spinner or the die.

EXAMPLE 3

What is each player's probability of winning the game in Explore? Which player is more likely to win?

Solution

Refer to the sample space in Example 1. There are 18 possible outcomes in the sample space. For the event "spinner player wins," there are 9 outcomes.

$$P(\text{spinner player wins}) = \frac{9}{18}, \text{ or } \frac{1}{2}$$

For the event "die player wins," there are 7 outcomes.

$$P(\text{die player wins}) = \frac{7}{18}$$

The spinner player is more likely to win. ◄

Since there are two possible ties, the $P(\text{a player wins})$ is $\frac{16}{18}$, or $\frac{8}{9}$.

The event "game ends in a tie" is the same as "no player wins." Because the sum of the probabilities of all outcomes in the sample space is 1, to find $P(\text{no player wins})$, you can subtract $P(\text{a player wins})$ from 1.

$$P(\text{no player wins}) = 1 - \frac{8}{9} = \frac{1}{9}$$

In general, the events A and *not A* are **complements** and

$$P(A) + P(\text{not } A) = 1, \text{ or}$$
$$P(A) = 1 - P(\text{not } A), \text{ or}$$
$$P(\text{not } A) = 1 - P(A)$$

EXAMPLE 4

A shop sells jackets in smooth leather, suede, or cotton. Each is available in four sizes (small, medium, large, and extra large) and three colors (brown, tan, and green). The store has one of each possible jacket. What is the probability that a jacket chosen at random is large and brown?

Solution

Use the fundamental counting principle to determine the number of outcomes in the event "choose a large brown jacket" and the total number of outcomes.

Number of Outcomes in the Event	Total Number of Outcomes
material • size • color	material • size • color
3 • 1 • 1 = 3	3 • 4 • 3 = 36

$$P(\text{large, brown}) = \frac{3}{36} = \frac{1}{12}$$ ◄

THINK BACK

What is the greatest value that can be found for the probability of any event? the least value? What do each of these values indicate?

COMMUNICATING ABOUT ALGEBRA

Use the information in Example 4. How could you determine the probability that a jacket chosen at random isn't a small cotton one?

Use the following information to solve Exercises 1–4. You will toss a coin and draw a card from a standard deck of 52 cards.

1. How many different outcomes are there in the sample space?

2. Show a table of ordered pairs for the event "tossing a head and drawing a red card." How many outcomes are in the event?

3. What is the probability of tossing a head and picking a red card?

4. What is the probability of tossing a tail and picking the ace of clubs?

BUSINESS INVENTORY   The Shirt Gallery has long- and short-sleeved shirts. They are available in blue, white, tan, yellow, green, and pink, as well as with diagonal, vertical, and horizontal stripes. Every shirt is available in ten different sizes.

5. What is the least number of shirts the store must carry in inventory to have one of each possible selection?

6. If the Shirt Gallery keeps the minimum number of shirts in stock, how many will be
   a. short-sleeved?
   b. blue?
   c. long-sleeved and yellow?
   d. short-sleeved, pink, with vertical stripes?

7. What is the probability that a shirt chosen at random will be green with vertical stripes?

PRACTICE

Use the following information to solve Exercises 1–5. A license plate in one state consists of two digits from 0 to 9, the first of which cannot be zero, followed by three letters.

1. If the two digits are known, how many different license plates are possible?

2. If the three letters are known, how many different license plates are possible?

3. How many different license plates are possible for the given conditions?

4. If you try to remember the license plate but can only recall the two numbers and first letter, what is the probability that you can guess the rest of the plate in one guess?

5. How many different plates can have the same letter used three times?

6. An ice-cream store advertises that it has 28 different flavors and 12 different toppings. How many different cups can be made with one flavor and one topping?

7. If the middle digit of a three-digit area code must be either 0 or 1, and the first digit cannot be 0, how many area codes are possible?

8. In Exercise 7, if you dial three digits at random, what is the probability that you will dial a three-digit sequence that is an actual area code?

9. A restaurant offers a complete meal that consists of three courses. If the choice is one each of six appetizers, eight main courses, and five desserts, how many different complete meals are possible?

10. In Exercise 9, what is the greatest number of main courses that the restaurant can remove from its menu and still allow at least 150 different meals to be chosen?

11. On a five-question true/false quiz, what is the propability of answering all five questions correctly by guessing?

EXTEND

For Exercises 12–14, decide whether the games described give each player an equal chance of winning. If not, name the player who has the advantage and explain why. Assume dice are labeled 1 to 6.

12. Two dice are rolled and the numbers are added. Player A wins if the sum is even; player B wins if the sum is odd.

13. Two dice are rolled and the numbers are multiplied. Player A wins if the product is even; player B wins if the product is odd.

14. Three coins are tossed. Player A wins if all three show either heads or tails. Player B wins if any one coin is different from the other two. For each win, Player A scores 3 points. For each win, Player B scores 1 point.

THINK CRITICALLY

Use the following information to solve Exercises 15–19. A husband and wife are planning to have four children. Assume that having a boy or a girl is equally likely. (*Hint:* List all possible outcomes.)

15. What is the probability that all four children will be boys?

16. What is the probability that all four children will be of the same sex?

17. What is the probability of having three girls and one boy? of having three boys and one girl?

18. What is the probability of having two boys and two girls?

19. What is the sum of all the probabilities in Exercises 16–18? Explain this result.

PROJECT *Connection*

Suppose you want to purchase an expensive item, such as a computer. If you do not have enough cash to pay for the item, you might consider *installment buying*. You pay part of the price as a *down payment* at the time of purchase and the remainder in equal monthly installments. Buying on an installment plan is a form of credit, and you should expect to pay extra for credit. Suppose, for example, you can buy a $1200 computer for a down payment of $240 and 12 monthly payments of $90 each.

1. Determine the total price you will pay for the computer. How much extra are you paying for an installment plan?

2. What percent of the list price of the computer does the extra cost in Question 1 represent? If you paid the total installment price in a single payment at the end of the year, that percent would be the interest rate. Since you make monthly payments, the *annual percentage rate* (APR) of interest is actually higher. Determine the APR using this formula.

$$\text{APR} = (24 \cdot \text{extra cost}) \div [(\text{list price})(\text{number of payments} + 1)]$$

The buyer for a retail store selects and purchases the merchandise to be sold by the store. Some of the decisions the buyer makes are based on probabilities.

Decision Making

Colors	Sizes	Styles
white	small	crewneck
grey	medium	hooded
black	large	
red	X large	
blue	XX large	
green	XXX large	
yellow		
pink		
purple		

1. The buyer for Ace Sports wants to purchase the line of sweatshirts described in the table at the left. The shirt manufacturer is offering a special introductory purchase price if the buyer purchases three of each possible shirt. How many shirts is this?

2. Suppose the buyer takes advantage of the special introductory offer. What is the probability that the first customer will select a large, blue, hooded sweatshirt?

3. What is the probability that the first sweatshirt sold will be a red crewneck sweatshirt?

4. A buyer relies on statistics from market trend reports and sales reports to predict which items are most likely to sell. Below are portions of the Ace Sports sales report for last year. The shirt manufacturer offers the same special rate for larger orders as long as the wholesale buyer purchases at least three of each shirt. Based on the statistics in the table below at the right, how many of each shirt would you recommend for the wholesale buyer?

Total Plain Sweatshirts Sold: 1500					
Styles		**Sizes**		**Colors**	
Crewneck:	1150	Small:	80	White:	191
Hooded:	350	Medium:	152	Grey:	296
		Large:	220	Black:	143
		X Large:	465	Red:	360
		XX Large:	435	Blue:	375
		XXX Large:	148	Green:	135

· · · CHAPTER REVIEW · · ·

VOCABULARY

Choose the word from the list at the right that completes each statement.

1. If an inequality is written with ≤ or ≥, the boundary line is shown as a(n) __?__.
   a. vertex

2. A(n) __?__ half-plane does not include the boundary line.
   b. sample space

3. In a probability experiment, the set of all possible outcomes is called the __?__.
   c. open half-plane

4. In linear programming, the maximum or minimum value of the objective function will occur at a(n) __?__ of the feasible region.
   d. solid line

Lesson 8.1 GRAPH LINEAR INEQUALITIES IN TWO VARIABLES   pages 373–378

- A **linear inequality** in two variables is an inequality of the form

  $$y < mx + b, y > mx + b, y \leq mx + b, \text{ or } y \geq mx + b,$$

  in which x and y are variables and m and b are constants. To determine whether an ordered pair is a solution of a linear inequality, substitute the values for the variables and evaluate the inequality.

- The graph of a linear equation separates the coordinate plane into three parts: points on the line, points above the line, and points below the line. Points above the line and below it form two **half-planes**. The line forms the **boundary** of each of the half-planes.

- The graph of a linear inequality in two variables includes all points in a half-plane. For inequalities with ≤ or ≥, the half-plane includes the boundary line and is a **closed half-plane**.

- For inequalities with < or >, the half-plane does not include the boundary line and is an **open half-plane**.

Tell whether the ordered pair is a solution of the linear inequality.

5. $y \leq 3x - 6$; $(1, 1)$ 6. $3y > x$; $(0, 0)$ 7. $2x + y > 4$; $(4, 6)$

Graph each linear inequality.

8. $y \leq x + 3$ 9. $3x + 6y \leq 12$ 10. $3x - 2y > 6$

Lesson 8.2 SOLVE SYSTEMS OF LINEAR INEQUALITIES BY GRAPHING   pages 379–385

- If you graph a **system of linear inequalities** on a coordinate plane, the region in which their graphs overlap contains the solution set of the system.

Graph each system of linear inequalities.

11. $\begin{cases} y \leq x + 1 \\ y \geq 2x - 2 \end{cases}$
12. $\begin{cases} 3y < x + 6 \\ y > -2x + 3 \end{cases}$
13. $\begin{cases} x - 4y \leq 4 \\ 3x - 2y \geq 0 \end{cases}$

Lesson 8.3 USE GRAPHING UTILITIES TO SOLVE SYSTEMS OF INEQUALITIES pages 386–387

• You can use a graphing utility to solve a system of inequalities.

Use a graphing utility to solve each set of inequalities.

14. $\begin{cases} y > 3 \\ y < x + 3 \end{cases}$

15. $\begin{cases} y < 3 \\ y > x + 3 \end{cases}$

16. $\begin{cases} y > 3 \\ y > x + 3 \end{cases}$

Lesson 8.4 LINEAR PROGRAMMING: THE OBJECTIVE FUNCTION pages 388–393

• **Linear programming** is a technique in which graphs of linear inequalities are used to find a maximum value for profit or a minimum value for costs. The inequalities represent **constraints**, or restrictions, that limit business activity.

• The region containing possible solutions is the **feasible region**. The expression to be maximized or minimized is called the **objective function**. A maximum or minimum value always occurs at a vertex of a feasible region.

• To avoid negative solutions, the inequalities $x \geq 0$ and $y \geq 0$ are included as constraints.

17. Find the maximum profit P for the objective function $P = 2x + 3y$ if P is subject to the following constraints. Name the point at which it occurs.

$$y \geq x + 5 \qquad 2x + y \leq 8 \qquad x \geq 0 \qquad y \geq 0$$

Lesson 8.5 USE LINEAR PROGRAMMING pages 394–397

• You can use linear programming to set up a model to determine the conditions under which costs will be minimized or profits maximized for a given set of constraints.

18. The Home Ec Club plans to sell two kinds of pies at the Health Fair, apple and blueberry. They sell apple pies for $3 each and blueberry pies for $2 each. Members usually sell equal numbers of apple and blueberry pies. They can make at least 20 pies but no more than 40. They can get only enough blueberries for 20 pies.

  a. What expression represents the profits P to be maximized?

  b. Write expressions for the constraints.

  c. Use a graph to find how many of each type of pie the club should sell to maximize profits.

Lesson 8.6 EXPLORE PROBABILITY: THEORETICAL PROBABILITY pages 398–403

• In a probability experiment, the set of all possible outcomes is called the **sample space**. The number of possible outcomes can be found by the **fundamental counting principle**. To apply the principle, find the product of the number of possible outcomes of each activity.

• The **theoretical probability** P for each event E in an experiment is the ratio

$$P(E) = \frac{\text{number of outcomes in } E}{\text{total number of outcomes in the sample space}}$$

19. You toss three number cubes. What is the total number of outcomes in the sample space?

CHAPTER ASSESSMENT

CHAPTER TEST

Tell whether the ordered pair is a solution of the linear inequality.

1. $y > 4x - 3$; $(4, 1)$ 2. $3x - y < 4$; $(2, 2)$

Graph each system of inequalities.

3. $\begin{cases} x \le 2 \\ y \ge -4 \end{cases}$ 4. $\begin{cases} y - x < 2 \\ y + 2x < 6 \end{cases}$

5. $\begin{cases} x + 3y \le -9 \\ x - y \ge 0 \end{cases}$ 6. $\begin{cases} y < 2x + 4 \\ y > -2x + 4 \end{cases}$

7. $\begin{cases} x - 3y \ge -1 \\ 3x - 4y \le 2 \end{cases}$ 8. $\begin{cases} x + 2y \le 3 \\ -2x - 4y \le 3 \end{cases}$

9. Carla plans to buy at least 16 CDs and books as graduation gifts. CDs will cost her an average of $12, and books will cost an average of $16. Carla wants to spend no more than $288.

   a. Write and graph a system of linear inequalities to represent the situation described above.

   b. Name three combinations of single CDs and albums that Carla could buy.

Graph each system of inequalities.

10. $\begin{cases} y - x \ge 0 \\ x + y \le 8 \\ x \ge 2 \end{cases}$ 11. $\begin{cases} x \ge 1 \\ y \le 1 \\ x - y \le 7 \end{cases}$

12. **WRITING MATHEMATICS** Write a paragraph to explain how you used the data in Question 9 to write the system of inequalities.

13. A restaurant offers a three-course meal. If there is a choice of seven appetizers, nine main courses, and seven desserts, how many different complete meals are there?

Use the graphs of $y = -2$, $y = x$, and $y = 10 - 2x$, shown at the right, to name the numbered region that contains the solution set of each system given.

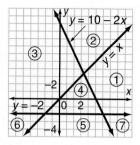

14. $\begin{cases} y > -2 \\ y < x \end{cases}$ 15. $\begin{cases} y < -2 \\ y < x \end{cases}$

16. $\begin{cases} y > x \\ y > -2 \end{cases}$ 17. $\begin{cases} y > x \\ y < -2 \end{cases}$

18. $\begin{cases} y < x \\ y < 10 - 2x \\ y > -2 \end{cases}$ 19. $\begin{cases} y > -2 \\ y > x \\ y < 10 - 2x \end{cases}$

For the given objective function and constraints, find the maximum profit or minimum cost and name the point at which it occurs.

20. $P = 5x + 3y$;
    $x + y \le 6$; $x - y \le 4$; $x \ge 0$; $y \ge 0$

21. $C = 2x + y$;
    $x + y \ge 6$; $x - y \ge 4$; $x \ge 0$; $y \ge 0$

22. Joe's Heavy Hauling has 8 five-ton trucks, 7 ten-ton trucks, and 12 drivers. Joe hauls 420 tons of cement per day. The five-ton trucks can make 8 trips a day. The ten-ton trucks can make six trips per day. The cost is $40 per day for a five-ton truck and $60 for a ten-ton truck, not including drivers' salaries.

    a. Write an equation for the objective function representing cost. Then write inequalities for the constraints. Graph the system.

    b. If all 12 drivers are to work, how many trucks of each type must be used to minimize the cost?

PERFORMANCE ASSESSMENT

QUILT MODELS Many quilt patterns are based on simple shapes such as squares and triangles.

1. Write the systems of inequalities that generate the green square and each of the blue triangles on this quilt pattern.

2. Find a quilt pattern or other folk art design that you like. Copy the design onto graph paper and draw coordinate axes. Use equations and systems of inequalities to describe as many lines and regions in the design as you can.

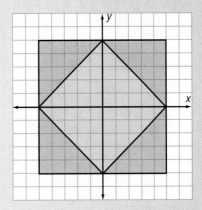

BAKE SALE Find recipes for two items you could prepare for a charity bake sale. The recipes should each require different amounts of two main ingredients, such as shortening and flour or eggs and flour. Make up numbers for how much of these ingredients you might have on hand. Then estimate how much they could sell for at the sale. Show how to write and solve a linear programming problem to determine how to maximize the profit from your baking.

PARALLELOGRAM PROBLEM Draw coordinate axes on a piece of graph paper. Shade a parallelogram that is not a rectangle and has an area of 48 square units. (Recall that area is the product of the base and the height.) Write the system of inequalities that describes the shaded area.

PROJECT ASSESSMENT

PROJECT *Connection* If your group has not already decided the format for your Teen Consumer Report, do so now. Plan a schedule for completing the report and determine responsibilities for each group member. Make any necessary arrangements for using school equipment such as video cameras or computers.

1. Think about how you can build student interest in attending your presentation. Let your target audience know why your project is important to them and when they should expect to read, watch, or listen to it.

2. Create a "What you have learned" quiz to distribute after your presentation.

3. Find out about consumer protection agencies or better business bureaus in your area. If possible, include a brief interview with a representative from one of these groups in your report. Ask about issues important to young people.

4. Invite all students to contribute to your *Tips for Teens* or alert you about misleading consumer practices. Update your audience regularly.

• • • CUMULATIVE REVIEW • • •

Graph each linear inequality.

1. $y \geq \dfrac{1}{2}x + 1$ **2.** $y < -2x + 3$

A bag contains 4 blue marbles, 3 red, 2 yellow, and 1 green marble. Use this bag of marbles for Questions 3–6.

3. Find $P(\text{yellow})$ as a fraction.

4. Find $P(\text{red})$ as a decimal.

5. Find $P(\text{orange})$.

6. WRITING MATHEMATICS Maria picked a red marble out of a bag, put it back, and then picked a blue marble. After 20 picks, she had chosen a red marble 8 times. She concluded that $P(\text{red}) = 0.4$. Is she correct? Explain.

7. Four hamburgers and 3 orders of fries cost $4.65. Five hamburgers and 2 orders of fries cost $4.85. How much does one hamburger and one order of fries cost?

8. The money a shopkeeper takes in varies directly with the number of customers that enter the shop. After 8 customers, the shopkeeper had $30. How many customers would he need in order to take in $135?

Use the line graph for Questions 9 and 10.

Miles Ridden on a Bicycle Trip

9. On which day did the riders ride 50 mi?

10. To the nearest mile, find the mean number of miles covered each day.

Solve each system of inequalities by graphing.

11. $\begin{cases} y > \dfrac{2}{3}x - 2 \\ y < -x + 1 \end{cases}$ **12.** $\begin{cases} 3x - 2y \leq 6 \\ x + y \geq 4 \end{cases}$

13. TECHNOLOGY Explain the importance of knowing which half-plane should be shaded when solving a system of inequalities on a graphing utility.

Solve each equation.

14. $n + 18 = 15$

15. $\dfrac{h}{-12} = -6$

16. $4(2d + 5) = 12$

17. $\dfrac{1}{4}x + 3 = 16 - \dfrac{2}{5}x$

18. A company makes a profit of $0.26 per unit sold. How many units does the company need to sell in order to earn a profit of at least $2000.00?

19. WRITING MATHEMATICS Describe the difference in appearance of the graph of $x > 3$ and $x + y > y + 3$.

20. Use Cramer's Rule to find the solution for x in the following system of equations.

$$\begin{cases} 3x - 7y = 13 \\ 5x + 2y = 8 \end{cases}$$

21. STANDARDIZED TESTS The point at which the two lines $2x - 5y = 22$ and $15y = 6x - 22$ intersect is

    A. $(11, 0)$
    B. $(6, -2)$
    C. $(16, 2)$
    D. They are the same line.
    E. They do not intersect—they are parallel.

QUANTITATIVE COMPARISON In each question compare the quantity in Column 1 with the quantity in Column 2. Select the letter of the correct answer from these choices:

A. The quantity in Column 1 is greater.
B. The quantity in Column 2 is greater.
C. The two quantities are equal.
D. The relationship cannot be determined by the information given.

Notes: In some questions, information which refers to one or both columns is centered over both columns. A symbol used in both columns has the same meaning in each column. All variables represent real numbers. Most figures are not drawn to scale.

Column 1	Column 2
1. the slope of the line through points, $(4, -2)$ and $(-2, 4)$	the slope of a horizontal line
2. the product of the coordinates of a point on the x-axis	the product of the coordinates of a point on the y-axis

3.
$$\begin{cases} 3b - a = 5 \\ 4b + 2a = 0 \end{cases}$$

$a + b$	ab

4. -3^2	$(-3)^2$
5. P(rolling a 7) with two number cubes	P(rolling a 10) with two number cubes

6.

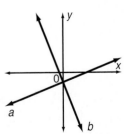

the absolute value of the slope of line a	the absolute value of the slope of line b

Column 1	Column 2

7. $x =$ amount of sales per week
$$\$1000 \leq x \leq \$2000$$

weekly salary of $\$150$ plus 8% of sales	weekly salary of $\$100$ plus 12% of sales
8. the product of the the slopes of two diagonal parallel lines	the product of the the slopes of two diagonal perpendicular lines

9.
$$y = x - 3$$

10.

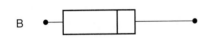

the median of boxplot A	the median of boxplot B

11. $a < b$

$2a$	b
12. the weakest possible correlation of two variables	the probability of an impossible event

13.
$$\begin{vmatrix} 3 & n \\ -2 & m \end{vmatrix} \qquad \begin{vmatrix} -2 & m \\ 3 & n \end{vmatrix}$$

14. $f(x) = x^2 - 3$

the least value in the domain	the least value in the range

15. $(5 - 3)^2$	$5^2 - 3^2$

9 Absolute Value and the Real Number System

Take a Look AHEAD

Make notes about things that look new.

- What different sets of numbers make up the real numbers? How are these sets of numbers related?
- In this chapter, several new properties of numbers and equality will be added to those you already know. Skim the lesson to identify the names of these properties. How might the name relate to the meaning of the property?

Make notes about things that look familiar.

- Recall your work with absolute value in Chapter 5. When is a compound statement containing and true? When is a compound statement containing or true?
- How is the mathematical usage of the term *dense* similar to its everyday usage? How is it different?

DATA Activity

Right Lane Closed Ahead!

Nobody enjoys being stuck in traffic, but road congestion is more than just an annoyance— it's a feature of daily life that costs individuals and businesses billions of dollars and contributes significantly to the serious problem of air pollution. The Federal Highway Administration compiles extensive data on road travel and delays. The results for selected cities are shown in the table on the opposite page.

SKILL FOCUS

- Add, subtract, multiply, and divide real numbers.
- Determine the mean of a set of data.
- Use ratios to compare.
- Use estimation.
- Construct graphs.

ALGEBRA WORKS

TRAFFIC

In this chapter, you will see how:

- **TIRE COMPANY ENGINEERS** use algebra to represent characteristics of ride, handling, and wear.
(Lesson 9.4, page 436)

- **LAW ENFORCEMENT OFFICERS** apply formulas to determine the speed at which a vehicle was traveling by its breaking distance.
(Lesson 9.5, page 444)

- **MARKET RESEARCH ANALYSTS** for a long-distance telephone company collect and organize data to attract customers.
(Lesson 9.6, page 451)

	Road Congestion		
City	Daily Vehicle Miles, thousands	Vehicle Hours of Delay	Delay and Final Cost, $ millions
New York, NY	80,920	1,512,740	6,040
Boston, MA	22,080	349,860	1,390
Cincinnati, OH	10,890	39,520	160
St. Louis, MO	18,720	138,450	540
Atlanta, GA	24,600	229,090	910
Miami, FL	8,350	222,270	870
Dallas, TX	22,650	241,530	980
Phoenix, AZ	7,050	179,560	700
Los Angeles, CA	106,680	1,752,200	7,000
Seattle, WA	18,200	255,630	1,020

Use the table above to answer the questions.

1. Determine the mean number of daily vehicle miles for the cities shown. Is the mean a useful statistic for this set of data?

2. Would it be correct to claim that the traffic delays in Phoenix are worse than those in New York? Explain.

3. Estimate the total delay and fuel cost for the cities shown. Describe the method you use.

4. Select one column from the table. Make a graph to display the data in that column.

5. **WORKING TOGETHER** Select a trip you can all take together, such as going to school or to the mall. Take the trip during a period of light traffic and again during heavy traffic. Make a graph to show how travel times compare.

411

PROJECT

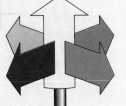

Which Way To Go?

Even when you have a destination in mind, there may be several different routes and methods of transportation that you can use to reach it. Traveling by car may be convenient and fast, but bicycling is good exercise and probably less expensive. Many people keep track of mileage, gasoline costs, and transportation time and use the data for travel decisions, household budgeting, or business purposes. In this project you will find out about people's travel habits, including your own.

PROJECT GOAL

To collect, analyze, and compare data about individual and group travel habits.

Getting Started

Work in groups.

1. Discuss methods of transportation used regularly by each person. Find out which places (other than school) are common destinations. Make a frequency table showing how often group members visit these destinations each week.

2. Obtain a detailed street map of your city, town, or county and a different colored pencil for each group member. Mark the location of your school. Have each member locate his or her home on the map and draw a straight line connecting home to school.

3. Have each member use the map scale to estimate the straight-line distance from home to school. Organize the information in a table or graph.

4. Discuss why the straight-line distance may not match the actual distance traveled to school. Plan how to find the actual distance. You will use this actual distance in a Project Connection.

PROJECT Connections

Lesson 9.2, page 423:
Determine an absolute value graph that shows the distance to and from home to school as a function of time.

Lesson 9.3, page 429:
Write absolute value equations using map distances and actual distances.

Lesson 9.5, page 443:
Plan and conduct a survey to learn more about people's travel habits.

Chapter Assessment, page 459:
Prepare travel profiles for the group and the community.

Internet Connection

www.swpco.com/
swpco/algebra1.html

412

9.1 Algebra Workshop
Absolute Value

Think Back/Working Together

Work with a partner. Recall that the absolute value of a number x is the distance between x and 0 on a number line. The absolute value of a number x is written $|x|$.

1. Evaluate each expression.
   a. $|4|$ **b.** $|-3.5|$ **c.** $|0|$ **d.** $\left|-\dfrac{1}{2}\right|$ **e.** $|-\pi|$

2. Use the information from Exercise 1 to complete the following definition of absolute value.

   $|x| = \blacksquare$ if $x \geq 0$ $|x| = \blacksquare$ if $x < 0$

3. Solve each equation. If there is no solution, explain.
   a. $|x| = 5$ **b.** $|x| = 0$ **c.** $|-x| = -2$ **d.** $|-x| = \dfrac{1}{2}$

Explore

You can use a graphing utility to graph absolute value equations. The equation $y = |x|$ is graphed at the right. A standard viewing window (x and y range from -10 to 10) with an x-scale and y-scale of 2 is used. The ABS (absolute value) function is used for $|x|$.

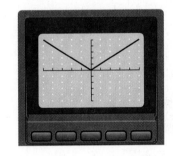

4. What letter of the alphabet best describes the shape of the graph?

5. If you can fold a drawing of a figure so that the part on one side coincides with the part on the other side, the figure is said to have **symmetry**. The fold is called a **line of symmetry**. Does the graph have a line of symmetry? Explain.

6. Why is the graph located only in Quadrants I and II?

7. Graph $y = |x|$ on your graphing utility. Use the TRACE feature to find the vertex of the graph.

You can add or subtract a positive number from the equation $y = |x|$ to obtain an equation of the form $y = |x| \pm c$, which means $y = |x| + c$ or $y = |x| - c$.

8. Graph $y = |x| + 3$ on the same set of axes (that is, without clearing your screen from Question 7). How is the graph of $y = |x| + 3$ similar to the graph of $y = |x|$? How is it different? Where is its vertex? What is its line of symmetry?

9. Graph $y = |x| - 4$ on the same set of axes. How is the graph of $y = |x| - 4$ similar to the graph of $y = |x|$? How is it different? Where is its vertex? What is its line of symmetry?

Algebra Workshop

PROBLEM SOLVING TIP

Recall that your calculator has a built-in order of operations. If you enter the equation $y = |x + 2|$ as ABS $x + 2$ instead of ABS$(x + 2)$, you will obtain the graph of $y = |x| + 2$.

COMMUNICATING ABOUT ALGEBRA

Reading from left to right in the equation $y = |x + 3| - 1$, the horizontal shift takes place first. Would the same graph result from entering $y = -1 + |x + 3|$? Discuss whether order matters when you perform these shifts.

10. Draw a conclusion about the type of shift that occurs when a positive number c is added to or subtracted from the right side of the equation $y = |x|$.

11. Clear your screen. Graph a new equation of the form $y = |x| \pm c$. Have your partner use your graph to identify the equation you entered. Then switch roles with your partner.

You can add or subtract a positive number from x within the absolute value bars to obtain an equation of the form $y = |x \pm b|$.

12. Clear your screen. Graph $y = |x + 2|$ and $y = |x|$ on the same set of axes. How is the graph of $y = |x + 2|$ similar to the graph of $y = |x|$? How is it different? Where is its vertex? What is its line of symmetry?

13. Clear your screen. Graph $y = |x - 3|$ and $y = |x|$ on the same set of axes. How is the graph of $y = |x - 3|$ similar to the graph of $y = |x|$? How is it different? Where is its vertex? What is its line of symmetry?

14. Draw a conclusion about the type of shift that occurs when a positive number b is added to or subtracted from x within the absolute value bars.

15. Clear your screen. Graph a new equation of the form $y = |x \pm b|$. Have your partner use your graph to identify the equation you entered. Then switch roles with your partner.

16. Predict how the graph of the equation $y = |x \pm b| \pm c$ will be shifted from the graph of $y = |x|$.

17. Clear your screen. Graph $y = |x + 3| - 1$ and $y = |x|$ on the same set of axes. How is the graph of $y = |x + 3| - 1$ similar to the graph of $y = |x|$? How is it different? Where is its vertex? What is its line of symmetry?

18. Clear your screen. Graph a new equation of the form $y = |x \pm b| \pm c$. Have your partner identify the equation you entered by studying the graph. Then switch roles with your partner. Where is the vertex of the graph of $y = |x + b| + c$?

Make Connections

- You have just learned that adding or subtracting a positive number c from the equation $y = |x|$ produces a vertical shift. Adding or subtracting a positive number b within the absolute value bars of the equation $y = |x|$ produces a horizontal shift.

Use the words *up*, *down*, *left*, and *right* (you can combine them) to indicate the direction in which the graph of each equation is shifted from the graph of $y = |x|$.

19. $y = |x - 2| - 4$ **20.** $y = |x + 5|$ **21.** $y = |x + 6| + 6$

Determine the equation of each graph. Then verify your answer using a graphing utility. Each tick mark is one unit.

22. 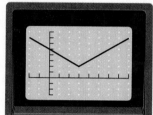 **23.** 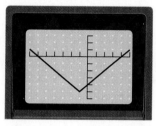 **24.**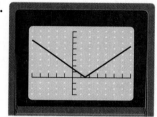

Write an equation of the form $y = |x \pm b| \pm c$ whose graph represents each of the following shifts from the graph of $y = |x|$.

25. 3 units down and 4 units left

26. 1.5 units up and 7.5 units right

27. 2.5 units down and 4.5 units right

28. 1.8 units up and 4.8 units left

Summarize

29. WRITING MATHEMATICS Write a paragraph that explains how to determine how much and in what direction the graph of $y = |x \pm b| \pm c$ is shifted from the graph of $y = |x|$.

30. THINKING CRITICALLY How is the graph of $y = a|x|$ different from the graph of $y = |x|$ if $a > 1$? if $0 < a < 1$?

31. THINKING CRITICALLY How is the graph of $y = -|x|$ different from the graph of $y = |x|$? Over what line is the graph of $y = -|x|$ a reflection of the graph of $y = |x|$?

32. GOING FURTHER Would you obtain the same graph by reflecting the graph of $y = |x|$ over the x-axis and then shifting it up 3 units as you would if you shifted it up 3 units and then reflected it over the x-axis? Does order matter when shifting and reflecting this absolute value graph?

Explore

1. On a coordinate plane draw a triangle in Quadrant I so that it does not touch either axis. Label the vertices of the triangle A, B, and C.

2. Fold the paper back along the y-axis. Trace your triangle in Quadrant II, labeling the vertices A', B', and C'.

3. Unfold the paper. What is the relationship of points A and A' to the y-axis? of B and B'? of C and C'?

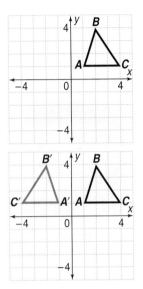

Build Understanding

The **absolute value function** is denoted by $f(x) = |x|$. The corresponding absolute value equation is $y = |x|$, where x is any real number and y is any number that is greater than or equal to 0.

The graphs of $y = x$ and $y = |x|$ are shown below.

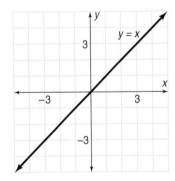
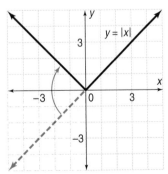

The graph of $y = x$ is a straight line that extends through Quadrants I and III. The graph of $y = |x|$ is V-shaped, has a vertex at $(0, 0)$, and opens upward in Quadrants I and II. The Quadrant III portion of the graph of $y = x$ has been "flipped" over the x-axis into Quadrant II. This flip is called a **reflection**, and the x-axis is called the **line of reflection**.

If you fold the graph of $y = |x|$ along the y-axis, the Quadrant I portion will fit exactly over the Quadrant II portion. The graph is said to have **line symmetry**, and the y-axis is the **line of symmetry**.

A **translation** of a graph is a slide or shift that produces the same graph in a new position. For example, when a number is added to or subtracted from the right side of the equation $y = |x|$, a *vertical shift* occurs.

EXAMPLE 1

Graph each equation. Identify its vertex and line of symmetry from the graph.

a. $y = |x| + 2$ **b.** $y = |x| - 3$

Solution

a.

x	y
-4	6
-2	4
0	2
2	4
4	6

b.

x	y
-4	1
-2	-1
0	-3
2	-1
4	1

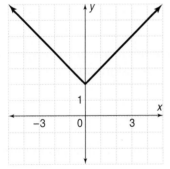

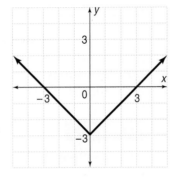

Vertex: $(0, 2)$
Line of symmetry: y-axis

Vertex: $(0, -3)$
Line of symmetry: y-axis   ◄

┌─ VERTICAL SHIFT ─────────────
The graph of $y = |x| + c$ is shifted c units *up* from the graph of $y = |x|$ when c is positive and c units *down* when c is negative.
└──────────────────────────────

When a number is added to or subtracted from x within the absolute value bars of the equation $y = |x|$, a *horizontal shift* occurs.

EXAMPLE 2

Graph each equation. Identify its vertex and line of symmetry from the graph.

a. $y = |x + 3|$ **b.** $y = |x - 1|$

Solution

a.

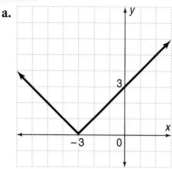

b.

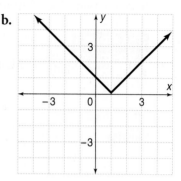

Vertex: $(-3, 0)$ Vertex: $(1, 0)$
Line of symmetry: $x = -3$ Line of symmetry: $x = 1$ ◄

— HORIZONTAL SHIFT —

The graph of $y = |x - b|$ is shifted b units to the *right* from the graph of $y = |x|$ when b is positive and b units to the *left* when b is negative.

The translations you have seen in this lesson are two types of **transformations**. A third type of transformation occurs in equations of the form $y = a|x|$.

EXAMPLE 3

Graph each equation. Tell how each graph differs from the graph of $y = |x|$.

a. $y = 2|x|$ **b.** $y = \frac{1}{2}|x|$ **c.** $y = -|x|$

Solution

a.

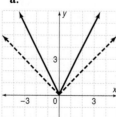

b.

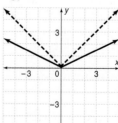

c.

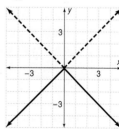

a. The graph of $y = 2|x|$ is V-shaped, but the rays that form the V are farther away from the x-axis than those of $y = |x|$.

b. The graph of $y = \frac{1}{2}|x|$ is V-shaped, but the rays that form the V are closer to the x–axis than those of $y = |x|$.

c. The graph of $y = -|x|$ is the reflection of the graph of $y = |x|$ over the x-axis. It is V-shaped and opens downward. ◀

COMMUNICATING ABOUT ALGEBRA

The graphs of the equations in Example 3 show different V-shapes. Explain what effect the value of a has on the shape of the graph $y = a|x|$.

All three types of transformations may occur in one equation of the form $y = a|x - b| + c$. You can determine the x-coordinate of the vertex of the graph by setting the expression within the absolute value bars equal to zero and solving for x.

$$x - b = 0$$
$$x = b \qquad \text{Solve for } x.$$

To determine the y-coordinate of the vertex, substitute this value of x into the original absolute value equation and solve for y.

If the vertex of an absolute value graph is at (b, c), then the equation of the line of symmetry is $x = b$.

EXAMPLE 4

Graph $y = 3|x + 2| + 1$ and indicate how it differs from the graph of $y = |x|$. Identify its vertex, line of symmetry, and shape.

Solution
The graph of $y = 3|x + 2| + 1$ has been shifted up one unit and to the left by two units. It is V-shaped, but the rays that form the V are farther away from the x-axis.

To find the x-coordinate of the vertex, set the expression inside the absolute value symbol equal to zero and solve for x.

$$x + 2 = 0$$
$$x = -2$$

To find the y-coordinate, substitute the x-coordinate into the original equation and solve for y.

$$y = 3|-2 + 2| + 1$$
$$y = 1$$

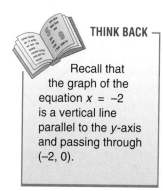

THINK BACK

Recall that the graph of the equation $x = -2$ is a vertical line parallel to the y-axis and passing through $(-2, 0)$.

So, the vertex is $(-2, 1)$, the line of symmetry is $x = -2$, and the graph is V-shaped with its sides farther from the x-axis than the graph of $y = |x|$. ◀

You can use the skills you developed in this lesson to determine the equation of an absolute value graph.

EXAMPLE 5

ENGINEERING A truss is a framework of beams that supports a bridge or roof. The triangular truss in the graph at the right can be represented by an absolute value equation. Which of the following equations represents the graph?

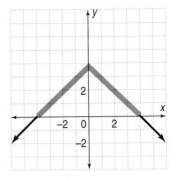

   a. $y = |x| + 4$ **b.** $y = -|x - 4|$

   c. $y = -|x| + 4$ **d.** $y = -|x + 4|$

Solution

The vertex of the graph is at $(0, 4)$. There is no horizontal shift. Only options a and c do not have a horizontal shift. The graph opens downward, so there must be a negative sign in front of the absolute value bars. The correct answer is c. Check by graphing the equation. ◄

TRY THESE

Determine the vertex and line of symmetry for the graph of each equation.

1. $y = |x - 2.5|$ **2.** $y = |x + 4|$ **3.** $y = |x| - 1.5$ **4.** $y = |x| + 2.5$

5. $y = |x + 4| - 9$ **6.** $y = |x - 3| - 3$ **7.** $y = -|x + 3| - 1$ **8.** $y = -|x - 1| + 2$

Tell how each graph differs from the graph of $y = |x|$.

9. $y = 4|x|$ **10.** $y = \frac{1}{4}|x|$ **11.** $y = -2|x|$ **12.** $y = -\frac{1}{2}|x|$

Match each equation with its graph.

13. $y = |x + 2|$ **14.** $y = |x| + 2$ **15.** $y = |x - 2|$ **16.** $y = |x| - 2$

  a.   **b.**   **c.**   **d.**

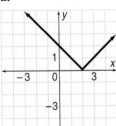 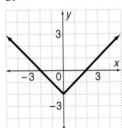 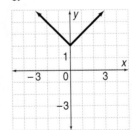 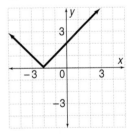

Graph each equation.

17. $y = |x - 2| + 3$ **18.** $y = |x + 1| - \frac{1}{2}$ **19.** $y = 2|x + 1.5|$ **20.** $y = -2|x - 1.5|$

21. WRITING MATHEMATICS Write an equation that represents the graph of the equation $y = |x|$ shifted 2 units to the right and 8 units downward.

22. WRITING MATHEMATICS Explain how the graph of $y = a|x - b| + c$ differs from the graph of $y = |x|$.

PRACTICE

Determine the vertex and the line of symmetry for the graph of each equation.

1. $y = |x| - 4$

2. $y = |x| + 5$

3. $y = |x - 1.5|$

4. $y = |x + 0.5|$

5. $y = |x - 4| - 3$

6. $y = |x + 3| - 3$

7. $y = -|x + 1| - 2$

8. $y = -|x - 3| + 2$

Explain how the graph of each equation differs from the graph of $y = |x|$.

9. $y = 3|x|$

10. $y = \frac{1}{3}|x|$

11. $y = -\frac{1}{5}|x|$

12. $y = -5|x|$

Match each equation with its graph.

13. $y = |x + 1|$

14. $y = |x| + 1$

15. $y = |x - 1|$

16. $y = |x| - 1$

a.

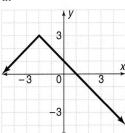

b.

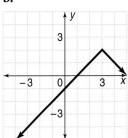

c.

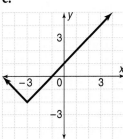

d.

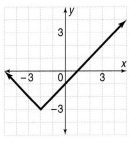

Match each equation with its graph.

17. $y = |x + 2| - 3$

18. $y = -|x - 3| + 2$

19. $y = -|x + 2| + 3$

20. $y = |x + 3| - 2$

a.

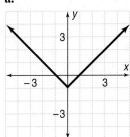

b.

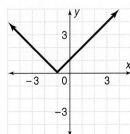

c.

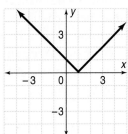

d.

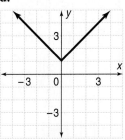

Graph each equation.

21. $y = |x - 3| + 2$

22. $y = |x + 2| - 2$

23. $y = 2|x - 2.5|$

24. $y = 0.5|x - 2| - 1.5$

25. $y = -2|x + 0.5|$

26. $y = 3|x + 3| + 3$

27. $y = 2|x - 2| - 2$

28. $y = -\frac{1}{2}|x + 2.5| - 1$

29. WRITING MATHEMATICS Write an equation of the form $y = a|x - b| + c$ that reflects the graph of $y = |x|$ over the x-axis and then moves it 5 units right and 2 units down.

30. WRITING MATHEMATICS The graph of the equation $y = a|x|$ has a right angle at its vertex when $a = 1$. For what values of a will the vertex angle be acute and when will it be obtuse? Explain your reasoning.

Write an equation that represents each graph. (*Hint:* Find the value of x when $y = 0$. Use that value to find a.)

31.

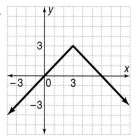

32.

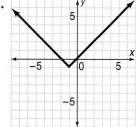

33. **FASHION DESIGN** The width of neckties that are considered fashionable changes frequently. As the width of a tie changes, the V at the bottom of the tie changes. Ties may be designed using a grid such as the one shown with three ties. Write an absolute value equation for the triangular shape at the bottom of each tie.

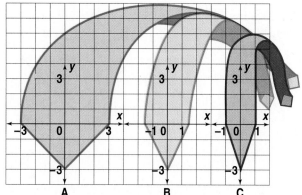

A   B   C

EXTEND

34. **TECHNOLOGY** Use a graphing utility to estimate the coordinates of the points at which the graphs of $y = |2x|$ and $y = -|2x| + 2$ intersect.

35. **TECHNOLOGY** Use a graphing utility to estimate the coordinates of the points at which the graphs of $y = |2x| - 1$ and $y = -|2x| + 3$ intersect.

INTERNATIONAL FLAGS The graphs of the absolute value equations in this lesson have vertical lines of symmetry of the form $x = r$, where r is a real number. If you graph $y = |x|$ and $y = -|x|$ on the same coordinate plane, the x- and y-axes are both lines of symmetry. Some figures have the line $y = x$ as a line of symmetry. Flags from six different countries are shown below. Imagine a flag is placed on a coordinate plane where the center of the flag is at $(0, 0)$. Determine the line or lines of symmetry for each flag.

36. Vietnam

37. Israel

38. Finland

39. Switzerland

40. Rwanda

41. Nigeria

GEOMETRY Copy each figure. Then draw and/or describe all the lines of symmetry.

42.   43.   44.   45.   46.

THINK CRITICALLY

47. Graph $y = -x$ and $y = |-x|$. What portion of the graph of $y = -x$ must be reflected to obtain the graph of $y = |-x|$? What is the line of reflection?

48. Graph $y = -x$ and $y = -|x|$. What portion of the graph of $y = -x$ must be reflected to obtain the graph of $y = -|x|$? What is the line of reflection?

49. The graph of $y = |2x|$ is the same as the graph of $y = 2|x|$. Is it also true that the graph of $y = |-2x|$ is the same as the graph of $y = -2|x|$? Verify your solution.

50. Do the graphs of $y = -3x$ and $y = |-3x|$ ever overlap? Write the equation of another graph, half of which overlaps the graph of $y = -3x$.

51. Write the equation of an absolute value function that intersects the graph of $y = |x|$ in exactly two points.

52. Write the equation of an absolute value function that does not intersect the graph of $y = -|x|$.

PROJECT *Connection*

In this activity, you will graph the distance between your home and school compared to the time you spend traveling.

1. Use the actual route, method of travel, and distance you determined earlier. Decide on the equal time intervals (such as every 5 min) you will use to record distances. Estimate the distance you have traveled at each time interval. Record each distance on your way to school and on your way home as a distance from home. Travel the same route each way. Record each time as total time you spend traveling so when you start home add to the amount of time you had previously.

2. Graph the data. Show time on the horizontal axis and distance from school on the vertical axis. Do the points for each part of the round trip lie on or near a straight line? If not, explain why.

3. Draw a line connecting your starting point to the school point. Draw another line connecting the school point to the home point again. Describe the graph. Compare your graph with those of other group members.

9.3 Absolute Value Equations

Explore

1. Use a straightedge and graph paper to make two identical number lines that extend from –5 to 5 in increments of 1 unit. Cut out the number lines as shown below. Label one number line as the "zero finder" and the other as the "solution finder." Obtain two pieces of uncooked spaghetti.

2. Use the number lines and spaghetti to solve the equation $|x - 2| = 3$ in the following manner.

   a. Set $x - 2$, the expression within the absolute value bars, equal to zero to obtain $x = 2$.

   b. Align the 2 on the second number line with the 0 on the zero finder.

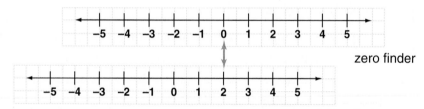

zero finder

solution finder

   c. Since you are looking for numbers that are 3 units from $|x - 2|$, place a piece of spaghetti vertically on 3 in the zero finder. Place another piece of spaghetti vertically on –3 on the zero finder.

   d. The pieces of spaghetti intersect the solution finder at $x = -1$ and $x = 5$. These values are the solutions to the equation $|x - 2| = 3$.

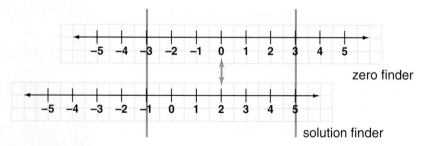

zero finder

solution finder

   e. Check the solutions by substituting in the original equation.

3. Work in the same way to solve $|x + 1| = 2$. What numbers do the pieces of spaghetti cross on the solution finder?

Build Understanding

In the previous lesson you found the line of symmetry for two-variable absolute value equations. In Explore you found the *point of symmetry* for one-variable absolute value equations by setting the absolute value expression equal to zero and finding equal distances from both sides of the point of symmetry to solve the original equation.

Recall that the absolute value of a number x is the distance between x and 0 on a number line. Since distance is positive, the absolute value of an expression must be positive. For example, the distance between 0 and 2 is 2, so $|2| = 2$. Since distance is positive, the distance between 0 and –2 is also 2, and $|-2| = 2$.

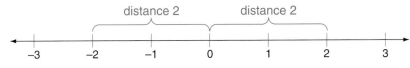

An algebraic way to state this is

$$|x| = x \text{ if } x \geq 0 \quad \text{and} \quad |x| = -x \text{ if } x < 0$$

A **disjunction** is two statements joined by the word *or*. A disjunction is true if at least one of the statements is true. The equation $|x| = 5$ is equivalent to the disjunction

$$x = 5 \text{ or } x = -5$$

EXAMPLE 1

Express as a disjunction: $|3x - 5| = 15$

Solution

$$3x - 5 = 15 \quad \text{or} \quad 3x - 5 = -15 \qquad \blacktriangleleft$$

To solve an absolute value equation, begin by writing it as a disjunction. Then solve the two resulting equations.

EXAMPLE 2

Solve and check: $|2x - 1| = 4$

Solution

$$
\begin{array}{llll}
2x - 1 = 4 & \text{or} & 2x - 1 = -4 & \text{Write a disjunction.} \\
2x = 5 & \text{or} & 2x = -3 & \text{Add 1 to both sides.} \\
x = 2.5 & \text{or} & x = -1.5 & \text{Divide both sides by 2.}
\end{array}
$$

To check, substitute both solutions into the original equation.

$$
\begin{array}{cc}
|2x - 1| = 4 & |2x - 1| = 4 \\
|2(2.5) - 1| \stackrel{?}{=} 4 & |2(-1.5) - 1| \stackrel{?}{=} 4 \\
4 = 4 \checkmark & 4 = 4 \checkmark
\end{array}
$$

The solutions are 2.5 and –1.5. $\qquad \blacktriangleleft$

COMMUNICATING ABOUT ALGEBRA

Describe how you could use the zero finder and solution finder to solve Example 2.

To solve some absolute value equations, you must first isolate the expression containing the absolute value symbol.

EXAMPLE 3

Solve and check: $2|2z + 4| - 5 = 11$

Solution

$$2|2z + 4| - 5 = 11$$
$$2|2z + 4| = 16 \qquad \text{Add 5 to both sides.}$$
$$|2z + 4| = 8 \qquad \text{Divide both sides by 2.}$$

$2z + 4 = 8 \quad \text{or} \quad 2z + 4 = -8 \qquad$ Write a disjunction.
$2z = 4 \quad \text{or} \qquad 2z = -12 \qquad$ Subtract 4 from both sides.
$z = 2 \quad \text{or} \qquad z = -6 \qquad$ Divide both sides by 2.

To check, substitute both solutions into the original equation.

$$2|2z + 4| - 5 = 11 \qquad\qquad 2|2z + 4| - 5 = 11$$
$$2|2(2) + 4| - 5 \stackrel{?}{=} 11 \qquad\qquad 2|2(-6) + 4| - 5 \stackrel{?}{=} 11$$
$$11 = 11 \checkmark \qquad\qquad\qquad 11 = 11 \checkmark$$

The solutions are 2 and -6.   ◄

Some absolute value equations have no solution. For example, the absolute value equation $|x + 5| = -3$ has no solution because the absolute value of an expression cannot be negative.

You can also use a graphing utility to solve an absolute value equation.

COMMUNICATING ABOUT ALGEBRA

Solve Example 4 in a different way by graphing three equations. Name the three equations. Which method do you prefer?

EXAMPLE 4

TECHNOLOGY Use a graphing utility to solve $|2x + 1| = 5$.

Solution
Graph $y_1 = |2x + 1|$ and $y_2 = 5$ on the same set of axes. Use the TRACE and ZOOM features to find the x-values of the two points of intersection, which are -3 and 2. These are the solutions to the equation $|2x + 1| = 5$.

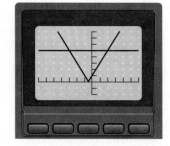

◄

You can use absolute value equations to model real world events.

EXAMPLE 5

RADIOLOGY A hospital ultrasound machine can measure with a 10% tolerance the weight of an unborn baby. This means that the weight x of the unborn baby will be measured as $x \pm 0.1x$. Suppose an unborn baby weighs 110 oz. Determine the maximum and minimum weights of the unborn baby.

Solution

Since the weight x of the baby is within 10% of 110 oz, the maximum and minimum weights must be $0.1(110) = 11$ oz above or below 110 oz. Write an absolute value equation for this situation.

$$|x - 110| = 11$$
$$x - 110 = 11 \quad \text{or} \quad x - 110 = -11 \quad \text{Write a disjunction.}$$
$$x = 121 \quad \text{or} \quad x = 99 \quad \text{Add 110 to both sides.}$$

The maximum and minimum weights of the baby are 121 oz and 99 oz, respectively. ◄

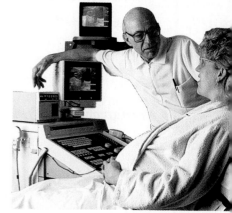

TRY THESE

Express each absolute value equation as a disjunction.

1. $|2x - 3| = 12$

2. $|3x - 2| = 10$

3. $|2x + 5| = 17$

Solve and check each equation. If an equation has no solution, write "no solution."

4. $|x - 2| = 6$

5. $|x - 3| = 9$

6. $|6 - x| = 4$

7. $|8 - x| = 2$

8. $|2x - 7| = 5$

9. $|2x - 9| = 11$

10. $3|x - 5| = 6$

11. $4|x - 3| = 16$

12. $2|3z + 1| - 3 = 9$

13. $2|4w - 1| + 2 = 12$

14. $-|x + 4| - 3 = 12$

15. $-|x + 5| - 5 = 5$

Use a graphing utility to solve each equation.

16. $|x + 1| = 5$

17. $|x + 3| = 2$

18. $|6 - x| = 2$

19. SCIENCE The volume of a liquid in a graduated cylinder is 30 mL. Write an equation that represents the volume if the cylinder is accurate to 5%.

20. HEALTH AND FITNESS A fitness study concluded that the optimal amount of body fat for a male is 15% ± 3% of total body weight. Determine the maximum and minimum optimal percentages of body fat for a male.

21. WRITING MATHEMATICS Write about a real world situation in your classroom involving a maximum and minimum. Write an absolute value equation to model the situation.

PRACTICE

Express each absolute value equation as a disjunction.

1. $|4x| = 12$

2. $|2x| = 10$

3. $|3x + 4| = 11$

4. $|2x + 1| = 8$

5. $|5x - 1.5| = 8.5$

6. $|6x - 4.1| = 9.1$

Solve and check each equation. If an equation has no solution, write "no solution."

7. $|2w| = 16$

8. $|-5t| = 15$

9. $\left|-\frac{1}{2}q\right| = -16$

10. $|x - 2| = 4$

11. $|8 - x| = 5$

12. $|7 - x| = 3$

Solve and check each equation. If an equation has no solution, write "no solution."

13. $|2x - 5| = 7$

14. $3|x - 4| = 10$

15. $8|3z + 1| - 6 = 18$

16. $3|2w + 1| + 3 = 12$

17. $-|x + 4| + 12 = 3$

18. $2|2x + 3| + 3 = 3$

19. $2|x + 3| + 3 = 9$

20. $4|x + 2| - 5 = 3$

21. $\frac{1}{2}|4x - 2| + 6 = 12$

Use a graphing utility to solve each equation.

22. $|x + 1| = 5$

23. $|x + 3| = 2$

24. $|6 - x| = 2$

25. $|5 - x| = 7$

26. $\frac{1}{2}|5y - 4| + 4 = 6$

27. $\frac{1}{2}|3y - 4| + 4 = 10$

28. SALARY AND BENEFITS A law firm pays a bonus of $10,000 to any associate who charges fees for 2,000 $\pm$ 50 hours per year. Determine maximum and minimum hours for which an associate can earn a $10,000 bonus.

29. VETERINARY MEDICINE Joline adopted a kitten from the local animal shelter. Her veterinarian estimated the kitten's age at 9 $\pm$ 3 mo. Determine the maximum and minimum ages of the kitten.

30. WRITING MATHEMATICS Describe the graph of an absolute value equation that has no solution. Provide such an equation and sketch its graph.

EXTEND

31. Some absolute value equations may have only one solution. What are the possible solutions of $|x - 5| = 5x$? Check your possible solutions in the original equation. Which possible solution must you reject? What is the solution?

Solve and check each equation.

32. $|x - 4| = 2x$

33. $|x - 3| = 4x$

34. $\frac{1}{4}|5 - 2x| = 2 - x$

ESTIMATION Estimate the solutions of each equation to the nearest integer. Check your solutions to see how accurate they are.

35. $|x - 2.1| = 4.1$

36. $|x - 2.9| = 9.9$

37. $|8 - 0.9x| = 4.9$

COLLEGE ADMISSION Havenhearst College accepts students with mathematics SAT scores of 700 $\pm$ 100 and high school grade-point averages of 3.5 $\pm$ 0.5.

38. Determine the maximum and minimum mathematics scores accepted at Havenhearst.

39. Determine the maximum and minimum high school grade-point averages accepted at Havenhearst.

The SKS Corporation established the following salary grade levels for middle management.

LEVEL	LOW	MIDDLE	HIGH
1	$ 18,000	$ 25,000	$ 32,000
2	$ 23,000	$ 31,000	$ 39,000
3	$ 36,000	$ 45,000	$ 54,000
4	$ 50,000	$ 61,000	$ 72,000

40. Determine the maximum and minimum salaries at each level.

THINK CRITICALLY

41. Write an absolute value equation that has two solutions, one of which is 0.

42. Write an absolute value equation that has only one solution, which cannot be 0.

43. Describe the graph of an absolute value equation with only one solution, which is not 0.

Solve each equation for *x*. Assume *a*, *b*, *c*, *d*, and *e* are positive numbers.

44. $|ax| = b$ **45.** $|cx - d| = e$ **46.** $-|x + c| = d$ **47.** $|x - 4| = 2x + c$

PROJECT *Connection*   Use the map on which you drew a straight line connecting your home and school.

1. First, make a copy of the map you drew. Let zero represent the location of your home. Mark off units on the line you drew so that the coordinate of the school is the actual distance from your home. Determine the midpoint of your distance line. Then write an absolute value equation with solutions that place you at school and at home. How did knowing the midpoint help you?

2. Write another absolute value equation using the distances on the map instead of actual distances.

3. How are your two equations related?

4. Compare your equations with those of other group members. How are the equations alike? Why? How do they differ? Why?

9.4 Absolute Value Inequalities

Explore

In the circle diagram at the right, the universal set U is $\{-4, -3, -2, -1, 0, 1, 2, 3, 4, 5, 6, 7\}$. Set A includes the integers greater than or equal to -2, and set B includes the integers less than or equal to 5. The shaded region represents the intersection of sets A and B.

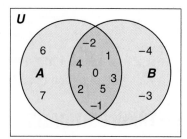

1. What numbers are included in set A? in set B? in set A but not in set B? in set B but not in set A? in the intersection of sets A and B?

Let x represent any integer. The graph below shows the solution set of the conjunction

$$x \geq -2 \quad and \quad x \leq 5$$

2. How many solutions are there?

3. Why are there solid dots at $x = -2$ and $x = 5$?

4. What would open dots indicate?

5. If x were not limited to integers, would the number of solutions be different? Explain.

Build Understanding

The double-shaded area in the circle diagram in Explore is where circles A and B overlap. The overlapping represents a *conjunction*. A **conjunction** is two statements connected by the word *and*.

The graph of $|x| < 4$ below contains points less than 4 units from 0. The solution is the conjunction

$$-4 < x \quad and \quad x < 4$$

which can be written $-4 < x < 4$.

The graph of $|x| > 4$ below contains points more than 4 units from 0. The solution is the disjunction

$$x < -4 \quad or \quad x > 4$$

These conjunctions and disjunctions are called **compound inequalities**.

You can solve absolute value inequalities of the form $|ax \pm b| < c$ or $|ax \pm b| > c$ by expressing each inequality as a compound inequality.

EXAMPLE 1

Solve $|x + 2| < 3$. Graph the solution.

Solution

Number Line Method
First, find the point of symmetry.

$$x + 2 = 0$$
$$x = -2$$

Align 0 with -2. Place spaghetti on 3 and -3 on the zero finder.

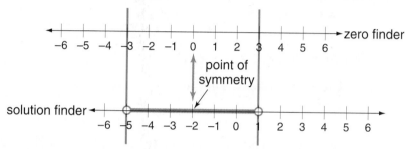

CHECK UNDERSTANDING

Example 1 shows a conjunction and Example 2 shows a disjunction. How can you tell which is which by looking at the inequality symbols?

The spaghetti intersects the solution finder at -5 and 1. Since the absolute value is "less than," the distance from the point of symmetry is less than 3. The solution is $-5 < x < 1$. Place open dots on -5 and 1. Connect them.

Algebraic Method
$|x + 2| < 3$ means $-3 < x + 2$ *and* $x + 2 < 3$.

$$-3 < x + 2 \quad \text{and} \quad x + 2 < 3$$
$$-5 < x \qquad\qquad\quad x < 1 \quad \text{Solve for } x.$$

The solution is $-5 < x < 1$. ◄

When the inequality is "greater than" or "greater than or equal to," the solution is a disjunction.

EXAMPLE 2

Solve $|2x - 3| > 5$. Graph the solution.

Solution

Use the algebraic method.
$|2x - 3| > 5$ means $2x - 3 < -5$ *or* $2x - 3 > 5$.

$$2x - 3 < -5 \quad \text{or} \quad 2x - 3 > 5$$
$$x < -1 \qquad\qquad\quad x > 4 \quad \text{Solve for } x.$$

The solution is $x < -1$ *or* $x > 4$.

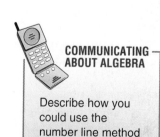

COMMUNICATING ABOUT ALGEBRA

Describe how you could use the number line method to solve Example 2.

Be sure to isolate the absolute value on one side of the inequality before you express it as a compound inequality.

EXAMPLE 3

Solve $2|4 - 2z| - 3 \leq 15$. Graph the solution.

Solution

$$2|4 - 2z| - 3 \leq 15$$
$$2|4 - 2z| \leq 18 \quad \text{Add 3 to both sides.}$$
$$|4 - 2z| \leq 9 \quad \text{Divide both sides by 2.}$$

$|4 - 2z| \leq 9$ means $-9 \leq 4 - 2z$ *and* $4 - 2z \leq 9$.

$-9 \leq 4 - 2z$	and	$4 - 2z \leq 9$	
$-13 \leq -2z$		$-2z \leq 5$	
$6.5 \geq z$		$z \geq -2.5$	Divide both sides by -2. Reverse the inequality.

The solution is $-2.5 \leq z \leq 6.5$.

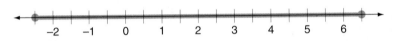

Some inequalities have no solution. For example, $|3x - 7| < -2$ has no solution because absolute value cannot be negative.

Absolute value inequalities are useful in representing real world data.

EXAMPLE 4

TRAVEL A private commuter airline charges customers a flat rate for all flights from its home airports to destinations that are between and including 100 mi and 130 mi away. Choose the graph that best represents this situation. Then write the absolute value equation that represents the situation.

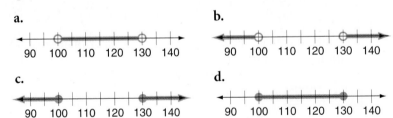

a.

b.

c.

d.

Solution

Eliminate graphs **b** and **c** because they do not indicate values between 100 and 130. Eliminate graph **a** because it does not include 100 and 130. The correct answer is **d**.

The number 115 represents the point of symmetry or the midpoint between the solutions to the inequality, 100 and 130. So, this situation can be modeled by the absolute value inequality $|x - 115| \leq 15$. ◄

CHECK UNDERSTANDING

In Example 3, why is the order of the inequality signs reversed in the last line of the solution?

COMMUNICATING ABOUT ALGEBRA

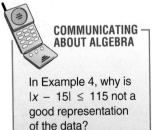

In Example 4, why is $|x - 15| \leq 115$ not a good representation of the data?

The following rules summarize the algebraic method for solving absolute value inequalities. You can replace < with ≤ and > with ≥ in these rules.

For all real numbers a, b, and c where $b \geq 0$ and $c \geq 0$,

If $|ax + b| < c$, then $-c < ax + b$ and $ax + b < c$.
If $|ax - b| < c$, then $-c < ax - b$ and $ax - b < c$.

If $|ax + b| > c$, then $ax + b < -c$ or $ax + b > c$.
If $|ax - b| > c$, then $ax - b < -c$ or $ax - b > c$.

You may prefer to solve an absolute value inequality using a number line. If the inequality is $|x - b| < c$, then its solutions are those values of x that are less than c units from the point of symmetry b. If the inequality is $|x - b| > c$, then its solutions are those values of x that are more than c units from b.

TRY THESE

Write a compound inequality that is equivalent to each absolute value inequality.

1. $|x + 3| < 6$ **2.** $|x - 4| > 7$ **3.** $|2x - 4| \geq 6$ **4.** $|3x + 5| \leq 9$

Use the number line method to solve each absolute value inequality. Graph the solution.

5. $|z + 3| < 5$ **6.** $|x - 4| \leq 3$ **7.** $|t - 5| \geq 2$ **8.** $|w + 2| > 4$

Match each equation with its graph.

9. $|x| \leq 5$ **10.** $|2 - x| > 1$ **11.** $|x + 2| \geq 1$ **12.** $|x - 3| < 2$

a.
0 1 2 3 4
b.
0 2 4
c.
-6 -4 -2 0 2 4 6
d.
-4   -2   0

Solve each inequality. For Exercises 13–16, also graph each solution.

13. $|2q - 7| \geq 1$ **14.** $|9x - 6| > 12$

15. $|3t + 5| < 7$ **16.** $|4x - 2| \leq 6$

17. $|2x| \geq 0$ **18.** $|2x - 8| < 0$

19. GRADING Ms. Hewlett indicated that those students with semester averages from 82 to 88 would receive a B. Write an absolute value inequality that represents this range.

20. WRITING MATHEMATICS Explain the similarities and differences between the graphs of $|x - b| < c$ and $|x - b| > c$.

21. WRITING MATHEMATICS What is the solution of $|x - 9| < -4$? Explain.

PRACTICE

Write a compound inequality that is equivalent to each absolute value inequality.

1. $|x + 1| < 5$
2. $|x + 2| < 3$
3. $|x - 4| > 6$
4. $|x - 3| > 5$

5. $|2x - 2| \leq 6$
6. $|3x - 3| \leq 15$
7. $|2x + 5| \geq 3$
8. $|3x + 5| \geq 13$

Solve each absolute value inequality. Graph the solution.

9. $|z + 1| < 3$
10. $|z + 3| < -7$
11. $|x - 5| > 2$
12. $|x - 4| > 2$

13. $|x - 0.5| \leq 3.5$
14. $|t - 1.5| \leq 4.5$
15. $|w + 2.5| \geq 3.5$
16. $|q + 3.5| \geq -2.5$

Match each equation with its graph.

17. $|x - 2| \leq 3$
18. $|3 - x| > 2$
19. $|x + 3| \geq 2$
20. $|x| < 3$

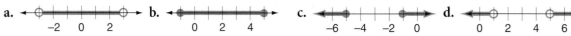

Solve each inequality. For Exercises 21–28, also graph each solution.

21. $|2q - 4| \geq 6$
22. $|3x - 9| > 6$
23. $|3t + 5| < 10$
24. $|4x - 2| \leq 10$

25. $|3z - 6| > 12$
26. $|3z - 1| \geq -7$
27. $|2q + 5| \leq 9$
28. $|5z - 10| < 5$

29. $|5x| > 0$
30. $|6x| \geq 0$
31. $|5x + 4| < 0$
32. $|2x - 9| < -5$

33. $\left|\frac{1}{2}x - 1\right| > 0$
34. $\left|\frac{1}{4}x - 2\right| > 0$
35. $\left|2 - \frac{1}{2}x\right| > -4$
36. $\left|4 - \frac{1}{2}x\right| \geq 2$

37. $|6 - 2x| < 12$
38. $|5 - 2x| \leq 7$
39. $|2x + 4| \leq 0$
40. $|3x + 9| \leq 0$

41. $2|1 + 2z| + 5 \leq 9$
42. $3|1 + 3x| - 6 < 12$
43. $4 - |5 - t| \geq 1$

44. $3 - |4 - t| > 1$
45. $\left|\frac{8 - 4x}{3}\right| < 2$
46. $\left|\frac{5 - 3x}{2}\right| > 4$

47. ARCADE GAMES One arcade game consists of a ramp with slots at one end that are labeled 1–6. To win at this game, you must roll 6 balls down the ramp to get a total score less than 11 or more than 31. Write an absolute value inequality that models the situation in which you would lose the game.

48. AUTOMOTIVE SALES An automobile salesman estimates that he will sell 30 ± 5 red cars next year. Write an absolute value inequality that models this estimate.

49. WRITING MATHEMATICS Write about a real world situation concerning food in which it is important to know the numbers within a specific range. Write an absolute value inequality to model this situation.

EXTEND

Some absolute value inequalities with the variable on each side of the equation have solutions that are single inequalities instead of compound inequalities. Solve each inequality.

50. $|x + 2| \leq x + 2$
51. $|x - 2| < 5x$
52. $|3 - x| > 3 - x$
53. $|5 - x| \geq 3x$

ESTIMATION Estimate the solutions to each inequality to the nearest integer. Check to see how accurate your estimates were.

54. $|x - 3.1| \leq 7.1$ **55.** $|x - 1.9| < 8.9$ **56.** $|7 - 0.9x| > 3.9$ **57.** $|5 - 1.1x| \geq 3.1$

HEMATOLOGY Lab technicians use the reference ranges below when measuring the blood chemistries of patients. A portion of the lab report for a patient is shown below.

58. Write an absolute value inequality that represents the normal range for glucose.

Chemzyme	Result	Normal Range
Glucose	110	70–116
Creatinine	1.3	0.7–1.4
GGTP	33	0–45

59. Write an absolute value inequality that represents the normal range for creatinine, a kidney function indicator.

60. Write an absolute value inequality that represents the normal range for GGTP, a liver function indicator.

THINK CRITICALLY

61. Write an absolute value inequality that has $x \neq 0$ as its solution.

62. Write an absolute value inequality that has no solution. (Do not use inequalities of the form $|x - b| < 0$ or $|x + b| < 0$.)

63. Use the following strategy for solving the inequality $|x| \geq c$. Explain your answer.

- Solve the inequality $|x| < c$.
- Conclude that the solutions of $|x| \geq c$ are the real numbers that do not satisfy $|x| < c$.

64. Under what circumstances, if any, are the following equations true?

a. $|a + b| = |a| + |b|$ **b.** $|a \cdot b| = |a| \cdot |b|$ **c.** $|a - b| = |a| - |b|$ **d.** $\left|\dfrac{a}{b}\right| = \dfrac{|a|}{|b|}$

MIXED REVIEW

65. STANDARDIZED TESTS Choose the value that correctly represents the expression in simplified form.

$$2(100 - 5^2)^2 - 3(8 + 2)^3$$

A. 14,250 **B.** 15,750 **C.** 8,250 **D.** −2,845

Express the equation of each line in slope-intercept form.

66. $3x - 7 = 4y + 3$ **67.** $5y - 5 = 2x + 11$

Solve each system of linear equations using substitution.

68. $\begin{cases} x + 3y = 15 \\ 4x = 24 \end{cases}$ **69.** $\begin{cases} x + 7y = 30 \\ -2y = -8 \end{cases}$

Solve each absolute value inequality algebraically.

70. $|2x + 6| > 15$ **71.** $|3x - 7| < 21$

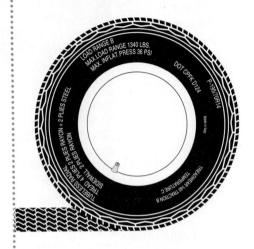

Career
Tire Company Engineer

Tire company engineers design tires that have specific characteristics of ride, handling, and wear. All the information about the tire is molded onto the tire sidewall.

The most common size tire in the United States is the P205/70R14. The "P" stands for passenger car. The 205 is the section width of the tire in millimeters. The 70 is the aspect ratio, which is a measure of the handling of the car. The lower the aspect ratio, the higher the performance. The "R" stands for radial construction, and the 14 is the diameter of the wheel in inches.

Decision Making

A car maker's 1996 sedan has P195/60R15 tires, and the same maker's turbo sedan has P205/50R16 tires.

1. Which tires have the larger section width?

2. Which tires have the higher aspect ratio?

3. Which tires have the higher performance?

4. How much greater is the radius of the turbo sedan tire than that of the regular model?

Another rating that is shown on a sidewall is the maximum load. A typical tire on a passenger car has a maximum load of 1100–1400 lb.

5. Express the maximum load on one tire as an absolute value inequality.

6. Express the maximum load on the four tires as an absolute value inequality.

7. A P205/70R14 tire has a sidewall that is 5.7 in. high. It is expected to last for approximately 60,000 mi. How many rotations will the tire make in its lifetime?

9.5 The Real Number System

Explore

For Questions 1–3, begin with a new number line. Consider the interval between 0 and 2.

1. Divide each unit interval between 0 and 2 into 2 equal parts. Locate and graph the numbers $\frac{1}{2}$, 1, $\frac{3}{2}$, and 2.

2. Divide each unit interval between 0 and 2 into 3 equal parts. What numbers can you locate and graph as a result?

3. Divide each unit interval between 0 and 2 into 4 equal parts. What numbers can you locate and graph as a result?

4. Suppose your number line is p units long and you divide each unit interval into q equal parts (p and q are both integers > 1). What numbers can you locate and graph as a result?

5. If you continue indefinitely, do you think you will eventually be able to locate and graph every point on the number line? Explain.

Build Understanding

You have worked with whole numbers, integers, and rational numbers throughout this book. Each of these groups of numbers forms a set of numbers. A **set** of numbers is any group of numbers with one or more common attributes.

Enclosing numbers in braces, { }, indicates that they form a set. You have studied sets of numbers, such as {0, 2, 4, 6}. The **elements** of this set are 0, 2, 4, and 6. The set is said to be **finite** because each element of the set can be listed. The set of all even numbers is **infinite** because its elements go on indefinitely, without limit (indicated by ellipsis points, . . .). The following are also infinite sets.

Natural or counting numbers: {1, 2, 3, 4, . . .}
Whole numbers: {0, 1, 2, 3, 4, . . .}
Integers: { . . . , −4, −3, −2, −1, 0, 1, 2, 3, 4, . . .}

Set A is a **subset** of a set B if every element of A is also an element of B. For example, the set of whole numbers is a subset of the set of integers because every whole number is also an integer.

Recall that a **rational number** is a number that can be expressed as $\frac{a}{b}$, where a and b are integers and b does not equal zero. A rational number can be expressed as either a terminating decimal such as 5.2 or a nonterminating repeating decimal such as $5.\overline{12}$.

CHECK UNDERSTANDING

Name the sets of numbers listed at the left that are subsets of the integers.

THINK BACK

Recall that the bar used in $5.\overline{12}$ is a symbol for the repeating decimal 5.12121212. . . .

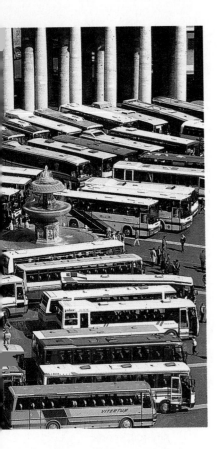

EXAMPLE 1

Write each of the following numbers as the quotient of two integers.

a. −6 **b.** 3.05 **c.** $4\frac{1}{2}$

Solution

a. $-6 = \dfrac{-6}{1}$ **b.** $3.05 = 3\dfrac{5}{100} = \dfrac{305}{100}$ **c.** $4\dfrac{1}{2} = \dfrac{9}{2}$ ◄

Example 1 demonstrates that the set of rational numbers includes integers, decimals, and fractions. You can also express a nonterminating repeating decimal as the quotient of two integers.

EXAMPLE 2

Write $0.\overline{57}$ as the quotient of two integers.

Solution

Let $n = 0.575757\ldots$ Note that two decimal places are repeated.

$$100n = 57.575757\ldots$$ Multiply both sides of the equation by 100 to move the decimal two places.

$$\underline{n = 0.575757\ldots}$$

$$99n = 57$$ Subtract the first equation from the second.

$$n = \dfrac{57}{99} = \dfrac{19}{33}$$ Solve for n. ◄

EXAMPLE 3

Express each quotient as a rational number in decimal form.

a. $\dfrac{7}{2000}$ **b.** $\dfrac{-56}{4}$ **c.** $\dfrac{323}{37}$

Solution

a. *Calculator method* Divide 7 by 2000. The result is 0.0035.

Mental math method Think of the denominator 2000 as $2 \cdot 1000$. First divide 7 by 2, then divide the result by 1000.

$$7 \div 2 = 3.5$$
$$3.5 \div 1000 = 0.0035$$

b. $-56 \div 4 = -14$

The integer −14 is already in decimal form because the decimal point is assumed to be to the right of the integer.

c. $323 \div 37 = 8.729729729\ldots$

To the right of the decimal point, the digits 729 repeat forever. Write the number using a bar over the first 729 to indicate which group of digits repeats. The result is $8.\overline{729}$. ◄

An **irrational number** cannot be expressed as the quotient of two integers. It is represented as a decimal that neither repeats nor terminates. Examples are $1.45445444544445\ldots$, $\sqrt{5}$ and π.

The diagram at the right shows the relationships between the subsets of real numbers. Notice that the rational numbers and irrational numbers have no common elements. The **real numbers** are composed of the rational and irrational numbers.

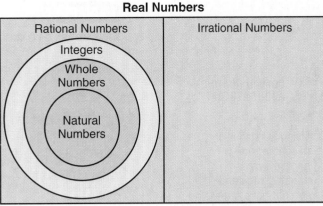

Real Numbers

Rational Numbers | Irrational Numbers

Integers
Whole Numbers
Natural Numbers

EXAMPLE 4

List all subsets of the real numbers to which each number belongs.

a. −3.2   **b.** 6   **c.** 0   **d.** $\sqrt{17}$

Solution

a. −3.2: rational numbers, real numbers

b. 6: natural numbers, whole numbers, integers, rational numbers, real numbers

c. 0: whole numbers, integers, rational numbers, real numbers

d. $\sqrt{17}$: irrational numbers, real numbers   ◄

The inverse process of squaring a number is finding its square root. Since 6^2 equals 36, 6 is a square root of 36. However, since $(-6)^2 = 36$, −6 is also a square root of 36. Every positive real number has one positive square root and one negative square root. The positive square root of a number k is called the **principal square root**. It is denoted $\sqrt{k}$. The symbol $\sqrt{}$ is called a **radical symbol**, the number under the radical symbol is called the **radicand**, and an expression such as $\sqrt{36}$ is called a **radical**. The negative square root of a number k is indicated by $-\sqrt{k}$.

The product and quotient properties of square roots can be used to find the square root of a number.

┌─ PRODUCT PROPERTY OF SQUARE ROOTS ─
│ For all real numbers a and b, where $a \geq 0$ and $b \geq 0$,
│ $$\sqrt{ab} = \sqrt{a} \cdot \sqrt{b}$$

┌─ QUOTIENT PROPERTY OF SQUARE ROOTS ─
│ For all real numbers a and b, where $a \geq 0$ and $b \geq 0$,
│ $$\sqrt{\frac{a}{b}} = \frac{\sqrt{a}}{\sqrt{b}}$$

ALGEBRA: WHO, WHERE, WHEN

Irrational numbers were first called *incommensurable* ("unmeasurable") *numbers*. The mathematician Pythagoras (580–500 B.C.) taught that all measurements could be expressed as ratios of whole numbers. The discovery that lengths such as the diagonal of a unit-square could not be expressed as a ratio led to the concept of irrational numbers.

CHECK UNDERSTANDING

Why are perfect squares always positive?

EXAMPLE 5

Determine each square root.

a. $\sqrt{1089}$ **b.** $-\sqrt{0.49}$ **c.** $\sqrt{-25}$ **d.** $\sqrt{\dfrac{324}{361}}$

Solution

a. $\sqrt{1089} = \sqrt{9 \cdot 121} = \sqrt{9} \cdot \sqrt{121} = 3 \cdot 11 = 33$

b. $-\sqrt{0.49} = -0.7$

c. $\sqrt{-25}$ has no real square root since there is no real number whose square is -25.

d. $\sqrt{\dfrac{324}{361}} = \dfrac{\sqrt{324}}{\sqrt{361}} = \dfrac{18}{19}$ ◀

The real numbers can be graphed on a real number line. You can use a calculator to find decimal approximations of irrational square roots.

EXAMPLE 6

Approximate the square root to the nearest hundredth using a calculator. Then graph each on the same number line.

a. $\sqrt{75}$ **b.** $\sqrt{224}$ **c.** $-\sqrt{21}$ **d.** $-\sqrt{0.004}$

Solution

a. $\sqrt{75} \approx 8.66$ **b.** $\sqrt{224} \approx 14.97$

c. $-\sqrt{21} \approx -4.58$ **d.** $-\sqrt{0.004} \approx -0.06$

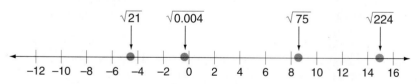

A number in front of a radical means that the radical is multiplied by that number. So $8\sqrt{5}$ means 8 times $\sqrt{5}$.

EXAMPLE 7

PHYSICS The formula $T = 2\pi\sqrt{\dfrac{L}{g}}$ represents the period of oscillation of a pendulum, where T is the period of oscillation, L is the length of the pendulum in centimeters and $g = 980$ cm/s^2, the force of gravity. Approximate the period of oscillation for a pendulum of length 1.25 m to the nearest hundredth of a second.

Solution

There are 100 cm in 1 m, so 1.25 m $=$ 125 cm.

$$T = 2\pi \sqrt{\frac{L}{g}}$$

$$T = 2\pi \sqrt{\frac{125}{980}} \approx 2.24$$

The period of oscillation is approximately 2.24. ◄

TRY THESE

1. Graph and label each of the following real numbers on the same number line.

   a. $-\dfrac{\pi}{2}$ **b.** $0.\overline{8}$ **c.** -2.8 **d.** $\sqrt{11}$ **e.** $-3\dfrac{1}{4}$

List the sets of the real number system to which each number belongs.

2. -4.4 **3.** $\sqrt{13}$ **4.** -15 **5.** $\dfrac{0}{13}$ **6.** $7\dfrac{1}{5}$

Write each as the quotient of two integers.

7. -3 **8.** 4.11 **9.** $5\dfrac{3}{7}$ **10.** 6.06 **11.** 0.01

12. $0.\overline{7}$ **13.** $0.\overline{35}$ **14.** $0.\overline{2}$ **15.** $0.\overline{123}$ **16.** $0.7\overline{53}$

Write each as a rational number in decimal form.

17. $\dfrac{9}{20}$ **18.** $\dfrac{451}{8}$ **19.** $\dfrac{-7}{5}$ **20.** $\dfrac{-336}{21}$ **21.** $\dfrac{27}{33}$

Determine each square root. Do not use a calculator.

22. $\sqrt{\dfrac{225}{256}}$ **23.** $\sqrt{-36}$ **24.** $-\sqrt{0.81}$ **25.** $\sqrt{2704}$ **26.** $\sqrt{3.61}$

Approximate the square root to the nearest hundredth using a calculator.

27. $\sqrt{82}$ **28.** $\sqrt{179}$

29. $-\sqrt{47}$ **30.** $-\sqrt{111}$

31. $\sqrt{0.88}$ **32.** $\sqrt{0.69}$

33. SPORTS In 1985, Dwight Evans had a batting average of 0.263. Show that this number can be expressed as the quotient of two integers. If he was at bat 617 times, how many hits did he get?

34. WRITING MATHEMATICS For which real numbers is $\sqrt{x} > x$?

PRACTICE

1. Graph and label each of the following real numbers on the same number line.

   a. $\dfrac{\pi}{3}$ b. $0.\overline{2}$ c. -1.6 d. $-\sqrt{15}$ e. $-1\dfrac{1}{4}$

List the sets of the real number system to which each number belongs.

2. -3.33 3. $\dfrac{0}{-9}$ 4. 99 5. $\sqrt{1.69}$ 6. $6\dfrac{1}{8}$

7. $-0.\overline{3}$ 8. $\sqrt{25}$ 9. $\sqrt{26}$ 10. $6.\overline{0}$ 11. $7\dfrac{1}{2}$

Write each as the quotient of two integers.

12. -4 13. 5.22 14. $10\dfrac{1}{6}$ 15. 0.078 16. $0.\overline{5}$

17. $0.\overline{12}$ 18. $-0.\overline{96}$ 19. $0.1\overline{67}$ 20. $0.\overline{779}$ 21. $0.1\overline{57}$

Write each as a rational number in decimal form.

22. $\dfrac{21}{50}$ 23. $\dfrac{756}{400}$ 24. $\dfrac{-17}{8}$ 25. $\dfrac{1664}{32}$ 26. $\dfrac{568}{1111}$

Determine each square root. Do not use a calculator.

27. $\sqrt{\dfrac{900}{529}}$ 28. $\sqrt{-144}$ 29. $-\sqrt{0.64}$ 30. $\sqrt{1764}$ 31. $\sqrt{2.89}$

Approximate the square root to the nearest hundredth using a calculator.

32. $\sqrt{77}$ 33. $\sqrt{151}$ 34. $-\sqrt{59}$ 35. $-\sqrt{166}$ 36. $\sqrt{0.67}$

37. $\sqrt{0.56}$ 38. $\sqrt{0.05}$ 39. $\sqrt{0.09}$ 40. $-\sqrt{1122}$ 41. $-\sqrt{2233}$

42. **PHYSICS** Galileo's formula states that the distance d, in meters, covered by a falling object can be calculated using $d = \dfrac{1}{2}gt^2$, where $g = 9.8$ m/s^2 and t represents time in seconds. If an object falls a distance of 100 m, for how many seconds is it falling?

43. **PHYSICS** Use Galileo's formula in Exercise 42 to determine how many seconds it takes for a ball to fall to the ground from the top of a table that is 100 cm high.

44. **WRITING MATHEMATICS** For what numbers is the square root of a real number a rational number?

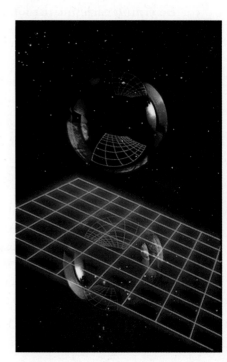

EXTEND

You know that the positive square root of a number k, denoted by $\sqrt{k}$, is called the principal square root. The principal cube root of a number k is denoted by $\sqrt[3]{k}$. Any real number k has exactly one real number cube root, the solution to the equation $x^3 = k$. It may be either positive or negative. You use the cube root function in your calculator to find the principal cube root of a real number k.

Determine each cube root to the nearest hundredth.

45. $\sqrt[3]{27}$ **46.** $\sqrt[3]{-27}$ **47.** $-\sqrt[3]{0.027}$ **48.** $\sqrt[3]{-990}$ **49.** $\sqrt[3]{75}$

50. GEOMETRY The area of a rectangle whose length is double its width is 16 cm². Find its length and its width to the nearest hundredth of a centimeter.

51. GEOMETRY The area A of a circle is given by the formula $A = \pi r^2$, where r is the radius. If $A = 122$ cm², find the radius of the circle to the nearest hundredth of a centimeter.

52. GEOMETRY The volume V of a sphere with radius r is given by the formula $V = \frac{4}{3}\pi r^3$. If $V = 905$ in.³, find the radius of the sphere to the nearest hundredth of an inch.

53. NUMBER THEORY Name three perfect squares between 80 and 130.

THINK CRITICALLY

Determine which number is the greatest in each group.

54. $\sqrt{13}$, $\sqrt{7}$, $\sqrt{18}$, or $\sqrt{21}$

55. $\sqrt{\dfrac{1}{13}}$, $\sqrt{\dfrac{1}{7}}$, $\sqrt{\dfrac{1}{18}}$, or $\sqrt{\dfrac{1}{21}}$

56. $\sqrt[3]{27}$, $\sqrt[3]{64}$, $\sqrt{64}$, or $\sqrt{27}$

57. $\sqrt[3]{0.027}$, $\sqrt{0.027}$, $\sqrt[3]{0.064}$, or $\sqrt{0.064}$

Tell whether each statement is *true* or *false*. If it is false, explain why.

58. $\sqrt{a + b} = \sqrt{a} + \sqrt{b}$

59. $\sqrt{a - b} = \sqrt{a} + \sqrt{b}$

60. Why are there no real square roots of negative real numbers?

61. When is a rational number a perfect square?

62. For what values of x is $\sqrt{3x - 6}$ real?

63. For what values of x is $\sqrt{4x + 12}$ real?

PROJECT Connection

In this activity, you will work with your group to plan and conduct a survey of people's transportation habits. Some examples of the type of information you may want to collect follow.

- What means of transportation do people use to go to work or school?
- What percent of people surveyed use public transportation?
- Do people drive alone or do they carpool?
- What percent of people work at home?
- What is the average travel time (in minutes) to work or school?

1. Each group member should try to survey at least ten people.

2. Tally the group results and discuss the conclusions that can be drawn from the data. Prepare tables and graphs to communicate the results.

3. Compare your results with other groups. Explore whether pooling all the data leads to different conclusions.

When law enforcement officers investigate accident scenes, they measure the length of a skid mark to determine how fast a car was traveling. The table below shows the information in a highway code.

Speed, mi/h	Reaction Distance, ft	Braking Distance, ft	Stopping Distance, ft
10	10	5	15
20	20	20	40
30	30	45	75
40	40	80	120
50	50	125	175
60	60	180	240

Decision Making

Use the table above to answer the following questions.

1. If the car's speed is x mi/h, what is the reaction distance in feet?

2. Verify that speed s is related to braking distance b by the formula $s = \sqrt{20b}$ using three ordered pairs (speed, braking distance).

3. Write an expression for stopping distance.

4. If speed doubles, by what factor does braking distance increase?

5. Over how many feet would a car traveling at 45 mi/h brake? What is its stopping distance?

Investigating officers use the formula $s = \sqrt{30fd}$ to estimate the speed s in miles per hour of a car that skids a distance of d ft. The variable f represents the coefficient of friction, which is determined by the type and condition of the road. The table below shows values of f for two road conditions, wet and dry.

	Concrete	Tar
Wet	0.4	0.5
Dry	0.8	1.0

6. At 65 mi/h, approximately how many feet would you skid on a wet tar road?

7. Suppose the car in Question 6 had been traveling at half the speed. By what factor could you multiply your answer to determine the skid distance?

8. The driver of the car involved in an accident claimed to be going 35 mi/h. The officer measured the skid to be 200 ft. Write up the report as if you were the officer.

9.6 Properties of Equality and Order

Explore

1. Use the two clues to graph the three numbers a, b, and c on a number line.

   - The graph of b is to the right of a.
   - The graph of a is to the right of c.

2. Where is the graph of b in relation to the graph of c?

3. What is the order of the numbers on the number line?

4. Express each clue using an inequality symbol. Then express your answer to Question 3 using inequality symbols.

5. Use the two clues to graph the three numbers d, e, and f.

   - The graph of d is to the left of f.
   - The graph of f is to the left of e.

6. Where is the graph of d in relation to the graph of e?

7. What is the order of the numbers on the number line?

8. Express each clue about d, e, and f using an inequality symbol. Then express your answer to Question 7 using inequality symbols.

Build Understanding

In the previous lesson, you graphed real numbers on a number line. Each real number is associated with exactly one point on the number line. Each point on the number line is associated with a unique real number. The real numbers are said to be **complete** because of this one-to-one correspondence between the real numbers and the points on the number line.

Number lines help when comparing real numbers. As Explore illustrates, one number can be less than a number, greater than a number, or equal to a number, just as your test score can be less than, greater than, or exactly the same as the test score of another person. These relationships can be summarized in the *trichotomy property*.

> **TRICHOTOMY PROPERTY**
>
> For all real numbers a and b, one and only one of the following is true:
>
> $a = b$ $a < b$ $a > b$

EXAMPLE 1

Graph the numbers $-\sqrt{2}, -\frac{4}{3}, -\frac{3}{4}, -1, -\sqrt{3}, -\sqrt{\frac{1}{4}}$, and $-\frac{1}{3}$ on a real number line. Then order them from least to greatest.

Solution

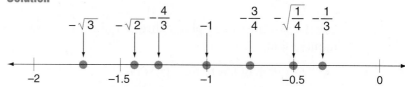

From the number line, list the numbers from least to greatest.

$$-\sqrt{3}, -\sqrt{2}, -\frac{4}{3}, -1, -\frac{3}{4}, -\sqrt{\frac{1}{4}}, -\frac{1}{3}$$ ◀

The real numbers are said to be **dense**. That is, no matter how close two real numbers appear to be, another real number can be found between them. For example, the numbers 0.4491, 0.4495, and 0.4499 are just three of the numbers between 0.449 and 0.45.

EXAMPLE 2

Name three numbers between 2.89 and 2.9.

Solution

The choices are endless; examples include 2.891, 2.8913, and 2.89999. ◀

You have studied and applied properties of real numbers throughout this book. In fact, you used some of them earlier in this chapter to solve absolute value equations and inequalities. These four *properties of equality* are true for all real numbers.

PROPERTIES OF EQUALITY

For all real numbers a, b, and c,

$a = a$	**Reflexive Property**
If $a = b$, then $b = a$.	**Symmetric Property**
If $a = b$ and $b = c$, then $a = c$.	**Transitive Property**
If $a = b$, then a may replace b or b may replace a in any statement.	**Substitution Property**

Mathematicians use the properties of equality to *prove* other statements. A **proof** is a logical sequence of statements that show another statement to be true. Properties, definitions, and already proven facts can be used as reasons for each statement in a proof.

CHECK UNDERSTANDING

Find three numbers located between 2.89 and 2.9 other than the numbers in Example 2. Take the smallest of these and find three numbers between 2.89 and it. Would you ever find a number n, such that $n > 2.89$, for which there is no number between n and 2.89?

EXAMPLE 3

Prove: For all real numbers x and y, if $x = y$, then $x - y = 0$.

Solution

Statements	Reasons
1. $x = y$	1. given
2. $x - y = x - y$	2. reflexive property
3. $x - y = y - y$	3. substitution property
4. $x - y = 0$	4. additive inverse property ◀

Properties of real numbers are used to interpret real world data.

EXAMPLE 4

THE STOCK MARKET The 1994 performance of the Dow Jones industrial average was mixed. The performance of 6 of the 30 companies that make up the 1994 index are displayed in the table.

Company	% Change
Woolworth	−40.9
AT&T	−4.3
Exxon	−3.8
McDonald's	2.6
Merck	10.9
Union Carbide	31.3

THINK BACK

Recall the following addition properties.

Commutative:
$a + b = b + a$

Associative:
$(a + b) + c = a + (b + c)$

Inverse:
$a + (-a) = 0$
and
$-a + a = 0$

Identity:
$a + 0 = a$
and
$0 + a = a$

a. Which two companies experienced the greatest percentage increases?

b. Which two companies experienced the greatest percentage decreases?

c. Which company experienced the greatest percentage change?

Solution
Order the percent change from least to greatest.

a. Merck and Union Carbide

b. Woolworth and AT&T

c. Woolworth ◀

TRY THESE

Graph each set of numbers on the real number line. Then order them from least to greatest.

1. $\sqrt{5}, 4, -3, 2, -1, 0$

2. $3, -3.5, 2.5, -\sqrt{2}, -2, 1$

3. $1.2, 1.8, \dfrac{5}{4}, \dfrac{3}{2}, \dfrac{3}{4}$

4. $-\sqrt{3}, -1.25, -1.5, -\dfrac{7}{4}, -2, -\dfrac{1}{2}$

Find three numbers between each pair of numbers.

5. 1.1 and $1.\overline{1}$

6. 0.17 and $0.1\overline{7}$

Determine which property of equality is demonstrated in each statement.

7. If $6 = 3 \cdot 2$, then $3 \cdot 2 = 6$.

8. If $24 \div 2 = 12$ and $12 = 4 \cdot 3$, then $24 \div 2 = 4 \cdot 3$.

9. $7 = 7$

10. $8 = 5 + 3$. Since $16 \div 8 = 2$, then $16 \div (5 + 3) = 2$.

11. Prove: For all real numbers a, b, and c, if $a = b$, then $ac = bc$.

Statements	Reasons
1. $a = b$	**1.** given
2. $ac = ac$	**2.** _____
3. $ac = bc$	**3.** _____

TRANSPORTATION Despite advertisements encouraging people to form carpools to drive to work and the increase in high occupancy lanes on the highways, the percentage of commuters traveling in carpools in representative New York counties and in the entire United States decreased from 1980 to 1990. This information is displayed in the chart to the right.

Commuter Poll, New York Counties and U.S. How Commuters Traveled, 1980–1990, by percent								
Means of Travel	**Westchester**		**Putnam**		**New York**		**U.S.**	
	1980	**1990**	**1980**	**1990**	**1980**	**1990**	**1980**	**1990**
Drive alone	54.6	61.16	64.8	77.43	54.3	55.7	64.4	73.2
Carpool	15.7	10.20	24	11.71	18.4	19.75	19.7	13.4
Bus	5	4.94	1.1	0.59	8.2	6.69	4.1	3
Rail	12.6	12.47	4.5	5.14	3.4	2.9	0.6	0.6
Bicycle	0.2	0.14	0.1	0.06	0.4	0.25	0.5	0.5
Walk	7.2	5.43	2.6	1.59	9.2	7.18	5.6	5.6
Other	4.7	5.71	2.9	3.48	6.1	7.52	5.2	3.7

12. Which mode of transportation experienced the greatest percentage point increase?

13. Which mode of transportation experienced the greatest percentage point decrease?

14. WRITING MATHEMATICS Write a paragraph in your own words explaining how to order $\sqrt{3}$, 1.7, $1\frac{71}{100}$, 1.732, and $1.7\overline{32}$.

PRACTICE

Graph each set of numbers on the real number line. Then order them from least to greatest.

1. $-\sqrt{6}, 5, -4, 3, -2, 0$

2. $5, -5.5, 4.5, -4, 3.5, \sqrt{20}$

3. $1.3, 1.8, \frac{6}{5}, \frac{8}{5}, \frac{19}{10}$

4. $-1.75, -1.5, -\frac{5}{4}, -1, -\frac{3}{4}$

Write each set of numbers in order from least to greatest.

5. $\frac{5}{8}, \frac{4}{7}, \frac{6}{9}$

6. $\frac{14}{17}, \frac{16}{19}, \frac{13}{16}$

7. $1\frac{6}{7}, \frac{15}{7}, 2.1$

8. $2\frac{5}{9}, 2\frac{2}{3}, 2.5$

9. $-0.151, -0.1501, -0.1511$

10. $-3.343, -3.434, -3.43$

Find three numbers between each pair of numbers.

11. 3.1 and 3.11

12. 5.7 and 5.75

13. 4.118 and 4.$\overline{118}$

14. 3.345 and 3.$\overline{345}$

Determine which property of equality is demonstrated in each statement.

15. $-3 = -3$

16. $7 = 9 - 2$; since $21 \div 7 = 3$, then $21 \div (9 - 2) = 3$.

17. If $4 = 8 \div 2$, then $8 \div 2 = 4$.

18. If $30 \cdot 2 = 60$ and $60 = 54 + 6$, then $30 \cdot 2 = 54 + 6$.

19. Prove: For all real numbers x, y, and z, if $x = y$, then $x + z = y + z$.

Statements	Reasons
1. $x = y$	**1.** given
2. $x + z = x + z$	**2.** _____
3. $x + z = y + z$	**3.** _____

20. Prove: For all real numbers x, y, and z, if $x + y = z$, then $x = z - y$.

Statements	Reasons
1. $x + y = z$	**1.** given
2. $x + y - y = z - y$	**2.** _____
3. $x + 0 = z - y$	**3.** _____
4. $x = z - y$	**4.** _____

ASTRONOMY The asteroids are approximately 30,000 pieces of rocky debris located between the orbits of Mars and Jupiter. The first asteroid was discovered in 1801 by Father Piazzi. Karl Friedrich Gauss calculated its orbit. Data for the first ten asteroids to be discovered are given in the chart.

21. Which asteroid has the least mean distance from the sun?

22. Which asteroid has the greatest mean distance from the sun?

23. Order the ten asteroids from shortest to longest orbital period.

24. **WRITING MATHEMATICS** Use a dictionary to find the definition of the word *symmetric*. Compare and contrast the meaning of this word as it is used in the properties of equality in this lesson and as it is used in Lessons 9.1 and 9.2 in relation to the graphs of absolute value equations.

Name	Mean Distance from Sun, millions of miles	Orbital Period, years
Ceres	257.0	4.60
Pallas	257.4	4.61
Juno	247.8	4.36
Vesta	219.3	3.63
Astraea	239.3	4.14
Hebe	225.2	3.78
Iris	221.4	3.68
Flora	204.4	3.27
Metis	221.7	3.69
Hygeia	222.6	5.59

EXTEND

The transitive property holds true for inequalities as well as equalities.

a. If $a < b$ and $b < c$, then $a < c$.
b. If $a > b$ and $b > c$, then $a > c$.

25. Provide a numerical example for part a of this property.

26. Provide a numerical example for part b.

27. Does the reflexive property hold for inequalities? Provide an example.

28. Does the symmetric property hold for inequalities? Provide an example.

29. Does the substitution property hold for inequalities? Provide an example.

TECHNOLOGY On some calculators, you can use the TEST menu to determine whether an inequality is true. If the statement is true, the calculator displays a 1. If the statement is false, the calculator displays a 0. Determine whether each of the following is true or false.

30. $0.889 > \dfrac{8}{9}$ **31.** $\dfrac{8}{9} < -0.889$ **32.** $0.11 < \dfrac{1}{9}$ **33.** $\dfrac{1}{4} < 0.252525$

THINK CRITICALLY

34. Can you replace the $<$ symbol with the $\leq$ symbol and the $>$ symbol with the $\geq$ symbol in the trichotomy property? Explain.

35. Write a property similar to the trichotomy property for $=$ and $\neq$.

36. Is the reflexive property true for $\neq$? Provide an example.

37. Is the symmetric property true for $\neq$? Provide an example.

38. Is the transitive property true for $\neq$? Provide an example.

39. Is the substitution property true for $\neq$? Provide an example.

40. Is $-x < x$ for all x? Explain. (Assume $x \neq 0$.)

MIXED REVIEW

Find the following measures of central tendency for these scores.

89, 92, 88, 77, 92, 83, 69, 98, 72, 95.

41. mean   **42.** mode   **43.** median   **44.** range

45. STANDARDIZED TESTS Which is the value of the function for $x = 8$?

$$f(x) = \dfrac{x^2 - x}{4}$$

A. 16.5   **B.** 14   **C.** 11.5   **D.** 18

Write each set of numbers in order from least to greatest.

46. $\dfrac{17}{19}, \dfrac{13}{16}, \dfrac{17}{21}$ **47.** $-0.161, 0.1601, -0.1611$

450   CHAPTER 9   **Absolute Value and the Real Number System**

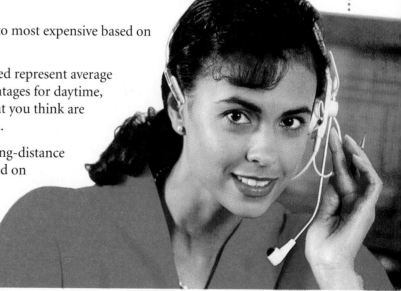

Career
Market Research Analyst

Telephone companies constantly seek ways to increase communication traffic on their lines. A market research analyst for Phone, Inc., a long-distance telephone company, analyzes the company's services and compares them to those of its competitors. To help the advertising department decide on an appropriate advertising campaign, the analyst may point out the areas in which the company has a price advantage over the competition.

Over 85% of the long-distance telephone market is controlled by three major companies, Phone, TelCo, and ABC, Inc. Each carrier has more than one plan. Although the three major carriers offer basic rates that are similar, the market researcher points out that Phone offers advantages for those people who have heavy night and weekend usage. The analyst compiles the following data based on 5% of calls made during the daytime, 25% in the evening, and 70% during night/weekend hours.

	10 calls/ 108 min	30 calls/ 318 min
PHONE Plan A	$17.53	$39.07
PHONE Plan B	$19.41	$45.35
TELCO Plan A	$19.98	$46.00
TELCO Plan B	$19.51	$46.66
ABC, INC Plan A	$21.08	$49.69
ABC, INC Plan B	$19.61	$45.35

Decision Making

1. Order the plans from least expensive to most expensive based on 10 calls/108 min.

2. Order the plans from least expensive to most expensive based on 30 calls/318 min.

3. Do you think the time percentages used represent average customer usage? If not, suggest percentages for daytime, evening, and night/weekend usage that you think are more typical and should be compared.

4. Name two types of discounts that a long-distance carrier can offer other than those based on the time of day of the call.

5. What other factors could affect a customer's choice of company and plan? Why might a customer be interested in rate history as well as current prices?

Problem Solving File

Use Multiple Representations

Many mathematical problems require different approaches, such as generating a table of numbers, making a graph, or writing an equation or a formula. Trying different approaches may help you understand a problem. Sometimes, one approach is more efficient than the others.

Problem

Government agencies take interest in earnings trends. For example, in 1970 the average weekly income of mining workers was $164; in 1990 the average weekly income for this group was $604. If the rate of change in the weekly income was about the same each year, how can the weekly income be represented as a function of the year? What is the predicted income for mining workers in the year 2000?

Explore the Problem

1. Use a graph. Let x represent time and use the last two digits of the year (so that 70 represents 1970 and 100 represents 2000). Using the given information, what two points can you plot? Use these points to graph the function. Describe your graph.

2. How can you find the yearly rate of change? What is this rate?

3. Use a table. Generate a table of weekly incomes for the period 1970–1980. Which was the first year that weekly income exceeded $300? Does the result agree with your graph above?

4. Use an equation. Use the information for 1970 and the slope-intercept method to write an equation that gives income as a function of the year. (*Hint:* Begin with $y = m(70) + b$.)

5. How could you use each of the methods to predict earnings for the year 2000? Carry out the work and compare your results.

6. WRITING MATHEMATICS Which method was most useful for making your prediction? Explain. What are some advantages of each method? Describe situations where each would be useful.

Investigate Further

Analysts for government agencies often compare earnings in several industries. Consider these data for workers in construction trades: in 1970 the average weekly income was $195; in 1990 it was $526. Assume a constant yearly rate of change.

7. **a.** Which group had the higher income in 1970, mining or construction workers?

    b. Which group had the higher income in 1990?

    c. What can you conclude about the yearly rate of change in income for construction workers as compared to mining workers? Explain.

8. Use the same set of axes as for the mining graph. Plot two points and draw the construction income graph. Describe the graph and explain what information you gain from it.

9. **a.** Find the yearly rate of change in construction income. Use this result to add a column for construction workers to the table you made for Question 3.

    b. Explain how you can determine from the table when the income for the two groups was approximately the same. Compare with your results using the graph.

10. **a.** Write an equation that determines the weekly construction income as a function of the year. Which method did you use?

    b. To show the graph of this equation on a graphing utility, what range values would you use?

11. How could you use elimination or substitution to determine when weekly income for both groups was equal? Why might you not want to use this approach?

12. Predict the income for construction workers in the year 2000. Which representation did you use? Explain how to use a different representation to check your result.

> ┌──────────────────
> │ **PROBLEM**
> │ **SOLVING PLAN**
> │
> │ • Understand
> │ • Plan
> │ • Solve
> │ • Examine
> └──────────────────

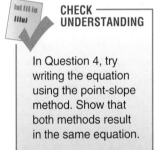

CHECK UNDERSTANDING

In Question 4, try writing the equation using the point-slope method. Show that both methods result in the same equation.

Apply the Strategy

13. ECONOMICS In 1970 the average weekly income of transportation workers was $133. In 1990 the weekly average was $442.

   a. Make a prediction about the slope of the transportation workers' graph and how this graph would look in comparison with the mining and construction workers' graphs. Then construct the graph for the transportation workers' incomes to check your prediction.

   b. What is the average rate of change in income per year for transportation workers over the 20-year period? Use this information to write an equation that determines their average weekly income as a function of the year.

   c. Predict the income for transportation workers in the year 2000.

14. DESIGN Leroy has red, blue, and green square tiles. He creates square designs according to these rules.

   • Outside corner tiles must be red.

   • Other outside tiles must be blue.

   • Inside tiles must be green.

corner
outside
inside

   a. A 3×3 design is shown. Color or label a 4×4 design. How many tiles of each color did you use?

   b. Make a table to show how many tiles of each color you need for a 2×2, a 3×3, . . . , a 10×10 design.

   c. On one set of axes, graph the number of each color tile as a function of side length. Does it make sense to connect the points? Explain.

   d. Which color increases most quickly? Describe the change for each of the other colors. Explain how either the table or the graph can be used to answer these questions.

   e. Write a formula for the number of tiles of each color in an $n \times n$ design.

   f. Find the number of each color tile for a 100×100 design. Which method is most useful this time?

   g. How many tiles are in a 100×100 design? Why? How can you check your answers to Question 14 f?

15. WRITING MATHEMATICS Write a paragraph about why it is useful to be able to use more than one method to solve a problem.

PROBLEM SOLVING TIP

Use the values of x and y for one of the years given to find the value of b in the equation for weekly income.

REVIEW PROBLEM SOLVING STRATEGIES

WHAT'S IT WORTH TO YOU?

1. Jan's Jewels is having a contest. Displayed in the window is a necklace of 33 pearls. A sign near the necklace gives the information at the right.

The middle pearl in this necklace is the most valuable. From one end of the necklace, the pearls have been arranged so that each successive pearl is worth $100 more than the preceding one, up to and including the center pearl. From the other end, the pearls increase in value by $150, up to and including the center pearl. The whole necklace is worth $65,000. What is the value of the center pearl?

Anyone who solves the problem gets a 33% discount on any purchase, including the necklace. (You didn't think Jan's would give the necklace away, did you?)

a. Can the pearls at each end have equal values? Explain.
b. Will you solve this problem using one or two variables?
c. How can you write two different expressions for the value of the center pearl?
d. What equation can you write for the value of the whole necklace?
e. What is the value of the center pearl? What is the value of the pearl at each end?
f. How much would a contest winner pay for the necklace?

CAN DO!

3. A case of soup cans holds 24 cans in four rows and six columns. Draw X's to show how to arrange 18 cans in the case so that there are an even number of cans in each row and each column. (You may want to use small circles to represent cans.)

FIND A PATTERN

2. Without listing them all, can you tell how many numbers are in this list?

2, 5, 8, 11, 14, . . . , 683

Explain your reasoning.

··· CHAPTER REVIEW ···

VOCABULARY

Choose the word from the list that completes each statement.

1. A(n) __?__ is a sentence formed by joining two open sentences by the word *or*.

2. The __?__ of a number is the distance between zero and the point representing the number on a number line.

3. A(n) __?__ is a number that neither repeats nor terminates.

4. A(n) __?__ is a number that can be expressed as the quotient of two integers.

5. A(n) __?__ is a sentence formed by joining two open sentences by the word *and*.

a. irrational number

b. conjunction

c. absolute value of a number

d. rational number

e. disjunction

Lesson 9.1   ABSOLUTE VALUE   pages 413–415

- Graphing utilities may be used to graph absolute value equations.

Use a graphing utility to graph each absolute value equation.

6. $y = |x| - 5$ 7. $y = |x| + 5$ 8. $y = |x - 5|$ 9. $y = |x + 5|$

Lesson 9.2   GRAPH ABSOLUTE VALUE FUNCTIONS   pages 416–423

- The graph of $y = |x - b| + c$ will be shifted upward or downward from the graph of $y = |x|$ depending on the value of c. It will be shifted left or right depending on the value of b.
- Determine the vertex and line of symmetry before you graph an absolute value function.

Determine the vertex and line of symmetry for the graph of each equation. Graph the absolute value function.

10. $y = |x + 4| - 1$ 11. $y = |x - 4| + 1$ 12. $y = 2|x - 2| + 3$ 13. $y = -\frac{1}{2}|x - 2| + 3$

Lesson 9.3   ABSOLUTE VALUE EQUATIONS   pages 424–429

- To solve an absolute value equation, express it as a disjunction. Then solve the two resulting equations.

Solve and check each equation. If an equation has no solution, write "no solution."

14. $|6x| = 36$ 15. $|x - 5| = 8$

16. $|3 + x| = 9$ 17. $-3|4g + 2| - 3 = 7$

---

456   CHAPTER 9   Absolute Value and the Real Number System

pages 430–436

- To solve an absolute value inequality, express it as either a disjunction or a conjunction. Then solve the two resulting inequalities.

Solve each inequality. Graph each solution.

18. $|x - 2| < 7$ **19.** $|z + 1| > 4$ **20.** $|3w + 6| \leq 12$ **21.** $|5 + 3x| \geq 0$

22. REPORT CARDS Miko's teacher told the class that any students with grade averages from 94 to 100 would receive an A. Write an absolute value inequality representing this range.

Lesson 9.5   THE REAL NUMBER SYSTEM
pages 437–444

- A **rational number** can be expressed as $\frac{a}{b}$, where a and b are integers and $b \neq 0$. It can be represented as either a terminating decimal or a nonterminating, repeating decimal. An **irrational number** cannot be expressed as the quotient of two integers and is represented as a decimal that neither terminates nor repeats.
- To determine the square root of a number, use the product property of square roots and the quotient property of square roots.

Write each as a rational number in decimal form.

23. $\frac{2}{11}$ **24.** $\frac{5}{9}$ **25.** $\frac{11}{20}$ **26.** $\frac{323}{19}$ **27.** $\frac{5}{18}$

Determine each square root if it is a real number.

28. $-\sqrt{0.49}$ **29.** $\sqrt{-36}$ **30.** $\sqrt{\dfrac{625}{961}}$ **31.** $\sqrt{4096}$

Lesson 9.6   PROPERTIES OF EQUALITY AND ORDER
pages 445–451

- The real number line can be used to order real numbers. The real numbers are **dense**; that is, between any two real numbers there is another real number.
- The properties of equality can be used to prove other statements.

Complete each of the following exercises.

32. Graph $-\dfrac{2}{3}, -\sqrt{2}, -\sqrt{\dfrac{1}{9}}, -2$, and $-\sqrt{5}$ on a number line. Order them from least to greatest.

33. Prove: For all real numbers a, b, and c, if $a = b$, then $a - c = b - c$.

Lesson 9.7   USE MULTIPLE REPRESENTATIONS
pages 452–455

- Using different methods to solve a problem can help you understand it better.

34. ENERGY CONSUMPTION In 1980 the total amount of energy consumed by the United States was 2364.4 million metric tons (in coal equivalents). In 1990 the amount was 2572.3 million metric tons. Construct a graph showing the rate of change in energy consumption from 1980 to 1990. Extend the graph to the year 2000 and give an approximate estimate of the energy consumption in that year.

CHAPTER ASSESSMENT

CHAPTER TEST

1. **WRITING MATHEMATICS** Write a paragraph explaining how the graph of $y = 3|x - 1| + 4$ differs from the graph of $y = |x|$.

Determine the vertex and the line of symmetry for the graph of each equation.

2. $y = -2|x|$

3. $y = |2x + 6| - 7$

4. $y = |x + 1| + 5$

5. $y = -6|x - 2| + 3$

6. **STANDARDIZED TESTS** Choose the equation that represents the graph of $y = |x|$ shifted 3 units left and 5 units downward.

   A. $y = |x - 3| - 5$ **B.** $y = |x + 5| + 3$

   C. $y = |x + 3| - 5$ **D.** $y = |x - 3| + 5$

Solve each equation. If an equation has no solution, write "no solution."

7. $|x + 3| = 2$

8. $|x - 8| = 3$

9. $3|4x - 1| - 7 = 5$

10. $4|3x + 2| + 6 = 2$

Solve each inequality. Graph each solution.

11. $|t + 7| > 2$

12. $|w - 5| < 1$

13. $|2x + 5| \geq 1$

14. $\frac{1}{2}|3 - 5x| > 0$

15. **STANDARDIZED TESTS** Which set lists the elements in order from least to greatest?

   A. $-\frac{6}{3}, \frac{7}{8}, -\sqrt{7}, \frac{5}{3}$ **B.** $-\sqrt{7}, -\frac{6}{3}, \frac{5}{3}, \frac{7}{8}$

   C. $-\sqrt{7}, \frac{5}{3}, -\frac{6}{3}, \frac{7}{8}$ **D.** $-\sqrt{7}, -\frac{6}{3}, \frac{7}{8}, \frac{5}{3}$

Solve the problem.

16. The birth rate per 1000 people in Japan was 17.2 in 1975 and 9.7 in 1992. Predict the approximate birth rate in the year 2000 if this trend continues.

17. **STANDARDIZED TESTS** Which item lists all the subsets of the real numbers to which each of the following numbers belong?

   I. natural   **II.** whole   **III.** integers
   IV. rational   **V.** irrational   **VI.** real

17. $-\frac{12}{4}$

   A. III, IV, and VI   **B.** I, II, III, IV, and VI

   C. IV, V, and VI   **D.** I, II, III, and VI

18. $\sqrt{21}$

   A. III, V, and VI   **B.** IV and VI

   C. V and VI   **D.** IV, V, and VI

Determine each square root if it is a real number.

19. $\sqrt{1.44}$ 20. $-\sqrt{121}$

21. $\sqrt{484}$ 22. $\sqrt{-1681}$

Write each as the quotient of two integers.

23. -4.17 24. 0.091

25. $0.\overline{8}$ 26. $0.\overline{16}$

27. **STANDARDIZED TESTS** Which of the following numbers represents the rational number $\frac{17}{6}$ in decimal form?

   A. 2.83   **B.** $2.\overline{83}$

   C. $2.8\overline{3}$ **D.** 2.8333

Solve the problem.

28. A clothing store pays each salesperson a yearly bonus of $1,000 if they achieve total sales of $75,000 $\pm$ $5,000 for the year. Write an inequality for the maximum and minimum sales for which a person can earn the bonus.

PERFORMANCE ASSESSMENT

USE THE REAL NUMBER LINE

Write absolute value equations in the form of $|ax + b| < c$ and $|ax + b| > c$ for positive, negative, and zero values for a and b and positive values for c. Show how to model solutions of the different absolute value equations using the real number line and strands of colored yarn and tape.

BUILD PERFECT SQUARES

Cut out 100 squares of construction paper each 1 in. × 1 in. Use the method shown to find the first 10 perfect squares.

☐ $1 = 1 \cdot 1 = 1^2 \rightarrow \sqrt{1} = 1$

⬚   2, not a square

⬚   3, not a square

⬚ $4 = 2 \cdot 2 = 2^2 \rightarrow \sqrt{4} = 2$

ABSOLUTE VALUE PATTERNS

Graph $y = |x|$ using a black marker. Then on the same coordinate grid, graph each of the following absolute value equations using the color listed.

$y = |x - 1|$	Red
$y = |x - 2|$	Blue
$y = |x - 3|$	Yellow
$y = |x - 4|$	Green
$y = |x - 5|$	Orange

Then on a separate coordinate grid, graph $y = |x|, y = \frac{1}{4}|x|, y = \frac{1}{3}|x|, y = \frac{1}{2}|x|, y = 2|x|, y = 3|x|,$ and $y = 4|x|$ using a different color for each. Graph other types of absolute value equations to make different patterns.

MAKE PREDICTIONS

Find out what the population of your town or city was in 1970 and what it is today. Make a prediction based on this information as to what the population might be in the year 2000 if the current trend continues.

PROJECT ASSESSMENT

PROJECT Connection

Use the graphs, equations, and results of your survey to prepare travel profiles of your group and the community.

1. As a group, discuss and agree upon the key ideas about travel habits that you wish to communicate. Plan how to arrange your material to support these ideas. Consider a large poster, bulletin board display, computer slide show, or video.

2. Research and include national statistics for some of the items in your survey. One source of information is the *Statistical Abstract of the United States*, a government publication available in most public libraries.

3. In a section of your presentation called "Into the Future," verbally or visually describe solutions to some of the transportation problems you have identified. Be creative!

··· CUMULATIVE REVIEW ···

Graph each equation.

1. $y = |x + 3|$ **2.** $y = |x| - 4$

3. $y = 2|x|$

Solve each inequality. Then graph the solution.

4. $|4x + 2| \geq 10$ **5.** $|2x - 5| < 7$

6. $12 - 3(6n + 5) + 4n < \frac{1}{2}(10n - 6)$

Solve each system of equations.

7. $\begin{cases} 9x - 5y = 22 \\ y = 3x - 8 \end{cases}$ **8.** $\begin{cases} 7x - 4y = 11 \\ 8y - 14x = -11 \end{cases}$

Use the two scatter plots for Questions 9–11.

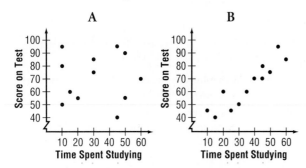

9. Which scatter plot shows the stronger correlation?

10. STANDARDIZED TESTS Consider statements I–IV as they pertain to scatter plots A and B.

   I. Scatter plot B shows a correlation close to 1.

   II. Scatter plot A shows a correlation close to −1.

   III. Scatter plot A shows a weak correlation.

   IV. Scatter plot B shows that the longer you study, the higher your test score can be.

Which statements are true?

   A. I and II   **B.** I and IV
   C. II and III   **D.** I, III, and IV

11. WRITING MATHEMATICS If you were the teacher, would you be pleased with the results in scatter plot A? Why or why not?

12. A concession stand sells hot dogs for $1.50 each and hamburgers for $2.00 each. It costs the owner of the stand $0.30 per hot dog and $0.60 per hamburger. For the big game, the owner decides that a total of 500 hot dogs and hamburgers should be enough, but does not want to spend more than $240 to purchase the food. How many hot dogs and hamburgers should the owner purchase in order to maximize profits?

Solve for the variable indicated in each formula.

13. $A = P + Prt$ for r

14. $SA = 2\pi r^2 + 2\pi rh$ for h

Identify the property illustrated.

15. If $c = a \cdot b$ and $a \cdot b = d$, then $c = d$.

16. $(10 + 13) + 24 = 10 + (13 + 24)$

17. $7 \cdot \frac{1}{7} = 1$ **18.** $x + 5 = x + 5$

19. WRITING MATHEMATICS Explain how the commutative properties and the symmetric property are similar and different.

Graph each inequality. Then determine whether $(2, -3)$ is a solution.

20. $4x + 2y \leq 7$ **21.** $3x - y < 5$

Compare the expressions. Use <, =, or >.

22. $(5 - 3)^2$ ▨ $5^2 - 3^2$

23. $-(-3)^3$ ▨ $-(-2)^6$

24. $|-5 - 7|$ ▨ $|-5| + |-7|$

··· STANDARDIZED TEST ···

STANDARD FIVE-CHOICE Select the best choice for each question.

1. The solution to $|3x - 2| \geq 5$ is

   A. $-5 \leq x \leq 5$
   B. $x \leq 3$ or $x \geq 7$
   C. $x \leq -1$ or $x \geq \dfrac{7}{3}$
   D. $-1 \leq x \leq \dfrac{7}{3}$
   E. $x \geq \dfrac{7}{3}$

2. If $A = 8$, $B = 4$, and $C = 6$, then which expression in BASIC will output 16?

   A. PRINT C / B * A
   B. PRINT A * B − C
   C. PRINT C * A / B
   D. PRINT C * B − A
   E. none of these

3. All of the following numbers are irrational except

   A. $\sqrt{5}$
   B. $1.833333\ldots$
   C. π
   D. $-\sqrt{8}$
   E. all are irrational

4. After three tests, Juan has an 86% average. What does he need to get on the next test to raise his average to 90%?

   A. 90%
   B. 94%
   C. 98%
   D. 100%
   E. he can't do it—he needs more than 100%

5. Which of the following cannot be the result of solving a linear system of equations?

   A. no solutions
   B. one solution
   C. two solutions
   D. infinite number of solutions
   E. all are possible

6. Which of the following equations can be solved by first subtracting 5 from both sides and then multiplying both sides by 3?

   A. $3x - 5 = 10$
   B. $3x + 5 = 14$
   C. $\dfrac{1}{3}x - 5 = 7$
   D. $\dfrac{1}{3}x + 5 = 11$
   E. none of these

7. Consider statements I–III as each applies to the correlation coefficient, r.

   I. $|r| \leq 1$
   II. $r = 0.6$ indicates a weak correlation
   III. $r = 0.9$ and $r = -0.9$ indicate the same strength of correlation

   Which statement is true?

   A. II only
   B. III only
   C. I and II
   D. I, II, and III
   E. none are true

8. All of the following functions have a domain of all real numbers except

   A. $y = \sqrt{x} + 2$
   B. $y = 5$
   C. $y = x^2 - 3$
   D. $y = |x - 4|$
   E. two of the above

9. Which graph shows the solution of $3(3 - 2x) > 4x - 1$?

   A.
   B.
   C.
   D.
   E.

10 Quadratic Functions and Equations

Take a Look AHEAD

Make notes about things that look new.

- Locate and copy the quadratic formula.
- Do all rectangles with the same perimeter have the same area? Given a rectangle with perimeter 40, how many different rectangles can you draw? Are the areas of any of your rectangles equal?

Make notes about things that look familiar.

- The Latin word "quadrare" means "to make a square." Why do you think the functions and equations in this chapter are called quadratics?
- When do you use the term "perfect square"? What is true about the square of any real number?

DATA Activity

Business Risk

Running a business can be exciting, but it also involves risk. For a business to grow and prosper, its owners and managers must understand their potential market, plan carefully, and always watch expenses. Business successes and failures also provide private and government analysts with important statistics for evaluating the condition of the national economy. The table on the next page gives data about businesses and failures during the 1980s and early 1990s.

SKILL FOCUS

▶ Construct and interpret a line graph.

▶ Compare real numbers.

▶ Add, subtract, multiply, and divide real numbers.

▶ Determine equivalent rates.

A BUSINESS INDUSTRY

In this chapter, you will see how:

- **PIZZA PARLOR OWNERS** use algebra to provide their customers with a better value for the money than their competitors do.
(Lesson 10.2, page 476)

- **AMUSEMENT PARK RIDE DESIGNERS** use algebra to create thrilling but safe rides.
(Lesson 10.4, page 489)

- **SATELLITE ANTENNA ENGINEERS** use parabolas to design effective satellite dish antennas.
(Lesson 10.5, page 497)

Business Failures					
Year	Total Number of Businesses, in 1000s	Number of Failures	Year	Total Number of Businesses, in 1000s	Number of Failures
1981	2,745	16,794	1987	6,004	61,111
1982	2,806	24,908	1988	5,804	57,098
1983	2,851	31,334	1989	7,694	50,361
1984	4,885	52,078	1990	8,038	60,747
1985	4,990	57,078	1991	8,218	88,140
1986	5,119	61,616	1992	8,805	96,857

Use the table to answer the following questions.

1. Make a line graph showing the total number of businesses for each year.

2. Use your graph to determine the two periods during which the greatest business growth occurred.

3. Is it accurate to say that more businesses failed in 1992 than in 1986? In what way is this statement misleading?

4. One method that business analysts use to make meaningful comparisons is to determine a rate of failure. In this situation, they might calculate the number of businesses that failed per 10,000 of the total number of businesses. Determine this rate to the nearest whole number for each year shown in the table.

5. **WORKING TOGETHER** Consult local business organizations and/or the Chamber of Commerce to find out how many businesses opened and how many closed during the last 2 years. Would you want to open a new business?

463

The Company Picture

The stock market offers people the opportunity to earn money on their money. Investors who know about industries, corporations, and the workings of the market itself can optimize their chances of making profitable choices. While there is never a guarantee of success in financial investing, it is no accident that most good decisions are based on solid research. In this project you will gain some experience in building knowledge about a potential investment.

PROJECT GOAL

To combine algebra and research skills to learn about an industry and the recent stock activity of corporations in that industry.

Getting Started

Work in small groups. Look over the Project Connections to familiarize yourself with the materials required for activities. You need to request information immediately so that it will be available for your corporate summary.

1. Select an industry that interests your group.

2. Use the stock market section of your newspaper to name several corporations in that industry. Each group member should be responsible for one corporation.

3. Learn how to read the stock market reports in the newspaper. Identify the listing for your corporation.

PROJECT Connections

Lesson 10.3, page 482:
Contact corporations to secure information and learn about stocks and stock markets.

Lesson 10.4, page 488:
Construct a special graph to display daily stock activity for a selected corporation.

Lesson 10.5, page 496:
Analyze corporate information, prepare a timeline, and explore models to fit data.

Chapter Assessment, page 511:
Groups present reports on industries, corporations, and stock markets, determine the success of imaginary investments, and speculate on the future.

Internet Connection

www.swpco.com/
swpco/algebra1.html

Think Back/Working Together

Work with a partner. Recall the slope-intercept form of a linear equation $y = mx + b$. The graph of this equation is a line with slope m and y-intercept b.

1. Copy and complete the table at the right for the equation $y = 2x - 5$.

2. **a.** How much does y change for every *unit* increase in x?

    b. What do you observe about this change?

    c. How did you determine y when $x = 25$?

3. How can you answer Question 2a without making a table?

4. Determine the slope of the line that contains the points $(10, 15)$ and $(25, 45)$.

x	y	Change in y
−2	−9	
		2
−1	−7	
		2
0	−5	
1		
2		
3		
4		
10		
25		

Explore

A **quadratic equation** is an equation of the form

$$y = ax^2 + bx + c$$

where a, b, and c are real numbers and $a \neq 0$.

5. Use a graphing utility to graph the equations $y_1 = 2x - 5$ and $y_2 = 2x^2 - 5$ on the same set of axes. Discuss the differences between the two equations and the two graphs.

6. What letter of the alphabet best describes the shape of the graph of $y_2 = 2x^2 - 5$?

7. Copy and complete the following table for the equation $y = 2x^2 - 5$.

x	y	Difference Column 1 (Change in y)	Difference Column 2 (Change in the change in y)
−4	27		
−3	13	14	
−2	3	10	4
−1			
0			
1			
2			
3			
4			

8. What do you notice about the rate of change in y per unit increase in x? (See Difference Column 1.)

9. Compare the table in Question 7 to the table in Explore. Why do you think the numbers in Difference Column 1 change in the table in Question 7?

10. What do you notice about the change in the change in y? (See Difference Column 2.)

Make Connections

COMMUNICATING ABOUT ALGEBRA

Discuss a possible relationship between the sign of a in $y = ax^2 + bx + c$ and the sign of the values in Difference Column 2. Use additional equations to test your hypothesis.

11. Work in a group of four students. Assign one of the following four quadratic equations to each member of your group, and complete a table for each equation similar to the table in Question 7. Use consecutive integer values of x from −4 to 5 in the x-column.

a. $y_1 = x^2 - x - 6$ **b.** $y_2 = x^2 - 2x - 3$

c. $y_3 = 3x^2 + x + 2$ **d.** $y_4 = -2x^2 + 4$

12. Compare your tables in Question 11. Do the values of y increase or decrease as x increases?

13. Based upon the results you obtained for Questions 7 and 11, draw conclusions about the behavior of the numbers in Difference Column 1 and in Difference Column 2.

14. Work with a partner. Use your graphing utility to graph each of the following equations on the same set of axes. Then draw a conclusion about the effect on the graph of $y = x^2$ of adding or subtracting a constant to the right side of the equation.

$$y_1 = x^2 \qquad\qquad y_3 = x^2 + 6$$
$$y_2 = x^2 + 3 \qquad\qquad y_4 = x^2 - 4$$

COMMUNICATING
ABOUT ALGEBRA

15. Use a graphing utility to graph a quadratic equation of the form $y = x^2 + c$. Have your partner identify the equation you entered by studying the graph. Then switch roles with your partner.

16. Use your graphing utility to graph each of the following sets of equations on the same coordinate plane. Then draw a conclusion about the effect on the graph $y = x^2$ of multiplying x^2 by a constant.

Set I: $y_1 = x^2$ Set II: $y_1 = x^2$
$\qquad y_2 = 3x^2 \qquad\qquad y_2 = -x^2$
$\qquad y_3 = 5x^2 \qquad\qquad y_3 = -3x^2$
$\qquad y_4 = \dfrac{1}{2}x^2 \qquad\qquad y_4 = -\dfrac{1}{2}x^2$

The constants and coefficients in equations are called *parameters* of the equations. Look in a dictionary to compare the various meanings of *parameter*. Discuss the similarities and differences in meaning you find. Explain how the presence of parameters in a quadratic equation can change the graph of the function.

17. Use a graphing utility to graph a quadratic equation of the form $y = ax^2$ where $a > 0$. Have your partner identify the equation you entered by studying the graph. Then switch roles with your partner.

Summarize

18. **WRITING MATHEMATICS** Explain the similarities and differences between linear functions and quadratic functions. Create examples to illustrate your points.

19. **MODELING** Graph $y = (x - 3)^2 - 5$. Imagine the curve as the track of a roller coaster. At what point does the track reach its lowest point?

20. **THINKING CRITICALLY** Will the graphs of $y_1 = x^2 + 3x + 4$ and $y_2 = x^2 + 3x + 6$ ever intersect? Explain.

21. **GOING FURTHER** Use your graphing utility to graph each of the following equations on the same set of axes. Then draw a conclusion about the effect on the graph of $y = x^2$ of adding a constant to x before squaring the quantity.

$$y_1 = x^2 \qquad\qquad y_3 = (x + 3)^2$$
$$y_2 = (x - 2)^2 \qquad\qquad y_4 = \left(x - \dfrac{1}{2}\right)^2$$

10.1   **Algebra Workshop: Explore Quadratic Functions**   **467**

Graphs of Quadratic Functions

Explore

1. Graph $y = x^2 - 6x + 8$ on a graphing utility. Find the vertex (highest or lowest point) of the equation.

2. Would you expect the graphs of
$$y = 2(x^2 - 6x + 8) = 2x^2 - 12x + 16 \quad \text{and}$$
$$y = 3(x^2 - 6x + 8) = 3x^2 - 18x + 24$$
to have the same vertex as the graph of $y = x^2 - 6x + 8$? Why?

3. Would you expect the vertices of the graphs of
$$y = 2x^2 - 12x + 16 \quad \text{and}$$
$$y = 3x^2 - 18x + 24$$
to have the same x-coordinate as the vertex of the graph of $y = x^2 - 6x + 8$? Why?

4. Verify your answers to Questions 2 and 3 using your graphing utility. Find the vertices of the graphs of
$$y = 2x^2 - 12x + 16 \quad \text{and}$$
$$y = 3x^2 - 18x + 24$$

5. Without graphing, predict the vertices of the graphs of
$$y = 4(x^2 - 6x + 8) = 4x^2 - 24x + 32 \quad \text{and}$$
$$y = -(x^2 - 6x + 8) = -x^2 + 6x - 8$$

Build Understanding

A **quadratic function** is a function of the form $f(x) = ax^2 + bx + c$ where a, b, and c are real numbers and $a \neq 0$. The corresponding quadratic equation is $y = ax^2 + bx + c$.

The graph of a quadratic equation is U-shaped and is called a **parabola**. If the leading coefficient a is positive, the parabola opens upward. If the leading coefficient a is negative, the parabola opens downward. The point where a parabola has its **maximum** (highest) point or **minimum** (lowest) point is called the **vertex** of the parabola.

EXAMPLE 1

Use a table of values for each equation to graph $y = x^2$ and $y = -x^2$.

Solution

Make a table of ordered pairs for each equation as shown at the top of page 469. Graph each point. Then connect the points with a smooth curve. Both graphs have vertices at $(0, 0)$. The graph of $y = x^2$ opens upward, so the vertex is the minimum point. The graph of $y = -x^2$ opens downward, so the vertex is the maximum point.

x	y
-3	9
-2	4
-1	1
0	0
1	1
2	4
3	9

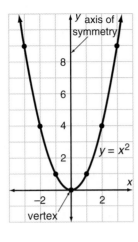

x	y
-3	-9
-2	-4
-1	-1
0	0
1	-1
2	-4
3	-9

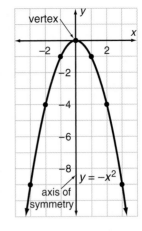

If you fold either graph along the y-axis, $x = 0$, one side fits exactly over the other. The graph is said to have **line symmetry**. For these two graphs the y-axis is the **axis of symmetry**. The vertex of a parabola is the point where the axis of symmetry intersects the graph. Each point on the graph has a corresponding **point of reflection** across the axis of symmetry. On the graph of $y = x^2$, the reflection of $(-3, 9)$ is $(3, 9)$.

When a quadratic equation is in the form $y = ax^2 + bx + c$, the x-value of the vertex is $x = \dfrac{-b}{2a}$. To find the y-value of the vertex, substitute the x-value into the original equation and solve for y. Use the x-value of the vertex to write the equation of the line of symmetry, $x = \dfrac{-b}{2a}$.

CHECK UNDERSTANDING

How do the graphs of $y = x^2$ and $y = -x^2$ appear to be related?

EXAMPLE 2

For the equation $y = 3x^2 - 24x + 40$,
a. Determine the coordinates of the vertex of the graph.
b. Write the equation for the axis of symmetry.
c. Determine whether the vertex is a maximum or a minimum.

CHECK UNDERSTANDING

Name three other points and their reflections that are on the graph of $y = x^2$.

Solution

a. For $y = 3x^2 - 24x + 40$, $a = 3$ and $b = -24$.

$$x = \frac{-b}{2a} \qquad \text{Equation of the line of symmetry.}$$

$$x = \frac{-(-24)}{2(3)} = 4 \qquad \text{Substitute and solve.}$$

Substitute 4 for x in the original equation and solve for y.

$$y = 3x^2 - 24x + 40$$
$$= 3(4)^2 - 24(4) + 40 = -8$$

The vertex is the point $(4, -8)$.

b. Write the equation of the line of symmetry, $x = 4$.

c. Since $a = 3$, the graph opens upward. The vertex is a minimum. ◄

COMMUNICATING ABOUT ALGEBRA

Study some of the graphs you drew earlier in the lesson. How is the x-coordinate of the vertex of a parabola related to the x-coordinates where the graph crosses the x-axis?

10.2   **Graphs of Quadratic Functions**   **469**

Recall that graphs of quadratic equations shift up or down, right or left, and are wide or narrow, depending on the equation.

EXAMPLE 3

Graph on the same pair of axes and compare the graphs.

a. $y = x^2$, $y = x^2 + 2$, and $y = x^2 - 3$

b. $y = x^2$, $y = (x + 2)^2$, and $y = (x - 3)^2$

c. $y = x^2$, $y = \frac{1}{2}x^2$, and $y = 2x^2$

Solution

a. The three graphs have the same shape, the same axis of symmetry, and open upward.

Graph of $y = x^2 + 2$
   Vertex: $(0, 2)$
   Axis of symmetry: y-axis, $x = 0$
   Shifted up 2 units.

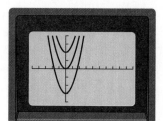

Graph of $y = x^2 - 3$
   Vertex: $(0, -3)$
   Axis of symmetry: y-axis, $x = 0$
   Shifted down 3 units.

b. The three graphs have the same shape and open upward.

Graph of $y = (x + 2)^2$
   Vertex: $(-2, 0)$
   Axis of symmetry: $x = -2$
   Shifted left 2 units.

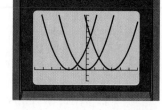

Graph of $y = (x - 3)^2$
   Vertex: $(3, 0)$
   Axis of symmetry: $x = 3$
   Shifted right 3 units.

c. The three graphs open upward, have the same vertex $(0, 0)$, and have the y-axis, $x = 0$, as the axis of symmetry.

Compared to the graph $y = x^2$, the graph of $y = \frac{1}{2}x^2$ is wider and the graph of $y = 2x^2$ is narrower. ◄

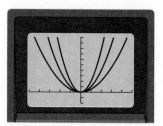

The graph of a quadratic equation of the form $y = ax^2 + bx + c$, in which $a \neq 1$, $b \neq 0$, and $c \neq 0$, exhibits all types of transformations from the graph of $y = x^2$.

EXAMPLE 4

Graph: $y = -2x^2 + 4x + 3$

Solution

Since a is negative, the graph opens downward. The x-coordinate of the vertex is $\dfrac{-4}{2(-2)} = 1$. The y-coordinate is $-2(1)^2 + 4(1) + 3 = 5$.
The vertex is at $(1, 5)$. The axis of symmetry is the line $x = 1$.

x	-2	-1	0	1	2	3	4
y	-13	-3	3	5	3	-3	-13

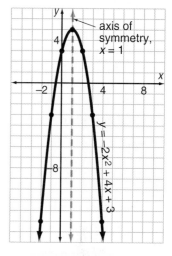

Knowing the coordinates of the vertex of a parabola is often useful in real world applications.

EXAMPLE 5

CIVIL ENGINEERING   Engineers building a highway must find a curve that will make a smooth roadbed connecting point A at an elevation of 1000 ft to point B 1500 ft away. The curve giving the smoothest roadbed is called a *transition curve*. It is a parabola modeled by the equation $y = 0.000038x^2 - 0.045x + 1000$. A storm drain is to be installed at the lowest point on the transition curve. Find the coordinates of the position of the storm drain to the nearest foot.

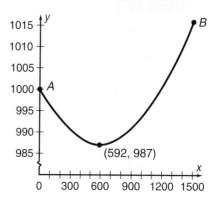

Solution

Find the vertex of $y = 0.000038x^2 - 0.045x + 1000$.
The x-coordinate of the vertex is

$$x = \frac{-b}{2a} = \frac{-(-0.045)}{2(0.000038)} = 592 \qquad \text{To the nearest foot}$$

Substitute the value for x in the equation to find the y-coordinate.

$$y = 0.000038(592)^2 - 0.045(592) + 1000$$
$$= 987 \qquad \text{To the nearest foot}$$

So, the coordinates of the storm drain are $(592, 987)$.

The graph is shown at the right. The y-coordinate of the vertex indicates that the low point of the curve elevation will be $1000 - 987$ or 13 ft below point A. The x-coordinate shows that the low point for the storm drain is 592 ft from point A. Point B is 908 ft from the vertex. Notice the vertex of the parabola is not at the midpoint of the distance A to B. ◄

Determine whether the graph of each equation opens upward or downward.

1. $y = -2x^2 + 10$

2. $y = 0.1x^2 - 6x + 9$

3. $y = 4(x - 3)^2 - 15$

Determine the vertex and axis of symmetry for the graph of each equation.

4. $y = 5x^2$

5. $y = -7x^2$

6. $y = 3x^2 + 4$

7. $y = -3x^2 + 15$

8. $y = 4x^2 - 2x + 15$

9. $y = -3x^2 - 5x - 9$

Tell how each graph differs from the graph of $y = x^2$.

10. $y = x^2 - 7$

11. $y = x^2 + 6$

12. $y = 4x^2$

13. $y = \frac{1}{4}x^2$

14. $y = (x - 5)^2$

15. $y = (x + 8)^2$

For Exercises 16–18, match the equation with its graph.

16. $y = 3x^2$

17. $y = x^2 + 3$

18. $y = (x - 3)^2$

a.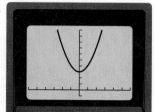

b.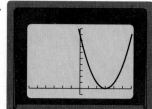

c.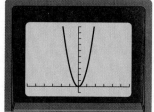

Graph each equation. Show and label the vertex and axis of symmetry for each.

19. $y = x^2 - 2x + 3$

20. $y = \frac{1}{4}x^2 - 3$

21. $y = -2x^2 - 4x - 2$

22. BUSINESS A business supply store found that pens that sell for x dollars have a profit in thousands of dollars modeled by the equation $y = -x^2 + 8x - 4$. What price will give the maximum profit, and what is the profit at that price?

23. WRITING MATHEMATICS Explain why the axis of symmetry of the graph of $y = ax^2 + bx + c$ is either the y-axis or parallel to the y-axis.

PRACTICE

Determine whether the graph of each equation opens upward or downward.

1. $y = 2x^2 + 10$

2. $y = -x^2 + 9x + 15$

3. $y = \frac{1}{2}x^2 + 3x + 12$

Determine the vertex and axis of symmetry for the graph of each equation.

4. $y = -66x^2$

5. $y = 82x^2$

6. $y = 7x^2 + 3$

7. $y = -2x^2 + 14$

8. $y = 3x^2 - 4x - 11$

9. $y = -5x^2 - 10x - 15$

10. $y = \frac{1}{2}x^2 - 6x - 2$

11. $y = -\frac{1}{2}x^2 - 4x + 14$

12. $y = \frac{1}{4}x^2 + 6x - 5$

Tell how each graph differs from the graph of $y = x^2$.

13. $y = x^2 - 10$ **14.** $y = x^2 + 16$ **15.** $y = 3x^2$

16. $y = \frac{1}{3}x^2$ **17.** $y = (x + 25)^2$ **18.** $y = (x - 18)^2$

For Exercises 19–21, match the equation with its graph.

19. $y = 4x^2$ **20.** $y = x^2 - 4$ **21.** $y = (x + 4)^2$

a. 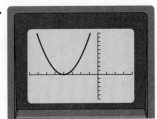   **b.** 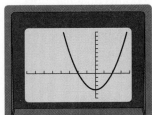   **c.**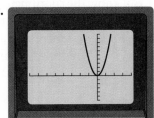

For Exercises 22–24, match the equation with its graph.

22. $y = 2x^2 + 4x - 1$ **23.** $y = -\frac{1}{2}x^2 - 2x + 1$ **24.** $y = -\frac{1}{4}x^2 + 2x + 1$

a. 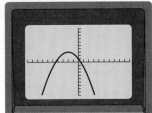   **b.** 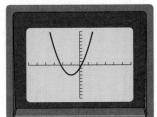   **c.**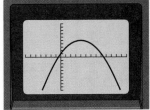

Graph each equation. Show and label the vertex and the axis of symmetry.

25. $y = x^2 + 2x - 1$ **26.** $y = 0.5x^2 - 1.5$ **27.** $y = -2x^2 + 8x - 8$

28. URBAN PLANNING The city council is considering lowering the city bus fare. They determined that the daily revenue from people riding the bus can be represented by $R(x) = -50x^2 + 200x + 19,800$ where x represents the number of $0.05 decreases off the current fare of $1.10. What fare will yield the highest revenue?

29. CIVIL ENGINEERING Engineers are designing a transition curve for a stretch of highway on a mountain from point A with an elevation of 100 ft to point B 1000 ft away. They want a lookout station at the highest point (summit) on the curve. The curve is modeled by

$$y = -0.00004x^2 + 0.025x + 100$$

Determine the coordinates of the summit to the nearest hundredth. Explain what the coordinates mean.

30. WRITING MATHEMATICS Ramiro says that if a parabola that opens upward and a parabola that opens downward are drawn on the same axes, they must intersect twice. Sultan says that the two parabolas do not have to intersect. Ariel says that they intersect 3 or 4 times, and Joelle says that they intersect only one time. Determine the different number of ways in which these parabolas can intersect. Provide examples and draw graphs.

EXTEND

If you know the coordinates of three points on a parabola with an equation of the form $y = ax^2 + bx + c$, you can solve for the values of a, b, and c. For example, if the points $(0, 5)$, $(1, 12)$, and $(-1, 4)$ are solutions, then

I. $a(0)^2 + b(0) + c = 5$
$$c = 5$$

II. $a(1)^2 + b(1) + c = 12$
$$a + b + c = 12$$

III. $a(-1)^2 + b(-1) + c = 4$
$$a - b + c = 4$$

Substitute 5 for c in Equations II and III and then add them.

$$
\begin{array}{ll}
a + b + 5 = 12 & \text{Equation II} \\
\underline{a - b + 5 = 4} & \text{Equation III} \\
2a + 10 = 16 & \\
2a = 6 & \\
a = 3 & \text{Solve for } a.
\end{array}
$$

Substitute $a = 3$ and $c = 5$ into Equation II: $3 + b + 5 = 12$. So, $b = 4$. Check these values in Equation III. The quadratic equation that has points $(0, 5)$, $(1, 12)$, and $(-1, 4)$ as solutions is $y = 3x^2 + 4x + 5$.

Determine an equation of the form $y = ax^2 + bx + c$, given the points that satisfy the equation.

31. $(0, 4), (1, 5), (-1, 9)$ **32.** $(0, 2), (1, -4), (-1, 12)$

33. $(-1, 1), (4, 26), (0, -2)$ **34.** $(4, 12), (-2, 9), (0, 6)$

PHYSICS Galileo found that the height h of an object t seconds after being released can be modeled by the equation $h = \frac{1}{2}at^2 + vt + s$ where a is the acceleration due to gravity, v is the upward speed of the object upon release, and s is the starting height of the object. On earth, a is approximately -32 ft/s² in customary units or -9.8 m/s² in metric units.

35. Ricky threw a ball straight up with a speed of 25 ft/s. His hand was 6 ft above the ground when he released the ball. Write the equation that represents the height of the ball in the air as a function of time.

36. Graph the equation you found in Exercise 35. What is the maximum height to the nearest hundredth of a foot that the ball attains?

37. CREDIT CARD TRENDS During each of the years 1990–1994, bank credit card offers can be approximated by the quadratic equation $C = 104.143x^2 - 19,026.286x + 869,356.657$ where C represents millions of bank credit card offers for the year and x represents the last two digits of the year. Use the equation to estimate the number of bank card offers made in 1994.

38. TECHNOLOGY Use a graphing utility to determine to the nearest hundredth the coordinates of the points at which the graphs of $y = 2x^2 - 7$ and $y = -2x^2 + 4$ intersect.

39. TECHNOLOGY Use a graphing utility to determine to the nearest hundredth the coordinates of the points at which the graphs of $y = x^2 - 5$ and $y = -3x^2 + 16$ intersect.

THINK CRITICALLY

40. Write an equation of the form $y = ax^2$ whose graph will intersect the graph of $y = -2x^2$ in exactly one point.

41. Find the vertex of the graph of $y = ax^2 + c$, $a \neq 0$.

42. Compare the effect that h has on the graph of the quadratic equation $y = (x + h)^2$ with the effect that b has on the graph of the absolute value equation $y = |x + b|$.

43. Compare the effect that c has on the graph of the quadratic equation $y = ax^2 + c$ with the effect that c has on the graph of the absolute value equation $y = a|x| + c$.

44. If the graph of $y_1 = ux^2 - 5$ is narrower than the graph of $y_2 = vx^2 - 5$, is $u > v$ or $u < v$?

45. Explain why the graphs of all equations of the form $y = ax^2$ have the same vertex and axis of symmetry.

MIXED REVIEW

Solve each equation.

46. $3x - 7 = 5x + 20$

47. $\frac{1}{2}(6x - 4) = 4(2x + 5)$

48. STANDARDIZED TESTS Choose the correct slope and y-intercept of the line

$3y - 4 = 6x$

   A. $6; -4$
   B. $-2; \frac{4}{3}$
   C. $2; \frac{4}{3}$
   D. $6; 4$

Compare the graph of each equation to the graph of $y = |x|$.

49. $y = 2|x - 1| + 3$

50. $y = \frac{1}{2}|x + 3| - 1$

Determine the vertex and the equation for the line of symmetry for the graph of each equation.

51. $y = 2x^2 - 8x - 17$

52. $y = -4x^2 + 12x - 22$

Career
Pizza Parlor Owner

Small businesses such as local butchers, hardware stores, shoe stores, and bakeries face tremendous competition from malls, department stores, discount stores, and warehouse stores. National chains are forcing many small downtown stores out of business.

The traditional "pizza parlor" faces competition from national chains and pizza delivery businesses. In the true spirit of competition, some pizza parlor owners offer better value than their competition.

The area of a pizza can be represented by $A = \pi r^2$, a quadratic equation in which r represents the radius of the pizza.

Decision Making

1. Pizzas are often priced according to diameter. Determine the area of a pizza with an 18-in. diameter to the nearest square inch. Use $\pi = 3.14$.

2. Express the area of the pizza as a function of its radius. Use a graphing utility to draw the graph of the resulting equation and describe it.

3. To promote the business, Rossi's pizza parlor owner, who regularly sells pies with 16-in. diameters, is now offering 20-in. pies for the same price. Determine the difference in the areas of the pizzas to the nearest square inch.

4. Best Pizza also sells 16-in. pies for the same price as Rossi's. It is now offering a free 10-in. pie with the purchase of a regular 16-in. pie. Their advertisements claim, "We give more—10 inches of free pizza to Rossi's 4 inches of free pizza!" Find the area of the free 10-in. pizza to the nearest square inch.

5. Is Best Pizza making an accurate claim? Explain why or why not.

6. Use the graph of $A = \pi r^2$ to show that adding 2 in. to the radius of a 14-in.-diameter pie does not add the same amount of pizza as adding 2 in. to the radius of an 18-in.-diameter pie.

Explore/Working Together

1. Guess and check until you find as many solutions as you can for each equation. (*Hint:* One equation has only one solution, one equation has no solutions, and two have two solutions.)

   a. $x^2 - 4x = 0$

   b. $x^2 - 25 = 0$

   c. $x^2 - 2x + 1 = 0$

   d. $x^2 + 1 = 0$

2. Use a graphing utility to graph $y = x^2 - 4x$. How many times does the graph cross the x-axis? Use the ZOOM and TRACE features to determine where the graph crosses the x-axis. How do these points compare to the answer for Question 1a?

3. Use a graphing utility to graph $y = x^2 - 25$. How many times does the graph cross the x-axis? Use the ZOOM and TRACE features to determine where the graph crosses the x-axis. How does this relate to the answer for Question 1b?

4. Use a graphing utility to graph $y = x^2 - 2x + 1$. How does this graph relate to the answer for Question 1c?

5. Use a graphing utility to graph $y = x^2 + 1$. How does this graph relate to the answer for Question 1c?

Build Understanding

You can use graphing to solve equations of the form $d = ax^2 + bx + c$. When $d = 0$, the equation is in **standard form**. The solutions, also called the **roots**, of the equation $0 = ax^2 + bx + c$ are the values of x where the graph of $y = ax^2 + bx + c$ crosses the x-axis ($y = 0$). These values are the **x-intercepts** of the graph.

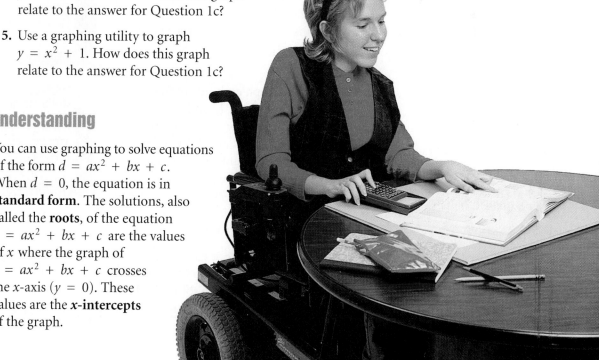

EXAMPLE 1

Solve $0 = x^2 - 2x - 8$ by finding the x-intercepts of its graph.

Solution

Graph the equation $y = x^2 - 2x - 8$
on a graphing utility. Use the ZOOM and
TRACE features to determine the
x-intercepts of the graph. The x-intercepts
are -2 and 4. So, the solutions are -2 and 4.

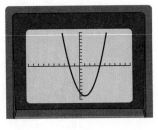

Check

Substitute the solutions into the original equation.

$$0 = x^2 - 2x - 8$$
$$0 \overset{?}{=} (-2)^2 - 2(-2) - 8$$
$$0 = 0 \checkmark$$

$$0 = x^2 - 2x - 8$$
$$0 \overset{?}{=} 4^2 - 2(4) - 8$$
$$0 = 0 \checkmark$$

◀

When $d \neq 0$, you can use your graphing utility to solve equations of
the form $d = ax^2 + bx + c$ using either of two methods.

EXAMPLE 2

Solve the equation $12 = x^2 - 4x + 15$ using methods A and B below.

Solution

Method A Rewrite $12 = x^2 - 4x + 15$
as $0 = x^2 - 4x + 3$. Graph the equation
$y = x^2 - 4x + 3$ on your graphing utility.
Use the ZOOM and TRACE features to
determine the x-intercepts of the graph.
The x-intercepts are 1 and 3. So, the
solutions are 1 and 3.

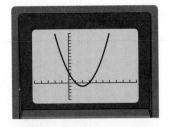

Method B Graph each side of the
equation. Graph $y = 12$ and
$y = x^2 - 4x + 15$ on the same set of
axes. Use the ZOOM and TRACE features
to determine the points of intersection of
the two equations. The points of
intersection are $(1, 12)$ and $(3, 12)$, so the
solutions are 1 and 3.

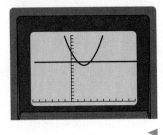

◀

CHECK UNDERSTANDING

In Example 2, what
equations would
you graph to solve
the equation
$0 = x^2 - 4x + 3$
using Method B?

As you saw in Explore, a quadratic equation may have one, two,
or no real solutions. If its graph crosses the x-axis at two points,
then the equation has two solutions. If its graph touches the x-axis
at only one point, then the equation has one solution. If its graph
neither touches nor crosses the x-axis at any point, then the
equation has no real solutions.

EXAMPLE 3

Graph each quadratic equation to decide how many solutions it has.

a. $0 = x^2 + 2x + 1$ **b.** $0 = x^2 + 2x - 3$ **c.** $0 = x^2 + 2x + 3$

Solution

a. one solution   **b.** two solutions   **c.** no solutions

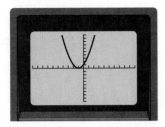 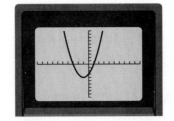 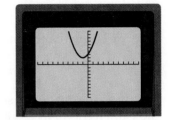

Recall that the height h of an object t seconds after being released is modeled by the equation $h = \frac{1}{2}at^2 + vt + s$ where a is the acceleration due to gravity, v is the upward speed of the object upon release, and s is the initial height. On Earth, a is approximately -32 ft/s^2.

EXAMPLE 4

PHYSICS A toy rocket is launched from ground level with an upward speed of 120 ft/s.

a. Write an equation that represents the height of the rocket as a function of time.

b. After how many seconds will the rocket be 150 ft above ground?

Solution

a. In this example, $a = -32$ ft/s, $v = 120$ ft/s, and $s = 0$.

$$h = \frac{1}{2}(-32)t^2 + 120t$$

$$h = -16t^2 + 120t$$

b. The equation $150 = -16t^2 + 120t$ represents the time at which the rocket is 150 ft above the ground. Rewrite the equation as $0 = -16t^2 + 120t - 150$, then graph it on your graphing utility. Use the TRACE and ZOOM features to determine the x-intercepts of the graph. The x-intercepts are 1.58 and 5.92 to the nearest hundredth.

x-scale: 1 y-scale: 10

The rocket is at a height of 150 ft on its way up at $t = 1.58$ s and again on its way down at $t = 5.92$ s.

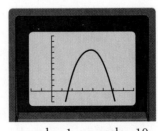

CHECK UNDERSTANDING

Can the vertex ever be the solution to the equation? Explain.

PROBLEM SOLVING TIP

To help you decide what viewing window (range) to use, first determine the coordinates of the vertex. Decide if the graph opens upward or downward. Then choose your window to include the vertex and surrounding points of the graph.

Rewrite each equation in the form $0 = ax^2 + bx + c$.

1. $7 = 3x^2 - 4x + 15$ **2.** $4 = 8 - 2x^2 + 5x$ **3.** $2 = 2 - 3x + \frac{1}{2}x^2$

Determine whether each equation is equivalent to $8 = 4x^2 - 6x - 2$.

4. $-8 = -4x^2 + 6x - 2$ **5.** $0 = 4x^2 - 6x - 10$ **6.** $10 = 4x^2 - 6x$

Solve each equation by finding the x-intercepts of its graph.

7. $0 = x^2 + 2x - 15$ **8.** $0 = -x^2 + 3x + 4$ **9.** $0 = x^2 - 10x + 25$

Rewrite each equation in standard form. Then solve each equation by finding the x-intercepts of its graph. Round decimals to the nearest hundredth.

10. $6 = x^2 + 4x - 6$ **11.** $8 = -x^2 + 6x + 1$ **12.** $-25 = -x^2 + x - 5$

Solve each equation by graphing $y = d$ and $y = ax^2 + bx + c$ on the same set of axes and then finding the x-coordinates of their points of intersection. Round decimals to the nearest hundredth.

13. $12 = x^2 - 2x - 3$ **14.** $-5 = -x^2 + 2x + 1$ **15.** $1 = 2x^2 - 2x - 8$

Use a graphing utility to determine the number of solutions to each equation.

16. $0 = 3x^2 - 6x + 2$ **17.** $0 = 4x^2 + 5x + 2$ **18.** $0 = x^2 - 8x + 16$

19. PHYSICS If the rocket discussed in Example 4 is launched from the moon, where the acceleration due to gravity is -5.32 ft/s^2, after how many seconds will it be at a height of 150 ft? Round to the nearest hundredth.

20. WRITING MATHEMATICS Explain how to use a graphing utility to find the solution(s) to a quadratic equation in standard form.

PRACTICE

Rewrite each equation in the form $0 = ax^2 + bx + c$.

1. $9 = 2x^2 - 5x + 12$ **2.** $3 = 6 - x^2 + 3x$ **3.** $5 = 5 - 5x + \frac{1}{4}x^2$

Determine whether each equation is equivalent to $5 = 3x^2 - 4x - 4$.

4. $9 = 3x^2 - 4x$ **5.** $0 = 3x^2 - 4x - 9$ **6.** $-5 = 3x^2 - 4x - 14$

Solve each equation by finding the x-intercepts of its graph. Round decimals to the nearest hundredth.

7. $0 = x^2 - 2x - 24$ **8.** $0 = x^2 - x - 30$ **9.** $0 = -x^2 - 3x + 10$

10. $0 = -x^2 + 4x + 12$ **11.** $0 = x^2 - 12x + 36$ **12.** $0 = x^2 + 6x + 9$

13. $0 = 2x^2 - 7x + 2$ **14.** $0 = 3x^2 - 8x - 2$ **15.** $0 = -4x^2 + 2x + 4$

Rewrite each equation in standard form. Then solve each equation by finding the x-intercepts of its graph. Round decimals to the nearest hundredth.

16. $4 = x^2 + 2x - 4$

17. $48 = x^2 + 8x - 17$

18. $9 = x^2 - 18x - 31$

19. $-10 = -x^2 + 2x - 6$

20. $2 = 3x^2 + 4x - 1$

21. $5 = 4x^2 - 2x + 1$

Solve each equation by graphing $y = d$ and $y = ax^2 + bx + c$ on the same set of axes and then finding the x-coordinates of their points of intersection. Round decimals to the nearest hundredth.

22. $5 = x^2 - 4x$

23. $-6 = -x^2 + 3x + 6$

24. $1 = 3x^2 - x - 4$

Use a graphing utility to determine the number of solutions to each equation.

25. $0 = 4x^2 - 12x + 9$

26. $0 = 3x^2 + x + 4$

27. $0 = x^2 - 8x + 14$

28. $0 = 4x^2 - x + 5$

29. $0 = x^2 + 5x - 3$

30. $0 = 9x^2 + 6x + 1$

31. BUSINESS TECHNOLOGY On January 17, 1995, the Prodigy on-line information service became the first consumer computer network to open its electronic portals to the Internet service known as the World Wide Web. The number (in trillions) of packets (bundles of data) traversing a major Internet conduit between May 1993 and December 1994 can be modeled with the equation $y = 0.052x^2 - 0.551x + 1.7$ where x is the month ($x = 1$ corresponds to May 1993, $x = 2$ corresponds to June 1993, . . . , and $x = 20$ corresponds to December 1994). In which month was the number of packets approximately 2 trillion?

32. PHYSICS On Mars the acceleration due to gravity is -12.2 ft/s². A ball is thrown from a point 5 ft above the ground straight upward at a speed of 20 ft/s. Use the formula for Example 4 to write the equation that represents the height of the ball as a function of time. Determine after how many seconds the ball will be 10 ft off the ground.

33. WRITING MATHEMATICS Describe the graph of $y = ax^2 + bx + c$ if the quadratic equation $0 = ax^2 + bx + c$ has no solutions.

EXTEND

TRAVEL PLEASE GO AWAY! travel agency offers a vacation package to Europe at a discounted rate. They have calculated that their per-person profit on the purchase can be modeled by the equation $y = -x^2 + 60x$ where x is the number of people who buy the package.

34. If their profit was $371 per person, how many people went to Europe?

35. If their profit was $704 per person, how many people went to Europe?

36. Determine the number of travelers that will provide the agency with the maximum profit per person. What is the maximum profit per person?

37. If more than a certain number of people accept this package, the agency will lose money. What is the maximum number of people that can use this package before the agency begins to lose money?

THINK CRITICALLY

38. Find a value of c such that the equation $y = x^2 - 4x + c$ will have only one solution.

39. Find two values of c such that the equation $y = x^2 - 4x + c$ will have no real solutions.

40. Write a quadratic equation that has two real solutions.

41. Write a quadratic equation that has two real solutions, one of which is a solution to the equation you wrote in Problem 43.

42. Graph to determine the solutions to the equation $x^2 - 6x + 5 = 0$. How do the solutions relate to the axis of symmetry?

MIXED REVIEW

43. STANDARDIZED TESTS Which is the slope-intercept form of the equation $9x + 3y = 12$?

A. $y = 3x + 4$ B. $y = 3x - 4$
C. $y = -3x + 4$ D. $y = 27x - 36$

Determine the number of solutions each system of equations has.

44. $\begin{cases} 2x + 3y = 6 \\ -12 + 4x = -6y \end{cases}$

45. $\begin{cases} 5x + 4y = 13 \\ 4x + 6y = 16 \end{cases}$

Demonstrate that each of the following numbers can be written as the quotient of two integers.

46. -8 **47.** 5.33 **48.** $-9\frac{1}{2}$ **49.** 0.095

Rewrite each equation in standard form. Solve each equation by finding the x-intercepts of its graph.

50. $-6 = x^2 + 5x$ **51.** $-2 = -x^2 + 3x + 1$

PROJECT *Connection*   In this activity, you will begin researching the industry and corporation you selected.

1. Obtain the address and/or telephone number of your corporation.

2. Contact each corporation immediately, by business letter or telephone. Explain the purpose of your project and request a copy of the Annual Report and any other available information (some companies have videos). Have the material sent to your home address. Send a thank you letter when you receive the information.

3. Research and summarize information on stocks and stock markets. You will need this information for the graph you will construct in the next Project Connection. Prepare a set of six to ten "attention-getting" facts using very large or very small numbers as they relate to market activity (for example, trading volume, total worth of corporations, amount of time for an electronic transaction). Consider using scientific notation for your data.

Solve Quadratic Equations Using Square Roots

Explore/Working Together

1. If you are the first person chosen by your teacher, select three cards from the deck in your teacher's hand. Assign the number on the first card to be a, the number on the second card to be b, and the number on the third card to be c in the equation $ax^2 + b = c$.

2. Write the resulting equation on the chalkboard.

3. Select a student to use mental math in solving the equation.

4. After that student provides you with an answer, solve the equation algebraically on the board.

5. Discuss whether the equation has a real number solution and, if so, discuss whether the solution is rational or irrational.

6. If the student you selected has the correct answer, repeat Steps 1–5. If that student does not have the correct answer, your teacher will select a new student to pick three cards.

7. After the class finishes the game, write the conditions under which the equation $ax^2 + b = c$ has no real solutions.

Build Understanding

Recall that every positive real number k has one positive and one negative square root. This fact is stated in the *square root property*.

> **SQUARE ROOT PROPERTY**
>
> If $x^2 = k$, then $x = \sqrt{k}$ or $x = -\sqrt{k}$ for any real number k, $k > 0$.
>
> If $k = 0$, then the equation $x^2 = 0$ has one solution, which is 0.

CHECK UNDERSTANDING

The square root property states that k in $\sqrt{k}$ must be greater than or equal to zero. Why is $\sqrt{k}$ not defined for $k < 0$?

To solve a quadratic equation of the form $ax^2 + b = c$, you can use the square root property. First solve for x^2 and then take the square root of both sides. This technique is called the **square root method** of solving quadratic equations.

Since $x^2 = 4$, $x = 2$ or $x = -2$. To indicate this you can write $x = \pm 2$. Therefore, when taking the square root of both sides of an equation, write $\pm\sqrt{}$ to indicate both the positive square root and the negative square root.

Recall the product and quotient properties of square roots.

For real numbers $a \geq 0$ and $b \geq 0$,

$$\sqrt{ab} = \sqrt{a} \cdot \sqrt{b}$$

For real numbers $a \geq 0$ and $b > 0$,

$$\sqrt{\dfrac{a}{b}} = \dfrac{\sqrt{a}}{\sqrt{b}}$$

PROBLEM SOLVING TIP

In Example 2, think of the expression $2x - 1$ as a "chunk." Then let $p = 2x - 1$. The equation becomes $4p^2 = 48$. After you find $p = \pm\sqrt{12}$, replace p with $2x - 1$ and solve the two equations $2x - 1 = \sqrt{12}$ or $2x - 1 = -\sqrt{12}$.

COMMUNICATING ABOUT ALGEBRA

Do all equations of the form $ax^2 + b = c$ where $c < 0$ have no real solutions?

EXAMPLE 1

Solve: $2x^2 + 5 = 167$

Solution

$$
\begin{aligned}
2x^2 + 5 &= 167 \\
2x^2 &= 162 \qquad &\text{Subtract 5 from both sides.} \\
x^2 &= 81 \qquad &\text{Divide both sides by 2.} \\
x &= \pm\sqrt{81} \qquad &\text{Take the square root of both sides.} \\
x &= \pm9 \qquad &\text{Simplify.}
\end{aligned}
$$

Check

Substitute both solutions in the original equation.

$$
\begin{aligned}
2x^2 + 5 &= 167 \\
2(9)^2 + 5 &\stackrel{?}{=} 167 \\
2(81) + 5 &\stackrel{?}{=} 167 \\
162 + 5 &\stackrel{?}{=} 167 \\
167 &= 167 \ \checkmark
\end{aligned}
\qquad\qquad
\begin{aligned}
2x^2 + 5 &= 167 \\
2(-9)^2 + 5 &\stackrel{?}{=} 167 \\
2(81) + 5 &\stackrel{?}{=} 167 \\
162 + 5 &\stackrel{?}{=} 167 \\
167 &= 167 \ \checkmark
\end{aligned}
$$

Therefore, the solutions are 9 and –9. ◀

You can also use the square root method to solve quadratic equations when an expression such as $(2x - 1)$ is squared.

EXAMPLE 2

Solve: $4(2x - 1)^2 = 48$

Solution

$$
\begin{aligned}
4(2x - 1)^2 &= 48 \\
(2x - 1)^2 &= 12 \qquad &\text{Divide both sides by 4.} \\
2x - 1 &= \pm\sqrt{12} \qquad &\text{Take the square root of both sides.}
\end{aligned}
$$

$$2x - 1 = 2\sqrt{3} \qquad \text{or} \qquad 2x - 1 = -2\sqrt{3}$$

$$x = \dfrac{1 + 2\sqrt{3}}{2} \qquad\qquad\qquad x = \dfrac{1 - 2\sqrt{3}}{2}$$

Check

Use your calculator or computer to determine a decimal approximation for each solution. Store each result in a memory.

$$(1 + 2\sqrt{3}) \div 2 \approx 2.232050808 \qquad \text{Store in memory A.}$$
$$(1 - 2\sqrt{3}) \div 2 \approx -1.232050808 \qquad \text{Store in memory B.}$$

Substitute each stored value in the original equation.

$$
\begin{aligned}
4(2x - 1)^2 &= 48 \\
4(2A - 1)^2 &\stackrel{?}{=} 48 \\
48 &= 48 \ \checkmark
\end{aligned}
\qquad\qquad
\begin{aligned}
4(2x - 1)^2 &= 48 \\
4(2B - 1)^2 &\stackrel{?}{=} 48 \\
48 &= 48 \ \checkmark
\end{aligned}
$$

The solutions to the nearest hundredth are 2.23 and –1.23. Note that if you round the solution before you store it in memory, you may only get a number close to 48. ◀

You can use the square root method to determine the time it takes for an object in free fall to descend a specific distance. Galileo found that the formula that models an object in free fall is $d = \frac{1}{2}at^2$ where d is the distance that the object has fallen, a is the acceleration due to gravity, and t is the elapsed time. The value of a varies depending on the location on Earth's surface. In the United States, the value of a is approximately -9.8 m/s^2.

EXAMPLE 3

PHYSICS A window washer accidentally dropped his watch from a top floor window 60 m above the ground. The acceleration due to gravity is approximately -9.8 m/s^2. In how many seconds will the watch hit the ground?

Solution

The watch falls 60 m downward, which can be represented by -60 m.

$$d = \frac{1}{2}at^2$$

$$-60 = \frac{1}{2}(-9.8)t^2 \quad \text{Substitute.}$$

$$-60 = -4.9t^2 \quad \text{Multiply.}$$

$$\frac{-60}{-4.9} = t^2 \quad \text{Solve for } t^2.$$

$$\pm\sqrt{\frac{-60}{-4.9}} = t \quad \text{Take the square root.}$$

$$\pm 3.50 \approx t$$

The watch will hit the ground in about 3.50 s.

TRY THESE

Solve each equation. Use the product and quotient properties to simplify the radicals.

1. $x^2 = 81$

2. $x^2 = 100$

3. $x^2 + 2 = 50$

4. $x^2 - 3 = 72$

5. $\frac{1}{2}x^2 + 3 = 23$

6. $\frac{1}{2}x^2 - 8 = 44$

7. $(x - 1)^2 = 25$

8. $(x - 1)^2 = 64$

9. $2(x + 1)^2 = 162$

10. $3(x - 4)^2 = 108$

11. $3\left(x + \frac{2}{3}\right)^2 = \frac{1}{3}$

12. $2\left(x + \frac{3}{4}\right)^2 = \frac{1}{8}$

Solve each equation. Round your answers to two decimal places.

13. $x^2 - 5 = -6$

14. $x^2 - 10 = -10$

15. $4x^2 - 3 = 18$

16. $2x^2 - 7 = 13$

17. $3x^2 - 4 = 16$

18. $3(x - 3)^2 = 17$

19. $5(x - 2)^2 = 12$

20. $6\left(x - \frac{1}{2}\right)^2 = 3$

21. $5(x + 0.5)^2 = 7.5$

22. **BANKING** The formula for determining the amount of money A that you will have at the end of 2 years if you deposit a principal of P in an account in which the interest rate is $r\%$ compounded annually is $A = P\left(1 + \dfrac{r}{100}\right)^2$. Determine the interest rate paid by this account if you deposit \$400 and it grows to \$441 in 2 years.

23. **WRITING MATHEMATICS** Explain in your own words the conditions under which a quadratic equation of the form $ax^2 + b = c$ will have two real number solutions, one real number solution, and no real number solutions.

PRACTICE

Solve each equation. Express radicals in simplest form.

1. $x^2 = 49$

2. $x^2 = 121$

3. $x^2 = \dfrac{25}{81}$

4. $x^2 - \dfrac{16}{64} = 0$

5. $x^2 = \dfrac{361}{484}$

6. $x^2 + 6 = 0$

7. $x^2 + 3 = 0$

8. $x^2 + 4 = 112$

9. $x^2 + 6 = 86$

10. $4x^2 - 5 = 283$

11. $3x^2 - 4 = 290$

12. $(x - 1)^2 = 64$

13. $(x - 4)^2 = 225$

14. $\dfrac{1}{2}x^2 + 4 = 158$

15. $\dfrac{1}{2}x^2 + 5 = 250$

16. $3(x + 5)^2 = 75$

17. $5(x + 6)^2 = 80$

18. $3(2x + 5)^2 = 300$

19. $2(3x + 4)^2 = 180.5$

20. $2\left(x - \dfrac{5}{6}\right)^2 = \dfrac{1}{18}$

21. $5(x + 0.3)^2 = 4.05$

22. **BANKING** If you invest \$1440 in a savings account in which the interest is compounded annually, and the principal grows to \$1690 in 2 years, what is the interest rate on the account? (*Hint:* Use the formula in Exercise 22 of Try These.)

Solve each equation. Round your answers to two decimal places.

23. $3x^2 - 5 = 16$

24. $4x^2 - 10 = 25$

25. $5x^2 + 9 = 38$

26. $3x^2 + 7 = 70$

27. $x^2 + 8 = 81$

28. $2(x - 3)^2 = 55$

29. $5(x - 6)^2 = 82$

30. $7(x - 0.6)^2 = 3$

31. $5(x - 0.8)^2 = 12$

32. **PHYSICS** Students conducting a physics experiment dropped an egg from a window 222 m above the ground. In how many seconds will the egg reach the ground? (*Hint:* Use the formula given for Example 3.)

33. **WRITING MATHEMATICS** Explain why one square root that is a solution to a quadratic equation in certain real world applications may be meaningless. Provide an example of such a situation.

34. **GEOMETRY** Alejandro is using posterboard to construct a cone for his geometry class. He has determined that the volume V should be 18π cm^3. He is ready to cut the circle for the base of the cone. Determine the radius of the circular base if the height h of the cone is 6 cm. $\left(\textit{Hint: Use the formula } V = \dfrac{1}{3}\pi r^2 h.\right)$

EXTEND

The square root method of solving a quadratic equation can be used to solve quadratic inequalities of the form $ax^2 + b < c$ or $ax^2 + b > c$. The following shows how you can solve a quadratic inequality such as $3x^2 + 7 < 82$.

a. Solve the corresponding quadratic equation.

$$3x^2 + 7 = 82$$
$$3x^2 = 75$$
$$x^2 = 25$$
$$x = \pm 5$$

b. Use the solutions to divide the real number line into three sections. Test a value from each section.

c. Check each of these sections by trying a test value in the section.

- Try $x = -7$ from Section I: $3(-7)^2 + 7 < 82$, which is false.
- Try $x = 0$ from Section II: $3(0)^2 + 7 < 82$, which is true.
- Try $x = 6$ from Section III: $3(6)^2 + 7 < 82$, which is false.

d. The solution is $-5 < x < 5$.

Use the above method to solve each inequality.

35. $x^2 > 144$

36. $x^2 < 225$

37. $2x^2 - 6 > 66$

38. $4x^2 - 8 < 392$

39. $4(x - 2)^2 < 16$

40. $3(x + 1)^2 > 147$

GEOMETRY The surface area S of a sphere can be calculated by using the formula $S = 4\pi r^2$ where r represents the radius of the sphere.

41. If the surface area of a sphere is 100 m^2, determine its radius to the nearest hundredth. Use 3.14 for π.

42. Write an equation for the surface area of an enclosed hemisphere. Then determine its radius to the nearest hundredth if the surface area of an enclosed hemisphere is 100 m^2. Solve using 3.14 for π.

43. GEOMETRY The surface area of a cube can be calculated using the formula $S = 6x^2$ where x is the length of a side. If the surface area of a cube is 350 m^2, determine the length of a side to the nearest hundredth.

THINK CRITICALLY

44. If the radius of a sphere is doubled, by what factor is the surface area multiplied?

45. If the side x of a cube is doubled, by what factor is the surface area multiplied?

46. If the height from which an object goes into free fall is doubled, by what factor is the time for the object to reach the ground multiplied?

47. Derive a formula that allows you to determine the amount of money A that you will have at the end of 4 years if you deposit P in an account in which the interest rate is $r\%$ compounded annually.

48. Derive a formula that allows you to determine the amount of money A that you will have at the end of 2 years if you deposit P in an account in which the interest rate is $r\%$ compounded semi-annually.

49. If you deposited $800 in an account that pays 8.1% simple interest per year for 2 years, will you have more or less money than you would have if you deposited $800 in an account that pays 8% compounded annually for 2 years? Explain.

50. In an equation of the form $ax^2 + b = c$, does c have to be a perfect square to solve the equation by the square root method? Explain.

MIXED REVIEW

51. STANDARDIZED TESTS One number cube having faces labeled 1, 2, 3, 4, 5, and 6 is rolled. What is the probability of rolling a number divisible by 3?

    A. $\dfrac{1}{2}$ **B.** $\dfrac{1}{6}$ **C.** $\dfrac{1}{4}$ **D.** $\dfrac{1}{3}$

Which property of real numbers is demonstrated below?

52. $2 \cdot \dfrac{1}{2} = 1$ **53.** $5(0) = 0$

Determine the vertex and the line of symmetry for the graph of each equation.

54. $y = |x - 5| + 1$ **55.** $y = |x + 4| - 2$

Solve each equation. Round your answer to the nearest hundredth.

56. $3(x + 3)^2 = 18$ **57.** $4(x - 5)^2 = 29$

PROJECT *Connection* In this activity, you will construct a special type of graph that gives information about the daily high, daily low, closing price, and the number of shares traded of a corporation's stock. Examine the sample graph shown.

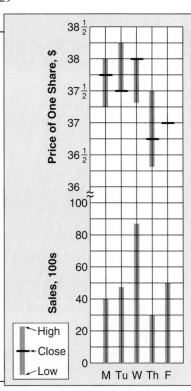

1. On a sheet of paper, draw and label the axes for the *Price of One Share* graph.

2. Label the middle of the vertical axis with today's closing price. This will allow space to graph the price if it rises or falls. Use intervals of $\dfrac{1}{4}$ dollar.

3. On another sheet of graph paper, draw and label the axes for the *Sales* graph. Do not set the range of numbers for the vertical axis. Use a table to record daily sales data. At the end of the project, examine your high and low sales figures and then decide how to label the vertical axis.

4. At the end of the project, combine or redraw the two graphs so that all the information is clearly displayed as shown at the right.

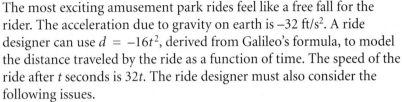

The most exciting amusement park rides feel like a free fall for the rider. The acceleration due to gravity on earth is –32 ft/s². A ride designer can use $d = -16t^2$, derived from Galileo's formula, to model the distance traveled by the ride as a function of time. The speed of the ride after t seconds is $32t$. The ride designer must also consider the following issues.

- How high should the ride be?
- How many seconds should the free fall last?
- What speed is too slow to be exciting?
- What speed is too fast to be safe?
- How much money will rides of different heights cost to build?
- How much distance should be allowed for the safe braking of the ride?

Decision Making

1. If a ride drops 36 ft, how many seconds is the rider in free fall?

2. Use your answer to Question 1 to determine the speed at which the ride is dropping when it has dropped 36 ft.

3. If a ride drops 72 ft, how many seconds is the rider in free fall?

4. Use your answer to Question 3 to determine the speed at which the ride is dropping when it has dropped 72 ft.

A speed of 60 mi/h is equivalent to 1 mi/min. There are 5280 ft in a mile.

5. Convert a speed of 60 mi/h to feet per second.

6. Determine the speed on the ride after 1 second and convert your answer to miles per hour.

7. Determine the speed on the ride after 2.5 seconds and convert your answer to miles per hour.

8. What do you think is a reasonable time for this ride?

10.5 The Quadratic Formula

Explore

1. Copy and complete the chart.

Use a graphing utility to find the solutions to each quadratic equation below.

a	b	c	$\dfrac{-b + \sqrt{b^2 - 4ac}}{2a}$	$\dfrac{-b - \sqrt{b^2 - 4ac}}{2a}$
1	−7	12		
1	5	6		
1	6	5		
1	−1	2		

2. $x^2 - 7x + 12 = 0$

3. $x^2 + 5x + 6 = 0$

4. $x^2 + 6x + 5 = 0$

5. $x^2 + x - 2 = 0$

6. What similarities do you notice between your answers to Question 1 and Questions 2–5?

7. What does your answer to Question 6 suggest about the expressions $\dfrac{-b + \sqrt{b^2 - 4ac}}{2a}$ and $\dfrac{-b - \sqrt{b^2 - 4ac}}{2a}$?

Build Understanding

The two algebraic expressions in Explore are formulas for finding the solutions to any quadratic equation of the form $ax^2 + bx + c = 0$, known as the *quadratic formula*.

QUADRATIC FORMULA

For a quadratic equation of the form $ax^2 + bx + c = 0$ where a, b, and c are real numbers and $a \neq 0$,

$$x = \frac{-b \pm \sqrt{b^2 - 4ac}}{2a}$$

EXAMPLE 1

Use the quadratic formula to solve $x^2 + 6x - 16 = 0$.

Solution

In $x^2 + 6x - 16 = 0$, $a = 1$, $b = 6$, and $c = -16$.

$$x = \frac{-b \pm \sqrt{b^2 - 4ac}}{2a}$$ Write the quadratic formula.

$$x = \frac{-6 \pm \sqrt{6^2 - 4(1)(-16)}}{2(1)}$$ Substitute known values.

$$x = \frac{-6 \pm \sqrt{36 + 64}}{2} \qquad \text{Simplify.}$$

$$x = \frac{-6 \pm \sqrt{100}}{2} \qquad \text{Simplify under the radical symbol.}$$

$$x = \frac{-6 \pm 10}{2} \qquad \text{Determine the value of the radicand.}$$

Check

Check your solutions in the original equation.

$$\begin{aligned} x^2 + 6x - 16 &= 0 \\ 2^2 + 6(2) - 16 &\overset{?}{=} 0 \\ 0 &= 0 \checkmark \end{aligned} \qquad \begin{aligned} x^2 + 6x - 16 &= 0 \\ (-8)^2 + 6(-8) - 16 &\overset{?}{=} 0 \\ 0 &= 0 \checkmark \end{aligned}$$

The solutions are $x = \dfrac{-6 + 10}{2} = 2$ and $x = \dfrac{-6 - 10}{2} = -8$. ◄

You must write a quadratic equation in the standard form $ax^2 + bx + c = 0$ before you apply the quadratic formula.

EXAMPLE 2

Use the quadratic formula to solve $3x^2 = -5 - 10x$.

Solution

Write $3x^2 = -5 - 10x$ in standard form as $3x^2 + 10x + 5 = 0$.
In $3x^2 + 10x + 5 = 0$, $a = 3$, $b = 10$ and $c = 5$.

$$x = \frac{-b \pm \sqrt{b^2 - 4ac}}{2a} \qquad \text{Write the quadratic formula.}$$

$$x = \frac{-10 \pm \sqrt{10^2 - 4(3)(5)}}{2(3)} \qquad \text{Substitute known values.}$$

$$x = \frac{-10 \pm \sqrt{100 - 60}}{6} \qquad \text{Simplify.}$$

$$x = \frac{-10 \pm \sqrt{40}}{6} \qquad \text{Simplify under the radical symbol.}$$

$$x = \frac{-10 \pm 2\sqrt{10}}{6} \qquad \text{Write } \sqrt{40} \text{ as } \sqrt{4} \cdot \sqrt{10} = 2\sqrt{10}.$$

$$x = \frac{-5 \pm \sqrt{10}}{3} \qquad \text{Divide each term by 2.}$$

The solutions are $x = \dfrac{-5 + \sqrt{10}}{3}$ and $x = \dfrac{-5 - \sqrt{10}}{3}$. Check your solutions in the original equation. ◄

In Lesson 10.3, you graphed equations to determine whether a quadratic equation has zero, one, or two real number solutions. Another method is to use $b^2 - 4ac$, the expression under the radical symbol in the quadratic formula, to determine the number of real number solutions. The expression $b^2 - 4ac$ is called the **discriminant**.

For a quadratic equation of the form $ax^2 + bx + c = 0$, the determinant is $b^2 - 4ac$.

- If $b^2 - 4ac > 0$, the equation has two real solutions.
- If $b^2 - 4ac = 0$, the equation has one real solution.
- If $b^2 - 4ac < 0$, the equation has no real solutions.

CHECK UNDERSTANDING

Describe the graphs of a quadratic function whose related equation has each of the following discriminants: greater than zero, less than zero, and equal to zero. Give examples.

EXAMPLE 3

Use the discriminant to determine the number of real solutions for each quadratic equation.

a. $x^2 - 4x + 4 = 0$

b. $x^2 - 4x + 5 = 0$

c. $x^2 - 4x - 2 = 0$

Solution

a. $b^2 - 4ac = (-4)^2 - 4(1)(4) = 16 - 16 = 0$
one real solution

b. $b^2 - 4ac = (-4)^2 - 4(1)(5) = 16 - 20 = -4$
no real solutions

c. $b^2 - 4ac = (-4)^2 - 4(1)(-2) = 16 - (-8) = 24$
two real solutions ◄

You can also use the discriminant to find values for one of the coefficients of a quadratic equation if you know the number of solutions the equation has.

EXAMPLE 4

Determine the values of k for which the equation $4x^2 + kx + 25 = 0$ has exactly one solution.

Solution

Since the equation has exactly one solution, $b^2 - 4ac = 0$ for $a = 4$, $b = k$, and $c = 25$.

$$b^2 - 4ac = 0$$

$$k^2 - 4(4)(25) = 0 \qquad a = 4, b = k, c = 25$$

$$k^2 - 400 = 0 \qquad \text{Solve for } k.$$

$$k^2 = 400$$

$$k = \pm 20$$

The equation has exactly one solution when $k = 20$ or $k = -20$. ◄

The quadratic formula can be used to solve real world problems.

EXAMPLE 5

BUSINESS Rig Outfitters makes equipment for oil companies. The projected revenue on pipe connectors is modeled by the function $R(x) = 20x^2 + 82x + 600$ where x is the number of connectors produced in hundreds. The cost to produce these items can be modeled by the function $C(x) = 4x^2 + 12x + 1296$. Assume that the company sells every connector it produces. The company determined that it made a profit of $300 on these connectors.

a. If profit is defined as revenue minus cost, write an equation that models profit.

b. Solve the equation in **a** to determine the number of connectors that the company sold.

Solution

a. $P(x) = R(x) - C(x)$

$$P(x) = 20x^2 + 82x + 600 - (4x^2 + 12x + 1296)$$

$$= 20x^2 - 4x^2 + 82x - 12x + 600 - 1296$$

$$= (20 - 4)x^2 + (82 - 12)x - 696$$

$$= 16x^2 + 70x - 696$$

Since $P(x) = 300$, $16x^2 + 70x - 696 = 300$ or $16x^2 + 70x - 996 = 0$.

b. Use the quadratic formula, with $a = 16$, $b = 70$, and $c = -996$.

$$x = \frac{-b \pm \sqrt{b^2 - 4ac}}{2a} \qquad \text{Quadratic formula.}$$

$$x = \frac{-70 \pm \sqrt{(70)^2 - 4(16)(-996)}}{2(16)}$$

$$x = \frac{-70 \pm 262}{32}$$

$$x = -10.375 \quad \text{or} \quad x = 6$$

Since x is in hundreds, the company sold 600 connectors. ◄

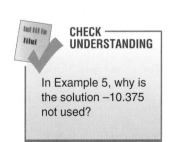

CHECK UNDERSTANDING

In Example 5, why is the solution -10.375 not used?

Determine the value of the discriminant for each equation.

1. $2x^2 + 4x + 9 = 0$ **2.** $-x^2 - 6x + 12 = 0$ **3.** $16x^2 + 16x + 4 = 0$

Use the discriminant to determine the number of real solutions to each quadratic equation.

4. $x^2 - 4x - 5 = 0$ **5.** $x^2 + 3x + 52 = 0$ **6.** $x^2 + 4x + 4 = 0$

7. $9x^2 + 6x + 1 = 0$ **8.** $6x^2 - 2x - 8 = 0$ **9.** $2x^2 + 2x + 4 = 0$

Use the quadratic formula to solve each equation. Round answer to the nearest hundredth where necessary.

10. $x^2 - 9x + 8 = 0$ **11.** $x^2 - 7x - 8 = 0$ **12.** $x^2 - 6x = -8$

13. $x^2 = 10 - 3x$ **14.** $4x^2 + 3x = 5$ **15.** $2x^2 + 4x = 7$

16. $3x^2 = 2x + 2$ **17.** $5x^2 - x - 3 = 0$ **18.** $2x^2 - 6x + 3 = 0$

Determine the values of k for which each equation has exactly one solution.

19. $x^2 + 12x + k = 0$ **20.** $kx^2 + 12x + 9 = 0$ **21.** $16x^2 + kx + 1 = 0$

22. PHYSICS A small rocket is fired from a height of 60 ft above ground level. Its height at any given time can be modeled by the function $h(t) = -16t^2 + 300t + 60$. Use the quadratic formula to determine the time, to the nearest second, at which the rocket will be 20 ft from the ground.

23. WRITING MATHEMATICS Explain in your own words how the value of the discriminant relates to the number of solutions that an equation has.

PRACTICE

Determine the value of the discriminant for each equation.

1. $3x^2 + 3x + 8 = 0$ **2.** $-x^2 - 8x + 6 = 0$ **3.** $25x^2 + 20x + 4 = 0$

Use the discriminant to determine the number of real solutions to each quadratic equation.

4. $x^2 + x + 1 = 0$ **5.** $x^2 + 4x - 6 = 0$ **6.** $x^2 + 10x + 25 = 0$

7. $36x^2 + 12x + 1 = 0$ **8.** $6x^2 - 3x - 5 = 0$ **9.** $2x^2 + 3x + 6 = 0$

Use the quadratic formula to solve each equation.

10. $x^2 - 10x + 9 = 0$

11. $x^2 - 6x - 7 = 0$

12. $x^2 - 9x + 18 = 0$

13. $x^2 - 11x + 30 = 0$

14. $x^2 + 2x = 35$

15. $x^2 - 7x = 30$

16. $x^2 = 4x + 45$

17. $x^2 = -4x + 21$

18. $6x^2 + x - 35 = 0$

19. BUSINESS The revenue function for SRB Company can be approximated by the function $R(p) = -16p^2 + 320p + 20{,}000$ where p represents the price of their product. If the revenue is \$10,000, use the quadratic formula to determine the price of their product to the nearest dollar.

Use the quadratic formula to solve each equation. Round decimal answers to the nearest hundredth.

20. $10x^2 - 3x = 1$

21. $14x^2 - 3x - 5 = 0$

22. $20x^2 = 4x + 24$

23. $x^2 + 5x - 2 = 0$

24. $x^2 + 4x = 1$

25. $2x^2 - 4x = 12$

26. $6x^2 - 6x - 7 = 0$

27. $3x^2 + 10x + 4 = 0$

28. $5x^2 + 12x - 2 = 0$

Determine the values of k for which each equation has exactly one solution.

29. $x^2 + 14x + k = 0$

30. $kx^2 - 30x + 25 = 0$

31. $25x^2 + kx + 16 = 0$

32. NUMBER THEORY Find a number whose square is 80 greater than twice the number.

33. WRITING MATHEMATICS Describe the graph of a quadratic equation whose discriminant is negative.

EXTEND

If s_1 and s_2 are the solutions of a quadratic equation of the form $ax^2 + bx + c = 0$ and $a = 1, a \neq 0$, then $x^2 - (s_1 + s_2)x + s_1s_2 = 0$. For example, if a quadratic equation has -4 and 3 as its solutions, then

$$x^2 - (s_1 + s_2)x + s_1s_2 = 0$$
$$x^2 - (-4 + 3)x + -4(3) = 0$$
$$x^2 + x - 12 = 0 \qquad \text{This equation has } -4 \text{ and } 3 \text{ as its solutions.}$$

Write a quadratic equation having the given solutions.

34. 5 and -6

35. 7 and -3

36. $\dfrac{1}{2}$ and $-\dfrac{1}{4}$

37. 7 and $-\dfrac{1}{2}$

38. 10.5 and -10.5

39. c and $-c$

NUMBER THEORY For each exercise, use the sum and the product of the numbers to write a quadratic equation. Then use the quadratic equation to solve for the numbers.

40. The sum of two numbers is 30 and their product is 216.

41. The sum of two numbers is -37 and their product is -650.

42. The sum of two numbers is -26 and their product is 153.

43. For what values of k are there two real solutions to the quadratic equation $x^2 + 6x + k = 0$?

44. For what values of k are there no real solutions to the quadratic equation $-x^2 - 5x + k = 0$?

Use the quadratric formula to solve each equation for x.

45. $(x - a)^2 - (x - a) - 5 = 0$ **46.** $(x - a)^2 - (x - a) = 6$

47. Describe the solutions to the equation $ax^2 + bx + c = 0$ if $\sqrt{b^2 - 4ac}$ is a perfect square.

48. Use the quadratic formula to determine where the graph of $y = 2x^2 - 7x - 4$ crosses the x-axis.

MIXED REVIEW

49. STANDARDIZED TESTS Which is the mean of the set of numbers?

   18, 44, 76, 6, 16

   A. 30   **B.** 160
   C. 32   **D.** 18

Determine whether each relation is a function.

50. $\{(3, 4), (4, 5), (5, 6)\}$ **51.** $\{(1, 3), (2, 5), (2, 7)\}$

Express each absolute value equation as a disjunction.

52. $|4 - 3x| = 21$ **53.** $\left| 2x - \dfrac{1}{2} \right| = \dfrac{1}{4}$

Use the quadratic formula to solve each equation.

54. $0 = 2x^2 - 37x + 105$ **55.** $12x^2 + 2x - 2 = 0$

PROJECT *Connection*   Read the material your corporation sent. Collect any other information you can find about your corporation in books, magazines, and other media.

1. Highlight important dates, people, and inventions that directly affected your corporation. Then find out about events that occurred in American history as your industry and corporation developed. Use the information to make a timeline that can include photographs and technical drawings.

2. Interpret at least one graph from your company's Annual Report. Consider using a different method to display the data.

3. Identify data in the Annual Report that might fit a linear or quadratic model. Use a graphing utility to determine the equation of the model. What predictions can you make for the corporation's future based on the equation?

"Satellite dish" antennas allow some homeowners to receive hundreds of television channels. Engineers use parabolas and quadratic equations to ensure that these antennas work correctly.

Geometrically, a *parabola* is the set of all points that are the same distance from a given line and a fixed point not on the line. This fixed point is called the *focus* of the parabola and the given line is called the *directrix*. Parabolic antennas reflect microwaves (represented by the parallel lines in the diagram) so that they converge at the focus of the parabola.

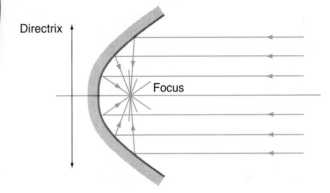

The structure of the satellite dish must support its weight and stand up to wind. If it bends, the incoming parallel waves will not be reflected correctly. Two factors in its construction are the area and circumference of the large circle created by the dish. To construct a particular antenna, an engineer begins with the parabola described by the function $f(x) = 0.04x^2$ on the domain -9 ft $\leq x \leq 9$ ft. The parabola is spun about its axis of symmetry to create a parabolic antenna.

Decision Making

1. Graph the parabola described by the function $f(x) = 0.04x^2$ with X-min: -9, X-max: 9, X-scl: 1, Y-min: -1, Y-max: 5, and Y-scl: 1.

2. Determine the circumference of the edge of the parabolic antenna.

3. Determine the area of the circle formed by the edge of the antenna.

4. Determine the depth of the antenna from its vertex to the plane of the circle formed by the edge.

5. If the parabola $g(x) = 0.05x^2$ were used to create a dish with the same diameter as $f(x)$ above, which would have greater depth?

Problem Solving File

Maximizing Area

Many problems involve determining shape for a given set of conditions. You can analyze such problems by using diagrams, tables, graphs, and equations.

Problem

Your community is building a children's playground. A local construction company has donated 200 ft of fencing for the project. The playground will be in the shape of a rectangle, having the greatest possible area for the given perimeter. If you use all of the available fencing, what should the length and width of the rectangle be?

Explore the Problem

Start by examining rectangles with whole-number dimensions and a constant perimeter of 200 ft.

CHECK UNDERSTANDING

Can you have a rectangle in this problem with a width of 100 ft? Explain.

1. Suppose the rectangle has a width of 10 ft. Explain how you determine the length of this rectangle. What is the area?

2. If you make the width 20 ft, then what must the length be? What is the area? Does it appear that the dimensions of the rectangle have an effect on area even when the perimeter is constant?

3. Write an equation that expresses the relationship between length and width for rectangles with a perimeter of 200 ft.

4. Copy and complete the table. Compare the table of widths, lengths, and areas below. Look for patterns.

w	l	A	w	l	A	w	l	A	w	l	A
5 ft			30 ft			55 ft			80 ft		
10 ft			35 ft			60 ft			85 ft		
15 ft			40 ft			65 ft			90 ft		
20 ft			45 ft			70 ft			95 ft		
25 ft			50 ft			75 ft			99 ft		

5. As widths increase, for what values do areas increase as well?

6. How would you predict the results for widths greater than 50 ft?

7. What is the most efficient shape for the playground?

8. Use the data from your table. Are there any rectangles whose dimensions are not shown in the table but which might have greater area? Make a graph of area as a function of width.

9. If you had 350 ft of fencing and wanted to enclose the maximum rectangular area, what would the dimensions and the area be? (Dimensions do not have to be whole numbers.)

PROBLEM
SOLVING PLAN

• Understand
• Plan
• Solve
• Examine

Investigate Further

For rectangles with a constant perimeter of 200 ft, you know that $2l + 2w = 200$ or $l + w = 100$. So, $l = 100 - w$. The area can be expressed as a function of width.

$$A = w(100 - w) = 100w - w^2 = -w^2 + 100w$$

10. What type of function is $A(w) = -w^2 + 100w$?

11. Graph this area function for $0 \le w \le 100$. How can knowing the shape and position of the graph of $B(w) = w^2$ help you to graph $A(w) = -w^2 + 100w$?

12. What are the coordinates of the vertex of the graph you drew above? Does this result agree with the solution you found by making a table?

CHECK
UNDERSTANDING

Use what you learned in Lesson 10.1 and your description in Question 11 to write an equivalent equation for $A(w)$.

Apply the Strategy

Draw a diagram to help you analyze each of the questions below.

13. Suppose you have 120 ft of fencing to enclose a rectangular area for a garden. You plan to use a house wall as one side of the garden.

   a. If the width is w, write an expression for the length.

   b. Write an expression for the area as a function of width.

   c. Use any strategy you choose. Find the dimensions of the rectangle having the maximum area and determine the area.

   d. How could you have used your fencing to make a square with one side being the wall? Determine the dimensions and the area for this square.

14. Suppose you had 120 ft of fencing and a house wall to use as one side. You plan to use some of the fencing to partition your garden into two sections, one for vegetables and one for flowers. The partition will make a right angle with the house wall.

   a. If the width of the rectangle is w, write expressions for the length and the area as a function of width.

   b. Find the dimensions of the rectangle with maximum area and find the area. If you wanted each section to have equal area, how would you partition the garden?

   c. If you wanted each section of the garden to be a square, what would be the dimension of the whole garden? What would be the area of each square section?

   d. Suppose you wanted the garden to be a square with the partition dividing it equally. What would be the area of each section?

15. Suppose you want to enclose a rectangular garden area of 400 ft^2. You want to know what dimensions will minimize the cost of fencing you will use. (Assume all dimensions are whole numbers.)

   a. How is this problem different than the others you have solved in this lesson?

   b. Will the dimensions of the rectangle make any difference as long as the rectangle has an area of 400 ft^2? Justify your answer with examples.

   c. Find the shape with the minimum perimeter by making a table for different lengths and widths.

   d. If the width of the rectangle is w, write an expression for the length in terms of w.

   e. Write an expression for the perimeter as a function of the width. Use a graphing utility to graph this function. Describe the graph and how you can use it to solve the problem.

16. WRITING MATHEMATICS Write a paragraph explaining the importance of drawing a diagram to help you solve problems.

REVIEW PROBLEM SOLVING STRATEGIES

WELL, WELL!

1. One day, Sharla was watching her friend Randy digging a well.

"How deep is the well so far?" Sharla asked.

"Guess," said Randy. "I'm exactly 5 ft 10 in. tall."

"How much deeper will you dig?" asked Sharla.

"I'm going to dig twice as deep as I've gone already," answered Randy, "and then my head will be twice as far below the ground as it is now above the ground."

Drawing a diagram may help you understand the relationships in the problem.

a. How many inches tall is Randy?

b. Let x represent the number of inches above ground Randy's head is at present. Represent the current depth of the well.

c. Using the expression you wrote in Question b, represent the final depth of the well in terms of x.

d. Have you used all the information in the problem? What other expression can you write?

e. How far above ground is Randy's head? What is the present depth of the well? How deep will the well be when completed?

REVA'S GLASSES

2. Reva took a small glass and filled it half full with pure orange juice. Then she took another glass, which was twice the size of the first glass, and filled it one-third full of pure orange juice. Then she filled each glass to capacity with water and poured the contents of both into a large pitcher. Then Reva figured out what part of the mixture was orange juice and what part was water. Can you figure it out? Use these hints if you need them.

a. What part of the total contents of the pitcher came from the smaller glass? What part of the total contents was the juice from the smaller glass?

b. What part of the total contents came from the larger glass? What part of the total contents was the juice from the larger glass?

Measurement Marks

3. Suppose you have an unmarked piece of wood that you know is 12 in. long.

— 12 in. —

You want to be able to measure all lengths from 1 to 12 in. using the piece of wood only once for each length you measure. What is the least number of marks you must make on the wood, and where would you place them? Explain.

Explore/Working Together

• An animal's teeth reveal much about its diet and way of life. Humans, like all mammals, have four kinds of specialized teeth: *incisors*, chisel-like teeth used for cutting; *canines* and *premolars*, for gripping and tearing; and *molars*, for crushing and breaking. Dentists and physical anthropologists describe the shape of the arrangement of human teeth as a *parabolic dental arcade*.

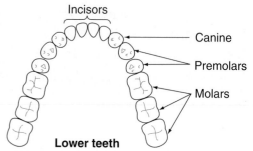

Lower teeth

You can find a quadratic equation that describes your dental arcade. Each member of your group will need a sheet of graph paper, plain paper, and a graphing utility that does quadratic regression.

1. Take an impression of your bite by biting the plain sheet of paper.

2. Draw a parabolic curve through the center of the impression as shown at the right.

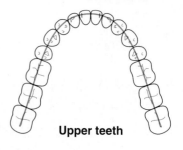

Upper teeth

3. Draw a set of axes on the graph paper. Using the *y*-axis as the axis of symmetry and (0, 0) as the vertex, trace the parabola.

4. Pick ten points on the parabola and estimate their coordinates, to the nearest tenth.

5. Use a graphing utility to draw a scatter plot of these ten points.

6. Use your graphing utility to find a quadratic equation that approximates the scatter plot.

7. Describe how well the parabola approximates the scatter plot.

8. Compare and discuss your work with other groups. In general, were the equations a good fit for the data?

Build Understanding

A linear model is of the form $y = ax + b$. One type of nonlinear model, the quadratic, is of the form $y = ax^2 + bx + c$. You can make a scatter plot to decide if a data set shows a linear or quadratic relationship.

EXAMPLE 1

Decide whether each data set shows a *linear relationship*, a *quadratic relationship*, or *neither*.

a. $(1.3, 1.4)$ $(2.5, 4.8)$ $(3, 7)$ $(0, 1)$ $(-1.5, 4.7)$ $(-2.2, 8)$
b. $(1.3, 5.5)$ $(2.3, 1.7)$ $(4, 7.7)$ $(0, 0)$ $(-1, -2.3)$ $(-2.5, 8.1)$
c. $(0, 2.7)$ $(1.3, 3.6)$ $(2.1, 3.9)$ $(3.0, 4.5)$ $(3.5, 5.1)$ $(4.5, 5.7)$

Solution

Make scatter plots of each data set. Examine the plots to see if the points appear to lie near a line or a parabola.

a.
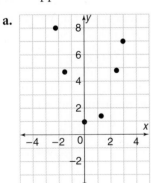

b.
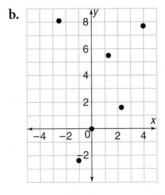

c.
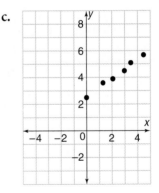

a. The points appear to lie near a parabola. The data show a quadratic relationship.

b. The points do not appear to lie near a line or a parabola. The data show neither a linear nor quadratic relationship.

c. The points lie near a line. The data show a linear relationship. ◄

Many nonlinear relationships in the real world can be approximated by a quadratic model. However, as the next example shows, you must be careful when rounding.

EXAMPLE 2

COMPUTER VIRUSES Businesses and individuals lose time and productivity when a virus infects their computers, making progress impossible. The table below shows financial losses caused by viruses during the years 1990–1994.

Year, last two digits	90	91	92	93	94
Loss, billions of dollars	0.1	0.3	0.7	1.4	2.7

a. Find the equation that best fits the data.

b. Check to see how well the curve fits the data for 1992.

c. Round the coefficients to the nearest hundredth and check to see how well the curve fits the data for 1992.

Solution

a. A scatter plot of the data shows that a linear model may not be the best one. The points seem to rise more rapidly in the later years and suggest the appearance of a parabola. Use a quadratic model and a graphing utility to find

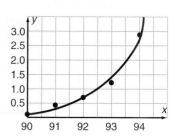

$$y = 0.1785714286x^2 - 32.22714286x + 1454.151429$$

b. Substitute 92 for x in the quadratic equation in **a** and find the value of y.

$$y = 0.1785714286(92)^2 - 32.22714286(92) + 1454.151429$$
$$= 0.6828575504$$

Subtract the predicted value from the actual value to find the error.

Error	=	Actual Value	−	Predicted Value

$$= 0.7 - 0.6828575504$$
$$= 0.0171424496$$

The percent error is $\dfrac{(0.0171424496)100}{0.7}$ or about 2.45%. Because the percent error is small, the quadratic model is a good fit.

c. With rounded coefficients, the equation is

$$y = 0.18x^2 - 32.23x + 1454.15$$

COMMUNICATING ABOUT ALGEBRA

For Example 2, discuss why an industry analyst might hypothesize that the financial loss due to viruses might increase more rapidly in the years following 1994.

Substitute 92 for x in the new equation and find the value of y.

$$y = 0.18(92)^2 - 32.23(92) + 1454.15$$
$$= 12.51$$

Find the error.

$$\text{error} = 0.7 - 12.51$$
$$= -11.81$$

The percent error is $\dfrac{(-11.81)100}{0.7}$ or about -1687%. The percent error is large. ◄

Why was the percent error in Example 2c so large? You can verify that rounding the coefficient of the x^2 term contributed most of the error. Rounding the coefficient of the x^2 term magnifies the errors from rounding, especially if the x^2 term is relatively large. Rounding coefficients can destroy the accuracy of the regression equation.

TRY THESE

Make a scatter plot for each set of data. Tell whether each data set shows a *linear relationship*, a *quadratic relationship*, or *neither*.

1. $(0, 0)$ $(5, 99.6)$ $(10, 148.9)$ $(12.5, 156)$ $(20, 100.5)$ $(25, 0.8)$

2. $(-2, -3.1)$ $(-1, -1.75)$ $(1, 0.85)$ $(3, 3.4)$ $(5, 6)$ $(8, 9.8)$

3. $(1, 2)$ $(2, 0)$ $(3, 5)$ $(4, 3)$ $(6, 4)$ $(9, 9)$ $(12, 2)$

4. $(-5, 6.2)$ $(-2, 1)$ $(0, 0)$ $(1, 0.3)$ $(3, 2.3)$ $(6, 9)$

EXPANSION OF WATER Density, or mass/unit volume, decreases as temperature increases for most liquids. For a certain range of temperature, the density of water increases as it is warmed from its frozen state at $0°C$ and then decreases as it is warmed further. The table shows values for the density of water at different temperatures.

Temperature, °C	Density, g/cm³
0	0.99986
1	0.99993
2	0.99997
3	0.99999
4	1.00000
5	0.99999
6	0.99997
7	0.99993
8	0.99987
9	0.99981

5. Make a scatter plot of the data. (*Hint:* Use intervals of 0.00005 from 0.99970 to 1.00000 on the vertical axis.) What type of relationship does the data show?

6. For which temperature interval does the density of water increase? What is the maximum density?

7. Using a graphing utility, a researcher determined that the equation $y = -0.000008x^2 + 0.000065x + 0.999867$ modeled the data in the table. What is the predicted value for $4°C$. What is the percent error?

8. Use the model to predict the density at $10°C$.

PRACTICE

SOLUBILITY You can make a chemical solution by dissolving a substance such as sugar in water. The amount of a substance that can be dissolved in a given liquid is a function of the temperature of the liquid. A chemist collected data on the solubility of several substances in water.

Temperature, °C	GRAMS OF SUBSTANCE PER 100 GRAMS OF WATER					
	0	20	40	60	80	100
Substance A	179	204	238	287	362	487
Substance B	28.0	34.2	40.1	45.8	51.3	56.3
Substance C	128	144	162	176	192	206
Substance D	13.9	31.6	61.3	116	167	245
Substance E	73	87.6	102	122	148	180

1. Make a separate scatter plot for each substance. For which substances does the temperature–solubility relationship appear to be linear? For which does the relationship appear quadratic? Indicate if there are any for which you are undecided.

2. Use a graphing utility to find the quadratic equation that best fits the data for substance E. Round coefficients to three decimal places.

3. Use your equation from Exercise 2 to predict values for 0°C and 100°C.

4. Use the actual values for 0° and 100° to compute the percent error. Do you think rounding the coefficients produced an acceptable model?

5. Use a graphing utility to find the linear equation that best fits the data for substance C. Round coefficients to three decimal places.

6. Compare the predicted and actual values for substance C at 60°C. Does the linear model appear to be a good fit?

7. WRITING MATHEMATICS Suppose you have collected a set of data that involves two variables. After making a scatter plot, you are still not sure whether a linear or quadratic model best fits the data. Explain what you might do to decide on a model.

EXTEND

You can construct and solve a system of equations to find an equation of a quadratic model. To determine the coefficients a, b, c, you will need three equations. Exercises 8–11 show the procedure.

8. Use the solubility data for substance A at 0°C, 40°C, and 80°C. Substitute the coordinates of each point into the equation $y = ax^2 + bx + c$. Write a system of three equations for a, b, c.

9. Solve the system for a, b, c. Do not round.

10. Use your results to write a quadratic equation of the form $y = ax^2 + bx + c$.

11. What values does your model predict for 60°C and 100°C? Do you think the equation is a good fitting model?

12. Use a graphing utility to find the quadratic equation that best fits the data. Round coefficients to three decimal places. What are the predicted values for 60°C and 100°C?

THINK CRITICALLY

Write *true* or *false*.

13. If two variables do not have a quadratic relationship, then they have a linear relationship.

14. A set of data involving two variables can have both a linear model and a quadratic model.

MIXED REVIEW

15. STANDARDIZED TESTS Identify the property illustrated.

$$9 + (13 + 26) = (9 + 13) + 26$$

   A. commutative property of addition

   B. identity property of addition

   C. inverse property of addition

   D. associative property of addition

16. Solve and graph: $-2m + 3 > -5$

Determine the slope of the line passing through each pair of points.

17. $(3, -9), (4, -12)$

18. $(4, -4), (9, 6)$

Solve each system of equations.

19. $\begin{cases} 25x - 10y = 51 \\ 5x + 15y = 60 \end{cases}$

20. $\begin{cases} m = 16 + n \\ -m - n = 0 \end{cases}$

Use the function $f(x) = x^2 + 6x + 8$.

21. Find the vertex.

22. Find the axis of symmetry.

· · · CHAPTER REVIEW · · ·

VOCABULARY

Choose the word from the list that completes each statement.

1. Every point on the graph of a function $y = ax^2 + bx + c$ has a point of reflection over the __?__ of the graph.

2. The intersection of a parabola and its axis of symmetry is called the __?__.

3. "The solutions of the equation $ax^2 + bx + c = 0$ can be expressed in terms of a, b, and c." This statement refers to the __?__.

4. The number of real solutions to a quadratic equation can be found by evaluating the __?__.

a. discriminant

b. line of symmetry

c. quadratic formula

d. vertex

Lesson 10.1   EXPLORE QUADRATIC FUNCTIONS   pages 465–467

- The graph of $y = x^2$ is translated vertically when a constant is added or subtracted, and horizontally when a constant is added to x before squaring the quantity. The graph of $y = ax^2$ is wider than $y = x^2$ for $-1 < a < 1$ and $a \neq 0$ and narrower for $a < -1$ or $a > 1$.

Match each in equation with its graph.

5. $y = x^2 + 4$ **6.** $y = 3x^2$ **7.** $y = (x - 4)^2$

a.
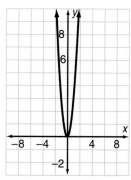

b.
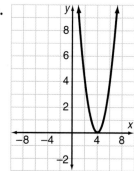

c.
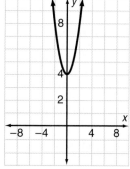

Lesson 10.2   GRAPHS OF QUADRATIC FUNCTIONS   pages 468–476

- The graph of $y = ax^2 + bx + c$ opens upward for $a > 0$ and downward for $a < 0$.
- The x-value of the vertex of the graph $y = ax^2 + bx + c$ is $-\dfrac{b}{2a}$. To find the y-value, substitute the x-value in the equation and solve for y.

Determine the coordinates of the vertex and the equation of the axis of symmetry for the graph of each equation.

8. $y = x^2 + 5$ **9.** $y = 2x^2 - 4x + 11$ **10.** $y = x^2 + 6x - 1$

Lesson 10.3 SOLVE QUADRATIC EQUATIONS BY GRAPHING pages 477–482

- To solve a quadratic equation of the form $ax^2 + bx + c = 0$, find the x-intercepts of the graph of $y = ax^2 + bx + c$.
- To solve a quadratic equation of the form $ax^2 + bx + c = d$, rewrite the equation as $ax^2 + bx + c - d = 0$. Alternatively, graph both $y_1 = ax^2 + bx + c$ and $y_2 = d$ and find the x-values of their intersection.

Solve each equation. Round decimals answers to the nearest hundredth.

11. $x^2 - 3x + 2 = 0$　　　　**12.** $x^2 - x = 20$　　　　**13.** $x^2 + 4x - 7 = 0$

Lesson 10.4 SOLVE QUADRATIC EQUATIONS USING SQUARE ROOTS pages 483–489

- To solve a quadratic equation of the form $ax^2 + b = c$, use the square root property.

Solve. Express answers in simplest radical form.

14. $x^2 - 49 = 0$　　　　**15.** $2x^2 = 64$　　　　**16.** $3x^2 - 1 = 47$

Lesson 10.5 THE QUADRATIC FORMULA pages 490–497

- To solve a quadratic equation of the form $ax^2 + bx + c = 0$, use the quadratic formula.
- Use the discriminant to determine the number of real solutions of a quadratic equation.

Solve. Round decimal answers to two decimal places.

17. $x^2 + 7x + 10 = 0$　　　　**18.** $4x^2 - 12x + 9 = 0$　　　　**19.** $x^2 - x = 5$

Determine the number of solutions.

20. $x^2 - 4x - 1 = 0$　　　　**21.** $x^2 + 3x + 6 = 0$　　　　**22.** $2x^2 + 7x = 9$

Lesson 10.6 MAXIMIZING AREA pages 498–501

- You can determine a maximum area for a given shape by solving a quadratic equation.

Solve.

23. A rectangular herb garden will be enclosed by 12 yards of low fencing. Determine the dimensions of the rectangle that will give the maximum area.

Lesson 10.7 EXPLORE STATISTICS: USING QUADRATIC MODELS pages 502–507

- A **scatter plot** may reveal that a set of data has a quadratic relationship. Use a graphing utility to find a quadratic equation that models the data.

Solve.

24. A set of data has a y-value of –5 when $x = 2$. A graphing utility gives the function $y = 0.286x^2 - 2.857x + 0.071$ as a model for this data. Determine the percent error for $x = 2$ between the actual and predicted values.

CHAPTER ASSESSMENT

CHAPTER TEST

Determine the coordinates of the vertex and the equation of the axis of symmetry of each equation.

1. $y = x^2 - 4x + 1$ **2.** $y = 2x^2 + 8x - 5$

3. $y = x^2 - 8x + 15$ **4.** $y = x^2 + 6x + 2$

Solve each equation. Express answers in simplest radical form.

5. $x^2 - 3x + 2 = 0$ **6.** $x^2 - 8x = -16$

7. $x^2 - 70 = 11$ **8.** $3x^2 - 12x = -4$

9. $x^2 - 2x - 48 = 0$ **10.** $x^2 + 6x - 9 = 0$

11. $x^2 + 6x + 9 = 15$ **12.** $x^2 + 49 = 14x$

13. $x^2 + 6 = 48$ **14.** $x^2 - 2x - 2 = 0$

15. STANDARDIZED TESTS Which is the equation of the graph shown?

A. $y = (x + 2)^2$

B. $y = x^2 - 4$

C. $y = (x - 2)^2$

D. $y = x^2 + 4$

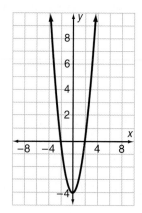

Solve each equation. Round decimal answers to the nearest hundredth.

16. $x^2 - 7x + 12 = 0$ **17.** $x^2 + 4x - 14 = 0$

18. $2x^2 + x = 10$ **19.** $3x^2 - 16 = 2x$

20. $x^2 = 11$ **21.** $2x^2 - 108 = 20$

22. STANDARDIZED TESTS Which of the following equations is equivalent to $x^2 - 3x + 4 = 0$?

A. $x^2 - 3x = 4$ **B.** $x^2 - 3x + 1 = -3$
C. $x^2 - 4 = 3x$ **D.** $x^2 - 3x - 2 = 2$

23. WRITING MATHEMATICS Write a paragraph explaining how to solve a quadratic equation of the form $ax^2 + bx + c = d$.

Use the discriminant to determine the number of real solutions of each equation.

24. $x^2 - 10x + 25 = 0$

25. $x^2 + 5x = 12$

26. $2x^2 - x + 23 = 9$

27. $x^2 + 3x - 5 = 7$

Solve each problem. Round decimal answers to the nearest hundredth.

28. The revenue y from ticket sales for a concert can be modeled by the equation $y = -5x^2 + 250x + 70$, where x is the price of the ticket. Find the ticket price that will give the maximum revenue.

29. Andrea is enclosing a rectangular area for a vegetable garden near the back of her house. One wall of the house will form one side of the garden. Andrea plans to use 50 ft of fencing. Determine the dimensions of the enclosed garden that will produce a maximum area.

30. The height of a rocket launched from the ground can be modeled by the equation $y = -16x^2 + 90x$, where x is the number of seconds after launching and y is the height in feet. Find the number of seconds it will take for the rocket to reach the ground after launching.

31. The equation $y = 0.198x^2 - 0.396x + 1.102$ models the following data: $(0, 1)$, $(3, 2)$, $(5, 4)$, $(6.7, 7)$, $(7.5, 9)$, $(8, 11)$.

   a. Find the predicted value of y for $x = 5$.
   b. Find the percent error for the predicted and actual values of y for $x = 5$.

PERFORMANCE ASSESSMENT

"WEIGHTLESS" EQUATIONS The equation $d = -9.8t^2$ can be used to determine the time t in seconds it takes for an object to reach the ground in free fall from a height d meters above the ground.

a. Use the equation to help you answer this question: If a brick and a nickel are dropped from a 10-meter height at the same moment, which object will hit the ground first?

b. Test your answer by performing an experiment. Hold a heavy book in one hand and a coin in the other. Drop them simultaneously from the same height. What do you notice?

c. Make a conjecture about your results.

USING A GRAPHING UTILITY Write three quadratic equations. Graph them with a graphing utility. Discuss with a partner how each graph relates to its equation. For each graph, determine the vertex and locate the axis of symmetry. Find the x-values for which the value of the function is 0.

DESIGN A POSTER Draw a parabola on posterboard. Label the vertex, the axis of symmetry, and the x-intercept or intercepts. On the poster, display the standard form of a quadratic equation, the quadratic formula, the algebraic expression for the discriminant, and a table that shows how to interpret the discriminant.

OUTLINING STEPS Make a list of steps you need to take to solve a quadratic equation. Give your list to a partner who will use it to solve an equation. Revise the steps if necessary. Then exchange roles with your partner.

PROJECT ASSESSMENT

PROJECT *Connection*

Work with your teacher to determine when a class Business Week can be held. Groups will organize the results of their research and take turns presenting an in-depth look at different industries and corporations and the working stock market. Here are some ideas for related activities.

1. Create a collage of logos of the corporations your group followed. See if your classmates can identify the corporations associated with each logo.

2. Pretend that each member of your group purchased 100 shares of stock in his or her chosen corporation. How much did each of you "invest" initially? What was the total value of your imaginary stock portfolio at the beginning and at the end of the project? Who came out ahead and who came out behind? Could you identify trends for the stock market or the industry in general? Were there reasons for individual corporate performance?

3. As a class, brainstorm new products or services for each industry studied. Speculate on the future of each industry and how it might be affected by technological advances.

CUMULATIVE REVIEW

Graph each equation. Give the coordinates of the vertex.

1. $y = x^2 - 6x + 9$ **2.** $y = 2x^2 + 8x - 10$

3. $y = -|x| + 3$ **4.** $y = 3x - 2$

Solve each equation by graphing. Round decimals to the nearest hundredth.

5. $x^2 - 3x + 1 = 7$

6. $5x^2 + 2x - 3 = 4x - 1$

Write three numbers that are between the two given numbers.

7. 5.06 and 5.07   **8.** $\frac{1}{2}$ and $\frac{3}{4}$

9. The Nut Hut sells cashews for $6.50 per pound and almonds for $2.25 per pound. Customers have suggested a mixture of cashews and almonds, so that the cost per pound would be $4.50. The owner agrees to make 20 pounds of the mixture on a trial basis. To the nearest tenth, how many pounds of each kind of nut will be needed to make the mixture?

10. Arrange the numbers in order from least to greatest.

$$-3.94, -\sqrt{17}, \frac{-11}{3}, -3.943$$

11. What percent of 5 is 6?

12. About 800 runners entered the 10 K race. It was determined that 28% of the runners finished the race in less than 32 minutes. How many runners was this?

Write the equation of each line described.

13. passes through $(3, 1)$ and $(-2, -4)$

14. passes through $(-4, 2)$ and $(4, 2)$

15. passes through $(5, -4)$ and parallel to the y-axis

Use the bar graph for Questions 16–18.

One Day's Activities

16. At which activity was 7 hours spent?

17. How much time was spent on homework?

18. WRITING MATHEMATICS What other type of graph might have been used to represent these data? Why?

Solve each quadratic equation. Use the product and quotient properties to simplify the radicals.

19. $3(x + 1)^2 = 24$ **20.** $9x^2 + 5 = 45$

21. Benny asked Jenny to try to guess the number he was thinking of. Jenny guessed 26. Benny said that Jenny was close, that she was only off by at most 7. In what range was Benny's number? Write an absolute value inequality and solve.

22. A deli offers five different kinds of lunch meat, three kinds of bread, and six different kinds of cheese. If customers can order one kind of lunch meat on a sandwich with one kind of cheese on one type of bread, how many different sandwiches can a customer order?

Solve each inequality. Then graph the solution.

23. $3x - 5 > -2$

24. $-3x \leq 9$ and $x + 2 < 5$

• • • STANDARDIZED TEST • • •

STUDENT PRODUCED ANSWERS Solve each question and on the answer grid write your answer at the top and fill in the ovals.

Notes: Mixed numbers such as $1\frac{1}{2}$ must be gridded as 1.5 or 3/2. Grid only one answer per question. If your answer is a decimal, enter the most accurate value the grid will accommodate.

1. $\begin{bmatrix} -12 & x+y \\ -3y & w \end{bmatrix} = \begin{bmatrix} -4x & z \\ 15 & 2z \end{bmatrix}$

   Find the value of w.

2. Solve the system of inequalities. Find the product of m and n.

   $$\begin{cases} 5m + 8n = 13 \\ 4m - 5n = 56 \end{cases}$$

3. Find the product of the integer solutions of $|x - 2| < 2$.

4. Find the value of the discriminant for the equation $x^2 + 8x = 5$.

5. If $f(x) = 4x + 5$ and $g(x) = 2x^2 - 7x + 3$, find $f(-2) + g(4)$.

6. A reporter asked the people attending the movie to rate the movie on a scale of 0 to 4. The results are shown in the table at the right.

Score	Frequency
0	6
1	15
2	25
3	42
4	12

   Find the mean rating score given by the people at the movie.

7. Solve the equation $-3x^2 + 10x - 5 = 0$. Grid the greater solution.

8. A store buys a certain item for $3.50. The store gives the item a selling price by marking it up 32%. Omitting the dollar sign, find the amount of the store's selling price for the item.

9. Find the slope of a line perpendicular to the line $2x - 9y = 14$.

10. Evaluate the expression $a + b \cdot a - b$ if $a = 3$ and $b = -4$.

11. A bag contains 4 red rubber balls, 3 blue rubber balls, and 1 black rubber ball. Azure gets to choose 3 balls and keep them. Azure's favorite color is blue. What is the probability that Azure will get the 3 blue balls with the three picks? Enter the answer as a fraction.

12. Solve the equation

    $$4n + \frac{1}{2}(5 - 2n) = 5(n + 2).$$

    Find $|n|$.

13. A football team has scored 31, 27, 35, 24, and 14 points in its five games. They still have two games to play. What is the least number of points the team needs to total in the two games so that they have an average of at least 28 points per game?

14. A homeowner wants to plant a vegetable garden and enclose it with 60 ft of fencing. The garden is to be adjacent to the house so it only has to be fenced on three sides. If the garden is to be rectangular, what is the greatest number of square feet the homeowner can enclose?

15. Two weeks ago, a worker was earning $100 per week. Last week, the worker's salary was reduced by 10%. This week, the worker's salary was increased by 10%. Omitting the dollar sign, find the worker's current weekly salary.

11 Polynomials and Exponents

Take a Look AHEAD

Make notes about things that look new.
- Look for places where you will be expected to multiply and divide terms with exponents in them.
- Locate two different pages where negative exponents appear. What will you be doing with negative exponents?

Make notes about things that look familiar.
- Find five different examples of subtraction in this chapter.
- What other operations are used in this chapter?

DATA Activity

Up, Up, and Away

Each of the seven aircraft in the Hall of Air Transportation at the National Air and Space Museum in Washington, D.C., played a role in the development of air transportation. In the early years of commercial aviation, most revenue came from mail routes flown by planes such as the Pitcairn PA-5 Mailwing. Although the 10-seat Boeing 247D expanded passenger travel, it was the Douglas DC-3 that changed the industry. With 21 seats, it was the first plane to make a profit transporting people.

SKILL FOCUS

- Add, subtract, multiply, and divide real numbers.
- Round decimal numbers.
- Determine ratios.
- Use scaled dimensions.
- Estimate metric and customary units.

ALGEBRA WORKS

HELP WANTED

Exploring FLIGHT

In this chapter, you will see how:

- **MACHINISTS** use equations to help them determine how much force is required by a metal punch machine to make rivet holes in sheets of metal.
(Lesson 11.2, page 526)

- **AERIAL PHOTOGRAPHERS** use equations to help them compare the distance between objects in a photograph to the actual distance between real objects on the ground.
(Lesson 11.4, page 537)

	Pitcairn PA-5 Mailwing	Boeing 247D	Douglas DC-3
Wingspan	10.05 m	22.55 m	28.95 m
Length	6.67 m	15.72 m	19.66 m
Height	2.83 m	3.70 m	5 m
Maximum Weight	1,139 kg	7,623 kg	11,430 kg

Use the table to answer the following questions.

1. How much greater is the wingspan of the Douglas DC-3 than that of the Boeing 247D?

2. What is the ratio of the Pitcairn's length to its height? Round to one decimal place.

3. If you were making a scale model of the Douglas DC-3 where 2 cm represents 0.5 m, how long would your model be? Round to one decimal place.

4. What is the wingspan of the Boeing 247D expressed in feet?

5. When expressed in customary units, would the Boeing's maximum weight be closer to 18,600 lb or 16,800 lb?

6. **WORKING TOGETHER** Research some of the highlights in the development of air transportation and prepare a timeline of major events in the history of flight.

Aspects of Flight

Flying, gliding, and soaring above mountaintops is easy for birds, but thousands of years passed from when humans started wondering about flight until they figured out how flight works. Building a machine that could lift itself and several hundred people off the ground and stay airborne for long distances was no easy task. During this project, you will explore different aspects of flight and make a paper airplane for a team competition.

PROJECT GOAL

To make a paper airplane that will fly the greatest distance or stay airborne the longest.

Getting Started

1. As a class, discuss rules for a paper airplane competition. For example, does the plane have to be folded from a single sheet of paper? Are there any size limitations? How many "extra" materials, such as rubber bands, tape, clips, string, and so on, is each team allowed to use?

2. Research ways of constructing paper airplanes. Many books are available on the subject. One is *Paper Airplane Book* published by the American Association for the Advancement of Science. Another is *Experimenting with Air and Flight* by O.H. Walker.

Internet Connection

www.swpco.com/
swpco/algebra1.html

PROJECT Connections

Lesson 11.2, page 525:
Learn how two different types of drag affect aircraft performance.

Lesson 11.4, page 536:
Experiment to locate a paper airplane's center of gravity.

Lesson 11.6, page 547:
Construct several model parachutes to determine design features that work best.

Lesson 11.8, page 557:
Find out what Bernoulli's principle is and how an aircraft is able to lift off.

Chapter Assessment, page 565:
Have the group airplane competition. Record and analyze data to determine design features that give best performance.

11.1 Algebra Workshop
Multiply and Divide Variables

Think Back/Working Together

To multiply integers using Algeblocks, use a Quadrant Mat and unit blocks. Work with a partner. Set up unit blocks on the Quadrant Mat as shown at the right.

1. What multiplication is shown?

2. Use unit blocks to complete the model. What is the product?

3. With blocks, make a model that shows that the product of two negative integers is positive.

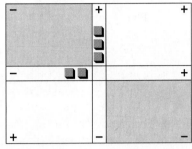

SPOTLIGHT ON LEARNING

WHAT? In this lesson you will learn
• to use Algeblocks to model multiplication and division of variables.

WHY? Algeblocks can help you to understand multiplication and division of variables useful in solving a variety of practical problems.

Explore

The Quadrant Mat can be used to model multiplication with variables.

4. What multiplication is shown at the right?

5. Use Algeblocks to complete the model for this product. Write an expression for the product.

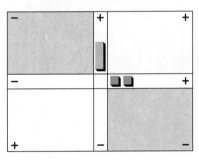

Write a multiplication sentence for each model.

6.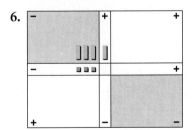

7.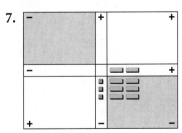

8.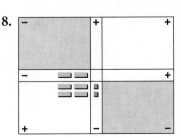

9.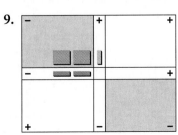

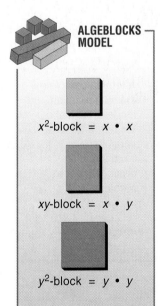

ALGEBLOCKS MODEL

x^2-block $= x \cdot x$

xy-block $= x \cdot y$

y^2-block $= y \cdot y$

Algebra Workshop

Remember, an exponent tells you how many times a number or variable is used as a factor. For example, 2^3 means $2 \cdot 2 \cdot 2$ and x^2 means $x \cdot x$.

10. Using unit blocks, make a square that is 3 units on each side. Write an expression with an exponent for the number of blocks in the square.

11. Using unit blocks, make a cube that is 2 units wide, 2 units long, and 2 units high. Write an expression with an exponent for the number of blocks in the cube.

12. Which is greater in value, 3^2 or 2^3? How do you know?

COMMUNICATING ABOUT ALGEBRA

After you complete the multiplication models, think about the numerical and variable parts of the factors and their product in each problem. With a partner, discuss the relationships you notice.

Use Algeblocks to complete the following products. Write an algebraic expression for the product.

13.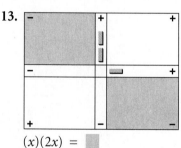

$(x)(2x) = $ ▢

14.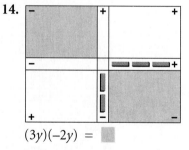

$(3y)(-2y) = $ ▢

15.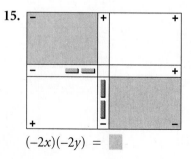

$(-2x)(-2y) = $ ▢

16.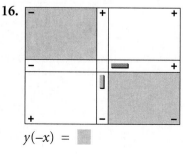

$y(-x) = $ ▢

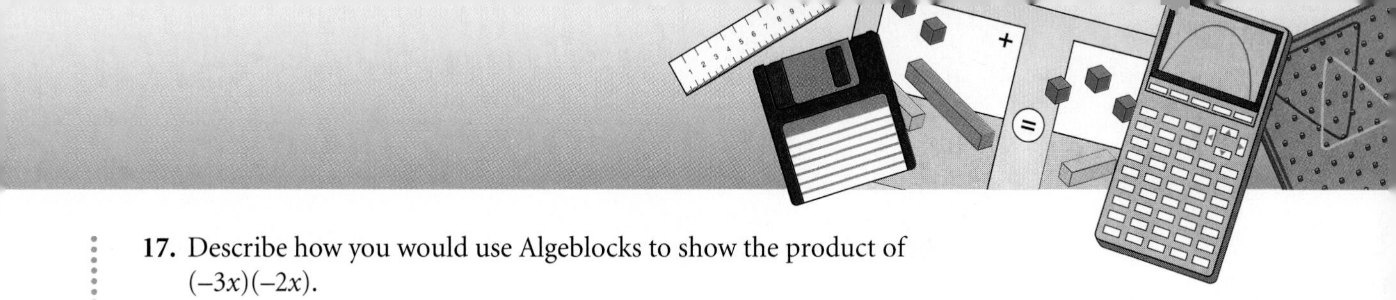

17. Describe how you would use Algeblocks to show the product of $(-3x)(-2x)$.

Find each product.

18. $(2x)(5x)$ **19.** $(-3x)(3x)$ **20.** $(-4y)(2y)$

21. $(3x)(-2y)$ **22.** $(-x)(2y)$ **23.** $(-y)(-2y)$

Since multiplication and division are inverse operations, you can use Algeblocks to show division.

24. The diagram shows a way of using Algeblocks to find the quotient for $-4xy \div 2x$. How will you complete the problem? What is the quotient?

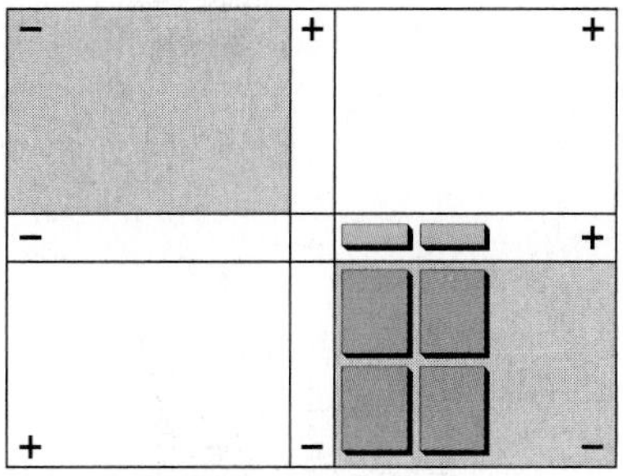

25. Describe how to use Algeblocks to find a quotient such as $2x^2 \div 2x$.

CHECK UNDERSTANDING

For Question 24, would the Algeblocks still work if you placed $-4xy$ in the *other* negative region? Draw a sketch showing your model.

Complete each division using Algeblocks. Write an algebraic expression for the quotient.

26.

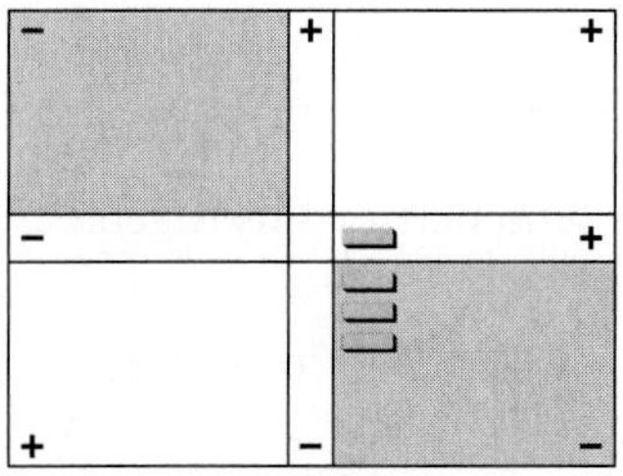

$-3x \div x = \square$

27.

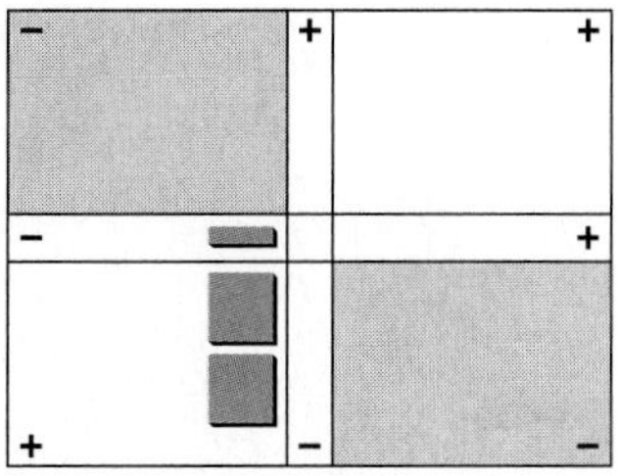

$2y^2 \div -y = \square$

28.

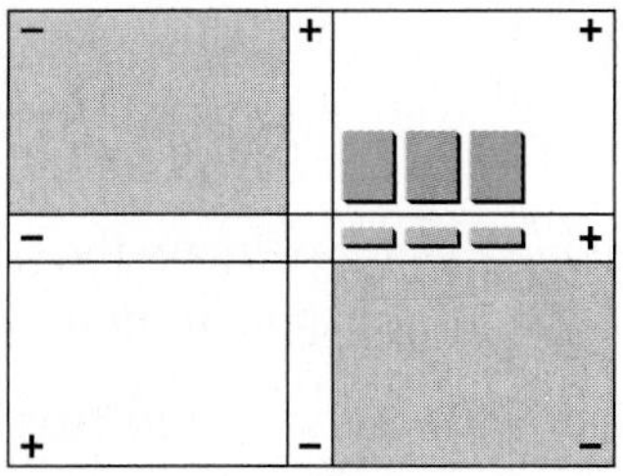

$3xy \div 3x = \square$

29.

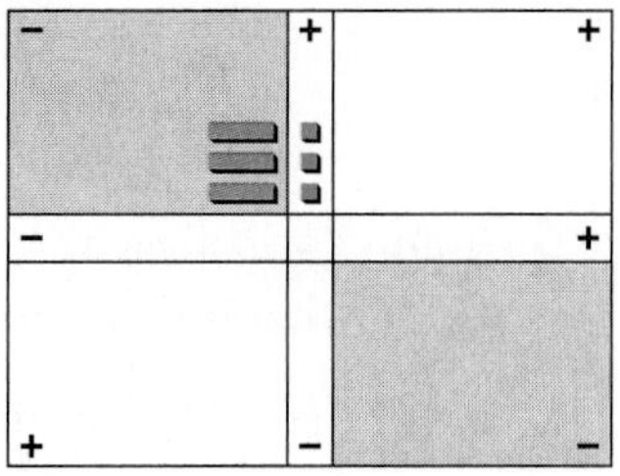

$-3y \div 3 = \square$

COMMUNICATING ABOUT ALGEBRA

Discuss the patterns you noticed in your division models. How do the numerical and variable parts of the dividend and divisor relate to the numerical and variable parts of the quotient?

Find each quotient.

30. $\dfrac{-6y}{3y}$ **31.** $\dfrac{2xy}{x}$ **32.** $\dfrac{-4x^2}{-2x}$

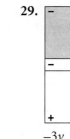

Algebra Workshop

Make Connections

33. What shape do you use to model $x(x)$? How many dimensions does the shape have? Write the algebraic expression for the product.

34. What shape would you use to model $x^2(x)$? How many dimensions does the shape have? Write the algebraic expression for the product.

35. Can you make a physical model for x^4? Explain.

36. How could you find the product of $2x$ and $3x^3$ without using Algeblocks? Justify your answer.

37. How could you find the quotient of $5x^6 \div 10x$ without using Algeblocks? Justify your answer.

Summarize

38. MODELING Write a multiplication sentence for this Algeblocks model.

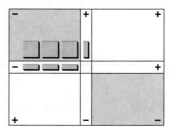

39. MODELING Write a division sentence for this Algeblocks model.

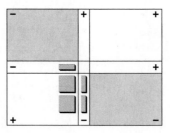

40. THINKING CRITICALLY When both factors contain the same variable, what is true about the variable and its exponent in the product?

41. THINKING CRITICALLY When the dividend and divisor have the same variable, what is true about the variable in the quotient?

42. GOING FURTHER Write an expression for the volume of the figure shown at the right.

11.2 Multiply Monomials

Explore/Working Together

Work with a partner.

1. Fold a blank sheet of paper in half. How many regions are formed by one fold?

2. Continue to fold the sheet in half. Copy and complete this table to record the number of regions formed from each additional fold.

Fold Number	1	2	3	4	5
Number of Regions	2	?	?	?	?

3. What do you observe from the table about the number of regions? How is the number of regions related to the number of folds?

4. Predict the number of regions that will be formed by 10 folds.

5. Suppose the paper is folded into thirds each time. Describe the pattern of the number of regions as the number of folds increases.

Build Understanding

A **monomial** is a single term that is a number, a variable, or the product of a number and one or more variables. A monomial can be made of any combination of numerical and variable factors. The variable must not appear in the denominator of a fraction or under a radical sign.

These expressions are monomials.

$$\frac{4}{5}x \qquad 3y^2 \qquad 17 \qquad -5x^2y^3 \qquad m \qquad \sqrt{2}xy$$

The following expressions are *not* monomials because they have more than one term, have a variable in the denominator, or have a variable under a radical sign.

$$x + 2 \qquad 3x + y + 5x \qquad \frac{5y}{2z} \qquad 3\sqrt{a}$$

The **degree of a variable** in a monomial is the number of times the variable occurs as a factor in the monomial. In the monomial $7x^2y^4$, the degree of x is 2 and the degree of y is 4.

The **degree of a monomial** is the sum of the degrees of all the variable factors. To find the degree of any monomial, add the degrees of the variables in the monomial. The degree of $7x^2y^4$ is $2 + 4 = 6$. The degree of a nonzero real number is zero. The monomial 0 has no degree.

SPOTLIGHT ON LEARNING

WHAT? In this lesson you will learn
- to recognize monomials.
- to multiply monomials using properties of exponents.

WHY? Multiplying monomials helps you solve problems involving gravity, electricity, and machines.

THINK BACK

In the algebraic expression $2x^5$, 2 is the coefficient, x is the base, and 5 is the exponent. The exponent indicates how many times the base is used as a factor.

ALGEBRA: WHO, WHERE, WHEN

Chu Shih-chieh, a Chinese mathematician who flourished about 1280–1303, wrote a book entitled *Ssu-yüan yü-chien* that deals with equations of degrees as high as fourteen.

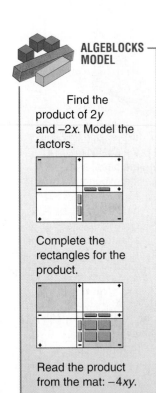

ALGEBLOCKS MODEL

Find the product of $2y$ and $-2x$. Model the factors.

Complete the rectangles for the product.

Read the product from the mat: $-4xy$.

EXAMPLE 1

Find the degree of each monomial.

a. $9x^6$ **b.** $-2y$ **c.** $5x^3y^4$ **d.** -7

Solution

a. The degree of $9x^6$ is 6, the exponent of the variable x.

b. The degree of $-2y$ is 1, since $-2y = -2y^1$. If a variable has no written exponent, its exponent is 1.

c. The degree of $5x^3y^4$ is 7. The degree of x is 3, the degree of y is 4, and the sum of 3 and 4 is 7.

d. The degree of -7 is 0. ◄

When monomials are multiplied, you can use the commutative and associative properties of multiplication to rearrange the factors.

EXAMPLE 2

Find the product of $-3x$ and $4xy$.

Solution

The product of $-3x$ and $4xy$ is the product of all five factors.

$$-3x(4xy) = -3 \cdot x \cdot 4 \cdot x \cdot y$$

You can use the commutative property to order the factors so that coefficients are together and like variables are together.

$$-3 \cdot x \cdot 4 \cdot x \cdot y = -3 \cdot 4 \cdot x \cdot x \cdot y$$

By the associative property, you can group numbers and variables.

$$(-3 \cdot 4) \cdot (x \cdot x) \cdot y = -12x^2y$$

Therefore, the product $-3x(4xy) = -12x^2y$. ◄

When monomials having the same base are multiplied, you can see a pattern.

$$x^2 \cdot x^3 = (x \cdot x)(x \cdot x \cdot x) = x^5$$
$$y^3 \cdot y^4 = (y \cdot y \cdot y)(y \cdot y \cdot y \cdot y) = y^7$$
$$a \cdot a^3 = a(a \cdot a \cdot a) = a^4$$

This pattern suggests the following property.

PROPERTY OF EXPONENTS: PRODUCT OF POWERS

For any real number a and all positive integers m and n,
$$a^m \cdot a^n = a^{m+n}$$

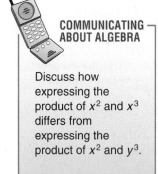

EXAMPLE 3

Find each product.

a. $x^5 \cdot x^7$ **b.** $3y^3(-2y^{10})$ **c.** $-4y^6(-x^3y^8)$

Solution

a. $x^5 \cdot x^7 = x^{5+7} = x^{12}$

b. $3y^3(-2y^{10}) = 3(-2)(y^3 \cdot y^{10}) = -6y^{13}$

c. $-4y^6(-x^3y^8) = -4(-1)(x^3)(y^6 \cdot y^8) = 4x^3y^{14}$ ◀

A number or a monomial in exponential form can be raised to a power. You could write the factors, then use exponents to write the product.

$$(2^2)^3 = (2^2)(2^2)(2^2) = (2 \cdot 2)(2 \cdot 2)(2 \cdot 2) = 2^6$$

$$(a^2)^3 = (a^2)(a^2)(a^2) = (a \cdot a)(a \cdot a)(a \cdot a) = a^6$$

Comparing the exponents in the original expression to the exponents in the final expression suggests the following property.

PROPERTY OF EXPONENTS: POWER OF A POWER

For any real number a and all positive integers m and n,
$$(a^m)^n = a^{mn}$$

EXAMPLE 4

Simplify.

a. $(2^3)^2$ **b.** $(x^7)^3$ **c.** $(y^4)^{10}$

Solution

a. $(2^3)^2 = 2^{3 \cdot 2} = 2^6 = 64$

b. $(x^7)^3 = x^{7 \cdot 3} = x^{21}$

c. $(y^4)^{10} = y^{4 \cdot 10} = y^{40}$ ◀

When a product is raised to a power, each factor is raised to that power.

$$(3x^2)^4 = (3x^2)(3x^2)(3x^2)(3x^2)$$

$$= (3 \cdot 3 \cdot 3 \cdot 3)(x^2 \cdot x^2 \cdot x^2 \cdot x^2) = 3^4 \cdot (x^2)^4 = 81x^8$$

This pattern suggests the following property.

PROPERTY OF EXPONENTS: POWER OF A PRODUCT

For any real numbers a and b and positive integer m,
$$(ab)^m = a^m b^m$$

PROBLEM SOLVING TIP

Use the exponent key on a calculator to find powers of numbers quickly.

EXAMPLE 5

Simplify.

  a. $(-2x^2)^5$ **b.** $(5xy^7)^3$ **c.** $-8(2x^3)^2$

Solution

  a. $(-2x^2)^5 = (-2)^5(x^2)^5$
$$= -32x^{10}$$

  b. $(5xy^7)^3 = (5)^3(x)^3(y^7)^3$
$$= 125x^3y^{21}$$

  c. $-8(2x^3)^2 = -8(2^2x^6) = -8(4x^6) = -32x^6$ ◀

TRY THESE

Find the degree of each monomial.

1. y^{10} **2.** $4b^2$ **3.** $-5m$ **4.** $2ab^3c^2$

Find each product.

5. $3a(-2a^3)$ **6.** $-2b^4(b^8)$ **7.** $-6c^3d(-3c^8d^2)$ **8.** $10ef^2(-ef^5)$

Simplify.

9. $(3^2)^3$ **10.** $(x^8)^4$ **11.** $(z^{14})^2$ **12.** $(4a^3)^2$

13. $(-3b)^3$ **14.** $(-2c^2d)^4$ **15.** $-(2m)^6$ **16.** $-3(x^3y)^4$

17. WRITING MATHEMATICS Make a chart summarizing the three properties of exponents presented in this lesson. Make up one example for each property.

PRACTICE

Find the degree of each monomial.

1. $3x^2$ **2.** $4a^3$ **3.** $-9c^8$ **4.** $-m^5$

5. $2x$ **6.** $-6de$ **7.** $3x^2y^4$ **8.** $-a^4b^5$

Find each product.

9. $a^3(a^4)$ **10.** $b^6(b^2)$ **11.** $2x^3(-5x^2)$ **12.** $8y^4(3y^4)$

13. $5g(-4g^9)$ **14.** $-6h^3(-7h)$ **15.** $2a^2b(9ab^5)$ **16.** $(-3p^5q^2)(-pq^7)$

17. $(-2m^3n^3)(7m^4n)$ **18.** $(-5x^6y^9)(3x^9y^6)$ **19.** $3xyz(-4x^2z^3)$ **20.** $(-8a^4b^2)(-9a^3b^2)$

Simplify.

21. $(a^3)^4$ **22.** $(b^2)^4$ **23.** $(3c^2)^3$ **24.** $(5d^7)^2$

25. $(6a^2b^3)^3$ **26.** $(-4x^9y^3)^2$ **27.** $(-3gh^7)^3$ **28.** $(-2j^4k)^5$

29. $(-2x^4yz^2)^3$ **30.** $-2(-3x^6y^{12}z^5)^4$ **31.** $-1(-4a^5b^3)^3$ **32.** $3(-2m^6n^5)^4$

33. FALLING OBJECTS The formula $d = |-16t^2|$ is used to find the distance d in feet that an object falls in t seconds. How far will a brick fall in 5 s?

34. GEOMETRY Express the volume of the box at the right in terms of x.

EXTEND

Simplify.

35. $(3x^4y)^2(-4xy^3)$

36. $(-ab^5)^5(7a^4b^2)$

37. $(p^3q^7)(-2p^2q^4)^3$

38. Write an expression for the result of n doublings of a quantity x.

39. ELECTRICITY Ohm's law is the formula $E = IR$ where E is the voltage in volts in an electrical circuit, I is the current in amperes of the circuit, and R is the resistance in ohms. The formula $W = EI$ can be used to find the power of the circuit in watts. Use these two formulas to write a formula for W in terms of I and R only.

40. PROBABILITY Write an expression for finding the probability of getting all heads on n tosses of a coin. (*Hint:* Look at the case for $n = 1$, 2, and 3 and find a pattern.)

THINK CRITICALLY

Find the value of n that makes each statement true.

41. $9x^n(7x^5) = 63x^8$

42. $(5a^nb^3)^2 = 25a^8b^6$

43. $(-q^3r^5)^n = -q^{15}r^{25}$

44. Does $(a^m)^n = (a^n)^m$? Justify your answer.

45. If $x^p(x^q) = x^{2p+1}$, write a formula for q in terms of p.

46. GEOMETRY The volume of a cube is $V = s^3$ where s is the length of a side. How much more volume does a cube with side of length $2x$ have than a cube with side of length x? Justify your answer.

PROJECT *Connection*   Aircraft must overcome the effects of drag in order to fly.

1. *Parasitic drag* is energy lost moving air around the aircraft's body. To model parasitic drag, hold a large piece of cardboard in front of you and run. Is it easier or harder to run with it? Why?

2. *Induced drag* is energy lost as the aircraft rises and air must be moved around the rising wing. One way to lower induced drag is to increase the *aspect ratio* defined as $A = \dfrac{b^2}{S}$ where b represents wing span and S represents wing area. If the wing area is held constant, what effect would increasing the wing span by 1 ft have on the aspect ratio?

3. The wing area S is defined as wing span times average chord length, or bc. Find another way to express the aspect ratio.

4. Using this new expression, explain how a designer could increase the aspect ratio.

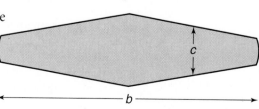

Career
Machinist

Machinists build and maintain machines used in mass production. They are involved in the manufacturing of most factory-made items, including airplanes. They must understand how many different kinds of machines work.

Airplanes are constructed from many sheets of metal. A metal punch machine is used to punch holes in a metal sheet. Two pieces of metal are joined together with rivets hammered through the holes. Usually rivet holes are punched in either a single row or a double row.

Machinists use an equation called *Pomeroy's formula* to find the power or force required by a metal punch machine to punch holes in a metal sheet. The formula is

$$P = \frac{t^2 d N}{3.78}$$

where P represents the power needed in horsepower (hp), t is the thickness of the metal in inches, d is the diameter of the hole in inches, and N is the number of holes to be punched at once.

Decision Making

1. Find the power needed to punch eight 2-in. diameter holes at the same time in a sheet of metal that is $\frac{1}{16}$-in. thick. Round to the nearest thousandth.

2. A machinist wants to punch as many holes as possible at one time. Solve Pomeroy's formula for N, the number of holes punched at once.

3. How many holes $1\frac{1}{2}$ in. in diameter can be punched at once using 0.1 hp on metal $\frac{3}{16}$-in. thick?

4. At a constant power, how can you increase the number of holes that can be punched at once?

11.3 Divide Monomials

Explore/Working Together

● Work with a partner.

1. Complete the table.

2. How does the exponent of the expression in the left column of the table change as you move down the table?

3. How does the expression in the right column of the table change as you move down the table?

Expression	Value
2^6	
2^5	
2^4	
2^3	
2^2	
2^1	

SPOTLIGHT ON LEARNING

WHAT? In this lesson you will learn
• to use properties of exponents to divide monomials.
• to use negative and zero exponents.

WHY? Dividing monomials helps you solve problems in geometry, demography, and astronomy.

4. Based on the pattern in the table, what should be the next entry below 2^1? What should its value be?

5. Continue the pattern by writing the next three entries and their values.

6. Rewrite each expression you wrote for Question 5 in the form $\frac{1}{2^n}$. What do you notice?

Build Understanding

● Dividing two monomials is similar to simplifying fractions to lowest terms. To write the fraction $\frac{15}{25}$ in lowest terms, a common factor of 5 is used to divide both numerator and denominator. So, $\frac{15}{25} = \frac{3}{5}$.

EXAMPLE 1

Simplify: $\frac{16x^5}{2x^2}$

Solution

$$\frac{16x^5}{2x^2} = \frac{2 \cdot 8 \cdot x \cdot x \cdot x \cdot x \cdot x}{2 \cdot x \cdot x}$$

$$= \frac{8 \cdot x \cdot x \cdot x}{1}$$

$$= 8x^3$$

Rewrite the variable terms without exponents.

Divide both numerator and denominator by their common factor, $2 \cdot x \cdot x$.

◀

The following examples suggest that when monomials with the same base are divided, the exponents of the bases can be subtracted.

$$\frac{y^4}{y} = \frac{y \cdot y \cdot y \cdot y}{y} = y^3 \qquad\qquad y^{4-1} = y^3$$

$$\frac{a^7}{a^3} = \frac{a \cdot a \cdot a \cdot a \cdot a \cdot a \cdot a}{a \cdot a \cdot a} = a^4 \qquad a^{7-3} = a^4$$

The pattern is summarized in the following property.

> **PROPERTY OF EXPONENTS: QUOTIENT RULE**
>
> For any real number a, $a \neq 0$, and positive integers m and n,
> $$\frac{a^m}{a^n} = a^{m-n}$$

EXAMPLE 2

Find each quotient.

a. $\dfrac{-10a^7b^4}{2ab}$ 　　　　　　　　 b. $\dfrac{4w^5z^7}{-10z^4}$

Solution

a. $\dfrac{-10a^7b^4}{2ab} = -5a^{7-1}b^{4-1} = -5a^6b^3$

b. $\dfrac{4w^5z^7}{-10z^4} = \dfrac{-2w^5z^3}{5}$　　◄

In Explore you saw that $2^0 = 1$. Any base raised to the zero power is 1. This is summarized in the following property.

> **ZERO PROPERTY OF EXPONENTS**
>
> For any real number a, $a \neq 0$,
> $$a^0 = 1$$

EXAMPLE 3

Simplify each expression.

a. $\dfrac{y^6}{y^6}$ 　　　　　 b. $\dfrac{-9x^5y^0}{3x}$ 　　　　　 c. $4^3 \cdot 6^0 - 3^4$

Solution

a. $\dfrac{y^6}{y^6} = y^{6-6} = y^0 = 1$ 　　　　 b. $\dfrac{-9x^5y^0}{3x} = \dfrac{-9x^{5-1}(1)}{3} = -3x^4$

c. $4^3 \cdot 6^0 - 3^4 = 64 \cdot 1 - 81 = 64 - 81 = -17$　　◄

ALGEBLOCKS MODEL

Find the quotient of $4xy$ and $-2x$. Model the divisor and the dividend.

Make the other dimension of the rectangle.

Read the quotient from the mat: $-2y$.

ALGEBRA: WHO, WHERE, WHEN

In 1484, Nicholas Chuquet, a French physician, wrote *Triparty en la science des nombres*, which contained an early form of exponential notation, including a notation for zero and negative exponents.

Also in Explore you saw that $2^{-1} = \frac{1}{2}$ and that $2^{-3} = \frac{1}{2^3}$. Any base raised to a negative power is the reciprocal of the base raised to the positive power. This is summarized in the following property.

┌─ **PROPERTY OF NEGATIVE EXPONENTS** ─────────────────────┐
For any real number a, $a \neq 0$, and any positive integer n,
$$a^{-n} = \frac{1}{a^n}$$
└──┘

COMMUNICATING ABOUT ALGEBRA

Write the reciprocal of a^n in two different ways. Show why each expression is the reciprocal of a^n. Compare and discuss the expressions you wrote with those of your classmates.

EXAMPLE 4

Simplify each expression. Write with positive exponents.

a. $-5a^{-3}$ **b.** $3^0 \cdot 5^{-2} \cdot 4^2$ **c.** $\dfrac{x^4 y^3}{x^7 y^2}$

Solution

a. $-5a^{-3} = \dfrac{-5}{a^3}$

b. $3^0 \cdot 5^{-2} \cdot 4^2 = 3^0 \cdot \dfrac{1}{5^2} \cdot 4^2$

$$= 1 \cdot \frac{1}{25} \cdot 16$$

$$= \frac{16}{25}$$

c. $\dfrac{x^4 y^3}{x^7 y^2} = x^{4-7} \cdot y^{3-2}$

$$= x^{-3} y$$

$$= \frac{y}{x^3}$$

◄

When a fraction is raised to a power, both the numerator and the denominator are raised to that power.

┌─ **PROPERTY OF EXPONENTS: POWER OF A QUOTIENT RULE** ─────┐
For all real numbers a and b, $b \neq 0$, and any positive integer m,
$$\left(\frac{a}{b}\right)^m = \frac{a^m}{b^m}$$
└──┘

THINK BACK

Remember:
$$\frac{1}{\frac{2}{3}} = \frac{3}{2}$$

EXAMPLE 5

Simplify each expression. Write with positive exponents.

a. $\left(\dfrac{3}{10}\right)^3$ **b.** $\left(\dfrac{5x^3}{2y}\right)^2$ **c.** $\left(\dfrac{2}{3}\right)^{-2}$

Solution

a. $\left(\dfrac{3}{10}\right)^3 = \dfrac{3^3}{10^3} = \dfrac{27}{1000}$

b. $\left(\dfrac{5x^3}{2y}\right)^2 = \dfrac{(5x^3)^2}{(2y)^2} = \dfrac{25x^6}{4y^2}$

c. $\left(\dfrac{2}{3}\right)^{-2} = \left(\dfrac{3}{2}\right)^2 = \dfrac{3^2}{2^2} = \dfrac{9}{4}$

◄

Simplify. Write with positive exponents.

1. $-\dfrac{-15x^7}{5x^5}$

2. $\dfrac{-24a^4}{-3a^3}$

3. $\dfrac{6a^{12}b^4}{2a^4b^2}$

4. $\dfrac{20p^9q}{-4p^3q}$

5. 100^0

6. w^0

7. 4^{-2}

8. 3^{-3}

9. $\dfrac{-3g^3h^2}{-9g^5h^4}$

10. $\dfrac{15km^4}{5k^2m^3}$

11. $\dfrac{6p^3q^4}{3p^3q^3}$

12. $\dfrac{-r^2s^5}{rs^6}$

13. $\left(\dfrac{4}{5}\right)^3$

14. $\left(\dfrac{-3}{4}\right)^{-2}$

15. $\left(\dfrac{2x^4}{y^3}\right)^5$

16. $\left(\dfrac{-3a}{2b^4}\right)^3$

17. MODELING Write a division sentence for the Algeblocks at the right.

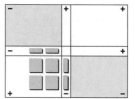

18. WRITING MATHEMATICS Write the rules of exponents presented in the lesson. Use words instead of symbols. Make up one example for each rule.

PRACTICE

Simplify. Write with positive exponents.

1. $\dfrac{-16a^7}{-2a^6}$

2. $\dfrac{9b^4}{-3b^2}$

3. $\dfrac{12a^6b^4}{-4a^5b}$

4. $\dfrac{-5c^4d^6}{7c^2d^5}$

5. $\dfrac{6x^0y^3}{y^2}$

6. $\dfrac{-7cd^0}{14c}$

7. $\dfrac{62yz^4}{14y^5z^4}$

8. $\dfrac{-2p^2q^7}{8p^4q}$

9. $2^4 \cdot 5 - 7^0$

10. $8^2 + 2 \cdot 5^0$

11. $9^0 \cdot 2^{-3} \cdot 6$

12. $5^{-2} \cdot 3^{-2} \cdot 4^0$

13. $\dfrac{-52x^7y}{-4x^3y^3}$

14. $\dfrac{35rs^4}{-7rs^{10}}$

15. $\dfrac{12s^2t^3}{-4st^4}$

16. $\dfrac{-8t^2v}{-10tv^4}$

17. $\left(\dfrac{2}{3}\right)^4$

18. $\left(\dfrac{1}{2}\right)^{-3}$

19. $\left(\dfrac{2c^4}{5d}\right)^2$

20. $\left(\dfrac{-3g^5}{2h^3}\right)^3$

21. GEOMETRY The volume of the rectangular solid shown is $4x^2y$. Find the height.

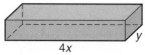

22. GEOMETRY The formula for the volume of a sphere is $V = \dfrac{4\pi r^3}{3}$ where r is the radius of the sphere. Write a formula for the volume that uses d, the diameter.

EXTEND

Simplify. Write with positive exponents.

23. $\dfrac{(2a^3b)^4(4ab^3)}{(3b)^2}$

24. $\dfrac{(4c^2)(-cd^4)^5}{8c^4}$

25. $\dfrac{(g^2h^4)^2(-2gh)^3}{(4g^3h^5)^2}$

26. $\dfrac{(3s^4t^5)^2(-st)^4}{(5s^4t^5)^2}$

27. $\left(\dfrac{2wx}{5w^4x^3}\right)^{-2}$

28. $\left(\dfrac{-y^2z^5}{3yz^3}\right)^{-3}$

29. WRITING MATHEMATICS Explain how you could use positive and negative exponents to evaluate $\dfrac{6}{K^2}$ when $K = \dfrac{2}{3}$.

DEMOGRAPHY The annual growth rate of the population of the Dominican Republic is 2.2%. The expression $p(1.022)^t$ will give an estimate of the population t years from now, if p is the present population.

30. Write an expression for the population t years ago.

31. Estimate the population two years ago if the population is now 7,800,000.

32. **ASTRONOMY** The force of gravitation between two objects is given by the formula $F = \dfrac{GmM}{r^2}$ where F is the force, G is a gravitational constant, m is the mass of one object, M is the mass of the other object, and r is the distance between the objects. Suppose that m is doubled, M is tripled, and r is doubled. Describe the effect on F. Explain your answer.

THINK CRITICALLY

Find a value of n that makes each statement true. Assume that none of the bases is 0.

33. $h^{-3} \cdot h^n = 1$

34. $(ab^3)^n = 1$

35. $\left(\dfrac{cd^6}{c^5d^n}\right)^2 = \dfrac{d^8}{c^8}$

36. If $\dfrac{p^a}{p^b} = p^2$, write a formula for a in terms of b.

37. Suppose $\dfrac{x^p}{x^q} = x^n$, $p > 0$, $q > 0$, and $n < 0$. Compare p and q. Justify your answer.

MIXED REVIEW

Solve.

38. $6x + 5 = 3x - 13$

39. $7y + 2(y - 1) = 4y + 18$

40. $2z = 9z + 28$

41. Graph these points on the coordinate plane.

    a. $(7, 0)$ **b.** $(0, -2)$ **c.** $(-1, -4)$ **d.** $(1, -3)$

42. Using the graph you made for Exercise 41, connect the points consecutively to form a closed figure. Is the figure a triangle? Justify your answer.

Solve.

43. $2x - 1 > -7$ **44.** $5 - 3x \geq 26$ **45.** $4x + 1 < -1$ **46.** $4 > 5x - 1$

47. **STANDARDIZED TESTS** Which system has the solution $(9, 2)$?

    A. $\begin{cases} x = 3y \\ x + 2y = -35 \end{cases}$ **B.** $\begin{cases} y = -2x \\ x - y = -27 \end{cases}$ **C.** $\begin{cases} x = 4y + 1 \\ 2x + y = 20 \end{cases}$ **D.** $\begin{cases} y = x + 3 \\ 2x - y = 2 \end{cases}$

Simplify. Write with positive exponents.

48. $\dfrac{-14b^9}{-7b^7}$ **49.** $\dfrac{-5x^6y^4}{x^8y^5}$ **50.** $\left(\dfrac{4a^2}{3b^3}\right)^3$ **51.** $\left(\dfrac{1}{3}\right)^{-4}$

11.4 Scientific Notation

Explore

1. Copy and complete the chart.

Exponential Expression										
10^5	10^4	10^3	10^2	10^1	10^0	10^{-1}	10^{-2}	10^{-3}	10^{-4}	10^{-5}
				10		0.1				
Value										

2. What patterns do you notice?

Build Understanding

Any number in decimal form, such as 12.456, 1,000,000,000, or 0.0083, is in **standard notation**. Large numbers, such as for the distance between planets, and small numbers, such as for the size of an atom, take up a great deal of space when written in standard notation. To conserve space, **scientific notation** is used to write such numbers. In scientific notation a number is written as a product of a number between 1 and 10 and a power of 10.

> **SCIENTIFIC NOTATION**
>
> **A number written in the form $m \times 10^n$ where $1 \leq m < 10$ and n is any integer.**

Example 1 shows the use of scientific notation for large numbers.

EXAMPLE 1

Write each number in scientific notation.

a. 8,000

b. 3,456,000

Solution

a. $8{,}000 = 8{,}000.$

8.000

$\overset{1\,2\,3}{}$

$= 8.000 \times 10^?$ Move the decimal point left so that it shows a number between 1 and 10.

$= 8 \times 10^3$ To find the exponent of 10, count the number of decimal places you moved the decimal point to the left.

b. $3{,}456{,}000 = 3.456000 \times 10^? = 3.456 \times 10^6$ ◄

Scientific notation is often used by astronomers to express distances in the universe.

EXAMPLE 2

ASTRONOMY The sun is approximately 93,000,000 mi from Earth. Write this distance in scientific notation.

Solution
$$93,000,000 = 9.3 \times 10^7$$

In scientific notation, the distance between Earth and the sun is 9.3×10^7 mi. ◄

Small numbers can also be written in scientific notation. The process is similar, except that the direction for moving the decimal point is reversed and negative exponents are used.

EXAMPLE 3

Write each number in scientific notation.

a. 0.0006

b. 0.000000792

Solution

a. $0.0006 = \underline{0006.} \times 10^?$ Move the decimal
point to the right of
$$= 6 \times 10^{-4}$$ the first nonzero digit.

b. $0.000000792 = \underline{0000007.92} \times 10^?$

$$= 7.92 \times 10^{-7}$$ ◄

CHECK UNDERSTANDING

To write 7,350,000 in scientific notation, would you use a positive number or a negative number for the exponent? to write 0.00894?

To write numbers in standard notation, reverse the process.

EXAMPLE 4

Write each number in standard notation.

a. 4.03×10^6 **b.** 5.2×10^{-5} **c.** 2.86×10^{-8}

Solution

a. $4.03 \times 10^6 = 4\,030000.$ Move the decimal point to
the right 6 places.
$$= 4,030,000$$

b. $5.2 \times 10^{-5} = .00005\,2$ Move the decimal point to
the left 5 places.
$$= 0.000052$$

c. $2.86 \times 10^{-8} = .00000002\,86$ Move the decimal point to
the left 8 places.
$$= 0.0000000286$$ ◄

CHECK UNDERSTANDING

When changing from scientific notation to standard notation, how do you know in which direction to move the decimal point?

Numbers written in scientific notation can be multiplied and divided using properties of exponents.

EXAMPLE 5

Find the product of 4.2×10^5 and 2.31×10^4.

Solution

Use the associative and commutative properties to rearrange the factors. Use the product of powers property for exponents to multiply the powers of 10.

$$(4.2 \times 2.31)(10^5 \times 10^4) = 9.702 \times 10^9 \qquad \blacktriangleleft$$

Sometimes, you must rewrite the number you get after computing so that your answer is in scientific notation.

EXAMPLE 6

ASTRONOMY The star Vega in the constellation Lyra is 23 light-years away from Earth. A *light-year* is the distance light travels in 1 year, 5.88×10^{12} mi. Find the distance from Earth to Vega in miles. Write the distance in scientific notation.

Solution

Find the product of 23 and 5.88×10^{12}.

$$
\begin{aligned}
23 \times 5.88 \times 10^{12} &= (23 \times 5.88) \times 10^{12} \\
&= 135.24 \times 10^{12}
\end{aligned}
$$

Rewrite 135.24 in scientific notation.

$$
\begin{aligned}
135.24 \times 10^{12} &= (1.3524 \times 10^2) \times 10^{12} \\
&= 1.3524 \times 10^{14}
\end{aligned}
$$

The star Vega is 1.3524×10^{14} mi from Earth. $\qquad \blacktriangleleft$

TRY THESE

Write each number in scientific notation.

1. 3,250,000,000   **2.** 109,000,000   **3.** 72,000,000   **4.** 923,000,000

5. 0.00315   **6.** 0.000072   **7.** 0.0000054   **8.** 0.000432

Evaluate. Write each result in scientific notation.

9. $(3.5 \times 10^4)(2.1 \times 10^5)$ **10.** $(1.4 \times 10^7)(4.5 \times 10^8)$

11. $\dfrac{9.6 \times 10^9}{1.2 \times 10^2}$ **12.** $\dfrac{8.4 \times 10^6}{2.1 \times 10^3}$

13. $(7.6 \times 10^4)^2$ **14.** $(6.8 \times 10^6)(9 \times 10^9)$

15. **ASTRONOMY** The mean distance of Venus to the sun is 67.2 million miles. Write this number in scientific notation.

16. **WRITING MATHEMATICS** Describe the steps used in writing 0.00062 in scientific notation.

PRACTICE

Write each number in standard notation.

1. 5.12×10^8 2. 9.7×10^{10} 3. 1.2×10^{-5} 4. 6.5×10^{-4}

Write each number in scientific notation.

5. 314,000   6. 4,300,000   7. 23,000,000   8. 610,000,000

9. 0.000415   10. 0.00103   11. 0.000008   12. 0.000032

Evaluate. Write each answer in scientific notation.

13. $(2.4 \times 10^5)(1.3 \times 10^9)$ 14. $(4 \times 10^6)(2.1 \times 10^8)$ 15. $\dfrac{9 \times 10^7}{3 \times 10^4}$

16. $\dfrac{7.4 \times 10^{10}}{2 \times 10^6}$ 17. $(2 \times 10^5)^2$ 18. $(3 \times 10^6)^2$

19. $(4.6 \times 10^4)(9 \times 10^{15})$ 20. $(8.1 \times 10^5)(7.5 \times 10^{11})$

Write the number in each problem in scientific notation.

21. **ASTRONOMY** The distance from Mercury to the sun is 57,900,000 mi.

22. **HISTORY** Archimedes wrote that the Greeks had traditional number names only for numbers through myriad myriads. (A myriad was equal to 10,000.) What was the largest number the Greeks could name?

23. **PHYSICS** The wave length of red light is 0.0000065 m.

24. **BIOLOGY** The number of hairs on the average human head is 1.5×10^5. There are approximately 5×10^9 people in the world. About how many human hairs are there in the world?

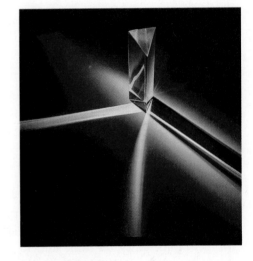

25. **CHEMISTRY** The cross-sectional area of a molecule is about 1.5×10^{-25} mm². What are the fewest number of molecules necessary to cover the head of a pin of diameter 2.25 mm?

EXTEND

Estimate. Write each estimate using scientific notation.

26. $(9.23 \times 10^7)^2$ 27. $(1.09 \times 10^9)^4$ 28. $\dfrac{2.03 \times 10^6}{3.98 \times 10^3}$

29. $\dfrac{3.13 \times 10^{10}}{8.6 \times 10^4}$ 30. $(7.35 \times 10^4)(8.19 \times 10^9)$ 31. $(4.7 \times 10^7)(8.3 \times 10^9)$

Use scientific notation to solve each problem.

32. **AERONAUTICS** The equatorial radius of Earth is about 4000 mi. How long will it take a caped superhero traveling at 600 mi/h to fly around Earth at a constant altitude of 100 mi directly over the equator?

33. **SPEED OF SOUND** The speed of sound through air varies with the air temperature. At 0°C, the speed of sound is 330 m/s. At 1000°C, it is 700 m/s. Find the number of meters per second increase for each degree Celsius.

34. **WRITING MATHEMATICS** Compare finding products and quotients using standard and scientific notation. What do you think are the advantages and disadvantages of each system?

THINK CRITICALLY

Find the value of p that will make each sentence true.

35. $4.23 \times 10^6 = 423 \times 10^p$

36. $9.1 \times 10^{-7} = 0.091 \times 10^p$

37. $5.3 \times 10^7 = p \times 10^6$

38. $6.7 \times 10^{-4} = p \times 10^{-8}$

39. If $(a \times 10^m)(b \times 10^n) = c \times 10^{m+n+1}$, in scientific notation, what must be true about the product, ab?

PROJECT *Connection*

1. Make a plane by folding an $8\frac{1}{2}$ by 11 in. sheet of paper as shown.

2. Use a ruler. Along the bottom of the plane, mark off and label every $\frac{1}{2}$ cm starting at the front.

3. Place a paper clip on the first mark, fly the plane, and record the distance. Repeat this test for each mark. For consistency, the same student should "pilot" the plane each time. Decide if you want one or more tests for each mark to collect reliable data.

4. Identify the two marks for which the plane seems to fly best, then test-fly with the clip between these marks to determine final placement. Graph and label your results.

5. When you have found the best spot for the paper clip, turn the plane upside down and hang it with a thread placed through the clip. How does the plane hang? You have just located the *center of gravity* of your plane. Discuss why this point is important when landing an aircraft.

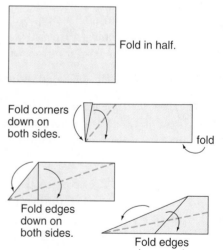

Fold in half.

Fold corners down on both sides.

fold

Fold edges down on both sides.

Fold edges down again.

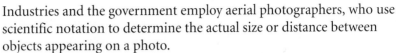

Industries and the government employ aerial photographers, who use scientific notation to determine the actual size or distance between objects appearing on a photo.

A photographer has taken an aerial photo and wants to determine the actual length of a road between two points for the photo caption. First he needs to determine the 1–1 scale factor.

The diagram shows the relationship between the actual length of the road AB and the length of the image of the road PQ on the photo. The camera lens is at C, the focal length f of the camera is 150 mm, and the height above ground H of the camera is 0.96 km.

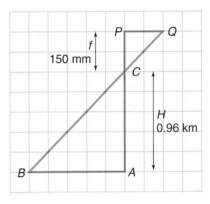

Decision Making

The ratio of the image length PQ to the actual length AB is the same as $\frac{f}{H}$. The ratio $\frac{f}{H}$ is called a 1–1 scale factor when the numerator and denominator are in the same units.

1. Change f, 150 mm, to meters.

2. Change H, 0.96 km, to meters.

3. Write the values you found for f and H in Questions 1 and 2 in scientific notation.

4. Find the 1–1 scale factor $\frac{f}{H}$ in scientific notation.

5. Use the proportion $\frac{PQ}{AB} = \frac{f}{H}$ to find the actual length of the road AB if the image length PQ is 225 mm.

Photo resolution is the smallest actual length whose image can be measured on the photograph. With current technology, it is possible to make measurements on a photograph to the nearest micron (10^{-6} m).

6. Using the proportion formula, find the resolution of the camera with the 1–1 scale factor you found in Question 4.

7. Name three posible applications for aerial photography.

11.5 Algebra Workshop
Add and Subtract Expressions

Think Back

1. Use Algeblocks and the Basic Mat to model $4x + 3y - 2x - y$. Combine like terms and remove any zero pairs, sketching each step. Write the resulting algebraic expression.

2. How many kinds of unlike terms were in the original expression in Question 1? How many terms did your algebraic expression for Question 1 have?

3. Model $x + 4 - 2x - 6$. Combine like terms and remove any zero pairs, sketching each step. Write the resulting algebraic expression.

4. How many kinds of unlike terms were in the original expression in Question 3? How many terms did your algebraic expression for Question 3 have?

5. What pattern do you notice in Questions 2 and 4?

6. Model $3x + 2 - y + 1 - 2x - y - x$. Combine like terms and remove any zero pairs, sketching each step. Write the resulting algebraic expression.

7. How many kinds of unlike terms were in the original expression in Question 5? How many terms did your algebraic expression for Question 5 have? Why are they different?

Explore

Work with a partner and Algeblocks.

8. Model the following addition.

$$(x^2 + 2x) + (2x^2 - 5x)$$

9. How many different kinds of blocks are on the mat? What do the different kinds of blocks represent?

10. Group blocks that are alike together. Remove any zero pairs from the mat. Write an algebraic expression for the blocks that are left.

Use Algeblocks to find each sum.

11. $(3x^2 - 2x + 4) + (5x - 3)$

12. $(x^3 + 4x + 3) + (-2x + 3)$

13. $(3xy + 5x - 3y) + (-2xy - 3x - 1)$

14. WRITING MATHEMATICS Make up an addition exercise that has the answer $3x + y - 5$.

15. Combine the expression $2x - 3$ with its opposite, $-2x + 3$, on the same mat. Find the sum. Explain your answer.

Use Algeblocks to model each of the following. First model the expression inside the parentheses. Then move each piece to the opposite side of the mat. Record the result.

16. $-(x + 2)$ **17.** $-(2xy - y)$ **18.** $-(x^2 + y^2)$ **19.** $-(-x + 3)$

20. Model $(4x - 2) - (2x - 3)$ by adding the opposite of $2x - 3$ to $4x - 2$. What is the result?

21. What is the opposite of $2x^2 + x - 3$? If you combined the original expression, $2x^2 + x - 3$, with its opposite, what would you find?

22. Model $(3x^2 + 4x + 2) - (2x^2 + x - 3)$. Remember to add the opposite of the expression being subtracted.

Find each difference.

23. $(3x - 2) - (x + 4)$ **24.** $(2xy + y) - (3xy - x)$

25. $(3x^2 + 4x - 1) - (2x^2 - 7)$ **26.** $(2y^2 - 3y) - (y^2 + 3y)$

Make Connections

27. Use the distributive property to show that $3x + 2x = 5x$.

Which of these expressions can be simplified by combining like terms? Write *yes* or *no*.

28. $4x^2 + 2x^2$ **29.** $2y + 2y^2$

30. $3xy + 2xy$ **31.** $3y^2 - 2xy$

32. Compare your answers to Questions 28–31 with those of other classmates. In your own words, define like terms.

33. Use the distributive property to show that $4 \times 10^2 + 2 \times 10^2$ equals 6×10^2. How is this expression like the expression in Question 28?

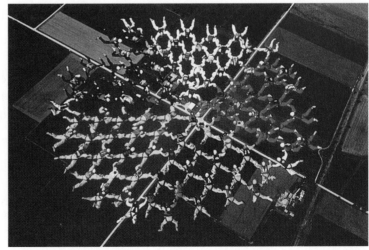

Algebra Workshop

The numbers below represent the total population of the world by region.

North America: 290,000,000   Asia: 3,392,000,000
Latin America: 470,000,000   Africa: 700,000,000
Europe: 728,000,000   Oceania: 28,000,000

34. Rewrite each number in scientific notation.

35. Note that the numbers you wrote for Question 34 do not have the same exponent for 10. How could you change the form of some of the numbers so that you can add them? (The new forms will not be in scientific notation.)

36. Add to find the total population of the world. Write the answer in scientific notation.

Summarize

37. MODELING The mat at the left shows Algeblocks for $3x + 2x$. The mat at the right shows $3x^2 + 2x$.

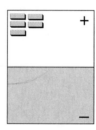
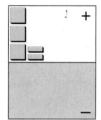

   a. Are the two mats different from one another?

   b. Use the mats to simplify $3x + 2x$. Can $3x^2 + 2x$ be simplified? Explain.

38. WRITING MATHEMATICS Write a paragraph describing how to add two algebraic expressions.

39. THINKING CRITICALLY How can you find the opposite of an algebraic expression? How is the opposite of an expression like the opposite of a number?

40. WRITING MATHEMATICS Write a paragraph describing how to do the subtraction $(6x^2 - y) - (x - y)$.

41. GOING FURTHER Which terms in the expression $3x^4 + 2x^3 + 2x + 5x^4 + 4x^3 + 3x + 12$ can be combined? Simplify the expression.

Explore

- Here is a mathematical trick you can play using the numbers in a calendar.

January 1997

S	M	T	W	T	F	S
			1	2	3	4
5	6	7	8	9	10	11
12	13	14	15	16	17	18
19	20	21	22	23	24	25
26	27	28	29	30	31	

1. A 3×3 arrangement of numbers has been chosen on the calendar at the right. Find the sum of the nine circled numbers.

2. Multiply 9 by the center number of the 3×3 arrangement. What do you notice?

3. Choose a different 3×3 arrangement on a different month in a calendar. Does the same relationship appear? Describe a shortcut for finding the sum of the nine numbers in a 3×3 arrangement for any month in a calendar.

4. The matrix at the right shows an algebraic expression for each number in any 3×3 arrangement on a calendar if the center number of the arrangement is x. Show that these expressions work for the dates circled in the calendar in Question 1.

$$\begin{bmatrix} x - 8 & x - 7 & x - 6 \\ x - 1 & x & x + 1 \\ x + 6 & x + 7 & x + 8 \end{bmatrix}$$

5. Use the algebraic expressions in the matrix to show that the sum of the nine expressions will always be $9x$. Explain how you found your answer.

Build Understanding

The expressions in the matrix in Question 4 of Explore are all *polynomials*. A **polynomial** is either a monomial or a sum or difference of monomials. Each monomial is a **term** of the polynomial. For example, the polynomial $x^2 + y^2 + 2xy + 5$ has four terms. Some polynomials have special names. A polynomial of two terms is a **binomial**. A polynomial of three terms is a **trinomial**.

The **degree of a polynomial** is the greatest degree of any of its terms.

EXAMPLE 1

Determine the degree of each polynomial.

a. $2x^2 + 5x + 3$

b. $x^2y^3 + 4y^4$

Solution

a. The degree of each term of the trinomial $2x^2 + 5x + 3$ is

$$2x^2 + 5x + 3$$
$$\uparrow \qquad \uparrow \qquad \uparrow$$
$$\text{degree:} \quad 2 \qquad 1 \qquad 0$$

Since the greatest degree of any of the terms is 2, the degree of the trinomial is 2.

b. Since the degree of x^2y^3 is 5 (the sum of the exponents of its variable factors) and the degree of $4y^4$ is 4, the degree of the binomial $x^2y^3 + 4y^4$ is 5.

Polynomials are usually arranged so that the degrees of one variable in the terms are either in ascending or descending order.

EXAMPLE 2

Arrange the terms of each polynomial in descending order of degree of the variable x.

a. $4 + 3x^2 + 7x + 5x^3$

b. $3x^2y^5 + 7xy^8 + 2x^3y^2$

Solution

For descending order, the term with the greatest degree should appear first.

a. $5x^3 + 3x^2 + 7x + 4$

b. $2x^3y^2 + 3x^2y^5 + 7xy^8$

Recall that the terms $4x^2$ and $-7x^2$ are **like terms** because they have the same variable with the same exponent. Like terms can be combined by using the distributive property.

$$4x^2 + (-7x^2) = [4 + (-7)]x^2$$
$$= -3x^2$$

Polynomials are added by combining their like terms.

EXAMPLE 3

Add the following polynomials.

$$(2x^2 + 3x - 2) + (x^2 - 3)$$

Solution

One way to find the sum is to use the commutative and associative properties to rearrange the terms of the two polynomials so that like terms are together.

$$(2x^2 + 3x - 2) + (x^2 - 3)$$
$$= (2x^2 + x^2) + 3x + [-2 + (-3)]$$

Then use the distributive property to combine like terms by adding their coefficients.

$$= (2 + 1)x^2 + 3x + [-2 + (-3)]$$
$$= (3)x^2 + 3x + (-5)$$
$$= 3x^2 + 3x - 5$$

Another way to find the sum is to write the polynomials in vertical form, with like terms aligned in columns.

$$\begin{array}{r} 2x^2 + 6x - 4 \\ + \ 2x^2 \qquad - 5 \\ \hline 4x^2 + 6x - 9 \end{array}$$

There is no x-term; leave a space.
Like terms are combined in each column.

The sum of the two polynomials is $4x^2 + 6x - 9$. ◀

Recall that you can subtract a number by adding its opposite.

$$a - b = a + (-b)$$

The opposite of a polynomial is the opposite of each of its terms.

EXAMPLE 4

Find the opposite of each polynomial.

a. $2x + 7$

b. $x^2 + 4x - 2$

Solution

a. The opposite of $2x + 7$ is $-(2x + 7) = -2x - 7$.

b. The opposite of $x^2 + 4x - 2$ is $-x^2 - 4x + 2$. ◀

To subtract two polynomials, add the opposite of the polynomial being subtracted.

CHECK UNDERSTANDING

Will the sum of two binomials always result in another binomial? Explain your answer.

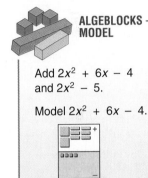

ALGEBLOCKS MODEL

Add $2x^2 + 6x - 4$ and $2x^2 - 5$.

Model $2x^2 + 6x - 4$.

Add $2x^2 - 5$.

Read the result: $4x^2 + 6x - 9$.

EXAMPLE 5

Subtract: $(5y^2 + 2y - 2) - (3y^2 + 7y - 8)$

Solution

Write the opposite of $3y^2 + 7y - 8$, then combine like terms.

$$\begin{aligned}
&(5y^2 + 2y - 2) - (3y^2 + 7y - 8)\\
=\ &5y^2 + 2y - 2 + (-3y^2 - 7y + 8)\\
=\ &5y^2 - 3y^2 + 2y - 7y - 2 + 8\\
=\ &(5 - 3)y^2 + (2 - 7)y + (-2 + 8)\\
=\ &2y^2 - 5y + 6
\end{aligned}$$

The difference is $2y^2 - 5y + 6$.

A polynomial is in **simplest form** when it contains no like terms.

EXAMPLE 6

GEOMETRY Find the perimeter of the triangle. Write the answer in simplest form.

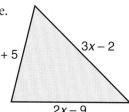

Solution

Find the perimeter by adding the lengths of each side.

$$\begin{array}{r}
x + 5\\
2x - 9\\
+\ 3x - 2\\
\hline
6x - 6
\end{array}$$

In simplest form, the perimeter of the triangle is $6x - 6$.

TRY THESE

Give the degree of each polynomial.

1. $2x^6 - 11x^4$ **2.** $-6y^2 + 2y + 9$ **3.** $9xy + 5x^2y$

Arrange each polynomial in descending order of degree of the variable x.

4. $4x + 8 + 2x^3 + 3x^2$ **5.** $6xy^3 - 2x^4y + 3x^2y^2 + 9$

Add.

6.
$$\begin{array}{r}
5x - 11\\
+\ 6x - 1\\
\hline
\end{array}$$

7.
$$\begin{array}{r}
-2x^2 + 6x - 9\\
+\ 4x^2 - 3x + 4\\
\hline
\end{array}$$

8. $8x + 2y + (7y - 8)$ **9.** $7x^2y - 3xy + (8y + 2x^2y + 4xy)$

Find the opposite of each polynomial.

10. $-3x + 9$ **11.** $8a^2 + 2a - 7$ **12.** $9b^4 - 1$ **13.** $a + 2b + c$

Subtract.

14. $7a - 2$
$\underline{-(4a + 1)}$

15. $3b^2 + 2b - 1$
$\underline{-(-2b^2 - 7b + 4)}$

16. $(6h + 2m) - (4h + 5m - 1)$

17. $(9p^2 + 5p + 3q) - (6q^2 + 7q)$

18. MODELING Show the sum of $(x^2 + 2x + 1) + (x^2 - 3x - 4)$ using Algeblocks.

19. WRITING MATHEMATICS In your own words, explain what like terms are. Tell how like terms are involved in the addition and subtraction of polynomials.

20. GEOMETRY Find the perimeter of the figure at the right and express it in simplest form.

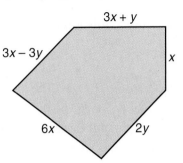

PRACTICE

Find the degree of each polynomial.

1. $10x^2 - 25$

2. $6x + 4x^3 + x^5$

3. $8a^2b - 2a^2b^2$

4. $11c^3 + 4cd^2 - 6d^7$

5. $9yz + 2y^2z^5 + 3z^3$

6. $100 - g^4$

Perform the indicated operation.

7. $4a - 3b + 5c$
$\underline{+ 8a + 3b - 7c}$

8. $5r + 3s + t$
$\underline{+ r + 9s + 7t}$

9. $-4p + q - 8$
$\underline{+ p + 5q + 17}$

10. $4x^2 - 8x$
$\underline{+ 9x^2 - 2x}$

11. $2a + 4b - c$
$\underline{-(6a + 3b + 5c)}$

12. $x - y + z$
$\underline{-(x + y - z)}$

13. $8y^4 + 7y^2 - 3$
$\underline{-(2y^4 - y^2 + 7)}$

14. $z^3 + 4z^2 - z$
$\underline{-(3z^3 - 2z^2 + 5z)}$

15. $(5x + 7y - 10) + (3y - 14)$

16. $(a^2 - 2ab + b^2) + (3a^2 - 6b^2)$

17. $(a + 2b) + (b - c) + (3a + 5c)$

18. $(4r^2 + 3r) + (5r^3 - r) + (r^3 + 4r)$

19. $(7d - 8d^2 + d^3) - (3d + d^2 - 4d^3)$

20. $(8k - 3k^2) - (9k + 2)$

21. $(8x^2 + 5xy - y^2) - (6x^2 - 3xy + y^2)$

22. $(4b^2 - 3b) - (7b + 6)$

23. GEOMETRY Write a polynomial for the area of the shaded region shown below.

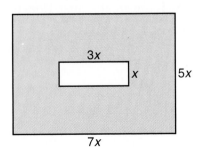

24. GEOMETRY Find the surface area of the rectangular prism shown below. Write the answer in simplest form.

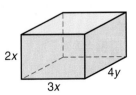

25. Subtract $4x + y$ from the sum of $x + 3y$ and $8x - 2y$.

26. Subtract $x^2 + x + 2$ from the sum of $2x + 3$ and $4x^2 + x - 7$.

Use a calculator to evaluate each polynomial for $x = 0.6$.

27. $2x^2 + 7x - 3.2$

28. $x^3 - 5x^2 + 4x - 7$

Evaluate each polynomial for $y = -2g$.

29. $y^2 + 2y - 3$

30. $2y^3 - 5y^2 + y - 3$

AERONAUTICAL DESIGN The total gross weight W_g of an airplane can be expressed by the polynomial $W_{st} + W_e + W_f + W_p$, where W_{st} is the structural weight of the empty plane less the weight of the engines, W_e is the weight of the engine or propulsion system, W_f is the weight of the fuel, and W_p is the weight of the payload (passengers and cargo). Use this information to answer Questions 31–32.

31. For a particular airplane with a particular engine, which of the terms in this polynomial are constants?

32. Suppose the total gross weight cannot exceed a certain limit. The amount of fuel determines the distance that the plane can travel. Assuming that W_g is at the limit, what is the effect on another of the variables if the amount of fuel and its weight W_f are increased?

DRIVING SAFETY The polynomial $x + \dfrac{x^2}{20}$ is the stopping distance of the car under ideal conditions in feet, after the brakes are applied. The variable x is the speed of the car in miles per hour before braking. Use this information to answer Questions 33–34.

33. Find the stopping distance if the car has been traveling at 40 mi/h.

34. A common formula is to leave one car length (approximately 20 ft) between your car and the car in front of you for each 10 mi/h of the speed at which you are traveling. Does this work based on the polynomial given here? Explain.

THINK CRITICALLY

Find the missing term.

35. $(3x^2 + 2x - 4) + (\boxed{} + 5x^2 - 9) = 8x^2 - 5x - 13$

36. $(x^2y + 5xy - 9y^2) - (3x^2y - 2y^2 + \boxed{}) = -2x^2y + 9xy - 7y^2$

37. Is the following *true* or *false*? Explain. The opposite of $a - b$ is $b - a$.

MIXED REVIEW

Evaluate. Round decimals to the nearest hundredth.

38. $z = \sqrt{121}$ **39.** $z = \sqrt{289}$ **40.** $z = \sqrt{47}$ **41.** $z = \sqrt{103}$

Solve.

42. A sales representative receives 4% commission on total sales. To earn $450, what amount of sales does she need?

43. The population of Briarwood decreased by 2.5% from last year. If the population last year was 20,000, what was is this year?

Write in standard notation.

44. 3.4×10^6 **45.** 1.02×10^8 **46.** 4.8×10^{-7} **47.** 3.03×10^{-5}

PROJECT Connection

You will be exploring different sizes and shapes of parachutes. You will need several sheets of paper, tape, thread, scissors, a ruler, a clothespin, a stopwatch, and a compass.

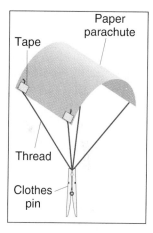

1. Cut out squares of paper with side lengths of 8 in., 10 in., 12 in., and 14 in. For each square, tape a piece of thread to each corner. The pieces of thread should be the same length and about double the length of one side of the square. Tape the pieces of thread together underneath the chute and clip the clothespin over them.

2. Drop each square parachute from a high place. (Stand on a chair or partway up a staircase and drop the chute. Be careful!) Time and record how long each parachute takes to hit the ground. For each chute, do at least three trials and average the times.

3. Repeat the experiment with circular chutes of the same area as the square chutes. Explain how you determine the radii of your circles.

4. Make a scatter plot for each data set, using different colors to represent the square chutes and the circular chutes.

5. From your graph, estimate the drop time of a 16-by-16-in. square parachute.

6. Estimate the drop time of a circular parachute with equal area.

7. Which parachute style works better, the square or the circular? Why do you think this is so?

8. What do you think would happen if you cut a small hole in the center of the chute? Test your prediction. Summarize your findings.

Think Back

● To use Algeblocks in multiplying two monomials, place the blocks for one factor along the horizontal axis and place the blocks for the other factor along the vertical axis of a Quadrant Mat.

1. What multiplication is shown by the Algeblocks at the right?

2. Use Algeblocks to model the multiplication. Read the product from the mat.

3. Write two factors whose product is $-2x^2$.

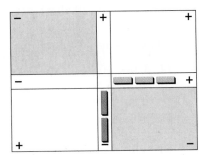

Explore

● Work with a partner. Use Algeblocks to multiply $2x(x - 2)$.

4. Place blocks for the first factor, $2x$, along the horizontal axis. Place blocks for the second factor, $x - 2$, along the vertical axis. Which regions will you use to model the multiplication?

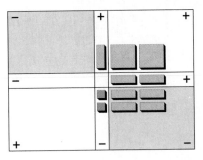

5. Model the multiplication. Write the product that matches each area.

6. The total product is the sum of the rectangular areas. Write a polynomial for the sum.

Write a polynomial for the product shown on each Quadrant Mat.

7.

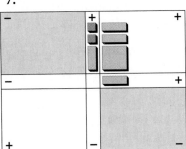

8.

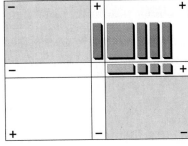

Use Algeblocks to model these products.

9. $2x(x + y)$

10. $y(y + 2)$

11. $-3x(2x + 1)$

12. $-2y(x - y)$

13. Use the distributive property to simplify $2x(x - 2)$. Explain how the product relates to the Algeblocks model completed in Question 5.

You can use Algeblocks to multiply two binomials.

Multiply: $(2x + 1)(x - 3)$

14. Place the first factor along the horizontal axis and the second along the vertical axis. Arrange the pieces as shown at the right.

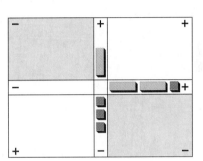

15. Model the multiplication. Notice that within the large rectangle are smaller rectangles made up of blocks that are the same size and shape. Write a monomial for each of these smaller rectangles.

16. Write a polynomial for the entire rectangular area. Can the polynomial be simplified? If so, how?

Use Algeblocks to model each product.

17. $(x + 2)(x + 3)$

18. $(x - 1)(x + 4)$

19. $(x + 1)(x - 4)$

20. $(x + 3)(x - 2)$

21. $(2y + 2)(y - 2)$

22. $(x + 1)(x + 2)$

23. The mat to the right shows Algeblocks that model the multiplication problem $(x + 1)(x - 2)$ and its product. Note the four circled rectangles.

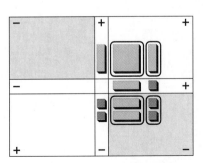

   a. Write the two factors that form each circled rectangle.

   b. Write a monomial for each circled rectangle.

   c. Write a simplified polynomial for the entire product.

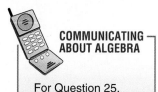

COMMUNICATING ABOUT ALGEBRA

For Question 25, how will your work look different if you multiply $(x - 3)(2x + 1)$? Will the product be the same?

Make Connections

24. Look at your answers in Question 23a. How does the distributive property of multiplication relate to these products?

25. In Questions 14–16, you used Algeblocks to find $(2x + 1)(x - 3)$. In using algebra to multiply two binomials, you use the distributive property twice to form the sum of four products.

$$(2x + 1)(x - 3) = (2x)(x - 3) + 1(x - 3)$$
$$= (2x)(x) + (2x)(-3) + (1)(x) + (1)(-3)$$

   a. Complete each product.
   b. Simplify the polynomial.

Use the distributive property to multiply.

26. $(x - 3)(x - 4)$ **27.** $(x + 7)(x - 5)$ **28.** $(2y - x)(y + 2x)$

You can check to see that the product of two binomials is correct by substituting values for the variable.

29. Substitute $x = 5$ in the expression $(x - 3)(x - 4)$. What value for the expression did you get?

30. Substitute $x = 5$ in the expression you got as an answer to Question 26. Is this the same value you got in Question 29? What does this tell you about your answer to Question 29?

Summarize

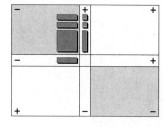

31. MODELING Write the two factors and the product for the model on the mat at the top left.

32. WRITING MATHEMATICS Describe how to multiply a polynomial by a monomial.

33. THINKING CRITICALLY If a polynomial in simplest form is multiplied by a monomial, how many terms will the product have?

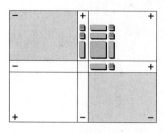

34. MODELING Write the two factors and the product for the model on the mat at the bottom left.

35. WRITING MATHEMATICS Write a paragraph describing how to use the distributive property to multiply two binomials.

36. GOING FURTHER Write out each step necessary to multiply $(x + 3)(x^2 - 2x + 5)$. Then find the total product.

11.8 Multiply Polynomials

Explore/Working Together

● Work with a partner.

Below are five arrangements of unit squares. While the first arrangement consists of only one square, the rest include more squares than you may notice at first.

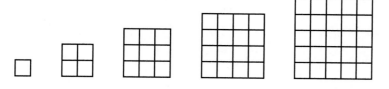

1. Study each arrangement of unit squares carefully. Then copy and complete this table in which n is the number of units of the side of the largest square.

	n				
	1	2	3	4	5
Number of 1 X 1 Squares					
Number of 2 X 2 Squares					
Number of 3 X 3 Squares					
Number of 4 X 4 Squares					
Number of 5 X 5 Squares					
Total Number of Squares					

A polynomial can model this problem.

2. In each polynomial below, substitute each value of n from the table to determine which polynomial will always give you the correct total number of squares. (More than one correct answer is possible.)

a. $\dfrac{n(n + 1)(2n + 1)}{6}$ b. $\dfrac{(n^2 + n)(2n + 1)}{6}$

c. $\dfrac{n^3 + 3n^2 + 2}{6}$ d. $\dfrac{2n^3 + 3n^2 + n}{6}$

3. More than one polynomial works for any given value of n. What does that tell you about those polynomials?

4. Use a polynomial model to determine the number of squares on a checkerboard, which is an 8×8 square.

Build Understanding

The distributive property of multiplication over addition can be used to multiply a polynomial by a monomial.

EXAMPLE 1

Find the product.

   a. $2x(x + 6)$ **b.** $y(y^2 - 3y + 4)$

Solution

   a. $2x(x + 6) = 2x(x) + 2x(6)$
                     $= 2x^2 + 12x$

   b. $y(y^2 - 3y + 4) = y(y^2) + y(-3y) + y(4)$
                                 $= y^3 + (-3y^2) + 4y$
                                 $= y^3 - 3y^2 + 4y$ ◀

The diagram at the right shows an area model for multiplying $x(x + 2)$.

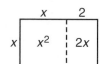

A similar diagram can model the product of two binomials. This rectangle has a width of $x + 2$ and a length of $x + 3$. Notice that the total area $(x + 2)(x + 3)$ consists of four smaller areas, $x^2, 3x, 2x,$ and 6. The total area is the simplified polynomial $x^2 + 5x + 6$.

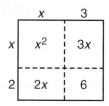

To multiply two binomials, use the distributive property twice.

EXAMPLE 2

Find the products.

   a. $(x + 4)(x - 3)$ **b.** $(2y - 1)(y - 5)$

Solution

   a. $(x + 4)(x - 3) = x(x - 3) + 4(x - 3)$
                         $= x(x) + x(-3) + 4(x) + 4(-3)$
                         $= x^2 + (-3x) + 4x + (-12)$
                         $= x^2 + x - 12$

   b. $(2y - 1)(y - 5) = 2y(y - 5) + (-1)(y - 5)$
                         $= 2y(y) + (2y)(-5) + (-1)(y) + (-1)(-5)$
                         $= 2y^2 + (-10y) + (-y) + 5$
                         $= 2y^2 - 11y + 5$ ◀

Any two polynomials can be multiplied using the distributive property. FOIL is a mnemonic device, or memory aid, for remembering which terms to multiply when multiplying two binomials.

COMMUNICATING ABOUT ALGEBRA

Show how you can multiply 43 by 25 by multiplying two binomials. (Use the expanded form of each number, 40 + 3 and 20 + 5.)

The following is an example of how **FOIL** works in finding the product of $w + 3$ and $w - 5$. Multiply the First terms, the Outer terms, the Inner terms, and the Last terms.

$$
\begin{aligned}
(w + 3)(w - 5) &= \overset{F}{w^2} - \overset{O}{5w} + \overset{I}{3w} - \overset{L}{15} \\
&= w^2 - 2w - 15 \quad \text{Combine like terms.}
\end{aligned}
$$

Sometimes you may prefer to multiply vertically.

EXAMPLE 3

Find the product of $x + 2$ and $x^2 + 4x - 3$.

Solution

Arrange the work vertically so that like terms are in the same column.

$$
\begin{array}{r}
x^2 + 4x - 3 \\
x + 2 \\
\hline
2x^2 + 8x - 6 \qquad 2(x^2 + 4x - 3) \\
x^3 + 4x^2 - 3x \qquad\;\; x(x^2 + 4x - 3) \\
\hline
x^3 + 6x^2 + 5x - 6 \qquad \text{Combine like terms.}
\end{array}
$$

The product is $x^3 + 6x^2 + 5x - 6$. ◄

CHECK UNDERSTANDING

How would you use an area model to multiply a binomial by a trinomial?

As you practice multiplying binomials, you will notice patterns. Sometimes the product of two binomials is not a trinomial but another binomial.

EXAMPLE 4

Find the products. Describe the pattern.

  a. $(x - 4)(x + 4)$ **b.** $(2x + 7)(2x - 7)$

Solution

  a. $(x - 4)(x + 4)$
    $= x(x + 4) + (-4)(x + 4)$
    $= x(x) + x(4) + (-4)(x) + (-4)(4)$
    $= x^2 + 4x + (-4x) + (-16) \qquad$ Opposite middle terms.
    $= x^2 - 16$

  b. $(2x + 7)(2x - 7)$
    $= 2x(2x - 7) + 7(2x - 7)$
    $= 2x(2x) + 2x(-7) + 7(2x) + 7(-7)$
    $= 4x^2 + (-14x) + 14x + (-49) \qquad$ Opposite middle terms.
    $= 4x^2 - 49$ ◄

CHECK UNDERSTANDING

Do you think $(3x + 2)(2 - 3x)$ fits the pattern described in Example 4? Explain.

The product in Example 4 reveals a pattern. When you multiply two binomials with identical terms, where one binomial is a sum and the other is a difference, the product is the difference of the squares of the first and second terms.

> **PRODUCT OF A SUM AND A DIFFERENCE**
>
> $$(a + b)(a - b) = a^2 - b^2$$

CHECK UNDERSTANDING

Do you think $(8x + 5)(8x^2 + 5)$ fits the pattern shown in Example 5? Explain.

Another pattern for multiplying binomials involves the square of a binomial.

EXAMPLE 5

Find the products. Describe the pattern.

a. $(y - 5)^2$ **b.** $(2x + 3)^2$

Solution

a. $(y - 5)^2$
$= (y - 5)(y - 5)$
$= y(y - 5) + (-5)(y - 5)$
$= y(y) + (y)(-5) + (-5)y + (-5)(-5)$ Identical middle terms.
$= y^2 + (-5y) + (-5y) + 25$
$= y^2 + (-10y) + 25$
$= y^2 - 10y + 25$

b. $(2x + 3)^2$
$= (2x + 3)(2x + 3)$
$= 2x(2x + 3) + 3(2x + 3)$ Identical middle terms.
$= 2x(2x) + 2x(3) + (3)2x + 3(3)$
$= 4x^2 + 6x + 6x + 9$
$= 4x^2 + 12x + 9$ ◀

The products in Example 5 reveal a pattern. When you square a binomial, the product is a trinomial whose first and last terms are the squares of the terms and whose middle term is twice the product of the two terms.

> **SQUARE OF A BINOMIAL**
>
> $$(a + b)^2 = a^2 + 2ab + b^2 \text{ and}$$
> $$(a - b)^2 = a^2 - 2ab + b^2$$

Multiplying binomials and polynomials can help solve many real world problems.

EXAMPLE 6

GEOMETRY The shaded area of the figure at the right is 63 ft^2. Solve for x.

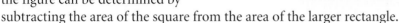

Solution

The area of the shaded region of the figure can be determined by subtracting the area of the square from the area of the larger rectangle.

area of rectangle $-$ area of square $= 63$

$$(x + 5)(x + 3) - x^2 = 63$$
$$(x^2 + 8x + 15) - x^2 = 63 \qquad \text{Multiply.}$$
$$8x + 15 = 63 \qquad \text{Combine like terms.}$$
$$x = 6 \qquad \text{Solve for } x.$$

So, the unknown length in the figure is 6 ft. ◄

TRY THESE

Multiply. Simplify each product.

1. $x(3x + 5)$
2. $c(c^2 + 9c)$
3. $-4(y + 2)$
4. $-3(2z - 7)$
5. $(x + 7)(x + 8)$
6. $(g + 4)(g - 6)$
7. $(2x + 1)(x - 4)$
8. $(z - 3)(2z - 3)$
9. $(a - 2)(a^2 - 3a + 2)$
10. $(k + 4)(k^2 - 4k + 1)$
11. $(a + 2)^2$
12. $(b - 6)^2$
13. $(x - 5)(x + 5)$
14. $(2y - 1)(2y + 1)$

15. MODELING What multiplication problem and product are modeled on the mat shown at the right?

16. WRITING MATHEMATICS Explain why $(x + 5)^2 \neq x^2 + 25$ for all values of x.

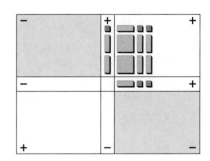

PRACTICE

Multiply. Simplify each product.

1. $x(x - 5)$
2. $2y(y - 3)$
3. $-5z(2z + 3)$
4. $-2k(k + 9)$
5. $-3m(2m - 4)$
6. $-7g(3g - 1)$
7. $2xy(x^2 - y^2)$
8. $3ab(a^2 - 2b^2)$
9. $(x + 2)(x + 5)$
10. $(y + 7)(y + 3)$
11. $(a - 6)(a - 4)$
12. $(b - 2)(b - 4)$
13. $(c - 5)(c + 3)$
14. $(d + 9h)(d + 4h)$

15. $(g^2 + 7)(g^2 - 2)$

16. $(h^3 + 8)(h^3 - 1)$

17. $(2m - 1)(m + 3)$

18. $(3n + 4)(n + 5)$

19. $(2p + q)(3p - 2q)$

20. $(4q + r)(5q + r)$

21. $(x - 2)(x + 2)$

22. $(s + 9)(s - 9)$

23. $(3r - 2)(3r + 2)$

24. $(4t + 5)(4t - 5)$

25. $(v - 6)^2$

26. $(g + 8)^2$

27. $(2w + 8)^2$

28. $(3x - 2)^2$

29. $(x + 3)(x^2 + 4x - 3)$

30. $(y + 5)(y^2 - 3y - 8)$

31. $(a - b)(a^2 - 2ab - 3b^2)$

32. $(p + 2)(p^2 - 5p - 1)$

33. GEOMETRY Find the area of the figure at the right.

34. COMPOUND INTEREST The formula for the amount of money A in an account that earns compound interest is $A = p(1 + r)^t$, where p is the principal (money invested), r is the rate of interest per time period, and t is the number of time periods.

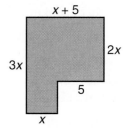

    a. Find A when $p = \$1000$, $r = 0.04$, and $t = 3$.

    b. Write the formula for A when $p = \$2000$ and $t = 2$. Write the formula as a polynomial in r without parentheses.

EXTEND

Simplify.

35. $(q - 2)^3$

36. $(a + 5)^3$

37. $(g + 3)(g - 1)^2$

38. $(k - 1)(k + 2)^2$

39. GEOMETRY Find x if the shaded area of the triangle measures 54 m^2.

40. GEOMETRY Find the value of x in the triangle at the right if the shaded area is 255 cm^2.

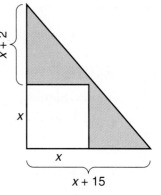

41. GEOMETRY A rectangle is 5 ft longer than it is wide. Its area is 66 ft^2. Find the width of the rectangle. (*Hint:* Guess and check your answer.)

42. HISTORY The relationship $(a + b)(a - b) = a^2 - b^2$ was known to astronomers of Mesopotamia (Iraq) by 2000 B.C. They could demonstrate this fact by a geometric model. Draw a geometrical model for this product.

43. Demography If p is the present population of a region and r is the percent increase of the population per year, an expression for the size of the population after y years is $p\left(1 + \dfrac{r}{100}\right)^y$. Find the population of Uganda two years from now if the present population is 19,800,000 and the rate of increase is 3%.

THINK CRITICALLY

Find the value of n that will make each statement true.

44. $(x + n)(x - 8) = x^2 - 12x + 32$

45. $(2y - 6)(y + n) = 2y^2 + 8y - 42$

46. Are there any values of a and b that make this statement true: $(a + b)^2 = a^2 + b^2$? Justify your answer.

47. If a and b are positive integers, which is greater, $(a + b)^2$ or $a^2 + b^2$? Justify your answer.

PROJECT *Connection*

Perform this experiment to find out about *lift* and the design of an airplane wing.

1. Cut a piece of paper 3 in. by 11 in. Fold it to make a 3 by $5\frac{1}{2}$-in. piece, but do not crease the fold. With ends together, slide one side of the folded paper so that it is $\frac{1}{2}$ in. shorter than the other end and tape it down. The top of the paper should have a slight curve, like a wing. Put your pencil through the folded end, curved side up. Blow across the paper. What happens?

You have just demonstrated *Bernoulli's principle* (discovered in 1738 by Swiss mathemetician Daniel Bernoulli), which states that as the speed of a gas or liquid increases, the pressure decreases. The velocity of the air passing over the curved wing increases to "catch up" with the air flowing below. Pressure decreases above the wing, lift is created, and the plane rises.

air

2. Bernoulli derived an equation for the pressure difference P, measured in newtons per square meter (N/m²):

$$P = \frac{1}{2}d(v_1^2 - v_2^2)$$

where d represents air density in km/m, v_1 represents velocity in m/s above wing, and v_2 represents velocity in m/s below wing. When an airplane is traveling at 192 km/h, $v_1 = 59.2$ m/s, $v_2 = 55.7$ m/s, and $d = 1.29$ km/m³. Determine P to one decimal place.

3. Show that Bernoulli's equation can be expressed as $P = \frac{1}{2}d(v_1 - v_2)(v_1 + v_2)$.

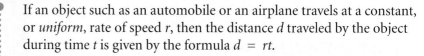

11.9 Problem Solving File

Use Variables

Rate, Time, and Distance Problems

If an object such as an automobile or an airplane travels at a constant, or *uniform*, rate of speed r, then the distance d traveled by the object during time t is given by the formula $d = rt$.

When you read a problem, note whether the situation involves motion in the same direction, motion in opposite directions, or a round trip.

Problem

Dan and Fran are truck drivers. Dan, averaging 55 mi/h, begins a 200-mi trip from their company's Chicago depot to Indianapolis at 7:00 A.M. Fran sets out from the Indianapolis depot at 8:00 A.M. the same day and travels to Chicago at 45 mi/h in the opposite direction on the same road. At what time will Dan and Fran pass each other?

Explore the Problem

1. Let t represent the amount of time Fran travels from 8:00 A.M. until the trucks pass each other. How long will Dan have been traveling when the trucks pass?

2. A table like the one below can help you to organize information about the problem. Complete the table.

	Rate, mi/h	x Time, h	= Distance, mi
Dan	55		
Fran	45	t	$45t$

3. A diagram can help you write an equation. Remember, the trucks are traveling toward each other.

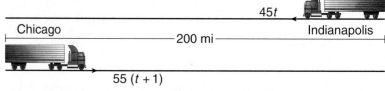

When the trucks pass, what will be true about the distance Dan has traveled from Chicago and the distance Fran has traveled from Indianapolis? Show this relationship algebraically.

4. Solve the equation and interpret your solution.

5. Check your result. (*Hint:* When the trucks pass, how far will each have traveled?)

Investigate Further

● When problems involve a round trip along the same route, you know that the distances each way are equal.

A jet traveled from New York to Mexico City at an average speed of 500 mi/h. Because of a very strong west-to-east tailwind, the average speed for the return trip was 600 mi/h. The return trip from Mexico City took 42 min less than the trip going. What is the air distance between New York and Mexico City?

> ┌─ **PROBLEM** ─
> **SOLVING PLAN**
>
> • Understand
> • Plan
> • Solve
> • Examine

6. Notice that the units do not match for all of the data. Identify what needs to be changed and then change it.

7. To find the distance, first find the time for each part of the trip. Let t represent the number of hours for the trip to Mexico City. How should you represent the time for the return trip?

8. Draw a diagram and complete the table.

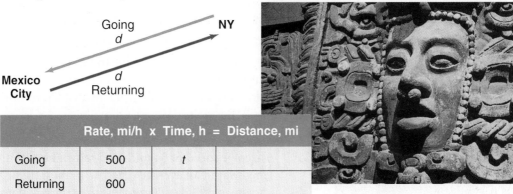

	Rate, mi/h	x Time, h	= Distance, mi
Going	500	t	
Returning	600		

9. You know that the distances are equal. Write an equation in which you can solve for t.

10. Solve the equation. Explain how to use your result to find the distance between the two cities.

11. How long was the return trip? Using the return time, find the distance.

12. **WRITING MATHEMATICS** Write a problem about your commute to and from school that requires the distance formula to solve. Then solve the problem.

> **PROBLEM** ─
> **SOLVING TIP**
>
> Draw a diagram as shown in Question 8.

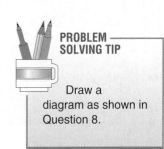

Apply the Strategy

HIKING Darrell and Ivan are spending a month at a summer camp in the mountains. One morning Darrell leaves camp and hikes along a mountain trail at 4 km/h. Thirty minutes later, his friend Ivan sets out along the same route, following after him at a rate of 6 km/h. How long will it take Ivan to catch up with Darrell?

13. Draw a diagram. What kind of motion will you show—in the same direction, in opposite directions, or a round trip?

14. Let t represent Ivan's hiking time. Is Darrell's time more or less than Ivan's? Represent Ivan's time.

15. When Ivan catches up to Darrell, what will be true about the distance each of them has walked? What equation fits the information? Solve your equation. Interpret and check your result.

16. **COMMUTING** Mr. and Mrs. Gonzalez leave home at the same time in separate cars. Mrs. Gonzalez travels north at a rate of speed that is 8 mi/h faster than that of her husband, who travels south. At the end of 3 h, Mr. and Mrs. Gonzalez's two cars are 300 mi apart. How fast was Mrs. Gonzalez traveling? How far south is Mr. Gonzalez after 3 h?

17. **RAIL TRAVEL** A passenger train averaging 62 mi/h begins the 355-mi trip from Glenville to Lintown at 12:00 noon. A freight train traveling 48 mi/h leaves Lintown at 2:00 P.M. the same day and travels to Glenville on an adjacent track. At what time will the two trains pass each other?

18. **RECREATION** The Padillas drive to Safari and Surf at an average speed of 54 mi/h and return home on the same highway at an average speed of 46 mi/h. If the trip to the theme park takes 20 min less than the trip home, how far is the park from the Padillas' home?

SPORTS Jason and Jackson are entered in the 26-mi Aeolian Marathon. Jason's average rate is 5 mi/h and Jackson's average rate is 8 mi/h. Both runners start at the same moment.

19. How far behind will the slower runner be when the faster one finishes the race?

20. If they start at 9:30 A.M., at what time will Jason and Jackson be 5.1 mi apart?

REVIEW PROBLEM SOLVING STRATEGIES

THE EMPTY ENVELOPE, PLEASE!

1. An assistant is sending copies of the sales meeting agenda to 27 of the company's representatives. After all the envelopes are addressed and sealed, the assistant discovers that one copy has fallen on the floor. Not wanting to open all the envelopes, the assistant decides to use a simple pan balance to determine which envelope is empty. What is the least number of weighings the assistant will need for 27 envelopes? Explain the method.

Swimmer's Puzzle

2. Each of nine swimmers has been assigned a lane for a 200-meter freestyle race. The leftmost lane is Lane 1. When a friend asks Joel in which lane he will be swimming, Joel replies, "The number of swimmers to the right of me multiplied by the number of swimmers to the left of me is 3 less than it would have been if my lane was 3 places to the right of where I will be." In which lane will Joel be swimming? Explain how you solved the problem. (*Hint:* You do not need algebra.)

SHOPPER'S SPECIAL

3. Bill's Bargain Basement is selling everything in the store at a 15% discount. Across the street, Good Buys is offering a discount of $1 on each $5 spent in the store. Where would you shop? Explain your shopping strategy. Use a graph to support your reasoning.

$1 DISCOUNT ON EACH $5 SPENT

SHOP HERE!
15% OFF ON EVERYTHING

· · · CHAPTER REVIEW · · ·

VOCABULARY

Choose the word from the list that completes each statement.

1. The __?__ is the greatest degree of any of its terms.

2. A base raised to a negative power is the __?__ of the base raised to a positive power.

3. The __?__ is the sum of the degrees of all the variable factors.

a. degree of a monomial

b. degree of a polynomial

c. reciprocal

Lesson 11.1   MULTIPLY AND DIVIDE VARIABLES   pages 517–520

- Algeblocks may be used to model multiplication and division of variables.

Use Algeblocks to model the multiplication or division.

4. $(-3y)(y)$

5. $3x^2 \div x$

6. $(-2x)(-3y)$

7. $-2xy \div (-2y)$

Lesson 11.2   MULTIPLY MONOMIALS   pages 521–526

- To multiply monomials, use the properties of exponents.

Simplify.

8. $z^4(z^2)$

9. $3a^4(-2a^3)$

10. $-4x^2y^5(3xy^2)$

11. $-2a^3bc(-3ab^3c^2)$

Lesson 11.3   DIVIDE MONOMIALS   pages 527–531

- To divide monomials use the zero property of exponents, the quotient rule, and the property of negative exponents.

Simplify. Write with positive exponents.

12. $\dfrac{-3x^5}{-x^3}$

13. $\dfrac{12a^8}{-2a^5}$

14. $\dfrac{4p^6q^5}{16pq^7}$

15. $\dfrac{5x^3y^0}{-x^2}$

16. $\left(\dfrac{3g^2}{2h^4}\right)^3$

Lesson 11.4   SCIENTIFIC NOTATION   pages 532–537

- A number is written in scientific notation when it is written as a product of a number greater than or equal to 1 and less than 10 and a power of 10.

Evaluate. Write the answers in scientific notation.

17. $(1.8 \cdot 10^6)(2.5 \cdot 10^4)$

18. $\dfrac{6.3 \cdot 10^8}{2 \cdot 10^5}$

19. $1{,}200{,}000{,}000 \cdot 2$

20. $0.000000023 \cdot 3$

Lesson 11.5   ADD AND SUBTRACT EXPRESSIONS   pages 538–540

- Algeblocks may be used to model addition and subtraction of algebraic expressions.

Use Algeblocks to model the addition or subtraction.

21. $(3x - y) + (2y - x)$

22. $(x^2 + y) + (-3x^2 + 2y)$

23. $(4x - 3) - (x + 2)$

24. $(3x + y) - (2y - x)$

Lesson 11.6   ADD AND SUBTRACT POLYNOMIALS   pages 541–547

- To add polynomials, combine like terms.
- To subtract a polynomial, add its opposite.

Perform the indicated operation.

25. $\begin{aligned} 4r + 2s - t \\ + 7r - 3s + 2t \end{aligned}$

26. $\begin{aligned} 7x^2 + 2x - 8 \\ -(4x^2 - 5x + 2) \end{aligned}$

27. $(3a + 2b) + (-a - 4b + 6c)$

28. $(4x^2 + xy - 6) - (x^2 - 2xy - 9)$

Lesson 11.7   PATTERNS IN MULTIPLYING POLYNOMIALS   pages 548–550

- Algeblocks may be used to model multiplication and division of variables.

Use Algeblocks to model the multiplication.

29. $(x + 2)(2x + 1)$

30. $(-x + 1)(x - 2)$

31. $(x - 1)(x + 3)$

Lesson 11.8   MULTIPLY POLYNOMIALS   pages 551–557

- To multiply polynomials use the distributive property, the FOIL method, an area model, or the vertical method.

Multiply. Simplify each product.

32. $k(k + 4)$

33. $-5x(3x - y)$

34. $2ab(3a^2 - 4b^2)$

35. $(a + 3)(a + 1)$

36. $(x - 2)(x + 4)$

37. $(2r + s)(r - s)$

38. $(y + 6)(y - 6)$

39. $(b + 2)^2$

40. $(y - 2)(y^2 + 3y - 4)$

Lesson 11.9   RATE, TIME, AND DISTANCE PROBLEMS   pages 558–561

- Use diagrams and charts to solve rate, time, and distance problems.

Draw a diagram and make a chart to solve.

41. At 10:30 A.M. Lizzie left Boston, driving at 54mi/h. At 10:50 A.M. José left Boston and followed the same route, driving at 60 mi/h. At what time did he overtake Lizzie?

CHAPTER ASSESSMENT

CHAPTER TEST

1. **STANDARDIZED TESTS** Which polynomials have a degree of 4?

   I. $6x^2 + y^3z$ **II.** x^6y **III.** $2x + y^4 + z$

   A. I and II   **B.** II and III

   C. I only   **D.** I and III

Simplify.

2. $g^2h(g^3h^4)$ 3. $(2r + s)^2$

4. How much greater is the expression $3x^2 - 2y$ than the expression $2x^2 - 3y$?

5. **WRITING MATHEMATICS** Write a paragraph explaining at least two different ways you can find the product $(x + 4)(2x - 1)$.

6. Which of the following expressions is *not* equivalent to $(-2x^3y)^2(-xy^3)$?

   A. $-4x^7y^5$ **B.** $(4x^6y^2)(-xy^3)$

   C. $(-2)^2(x^3)^2(y^2)(-x)^3(y^3)$

   D. $(-2)^2(x^3)^2(y^2)(-xy^3)$

7. Express the surface area of the rectangular prism as an algebraic expression in simplest form.

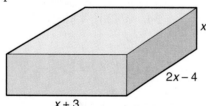

x

$2x - 4$

$x + 3$

8. **STANDARDIZED TESTS** Which of the following is the scientific notation for 0.000000215?

   A. 21.5×10^{-8} **B.** 2.15×10^{-7}

   C. 21.5×10^8 **D.** 2.15×10^7

9. Write 43 billion in scientific notation.

10. Evaluate. Write the answers in scientific notation.

    a. $(1.4 \times 10^6)(2.7 \times 10^8)$

    b. $(7.3 \times 10^5)(4.6 \times 10^4)$

    c. $(3 \times 10^7)^2$ **d.** $\dfrac{8 \times 10^9}{4 \times 10^3}$

11. Which expressions are equivalent to $\left(\dfrac{2x^3y}{-xy^4}\right)^2$?

    I. $4\left(\dfrac{x^3y}{-xy^4}\right)^2$ **II.** $-4\left(\dfrac{x^3y}{-xy^4}\right)^2$

    III. $\dfrac{4x^6y^2}{-x^2y^8}$ **IV.** $\dfrac{4x^4}{y^6}$ **V.** $4x^4y^{-6}$

    A. I and IV only   **B.** I, IV, and V

    C. IV and V only   **D.** I and III

Solve each problem.

12. A rectangle is four times as long as it is wide. Its width is $x + 1$. Express its area as an algebraic expression in simplest form.

13. Express the area of a rectangle whose length is three times its width as an algebraic expression in simplest form. If the area is 48 cm², what are the dimensions of the rectangle?

14. The volume of a rectangular prism is $8x^2y^2$. Find the width.

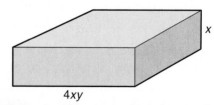

x

$4xy$

15. Joelle begins the 465 mi drive from Atlanta to Tampa at 6:30 A.M., driving at an average speed of 58 mi/h. Jason leaves Tampa at 8:00 A.M. the same day and travels to Atlanta on the opposite side of the same route. He drives at an average speed of 50 mi/h. At what time will the two drivers meet?

PERFORMANCE ASSESSMENT

USE ALGEBLOCKS Pick a variety of algebraic expressions that show addition and subtraction of polynomials as presented in this chapter. Show how to model each operation with Algeblocks. Then ask a partner to show the steps for completing each model. Check that each step is correct.

USE ALGEBLOCKS Use Algeblocks to model multiplying and dividing various polynomials. Have a partner write the algebraic expression to represent the multiplication or division and then show the steps for finding the product or quotient. Check that each step is correct.

USE GRID PAPER Let each unit square on grid paper represent one unit, a strip of 5 unit squares represent x, and a 5×5 square represent x^2. The diagram shows that $x^2 + 2x$ is the area of the rectangle with width x and length $x + 2$. What happens if you substitute 5 for x? Draw pictures of rectangles to represent products of polynomials. Use different values for x. Have a partner express the width, length, and area as algebraic expressions and confirm that the area of each rectangle is the product of the length and width.

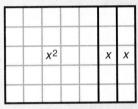

CALCULATE COMPOUND INTEREST Investigate the different savings plans that various banks in your community offer. Use the formula $A = p(1 + r)^t$ to decide which plan would be best for you. Remember that A represents the amount of money in the account, p is the principal or the amount of money invested, r is the rate of interest per time period, and t is the number of time periods.

PROJECT ASSESSMENT

PROJECT *Connection* Have a two-part competition to determine which group has designed the paper airplane that will either fly the greatest distance or stay airborne the longest. Hold the competition in a large indoor area such as a gymnasium. Have each group introduce their entry with a short talk that explains the aerodynamic principles they applied to their design.

1. Decide how many trials each entry will be allowed. Will the results be averaged or will only the best distance or time be used?

2. As a class, record all data on the distances and times of each plane. Then have each group choose a different way to organize and present the data visually. For example, one group might make a bar graph of distances, another group might make a circle graph showing the percent of entries with flight times in certain intervals, and another group could make a scatterplot to see if there is a relationship between some aspect of the plane, such as wingspan, and the flight distance or time.

3. Interested students can extend the project by building more durable and sophisticated models.

CUMULATIVE REVIEW

Simplify each expression.

1. $y^3 \cdot y^5 \cdot y^2$

2. $\dfrac{k^{16}}{k^8}$

3. $(t^5)^8$

4. $(3m^4n)^3$

5. $\dfrac{x^7}{x^{10}}$

6. $\left(\dfrac{-2a^5}{b^2}\right)^3$

Use the quadratic formula to solve each equation. Round answers to the nearest hundredth.

7. $2x^2 - 5x = 11$

8. $\dfrac{1}{2}x^2 + 6 = 13x$

Solve each compound inequality. Then graph the solution.

9. $4n - 3 < 2n + 9$ and $1 - 2n < 5$

10. $6(2w - 1) \le 3(3w - 5)$ or $w - 25 \ge -23$

11. $\dfrac{1}{4}z - 3 \le 1 - \dfrac{3}{4}z$ and $2z - 2.7 > 5.3$

Add or subtract.

12. $(4x^2 + 2x - 9) + (x^3 + x - 9) + (x^2 - 18)$

13. $(5v^3 - 7v^2 + v - 4) - (8v^3 - 7v^2 - 3v + 10)$

Write each decimal as the quotient of two integers.

14. $0.\overline{36}$

15. 8.75

16. $2.1\overline{6}$

17. Determine an equation of the form $y = ax^2 + bx + c$ that passes through the points $(1, 4)$, $(-2, 25)$, and $(7, 16)$.

18. Find the three greatest consecutive integers whose sum does not exceed 70.

Find each product.

19. $(x + 3)(x - 7)$

20. $(3t - 4)(3t + 4)$

21. $(2z + 5)(2z + 5)$

22. $-4mn^2(3m^4n + 2n^3)$

23. $(x - 3)(2x^2 + 7x - 4)$

24. **WRITING MATHEMATICS** Explain why FOIL is only applicable when multiplying two binomials.

25. A farmer wants to enclose a rectangular area for some animals, using the side of a barn as one side. He has 80 feet of fencing available for the other three sides. What dimensions should the farmer make the enclosed area if he wants to maximize the space available?

Identify the property illustrated.

26. $8 + (-8) = 0$

27. If $7 \cdot 5 = 35$, then $35 = 7 \cdot 5$

28. $(82 + 75) + 54 = 54 + (82 + 75)$

29. **STANDARDIZED TESTS** Which of the following expressions is equal to $2x^2 - x - 3$?

   I. $(2x + 3)(x - 1)$
  II. $(x^3 + 4x - 5) - (x^3 - 2x^2 + 5x - 2)$
 III. $(x^2 + 3x + 1) + (x^2 - 2x - 4)$

  A. I only   **B.** II only
  C. I and II   **D.** I, II, and III

30. **WRITING MATHEMATICS** Jorge and Consuela both make the same amount of money per month. One month, Jorge recieved a 10% raise while Consuela's salary was reduced by 10%. The next month Consuela recieved a 10% raise while Jorge's salary was reduced by 10%. Does Jorge still earn the same amount as Consuela? How does their current salary compare to their salary two months ago? Explain.

Write each number in scientific notation.

31. $40,500,000$

32. 0.0000000032

33. $(4.5 \cdot 10^6)(5.2 \cdot 10^6)$

STANDARDIZED TEST

STANDARD FIVE-CHOICE Select the best choice for each question.

1. A boy starts out from home at 10:00 A.M. walking east at a rate of 4 miles per hour. His sister leaves from home at 10:30 A.M. running at 7 miles per hour, following the same route. At what time will the sister catch up to the brother?

   A. 40 minutes
   B. 10:40 A.M.
   C. 11:00 A.M.
   D. 11:10 A.M.
   E. 11:40 A.M.

2. The slope of the line through points (a, b) and (c, d) is

   A. $\dfrac{a - c}{b - d}$ **B.** $\dfrac{d - b}{a - c}$ **C.** $\dfrac{d - b}{c - a}$

   D. $\dfrac{b - d}{c - a}$ **E.** $\dfrac{|b - d|}{|a - c|}$

3. The quadratic equation $x^2 - bx - c = 0$ will have how many real solutions if $b > 0$ and $c > 0$?

   A. cannot be determined
   B. 0
   C. 1
   D. 2
   E. infinite

4. The vertex of the graph of $y = -2|x - 3| - 4$ is located at

   A. $(-2, -4)$ **B.** $(-3, -4)$ **C.** $(3, -4)$
   D. $(-2, 3)$ **E.** $(-4, 3)$

5. A football team scored these point totals during the season:

   42, 27, 10, 13, 21, 16, 35, 7, 14, 24, 38

   $Q_3 - Q_1 =$

   A. 22   **B.** 35   **C.** 21
   D. 17.5   **E.** Q_2

6. Which of the following products simplifies to be $n^2 + 5n - 6$?

   A. $(n + 3)(n - 2)$
   B. $(n - 5)(n - 1)$
   C. $(n - 6)(n + 1)$
   D. $(n - 3)(n + 2)$
   E. $(n - 1)(n + 6)$

7. An item is priced at $42.50. A sale offers 20% off for this particular item. What is the item's sale price?

   A. $8.50
   B. $34.00
   C. $51.00
   D. $53.13
   E. $212.50

8. Consider statements I–III as each applies to the points in Quadrant III.

   I. The product of the coordinates is positive.
   II. Some of the points may lie on the line $y = -x$.
   III. Some of the points may lie on the line $x + y = 4$.

   Which statement is true?

   A. I only   **B.** II only   **C.** III only
   D. II and III   **E.** none are true

9. Six oranges and five apples cost $1.35. Five oranges and six apples cost $1.40. What is the difference in price of one apple and one orange?

   A. $0.25   **B.** $0.05   **C.** $0.10
   D. $0.15   **E.** no difference

10. Multiply $(2.5 \cdot 10^8)(8.4 \cdot 10^{-3})$. In scientific notation, the product is

    A. $21 \cdot 10^5$ **B.** $2.1 \cdot 10^4$ **C.** $2.1 \cdot 10^6$
    D. $21 \cdot 10^{11}$ **E.** $2.1 \cdot 10^{12}$

12 Polynomials and Factoring

Take a Look AHEAD

Make notes about things that look new.

- Find an example to illustrate the difference between a common factor of two monomials and the greatest common factor of the same monomials.
- How do you apply the Distributive Property to multiply? How do you apply the Distributive Property to factor?

Make notes about things that look familiar.

- In Chapter 11, you learned to recognize special cases of polynomial multiplication. Identify examples of these special products in this chapter.
- How do you think factoring a polynomial will be similar to factoring an integer? How do you think the factoring process might be different?

DATA Activity

Where Do the Waters Go?

A river and its *tributaries*, smaller channels such as brooks or streams that merge into the river, form a *river system*. All the water in a river system eventually flows to one location—the main river. The area drained by a river system is called the system's *drainage basin*. As you can see from the table on the next page, the Mississippi River has the largest drainage area in the United States. Most of the rain that falls between the Rocky Mountains in the West to the Appalachian Mountains in the East drains into the Mississippi.

SKILL FOCUS

- Add, subtract, multiply, and divide and compare real numbers.
- Determine the range of a data set.
- Solve percent problems.
- Use scientific notation.
- Construct a boxplot.

U.S. RIVERS

In this chapter, you will see how

- **HYDROLOGISTS** create mathematical models of river activity using velocity and flow rate data.
(Lesson 12.2, page 581)

- **CIVIL ENGINEERS** design dams to provide both flood control and low-cost electrical power.
(Lesson 12.4, page 590)

- **WATER RESOURCES TECHNICIANS** determine if the volume of water in a river is increasing toward flood levels.
(Lesson 12.6, page 601)

Drainage Areas of U.S. Rivers	
River	**Drainage Area, 1000s of mi²**
Mississippi	1150
Ohio	203
Columbia	258
Yukon	328
Missouri	529
Mobile	44.6
Snake	108
Tennessee	40.9
Atchafalaya	95.1
Susquehanna	27.2

Use the table at the right to answer the following questions.

1. What is the range of the drainage areas for the rivers shown?

2. Which river has a drainage area about four times the size of the drainage area of the Susquehanna River?

3. The drainage area of the Mississippi is about 40% of the total land area of the United States excluding Hawaii and Alaska. Determine the approximate total land area of the continental United States. Write your answer using scientific notation.

4. Consider the value for the Mississippi as an outlier. Make a boxplot of the data for the other rivers.

5. **WORKING TOGETHER** Research the size of the drainage areas of several major rivers of the world, such as the Amazon or the Nile. Write word problems that relate the size of the drainage area of a world river to the size of the basin of a United States river. Exchange problems with other groups and solve.

PROJECT

How Rapid Is Your River?

You've probably seen pictures of white-water boating and perhaps even participated in this increasingly popular sport. The area of the United States between the Rocky Mountains and the Pacific Ocean is known for some of the best white-water rivers in the world, although good white-water boating is found throughout the country. In this project, you will explore some of the river conditions that lead to crashing rapids and thrilling rides.

PROJECT GOAL

To construct a model of a white-water river.

Getting Started

Work with a group to build the river model.

1. If possible, borrow a *stream table* from the science department. You will also need large pails, rubber tubing, bricks or wooden blocks, sand to fill the stream table, a toy boat, a stopwatch, other materials such as rocks and clay to create special features in your model, lots of paper towels, and an approved working area. Consult a teacher or an earth science textbook for instructions on how to set up this equipment.

2. Research to find out which river characteristics affect the water flow. How do deep rivers compare with shallow ones? Does water flow faster in narrow channels or wide ones? How does a change in elevation affect water flow? How do different types of obstacles influence the water course?

3. Begin a glossary or special vocabulary list associated with rivers and white-water boating.

PROJECT Connections

Lesson 12.2, page 580:
Find out how to model backwaters and eddies.
Lesson 12.4, page 589:
Find out how to model souse holes.
Lesson 12.5, page 594:
Determine the gradient of the river model and explore how gradient affects velocity.
Chapter Assessment, page 609:
Demonstrate, evaluate, and compare each group's white-water river model.

Internet Connection

www.swpco.com/
swpco/algebra1.html

12.1 Algebra Workshop
Explore Patterns in Factoring

Think Back

- Algeblocks can be used to model multiplication of polynomials. The mat at the right shows the multiplication of two binomials.

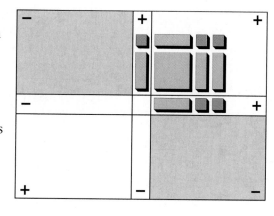

  1. What two binomials are multiplied?

  2. Describe how Algeblocks are used to form the product.

  3. Read the product from the mat.

SPOTLIGHT ON LEARNING

WHAT? In this lesson you will learn
- to find the two binomial factors of a trinomial.
- to recognize special factoring patterns.

WHY? Algeblocks can help you to model patterns in factoring polynomials so you can solve problems in geometry, physics, and industry.

Explore

- Work with a partner and a set of Algeblocks.

  4. Make a group of 8 unit blocks and another group of 12 unit blocks. For each group, form rectangles that have the length of one side in common. What is the greatest common length the two rectangles can have?

  5. Model each of the monomials $2x^2$ and $4x$.

  6. For each monomial in Question 5, form rectangles that have the length of one side in common. What is the greatest common length of the two rectangles?

  7. Model the polynomial $2x^2 + 4x$ in one large rectangle in Quadrant I on a Quadrant Mat. Use the greatest common length that you found in Question 6 as the length of one side. Model that length on one axis.

  8. Show the other length of the polynomial on the other axis. Read the dimensions from the mat.

  9. Write $2x^2 + 4x$ as the product of a monomial and a binomial.

  10. Explain why you think the product you wrote for Question 9 is called the factored form of $2x^2 + 4x$.

For each Algeblocks model, write the polynomial and its factored form.

11.

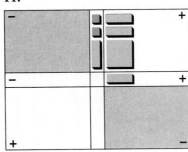

12.

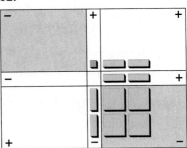

The trinomial $x^2 + 5x + 6$ can be written as the product of two binomials. Use Algeblocks to find the two binomial factors.

13. Place Algeblocks for $x^2 + 5x + 6$ in Quadrant I on a Quadrant Mat. Arrange the blocks to form a rectangle. Draw a diagram of your work.

14. Place blocks along the horizontal and vertical axes of the rectangle you formed in Question 13. Read the two binomial factors of $x^2 + 5x + 6$.

Use Algeblocks to write the two binomial factors for each trinomial.

15. $x^2 + 4x + 3$ **16.** $y^2 + 6y + 8$ **17.** $y^2 + 9y + 8$

18. What pattern do you notice for the terms of each trinomial in Questions 15–17?

19. Place Algeblocks for the trinomial $x^2 - 4x + 3$ on a Quadrant Mat. Arrange the blocks to form rectangles. (*Hint:* Use all four quadrants.) Place blocks for the dimensions of the rectangles on the horizontal and vertical axes. Draw a diagram to show your work.

20. Read the two binomial factors from the mat.

21. Compare your diagram for Question 19 with those of your classmates. Describe any differences you see. Did the differences affect the answer to Question 20?

Use Algeblocks to write the two binomial factors for each trinomial.

22. $x^2 - 2x + 1$ **23.** $x^2 - 7x + 6$ **24.** $y^2 - 3y + 2$

25. What pattern do you notice for the terms of each trinomial in Questions 22–24?

**CHECK —
UNDERSTANDING**

In Question 20, explain where to look on the mat when naming the terms of each binomial.

26. Place Algeblocks for the trinomial $x^2 + 3x - 4$ on a Quadrant Mat. Arrange the blocks to form rectangles. (*Hint:* You may use zero pairs to complete the rectangles.) Place blocks for the dimensions of the rectangles on the horizontal and vertical axes. Draw a diagram to show your work.

27. Read the two binomial factors from the mat.

Use Algeblocks to write the two binomial factors for each trinomial.

28. $y^2 + y - 2$ **29.** $x^2 + 2x - 3$ **30.** $y^2 + 2y - 8$

31. What pattern do you notice for the terms of each trinomial in Questions 28–30?

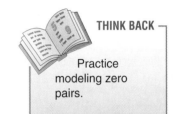

THINK BACK

Practice modeling zero pairs.

32. Place Algeblocks for the trinomial $x^2 - 3x - 4$ on a Quadrant Mat. Arrange the blocks to form rectangles. Place blocks for the dimensions of the rectangles on the horizontal and vertical axes. Draw a diagram to show your work.

33. Read the two binomial factors from the mat.

Use Algeblocks to write the two binomial factors for each trinomial.

34. $x^2 - x - 6$ **35.** $y^2 - 4y - 5$ **36.** $x^2 - 3x - 10$

37. What pattern do you notice for the terms of each trinomial in Questions 34–36?

Make Connections

- There are four different patterns that you should be aware of when factoring a trinomial into two binomials. The table below shows the patterns and an example of each.

	Trinomial Pattern	Example	Binomial Factors
1.	$x^2 + bx + c$	$x^2 + 5x + 6$	$(x + 3)(x + 2)$
2.	$x^2 - bx + c$	$x^2 - 5x + 6$	$(x - 3)(x - 2)$
3.	$x^2 + bx - c$	$x^2 + x - 6$	$(x + 3)(x - 2)$
4.	$x^2 - bx - c$	$x^2 - x - 6$	$(x - 3)(x + 2)$

38. How are all four trinomials the same?

39. How are the trinomials for Patterns 1 and 2 similar? How are they different?

40. How are the trinomials for Patterns 3 and 4 similar? How are they different?

Match each trinomial with its binomial factors.

41. $x^2 - 3x - 28$

42. $x^2 + 11x + 28$

43. $x^2 + 3x - 28$

44. $x^2 - 11x + 28$

a. $(x + 7)(x + 4)$

b. $(x - 7)(x - 4)$

c. $(x - 7)(x + 4)$

d. $(x + 7)(x - 4)$

45. Describe how the patterns helped you to determine the binomial factors in Questions 41–44.

Summarize

- **46.** MODELING Use Algeblocks to find the factored form of $3x^2 - 6x$. Sketch each step.

- **47.** WRITING MATHEMATICS Write a paragraph describing how to find two binomial factors for a trinomial in which the coefficient of the x^2 term is 1.

- **48.** GOING FURTHER Find the binomial factors of the trinomial $2x^2 + 7x + 3$. Use Algeblocks if you wish.

12.2 Factor Polynomials

Explore

Jimmy wants to make two rectangular side-by-side enclosures for his pets. His rabbits need an area of 80 ft². The two puppies should have a 24-ft² area. Jimmy wants the enclosures to have one dimension in common so that one complete side of each pen shares the same fence.

1. List all the different ways that Jimmy can design rectangular enclosures that have a common dimension. Sketch the possible enclosures on graph paper to be sure your list is complete.

2. Find the total fencing needed for each design you listed.

3. Which design uses the least fencing?

4. What is the relationship between the possible common dimensions you found and the two given areas?

SPOTLIGHT ON LEARNING

WHAT? In this lesson you will learn
- to factor polynomials whose terms have a common monomial factor.
- to factor a trinomial with a quadratic coefficient of 1.
- to factor a trinomial with an integral quadratic coefficient greater than 1.

WHY? Factoring polynomials can help you solve problems involving geometry, physics, and hydraulics.

Build Understanding

In Explore you saw several common dimensions or factors for two numbers. The terms of polynomials can also have common factors.

A polynomial is **factored**, or in **factored form**, when it is expressed as a product of other polynomials. Factoring is an important technique for many applications in advanced mathematics, science, and engineering. Factoring allows you to rewrite a complicated expression as the product of simpler expressions. To factor out a common monomial from the terms of a polynomial, you can use the distributive property in reverse.

THINK BACK

The distributive property is
$a(b + c) = ab + bc$.
So, by the reflexive property,
$ab + bc = a(b + c)$.

When factoring out a monomial from a polynomial, list all the factors of each term of the polynomial to find the greatest common factor (GCF) of each term. The GCF allows you to factor a polynomial so that no common factors remain.

EXAMPLE 1

Factor $2x^3y + 6x^2 + 8xy$.

Solution

List all factors of each term.

$$2x^3y \qquad + \qquad 6x^2 \qquad + \qquad 8xy$$
$$2 \cdot x \cdot x \cdot x \cdot y \qquad 2 \cdot 3 \cdot x \cdot x \qquad 2 \cdot 2 \cdot 2 \cdot x \cdot y$$

Common factors of all three terms are 2 and x. So, the GCF is $2x$.

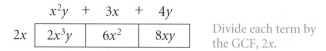

	x^2y +	$3x$ +	$4y$
$2x$	$2x^3y$	$6x^2$	$8xy$

Divide each term by the GCF, $2x$.

So, the factored form of $2x^3y + 6x^2 + 8xy$ is $2x(x^2y + 3x + 4y)$. ◀

Factoring a trinomial into two binomials is the reverse of multiplying two binomials whose product is a trinomial. Patterns you notice in multiplying two binomials will help you in factoring trinomials.

To factor a trinomial, you can use multiplication or you can use patterns of products of binomials, as in the next example.

EXAMPLE 2

Factor each trinomial.

a. $x^2 + 7x + 10$ **b.** $x^2 - 6x + 8$

Solution

a. *Multiplication diagram method* Use the multiplication diagram you used to multiply two polynomials. Fill in the products and terms you know for the product $x^2 + 7x + 10$.

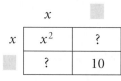

	x	
x	x^2	?
	?	10

The monomials that replace both ? must total $7x$, the x-term.

	x	$+5$
x	x^2	$+5x$
$+2$	$+2x$	10

Look for two factors having a sum of 7 and a product of 10. The factors 2 and 5 work.

The factored form of $x^2 + 7x + 10$ is $(x + 5)(x + 2)$.

ALGEBLOCKS MODEL

Factor the trinomial $x^2 + 7x + 10$. Place Algeblocks to model $x^2 + 7x + 10$ in Quadrant I. Arrange the blocks to form a rectangle.

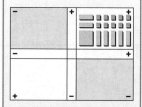

Place blocks along the horizontal and vertical axes to match the sides of the rectangle you formed.

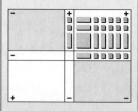

Read the two binomial factors of $x^2 + 7x + 10$ from the horizontal and vertical axes of the rectangle.

Patterns of products method $x^2 + 7x + 10 = (x + \blacksquare)(x + \blacksquare)$

Factors of 10	Sum of the Factors
1, 10	11
−1, −10	−11
2, 5	7
−2, −5	−7

List all pairs of factors of 10. Then find the sum of each pair of factors. Look for the pair of factors whose sum is +7.

Since $2 + 5 = 7$, use 2 and 5 to complete the binomial factors.

So, $x^2 + 7x + 10$ factored is $(x + 2)(x + 5)$.

Multiply the factors to see that the product results in the original trinomial.

b. *Patterns of products method* $x^2 − 6x + 8 = (x + \blacksquare)(x + \blacksquare)$

Factors of 8	Sum of the Factors
1, 8	9
−1, −8	−9
2, 4	6
−2, −4	−6

Factors −2 and −4 have a sum of −6.

So, $x^2 − 6x + 8$ factored is $(x − 2)(x − 4)$.

Check by multiplying the factors.   ◀

Example 3 shows that you can use the patterns of products method to factor a trinomial with a negative constant term. Also try using the multiplication diagram to see which method you prefer.

EXAMPLE 3

Factor each trinomial.

a. $x^2 − 4x − 5$ **b.** $x^2 + 2x − 15$

Solution

a. $x^2 − 4x − 5 = (x + \blacksquare)(x + \blacksquare)$

Factors of −5	Sum of the Factors
1, −5	−4
−1, 5	4

Factors 1 and −5 have a sum of −4.

So, $x^2 − 4x − 5$ factored is $(x + 1)(x − 5)$.

b. $x^2 + 2x − 15 = (x + \blacksquare)(x + \blacksquare)$

Factors of −15	Sum of the Factors
1, −15	−14
−1, 15	14
3, −5	−2
−3, 5	2

Factors −3 and 5 have a sum of 2.

So, $x^2 + 2x − 15$ factored is $(x − 3)(x + 5)$.   ◀

CHECK UNDERSTANDING

In Example 2b, why weren't the factors −1, 8 or −2, 4 considered as possibilities?

ALGEBLOCKS MODEL

Factor the trinomial $x^2 − 3x − 10$. Place Algeblocks to model $x^2 − 3x − 10$. Arrange the blocks to form rectangles. Use zero pairs to complete your rectangles if necessary.

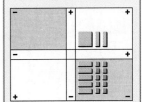

Place blocks along the horizontal and vertical axes to match the sides of the rectangle you formed.

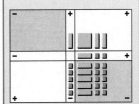

Read the two binomial factors of $x^2 − 3x − 10$ from the horizontal and vertical axes of the rectangle.

So far you have seen how to factor trinomials whose **leading coefficient**, the coefficient of x^2, is 1. The following example shows how you can factor polynomials whose leading coefficient is greater than 1. Whether you use the multiplication diagram or the patterns of products method, you will need to guess and check to determine the factors.

EXAMPLE 4

GEOMETRY Find the dimensions of a rectangle in terms of x if the area of the rectangle is $2x^2 + 7x + 5$.

Solution

$$2x^2 + 7x + 5 = (2x + \blacksquare)(x + \blacksquare)$$

The factors of $+5$ are $1, 5$ and $-1, -5$.

Check the product of every possible combination of binomial factors.

Trial Factors	Product	
$(2x + 1)(x + 5)$	$2x^2 + 11x + 5$	
$(2x - 1)(x - 5)$	$2x^2 - 11x + 5$	
$(2x + 5)(x + 1)$	$2x^2 + 7x + 5$	This is the product you want.
$(2x - 5)(x - 1)$	$2x^2 - 7x + 5$	

The required dimensions of the rectangle are $(2x + 5)$ and $(x + 1)$. ◄

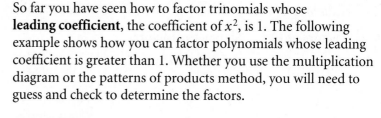

PROBLEM SOLVING TIP

Guess and check is the process of listing all possible answers and then trying each answer one by one until the correct one is found.

TRY THESE

Factor.

1. $9x^2 + 3x$

2. $15y^4z + 10y^2z^2 - 20yz$

3. $2y^3 - 16y^2$

4. **WRITING MATHEMATICS** Write a paragraph explaining how you would determine the binomial factors of $x^2 - 4x + 3$.

Factor each trinomial.

5. $x^2 + 5x + 6$

6. $x^2 + 6x + 8$

7. $x^2 - 4x + 3$

8. $x^2 - 11x + 30$

9. $x^2 - 2x - 8$

10. $x^2 - x - 12$

11. $x^2 + 3x - 28$

12. $x^2 + 5x - 14$

13. $2x^2 + 5x + 3$

14. **GEOMETRY** Determine the dimensions of the rectangle at the right.

15. **MODELING** Use Algeblocks to show that $x^2 - 3x + 2 = (x - 2)(x - 1)$.

$$A = x^2 + 7x + 6$$

PRACTICE

Factor.

1. $2a^4 + 8a$

2. $7b^3 + 21b$

3. $8ab^2 - 12a^2b^3$

4. $10c^3d^2 - 15cd^3$

5. $6c^2 - 9d^2$

6. $15f - 20g^2$

7. $4x^3 - 2x^2 + 14x$

8. $3y^4 + 9y^2 - 15$

9. $2z^3 + 3z^2 + 4z$

10. $9mn - 3m^2 + 4mn^2$

11. $8abc^2 - 4b^2c + 12a^2bc$

12. $6x^2yz + 2xy^2z - 4xyz$

13. $12a^4b^3c^2 - 4a^3bc^2 + 8a^2c - 16ab$

14. $9x^3yz^2 - 6x^2yz^2 + 12xyz^2 - 21yz^2$

Factor each trinomial.

15. $x^2 + 5x + 4$

16. $y^2 + 5y + 6$

17. $z^2 + 8z + 15$

18. $y^2 - 4y + 4$

19. $x^2 - 6x + 9$

20. $x^2 - 10x + 9$

21. $z^2 - 10z + 9$

22. $z^2 - 11z + 28$

23. $y^2 + 7y - 8$

24. $x^2 + x - 6$

25. $y^2 + 11y - 12$

26. $a^2 + 6a - 7$

27. $b^2 + 3b - 4$

28. $x^2 - x - 12$

29. $x^2 - 2x - 35$

30. $n^2 - 4n - 12$

31. $c^2 - 3c - 18$

32. $z^2 - 6z - 7$

33. $5x^2 + 12x + 7$

34. $2a^2 + 13a - 7$

35. $2b^2 + 5b - 3$

36. **GEOMETRY** Determine the area of the shaded region at the right. Express your answer in factored form.

37. **GEOMETRY** Determine the dimensions, in terms of x, of a rectangle if its area is $3x^2 + 13x - 10$.

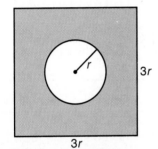
$3r$
$3r$

EXTEND

Find the missing factor.

38. $a^{n+4} + a^n = a^n(\quad)$

39. $6b^{2n} + 15b^{2n+2} = (\quad)(2 + 5b^2)$

40. **GEOMETRY** The area of a rectangle in square feet is $x^2 + 13x + 36$. How much longer is the length than the width of the rectangle?

41. **PHYSICS** An expression that is used in connection with certain atomic particles is $Z\left(\frac{1}{2}\right) + N\left(-\frac{1}{2}\right)$, where Z is the number of protons and N is the number of neutrons in the nucleus. Factor this expression.

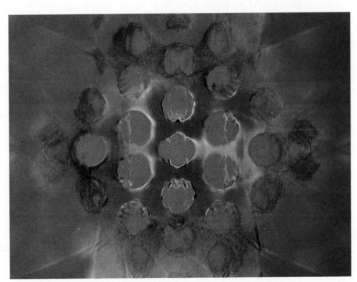

42. MACHINERY A hydraulic cylinder uses water power to turn a piston. The formula that applies to one type of hydraulic cylinder is

$$q = \frac{v\pi D^2 - v\pi d^2}{4}$$

where q is the flow rate in cubic inches per second, D is the diameter of the cylinder in inches, d is the diameter of the piston rod in inches, and v is the velocity of the piston in inches per second.

a. Rewrite the formula in factored form.

b. The velocity of the piston is 35 in./s, the diameter of the cylinder is 3 in., and the diameter of the rod is 0.75 in. Find the flow rate to the nearest whole number.

THINK CRITICALLY

43. What are all the possible values of b that make $x^2 + bx + 12$ factorable?

44. In the equation $x^2 + bx + 24 = (x + 3)(x + s)$, find all possible values of s and b.

PROJECT Connection Two features to include in your river model are *backwaters* and *eddies*. Both are the result of obstacles such as large rocks.

1. Look at the drawing of a backwater in Figure 1. In which direction is the main current A flowing? In which direction is the small current B behind the rock flowing?

2. In Figure 1, the water at B is higher than the water at A. Because the water has a lot of speed as it goes from A to B, it continues to flow even as it moves upstream. The water behind the rock has only a little speed, so it flows as expected from high to low water. To construct a backwater, you need to give a portion of your model river a slight rise.

3. Eddies are similar to backwaters, but water flows along the boundary layer of the rock. Figure 2 shows the water flow. To construct an eddy, you need a rapid river flow.

Figure 1

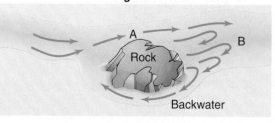

Figure 2

A hydrologist is a scientist who studies the distribution, circulation, and physical properties of underground and surface waters. Hydrologists are hired to observe and predict changes in many aspects of rivers, including the velocity of the river and its flow rate.

The Ohio River is the major eastern tributary of the Mississippi River. The word *Ohio* in Iroquois means "bright," "shining," or "great."

The table shows readings for the velocity and flow rate for four days on the Ohio River in Cincinnati.

Ohio River, Velocity and Flow Rate in Cincinnati				
Day	**1**	**2**	**3**	**4**
Velocity, mi/h	1.13	1.08	1.03	0.97
Flow rate, ft³/s	52.8	50.0	46.7	43.7

A mathematical model of data is an equation that will provide a good estimate for one variable based on another variable. For the table, look at various models that relate x, the velocity of the river in miles per hour, to y, the flow rate in cubic feet per second.

A graphing utility can take the data points and give the following regression equations (coefficients rounded to the nearest tenth).

Linear regression equation: $y = 57.6x - 12.3$
Quadratic regression equation: $y = 35.8x^2 - 17.4x + 26.9$

Decision Making

Complete the tables. Round to the nearest hundredth.

1.

Linear Model				
x	1.13	1.08	1.03	0.97
$y = 57.6x - 12.3$				

2.

Quadratic Model				
x	1.13	1.08	1.03	0.97
$y = 35.8x^2 - 17.4x + 26.9$				

3. Which model do you think is best for the given data? Explain.

4. What information would the hydrologist look for if he or she wanted to improve the mathematical model for the river at this point?

12.3 Factoring Special Products

Explore/Working Together

Work with a partner.

In ancient Iraq (about 2000 B.C.) the temple library had extensive tables of squares of whole numbers inscribed on clay tablets. It is thought that one way these tables were used was for multiplying two numbers. To find the product of two even numbers or two odd numbers, the Babylonians would compute the difference between the squares of two related numbers in the table. Then, they could multiply by subtracting! To find the product of 37 and 25, they would compute the difference between the square of 31 and the square of 6.

1. Show that $(37)(25) = 31^2 - 6^2$.

2. Show that $(36)(28) = 32^2 - 4^2$.

3. Explain how this method worked. (*Hint:* Find a relationship between each number that is squared and the original factors.)

4. Use this method to find the product $(51)(35)$.

Build Understanding

THINK BACK

Recall that $a - b = a + (-b)$.

Recall from Lesson 11.8 that the product of the sum and difference of the same two terms equals the difference between the squares of those terms.

$$(a + b)(a - b) = a^2 - b^2$$

By reversing the pattern, you can factor the difference of two squares.

EXAMPLE 1

Factor each expression if factorable.

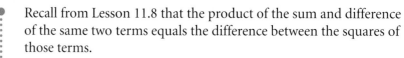

a. $x^2 - 36$ **b.** $16a^2 - 49b^2$ **c.** $9y - 100$ **d.** $h^2 + 25$

Solution

Determine whether the polynomial is the difference of two squares. If it is, write it as the product of a sum and a difference.

a. $x^2 - 36$ is the difference of two squares, x^2 and 6^2.

So, $x^2 - 36 = (x + 6)(x - 6)$.

b. $16a^2 - 49b^2$ is the difference of two squares, $(4a)^2$ and $(7b)^2$.

So, $16a^2 - 49b^2 = (4a + 7b)(4a - 7b)$.

c. $9y - 100$ is *not* the difference of two squares. The two terms have no common factors, so the polynomial cannot be factored.

d. $h^2 + 25$ is the sum, rather than the difference, of two squares. The two terms have no common factors, so the polynomial cannot be factored. ◄

Recall from Lesson 11.8 that the square of a binomial has special patterns.

$$(a + b)^2 = a^2 + 2ab + b^2 \text{ and } (a - b)^2 = a^2 - 2ab + b^2$$

Notice that the middle term in each trinomial is twice the product of the square roots of the first and last terms of the trinomial.

When you recognize either pattern in a trinomial, you can factor the trinomial into the square of a binomial.

EXAMPLE 2

Factor each polynomial.

a. $x^2 + 16x + 64$ **b.** $9y^2 - 30yz + 25z^2$

Solution

Test to see that the expression is a perfect square trinomial. First determine if the first and last terms are perfect squares. Then determine if the middle term is twice the product of the *square roots* of the first and last terms. Observe the sign of the middle term.

a. $x^2 + 16x + 64$

The first and last terms, x^2 and 64, are perfect squares.

The square root of the first term is x. The square root of the last term is 8. Twice the product of 8 and x is $16x$.

Therefore, $x^2 + 16x + 64$ is a perfect square trinomial.

Since the middle term of the trinomial is positive, the trinomial has the pattern $a^2 + 2ab + b^2 = (a + b)^2$.

So, $x^2 + 16x + 64 = (x + 8)^2$.

b. $9y^2 - 30yz + 25z^2$

The first and last terms, $9y^2$ and $25z^2$, are perfect squares.

The square root of the first term is $3y$. The square root of the last term is $5z$. Twice the product of $3y$ and $5z$ is $30yz$.

Therefore, $9y^2 - 30yz + 25z^2$ is a perfect square trinomial.

The trinomial fits the pattern $a^2 - 2ab + b^2 = (a - b)^2$.

So, $9y^2 - 30yz + 25z^2 = (3y - 5z)^2$. ◄

ALGEBLOCKS MODEL

Factor the difference of two squares, $x^2 - 4$.

Step 1: Form rectangles to represent the terms in the expression. Use zero pairs where necessary.

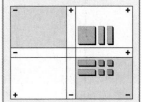

Step 2: Place Algeblocks in the axes to match the edges of the rectangles.

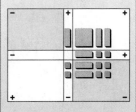

Read the factors from the mat.

CHECK UNDERSTANDING

Write a perfect square trinomial. Use Algeblocks to show that the model for your trinomial forms a square.

Some geometry problems can be solved by factoring.

EXAMPLE 3

MODELING Find the dimensions in terms of x of a rectangle that has the same area as the shaded region at the right.

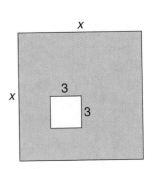

Solution

Write a polynomial for the area of the shaded region.

$$area = x^2 - 9$$

Factor $x^2 - 9$ as the difference of two squares.

$$x^2 - 9 = (x + 3)(x - 3)$$

Therefore, a rectangle with the same area as $x^2 - 9$ has dimensions $x + 3$ and $x - 3$.

TRY THESE

Factor if possible.

1. $a^2 - 16$

2. $b^2 - 81$

3. $c^4 + 49$

4. $d^4 - 36$

5. $9e^4 - 16f^2$

6. $25g^2 - 21h^2$

Determine if the trinomial is a perfect square. Factor if possible.

7. $a^2 - 10a + 25$

8. $b^2 + 14b + 49$

9. $c^4 - 4c^2 + 4$

10. $9y^2 + 6y + 4$

11. MODELING For the Algeblocks diagram below, write the trinomial and its factors.

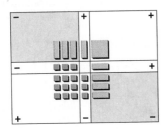

12. GEOMETRY Write the area of the shaded region as a polynomial. Factor if possible.

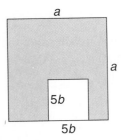

13. WRITING MATHEMATICS Write a paragraph describing a perfect square trinomial.

PRACTICE

Factor if possible.

1. $a^2 - 16a + 64$

2. $b^2 + 2b + 1$

3. $c^2 + 6c + 9$

4. $d^2 - 7d + 49$

5. $e^2 - 20e + 100$

6. $f^4 + 10f^2 + 25$

7. $81h^2 + 36h + 4$

8. $25j^2 - 15j + 9$

9. $100p^2 + 60pq + 9q^2$

10. $t^2 - 16$

11. $v^2 - 121$

12. $x^2 + 81$

13. $y^2 - 1$

14. $4z^2 - 49$

15. $9a^2 - 12$

16. $25b^2 - 64$

17. $36c^4 - d^2$

18. $16p^2 - 21$

19. GEOMETRY Find the dimensions of a rectangle that has the same area as the shaded region in the figure at the right.

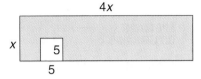

EXTEND

Factor if possible.

20. $g^6 + 12g^3 + 36$

21. $16r^4 - 8r^2s + s^2$

22. $9g^{10} - 100h^8$

23. $400b^2 + 60b + 9$

24. $900y^2 + 361$

25. $121x^2 - 289$

26. GRAPHING Recall from Lesson 10.2 that the graph of a function of the form $y = (x - a)^2$ is the same as the graph of $y = x^2$ moved a units to the right, as shown in the graph at the right. Describe the graph of $y = x^2 - 6x + 9$.

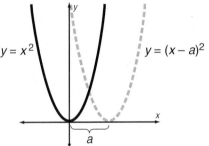

27. COMPUTER SECURITY Hard-to-crack codes are essential for computer security. Codes have been developed based on 200-digit numbers that are products of two prime numbers, which are difficult and time-consuming to determine. One time-saving approach is to find two numbers x and y such that $x^2 - y^2 = N$, the number to be factored.

 a. Determine two possible pairs of values of x and y for 24.

 b. Use these values of x and y to name two pairs of factors for 24.

THINK CRITICALLY

Factor if possible.

28. $(x + y)^2 - z^2$

29. $(a + b)^2 - (c + d)^2$

30. $(p^2 - 2pq + q^2) - s^2$

Find the value of N that makes the statement true.

31. $4x^2 - 3Ny^4 = (2x + 9y^2)(2x - 9y^2)$

32. $16a^2 + 5Nb = (4a + 5b^3)(4a - 5b^3)$

MIXED REVIEW

Graph the solution.

33. $-3 < x \leq 2$

34. $x < -1$ or $x > 2$

35. $0 \leq x \leq 3$

36. STANDARDIZED TESTS Which is the equation of the line through the points $(-2, 4)$ and $(3, -6)$?

 A. $y = 2x$

 B. $y = -2x$

 C. $5y = 2x$

 D. $5y = -2x$

Solve.

37. $\begin{cases} y = 2x \\ x + y = 24 \end{cases}$

38. $\begin{cases} y = -x \\ 2x - 3y = 25 \end{cases}$

39. $\begin{cases} y = 2x + 1 \\ 2x + 3y = 11 \end{cases}$

Explore

● A rectangle with length x and width y will be expanded by 2 units in one dimension and 3 units in the other dimension.

1. Draw a diagram of the original rectangle and of the two possible expanded rectangles.

2. Compute the area of all three rectangles.

3. Are the areas of the two expanded rectangles the same? Explain.

4. If $y > x$, which of the expanded areas is greater? Justify your answer.

Build Understanding

● Recall from Lesson 12.2 that a polynomial can have a monomial as a factor of each of its terms. A polynomial can also have a binomial as a common factor. You can use the distributive property to factor out a common binomial just as you would use it to factor out a common monomial.

EXAMPLE 1

Factor each polynomial.
  a. $y(a + b) + 3(a + b)$ **b.** $7(x^2 + 1) - y(x^2 + 1)$

Solution

a. The binomial $a + b$ is a common factor of both terms.

$$y(a + b) + 3(a + b) = (y + 3)(a + b)$$

b. The binomial $x^2 + 1$ is a common factor of both terms.

$$7(x^2 + 1) - y(x^2 + 1) = (7 - y)(x^2 + 1)$$ ◀

Look for common factors. Sometimes you will regroup terms.

EXAMPLE 2

Factor each polynomial.

a. $3ac + 6ab + 4c + 8b$ **b.** $4x^2 - 8y + 8xy - 4x$

Solution

a. The only common factor of the four terms in
$3ac + 6ab + 4c + 8b$ is 1. Try to factor by grouping.

$$3ac + 6ab + 4c + 8b$$
$$= (3ac + 6ab) + (4c + 8b) \quad \text{Group the terms.}$$
$$= 3a(c + 2b) + 4(c + 2b) \quad \text{Factor each group.}$$
$$= (3a + 4)(c + 2b) \quad \text{Factor out the binomial.}$$

So, $3ac + 6ab + 4c + 8b$ in factored form is $(3a + 4)(c + 2b)$.

Check by multiplying the binomials to see that the product is
$3a + 6ab + 4c + b$.

b. Regroup the terms of the polynomial $4x^2 - 8y + 8xy - 4x$.

$$4x^2 - 8y + 8xy - 4x$$
$$= 4x^2 - 4x + 8xy - 8y \quad \text{Rearrange the polynomial.}$$
$$= (4x^2 - 4x) + (8xy - 8y) \quad \text{Group the polynomial.}$$
$$= 4x(x - 1) + 8y(x - 1) \quad \text{Factor each group.}$$
$$= (4x + 8y)(x - 1) \quad \text{Factor out the binomial.}$$
$$= 4(x + 2y)(x - 1) \quad \text{Factor out 4 from the first binomial.}$$

In factored form $4x^2 - 8y + 8xy - 4x$ is $4(x + 2y)(x - 1)$. ◄

 CHECK UNDERSTANDING

Describe how you can factor the two expressions in Examples 2a and 2b by starting with a different grouping of terms. Compare the results.

It is helpful to recognize polynomials that are *additive inverses*, or *opposites*. Recall that multiplying x by -1 produces its opposite, $-x$. The opposite of a polynomial is one in which each term is the opposite of the corresponding term in the original polynomial. For example, the opposite of $x - y$ is $-1(x - y) = -x + y$. By the commutative property, $-x + y = y - x$. Therefore, $y - x$ is the opposite of $x - y$.

EXAMPLE 3

Factor: $y^2 - 7y + 21x - 3xy$

Solution

Group terms that have a common factor. Then factor each group.

$$y^2 - 7y + 21x - 3xy = y(y - 7) + 3x(7 - y)$$

Since $y - 7$ and $7 - y$ are opposites, you can replace
$7 - y$ with $-1(y - 7)$.

$$y^2 - 7y + 21x - 3xy = y(y - 7) + 3x(-1)(y - 7)$$
$$= y(y - 7) - 3x(y - 7)$$
$$= (y - 3x)(y - 7)$$ ◄

 PROBLEM SOLVING TIP

You can model factor-by-grouping problems using Algeblocks.

You may also group terms in special patterns to factor. You may have to factor twice, as in the following example, where you see the pattern of a perfect square trinomial as well as of a difference of two squares.

EXAMPLE 4

CERAMIC TILING The area of a square ceramic tile is $4x^2 + 12x + 9$ and the area of the circle is $25y^2$. Find the area of the shaded region in factored form.

Solution

The area of the shaded region = (area of square) − (area of circle). Notice that the area of the square, $4x^2 + 12x + 9$, is a perfect square trinomial and that the area of the circle, $25y^2$, is a perfect square.

$$A = 4x^2 + 12x + 9 - 25y^2$$

$$A = (2x + 3)^2 - (5y)^2$$ Factor the trinomial.

$$A = (2x + 3 + 5y)(2x + 3 - 5y)$$ Factor the difference of two squares.

So, in factored form the area of the shaded region is $(2x + 3 + 5y)(2x + 3 - 5y)$. ◀

PROBLEM SOLVING TIP

In Example 4, you might have begun with the grouping $(4x^2 - 25y^2) + (12x + 9)$ and factored to get $(2x + 5y)(2x - 5y) + 3(4x + 3)$, which cannot be factored. You may need to try several approaches to factoring before you find one that works.

TRY THESE

Factor.

1. $5(c + d) + 7(c + d)$

2. $13(f^2 + 8) - 9(f^2 + 8)$

3. $g(b + 3) - 4(b + 3)$

4. $xz + 10x + yz + 10y$

5. $2h - 2k + gh - gk$

6. $x^2 - x + xy - y$

7. $y^4 - 2y^3 + 3y - 6$

8. $3a - 3b + ab - a^2$

9. $2wz - w + 3 - 6z$

10. GEOMETRY Find the dimensions of a rectangle with area $ab - 3a + 5b - 15$.

11. MODELING Use Algeblocks to find the factors of $xy - 2x + y - 2$.

12. WRITING MATHEMATICS Can $ax - bx - ay + by$ be factored? Write a paragraph explaining how you know.

PRACTICE

Factor.

1. $5(x + 1) + w(x + 1)$

2. $z(y - 3) + 2(y - 3)$

3. $xy + 5x + 2y + 10$

4. $ab + 7a + 4b + 28$

5. $xy + 2x - 7y - 14$

6. $ab - 3a + 9b - 27$

7. $ps - 2pt + qs - 2qt$

8. $mw - mx - nw + nx$

9. $12ab + 15a + 4b + 5$

10. $2xy - 8x + 3y - 12$

11. $3wz + 12w - z - 4$

12. $cd - 8c - 3d + 24$

13. GEOMETRY Find the dimensions of a rectangle if its area is $mn - 4m + 2n - 8$.

Factor.

14. $2fg + 4f - 7g - 14$ **15.** $yx - 2y + 8 - 4x$ **16.** $gh - g + 3 - 3h$

17. $pq - 7p + 35 - 5q$ **18.** $st - 3s + 18 - 6t$ **19.** $xy + 3x - 4y - 12$

20. **GEOMETRY** Find the area of the shaded region shown at the right. The area of the square is $4x^2 + 4x + 1$ and the area of the circle is $36y^2$. Write the area in factored form.

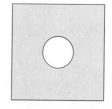

EXTEND

Factor by grouping.

21. $ax + bx + cx + 2a + 2b + 2c$ **22.** $xw + 2yw + 3zw - 4x - 8y - 12z$

23. $ap + aq - ar - bp - bq + br$ **24.** $x^2 - cx - ax + ac - bx + bc$

25. **GEOMETRY** Find the perimeter of a rectangle with an area of $xy - 5x + 4y - 20$.

26. **INSTALLMENT CAR LOANS** The formula for determining a monthly car payment R is $R = \dfrac{P + Pnr}{12n}$ where P is the loan amount, r is the annual interest rate, and n is the number of years of the loan. Solve this formula for P.

27. **HISTORY** A Hindu mathematician, Brahmagupta, who lived about A.D. 628, did much work in algebra and developed a system of notation. In his notation, the expression $xy + 4x + 2y + 8$ would be written as *ya ka bha ya* 4 *ka* 2 *ru* 8. The syllable *ya* stands for the first two letters of "the first unknown," *ka* stands for the first two letters of "the second unknown," *bha* stands for "product," and *ru* stands for "pure number."
   a. Factor this expression in our notation.
   b. How do you think Brahmagupta would write $3xy + 9y$?

THINK CRITICALLY

Replace N with a number or monomial that will make the polynomial completely factorable. Then, factor the polynomial.

28. $2jk - 7j + N - 14$ **29.** $ab^2 + N + 3b^2 - 48$ **30.** $25q^2 + N + 16 - p^2$

PROJECT *Connection*   Another feature to include in your river model is a *hole* or *souse hole*. A hole occurs when there is a large drop behind a submerged obstruction.

As shown in the figure at the right, the fast water flowing over the obstacle collides with the slower, deeper water behind it and a wave, called a *roller*, forms. This wave can overflow into your boat or cause it to turn over. Create a hole in your model river using a submerged obstruction.

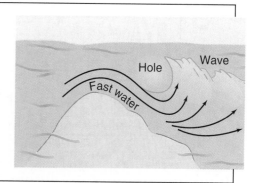

Civil engineers design and supervise the construction of roads, airports, tunnels, bridges, and dams. They often must solve problems, such as the one involving the Colorado River Basin in the early 1900s.

The Colorado River begins in Colorado and extends to California. The river used to flood low-lying areas in late spring and early summer. During other periods of the year, the river flow was so low that there was not enough water for irrigation or livestock. To alleviate this pattern of flood and drought, the Hoover Dam was built.

The dam, which is the highest concrete arch dam in the United States, is 726 feet high and contains 3.25 million cubic yards of concrete. It provides flood control; irrigation water; water for domestic, industrial, and municipal use; and low-cost electrical power for Arizona, Nevada, and southern California.

The energy generated by the dam is measured in kilowatt capacity. This table shows how the capacity has changed over the years.

Kilowatt Capacity—Hoover Dam				
Year	1952	1961	1972	1989	1992
Kilowatts	1,249,800	1,344,800	1,407,300	1,920,000	2,217,000

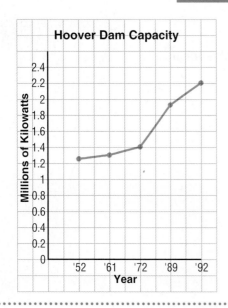

Decision Making

1. When the turbines were updated in 1972, what was the percent increase of kilowatt capacity from 1952?

2. What is the total percent increase in kilowatt capacity from 1952 to 1992?

Suppose the data in the table are graphed as shown at the left.

3. What does this graph seem to indicate about the future capacity of the dam?

4. Do you think this graph is a good representation for this data? Give reasons for your answer.

Explore/Working Together

● Work in a small group. Each group should have a calculator.

SPOTLIGHT
ON LEARNING

WHAT? In this lesson you will learn
• to perform two or more types of factoring on the same polynomial.

WHY?
Understanding how to factor polynomials completely can help you solve problems in geometry and metalwork.

1. Recall that the prime factorization of a number shows the number as a product of only prime numbers. For example, the prime factorization of 12 is 2 • 2 • 3, or 2^2 • 3. Each group member should work independently with paper and pencil to find the prime factorization of 756.

2. Compare your work with the work of other group members. Check with a calculator that each factorization is correct. Describe each person's method of factoring.

3. Which method appears to be the most efficient?

Build Understanding

● In this chapter, you have seen several ways of factoring polynomials.

Factor out a common monomial from a polynomial.

$$3x^3 - 12x^2 + 6x = 3x(x^2 - 4x + 2)$$

Factor a trinomial, including a perfect square binomial, as the product of two binomials.

$$x^2 + 4x + 3 = (x + 3)(x + 1)$$
$$4x^2 - 28x + 49 = (2x - 7)^2$$

Factor the difference of two squares as the product of two binomials.

$$16x^2 - 25 = (4x + 5)(4x - 5)$$

A polynomial is not considered "factored" until it is factored completely. A polynomial is **factored completely** only when it is expressed as the product of a monomial and one or more polynomial expressions that cannot be factored further. Some polynomials must be factored more than once before they are factored completely. You should check after each factoring step to see whether there are any expressions that can be factored further.

COMMUNICATING ABOUT ALGEBRA

Review by making a list of all the types of factoring. Order them based on what to look for first when factoring completely.

EXAMPLE 1

Factor: $2x^2 - 14x + 24$

Solution

First, look for common monomial factors. Then try to factor the resulting polynomial.

$$2x^2 - 14x + 24$$
$$= 2(x^2 - 7x + 12) \qquad \text{Factor out 2, the GCF.}$$
$$= 2(x - 3)(x - 4) \qquad \text{Factor the trinomial.}$$

Therefore, $2x^2 - 14x + 24 = 2(x - 4)(x - 3)$. ◄

Check by multiplying all the factors. Sometimes you need to factor more than twice.

EXAMPLE 2

Factor: $5x^2y + 20x^2 - 45y - 180$

Solution

$$5x^2y + 20x^2 - 45y - 180 \qquad \text{The GCF is 5.}$$
$$= 5(x^2y + 4x^2 - 9y - 36) \qquad \text{Factor out 5.}$$
$$= 5[(x^2y + 4x^2) - (9y + 36)] \qquad \text{Group.}$$
$$= 5[(x^2(y + 4) - 9(y + 4)] \qquad \text{Factor each group.}$$
$$= 5(y + 4)(x^2 - 9) \qquad \text{Factor out } y + 4.$$
$$= 5(y + 4)(x + 3)(x - 3) \qquad \text{Factor } x^2 - 9.$$

Since no other terms can be factored,

$$5x^2y + 20x^2 - 45y - 180 = 5(y + 4)(x + 3)(x - 3) \qquad ◄$$

Recall that the volume of a rectangular prism is the product of its length, width, and height.

EXAMPLE 3

GEOMETRY The volume of a rectangular prism is $3ab^2 - 6ab - 45a$. Find the dimensions of the prism in terms of a and b.

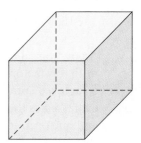

Solution

$$3ab^2 - 6ab - 45a$$
$$= 3a(b^2 - 2b - 15) \qquad \text{Factor out the GCF, } 3a.$$
$$= 3a(b - 5)(b + 3) \qquad \text{Factor the trinomial.}$$

The three dimensions of the rectangular prism are $3a$, $b - 5$, and $b + 3$. ◄

Factor.

1. $4x^2 + 8x - 32$

2. $3y^2 - 12y + 9$

3. $5x^2 - 20$

4. $7a^2 - 63$

5. $xy^2 - 16x + 2y^2 - 32$

6. $ab^2 - a + 9b^2 - 9$

7. GEOMETRY Find the dimensions of a box if its volume is $ab^2 + 13ab + 40a$.

8. WRITING MATHEMATICS Write a paragraph describing how you know when a polynomial has been factored completely.

9. MODELING Use Algeblocks to build a three-dimensional model of a rectangular prism with a base equal to $x^2 + 2x + 1$ and a height of 3. How do the dimensions of the prism relate to the polynomial $3x^2 + 6x + 3$?

PRACTICE

Factor.

1. $2x^2 + 24x + 70$

2. $3y^2 + 21y + 36$

3. $5z^2 - 15z - 90$

4. $2a^2 - 4a - 160$

5. $4bc^2 + 12bc - 40b$

6. $6gh^4 + 18gh^2 - 168g$

7. $3x^2 - 75$

8. $4m^2 - 144$

9. $4x^2 + 24x + 36$

10. $8y^2 - 160y + 800$

11. $5p^2q^2 - 500$

12. $3r^2s^4 - 147$

13. $2xy^2 - 32x$

14. $3a^3b^4 - 192a^3$

15. $2xy^2 - 2x + 4y^2 - 4$

16. $5ab^2 - 20a + 30b^2 - 120$

17. $4cd^2 - 4c - 12d^2 + 12$

18. GEOMETRY Find the dimensions of a rectangular prism with the volume $3x^2 - 243$.

EXTEND

Factor.

19. $ab^2 + 8ab + 12a + 3b^2 + 24b + 36$

20. $xy^2 - 12xy + 36x + 4y^2 - 48y + 144$

21. $6pq^2 - 54p - 12q^2 + 108$

22. $xy^2 + 7xy + 12x + 2y^2 + 14y + 24$

23. $ab^2 + 7ab + 10a + b^2 + 7b + 10$

24. $ac^2 + 5ac + 6a - bc^2 - 5bc - 6b$

25. $m^2n^2 + 9m^2n + 20m^2 - 4n^2 - 36n - 80$

26. $a^4b^2 + 2a^4b + a^4 - 9b^2 - 18b - 9$

27. METALWORK The formula for the volume of a hollow cylinder is $V = \pi R^2 h - \pi r^2 h$ where R is the radius of the cylinder, r is the radius of the hollow, and h is the height.

    a. Write the formula in factored form.

    b. The weight of aluminum is 168.5 lb/ft^3. Find the weight, to the nearest tenth of a pound, of a hollow aluminum cylinder if $R = 3$ ft, $r = 2.5$ ft, and $h = 4$ ft. Use 3.14 for π.

The following formulas can be used to factor some third-degree polynomials.

$$x^3 - a^3 = (x - a)(x^2 + ax + a^2) \quad \text{and} \quad x^3 + a^3 = (x + a)(x^2 - ax + a^2)$$

Factor the given polynomial.

28. $x^3 + 2^3$ **29.** $x^3 - 1$ **30.** $x^3 + 64$

31. $y^6 - 2^6$ (*Hint:* Let $y^3 = u$ and $2^3 = v$.)

THINK CRITICALLY

32. Show that $a^3 + b^3 = (a + b)(a^2 - ab + b^2)$.

33. Use the pattern from Exercise 32 to factor $16x^3 + 54y^3$ completely.

Find a value of n that makes the statement true.

34. $pq^2 - 14pq - 3np + 2q^2 - 28q - 30 = (p + 2)(q + 1)(q - 3n)$

35. $8r^{5n-1} - 50s^2 = 2(2r^{2n+3} + 5s)(2r^{2n+3} - 5s)$

MIXED REVIEW

36. STANDARDIZED TESTS   28% of what number is 42?

    A. 1.5   **B.** 150   **C.** 87.5   **D.** 11.76

Factor.

37. $xy - 5x + 2y - 10$ **38.** $2ab - 2a + 3b - 3$ **39.** $4y^2z + 4y^2 - 9z - 9$

PROJECT *onnection*   A river's *gradient*—how much it drops in vertical feet per mile—is a significant indicator of how rough the waterflow is likely to be. Rivers with a gradient of less than 20 feet per mile usually only have mild rapids, while those with a gradient over 50 feet per mile provide a dangerous ride.

1. Raise one end of your model using pieces of wood or other materials. Determine the average gradient or vertical drop in inches per foot along the entire course using the formula

$$g = \frac{\text{start elevation} - \text{end elevation (in inches)}}{\text{total length (in feet)}}$$

2. Sail a toy boat or piece of cork down the course. Use a stopwatch to time the trip. Then use the formula $d = rt$ to determine the speed.

3. Raise your model higher using another piece of wood. Determine the average gradient. Predict the time and speed for the course now. Then sail the boat again and check your predictions.

12.6 Solve Quadratic Equations by Factoring

Explore

1. Use a graphing utility to graph the function $y = x^2 + 5x - 6$. Set your calculator so that integers are displayed, and use the TRACE feature to locate the values of x for which $y = 0$. Substitute the values in the polynomial to check.

2. Write the polynomial $x^2 + 5x - 6$ in factored form.

3. Describe how the factored form relates to the values of x for which $y = 0$.

4. Based on the example above, find the values of x for which $y = 0$ in the function $y = x^2 + 3x + 2$. Check these values by using the graphing utility or by substituting in the polynomial.

Build Understanding

You know that any time you multiply a number by zero, the product is zero. Therefore, if you know that a product is zero, then you can be sure that at least one of its factors is zero. This is stated in the **zero product property**.

> ### THE ZERO PRODUCT PROPERTY
> For any real numbers a and b, if $ab = 0$, then $a = 0$ or $b = 0$, or both $a = 0$ and $b = 0$.

The zero product property can be used in solving quadratic equations.

EXAMPLE 1

Find all the real solutions to each equation.
  a. $y(2y - 3) = 0$ **b.** $(z + 2)(z - 5) = 0$

Solution

Use the zero product property.

a. $y(2y - 3) = 0$

$\quad y = 0 \quad$ or $\quad\quad 2y - 3 = 0$

$\quad y = 0 \quad$ or $\quad 2y - 3 + 3 = 0 + 3$

$\quad y = 0 \quad$ or $\quad\quad\quad \dfrac{2y}{2} = \dfrac{3}{2}$

$\quad y = 0 \quad$ or $\quad\quad\quad\quad y = \dfrac{3}{2}$

So, the solutions are 0 and $\dfrac{3}{2}$.

Check

$$y = 0$$
$$0(2 \cdot 0 - 3) \stackrel{?}{=} 0$$
$$0 = 0 \checkmark$$

$$y = \frac{3}{2}$$
$$\frac{3}{2}\left(2 \cdot \frac{3}{2} - 3\right) \stackrel{?}{=} 0$$
$$\frac{3}{2}(3 - 3) \stackrel{?}{=} 0$$
$$\frac{3}{2} \cdot 0 \stackrel{?}{=} 0$$
$$0 = 0 \checkmark$$

b.
$$(z + 2)(z - 5) = 0$$
$$(z + 2) = 0 \quad \text{or} \quad (z - 5) = 0$$
$$z + 2 - 2 = 0 - 2 \quad \text{or} \quad z - 5 + 5 = 0 + 5$$
$$z = -2 \quad \text{or} \quad z = 5$$

So, the solutions are –2 and 5.

Check

$$z = -2$$
$$(z + 2)(z - 5) \stackrel{?}{=} 0$$
$$(-2 + 2)(-2 - 5) = 0$$
$$0(-7) = 0$$
$$0 = 0 \checkmark$$

$$z = 5$$
$$(z + 2)(z - 5) \stackrel{?}{=} 0$$
$$(5 + 2)(5 - 5) = 0$$
$$7 \cdot 0 = 0$$
$$0 = 0 \checkmark$$ ◀

Some quadratic equations can be solved by factoring and using the zero product rule.

EXAMPLE 2

Solve each equation.

a. $x^2 - 7x = 0$ **b.** $x^2 - 9x + 20 = 0$

Solution

a. $x^2 - 7x = 0$
$x(x - 7) = 0$ Factor out the GCF.
$x = 0 \quad \text{or} \quad x - 7 = 0$
$x = 0 \quad \text{or} \quad x = 7$

So, the solutions are 0 and 7. Check each solution.

b. $x^2 - 9x + 20 = 0$
$(x - 4)(x - 5) = 0$ Factor into two binomials.
$x - 4 = 0 \quad \text{or} \quad x - 5 = 0$
$x = 4 \quad \text{or} \quad x = 5$

So, the solutions are 4 and 5. Check each solution. ◀

It is important to remember that when you use the factoring method to solve quadratic equations, you are also using the zero product property. The factored polynomial *must* equal zero.

EXAMPLE 3

Solve each equation.

  a. $a^2 = 8a$ **b.** $b^2 - 2b - 1 = 7$

Solution

a. $a^2 = 8a$

     $a^2 - 8a = 0$ Make one side of the equation equal to zero.

     $a(a - 8) = 0$ Factor and solve.

      $a = 0$ or $a - 8 = 0$

      $a = 0$ or $a = 8$

The solutions are 0 and 8. Check the solutions.

b. $b^2 - 2b - 1 = 7$

   $b^2 - 2b - 1 - 7 = 0$ Make one side of the equation equal to zero.

     $b^2 - 2b - 8 = 0$

   $(b - 4)(b + 2) = 0$ Factor and solve.

     $b - 4 = 0$ or $b + 2 = 0$

        $b = 4$ or $b = -2$

The solutions are 4 and –2. Check the solutions. ◄

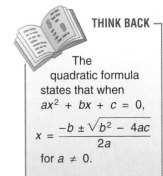

THINK BACK

The quadratic formula states that when $ax^2 + bx + c = 0$,

$$x = \frac{-b \pm \sqrt{b^2 - 4ac}}{2a}$$

for $a \neq 0$.

Remember that not all polynomials are factorable. You can use the *discriminant* of the quadratic formula to determine whether the polynomial is factorable. The **discriminant** is the expression under the radical sign of the quadratic formula, $b^2 - 4ac$.

   If $b^2 - 4ac < 0$, the equation has no real solution.

   If $b^2 - 4ac = 0$, there is one, rational solution.

   If $b^2 - 4ac > 0$, there are two real solutions.

Also, if $b^2 - 4ac > 0$ and is not a perfect square, then the solutions of the equation are irrational. If the discriminant is greater than zero and is a perfect square, then the solutions of the equation are rational numbers.

Therefore, for any polynomial $ax^2 + bx + c$, if $b^2 - 4ac \geq 0$ and is a perfect square, then the polynomial is factorable.

COMMUNICATING ABOUT ALGEBRA

Discuss why a polynomial is factorable only when the discriminant is a perfect square. You may want to start with two equations of the form $x = p$ and $x = q$ where p and q are rational numbers, and work backward to a polynomial.

EXAMPLE 4

Use the discriminant to determine whether each polynomial is factorable. Then, factor if possible.

a. $2x^2 + x - 6$ **b.** $x^2 + 10x + 2$

Solution

For each polynomial, identify a, b, and c. Then determine the value of $b^2 - 4ac$.

a. $2x^2 + x - 6$ $\qquad a = 2, b = 1, c = -6$

$$b^2 - 4ac = (1)^2 - 4(2)(-6)$$
$$= 1 + 48$$
$$= 49$$

Since 49 is a perfect square, the polynomial is factorable. By trial and error, you can find the factors.

$$2x^2 + x - 6 = (2x - 3)(x + 2)$$

b. $x^2 + 10x + 2$ $\qquad a = 1, b = 10, c = 2$

$$b^2 - 4ac = 10^2 - 4(1)(2)$$
$$= 100 - 8$$
$$= 92$$

Since 92 is not a perfect square, the polynomial cannot be factored. ◄

Solving quadratic equations by factoring is often useful in solving problems about area.

EXAMPLE 5

ARCHITECTURE The length of a rectangle is 5 m more than its width. Find the dimensions of the rectangle if its area is 84 m².

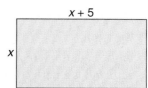

Solution

Let x represent the width of the rectangle. Let $(x + 5)$ represent the length.

$$(\text{length})(\text{width}) = \text{area}$$
$$(x + 5)x = 84$$
$$x^2 + 5x = 84 \quad \text{Multiply.}$$
$$x^2 + 5x - 84 = 0 \quad \text{Make one side of the equation equal to zero.}$$
$$(x + 12)(x - 7) = 0 \quad \text{Factor.}$$
$$x + 12 = 0 \quad \text{or} \quad x - 7 = 0$$
$$x = -12 \quad \text{or} \quad x = 7$$

Since the width cannot be −12, use the positive solution, 7. The width is 7. The length is 7 + 5, or 12. ◄

PROBLEM SOLVING TIP

Draw a diagram that helps you understand the problem.

Solve each equation.

1. $x(x + 3) = 0$

2. $y(y - 9) = 0$

3. $(x + 5)(x - 1) = 0$

4. $(x + 1)(x - 2) = 0$

5. $x^2 - 6x = 0$

6. $y^2 + 2y = 0$

7. $x^2 + 5x - 14 = 0$

8. $y^2 - 4y + 3 = 0$

9. $z^2 - z = 12$

10. $x^2 + 4x = 12$

11. $y^2 + 6y = -8$

12. $z^2 + 2z = 80$

13. GEOMETRY The length of a rectangle is 1 yd more than twice its width. The area is 55 yd^2. Find the dimensions of the rectangle.

Determine whether the polynomial is factorable. Then factor, if possible.

14. $x^2 + 13x + 22$

15. $x^2 - 3x - 24$

16. $x^2 - 20x + 36$

17. WRITING MATHEMATICS Write a paragraph explaining how some quadratic equations can be solved by factoring. Include directions for determining whether or not factoring can be used to solve a particular quadratic equation.

PRACTICE

Solve each equation.

1. $a(a + 1) = 0$

2. $b(b - 10) = 0$

3. $(c + 3)(c - 8) = 0$

4. $(d - 4)(d - 9) = 0$

5. $x^2 - 5x = 0$

6. $y^2 + 7y = 0$

7. $x^2 - 3x - 70 = 0$

8. $x^2 + 4x - 45 = 0$

9. $x^2 + 11x + 28 = 0$

10. $x^2 - 15x + 44 = 0$

11. $x^2 + 3x = 18$

12. $x^2 - 2x = 63$

13. $y^2 - 14 = 5y$

14. $z^2 + 10 = 11z$

15. $2x^2 + 7x = -3$

16. $3x^2 + 14x = 5$

17. $x^2 - x = 3x + 12$

18. $y^2 + 3y - 2 = y + 1$

19. GEOMETRY The length of a rectangle is 7 m longer than its width. The area is 18 m^2. Find the dimensions of the rectangle.

Determine whether the polynomial is factorable. Then factor, if possible.

20. $x^2 - 2x + 5$

21. $x^2 - 12x + 32$

22. $x^2 - 13x - 48$

23. $x^2 - x - 90$

24. $2x^2 + 13x + 15$

25. $2x^2 + x + 3$

26. $x^2 - 5x - 84$

27. $x^2 - x + 4$

28. $x^2 - 17x + 66$

29. NUMBER THEORY The product of two consecutive integers is 132. What are the integers?

30. NUMBER THEORY The product of two consecutive odd integers is 143. What are the integers?

EXTEND

Solve each equation.

31. $(x + 1)^2 + (x - 5)^2 = 20$

32. $(x + 3)^2 - 2(x + 1) = 3$

33. GEOMETRY The perimeter of a garden must be 40 m. The area of the garden must be 96 m^2. Find the length and width.

34. NUMBER THEORY Find two consecutive even integers such that the square of the larger less five times the smaller is 60.

35. CHINESE ARCHITECTURE The floor plan of Toshodaiji, an eighth-century Japanese monastery, built under Chinese influence and supervision, is a rectangle. Its length is 1 unit less than twice its width, and the floor area is 28 square units. Find the dimensions in units.

THINK CRITICALLY

Write a quadratic equation that has the given pair of solutions. Make all coefficients integers.

36. −2 and 8

37. 0 and −12

38. $\frac{1}{2}$ and −3

39. $\frac{2}{3}$ and −1

Find all integers, *n*, that make the polynomial factorable.

40. $2x^2 + nx - 10$

41. $nx^2 + 2x - 4$

MIXED REVIEW

Solve.

42. $7x + 8 = 3x - 20$

43. $4(x - 3) = 3x + 11$

44. $5x + 7 + 2x = 11 + 4x$

Find the range of the function given the domain {−2, 0, 3}.

45. $y = -3x$

46. $y = x + 5$

47. $y = x^2$

48. STANDARDIZED TESTS Which is the solution of the inequality $2 - x > 7$?

    A. $x > 9$ **B.** $x > -5$ **C.** $x < -9$ **D.** $x < -5$

Simplify.

49. $(3x^2 + 4x - 2) + (4x^2 - 3)$

50. $(4ab - 3a + b) - (2ab + a - 6b)$

The Mississippi River is an important waterway to the Gulf of Mexico and the Atlantic Ocean. *Mississippi* means "great river" in the language of the Ojibway, Native Americans of the Great Lakes region. Like all rivers, the Mississippi can be dangerous. In the summer of 1993, the Mississippi River overflowed, causing loss of life, property, and crops.

A water resources technician gathers data and makes computations for water projects, including flood control. These professionals pay close attention to river height as an indicator of flood potential. Even a small rise in a river can mean an increase in water volume of millions of gallons. The volume is determined by multiplying the length, width, and depth of the river ($V = lwd$). The table below models the change in volume when width, depth, or both are increased.

Length	Width	Depth	Volume	Change from Original
l	w	d	lwd	0
l	$w + x$	d	$lwd + xld$	$+ xld$
l	w	$d + x$	$lwd + lwx$	$+ lwx$
l	$w + x$	$d + x$	$lwd + lwx + lxd + lx^2$	$+ lwx + lxd + lx^2$

Decision Making

Use the table to answer these questions.

1. The last line of the table shows the change in volume when both the width and the depth of the river are increased by the same amount. Rewrite this polynomial in factored form.

2. Consider a stretch of the Mississippi that is $\frac{1}{2}$ mi wide. The depth of the river increases an average of $\frac{1}{2}$ ft along a 1-mi stretch. By how much does the volume of water (cubic feet) in the river increase along that stretch? [1 mi = 5280 ft]

3. Change your answer to Question 2, which is in cubic feet, to gallons. Round to the nearest whole gallon. [1 gal ≈ 0.13 ft³]

4. Compute the approximate increase in gallons of water for the same stretch as in Question 2, if the depth increases an average of $1\frac{1}{2}$ ft along that stretch. (Round to the nearest whole gallon.)

5. How many more gallons of water are in that 1-mi stretch when the river's depth is increased by $1\frac{1}{2}$ ft than when it increases by $\frac{1}{2}$ ft?

6. Write a paragraph explaining the importance of a water resources technician, based on what you've learned from these exercises.

Problems Involving Quadratic Equations

In applied situations, the mathematical model often results in a quadratic equation. If the quadratic equation has two real solutions, it may be that only one of them makes sense in the problem. For example, a negative number cannot represent a length or elapsed time. Reject solutions that do not fit the physical situation.

Problem

A jewelry designer sketched a rectangular belt buckle 80 mm long and 60 mm wide. The belt buckle has a uniform strip of bronze around a smaller rectangle of silver. How wide should the bronze strip be if the area of the silver rectangle is to be half the area of the belt buckle?

Explore the Problem

Begin by drawing a diagram to help you analyze the problem.

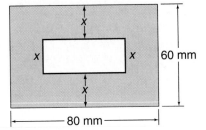

1. What will x represent?

2. In terms of x, what are the dimensions of the silver rectangle?

3. Express the area of the silver rectangle in terms of x.

4. Write a quadratic equation that relates the areas of the two metals.

5. Solve the quadratic equation by factoring.

6. Do both solutions make sense in the problem? Explain. What is the width of the bronze strip?

7. How can you check your solution?

Investigate Further

Cristina was curious to know how long her cousins Gustavo and Gloria have lived in Minneapolis. She asked her Uncle Hector, a skilled riddler. He replied, "Gustavo has lived in Minneapolis three years more than twice as long as Gloria has lived there. If you square the number of years each of them has lived in Minneapolis, the sum of the squares is 194." How long has each person lived in Minneapolis?

8. If y represents the number of years Gloria has lived in Minneapolis, what expression represents the number of years Gustavo has lived there?

9. What information from the problem can you use to write an equation? Show your equation.

10. Can you solve the equation by factoring? How do you know?

11. Explain how the solutions you find by factoring relate to the problem.

12. How long has each person lived in Minneapolis? Check your answer.

13. Suppose Uncle Hector had responded, "Find two real numbers such that one number is 3 more than twice the second, and the sum of their squares is 194." Would your answer be different? Explain.

14. WRITING MATHEMATICS Think about different problems that could lead to the equation $x(x + 1) = 72$. Solve the equation by factoring and consider the values you obtain. Then write one problem for which both values would make sense and another problem for which only one value makes sense.

PROBLEM
SOLVING PLAN

- Understand
- Plan
- Solve
- Examine

Apply the Strategy

15. HOME FURNISHINGS If the units are omitted, the sum of the area and perimeter of a square rug is 60.

   a. What equation can you use to find the length of one side of the rug?

   b. What is the length, assuming the units are feet?

16. RECREATION The area of a rectangular park is 1750 m². The park is 15 m longer than it is wide. What are the dimensions of the park?

17. MAGAZINE LAYOUTS The perimeter of a rectangular photograph is 32 in. and the area is 63 in.² The photograph is positioned on a magazine page so that the only remaining space on the page is a $1\frac{1}{2}$-in. strip all around the photograph. (Assume length is the greater dimension.)

   a. What are the dimensions of the photograph?

   b. What are the dimensions of the magazine page?

18. **AGES** Yoshi is 1 year younger than Keiko. The product of their ages is 756. How old is each person?

19. **NUMBER THEORY** Find two consecutive integers whose product is 1560.

20. **NUMBER THEORY** A positive integer is increased by 10. If the square of the resulting larger integer is nine times the square of the smaller integer, find the smaller integer.

21. **ART** A rectangular painting, with its wooden frame, measures 24 in. by 31 in. The width of the frame is equal all around. Without the frame, the painting has an area of 450 in.^{2}

 a. How wide is the frame?

 b. What are the dimensions of the painting?

22. **METALWORK** A piece of wire 56 in. long is cut into two pieces, each of which is formed into a square. If the sum of the areas of the squares is 106 in.^{2}, how long were the pieces of wire?

23. **GARDENING** A circular garden covers an area of 400π ft^2. A paved path of constant width is to surround the garden. If the total area of the garden and path is 676π ft^2, how wide is the path?

24. **PACKAGING** A square piece of cardboard has edges of 10 in. in length. A square piece is cut from each of the corners, and the remaining sides are folded up to form an open box. If the area of the bottom of the box is equal to the sum of the areas of the sides of the box, what is the length of the sides of the squares that were cut?

REVIEW PROBLEM SOLVING STRATEGIES

CRAZY CARDS

1. A box contains three cards. One card is green on both sides, one card is yellow on both sides, and one card is green on one side and yellow on the other. You pick a card without looking and place it on the table. The card has a green side showing. What is the probability that the side not showing is also green? Before you (incorrectly) answer $\frac{1}{2}$, use the steps below to analyze the problem.

a. Represent the cards as shown and consider the set of possible outcomes when a card is drawn and placed on the table. Use ordered pairs such as (G_1, G_2) where the first letter denotes the side placed face up and the second letter denotes the side not showing. How many ordered pairs are in the sample space?

b. Since you know a green side is showing, what is the possible sample space now?

c. So, if the side showing is green, what is the probability that the other side is also green?

d. Now try this problem. Mrs. Long is the mother of two children. One of these children is a boy. What is the probability that they are both boys?

CUT CAREFULLY

2. Suppose you have two cords. One cord is 14 in. long and the other cord is 19 in. long.

a. How can you cut one of the cords into two pieces so that the length of one of the three pieces of cord you now have will be the average of the lengths of the other two? What relationship to the original lengths do you notice?

b. Generalize your results from 2a for any two cords. How can you cut one cord into two pieces so that the length of one of the three pieces is the average of the lengths of the other two? Explain.

FREQUENT FLIER

3. A northbound train starts traveling at the same time as a southbound train. The two trains, initially 150 kilometers apart, head toward each other on parallel tracks. The northbound train averages 33 km/h, and the southbound train averages 42 km/h. A bird is flying back and forth between the two trains at 8 km/h, leaving the northbound train just as it starts out. How far has the bird flown when the two trains pass? Explain your reasoning. (*Hint:* What formula can you use to determine the bird's distance? What do you know? What must you find?)

CHAPTER REVIEW

VOCABULARY

Choose the word from the list that completes each statement.

1. The __?__ states that if a product is zero, then at least one of its factors is zero.

2. The __?__ is the number multiplying the first term in a polynomial written in standard form.

3. A __?__ is a polynomial that is the square of a binomial.

4. For a quadratic equation $ax^2 + bx + c$, the __?__ is the expression $b^2 - 4ac$.

a. leading coefficient

b. discriminant

c. zero product property

d. perfect square trinomial

Lesson 12.1   EXPLORE PATTERNS IN FACTORING   pages 571–574

● Algeblocks may be used to model polynomials and find their factors.

Use Algeblocks to write the two binomial factors for each trinomial.

5. $x^2 + 4x + 3$

6. $x^2 - 5x + 6$

7. $x^2 + 2x - 8$

8. $x^2 - 4x - 5$

Lesson 12.2   FACTOR POLYNOMIALS   pages 575–581

● To factor out a monomial from a polynomial, find the greatest common factor and use the distributive property in reverse. To factor a trinomial, use a multiplication diagram or patterns of products.

Factor each polynomial.

9. $15d + 25d^2$

10. $9xyz^2 - 3y^2z + 6x^2yz$

11. $z^2 + 11z + 24$

12. $y^2 + 8y - 9$

13. Find the dimensions in terms of x of a rectangle if its area is $2x^2 + x - 15$.

Lesson 12.3   FACTORING SPECIAL PRODUCTS   pages 582–585

● Use patterns to factor the difference of two squares and perfect square trinomials.

Factor if possible.

14. $c^2 + 8c + 16$

15. $x^2 - x + 1$

16. $r^2 - 6r + 9$

17. $w^2 - 100$

18. Find the dimensions of a rectangle that has the same area as the shaded region.

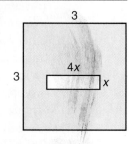

- To factor polynomials by grouping, use the distributive property.

Factor.

19. $4(x - 3) + y(x - 3)$

20. $ab + 2a + 4b + 8$

21. $pr - ps - qr + qs$

22. $cd + 5c - 5 - d$

23. Find the dimensions of a rectangle in terms of a and b if its area is $ab - 5a + 3b - 15$.

- To factor completely, you may need to perform two or more types of factoring on the same polynomial.

Factor completely.

24. $3x^2 + 27x + 42$

25. $2ab^4 - 8ab^2 - 90a$

26. $6y^2 - 36y + 54$

27. $3xy^2 - 12x$

28. Find the dimensions of a rectangular prism with the volume $4x^2 - 144$.

- Use the zero property to solve quadratic equations by factoring.

Solve each equation.

29. $k(k + 5)$

30. $n(n - 9)$

31. $(y + 7)(y - 10)$

32. $x^2 + 9x + 18$

33. The product of two consecutive even integers is 224. What are the integers?

- You can use quadratic equations to solve problems, but you may have to reject a solution that is not reasonable.

Solve each problem.

34. A rectangular flag 30 in. by 20 in. has a partial border of equal width covering two of its sides as shown. Find the width of the border if the total area covered by the border is 264 in.^{2}

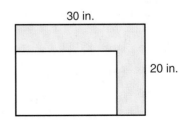

35. The difference of two positive numbers is 4. The sum of their squares is 170. What are the numbers?

CHAPTER ASSESSMENT

CHAPTER TEST

Factor completely.

1. $2g^3h^2 - 8gh$ **2.** $6a^3b^2 - 3a^2b + 18ab^2$

3. $4ab - 12b + 4a$ **4.** $3x^3y - 12xy^3 + 6x^2y^2$

5. WRITING MATHEMATICS Write a paragraph explaining how you can factor the polynomial $3x^3 + 9x^2 + 6x$.

6. The product of two consecutive odd whole numbers is 195. What are the numbers?

7. What is the difference between the expressions $4r^2$ and $16s^2$? Write your answer in factored form.

Factor completely.

8. $x^2 - 81$ **9.** $x^3 - 16x$

10. $50x^2 - 32$ **11.** $x^2 - 16x + 64$

12. $3z^2 - 2z - 1$ **13.** $2x^2 - 28x + 98$

14. $a^4 - 6a + 5$ **15.** $2b^6c^6 - 7a^3b^3 - 4$

16. STANDARDIZED TESTS Which expression does not have a difference of squares as at least one of its factors?

  A. $x^3 - 16x$ **B.** $4x^2 - 16$

  C. $2x^2 - 4$ **D.** $2x^3 - 32x$

17. STANDARDIZED TESTS Which expressions are factorable?

  I. $3x^2 + 13x - 10$

  II. $2x^2 + 11x + 14$

  III. $4x^2 - 4x - 15$

  IV. $2x^2 + 19x + 24$

  A. I, II, and III only **B.** I and IV only

  C. I, II, III, and IV **D.** I and III only

Solve each equation.

18. $r(r + 6) = 0$

19. $16x^2 - 9 = 0$

20. $y^2 - 10y - 56 = 0$

21. $x^2 + x - 0.75 = 0$

22. $k(k + 4) - 3(k + 4) = 0$

23. $n^2 - 7n + 10 = 0$

24. $y^2 - 6y + 4 = 20$

Solve each problem.

25. Find the area of the aqua region at the right and write it in factored form.

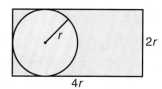

26. Find the dimensions of a rectangular prism with the volume $x^3 - 36x + 3x^2 - 108$.

27. The length of a rectangle is 7 cm longer than its width. The area is 120 cm². Find the dimensions of the rectangle.

28. The perimeter of a rectangle is 38 m and its area is 84 m². What are the dimensions of the rectangle?

29. A photograph is framed so that a uniform border surrounds the photograph. The entire framed photograph is 8 in. by 10 in. The area of the photograph is 32 in.² less than the area of the entire framed picture. How wide is the border?

30. Find the dimensions of a rectangle that has the same area as the shaded region at the right.

PERFORMANCE ASSESSMENT

MAKE A FLAG Design a rectangular flag 12 in. by 16 in. with a uniform border surrounding a smaller rectangle inside. Make the area of the smaller rectangle a ratio of the area of the entire flag. For example, try $\frac{1}{4}$ the area, $\frac{1}{3}$ the area, or $\frac{1}{2}$ the area of the entire flag for the area of the smaller rectangle. Use a quadratic equation and factoring to help you find measurements that will work. You may wish to design flags of different sizes with similar conditions.

USE MULTIPLICATION DIAGRAMS Select a variety of polynomials from this chapter. Draw multiplication diagrams and charts to find the factors as shown in the example for $x^2 + 2x - 8$. Draw a multiplication diagram and a chart.

	x	?
x	x^2	$?x$
?	$?x$	-8

Product, -8	Sum
$1, -8$	-7
$-1, 8$	7
$2, -4$	-2
$-2, 4$	2

	x	-2
x	x^2	$-2x$
$+4$	$4x$	-8

USE ALGEBLOCKS Use Algeblocks to build area models to determine factors of some polynomials. Select polynomials from this chapter that have all positive signs. Have a partner determine the factors. Follow the example below, a model of $x^2 + 4x + 3$. Remember the factors from the sides of the large rectangle are $(x + 3)$ and $(x + 1)$.

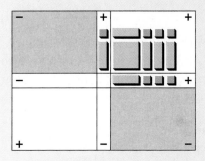

USE ALGEBLOCKS Pick a variety of polynomials from this chapter. Show how to model each polynomial with Algeblocks. Then ask a partner to show the steps for determining the factors of each polynomial. Check that each step is correct.

PROJECT ASSESSMENT

PROJECT *Connection* When each group has completed work on their river model, arrange a day for demonstrations.

1. Each group should name their water course and announce it when they begin the demonstration. Observers should take notes to be used later for comparison and evaluation of how well each group achieved its goal.

2. Each group should demonstrate three boat runs. These runs should be timed and averaged. The class can display the results in a bar graph and compare fastest and slowest times.

3. Geologists can estimate the age of a river based on its characteristics. Research and report on how they do so.

··· CUMULATIVE REVIEW ···

Factor each polynomial.

1. $12x^2y + 8xy^3$
2. $h^2 + 13h + 30$
3. $n^2 - 9n + 20$
4. $g^2 + 10g - 56$

Simplify each expression.

5. $\dfrac{x^8 \cdot x^7}{x^5}$
6. $(x^3y)^3(xy^4)^2$

7. A chemist has 3 liters of a 24% alcohol solution. She wants to add some pure alcohol to the solution to bring it up to 28%. How much pure alcohol should the chemist add?

Use the functions $f(x) = 5x - 2$ and $g(x) = 2x^2 + x - 1$ for Questions 8–11.

8. $f(-2) + g(-2)$
9. $\dfrac{f(4)}{g(2)}$
10. $f(x) + g(x)$
11. $g(x) - f(x)$

Factor each polynomial.

12. $2x^2 + 7x + 3$
13. $r^2 - 14r + 49$
14. $9t^2 - 4$
15. $16z^2 + 72z + 81$

Use the matrices for Questions 16–18.

$$A = \begin{bmatrix} 4 & -2 \\ -3 & 7 \end{bmatrix} \quad B = \begin{bmatrix} -5 & 4 \\ 3 & -9 \end{bmatrix}$$

16. $A + B$
17. $B - A$
18. (determinant of A) − (determinant of B)

19. Ed's Car Rental charges $24.50 per day plus $0.18 a mile. Ralph's Rental charges $27.00 per day plus $0.16 per mile. Alice prefers to rent from Ralph's. For a one-day rental, how many miles would Alice have to travel so that Ralph's Rental would have the lesser charge?

Solve and graph each compound inequality.

20. $2x < -6$ or $x - 5 \geq -2$
21. $-7 < 2x + 1 < 9$

Factor each polynomial completely.

22. $2x^2 + 6xy + 3x + 9y$
23. $x^3 - x^2 - 25x + 25$
24. $2p^4 - 8p^3 - 24p^2$
25. $2z^4 - 32$

Use the circle graph for Questions 26–28.

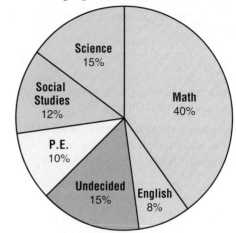

26. If 280 students gave their opinion, how many named Science as their favorite subject?

27. If 45 students said that Social Studies was their favorite, how many students were part of the survey?

28. **WRITING MATHEMATICS** Joe concluded that 40% of all the students at the school liked Mathematics best. Do you agree with Joe's conclusion? What questions might you like to ask Joe?

Solve each equation by factoring.

29. $x^2 + 8x = 20$
30. $3x^2 + 2 = 7x$
31. $y^3 - y^2 - 2y = 0$
32. $2n^2 + 12n = -18$

33. **TECHNOLOGY** Fred was having trouble factoring the trinomial $x^2 - 4x - 45$. Wilma came by and suggested to Fred that he graph the trinomial on his graphing calculator. Fred was able to figure it out right away. How did Wilma's idea help?

QUANTITATIVE COMPARISON In each question compare the quantity in Column 1 with the quantity in Column 2. Select the letter of the correct answer from these choices:

A. The quantity in Column 1 is greater.
B. The quantity in Column 2 is greater.
C. The two quantities are equal.
D. The relationship cannot be determined by the information given.

Notes: In some questions, information which refers to one or both columns is centered over both columns. A symbol used in both columns has the same meaning in each column. All variables represent real numbers. Most figures are not drawn to scale.

Column 1	Column 2

1.

Area = $2x^2 + 7x + 6$

length of the longer side	twice the length of the shorter side

2. the line $y = mx + b$, with $m < 0$

slope of a line parallel to the one given	slope of a line perpendicular to the one given

3. product of the coordinates of the vertex of the graph of $y = 2|x - 2| - 2$ | product of the coordinates of the vertex of the graph of $y = 2|x + 2| + 2$

4. $a < 0$

$a^5 \cdot a^3$	$(a^5)^3$

5. P(even number) rolling one number cube | P(tails) with one toss of a coin

Column 1	Column 2

6. $(8 \div 4) \div 2$ $\qquad$ $8 \div (4 \div 2)$

7. $-3x < 12$

x	4

8. A total of 20 nickels and dimes amounts to $1.45.

number of nickels	number of dimes

9.

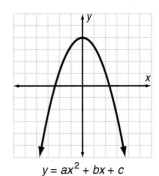

$y = ax^2 + bx + c$

a	c

10. 83% $\qquad$ $\dfrac{5}{6}$

11. $|x| < 7, |y| > 7$

x	y

12. m and n are integers, $a > 0$

a^{m-n}	$\dfrac{1}{a^{n-m}}$

13. $x \neq 0, y \neq 0$

number of terms when $(x + y)(x - y)$ is multiplied	number of terms when $(x + y)(x + y)$ is multiplied

14. $4(x + 5) < 12, \; 3(2 - n) < 12$

x	n

13 Geometry and Radical Expressions

Take a Look AHEAD

Make notes about things that look new.

- List the terms that are used to classify triangles according to the lengths of their sides. Make another list of terms that are used to classify triangles according to the measures of their angles.
- Identify the special ratios that are defined for corresponding sides of right triangles.

Make notes about things that look familiar.

- What does the radical symbol mean? What is the expression under the radical symbol called?
- Find a formula from this chapter that uses a radical symbol.
- In everyday language, when you describe two objects or ideas as being "similar," do you mean they are exactly alike? Use the word "similar" in a sentence. Find a sentence in this chapter that uses the word similar. How do you think their meanings compare?

DATA Activity

The Family Car

When Henry Ford began production of the Model T in 1908, the automobile's price was $850. By 1916 the price had dropped to $400 due to Ford's use of mass production methods. Since then, the cost of the average new car sold in the United States has risen significantly, and according to the Commerce Department, family income has not kept pace. The table on the next page shows average new car cost, in terms of the number of weeks' earnings of a median-income family. In addition to sticker price, costs include safety and pollution control equipment, dealer fees, registration fee, and taxes.

SKILL FOCUS

- Add, subtract, multiply, and divide real numbers.
- Interpret graphs.
- Write and solve an equation.
- Interpret the median of a data set.

TRAVEL and TRANSPORTATION

In this chapter, you will see how:

- **ROAD DEVELOPERS** apply the Pythagorean theorem to check that road intersections are perpendicular.
  (Lesson 13.2, page 624)

- **HIGHWAY ENGINEERS** use a formula involving a radical to design safe curves for freeway on-ramps.
  (Lesson 13.3, page 630)

- **AIR TRAFFIC CONTROLLERS** use a three-dimensional coordinate system when determining the distance between two airplanes.
  (Lesson 13.6, page 647)

NEW CAR COST IN WEEK'S EARNINGS

Use the table to answer the following questions.

1. In 1983, how many weeks' earnings would a median-income family have had to spend to purchase a new car?

2. How many more weeks' earnings were needed to purchase a new car in 1991 than in 1971?

3. Suppose the average new car price in 1994 was $21,800. Write and solve an equation to determine the figure that was used as the median family income for that year.

4. If a family's income is below the median level, will the number of weeks' earnings required to purchase a new car be more or less than the weeks shown on the graph?

5. **WORKING TOGETHER** As a group, discuss possible reasons for the decreasing trend shown in the 1970s. Try to confirm your ideas by doing some research.

You "Auto" Find Out

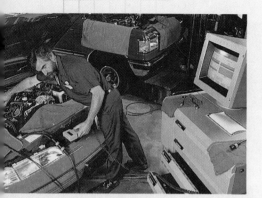

They cost many thousands of dollars to purchase, insure, and maintain, yet Americans own about 200,000,000 of them. People spend several hours each day inside them, possibly eating, listening to music, or talking on the telephone. No, it's not your home, it's your automobile, and it's the object under investigation in this project.

PROJECT GOAL

To analyze information relating to automobiles.

Getting Started

Work in groups.

1. Each group member should offer the name, make, and model year of a car for which he or she will collect information. This car might belong to a family member or a friend. Explain to the owners that you will need their help as you work on your project for the next few weeks.

2. Make a chart showing the cars to be analyzed by the group. List the options for each car, such as air conditioning, power windows, and stereo.

PROJECT Connections

Lesson 13.2, page 623:
Analyze data about features and prices of used cars.
Lesson 13.4, page 635:
Determine the maximum length of a package that will fit in a car's trunk.
Lesson 13.6, page 646:
Plan a triangular travel route, then determine distances and gasoline expense.
Chapter Assessment, page 669:
Use the information from the project to design classified advertisements for the used cars studied.

Internet Connection

www.swpco.com/
swpco/algebra1.html

3. For the next two weeks, each group member should read the automotive classified section of one or more newspapers to find advertisements for cars of the same make and model year for sale. Collect the advertisements and bring them to the next group meeting. Some students should go to a car dealer and obtain prices.

4. Make a list of the abbreviations used in the advertisements. Explain the meaning of each abbreviation.

Think Back/Working Together

- Work with a partner.

1. Explain how this figure shows that $3^2 = 9$.

2. Explain how this figure shows that $\sqrt{4} = 2$.

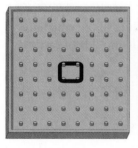 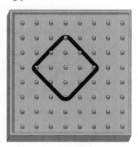

Find the area of each square. You may have to divide a square into smaller figures, calculate the area of each smaller figure, and then add to find the total area.

3.

4.

5.

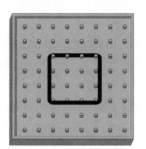 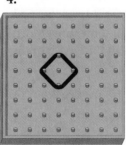 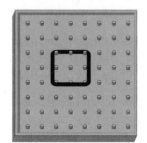

THINK BACK

A right triangle has a 90° angle. The side opposite the 90° angle is called the *hypotenuse*. The other two sides are called *legs*.

Explore

6. Using graph paper, cut out three squares: 3×3, 4×4, and 5×5. Arrange the squares as shown at the right. Notice the triangle formed by this placement of the squares is a right triangle.

7. Find the area of each square in the figure.

Repeat Questions 6 and 7 for triangles with the following dimensions. Make a table of your results.

8. 5–12–13   **9.** 6–8–10   **10.** 8–15–17

11. Repeat Questions 6 and 7 for a triangle with dimensions of your choosing.

Algebra Workshop

Make Connections

ALGEBRA: WHO, WHERE, WHEN

Pythagoras (6th century B.C.) founded a secret society for the study of mathematics, philosophy, and natural science. So closely did the Pythagoreans guard their secrets that, according to legend, one member of the society, Hippasus, was drowned at sea for revealing the existence of irrational numbers.

The Pythagorean theorem, named after Pythagoras, was discovered at least 3000 years ago. Pythagoras made the first systematic study of the relationship stated in the theorem.

12. Examine the results of your exploration. For each triangle, what is the relationship between the area of the largest square and the areas of the two smaller squares?

13. Suppose that you built squares on each leg of a $7-24-25$ right triangle. Predict the areas of the squares. Explain how you made your prediction.

14. Complete the following statement: In a right triangle, the square of the length of the hypotenuse is equal to ___?___ the squares of the lengths of the two legs.

The relationship between the sides of a right triangle is known as the *Pythagorean theorem*.

PYTHAGOREAN THEOREM

If a right triangle has lengths a and b, and a hypotenuse of length c, then
$$a^2 + b^2 = c^2$$

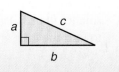

In Exercises 15–20, substitute known values in the equation $a^2 + b^2 = c^2$. Find the length of the missing side of each right triangle.

	Leg 1	Leg 2	Hypotenuse
15.	3	4	
16.	5	12	
17.		12	15
18.	8		17
19.		40	41
20.	m	n	

Suppose you know the length of the sides of a triangle. Substitute the lengths in the equation $a^2 + b^2 = c^2$. If this equation is true, then the triangle is a right triangle. This rule is called the *converse* of the Pythagorean theorem. So, you can use the converse of the Pythagorean theorem to determine if any triangle is a right triangle when you know the lengths of the sides.

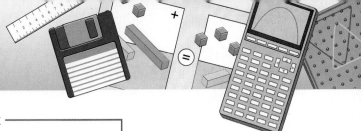

CONVERSE OF THE PYTHAGOREAN THEOREM

Let a, b, and c represent the lengths of the sides of a triangle. If $a^2 + b^2 = c^2$, then the triangle is a right triangle.

The lengths of the sides of a triangle are given. State whether the triangle is a right triangle.

	Side 1	Side 2	Side 3
21.	6	8	10
22.	4	6	7
23.	2.5	6	6.5
24.	12	35	37
25.	14	22	26

Summarize

26. MODELING Explain how the figure at the right demonstrates the Pythagorean theorem as it relates to the shaded triangle.

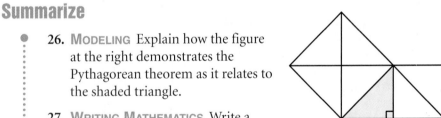

27. WRITING MATHEMATICS Write a word problem that can be solved using the Pythagorean theorem. Then solve the problem.

28. THINKING CRITICALLY Four right triangles with sides measuring a, b, and c are arranged to form a square as shown. Write the area of the large square as the sum of the areas of the four right triangles and smaller square. Set this sum equal to the area of the large square, $(a + b)^2$, and show that your results confirm the Pythagorean theorem.

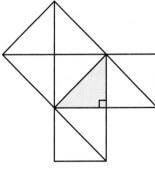

29. GOING FURTHER A set of numbers like {3, 4, 5} or {5, 12, 13} which satisfy the Pythagorean theorem is called a **Pythagorean triple**. Explore whether multiples (including fractional multiples) of Pythagorean triples are also Pythagorean triples. Describe your work and summarize your results.

Explore

Square the measures of each side of each triangle below. Copy and complete the table.

1.

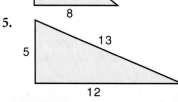

2.

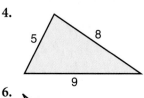

3.

4.

5.

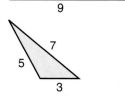

6.

	Square of a short side (a^2)	Square of a short side (b^2)	Square of the longest side (c^2)
1.			
2.			
3.			
4.			
5.			
6.			

7. Measure the angles of the triangles. Which of the triangles has a right angle? Which have an obtuse angle? Which have neither a right nor an obtuse angle?

In the Pythagorean theorem $a^2 + b^2 = c^2$, c is the hypotenuse, which is always the longest side of a right triangle.

8. For which triangles is $a^2 + b^2 = c^2$? What do these triangles have in common?

9. For which triangles is $a^2 + b^2 < c^2$? What do these triangles have in common?

10. For which triangles is $a^2 + b^2 > c^2$? What do these triangles have in common?

Build Understanding

Triangles can be classified according to the measures of their angles.

An **acute triangle** has three acute angles.

A **right triangle** has one right angle.

An **obtuse triangle** has one obtuse angle.

Triangles can be classified according to the length of their sides.

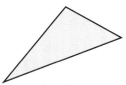

A **scalene triangle** has no sides equal in length.

An **isosceles triangle** has at least two sides equal in length.

An **equilateral triangle** has three sides equal in length.

In an isosceles triangle, the measures of the angles opposite the equal sides are equal. In an equilateral triangle, the measures of all three angles are equal.

Triangle *PQR* can be written △*PQR*.

EXAMPLE 1

Classify each triangle according to its side lengths and angle measures.

a.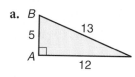

b.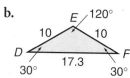

c.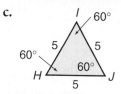

Solution

a. △*ABC* has no sides equal and contains a right angle. It is scalene and right.

b. △*DEF* has two equal sides and one obtuse angle. It is isosceles and obtuse.

c. △*HIJ* has three equal sides and three equal angles. It is equilateral and acute. ◀

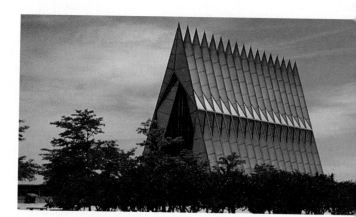

For triangles having side c as the longest side, the following are true.

> **PYTHAGOREAN RELATIONSHIPS**
>
> If $a^2 + b^2 > c^2$, then the triangle is acute.
> If $a^2 + b^2 = c^2$, then the triangle is right.
> If $a^2 + b^2 < c^2$, then the triangle is obtuse.

EXAMPLE 2

Use the Pythagorean relationships to classify a triangle having side lengths 9, 17, and 24.

Solution

Let $a = 9$, $b = 17$, and $c = 24$. Then $a^2 = 81$, $b^2 = 289$, and $c^2 = 576$. Since $81 + 289 < 576$, the triangle is obtuse. ◀

For any triangle, regardless of how the triangle is classified by side lengths and angle measures, the following is true.

> **SUM OF THE ANGLES OF A TRIANGLE**
>
> The sum of the measures of the angles of a triangle is 180°.

Angle Q can be written $\angle Q$. The measure of $\angle Q$ can be written as $m\angle Q$.

EXAMPLE 3

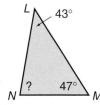

CITY PLANNING Roy is surveying a triangular plot of land for a city park. He determined that $m\angle L = 43°$ and $m\angle M = 47°$. Find $m\angle N$ and classify the triangle.

Solution

$$\begin{aligned}
m\angle L + m\angle M + m\angle N &= 180 \\
43 + 47 + m\angle N &= 180 \\
90 + m\angle N &= 180 \\
m\angle N &= 90
\end{aligned}$$

Sum of the measures of the angles is 180°.

The measure of $\angle N$ is 90°. So, $\triangle LMN$ is a right triangle. ◀

As you could see in Example 3, $\triangle LMN$ is a right triangle and the sum of its two acute angles is 90°. Another way to say this is that the two acute angles are *complements* of each other.

> **COMPLEMENTARY ANGLES**
>
> A pair of angles is *complementary* if their sum is 90°.
> Each angle is said to be the *complement* of the other.

Another special pair of angles are *supplements* of each other.

> **SUPPLEMENTARY ANGLES**
>
> A pair of angles is *supplementary* if their sum is 180°.
> Each angle is said to be a *supplement* of the other.

EXAMPLE 4

The measure of an angle is 60° less than its supplement. Determine the measure of the two angles.

Solution

Let x equal the measure of the angle. Then $180 - x$ equals the measure of its supplement.

$$x = (180 - x) - 60$$
$$x = 120 - x$$
$$2x = 120$$
$$x = 60$$

The measure of the angle is 60°. The measure of its complement is 120°. ◄

TRY THESE

Classify each triangle according to the lengths of its sides and the measures of its angles.

1.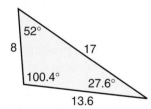

2.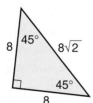

3.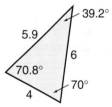

Determine the measure of the third angle for each triangle.

4. m∠C = 42, m∠D = 81

5. m∠C = 79, m∠D = 34

6. m∠C = 108, m∠D = 29

7. m∠C = 44, m∠D = 46

8. In isosceles triangle *RST*, m∠R = 34, and m∠S = m∠T. Determine the measures of angles *S* and *T*.

9. In right triangle *NQR*, one acute angle is 6° greater than three times the measure of its complement. Determine the measure of each angle.

Use the Pythagorean relationship to classify each triangle .

10. 45, 108, 117

11. 39, 80, 90

12. 15, 23, 24

13. 80, 192, 208

14. **ELECTRONICS** A guy wire attached to the top of a 200-ft radio antenna is bolted to the ground 75 ft from the base of the tower. Find the length of the guy wire to the nearest tenth.

15. **HOME SAFETY** Safety experts recommend that the bottom of a 12-ft ladder should be at least 4 ft from the base of the wall. How high can the ladder safely reach?

16. **WRITING MATHEMATICS** Assume you want to build a rectangular picture frame and hold it steady with a diagonal brace. Describe at least two ways by which you could test to see whether you have a true rectangle.

PRACTICE

Classify each triangle as either *acute*, *right*, or *obtuse* and as either *scalene*, *isosceles*, or *equilateral*.

1.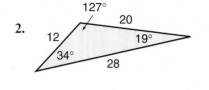

2.

3.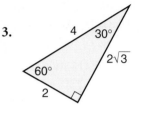

4. In isoceles triangle *ABC*, if the measure of ∠*A* is 92°, determine the measures of ∠*B* and ∠*C*.

5. In isosceles triangle *DEF*, the measure of each angle is a whole number of degrees. If one angle measures 35°, what are the measures of the other two angles?

6. The measures of angles *G* and *H* are equal. If the measure of ∠*I* is twice the measure of each of the other angles, determine each angle and describe the triangle.

7. **WRITING MATHEMATICS** The measures of the angles of a triangle can be three consecutive integers. They can also be three consecutive even integers. Can the measures of the angles of a triangle be three consecutive odd integers? Explain your reasoning.

EXTEND

A SPECIAL RIGHT TRIANGLE In right triangle *ABC* with 30° and 60° angles, the ratio of the side lengths is $1 : \sqrt{3} : 2$.

8. Name the sides of △*ABC* in order from shortest to longest.

9. **SHADOWS** Side $\overline{AB}$ represents a tree and $\overline{BC}$ the shadow it casts at a particular time of day. If the length of the shadow is 50.2 ft, how tall is the tree to the nearest tenth of a foot?

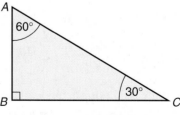

10. In △*PRS*, the measure of ∠*P* is 10° less than two-thirds the measure of ∠*R*. The measure of ∠*S* is 3° greater than twice the measure of ∠*R*. Find the measures of the angles.

11. **PACKING** Can a 66.5-in. fishing rod be packed in the carton shown at the right without bending the rod? Explain your reasoning.

THINK CRITICALLY

12. **PYTHAGOREAN TRIPLES** Pythagorean triples are integers a, b, and c such that $a^2 + b^2 = c^2$. They can be generated using a formula known to the ancient Greeks. Where x is any odd whole number, the following numbers will always be a Pythagorean triple.

$$x, \quad \frac{x^2 - 1}{2}, \quad \frac{x^2 + 1}{2}$$

Verify the rule by applying the Pythagorean theorem to the three expressions. Assume that $\frac{(x^2 + 1)}{2}$ is the greatest number.

CREATING TRIANGLES In Exercises 13–15, make up side lengths for triangles that meet the given descriptions. (If necessary, round to the nearest hundredth.) In each case, explain how you know that you are correct.

13. obtuse and scalene
14. acute and scalene
15. acute and isosceles

MIXED REVIEW

Evaluate each expression.

16. $|12|$
17. $-|17|$
18. $|-41|$

19. **STANDARDIZED TESTS** 30 is

    A. 25% of 150   **B.** 20% of 150   **C.** 150% of 20   **D.** 150% of 25

Solve each system using the elimination method.

20. $\begin{cases} 3x - y \\ 2x + 3y = 16 \end{cases}$
21. $\begin{cases} x - y = 1 \\ 2x - y = 8 \end{cases}$
22. $\begin{cases} 4x + 3y = 3 \\ x - y = \frac{1}{6} \end{cases}$

Factor.

23. $3x^3 - 48x$
24. $4x^2y - 4xy - 48y$
25. $x^4 - 81$

PROJECT *Connection*   In this activity, you will make some generalizations about the used cars you are studying.

1. Assemble all the classified advertisements. Determine the mode option listed in them.

2. Sort the advertisements by make and model year. For each specific type of car, determine the mean price and mode price of the used cars offered for sale. If these prices differ, give reasons why.

3. If the owner will cooperate, find out the original purchase price of the car. (If not, you may be able to find out the original suggested retail price of the car as equipped.) Use the average selling price of similar used cars to determine the percent decrease from the original price. Compare and discuss the percents you obtain for different models.

Career
Road Developer

To be certain that road intersections are perpendicular, road developers use a procedure based on the Pythagorean theorem to create small scale layout documents to model a road system. You can model the procedure using string, tape, and a centimeter ruler.

1. Begin by taping both ends of a piece of string to your work surface. This string represents the first road. Make two marks on the string to represent two points A and B on the first road. You can do this with small additional pieces of tape.

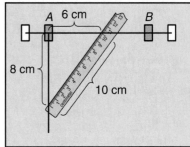

2. Tape one end of a second string at Point A on the first road and mark off 8 cm in the general direction of a right angle. Measure exactly 6 cm from Point A along the first road and mark the point. From that point, lay the end of a centimeter ruler toward the far end of the second road. Stretch the second string to form a triangle with the first road and the ruler. Tape the second string at the point where the 10 cm mark on the ruler meets the 8 cm mark of the second road. The second road must pass through this point.

3. Continue by repeating the process for each of the four corners where the roads are to intersect and form a right angle, by using a Pythagorean triple.

Decision Making

4. You can check the accuracy of your construction by measuring the diagonals of the rectangle you created. What should you look for? How will these measurements show whether you have constructed roads that meet at right angles?

13.3 Multiply and Divide Radicals

Explore

- In the figure, dots are 1 unit apart horizontally and vertically.

  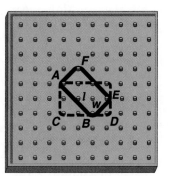

  1. $\triangle ABC$ is a right triangle with a hypotenuse of length l. Use the Pythagorean theorem to determine l.

  2. Find w, the hypotenuse of $\triangle BDE$.

  3. Find the area of rectangle $AFEB$ by counting the number of unit squares covered by the rectangle.

  4. How does this exploration show that $\sqrt{8} \cdot \sqrt{2} = \sqrt{16}$?

Build Understanding

- You have learned that the square root of a perfect square can be written in a simpler form, such as $\sqrt{36} = 6$. A **radical expression** (an expression containing a radical) is in **simplest form** if

  1. the **radicand** (the expression under the radical symbol) contains no perfect-square factors other than 1

  2. the radicand contains no fractions

  3. no denominator contains a radical

Recall from Lesson 9.5 that you can use the following properties to simplify radical expressions.

> **PROPERTIES OF SQUARE ROOTS**
>
> For all real numbers a and b, where $a \geq 0$ and $b \geq 0$,
>
> $$\sqrt{ab} = \sqrt{a} \cdot \sqrt{b} \qquad \textsc{Product Property}$$
>
> For all real numbers a and b, where $a \geq 0$ and $b > 0$,
>
> $$\sqrt{\frac{a}{b}} = \frac{\sqrt{a}}{\sqrt{b}} \qquad \textsc{Quotient Property}$$

Recall that $\sqrt{}$ stands for the *principal*, or positive, square root of a positive number. Thus, $\sqrt{x^2} = x$ only if $x \geq 0$. In other words, $\sqrt{x^2} = |x|$. You can use this fact to simplify radical expressions.

To solve Example 1a mentally, look for a factor of the radicand that is a perfect square.

$$54 = \boxed{9} \cdot 6, \text{ so}$$
$$\sqrt{54} = \sqrt{9} \cdot \sqrt{6}$$
$$= 3\sqrt{6}$$

CHECK UNDERSTANDING

In Example 1a, why is $\sqrt{3 \cdot 3 \cdot 3 \cdot 2}$ rewritten as $\sqrt{3^2} \cdot \sqrt{3 \cdot 2}$ rather than $\sqrt{3^3} \cdot \sqrt{2}$?

EXAMPLE 1

Simplify.

a. $\sqrt{54}$ **b.** $\sqrt{98x^2y}$

Solution

a. $\sqrt{54} = \sqrt{3 \cdot 3 \cdot 3 \cdot 2}$ Write the prime factorization of 54.
$\phantom{\sqrt{54}} = \sqrt{3^2} \cdot \sqrt{3 \cdot 2}$ Use the product property of square roots.
$\phantom{\sqrt{54}} = 3\sqrt{6}$

b. $\sqrt{98x^2y} = \sqrt{7 \cdot 7 \cdot 2 \cdot x^2 \cdot y}$ Write the prime factorization of 98.
$\phantom{\sqrt{98x^2y}} = \sqrt{7^2} \cdot \sqrt{x^2} \cdot \sqrt{2y}$ Product property of square roots.
$\phantom{\sqrt{98x^2y}} = 7|x|\sqrt{2y}$ ◄

In Example 1, the product property of square roots was used to simplify radical expressions. The product property can also be used to multiply radical expressions.

EXAMPLE 2

Multiply and simplify.

a. $-\sqrt{6} \cdot \sqrt{15}$ **b.** $\sqrt{5y}\left(8 - \sqrt{35y}\right)$

Solution

a. $-\sqrt{6} \cdot \sqrt{15}$
$= -\sqrt{6 \cdot 15}$ Product property of square roots.
$= -\sqrt{2 \cdot 3 \cdot 3 \cdot 5}$ Write the prime factorizations.
$= -\sqrt{3^2} \cdot \sqrt{2 \cdot 5}$ Commutative property and product property of square roots.
$= -3\sqrt{10}$

b. $\sqrt{5y}\left(8 - \sqrt{35y}\right)$
$= 8\sqrt{5y} - \sqrt{5y} \cdot \sqrt{35y}$ Distributive property.
$= 8\sqrt{5y} - \sqrt{5y \cdot 35y}$
$= 8\sqrt{5y} - \sqrt{5 \cdot 5 \cdot 7 \cdot y^2}$ Factor.
$= 8\sqrt{5y} - \sqrt{5^2 \cdot y^2 \cdot 7}$
$= 8\sqrt{5y} - 5y\sqrt{7}$ ◄

Notice that in Example 2b it is not necessary to write $|y|$ for $\sqrt{y^2}$. Since both $5y$ and $35y$ appear as radicands in the original problem, y must be greater than or equal to zero. Therefore, $\sqrt{y^2} = y$.

The quotient property of square roots can also be used to simplify radical expressions. Remember that in a radical expression in simplest form, no denominator should contain a radical.

EXAMPLE 3

Simplify.

a. $\dfrac{\sqrt{90}}{\sqrt{5}}$

b. $\sqrt{\dfrac{1}{2}}$

Solution

a. $\dfrac{\sqrt{90}}{\sqrt{5}} = \sqrt{\dfrac{90}{5}}$ Quotient property of square roots.

$\quad\quad = \sqrt{\dfrac{2 \cdot 3 \cdot 3 \cdot 5}{5}}$ Factor.

$\quad\quad = \sqrt{2 \cdot 3 \cdot 3}$ Divide by the common factor, 5.

$\quad\quad = 3\sqrt{2}$

b. $\sqrt{\dfrac{1}{2}} = \dfrac{\sqrt{1}}{\sqrt{2}}$ Quotient property of square roots.

$\quad\quad = \dfrac{1}{\sqrt{2}} \cdot \dfrac{\sqrt{2}}{\sqrt{2}}$ Multiply by $\dfrac{\sqrt{2}}{\sqrt{2}}$.

$\quad\quad = \dfrac{\sqrt{2}}{\sqrt{2 \cdot 2}}$ Product property.

$\quad\quad = \dfrac{\sqrt{2}}{2}$

CHECK UNDERSTANDING

What property assures that you can multiply the fraction $\dfrac{1}{\sqrt{2}}$ by $\dfrac{\sqrt{2}}{\sqrt{2}}$ without changing the value of the fraction?

Notice that in Example 3b, the radical $\sqrt{2}$ was eliminated from $\dfrac{1}{\sqrt{2}}$ by multiplying the expression by $\dfrac{\sqrt{2}}{\sqrt{2}}$, which is equal to 1, and applying the product property. The process of eliminating the radical from the denominator is called **rationalizing the denominator**.

EXAMPLE 4

PHYSICS The formula $T = 2\pi\sqrt{\dfrac{L}{32}}$ gives the length of time T in seconds that it takes a pendulum of length L in feet to complete one swing. Approximate T for a pendulum 3 in. long.

Solution

$L = 3 \text{ in.} = 0.25 \text{ ft}$

$\quad T = 2\pi\sqrt{\dfrac{L}{32}} = 2\pi\sqrt{\dfrac{0.25}{32}}$ Substitute 0.25 for L.

$\quad\quad \approx 0.5553603673$ Use a calculator.

$\quad\quad \approx 0.56$ Round.

The time is about 0.56 seconds.

ALGEBRA: WHO, WHERE, WHEN

The Italian physicist Galileo discovered that the time for the swing of a pendulum depended on the pendulum length, not on the size of the arc. Simple pendulum clocks work on the principle that when a longer pendulum is used, the clock runs more slowly than it does with a shorter one.

Simplify.

1. $\sqrt{50}$ **2.** $\sqrt{27k^2}$ **3.** $2\sqrt{6} \cdot \sqrt{8}$ **4.** $\sqrt{2}(\sqrt{3} - \sqrt{8})$

5. $\dfrac{\sqrt{60}}{\sqrt{5}}$ **6.** $\dfrac{3}{\sqrt{3}}$ **7.** $\sqrt{\dfrac{5}{6}}$ **8.** $\dfrac{4\sqrt{3}}{\sqrt{8}}$

9. GEOGRAPHY The formula $d = \sqrt{12h}$ can be used to approximate the distance d to Earth's horizon in kilometers from a point h meters above Earth's surface. Find and simplify a radical expression for the distance to the horizon from an elevation of 375 m.

10. SPORTS A baseball "diamond" is really a square measuring 90 ft on a side. Determine a radical expression for the distance from home plate directly to second base.

11. WRITING MATHEMATICS Determine the prime factors of 30 and explain why $\sqrt{30}$ is in simplest form.

PRACTICE

Simplify.

1. $\sqrt{18}$ **2.** $\sqrt{40}$ **3.** $\sqrt{300}$ **4.** $\sqrt{147}$

5. $\sqrt{9n^2}$ **6.** $\sqrt{50p^2}$ **7.** $3\sqrt{32c^3}$ **8.** $-5\sqrt{24x^3y}$

9. $\sqrt{3} \cdot \sqrt{3}$ **10.** $\sqrt{7} \cdot \sqrt{7}$ **11.** $5\sqrt{20} \cdot \sqrt{5}$ **12.** $-3\sqrt{18} \cdot \sqrt{2}$

13. $6(2 - \sqrt{3})$ **14.** $\sqrt{5}(5 + \sqrt{3})$ **15.** $\sqrt{12}(2\sqrt{3} - \sqrt{5})$ **16.** $\dfrac{\sqrt{30}}{\sqrt{6}}$

17. $\dfrac{\sqrt{72}}{\sqrt{8}}$ **18.** $\dfrac{\sqrt{28a^2}}{\sqrt{8}}$ **19.** $\dfrac{2}{\sqrt{5}}$ **20.** $\dfrac{6}{\sqrt{2}}$

Determine the area of each figure.

21.

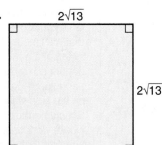

$2\sqrt{13}$
$2\sqrt{13}$

22.

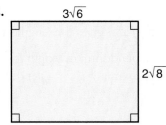

$3\sqrt{6}$
$2\sqrt{8}$

23.

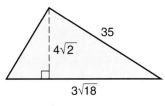

35
$4\sqrt{2}$
$3\sqrt{18}$

24. CHEMISTRY If two gases have masses of m_1 and m_2 respectively, the formula $\dfrac{r_1}{r_2} = \dfrac{\sqrt{m_1}}{\sqrt{m_2}}$ gives the ratio of the rates, r_1 and r_2, at which the gases diffuse (spread out to fill a space). Find and simplify the radical expression $\dfrac{r_1}{r_2}$ for two gases with $m_1 = 3$ mg and $m_2 = 2$ mg.

25. WRITING MATHEMATICS For each of the rules for simplification of radicals, give a radical and show how to simplify it.

EXTEND

26. **GEOMETRY** An equilateral triangle has sides measuring 12 cm. Find and simplify a radical expression for the height of the triangle.

27. **GEOMETRY** Find and simplify a radical expression for the height of an equilateral triangle with sides measuring s units.

If two radical expressions consist of the *sum* and *difference* of the same two terms, they are called **conjugates** of one another. The following expressions are conjugates.

$$\sqrt{2} + \sqrt{3} \text{ and } \sqrt{2} - \sqrt{3} \qquad 5 + 7\sqrt{13} \text{ and } 5 - 7\sqrt{13}$$

Give the conjugate of each radical expression. Then find the product of each pair of conjugates.

28. $4 + \sqrt{2}$ 29. $\sqrt{5} - \sqrt{3}$ 30. $2\sqrt{6} - 3\sqrt{2}$ 31. $\sqrt{10} + 4$

To rationalize a denominator that is a binomial, multiply the numerator and denominator by the conjugate of the binomial.

Simplify.

32. $\dfrac{\sqrt{3} + \sqrt{2}}{\sqrt{3} - \sqrt{2}}$ 33. $\dfrac{1 + \sqrt{2}}{3 - \sqrt{3}}$ 34. $\dfrac{1 + \sqrt{3}}{2 + \sqrt{5}}$

THINK CRITICALLY

35. Find two numbers such that the square root of their sum is 10 and the square root of their product is 48.

36. Find the number such that the square root of the product of 7 and a number divided by the square root of the product of 28 and the number squared is $\dfrac{1}{2}$.

37. Suppose that m and n are rational numbers. Is the product $(m + \sqrt{n})(m - \sqrt{n})$ also a rational number? Explain.

MIXED REVIEW

Solve each equation for x.

38. $3x - 2 = 2x + 7$ 39. $6 - 3x = -5x + 2$ 40. $5(2x - 5) = -(x - 8)$

Find the slope and y-intercept of the graph of each equation.

41. $y = -4x + 3$ 42. $6x + 2y = 11$ 43. $3(y - 4) = 3x$

44. **STANDARDIZED TESTS** In simplest form $\sqrt{20n^2}$ is

   A. $\pm 2n\sqrt{5}$ **B.** $|2n|\sqrt{5}$

   C. $2n\sqrt{5}$ **D.** $2|n|\sqrt{5}$

Career
Highway Engineer

The United States, with 3.9 million miles of graded roads, has the most extensive highway system in the world. Within the U.S. highway system, Texas has the greatest number of miles of roads of any state, totalling about 306,000 mi. Highway engineers have the job of designing and building this massive network of highways and seeing to it that the roads are maintained in safe condition for the millions of commuters, truckers, tourists, and leisure drivers who use them.

Decision Making

1. What percent of U.S. highways are in Texas?

2. The earth's circumference is about 24,900 mi. If all the roads in the United States were laid end to end, how many times would they reach around the earth?

In designing a curve for a freeway on-ramp, highway engineers must balance efficiency against safety. A short, sharp curve puts drivers on the freeway quickly but can be dangerous. If the curve is too sharp, a careless driver might drive off the edge of the road. A wide curve is safer but takes longer to drive. The formula for the relationship between the radius r in ft of an unbanked curve and the maximum velocity v in miles per hour at which a car can go around the curve without skidding is $v = \sqrt{2.5r}$.

3. As curve radius increases, does maximum safe velocity increase or decrease? Explain.

4. Determine the maximum safe velocities achievable on curves of radii that are multiples of 10 ft, up to 100 ft.

5. Graph the equation $v = \sqrt{2.5r}$ for radii ranging from 0 ft to 2000 ft.

6. An unbanked curve with a radius of 4.1 mi would be needed to permit a no-skid turn by a car traveling at the fastest speed ever achieved at the Indianapolis 500 auto race. How fast was the race car moving?

13.4 Add and Subtract Radicals

Explore

In the figure below, dots are 1 unit apart horizontally and vertically.

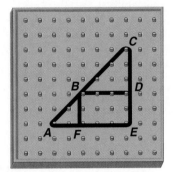

1. For $\triangle AFB$, use the Pythagorean theorem to find and simplify a radical expression for AB.

2. For $\triangle BDC$, use the Pythagorean theorem to find and simplify a radical expression for BC.

3. For $\triangle AEC$, use the Pythagorean theorem to find and simplify a radical expression for AC.

4. How does this exploration show that $2\sqrt{2} + 3\sqrt{2} = 5\sqrt{2}$?

SPOTLIGHT ON LEARNING

WHAT? In this lesson you will learn
• to add and subtract radical expressions.

WHY? Knowing how to add and subtract radical expressions can help you solve problems in physics, traffic safety, geography, and navigation.

Build Understanding

Radical expressions with the same radicand are **like radicals**.

Like Radicals	Unlike Radicals
$2\sqrt{5}$ and $-3\sqrt{5}$	$\sqrt{6}$ and $4\sqrt{7}$
$3\sqrt{ab}$ and $12\sqrt{ab}$	$-5\sqrt{x}$ and $3\sqrt{y}$

You can combine radical expressions that are like radicals just as you can combine like terms in algebraic expressions.

Like Terms	Like Radicals
$8x - 3x = 5x$	$8\sqrt{2} - 3\sqrt{2} = 5\sqrt{2}$

Use the distributive property to add or subtract like radicals.

EXAMPLE 1

Simplify: $5\sqrt{3} + 7\sqrt{3} + 4\sqrt{6} - 2\sqrt{3}$

Solution

The terms containing $\sqrt{3}$ are like radicals.

$5\sqrt{3} + 7\sqrt{3} + 4\sqrt{6} - 2\sqrt{3}$

$= 5\sqrt{3} + 7\sqrt{3} - 2\sqrt{3} + 4\sqrt{6}$ Commutative property.

$= (5 + 7 - 2)\sqrt{3} + 4\sqrt{6}$ Distributive property.

$= 10\sqrt{3} + 4\sqrt{6}$ ◀

Sometimes you will need to simplify radicals first to see whether any can be combined.

EXAMPLE 2

Add: $2\sqrt{50} + 3\sqrt{18}$

Solution

$$\begin{aligned}
&\quad 2\sqrt{50} + 3\sqrt{18} \\
&= 2\sqrt{5^2 \cdot 2} + 3\sqrt{3^2 \cdot 2} &&\text{Factor.} \\
&= 2 \cdot 5\sqrt{2} + 3 \cdot 3\sqrt{2} &&\text{Simplify.} \\
&= 10\sqrt{2} + 9\sqrt{2} &&\text{Multiply.} \\
&= 19\sqrt{2} &&\text{Add.}
\end{aligned}$$

◀

Radical expressions with variables in the radicand are added and subtracted in the same way as radicals having only real numbers in the radicand.

CHECK UNDERSTANDING

In Example 3, why is it not necessary to write the solution as $4|x|\sqrt{5x}$?

EXAMPLE 3

Subtract: $2\sqrt{45x^3} - \sqrt{20x^3}$

Solution

$$\begin{aligned}
&\quad 2\sqrt{45x^3} - \sqrt{20x^3} \\
&= 2\sqrt{3^2 \cdot 5 \cdot x^2 \cdot x} - \sqrt{2^2 \cdot 5 \cdot x^2 \cdot x} \\
&= 2 \cdot 3x\sqrt{5x} - 2x\sqrt{5x} \\
&= 6x\sqrt{5x} - 2x\sqrt{5x} \\
&= 4x\sqrt{5x}
\end{aligned}$$

◀

Real world problems involving square roots may require addition or subtraction of radicals.

EXAMPLE 4

PHYSICS The time t in seconds that it takes an object to fall d feet is given by $t = \dfrac{\sqrt{d}}{4}$. Two packages of equipment for Antarctic researchers are dropped simultaneously from two planes, one from an altitude of 4000 ft, the other from an altitude of 2560 ft. How long will one package be on the ground before the other one lands? Assume that air resistance is not a factor.

Solution

$$t_1 = \frac{\sqrt{4000}}{4} \qquad t_2 = \frac{\sqrt{2560}}{4}$$

difference in times: $t_1 - t_2$

$$t_1 - t_2$$

$$= \frac{\sqrt{4000}}{4} - \frac{\sqrt{2560}}{4}$$

$$= \frac{\sqrt{2^4 \cdot 5^2 \cdot 2 \cdot 5}}{4} - \frac{\sqrt{2^8 \cdot 2 \cdot 5}}{4}$$

$$= \frac{20\sqrt{10}}{4} - \frac{16\sqrt{10}}{4}$$

$$= 5\sqrt{10} - 4\sqrt{10}$$

$$= \sqrt{10}$$

The first package will land $\sqrt{10}$ seconds, or about 3.2 seconds, before the second package lands. ◀

TRY THESE

Simplify.

1. $5\sqrt{7} + 2\sqrt{7}$

2. $12\sqrt{5} - \sqrt{5}$

3. $10\sqrt{2} + 9\sqrt{2}$

4. $\sqrt{24} - \sqrt{6}$

5. $2\sqrt{10} - 3\sqrt{40} + 4\sqrt{5}$

6. $3\sqrt{b} - 5\sqrt{b}$

7. $7\sqrt{45} + 3\sqrt{20}$

8. $\sqrt{8x^2} + \sqrt{2x^2}$

9. $\sqrt{\frac{1}{2}} - \sqrt{\frac{1}{8}}$

10. GEOMETRY A rectangle is $\sqrt{54}$ in. long and $\sqrt{24}$ in. wide. Find the perimeter of the rectangle.

11. WRITING MATHEMATICS Write a paragraph that compares simplifying $7x^2 - 6x + 4x^2$ to simplifying $7\sqrt{6} - 6\sqrt{2} + 4\sqrt{6}$.

PRACTICE

Simplify.

1. $4\sqrt{3} + 5\sqrt{3}$

2. $8\sqrt{7} + 7\sqrt{7}$

3. $13\sqrt{5} - 5\sqrt{5}$

4. $\sqrt{10} + 3\sqrt{10} - \sqrt{5}$

5. $14\sqrt{3} + 6\sqrt{2} - 11\sqrt{3}$

6. $\sqrt{20} + \sqrt{80} - \sqrt{45}$

7. $11\sqrt{h} + 5\sqrt{h}$

8. $\sqrt{128y} - \sqrt{2y}$

9. $5\sqrt{18x} + 2\sqrt{8x}$

10. $2\sqrt{3} + 3\sqrt{12}$

11. $5\sqrt{50} - 4\sqrt{32}$

12. $\sqrt{200c^2} - \sqrt{98c^2}$

13. $\sqrt{8n + 8} + \sqrt{2n + 2}$

14. $\sqrt{2} + \sqrt{\frac{1}{2}}$

15. $\sqrt{\frac{2}{3}} - \sqrt{\frac{1}{6}}$

Determine the area and perimeter of each figure.

16.

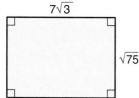

$7\sqrt{3}$

$\sqrt{75}$

17.

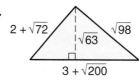

$2 + \sqrt{72}$ $\sqrt{63}$ $\sqrt{98}$

$3 + \sqrt{200}$

18.

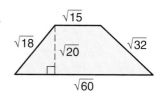

$\sqrt{15}$

$\sqrt{18}$ $\sqrt{20}$ $\sqrt{32}$

$\sqrt{60}$

19. Traffic Safety Police investigators can use the formula $s = 2\sqrt{5L}$ to approximate the rate of speed s in miles per hour of a car that leaves skids marks of length L in feet after the driver slams on the brakes to stop. Skid marks at a two-car accident measured 230.4 ft and 160 ft. Find the difference in speed of the two cars involved in the accident.

20. Geography The formula $d = \sqrt{12h}$ can be used to approximate the distance d in kilometers to Earth's horizon from a point h meters above Earth's surface. How much farther could you see from the top of New York City's Chrysler Bulding (height, 320 m) than you could see from the top of Cityspire (height, 245 m)? Assume that the views to the horizon are unobstructed.

21. Physics The formula $T = 2\pi\sqrt{\dfrac{L}{32}}$ gives the length of time T in seconds that it takes a pendulum of length L in feet to complete one swing. Two pendulums, 4 ft and 1 ft in length respectively, begin a swing simultaneously.

a. Which pendulum completes one swing first?

b. How much less time does the faster pendulum take to complete one swing than the slower one takes?

22. Writing Mathematics Explain how you can use a calculator to confirm that two radical expressions are equal. Provide an example to illustrate your explanation.

EXTEND

23. Navigation To take advantage of the wind, a sailor boating from Portsmouth to Lakeview sailed south 3 mi and east 3 mi, then south 4 mi and east 4 mi, and finally south 7 mi and east 7 mi.

a. Find the shortest distance from Portsmouth to Lakeview.

b. How much farther was it from Portsmouth to Lakeview by the route the sailor took?

Decide whether the given value of x is a solution of the equation.

24. $x^2 + 4x + 2 = 0; x = -2 + \sqrt{2}$

25. $x^2 - 4x - 7 = 0; x = 2 - \sqrt{11}$

26. $x^2 - 3x - 5 = 0; x = 1 + \sqrt{6}$

27. $x^2 - 10x + 5 = 0; x = 5 + 2\sqrt{5}$

28. **a.** Simplify: $\sqrt{x} + \sqrt{\dfrac{1}{x}}$

b. Use your results to find $\sqrt{5} + \sqrt{\dfrac{1}{5}}$.

THINK CRITICALLY

29. **a.** Use a calculator to evaluate the radical expressions $\sqrt{5} - 1$ and $\sqrt{6 - 2\sqrt{5}}$. What do you notice about your results?

b. Show that $\sqrt{5} - 1 = \sqrt{6 - 2\sqrt{5}}$ by squaring $\sqrt{5} - 1$.

c. Simplify: $\sqrt{10 + 4\sqrt{6}}$

d. Simplify: $\sqrt{7 + 4\sqrt{3}}$

MIXED REVIEW

Evaluate each expression for the given value(s) of the variables.

30. $3x - 4y$, for $x = -2$ and $y = 5$

31. $5a^2 + 7a$, for $a = -3$

32. $6(x - 2y) + 5(x + y)$, for $x = 2$ and $y = -2$

33. $m^3 - m^2 + 5m - 2$, for $m = -1$

Evaluate the function $f(x) = 4x - 5$ for the given values of x.

34. $f(3)$ **35.** $f(-5.5)$ **36.** $f\left(\frac{1}{2}\right)$ **37.** $f(x + 1)$

Solve each inequality.

38. $|x| < 3$ **39.** $|x| \geq 6$

40. STANDARDIZED TESTS The product of $(3 - \sqrt{5})(5 + 2\sqrt{5})$ is

A. $5 + 11\sqrt{5}$ **B.** $15 - \sqrt{5}$

C. $5 + \sqrt{5}$ **D.** $5 - 11\sqrt{5}$

Simplify.

41. $9\sqrt{7} - 3\sqrt{7}$ **42.** $\sqrt{18} - \sqrt{8}$ **43.** $\sqrt{27} + 5\sqrt{3}$ **44.** $5\sqrt{x} + 2\sqrt{9x}$

PROJECT *Connection*   In this activity, you will investigate what items can fit in your car's trunk compartment.

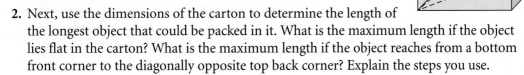

1. Obtain several large, empty cartons (such as those used for appliances). Experiment to determine the largest carton that will fit in the trunk with the hood securely closed. Then measure the dimensions of the carton (to the nearest tenth of a foot) and calculate the volume.

2. Next, use the dimensions of the carton to determine the length of the longest object that could be packed in it. What is the maximum length if the object lies flat in the carton? What is the maximum length if the object reaches from a bottom front corner to the diagonally opposite top back corner? Explain the steps you use.

13.5 Solve Radical Equations

Explore/Working Together

● You and a partner will solve the equation $x = \sqrt{5x + 6}$ using two different methods.

1. **a.** Use a graphing utility. How can you solve

$$x = \sqrt{5x + 6}$$

by graphing a system of equations?

   b. Try your method and state the solution you determine. Remember to check each solution in the original equation.

2. **a.** Now solve $x = \sqrt{5x + 6}$ algebraically. How can you eliminate the radical on the right side?

   b. What type of equation do you obtain? How can you solve this equation?

   c. Check each solution in the original equation. What do you notice?

Solve and check each equation using both methods. Describe any patterns you notice.

3. $x = \sqrt{x + 2}$	**4.** $\sqrt{10 - 3x} = x$
5. $3x = \sqrt{3 - 6x}$	**6.** $2x = \sqrt{4x + 15}$

SPOTLIGHT ON LEARNING

WHAT? In this lesson you will learn
• to solve equations containing radical expressions.

WHY? Knowing how to solve radical equations can help you solve problems in plumbing, auto mechanics, and geometry.

Build Understanding

● An equation having a radical with a variable in the radicand, such as those you worked with in Explore, is a **radical equation**. The following are other examples of radical equations.

$$\sqrt{x} = 6 \qquad\qquad x + 2 = \sqrt{7x + 4}$$

As you discovered in Explore, you can use the following principle to solve radical equations.

> **PRINCIPLE OF SQUARING**
>
> **If the equation $a = b$ is true, then the equation $a^2 = b^2$ is also true.**

To solve a radical equation in the form $\sqrt{a} = b$, first square both sides. Then solve the resulting equation .

EXAMPLE 1

Solve and check: $\sqrt{2x - 1} = x$

Solution

$$\sqrt{2x - 1} = x$$
$$\left(\sqrt{2x - 1}\right)^2 = x^2 \quad \text{Square both sides.}$$
$$2x - 1 = x^2$$
$$x^2 - 2x + 1 = 0$$
$$(x - 1)(x - 1) = 0$$
$$x = 1 \quad \text{Solve for } x.$$

Check

$$\sqrt{2x - 1} = x$$
$$\sqrt{2(1) - 1} \overset{?}{=} 1$$
$$\sqrt{1} \overset{?}{=} 1$$
$$1 = 1 \checkmark$$

The solution is $x = 1$.

◀

CHECK UNDERSTANDING

In the solution to Example 1, why does $\left(\sqrt{2x - 1}\right)^2 = 2x - 1$?

To solve a radical equation of the form $\sqrt{a} + d = b$, first isolate the radical on one side of the equation. Then square both sides and solve the resulting equation. Because it is possible to introduce extraneous solutions when you square both sides of a radical equation, be sure to check all apparent solutions to see if they are actual solutions.

EXAMPLE 2

Solve and check: $\sqrt{2x - 1} - x = -2$

Solution

$$\sqrt{2x - 1} - x = -2$$
$$\sqrt{2x - 1} = x - 2 \qquad \text{Isolate the radical.}$$
$$\left(\sqrt{2x - 1}\right)^2 = (x - 2)^2 \qquad \text{Square both sides of the equation.}$$
$$2x - 1 = x^2 - 4x + 4 \qquad \text{Square the binomial.}$$
$$0 = x^2 - 6x + 5 \qquad \text{Write the quadratic in standard form.}$$
$$0 = (x - 1)(x - 5) \qquad \text{Factor.}$$
$$x - 1 = 0 \quad \text{or} \quad x - 5 = 0 \qquad \text{Use the zero product property.}$$
$$x = 1 \quad \text{or} \qquad x = 5$$

Check

$x = 1$	$x = 5$
$\sqrt{2x - 1} - x = -2$	$\sqrt{2x - 1} - x = -2$
$\sqrt{2(1) - 1} - 1 \overset{?}{=} -2$	$\sqrt{2(5) - 1} - 5 \overset{?}{=} -2$
$\sqrt{1} - 1 \overset{?}{=} -2$	$\sqrt{9} - 5 \overset{?}{=} -2$
$0 \neq -2$	$-2 = -2 \checkmark$

The solution is $x = 5$; $x = 1$ is an extraneous solution.

◀

COMMUNICATING ABOUT ALGEBRA

Explain to another student why it is easier to solve the equation in Example 2 by first isolating the radical than it would be to square both sides of the given equation.

THINK BACK

The zero product property states that if the product of two quantities is zero, one or the other quantity, or both, must equal zero.

When solving a radical equation with a fraction, remember to multiply both sides by the denominator to clear the fractions.

EXAMPLE 3

Solve and check: $\dfrac{1}{2}x = \sqrt{2x - 3}$

Solution

$$\frac{1}{2}x = \sqrt{2x - 3}$$

$$\frac{x^2}{4} = 2x - 3 \qquad \text{Square both sides.}$$

$$x^2 = 8x - 12 \qquad \text{Multiply both sides by 4.}$$

$$x^2 - 8x + 12 = 0$$

$$(x - 6)(x - 2) = 0$$

$$x - 6 = 0 \quad \text{or} \quad x - 2 = 0$$

$$x = 6 \quad \text{or} \qquad x = 2$$

The solutions are 6 and 2 since both check in the original equation. ◀

When there are two different radicals, write the equation with one radical on each side before squaring both sides of the equation.

EXAMPLE 4

Solve and check: $2\sqrt{x - 9} - \sqrt{2x} = 0$

Solution

$$2\sqrt{x - 9} - \sqrt{2x} = 0$$

$$2\sqrt{x - 9} = \sqrt{2x} \qquad \text{Add } \sqrt{2x} \text{ to both sides.}$$

$$\left(2\sqrt{x - 9}\right)^2 = \left(\sqrt{2x}\right)^2 \qquad \text{Square both sides.}$$

$$4(x - 9) = 2x$$

$$4x - 36 = 2x \qquad \text{Use the distributive property.}$$

$$2x = 36$$

$$x = 18 \qquad \text{Solve for } x.$$

Check

$$2\sqrt{x - 9} - \sqrt{2x} = 0$$

$$2\sqrt{18 - 9} - \sqrt{2 \cdot 18} \overset{?}{=} 0 \qquad \text{Substitute 18 for } x.$$

$$2\sqrt{9} - \sqrt{36} \overset{?}{=} 0$$

$$0 = 0$$

The solution is $x = 18$. ◀

PROBLEM SOLVING TIP

Remember that when you square both sides of the equation in Example 3, you must square not only the radical $\sqrt{x - 9}$, but also square the coefficient, 2, as well.

Real world problems may involve radical equations.

EXAMPLE 5

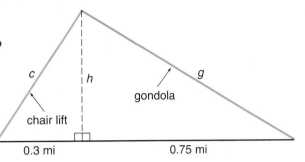

Skiing Both a chair lift and a gondola are used to transport skiers to the top of Triangle Mountain. The length of the gondola is 1.7 times the length of the chair lift cable, c. Use this information to determine h, the difference in elevation between the base of the mountain and the top.

Solution

By the Pythagorean theorem,

$$\sqrt{h^2 + (0.3)^2} = c \qquad \textbf{chair-lift triangle}$$

$$\sqrt{h^2 + (0.75)^2} = g \qquad \textbf{gondola triangle}$$

$$g = 1.7c \qquad \text{Write an equation.}$$

$$\sqrt{h^2 + (0.75)^2} = 1.7\sqrt{h^2 + (0.3)^2} \qquad \text{Substitute.}$$

$$\sqrt{h^2 + 0.5625} = 1.7\sqrt{h^2 + 0.09} \qquad \text{Simplify the radicands.}$$

$$\left(\sqrt{h^2 + 0.5625}\right)^2 = \left(1.7\sqrt{h^2 + 0.09}\right)^2 \qquad \text{Square both sides.}$$

$$h^2 + 0.5625 = 2.89(h^2 + 0.09)$$

$$h^2 + 0.5625 = 2.89h^2 + 0.2601 \qquad \text{Use the distributive property.}$$

$$0.3024 = 1.89h^2$$

$$0.16 = h^2$$

$$0.4 = h \qquad \text{Solve for } h.$$

From the base of the mountain to the top, the elevation is 0.4 mi. Check the solution. ◀

TRY THESE

Solve and check.

1. $\sqrt{x} = 3$

2. $7 = \sqrt{x}$

3. $\sqrt{x - 3} = 5$

4. $\sqrt{2x - 5} = 11$

5. $\sqrt{3x} = 6$

6. $\sqrt{-5x} = 10$

7. $7\sqrt{x + 3} - 6 = 22$

8. $-3\sqrt{x - 5} + 15 = -6$

9. $\sqrt{x + 7} = x - 5$

10. $x - 5 = \sqrt{18 - 2x}$

11. $\sqrt{5x - 9} - \sqrt{3x + 1} = 0$

12. $\sqrt{5x - 4} - 3\sqrt{x - 4} = 0$

13. **Geometry** The formula $r = \sqrt{\dfrac{A}{\pi}}$ determines the radius of a circle with area A.
Use the formula to write an equation to find the area of a circle with radius 11 cm. Then solve the equation.

14. **Writing Mathematics** Explain how solving $2\sqrt{x - 1} + 5 = 15$ is similar to solving $2w + 5 = 15$.

PRACTICE

Write the square of each expression.

1. 9   **2.** −5   **3.** 1.2   **4.** −8.5

5. $\sqrt{6}$ **6.** $\sqrt{27.13}$ **7.** $\sqrt{x-5}$ **8.** $\sqrt{3x^2+6}$

Solve and check.

9. $\sqrt{a}=2$ **10.** $-\sqrt{x}=-11$ **11.** $\sqrt{j}-9=0$

12. $\sqrt{m}-4=0$ **13.** $5\sqrt{p}=30$ **14.** $-28=-4\sqrt{x}$

15. $\sqrt{x}+8=11$ **16.** $\sqrt{n}-3=7$ **17.** $\sqrt{c+6}=8$

18. $\sqrt{s-9}=5$ **19.** $\sqrt{2d}=12$ **20.** $-\sqrt{5x}=-10$

21. $\sqrt{2k-5}=7$ **22.** $\sqrt{4x+4}=8$ **23.** $\sqrt{3t-5}=5$

24. $\sqrt{6y-11}=11$ **25.** $\sqrt{3h}=\dfrac{1}{3}$ **26.** $\sqrt{5x-1}=\dfrac{1}{2}$

27. $\sqrt{z+18}=z-2$ **28.** $\sqrt{x+5}=1-x$ **29.** $g+1=\sqrt{1-2g}$

30. $\sqrt{x+7}=\sqrt{3x-19}$ **31.** $\sqrt{5x+2}=\sqrt{-2x+23}$

32. $2\sqrt{x+5}=\sqrt{6x-2}$ **33.** $2\sqrt{6x+1}=5\sqrt{x}$

34. $3\sqrt{5x-4}-\sqrt{9x+9}=0$ **35.** $4\sqrt{x-1}-\sqrt{13x+14}=0$

36. PLUMBING A circular pipe has an inner radius of r and a cross-sectional area of A. The outer radius R of the pipe is given by $R=\sqrt{\dfrac{A}{\pi}+r^2}$.

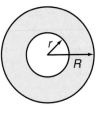

Use the formula to write an equation to find the inner radius of apipe with an outer radius of 2 cm and across-sectional area of 1.75π cm². Then solve the equation.

37. AUTO MECHANICS An engine piston is acted on by a force F producing pressure P. The diameter d of the piston is given by $d=\sqrt{\dfrac{5F}{4P}}$. Use the formula to write an equation you can use to find the force in pounds that produces a pressure of 80 lb/in.² on a piston with a diameter of 3 in. Then solve the equation.

38. WRITING MATHEMATICS Explain why it is important to check solutions to a radical equation. Provide examples to illustrate your explanation.

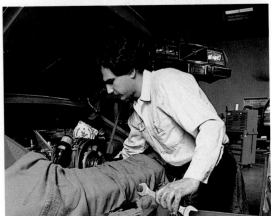

After squaring both sides of a radical equation, the equation may still contain a radical. To solve, isolate the remaining radical and repeat the process.

Solve and check.

39. $\sqrt{x + 7} = \sqrt{x - 1} + 2$

40. $5 - \sqrt{x} = \sqrt{x - 5}$

41. $\sqrt{x} + \sqrt{x + 1} = 2$

42. $\sqrt{x + 4} = 1 + \sqrt{x}$

43. The perimeter of the rectangle is 42. Find the value of x.

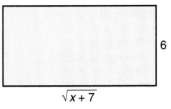

6

$\sqrt{x + 7}$

GEOMETRY A number b is the **geometric mean** between two numbers m and n if $\dfrac{m}{b} = \dfrac{b}{n}$. The geometric mean of two numbers m and n is $\sqrt{mn}$. Determine n, given m and the geometric mean of m and n.

44. $m = 5$, geometric mean $= 10$

45. $m = 3$, geometric mean $= 6$

46. $m = 16$, geometric mean $= 12$

47. $m = 20$, geometric mean $= 30$

THINK CRITICALLY

48. Graph $y_1 = \sqrt{x}$, $y_2 = \sqrt{x} + 2$, $y_3 = \sqrt{x} - 3$ on the same set of axes. How is the graph of y_2 related to the graph of y_1? How is the graph of y_3 related to the graph of y_1?

49. Graph $y_1 = \sqrt{x}$, $y_2 = \sqrt{x + 2}$, $y_3 = \sqrt{x - 3}$ on the same set of axes. How is the graph of y_2 related to the graph of y_1? How is the graph of y_3 related to the graph of y_1?

50. The graph of a certain function is translated 4 units up and 1 unit to the right from the graph of $y = \sqrt{x}$. Write the equation of the function. Check your answer by comparing the graph to the graph of $y = \sqrt{x}$.

51. When 8 times a number is decreased by 3, the square root of the result is 9. Find the number.

MIXED REVIEW

Find the mean, median, and mode of each set of data.

52. $\{9, 5, 4, 9, 8\}$

53. $\{42, 48, 29, 37, 66, 28, 37, 45\}$

Solve each inequality.

54. $4x - 5 \le 23$

55. $2x + 6 > 3x - 5$

56. $7(3 - x) < -5(x - 5)$

Solve each system by graphing.

57. $\begin{cases} y = 2x - 1 \\ y = -x + 5 \end{cases}$

58. $\begin{cases} y = -2x + 2 \\ y = x + 5 \end{cases}$

Solve each equation.

59. $\sqrt{3x - 2} = 5$

60. $\sqrt{x^2 + 2} = \sqrt{x^2 + x}$

61. $\sqrt{x + 2} - x = -4$

13.6 The Distance and Midpoint Formulas

Explore

Graph to determine the distance between each point.

1. $(2, 3)$ and $(8, 3)$ **2.** $(-7, -3)$ and $(0, -3)$

3. $(-4, 5)$ and $(5, 5)$ **4.** $(17, -1)$ and $(-5, -1)$

5. What is the distance between the points (a, c) and (b, c)?

Graph to determine the distance between each point.

6. $(4, 1)$ and $(4, 11)$ **7.** $(0, -8)$ and $(0, -12)$

8. $(-2, 9)$ and $(-2, -5)$ **9.** $(1, -7)$ and $(1, 2)$

10. What is the distance between the points (a, b) and (a, c)?

11. Graph the points $(4, -1)$, $(10, 7)$, and $(4, 7)$. Describe the kind of triangle formed. Determine the length of each side of the triangle.

Build Understanding

As you saw in Explore, you can determine the distance between two points in the coordinate plane. If the two points are on a vertical line, you can subtract the y-values. If the two points are on a horizontal line, you can subtract the x-values. If the two points are not on a vertical or horizontal line, you can think of the two points as endpoints of the hypotenuse of a right triangle. Then you can use the Pythagorean theorem to determine the distance.

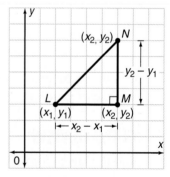

The *distance formula* combines all the steps you used in Explore to determine the distance between two points in the coordinate plane.

> **DISTANCE FORMULA**
>
> If d is the distance between the two points (x_1, y_1) and (x_2, y_2) in the coordinate plane, then
> $$d = \sqrt{(x_2 - x_1)^2 + (y_2 - y_1)^2}$$

SPOTLIGHT ON LEARNING

WHAT? In this lesson you will learn
- to use the Pythagorean theorem to find the distance between two points in the coordinate plane.

WHY? Knowing the distance and midpoint formulas can help you solve problems in geometry and navigation.

COMMUNICATING ABOUT ALGEBRA

State the distance formula in your own words.

13.6 The Distance and Midpoint Formulas

Explore

Graph to determine the distance between each point.

1. $(2, 3)$ and $(8, 3)$ **2.** $(-7, -3)$ and $(0, -3)$

3. $(-4, 5)$ and $(5, 5)$ **4.** $(17, -1)$ and $(-5, -1)$

5. What is the distance between the points (a, c) and (b, c)?

Graph to determine the distance between each point.

6. $(4, 1)$ and $(4, 11)$ **7.** $(0, -8)$ and $(0, -12)$

8. $(-2, 9)$ and $(-2, -5)$ **9.** $(1, -7)$ and $(1, 2)$

10. What is the distance between the points (a, b) and (a, c)?

11. Graph the points $(4, -1)$, $(10, 7)$, and $(4, 7)$. Describe the kind of triangle formed. Determine the length of each side of the triangle.

Build Understanding

As you saw in Explore, you can determine the distance between two points in the coordinate plane. If the two points are on a vertical line, you can subtract the y-values. If the two points are on a horizontal line, you can subtract the x-values. If the two points are not on a vertical or horizontal line, you can think of the two points as endpoints of the hypotenuse of a right triangle. Then you can use the Pythagorean theorem to determine the distance.

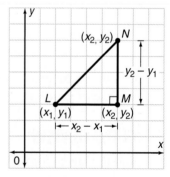

The *distance formula* combines all the steps you used in Explore to determine the distance between two points in the coordinate plane.

> **DISTANCE FORMULA**
>
> If d is the distance between the two points (x_1, y_1) and (x_2, y_2) in the coordinate plane, then
> $$d = \sqrt{(x_2 - x_1)^2 + (y_2 - y_1)^2}$$

SPOTLIGHT ON LEARNING

WHAT? In this lesson you will learn
- to use the Pythagorean theorem to find the distance between two points in the coordinate plane.

WHY? Knowing the distance and midpoint formulas can help you solve problems in geometry and navigation.

COMMUNICATING ABOUT ALGEBRA

State the distance formula in your own words.

EXAMPLE 1

Use the distance formula to determine the distance between the points $(2, -4)$ and $(14, 1)$.

Solution

Let $(x_1, y_1) = (2, -4)$ and let $(x_2, y_2) = (14, 1)$. Then $x_1 = 2$, $x_2 = 14$, $y_1 = -4$, and $y_2 = 1$.

$$\begin{aligned} d &= \sqrt{(x_2 - x_1)^2 + (y_2 - y_1)^2} & \text{Distance formula.} \\ &= \sqrt{(14 - 2)^2 + (1 - (-4))^2} & \text{Substitute.} \\ &= \sqrt{12^2 + 5^2} & \text{Subtract.} \\ &= \sqrt{144 + 25} & \text{Square the terms.} \\ &= \sqrt{169} & \text{Add.} \\ &= 13 & \text{Simplify.} \end{aligned}$$

With a calculator, enter the expression after you have substituted. Be sure to include parentheses.

$$\sqrt{((14 - 2)^2 + (1 - (-4))^2)} = 13$$

The distance between the points is 13 units. ◄

The **midpoint** of a line segment is the point that is equidistant from the endpoints of the segment. You can find the coordinates of a midpoint by finding the mean of the coordinates of the endpoints.

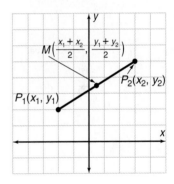

MIDPOINT FORMULA

The *midpoint M* of the line segment whose endpoints are $P_1(x_1, y_1)$ and $P_2(x_2, y_2)$ is

$$M\left(\frac{x_1 + x_2}{2}, \frac{y_1 + y_2}{2}\right)$$

EXAMPLE 2

Determine the midpoint of the segment with endpoints $(-3, 5)$ and $(11, -1)$.

Solution

Let $(x_1, y_1) = (-3, 5)$ and let $(x_2, y_2) = (11, -1)$. Then $x_1 = -3$, $x_2 = 11$, $y_1 = 5$, and $y_2 = -1$.

$$\begin{aligned} &\left(\frac{x_1 + x_2}{2}, \frac{y_1 + y_2}{2}\right) & \text{Midpoint formula.} \\ &= \left(\frac{-3 + 11}{2}, \frac{5 + (-1)}{2}\right) \\ &= (4, 2) & \textbf{The midpoint of the segment is } (4, 2). \blacktriangleleft \end{aligned}$$

To solve real world problems involving the distance formula, you may need to solve a radical equation.

EXAMPLE 3

A coordinate grid of Valley Park is drawn using Valley Creek as the positive x-axis and Park Road as the positive y-axis. (Units are in miles.) Park officials are laying out a straight-line race course from the Visitor Center $(1, 1.4)$ to a point on Valley Creek $(x, 0)$. If the course must be exactly 5 mi long, where will it intersect Valley Creek?

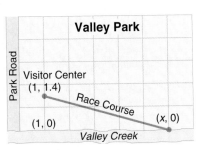

Valley Park

Park Road

Visitor Center
(1, 1.4)

Race Course

(1, 0)

(x, 0)

Valley Creek

Solution

Let $d = 5, x_2 = 1, x_1 = x, y_2 = 1.4$, and $y_1 = 0$.

$$d = \sqrt{(x_2 - x_1)^2 + (y_2 - y_1)^2}$$ Distance formula.

$$5 = \sqrt{(1 - x)^2 + (1.4 - 0)^2}$$ Substitute.

$$25 = (1 - x)^2 + (1.4 - 0)^2$$ Square both sides of the equation.

$$25 = (1 - 2x + x^2) + 1.96$$

$$0 = x^2 - 2x - 22.04$$ Write the equation in standard form.

$$x = \frac{2 \pm \sqrt{(-2)^2 - 4(1)(-22.04)}}{2(1)}$$ Use the quadratic formula.

$$x = \frac{2 \pm \sqrt{92.16}}{2}$$ Simplify.

$$x = 5.8 \quad \text{or} \quad x = -3.8$$ To the nearest tenth of a mile.

The race course intersects Valley Creek at $(5.8, 0)$ or $(-3.8, 0)$. Since $(-3.8, 0)$ lies outside the park boundaries, the intersection point is $(5.8, 0)$.

TRY THESE

Determine the distance between the points. Write radicals in simplest form.

1. $(5, -2)$ and $(-5, -2)$ **2.** $(-7, 4)$ and $(-7, -2)$ **3.** $(3, 7)$ and $(8, -5)$

4. $(5, 9)$ and $(-1, 6)$ **5.** $(-6, -10)$ and $(9, -2)$ **6.** $(\sqrt{2}, \sqrt{6})$ and $(0, 0)$

Determine the midpoint of the segment joining the two points.

7. $(2, 5)$ and $(6, 1)$ **8.** $(-7, 11)$ and $(-1, 3)$ **9.** $(3, 7)$ and $(14, 20)$

10. **GEOMETRY** Determine the lengths of the sides of the triangle formed by connecting the points (5, 1), (1, –2), and (2, –3). Then state whether the triangle is *scalene*, *isosceles*, or *equilateral*.

11. **WRITING MATHEMATICS** State whether you agree or disagree with this statement: the distance formula is an application of the Pythagorean theorem. Explain your reasoning.

PRACTICE

Determine the distance between the points. Write radicals in simplest form.

1. (12, 10) and (6, 2)

2. (9, 7) and (–6, 15)

3. (–2, –3) and (0, –7)

4. (0, 0) and (5, 6)

5. (–8, 6) and (–1, 7)

6. (5, 4) and (13, 8)

7. (–9, –15) and (–2, 9)

8. (4, 4.5) and (0, 12)

9. $\left(2\frac{1}{2}, 4\frac{1}{2}\right)$ and $\left(-6\frac{1}{2}, 44\frac{1}{2}\right)$

Determine the midpoint of the segment joining the two points.

10. (5, 6) and (3, 8)

11. (–1, –3) and (7, –11)

12. (8, 12) and (3, –2)

13. (–7, –9) and (2, –13)

14. (–3, 0) and (2, 0)

15. (m, n) and $(m, -n)$

GEOMETRY For Exercises 16 and 17, state whether the triangle connecting the points is *scalene, isosceles,* or *equilateral.*

16. (3, 4), (8, –1), (10, 3)

17. $(-4, 0), (0, 4\sqrt{3}), (4, 0)$

18. **GEOMETRY** The endpoints of a diameter of a circle are (–3, 6) and (9, 3). Find the coordinates of the center of the circle and the length of its radius.

19. **WRITING MATHEMATICS** To find the distance between (4, 2) and (6, –1), must you use 4 for x_1? Explain.

EXTEND

20. **NAVIGATION** A tanker at (630, 21) on a coordinate grid of the Atlantic Ocean sets a straight-line course for Libreville, Gabon, on the equator 29 mi away. Find the coordinates of Libreville. (Units are in miles.)

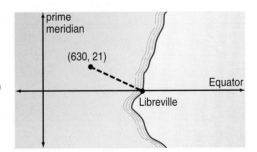

21. Use the distance formula to verify that the midpoint of the segment joining the points $A(2, 5)$ and $B(12, –1)$ is $M(7, 2)$.

22. Determine the point on the *x*-axis that is equidistant from the points (–4, 5) and (3, 2).

23. The point (3, a) is 5 units from the point (7, 9). Find all values of a.

24. The point (b, 12) is 17 units from the point (2, 4). Find all values of b.

THINK CRITICALLY

25. Use these steps to find the distance from the point $(1, 4)$ to the line $y = -\frac{1}{2}x - 3$.

    a. Find the slope of the line through $(1, 4)$ that is perpendicular to the given line.
    b. Find the equation of the line through $(1, 4)$ that is perpendicular to the given line.
    c. Find the point of intersection of the line you have found and the given line.
    d. Find the distance from the point $(1, 4)$ to the point of intersection.

26. A circle with its center at the origin has a radius of 3. Write the equation of the circle by using the distance formula to express the distance between the origin and any point (x, y) that is on the circle.

27. An ellipse is an oval or elongated circle. Each point (x, y) on a certain ellipse has the property that the sum of its distances from the points $(-4, 0)$ and $(4, 0)$ equals 12. Use the distance formula to write the equation of the ellipse.

MIXED REVIEW

Evaluate each expression.

28. $6 + 3 \cdot 2$

29. $36 \div 18 \div 2$

30. $20 - 10 \div 5$

31. $24 - 19 - 3$

32. $24 - (19 - 3)$

33. $5 + 2^2$

34. STANDARDIZED TESTS The equation of a vertical line passing through the point $(5, 7)$ is

    A. $y = 7$ **B.** $y = -7$ **C.** $x = -5$ **D.** $x = 5$

Solve by using the quadratic formula. Write answers in simplest radical form.

35. $x^2 - 10x + 22 = 0$

36. $3x^2 - 4x - 2 = 0$

37. $4x^2 + 20x + 23 = 0$

Determine the distance between the points.

38. $(-5, 0)$ and $(0, -5)$

39. $(9, 3)$ and $(-7, -9)$

40. $(6, -3)$ and $(4, 2)$

PROJECT *onnection* For this activity, you will need to know the average number of miles each car travels per gallon of gasoline (mpg). The owner may have a reliable value from past experience; otherwise, plan a test drive to collect the data you to need calculate the value.

  1. Obtain a road map of the United States. Select three major cities that seem to form a right triangle when segments are drawn to connect them. Use the map scale to determine the distances represented by the legs of the triangle. Then use the Pythagorean theorem to calculate the longest distance. Compare your result with the distance you obtain by measuring and using the map scale. Also, look up the distances between your cities in a reference book. How can you account for any differences?

  2. Use each car's mpg, determined earlier. How many gallons of gas would be necessary to drive from your hometown to the closest city on the triangle, travel on the perimeter of the triangle to visit the other two cities, and return home? Use an average per-gallon price to determine the gasoline expense for the trip.

In 1992, a record 473,305,000 passengers flew on commercial airlines in the United States, accumulating a grand total of 478 *billion* miles in the air. Chicago's O'Hare Airport was the world's busiest, with more than 64 million passengers. Every day about 2300 planes take off or land at O'Hare.

Given so much daily traffic, the job of air traffic controller is certainly challenging. At any given moment a controller may be keeping track of a dozen or more planes traveling at different altitudes, speeds, and directions. Because of this, a keen ability to visualize three-dimensional relationships is a prerequisite for controllers.

Decision Making

1. Determine the mean number of miles flown by each airline passenger in 1992.

2. To determine the distance between two planes in the air, a three-coordinate system must be employed using ordered triplets (x, y, z). Each plane's position is measured in relation not only to a horizontal x-axis and a vertical y-axis, but also to a depth-measuring z-axis perpendicular to the other two axes. Place the origin $(0, 0, 0)$ of the three axes at the O'Hare Airport control tower and measure units in miles. Where is a plane with position coordinates $(8, 2, 5)$?

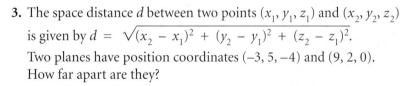

3. The space distance d between two points (x_1, y_1, z_1) and (x_2, y_2, z_2) is given by $d = \sqrt{(x_2 - x_1)^2 + (y_2 - y_1)^2 + (z_2 - z_1)^2}$. Two planes have position coordinates $(-3, 5, -4)$ and $(9, 2, 0)$. How far apart are they?

4. A passenger jet is midway between two private aircraft with position coordinates $(6, 2, -13)$ and $(1, -8, -5)$. Find the coordinates of the jet. Explain how you found your answer.

Similar Triangles

Think Back

● The word *similar* is used in everyday language to describe objects or ideas that are alike in some way. In mathematics, figures are said to be **similar** if they have the same shape, but not necessarily the same size.

1. Name two objects which meet the everyday definition of the word *similar*, but not necessarily the mathematical definition.

2. Name two objects that meet the mathematical definition of similarity.

3. Examine each pair of triangles below. Decide whether the triangles in each pair are mathematically similar.

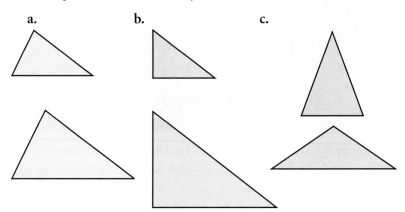

a. b. c.

Explore

● You can draw **similar triangles** using a projection point.

4. On a sheet of paper, draw any triangle *ABC* approximately in the center of the paper. Select a point *P* somewhere outside the triangle, as shown in the figure.

5. Measure the length of $\overline{PA}$ and draw $\overline{PA'}$ so that $\overline{PA'}$ passes through point *A* and is exactly twice as long as $\overline{PA}$.

6. Do the same with $\overline{PB}$ and $\overline{PC}$, making sure that $\overline{PB'}$ is twice the length of $\overline{PB}$ and $\overline{PC'}$ is twice the length of $\overline{PC}$.

7. Connect point *A'* to *B'*, *B'* to *C'*, and *C'* to *A'* to form $\triangle A'B'C'$.

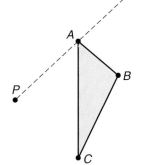

8. Use a ruler and protractor to measure the side lengths and angle measures of each triangle. Copy the table and enter your data.

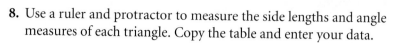

Side Lengths						Angle Measures					
$\overline{AB}$	$\overline{A'B'}$	$\overline{BC}$	$\overline{B'C'}$	$\overline{AC}$	$\overline{A'C'}$	$\angle A$	$\angle A'$	$\angle B$	$\angle B'$	$\angle C$	$\angle C'$

9. Compare your data with those of another student. What patterns do you notice?

10. Would the side and angle relationships be the same if you had begun with a different triangle? Draw a new triangle, choose a projection point, and test your prediction.

11. You can draw similar triangles by reducing, rather than enlarging, the original triangle. Draw another triangle, *ABC*, and choose a projection point *P* some distance from the triangle. Draw $\overline{PA}$ and measure its length. Locate the midpoint *A'* so that $\overline{PA'}$ is exactly half of $\overline{PA}$. Repeat to determine points *B'* and *C'*. Draw △ *A'B'C'*. How does this pair of triangles compare to previous pairs of similar triangles you have drawn?

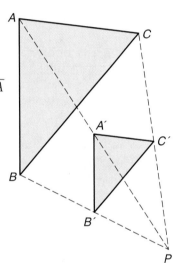

Make Connections

Triangle *JKM* and triangle *RST* are similar. Similar triangles have corresponding angles and sides. Triangles are similar if the measures of *corresponding angles* are equal and the ratios of *corresponding sides* form proportions. You can use this definition to find unknown measures in pairs of similar triangles or to determine whether a pair of triangles is similar.

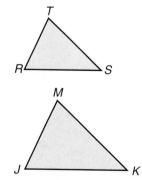

THINK BACK

A proportion is an equation that states that two ratios are equal.

Corresponding Angles	**Corresponding Sides**
$\angle J$ and $\angle R$	$\overline{JK}$ and $\overline{RS}$
$\angle K$ and $\angle S$	$\overline{KM}$ and $\overline{ST}$
$\angle M$ and $\angle T$	$\overline{MJ}$ and $\overline{TR}$

In the figure below, $\triangle ABC$ is similar to $\triangle DEF$. To determine the lengths of $\overline{BC}$ and $\overline{DE}$, use corresponding sides to write and solve proportions.

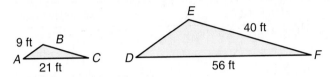

length of $\overline{DE}$: $\dfrac{AC}{DF} = \dfrac{AB}{DE}$ length of $\overline{BC}$: $\dfrac{BC}{EF} = \dfrac{AC}{DF}$

$$\dfrac{21}{56} = \dfrac{9}{DE}$$ $$\dfrac{BC}{40} = \dfrac{21}{56}$$

$$21DE = 504$$ $$56BC = 840$$
$$DE = 24$$ $$BC = 15$$

The length of $\overline{DE}$ is 24 ft.   The length of $\overline{BC}$ is 15 ft.

12. Each pair of triangles is similar. Find all unknown side lengths and angle measures.

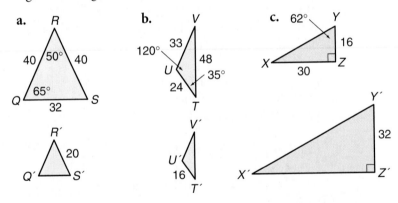

13. Use the given measures to determine whether each pair of triangles is similar.

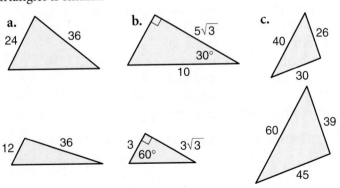

In Explore, you made enlargements and one reduction. In each case you created similar triangles. By measuring the lengths of one pair of corresponding sides of similar triangles and writing their ratio in simplest form, you can determine the **scale factor**, or multiplier, by which the original triangle is either enlarged or reduced.

14. Find the scale factor for each pair of similar triangles. For both pairs, assume that the smaller triangle is the original. All measures are given in millimeters.

a.
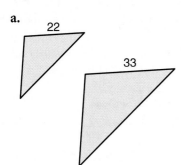

b.
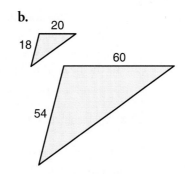

15. Draw a pair of similar triangles with a scale factor of your choosing. Exchange papers with a partner and have your partner determine the scale factor you used.

Summarize

16. WRITING MATHEMATICS Two sides of a triangle are 20 units and 34 units. André and Ana each tried to draw a similar triangle by taking $\frac{1}{2}$ of 20 and 34, respectively. Their triangles are shown below. Who drew the correct similar triangle? Explain.

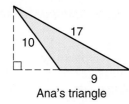
Ana's triangle

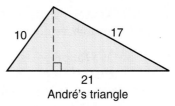
André's triangle

17. How many similar triangles can you find in the figure shown at the right? Explain your reasoning.

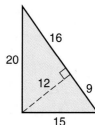

18. GOING FURTHER Triangles ABC and ADE are similar. The measures are as shown at the right. Determine the length of segments $\overline{BC}$ and $\overline{DB}$.

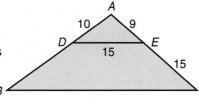

13.8 Sine, Cosine, and Tangent

Explore

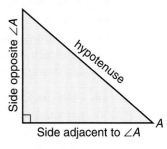

• Examine the right triangle at the right. For either of the two acute angles, you can identify the side opposite the angle, the side adjacent to the angle, and the hypotenuse. In the figure, the sides are described in relation to $\angle A$.

1. Work independently. Using a protractor and straightedge, draw right triangle XYZ so that $\angle X$ has a measure of 30°, side XY is adjacent to $\angle X$, and $\angle Y$ is a right angle. Measure each side length to the nearest millimeter. Copy the table below. Enter the results.

Lengths of Sides		
XY (adjacent to $\angle X$)	YZ (opposite $\angle X$)	XZ (hypotenuse)
_____ mm	_____ mm	_____ mm

2. Use the side lengths of your triangle to determine each of the following ratios. Copy the table below and enter the results. Round each ratio to the nearest thousandth.

Ratios of the Lengths of Sides		
$\dfrac{XY \text{ (adjacent)}}{XZ \text{ (hypotenuse)}}$	$\dfrac{YZ \text{ (opposite)}}{XZ \text{ (hypotenuse)}}$	$\dfrac{YZ \text{ (opposite)}}{XY \text{ (adjacent)}}$
_____	_____	_____

3. Compare results with those of others in your group. What do you notice about the ratios of corresponding sides?

4. What is the relationship between the triangles that members of the group have drawn? Explain.

5. Draw another right triangle, one that has a 45° angle. Measure side lengths. Copy and complete tables like those above for this triangle. What do you predict will be the relationship between your results and others' in your group?

6. In the right triangle with a 45° angle, what do you notice about the ratios of $\dfrac{\text{length of side opposite the angle}}{\text{length of hypotenuse}}$ and $\dfrac{\text{length of side adjacent to the angle}}{\text{length of hypotenuse}}$? Explain.

Build Understanding

● The word **trigonometry** comes from a Greek term meaning "measure of triangles." For any right triangle, there are three common **trigonometric ratios** of the lengths of the sides of the triangle. These ratios, called the **sine**, **cosine**, and **tangent**, are the same for all angles of equal measure in right triangles, even though the lengths of the sides of the triangles may be different.

For any right triangle, such as triangle *ABC* shown at the right, the following relationships apply to acute angle *B*.

TRIGONOMETRIC RATIOS	
Ratio	**Description**
sin *B* (read "sine of ∠*B*")	$\dfrac{\text{length of side opposite } \angle B}{\text{length of hypotenuse}}$
cos *B* (read "cosine of ∠*B*")	$\dfrac{\text{length of side adjacent to } \angle B}{\text{length of hypotenuse}}$
tan *B* (read "tangent of ∠*B*")	$\dfrac{\text{length of side opposite } \angle B}{\text{length of side adjacent to } \angle B}$

CHECK UNDERSTANDING

For right triangle *ABC* shown in the figure, describe the sine, cosine, and tangent ratios for angle *A*.

EXAMPLE 1

Determine the sine, cosine, and tangent ratios for each acute angle in △*QRS*, shown at the right. Round to the nearest thousandth.

Solution

$$\sin Q = \frac{\text{opposite}}{\text{hypotenuse}} = \frac{20}{29} \approx 0.690$$

$$\cos Q = \frac{\text{adjacent}}{\text{hypotenuse}} = \frac{21}{29} \approx 0.724$$

$$\tan Q = \frac{\text{opposite}}{\text{adjacent}} = \frac{20}{21} \approx 0.952$$

$$\sin R = \frac{\text{opposite}}{\text{hypotenuse}} = \frac{21}{29} \approx 0.724$$

$$\cos R = \frac{\text{adjacent}}{\text{hypotenuse}} = \frac{20}{29} \approx 0.690$$

$$\tan R = \frac{\text{opposite}}{\text{adjacent}} = \frac{21}{20} = 1.05 \quad ◀$$

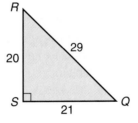

COMMUNICATING ABOUT ALGEBRA

For the triangle in Example 1, the values of cos *Q* and sin *R* are equal and the values of cos *R* and sin *Q* are equal. Do you think this equality will be true for all right triangles? Explain.

You can use a calculator to find sine, cosine, and tangent values given an angle value. You can also use a calculator to find the angle value when given a sine, cosine, or tangent value.

EXAMPLE 2

Use a graphing calculator to compute.

a. Find the value of sin 74° to the nearest ten thousandth.

b. Given cos B = 0.5987. Find the measure of $\angle B$ to the nearest degree.

Solution

a. Enter: SIN 74   The result is 0.9612616959.
To the nearest ten thousandth, the value of sin 74° is 0.9613.

b. Enter: 2nd   COS⁻¹ 0.5987   The result is 53.22315137.
To the nearest ten thousandth, the measure of $\angle B$ is 53. ◀

If you know the lengths of two sides of a right triangle, you can use a trigonometric ratio to find the measure of an acute angle of the triangle.

EXAMPLE 3

Determine the measure of $\angle X$. Round to the nearest degree.

Solution
Because you have information about the side opposite and the side adjacent to $\angle X$, use the tangent ratio.

$$\tan X = \frac{\text{opposite}}{\text{adjacent}}$$

$$\tan X = \frac{14}{10} \qquad \text{Use a calculator.}$$

$$x \approx 54.46232221 \qquad 2^{\text{nd}} \text{TAN}^{-1} (14 \div 10)$$

To the nearest degree, the measure of $\angle X$ is 54°. ◀

Trigonometric ratios enable you to calculate distances or lengths that would otherwise be difficult to measure directly.

EXAMPLE 4

DRIVING The figure shows a truck at the top of an inclined road. Suppose that the brakes on the truck suddenly begin to fail. Over what distance must the driver keep control of the truck until the road levels out?

350 ft   6°

Solution

Suppose the measure of the angle is 6° and you know the measure of the side opposite. Use the sine ratio to find the measure of the hypotenuse.

$$\sin 6° = \frac{\text{opposite}}{\text{hypotenuse}}$$

$$\sin 6° = \frac{350}{\text{hypotenuse}}$$

$$\text{hypotenuse} \approx \frac{350}{\sin 6°}$$

$$\text{hypotenuse} \approx 3348.370282$$

To the nearest foot, the driver must keep control of the truck for a distance of 3348 ft before the road levels out.

EXAMPLE 5

CONSTRUCTION A 32-ft antenna on a school building is attached to the roof and held in place by a wire from the top of the antenna to a roof 18 ft below the top of the building. To the nearest degree, what angle does the wire make with the roof to which it is attached?

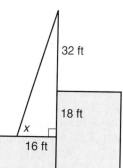

32 ft

18 ft

x

16 ft

Solution

The information includes the side opposite and the side adjacent to ∠*x*. Use the tangent ratio.

$$\tan x = \frac{(18 + 32)}{16}$$

$$\tan x = \frac{50}{16} \qquad \text{Use a calculator.}$$

$$x \approx 72.2553287$$

To the nearest degree, the antenna wire makes a 72° angle with the roof to which it is attached.

TRY THESE

For triangle *ABC*, name the trigonometric ratio or ratios represented by the given fraction. Choose from *sine*, *cosine*, or *tangent*.

1. $\dfrac{7}{25}$

2. $\dfrac{24}{25}$

3. $\dfrac{24}{7}$

4. $\dfrac{7}{24}$

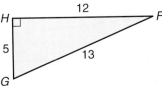

For triangle *FGH*, determine the value of each trigonometric ratio named. Write the ratio as a fraction.

5. tan *F*

6. cos *G*

7. sin *G*

8. sin *F*

9. tan *G*

10. cos *F*

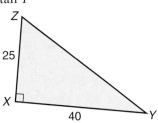

Determine the specified trigonometric ratio to the nearest ten thousandth.

11. sin *X*

12. tan *Y*

13. cos *B*

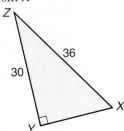

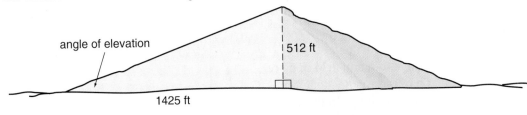

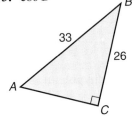

14. **ENGINEERING** What is the angle of elevation of the mountain in the figure below?

angle of elevation

512 ft

1425 ft

15. **WRITING MATHEMATICS** Is it true that in any right triangle, the tangents of the acute angles are reciprocals? Give several examples to support your answer.

PRACTICE

Determine the specified trigonometric ratio to the nearest ten thousandth.

1. tan *W*

2. cos *R*

3. sin *V*

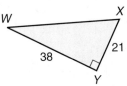

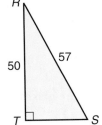

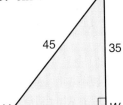

Determine the measure of each acute angle to the nearest degree.

4.
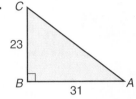

5.
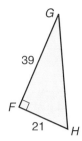

6.
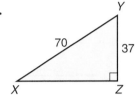

7.
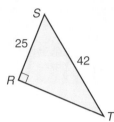

8.
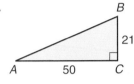

9.
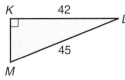

For each trigonometric ratio, determine the measure of the angle named.

10. $\sin C = 0.8746$

11. $\tan D = 0.7002$

12. $\cos T = 0.7193$

13. $\tan G = 19.0811$

14. $\cos Y = 0.8988$

15. $\sin H = 0.2079$

HOME REPAIRS When you lean a ladder against a wall, it may be unstable at certain angles. Some home repair manuals recommend that the measure of the angle that an unsecured ladder makes with the ground should not be more than 75°.

16. The base of an 8-m ladder is placed 1.5 m from a wall. Is the ladder safe?

17. An electrician places a ladder that is 9 m long against the outside wall of a house. At what distance should the base of the ladder be placed from the wall so that the ladder is safe to climb? Round to one decimal place.

18. **SURVEYING** As shown in the figure at the right, a surveyor finds that the distance from point M to point N is 350 m. The measure of the angle at point M is 43° To the nearest meter, what is the distance across the lake?

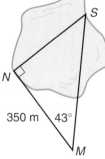

19. **WRITING MATHEMATICS** Using your calculator, a protractor, and a straightedge, draw right triangle RST in which the tangent value of $\angle R$ is 1.1503 (rounded to four decimal places) and $\angle T$ is a right angle. Describe the steps you used.

EXTEND

20. **STEEP STREETS** Filbert Street and 22nd Street in San Francisco are two of the world's steepest streets. Each has a slope of 0.3154. To the nearest degree, at what angle do these streets rise?

21. **HIGHWAY DRIVING** For a road with a 10% grade, the road rises 10 ft for every 100 ft of horizontal distance. In driving from point C to point D, a distance of 0.5 mi, what will your altitude be at point D if point C is at sea level? (*Hint:* First use slope to find the measure of $\angle C$ to the nearest tenth of a degree.)

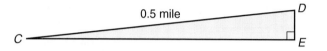

0.5 mile

THINK CRITICALLY

22. **SUNDIALS** Ancient Egyptians used the vertical shadow of a stick placed horizontally on the east side of a building to estimate the time of day. They kept a table of shadow lengths and hours of the morning. Describe what happened as the sun rose higher from early morning until noon. How is this an early use of the tangent concept?

23. Which of the following is a true equality?

$$\tan X = \frac{\sin X}{\cos X} \quad \text{or} \quad \tan X = \frac{\cos X}{\sin X}$$

Verify your answer by substituting the ratio for sine and cosine in each fraction and then simplifying the fraction.

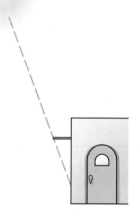

MIXED REVIEW

Name the degree of each polynomial.

24. $5y$

25. abc

26. $a^2b - 3$

27. $-a^2 - 3b^4c^2 + a^4$

28. $\dfrac{5w^2}{3} + \dfrac{w}{2}$

29. $-2y^4 - 3x^2y^2$

Solve each equation.

30. $2x^2 - 11x - 21 = 0$

31. $17x^2 - 7 = 418$

32. $x^2 - 7 = 4$

33. $x^2 + 7x = -6$

34. The ratio of two numbers is 5 : 4. Their product is 980. Find the numbers.

35. **STANDARDIZED TEST** The solution to the equation $2\sqrt{x + 3} = x$ is

   A. 3 and −4   **B.** 6   **C.** −6   **D.** 6 and −2

13.9 Algebra Workshop
Explore Sine and Cosine Graphs

Think Back

● Find each value. Round to hundredths.

1. sin 47°   **2.** cos 81°   **3.** sin 9°   **4.** cos 20°

Find the measure of the acute angle. Round to tenths.

5. sin x = 0.62   **6.** cos y = 0.22

7. sin z = 0.09   **8.** cos w = 0.91

9. Is y = sin x a function? Is y = cos x a function? Explain.

Explore

● You can graph the sine and cosine functions for an angle of any measure. The angle need not be acute or even less than 180°. The graphs of these functions show interesting patterns.

10. Copy the table and, using a graphing calculator, determine the sine for each angle. Round to hundredths where necessary.

Angle Measure	Sine	Angle Measure	Sine	Angle Measure	Sine
0°	____	135°	____	270°	____
30°	____	150°	____	300°	____
45°	____	180°	____	315°	____
60°	____	210°	____	330°	____
90°	____	225°	____	360°	____
120°	____	240°	____		

11. Within what range of degrees is the value of the sine positive? For what range is it negative?

12. For any angle between 0° and 360°, what is the greatest sine value? Which is the least sine value?

13. In graphing the function y = sin x, which variable will be independent and which will be dependent?

14. On graph paper, draw and label a set of axes for which the x-values range from 0° to 360° and the y-values range from −1 to 1. Using the table in Question 10, graph the data.

SPOTLIGHT ON LEARNING

WHAT? In this lesson you will learn
- to graph the functions y = sin x and y = cos x.
- to recognize how different parameters affect the appearance of these graphs.

WHY? Graphs of sine and cosine functions can help you solve problems in scientific experiments.

CHECK UNDERSTANDING

Does the graph of y = sin x pass the vertical line test? What does this mean?

PROBLEM SOLVING TIP

Make sure your graphing utility is in DEG mode.

15. On your graphing utility, set the range values as in Question 14. Use an x scale of 90 and a y scale of 0.5. Graph $y = \sin x$. How do the two graphs compare?

16. Change the x_{max} range value to 720. Predict how the graph will look. Then graph the function and check your prediction.

The graph of the sine function is called a **sine curve**. The horizontal distance from any point on the graph to that point where the graph begins to repeat is called the **period** of the function. The period of the sine function is 360°.

The **amplitude** of the function is the distance from the middle line of the graph to either the **crest** (the maximum value) or **trough** (minimum value). The middle line for the graph of $y = \sin x$ is the x-axis. The amplitude of $y = \sin x$ is 1.

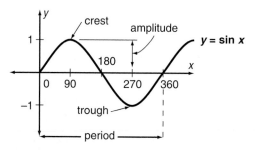

The function $y = \cos x$ is also a periodic function.

17. Graph the function of $y = \cos x$ for angles from 0° through 720°. Compare the period and amplitude to the graph of $y = \sin x$. You may wish to graph the two functions on the same screen.

18. For what values of x is $\sin x$ equal to 0? For what values of x is $\cos x$ equal to 0?

Make Connections

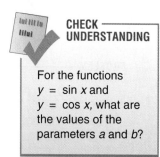

CHECK UNDERSTANDING

For the functions $y = \sin x$ and $y = \cos x$, what are the values of the parameters a and b?

In earlier lessons you used the graphs of $y = |x|$ and $y = x^2$ as references and explored the effect of different parameters. Now, you will use the graphs of $y = \sin x$ and $y = \cos x$ as references to compare graphs of general functions $y = a \sin bx$ and $y = a \cos bx$.

19. Clear the graph screen. Graph the function of $y = 2 \sin x$ for x values from 0° through 720°. Compare the period and amplitude of $y = 2 \sin x$ with those of $y = \sin x$.

20. Graph the function of $y = \sin 2x$ for values of x from 0° through 360°. Compare the period and amplitude of $y = \sin 2x$ with those of $y = \sin x$.

21. Predict the period and amplitude of the function $y = 3 \sin 3x$. Check your answer by graphing the function.

22. Graph the function $y = 2 \cos x$ for values of x from $0°$ through $720°$. Compare the period and amplitude of this graph with those of $y = \cos x$.

23. What function of the form $y = a \cos bx$ will have a period of $90°$ and an amplitude of 4? Check your answer by graphing the function.

Identify the period and amplitude for each function. Check your answer by graphing the function.

24. $y = \dfrac{1}{2} \sin 2x$ **25.** $y = 4 \cos \dfrac{1}{2}x$ **26.** $y = -3 \sin x$

27. $y = 2 \cos 6x$ **28.** $y = 3 \sin \dfrac{1}{3}x$ **29.** $y = 5 \cos 5x$

Summarize

30. WRITING MATHEMATICS   Explain how you determine the period and amplitude of the functions $y = a \sin bx$ and $y = a \cos bx$. Try to write general formulas that can be used to find the period and amplitude. Remember that the period and amplitude should always be positive numbers.

31. MODELING   The graph below shows temperature fluctuations during a controlled experiment in a chemistry laboratory. What is the period of the curve?

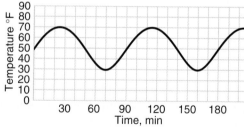

32. THINKING CRITICALLY   Explain how you can use a graphing utility to find the values of x for the range $0 \le x \le 360°$ that are solutions to the equation $\sin x = \cos x$. Use your method to solve the equation.

33. GOING FURTHER   Graph the function $y = \sin x + 2$ and compare it with $y = \sin x$. Then graph the function $y = \cos x - 4$ and compare it with $y = \cos x$. (*Hint*: Use a line of symmetry to determine the amplitude of the altered function.)

Problem Solving File

Use Measures Indirectly

In some situations a measurement is needed that you cannot directly measure by hand. However, you can determine the measure indirectly by using the Pythagorean theorem, trigonometric ratios, or proportions in similar figures. For some problems, you may need to use more than one method of indirect measurement.

Problem

An electric power utility wants to run a cable between the summits of two mountains, from point C to point D. What is the distance from point C to point D?

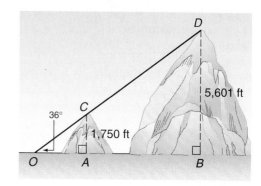

Explore the Problem

1. Which method of indirect measurement can you use to determine length OA?

2. What equation can you write?

3. Determine the length OA to the nearest foot.

4. What choice of methods do you have for finding the length of side OC of triangle OAC?

5. Find the length OC to the nearest foot using one of these methods.

6. With the information you have so far, how can you determine the length of $\overline{OD}$? How will this help you?

7. Use the plan you explained in Question 6. Determine the lengths of OD and CD to the nearest foot.

8. What if you also wanted to determine the distance from point A to point B. Describe a method you could use and carry out the steps.

9. **WRITING MATHEMATICS** Given a choice of methods for finding a particular length, which method did you choose? Why?

Investigate Further

● **CONSTRUCTION** The wire supporting a radio transmitting tower touches the top of a 10-ft pole as shown in the diagram. Right triangles *ABC* and *ADE* are similar. Use this information to determine the height *h* of the tower.

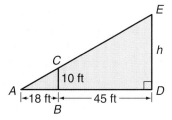

10. Describe how you could use principles of similar triangles to solve the problem.

11. Describe how you could use trigonometry to solve the problem.

12. Choose the method you prefer and carry out the steps to determine the height of the tower.

13. Find a clasmate who solved the problem differently than you did. Compare results and discuss the advantages and disadvantages of each method.

Apply the Strategy

● 14. **SURVEYING** In the figure, triangles *JKL* and *MNL* are similar.

   a. To the nearest yard, find the distance *x* across the lake.

   b. Could you use trigonometry to solve this problem? Why or why not?

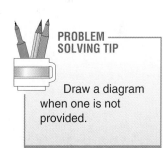

PROBLEM SOLVING TIP

Draw a diagram when one is not provided.

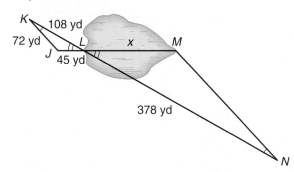

15. **TRAVEL** Two trains leave from the same station at the same time. One travels east averaging 50 mi/h. The other travels south averaging 40 mi/h. To the nearest mile, how far apart are the trains after 2 h?

16. CONSTRUCTION A vertical supporting beam divides the cross section of a roof into two right triangles of the same shape and size. If the slope of the roof is 32°, what is the length of the supporting beam *b*, to the nearest foot?

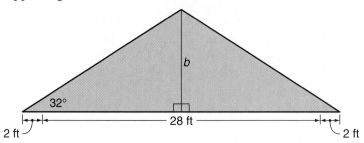

17. GEOMETRY Refer to rectangle *ABCD*.

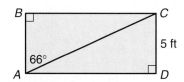

a. Find the length of the diagonal to the nearest foot.

b. Find the area of the rectangle to the nearest tenth of a square foot.

18. SURVEYING Find, to the nearest tenth of a foot, the height *h* of the antenna on top of the building shown in the figure below.

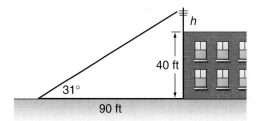

19. MOVING Some movers positioned a carton such that rectangle *RSTV* represents one side of the carton leaning against wall *TU* at point *T*. Triangle *RQV* is similar to triangle *VUT*. What is the distance from the corner *R* to the floor at *Q* to the nearest tenth of an inch?

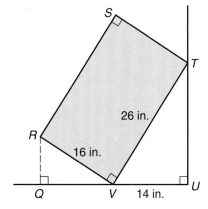

REVIEW PROBLEM SOLVING STRATEGIES

ALGEBRA IS GREAT

1. The words that make up the triangle range from 2 through 15 letters. Let a, b, c and d be four different numbers. Choose four different words that are a, b, c, or d letters long so that $a^2 = bd$ and $ad = b^2c$. You can use guess and check or use the hint to help you solve the problem algebraically. (*Hint*: if $a^2 = bd$, then $a^2(ad) = bd(b^2c)$ or $a^3d = b^3dc$. Dividing by d, $a^3 = b^3c$. What does this imply about c? See what equations result from substituting the value you get for c and use logic.)

```
A M
L E T
G O O D
E X T R A
B E H I N D
R O O S T E R
A M I C A B L E
I N T E L L E C T
S T A T I O N A R Y
G R A S S H O P P E R
R E O R G A N I Z I N G
E X P O N E N T I A L L Y
A S T R O B I O L O G I S T
T O N S I L L E C T O M I E S
```

A TAXING PROBLEM

2. The country of Santa Taxita has an unusual tax plan. The tax rate is equal to the number of thousands of dollars of your income—that is, the rate is 3% on an income of $3,000, 14% on $14,000, and a maximum of 100% on income of $100,000 or more.

   a. How much money will you have left if your income is $18,000? Is there any other income for which you would be left with the same amount?

   b. Express the amount A you have left as a function of x, the number of thousands of dollars of income.

   c. At what income do you have the most money after taxes in Santa Taxita? Explain how you determined this amount.

The Power of Two

3. Without adding all the terms, how can you find the sum?

$$\frac{1}{2} + \frac{1}{4} + \frac{1}{8} + \frac{1}{16} + \ldots + \frac{1}{1024} = \square$$

Here are some questions to guide your thinking.

   a. What is the sum of the first two terms?

   b. How is the denominator of the sum of the first two terms related to the second addend?

   c. How is the numerator of the sum of the first two terms related to the denominator?

   d. What is the sum of the first three terms?

   e. What patterns do you see in how the addends are related to the sum?

   f. Without adding, find the sum of the original expression.

   g. Express the pattern you discovered as a general statement.

   h. How many terms are there in the series of addends that has the sum $\frac{32,767}{32,768}$? Write the last addend.

· · · CHAPTER REVIEW · · ·

VOCABULARY

Choose the word from the list that completes each statement.

1. In a right triangle, the ratio of the side opposite an acute angle to the hypotenuse is called the __?__ of the angle.

2. The __?__ is the expression under the radical sign.

3. A __?__ has no sides equal in length.

a. scalene triangle

b. sine

c. radicand

Lessons 13.1, 13.2, and 13.7   TRIANGLES; THE PYTHAGOREAN THEOREM   pages 615–624, 648–651

- Use the properties of triangles and Pythagorean relationships to classify triangles and to find measures of sides. Triangles are *similar* if the measures of corresponding angles are equal and the ratios of corresponding sides form proportions.

Classify each triangle according to its side lengths.

4. 11, 16, 32   5. 21, 28, 35   6. 12, 15, 19

For each pair of similar triangles find all unknown side lengths and angle measures.

7.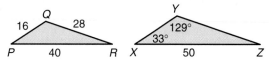

8.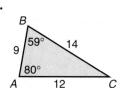

Lessons 13.3 and 13.4   MULTIPLY, DIVIDE, ADD, AND SUBTRACT RADICALS   pages 625–635

- To multiply and divide radicals, use the product and quotient properties of square roots. Use the distributive property to add or subtract like radicals.

Simplify.

9. $\sqrt{28}$ 10. $\sqrt{25y^2}$ 11. $2\sqrt{2} \cdot \sqrt{32}$ 12. $\sqrt{3}(4 + \sqrt{15})$ 13. $\dfrac{\sqrt{48}}{\sqrt{6}}$

14. $\sqrt{13} + 4\sqrt{13}$ 15. $4\sqrt{5} - 3\sqrt{2} + 2\sqrt{5}$ 16. $\sqrt{128a} + 3\sqrt{8a}$ 17. $\sqrt{300r^2} - \sqrt{48r^2}$

Lesson 13.5   SOLVE RADICAL EQUATIONS   pages 636–641

- To solve a radical equation, isolate radicals and square both sides of the equation.

Solve and check.

18. $3\sqrt{m} = 15$ 19. $\sqrt{18k} - 4 = 2$ 20. $\sqrt{3s + 18} = s$ 21. $8\sqrt{x - 1} = 6\sqrt{3x - 14}$

Lesson 13.6 THE DISTANCE AND MIDPOINT FORMULAS

pages 642–647

- To determine the distance between two points (x_1, y_1) and (x_2, y_2), use the distance formula.
- To determine the midpoint between two points (x_1, y_1) and (x_2, y_2), use the midpoint formula.

Determine the distance between the two points. Then determine the midpoint of the segment joining them. Write radicals in simplest form.

22. $(-12, 3)$ and $(2, 5)$

23. $(8, 1)$ and $(7, -1)$

Lesson 13.8 SINE, COSINE, AND TANGENT

pages 652–658

- Use the trigonometric ratios of sine, cosine, and tangent to determine the measure of an acute angle in a right triangle or to calculate distances or lengths indirectly.

Find the specified trigonometric ratio to the nearest ten thousandth.

24. $\sin J$

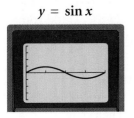

Determine the measure of each acute angle to the nearest degree.

25.

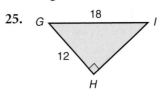

Lesson 13.9 EXPLORE SINE AND COSINE GRAPHS

pages 659–661

- Use the graphs of $y = \sin x$ and $y = \cos x$ as basic references to graph functions of the form $y = a \sin bx$ and $y = a \cos bx$.

Match each equation to its graph. Each graph shows values for x between $0°$ and $360°$ and values for y between -4 and 4.

$y = \sin x$

a.

b.

c.

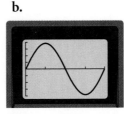

26. $y = \sin 2x$

27. $y = 4 \sin x$

28. $y = \sin 4x$

Lesson 13.10 USE MEASURES INDIRECTLY

pages 662–665

- Indirect measurement can be used to solve problems.

29. To find the distance from the ground to the place on a pole where a support wire is connected, Rob has placed a 3-ft stick at right angles to the ground so that it just touches the wire. The distance BE is 3.75 ft and the distance EC is 1.25 ft. What is the distance AB? What is the length of the wire?

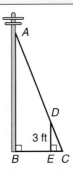

CHAPTER TEST

For the side lengths given, classify each triangle as *acute*, *right*, or *obtuse*.

1. 18, 24, 25
2. 15, 20, 25
3. 9, 40, 42

Determine the measure of the third angle of each triangle.

4. 32°, 50°
5. 48°, 110°
6. 63°, 99°

Simplify each radical expression.

7. $\sqrt{12}$
8. $\dfrac{3\sqrt{6}}{\sqrt{3}}$
9. $\sqrt{\dfrac{4}{3}}$
10. $\sqrt{9x^3}$
11. $\sqrt{8n^2}$
12. $\sqrt{\dfrac{x}{3}}$
13. $5\sqrt{50} - 8\sqrt{32}$
14. $4\sqrt{72a} - 3\sqrt{98a}$
15. $\sqrt{48x} - \sqrt{27x}$
16. $\sqrt{5p} + 3\sqrt{45p^3}$
17. $\sqrt{12x + 12} + \sqrt{27x + 27}$

18. **WRITING MATHEMATICS** Write a paragraph explaining how the distributive property is used to add like radicals. Use an example to illustrate.

19. **STANDARDIZED TESTS** Which of the following radical expressions does *not* simplify to $4\sqrt{3}$?

   A. $3\sqrt{12} - 2\sqrt{3}$
   B. $2\sqrt{75} + 2\sqrt{27}$
   C. $4\sqrt{27} - 2\sqrt{48}$
   D. $\sqrt{3} + \sqrt{27}$

Solve each radical equation.

20. $\sqrt{3x} = -6$
21. $3\sqrt{x - 5} - 7 = 5$
22. $\sqrt{4r + 5} + r = 0$
23. $\sqrt{3n - 2} - n = 0$
24. $\sqrt{3x + 1} = x - 3$
25. $1 + 6\sqrt{y - 9} = y$

26. Determine the distance between the points $(-3, -2)$ and $(5, 4)$ and the midpoint of the line segment connecting them.

27. Determine whether the two triangles shown are similar.

For $\triangle ABC$, find each trigonometric ratio.

28. $\sin A$
29. $\cos C$
30. $\tan C$
31. $\sin C$

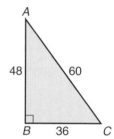

Solve each problem.

32. The measure of an angle is 48° less than its complement. Determine the measure of the angle and its complement.

33. The formula $d = \sqrt{12h}$ can be used to approximate the distance d to Earth's horizon in kilometers from a point h meters above the Earth's surface. How much farther could you see from a height of 525 m than you could from a height of 336 m?

34. Determine the lengths of the sides of the triangle formed by connecting the points $(-2, 4)$, $(4, 6)$, and $(2, -2)$. Then state whether the triangle is *scalene*, *isosceles*, or *equilateral*.

35. Determine the width of the river given that $\triangle ABC$ is similar to $\triangle EDC$.

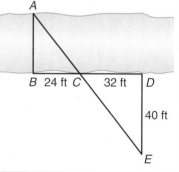

PERFORMANCE ASSESSMENT

DETERMINE HEIGHTS Work with a partner. Select a tall object such as a tree or building. Follow these steps to determine the height of the object using similar triangles.

- Select a point some distance from the object from which to sight a triangle and name it point *C*.
- Name the top of the object *A* and the bottom of the object *B*.

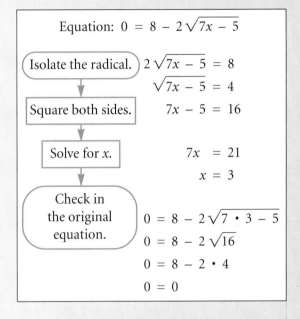

- Measure the distance from the sight point *C* to base *E*.
- While you look from the ground at sight point *C*, stand so that the sight line from the ground point to the top of the object grazes the top of your partner's head *B*. Make sure your partner stands so that he or she forms the same angle to the ground as that of the object. Then two similar triangles will be formed.
- Measure the distance *CD* to your partner.
- Determine the approximate height of the object.

USE A FLOWCHART Write a radical equation in the form $0 = a \pm b \sqrt{cx \pm d}$, where *a*, *b*, *c*, and *d* are real numbers greater than 0. Have a partner use the flowchart below to solve the radical equation. Check that each step is correct.

Equation: $0 = 8 - 2\sqrt{7x - 5}$

Isolate the radical. → $2\sqrt{7x - 5} = 8$
$$\sqrt{7x - 5} = 4$$

Square both sides. → $7x - 5 = 16$

Solve for *x*. → $7x = 21$
$$x = 3$$

Check in the original equation. →
$$0 = 8 - 2\sqrt{7 \cdot 3 - 5}$$
$$0 = 8 - 2\sqrt{16}$$
$$0 = 8 - 2 \cdot 4$$
$$0 = 0$$

USE GRAPH PAPER Draw a right isosceles triangle *ACE* with side lengths of your own choosing on graph paper. Have a partner show at least two different ways to draw two right isosceles triangles *ABF* and *BCD* to show that $AB + BC = AC$. Check that each step is correct.

USE GRAPH PAPER Draw a right triangle on graph paper. Have a partner draw a similar right triangle either inside your triangle or containing it. Determine the scale factor your partner used. Show that it is true for each side of the pair of triangles.

PROJECT ASSESSMENT

 It's time to sell your used cars. Assemble all the data and visuals you have collected during the project.

1. As a group, discuss each car you have studied. Decide on a fair price for the car and what features of the car you wish to highlight in an advertisement.

2. Have each group member work on a specific advertisement. He or she should use the correct abbreviations when developing the copy.

3. Contact one or more local newspapers to find out their automotive classified rates. Use these rates to determine the cost of running each advertisement for three days.

4. Compare your advertisements with those of other groups. Were some more attention-getting than others? Why?

···CUMULATIVE REVIEW···

Simplify each radical expression.

1. $\sqrt{300}$ 2. $\sqrt{\dfrac{4}{5}}$

3. $\sqrt{15} \cdot \sqrt{5}$ 4. $\dfrac{\sqrt{24}}{\sqrt{3}}$

5. Find two integers that differ by three and whose product is 54.

Solve each equation. Round to the nearest hundredth.

6. $x^2 + 29x - 30 = 0$

7. $4n - (5n - 8) = 8 - n$

8. $x^2 - 11x + 25 = 0$

9. $\dfrac{3}{4}d - 12 = 3 - \dfrac{1}{2}d$

10. **WRITING MATHEMATICS** You have learned three ways to solve quadratic equations: by factoring, by taking square roots, and using the quadratic formula. If you could choose only one of these methods for solving a particular quadratic equation, which would you choose? Why?

Simplify each radical expression.

11. $\sqrt{48} + \sqrt{27} - \sqrt{75}$

12. $3\sqrt{24} + 2\sqrt{45} - \sqrt{150} + 6\sqrt{20}$

13. **STANDARDIZED TESTS** The area of a rectangle is $2x^2 + 9x + 10$. What is the perimeter of the rectangle?

   A. $6x + 14$
   B. $3x + 7$
   C. $2x^2 + 9x + 10$
   D. 9
   E. cannot be determined

Simplify each expression.

14. $\dfrac{(2x^2)^3}{(4x^5)^2}$ 15. $\dfrac{(a^2b^{-3})^6}{(a^3b^{-4})^4}$

Write the equation of each line described, in standard form.

16. passes through $(4, -1)$ and parallel to the line $y = 2x - 3$

17. passes through $(-4, 5)$ and perpendicular to the line $3x - 2y = 8$

18. passes through $(3, -2)$ and perpendicular to the x axis

Solve each equation.

19. $\sqrt{2x + 1} + 4 = 9$ 20. $\sqrt{x + 7} + 5 = x$

Solve each system by graphing.

21. $\begin{cases} x - y = 1 \\ 2x + y = 5 \end{cases}$ 22. $\begin{cases} 3x + 2y < 8 \\ 2x - y < 1 \end{cases}$

23. Find the length of the segment whose endpoints are $(3, -5)$ and $(-3, -1)$.

24. Find the midpoint of the segment whose endpoints are $(10, 3)$ and $(2, -5)$.

25. **TECHNOLOGY** Teri used the distance formula and came up with $\sqrt{36 + 16}$. She entered $\boxed{\sqrt{}}\ 36\ \boxed{+}\ 16$ on her graphing calculator, but got an answer of 22, which she knew was not correct. Why did this happen, and how can Teri enter the calculation correctly?

STANDARDIZED TESTS Determine whether the quantity in Column 1 is greater than, less than, or equal to the quantity in Column 2, or whether the relationship cannot be determined.

	Column 1	Column 2
26.	the discriminant of $2x^2 + 7x - 5 = 0$	the discriminant of $2x^2 - 7x - 5 = 0$
27.	the hypotenuse of a triangle with sides 10 and 24	the hypotenuse of a triangle with sides 15 and 20

··· STANDARDIZED TEST ···

STANDARD FIVE-CHOICE Select the best choice for each question.

1. Consider the expressions numbered I–IV.

   I. $2x^4 + 5$
   II. $x^4 + x^2y^2 + y^4$
   III. $4x^3 + 2x$
   IV. $7y^2 - 2y^4$

   Which of the above is a fourth degree binomial?
   A. IV only   **B.** II and III   **C.** I and IV
   D. I and II   **E.** I, III, and IV

2. When factored correctly, the polynomial $x^2 - 13x - 30$ can be written as

   A. $(x - 10)(x - 3)$
   B. $(x - 10)(x + 3)$
   C. $(x - 15)(x - 2)$
   D. $(x + 15)(x - 2)$
   E. $(x - 15)(x + 2)$

3.

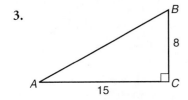

   $\cos A =$

   A. cannot be determined
   B. $\dfrac{8}{17}$ **C.** $\dfrac{15}{17}$ **D.** $\dfrac{15}{23}$ **E.** $\dfrac{8}{15}$

4. $\dfrac{(-1)^{16}(-2)^5}{-4^2 - (-4)^2}$

   A. 1   **B.** undefined   **C.** 0
   D. −32   **E.** 3

5. Darrell takes a $50 bill to the bank. He asks for change in bills. The teller returns 30 bills, all of which are $5's and $1's. How many $5's did Darrell get?

   A. 1   **B.** 3   **C.** 5   **D.** 7   **E.** 9

6. What is the maximum area that can be enclosed by a rectangular region with 40 ft of fencing?

   A. 10 ft²   **B.** 20 ft²   **C.** 40 ft²
   D. 100 ft²   **E.** 160 ft²

7. Which of the following numbers is between 5 and 6?

   I. $\sqrt{31}$ **II.** 35%
   III. 6.010010001...   **IV.** $\dfrac{23}{4}$

   A. I only   **B.** II and IV   **C.** I and IV
   D. II and III   **E.** I, II, and IV

8. A person is reading from a book. The sum of the two facing page numbers is 351. If each chapter in the book is 40 pages long, in what chapter is the person reading?

   A. Chapter 4   **B.** Chapter 5   **C.** Chapter 6
   D. Chapter 8   **E.** Chapter 9

9. If $f(x) = \dfrac{2}{3}x - 5$ and $g(x) = 2x^2 + 3x - 5$, find $f(6) + g(-2)$.

   A. −20   **B.** −4   **C.** −2
   D. 2   **E.** 10

10. The midpoint of a segment is $(1, -3)$. If one endpoint of the segment is located at $(4, -10)$, where is the other endpoint located?

    A. $(2.5, -6.5)$ **B.** $(1.5, -3.5)$ **C.** $(7, -17)$
    D. $(-2, 4)$ **E.** the origin

11. An item had an original selling price of $24.00. First, it was marked $\dfrac{1}{3}$ off. Then it was further reduced, so that a customer only had to pay $12.00 for the item. What percent was the second markdown?

    A. 50%   **B.** 25%   **C.** $58\dfrac{1}{3}$%
    D. $33\dfrac{1}{3}$%   **E.** 40%

Take a Look
AHEAD

Make notes about things that look familiar.

- Find a lesson on multiplication and division. What will be multiplied and divided?
- Describe the equations you find in this chapter.

Make notes about things that look new.

- Find graphs that are different from other graphs that you have seen in this book. Describe the differences.
- What is "dimensional analysis" and why is it a useful problem solving strategy?

DATA*Activity*

Technology and Agribusiness

Technology has greatly reduced the farmer's workload. Huge plows and combines make planting and harvesting easier tasks. Automated systems may feed animals or distribute water, pesticides, and fertilizer to fields. However, the expense of such equipment is feasible only for a large-scale operation. Traditional family-owned farms are gradually disappearing, replaced by giant corporate-owned farms called *agribusinesses*.

Today's farmer is more likely to use a computer than go to the market to find out the latest product prices. The table on the next page shows farm product prices for a week in July 1995.

- ▶ Add, subtract, multiply, and divide real numbers.
- ▶ Determine percent decrease.
- ▶ Use probability.
- ▶ Read a table.

AGRIBUSINESS

In this chapter, you will see how:

- **AGRIBUSINESS MANAGERS** work with several variables to determine break-even points, amount of profit, and sales goals.
(Lesson 14.1, page 678)

- **REFRIGERATION SYSTEM ENGINEERS** use direct and inverse relationships when designing cooling systems.
(Lesson 14.2, page 685)

- **CATTLE FARMERS** use rational expressions to select the most cost-effective feed mixture for their animals.
(Lesson 14.3, page 691)

Farm Product Prices, grain prices per bushel					
Product	Mon	Tue	Wed	Thur	Fri
Eggs (large white doz.)	0.83	0.85	0.85	0.86	0.86
Corn (No. 2 yellow)	2.94	2.98	$2.96\frac{1}{2}$	$2.92\frac{1}{2}$	2.91
Soybeans (No. 1 yellow)	$6.25\frac{1}{2}$	$6.30\frac{1}{4}$	$6.31\frac{1}{4}$	$6.22\frac{1}{2}$	$6.21\frac{1}{2}$
Wheat (No. 2 soft)	4.73	$4.56\frac{1}{4}$	$4.55\frac{1}{2}$	$4.52\frac{1}{2}$	4.54
Wheat (No. 2 dark)	$5.85\frac{1}{2}$	5.76	$5.74\frac{1}{2}$	5.69	$5.60\frac{3}{4}$
Wheat (No. 2 hard)	$5.13\frac{1}{2}$	$4.97\frac{1}{2}$	4.94	4.84	$4.85\frac{1}{2}$
Oats (No. 2 heavy)	2.09	2.08	2.09	$2.07\frac{1}{2}$	2.06

Use the table to answer the following questions.

1. What was the change in the price of No. 2 dark wheat from Monday to Friday?

2. A farmer sold 10,000 bushels of No. 2 yellow corn on Tuesday. How much money did the farmer receive? How much more is this than if the sale had taken place on Thursday?

3. What was the percent decrease in the price of No. 2 soft wheat from Monday to Friday? Round to the nearest percent.

4. If a farmer sold one of the listed products on Wednesday, what is the probability the farmer received the best price of the week?

5. **WORKING TOGETHER** Research and report on how farm product prices have varied over the last 40 years. For which products have prices risen? Which product prices have declined? Use several graphs to present your findings.

673

PROJECT

Specialize in Silos

Farmers use silos to store livestock feed, grain, and other products. Early silos were just pits covered with boards. In 1873, Fred Hatch, an Illinois farmer, built the first above-ground silo. Because of its square shape, feed could not be packed tightly enough to prevent air pockets that lead to spoilage. In 1882, Franklin King, a Wisconsin agricultural scientist, constructed a round silo that proved to be airtight. Today, tall, cylindrical, glass-lined, steel silos provide year-round storage.

PROJECT GOAL

To investigate the storage capacity of a silo.

Getting Started

Suppose a farmer has only 93.5 yd² of material to use for the lateral sides of a silo. How can the farmer maximize the storage capacity?

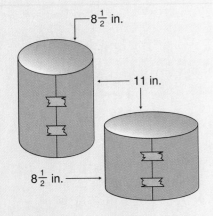

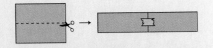

Internet Connection
www.swpco.com/ swpco/algebra1.html

1. Why is an $8\frac{1}{2}$ in. by 11 in. sheet of paper convenient for solving this problem?

2. First roll and tape the paper lengthwise, then tape the paper widthwise. Be careful not to overlap the edges. Which "silo" holds more? (In each case, one dimension of the paper will be the height h and the other will be the circumference C of the base.) Stand each cylinder on a tabletop, then fill with macaroni, popcorn, beans, or other similar material to determine volume. Record your results.

3. Continue investigating by cutting your paper into strips and taping the strips to build wider and taller silos. How do the height and the circumference of the base change?

4. Make a table showing the circumference of the base, the height, and the volume for each silo.

PROJECT *Connections*

Lesson 14.2, page 684:
Determine the equation of a variation that relates to the construction of the silos.
Lesson 14.3, page 690:
Explore volume as a function of the circumference of the base of the silo.
Lesson 14.8, page 717:
Use dry matter tonnage to predict silo capacities for different moisture levels of stored product.
Chapter Assessment, page 725:
Use roleplaying to present results and recommendations of project investigation.

14.1 Algebra Workshop
Inverse Variation

Think Back/Working Together

● Work with a partner. Recall that a direct variation is a relation where y varies directly as x and can be written in the form $y = kx, k \neq 0$.

SPOTLIGHT ON LEARNING

WHAT? In this lesson you will learn
- to use area models to explore inverse variation.
- to recognize the equation of an inverse variation.
- to explore the graph of an inverse variation.

WHY? Writing and solving equations of inverse variation can help you solve problems in geometry, agriculture, and business.

1. Copy and complete the table below for a car that travels at an average speed of 50 mi/h.

Number of Hours	1	2	3	4	5	x
Miles Traveled	50					

2. Write an equation that represents the information from the table above, letting x represent number of hours and y represent miles traveled. Is it an example of direct variation?

3. Use a graphing utility to graph the equation you wrote in Question 2. Set an appropriate range for the viewing window.

Explore/Working Together

● Work with a partner.

4. Use a grid similar to the one at the right to draw the different rectangles with lengths and widths that are whole numbers and having an area of 12 square units.

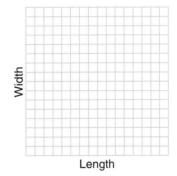

5. Use these rectangles to complete a table of the pairs of factors of 12.

Length					
Width					

6. Write an equation that represents the width of the rectangle, letting x represent the length and y represent the width. As x increases, what happens to y? As x decreases, what happens to y? Is this a direct variation?

7. Use a graphing utility to graph the equation you wrote in Question 6. Use a standard viewing window.

8. Trace along the portion of the graph in Quadrant I to investigate ordered pairs, other than those in your table, that satisfy the equation. Do these values make sense in this situation? Explain.

9. Trace along the portion of the graph in Quadrant III. Do these values make sense in this situation? Explain.

10. Can you find a point on the curve for which x equals zero? Why or why not? Would $x = 0$ make sense in this situation? Explain.

The equation you wrote in Question 6 is of the form

$$y = \frac{k}{x}, k \neq 0, x \neq 0$$

It shows an *inverse variation*. The variable y varies inversely as x, or y is inversely proportional to x.

Make Connections

11. Use the equation you wrote in Question 6. Copy and complete Tables 1 and 2 by finding the width that corresponds to each length.

Table 1

Length, x	0.01	0.02	0.05	0.08	0.1	0.2	0.5	0.8	1
Width, y									

Table 2

Length, x	12	24	36	48	60	72	84	96	100	120
Width, y										

12. Compare the change in the x-values to the change in the y-values in Table 1. What do you notice?

13. Compare the change in the x-values to the change in the y-values in Table 2. What do you notice?

14. What do you notice about the rate of change in y as x changes from 0.01 to 120? How does this compare to a direct variation?

15. What do you think happens to the values of y as the values of x get closer and closer to zero?

16. What do you think happens to the values of y as the values of x increase infinitely?

ALGEBRA: WHO, WHERE, WHEN

American biologist Rachel Louise Carson (1907–1964) noticed that the thickness of egg shells of predatory birds varied indirectly as the exposure of the parent birds to DDT. This threatened the survival of the species. Her discovery launched ecology awareness as we know it today.

17. Compare the two groups of graphs below. Describe the differences between the graphs of equations having direct and inverse variations.

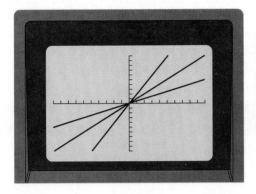

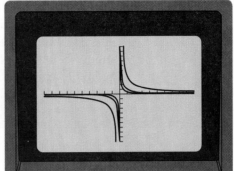

Direct: $y = x$, $y = 2x$, $y = 0.5x$ Inverse: $y = \dfrac{1}{x}$, $y = \dfrac{4}{x}$, $y = \dfrac{0.5}{x}$

Summarize

18. WRITING MATHEMATICS Summarize the characteristics of an inverse variation. Include the following information.

    a. the general form of the equation
    b. the shape of the graph
    c. the value of y when x equals zero
    d. the values of y when x gets infinitely small
    e. the values of y when x gets infinitely large
    f. how to find k given ordered pairs of x and y

19. THINKING CRITICALLY Will the graphs of $y = \dfrac{1}{x}$ and $y = \dfrac{10}{x}$ ever intersect? Explain your answer.

20. GOING FURTHER The graph of an inverse variation $xy = k$ is called a *hyperbola*. Use a graphing utility to explore the hyperbolas that correspond to different values of k. When k is positive, in which quadrants are the parts or *branches* of the hyperbola? What happens when k is negative?

21. GOING FURTHER Use a graphing utility to graph equations of the form $y = \dfrac{k}{x^2}$. Comment on the graphs of variations in which y varies inversely as the square of x.

Career
Agribusiness Manager

An agribusiness manager supervises the welfare of farm animals or crops in the interest of profit. The manager of Agri-Feed wants to project profits for the upcoming selling period. She must consider fixed costs such as buildings, equipment, and taxes, and variable costs such as labor and utilities. She must also determine the break-even point when all costs are covered and profit can begin. The break-even point occurs when the profit is zero.

TONS OF FEED PRODUCED	TOTAL REVENUE (P = $174/ton)	VARIABLE COST (C = $167/ton)	FIXED COSTS	TOTAL COSTS	TOTAL PROFITS
(Tons)	Tons x $174 (Dollars)	Tons x $167 (Dollars)	(Dollars)	Variable Costs Fixed Costs (Dollars)	Total Revenue Total Costs (Dollars)
0	0	0	12,500	12,500	−12,500
1000	174,000	167,000	12,500	179,500	−5500
2000	348,000	334,000	12,500	346,000	1500

Decision Making

Determining that the fixed costs are $12,500 and that the variable costs are $167 per ton, the manager produced the table shown above.

Let t represent the number of tons produced. Recall from Lesson 7.3 that the break-even point occurs when

$$\text{Revenue} = \text{Fixed Costs} + \text{Variable Costs}$$
$$174t = 12{,}500 + 167t$$
$$t = \frac{12{,}500}{174 - 167} \approx 1785.7$$

Agri-Feed must sell approximately 1785.7 tons of feed to break even. The solution shows that the break-even point occurs when

$$t = \frac{\text{Fixed Costs}}{\text{Selling Price per ton} - \text{Variable Cost per ton}}$$

So, the break-even tonnage varies inversely as the difference of the selling price per ton and the variable costs per ton.

1. What happens to the break-even point t when the selling price per ton goes up and the variable costs per ton stay the same?

2. If Agri-Feed raises the price of feed to $175 per ton, determine the new break-even point.

3. If the selling price is $174 per ton, the variable costs are $167 per ton, and Agri-Feed sells 2250 tons of feed, what is their profit? How many tons of feed must they sell to make a profit of $6500?

14.2 Direct, Inverse, and Joint Variation

Explore

In Lesson 14.1 you examined the difference between direct and inverse variation.

> **DIRECT VARIATION**
>
> When y varies directly as x and k is the constant of variation, an equation can be written in the form
> $$y = kx, k \neq 0$$

> **INVERSE VARIATION**
>
> When y varies inversely as x and k is the constant of variation, an equation can be written in the form
> $$y = \frac{k}{x}, k \neq 0$$

SPOTLIGHT ON LEARNING

WHAT? In this lesson you will learn
- to use direct variation, inverse variation, and joint variation.
- to apply the different forms of variation in real world settings.

WHY? Direct, inverse, and joint variation can help you solve problems about travel, electricity, and coordinate geometry.

Consider the equation $C = \pi d$ for the circumference of a circle with diameter d.

1. In the formula for the circumference of a circle, what happens to C as d increases? Which type of variation does this show?

2. For $C = \pi d$, what is the constant of variation? Solve the equation for the constant of variation.

3. Write the equation for the diameter d in terms of the radius r. As r increases, what happens to d? Does d vary directly as r?

4. If the circumference of a circle varies directly as the diameter, does it also vary directly as the radius? Explain.

5. Which would get you 4 mi from school in the least amount of time, walking or riding in a car? Which mode of transportation has the greater rate of speed? How does speed relate to time? Write an equation that relates speed and time to the distance of 4 mi.

THINK BACK

A radius is a segment from the center of a circle to any point on the circle.

A diameter is a segment through the center of the circle which touches the circle at two points.

Build Understanding

Explore describes two different kinds of variations, direct and inverse. Notice that when you solve for the constant of variation k in an equation of direct variation, $k = \frac{y}{x}$. When you solve for the constant of variation k in an equation of inverse variation, $k = xy$.

EXAMPLE 1

TRAVEL José recorded the time it took him to travel from home to the state capital driving at several different speeds. Do the results suggest that the time varies inversely as the speed?

Speed, s (mi/h)	60	50	40	30
Time, t (h)	2	2.4	3	4

Solution

If the time varies inversely as the speed, then the product st will be a constant. Calculate st for the given data.

$$s \cdot t$$
$$60 \cdot 2 \ = 120$$
$$50 \cdot 2.4 = 120$$
$$40 \cdot 3 \ = 120$$
$$30 \cdot 4 \ = 120$$

In each case, $st = 120$, which suggests that for any given distance, time varies inversely as the speed.

Proportions are useful when solving problems involving inverse variation. If the ordered pairs (x_1, y_1) and (x_2, y_2) satisfy the equation $xy = k$, then $x_1y_1 = k$ and $x_2y_2 = k$. Therefore,

$$x_1y_1 = x_2y_2$$
$$\frac{x_1}{x_1} = \frac{y_1}{y_2}$$

EXAMPLE 2

If y varies inversely as x, and $y = 2$ when $x = 50$, determine the value of y when $x = 200$. You can use the proportion method or the equation method.

COMMUNICATING ABOUT ALGEBRA

Can you use the proportion

$$\frac{x_1}{x_2} = \frac{y_2}{y_1}$$

to solve a direct variation? If so, explain why. If not, what proportion can you use?

Proportion Method

$$\frac{50}{200} = \frac{y}{2} \qquad \frac{x_1}{x_2} = \frac{y_2}{y_1}$$

$$\frac{50(2)}{200} = y$$

$$\frac{1}{2} = y$$

Equation Method

$$xy = k$$
$$50(2) = k \qquad \text{Substitute } x = 50, y = 2.$$
$$100 = k$$

$$xy = k$$
$$200y = 100 \qquad \text{Substitute 200 for } x \text{ and 100 for } k.$$
$$y = \frac{100}{200}$$
$$y = \frac{1}{2}$$ ◀

Sometimes one variable varies directly as the product of two or more variables. This is called *joint variation*.

> ┌─ **JOINT VARIATION** ─────────
> When *y* varies directly as the product of *w* and *x*, and *k* is the constant of variation, an equation can be written in the form
> $$y = kwx, k \neq 0$$

CHECK UNDERSTANDING

The formula for the area of a triangle, $A = \frac{1}{2}bh$, shows joint variation. The area increases as either the base or the height increases.

EXAMPLE 3

If *y* varies jointly as *x* and *z*, and $y = 144$ when $x = 3$ and $z = 12$, find *y* when $x = 5$ and $z = 11$.

Solution

$$y = kxz \qquad \text{Write the equation.}$$
$$144 = k(3)(12) \qquad \text{Substitute } x = 3 \text{ and } z = 12.$$

$$\frac{144}{(3)(4)} = k \qquad \text{Solve for } k.$$

$$4 = k \qquad \text{Simplify.}$$

$$y = kxz \qquad \text{Write the equation.}$$
$$y = 4(5)(11) \qquad \text{Substitute } k = 4, x = 5, \text{ and } z = 11.$$
$$y = 220 \qquad \text{Simplify.}$$ ◀

State whether each equation represents *direct variation*, *inverse variation*, or *joint variation*. Also, state the constant of variation.

1. $s = 2qr$ **2.** $d = 2r$ **3.** $xy = 10$

Determine whether the data in each table suggests that y varies inversely as x. If it does, state the constant of variation.

4.

x	1	2	4	8	16
y	8	4	2	1	$\dfrac{1}{2}$

5.

x	2	3	4	5	6
y	30	20	15	12	10

Write an equation for each variation using k as the constant of variation.

6. w varies inversely as x

7. c varies inversely as the square of d

8. f varies jointly as g and h

In Exercises 9–10, y varies inversely as x.

9. If $y = 10$ when $x = 50$, find y when $x = 20$.

10. If $y = 8$ when $x = 12$, find x when $y = 6$.

In Exercises 11–12, y varies jointly as x and z.

11. If $y = 30$ when $x = 3$ and $z = 5$, find y when $x = 12$ and $z = \dfrac{1}{2}$.

12. If $y = 75$ when $x = 25$ and $z = 6$, find y when $x = 30$ and $z = 4$.

13. SIMPLE MACHINES The force F needed to pry open a crate varies inversely as the length l of the crowbar used. When the length is 2 m, the force needed is 12 N (newtons). What force would be needed if the crowbar was 1.6 m long?

14. ELECTRICITY The energy an appliance uses varies directly as the time it is in operation. A stereo that operates for 21 h a week uses 2.3 kWh (kilowatt-hours) of electrical energy. How much energy would the stereo use for 80 h of operation? Round to two decimal places.

15. WRITING MATHEMATICS The area of a triangle is 24 cm². Write a paragraph that describes how its base varies when the height increases or decreases. Determine the constant of variation and include that in your discussion.

PRACTICE

State whether each equation represents *direct variation*, *inverse variation*, or *joint variation*. State the constant of variation.

1. $a = 9.5b$

2. $c = \dfrac{1}{2}de$

3. $lw = 72$

4. $g = 4ef$

5. $u = 0.14v$

6. $q = \dfrac{16}{r}$

In Exercises 7–10, y varies inversely as x.

7. If $y = 12$ when $x = 60$, find y when $x = 15$.

8. If $y = 15$ when $x = 20$, find y when $x = 30$.

9. If $y = 5$ when $x = 14$, find x when $y = 10$.

10. If $y = 9$ when $x = 50$, find x when $y = 15$.

In Exercises 11–14, y varies jointly as x and z.

11. If $y = 45$ when $x = 9$ and $z = 6$, find y when $x = 10$ and $z = 6$.

12. If $y = 75$ when $x = 25$ and $z = 6$, find y when $x = 30$ and $z = 4$.

13. If $y = 600$ when $x = 50$ and $z = 24$, find y when $x = 50$ and $z = 84$.

14. If $y = 100$ when $x = 10$ and $z = 40$, find y when $x = 28$ and $z = 14$.

GEOMETRY Write an equation for each variation with k as the constant of variation.

15. The diameter of a circle varies directly as the radius of the circle.

16. The area of a triangle varies jointly as its base and height.

17. In a rectangle with a constant area, the width varies inversely as the length.

18. WRITING MATHEMATICS Write an equation such that y varies inversely as x. Describe a situation about agriculture for your equation.

EXTEND

In some cases, one quantity varies directly or inversely as the square of another.
For example, the area of a circle varies directly as the square of the radius, as shown in the formula $A = \pi r^2$. This function is an example of **quadratic direct variation**.

19. Assume that y varies directly as x^2. When $x = 7$, $y = 245$. Find y when $x = 11$.

20. Assume that y varies inversely as x^2. When $x = 9$, $y = 16$. Find y when $x = 6$.

21. BIOLOGY The number of organisms in a culture varies directly as the square of the time the culture has been growing. A culture growing for 25 min has 5,000 organisms. About how long will it take for there to be 10,000 organisms?

THINK CRITICALLY

COORDINATE GEOMETRY The graph of an inverse variation of the form $y = \dfrac{k}{x}$ or $xy = k$, $x \neq 0$, $k \neq 0$ is a *hyperbola*. The graphs of $xy = 1$, $xy = 4$, and $xy = 10$ are displayed at the right.

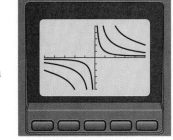

22. In which quadrants is the graph of $xy = k$ located if k is greater than zero? if k is less than zero?

23. What are the lines of symmetry of the graph of $xy = k$?

24. Determine the Quadrant I point on the graph of $xy = 1$ that is closest to the origin. (*Hint:* On what line does the point closest to the origin lie? What is the relationship of the x- and y-coordinates of this point?)

25. Determine the Quadrant I point on the graph of $xy = 4$ that is closest to the origin.

26. Determine the Quadrant I point on the graph of $xy = 10$ that is closest to the origin.

MIXED REVIEW

27. **STANDARDIZED TESTS** An 80-mL solution of water and acid contains 30% acid. How much water must you add to make it a 15% acid solution?

    A. 40 mL   **B.** 80 mL   **C.** 24 mL   **D.** 160 mL

Express the equation of each line in slope-intercept form.

28. $6x + 3y - 6 = 12$ 29. $2x + 15 = 3y + 9$

Use the algebraic method to solve each absolute value inequality.

30. $|2x - 6| > 12$ 31. $|4x + 8| < 16$

In Exercises 32–33, y varies inversely as x.

32. If $y = 15$ when $x = 5$, find y when $x = 25$.

33. If $y = 45$ when $x = 5$, find x when $y = 9$.

PROJECT Connection   Use the data collected during the Getting Started activity on page 674.

1. Determine the product of each pair of values for the height and circumference of the base. What do you notice?

2. Explain why this happens, in terms of the physical aspects of the situation.

3. Express the height h as a function of the circumference C of the base.

4. Does the equation you wrote in Question 3 represent a variation and, if so, what type of variation?

Some branches of agribusiness require the use of refrigeration. Most refrigerators use a compression system that operates on two principles. First, as liquid evaporates, it absorbs heat from the surrounding air. Second, the temperature T of an enclosed gas varies jointly with the volume V and pressure P; that is, $T = kPV$. The diagram below shows the refrigeration process.

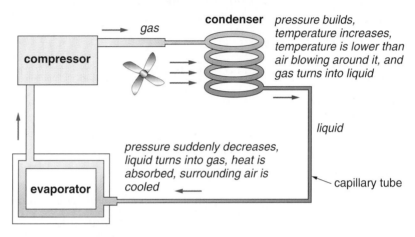

condenser *pressure builds, temperature increases, temperature is lower than air blowing around it, and gas turns into liquid*

gas

compressor

liquid

pressure suddenly decreases, liquid turns into gas, heat is absorbed, surrounding air is cooled

evaporator

capillary tube

Decision Making

The engineer who designs the cooling system must choose the diameter of the capillary tube. For a given compressor and refrigerant, different diameters require different lengths of tubing. For example, the table shows the dimensions required to maintain high temperatures using a particular compressor and refrigerant.

Inside Diameter, in.	Length of Tubing, ft
0.031	1.1
0.036	2.2
0.040	3.5
0.042	4.5
0.049	9.0
0.055	15.0

1. Use the relationship $T = kPV$ to explain why a larger diameter requires more tubing to achieve the same temperature results.

One size capillary tube that is often used for refrigerant R-12 has an outside diameter of 0.114 in. and an inside diameter of 0.049 in.

2. Determine the circumference of the inside and outside of the tube.

3. If temperature is held constant, the volume of a gas varies inversely as the pressure. When the volume is 120 in.³, the pressure is 30 lb/in.³ If the pressure changes to 20 lb/in.³, what is the volume?

4. If pressure is held constant, the volume of a gas varies directly with its temperature (in degrees Kelvin). If 10 m³ of a gas is kept under constant pressure while its temperature drops from 405°K to 345°K, approximately what is its new volume?

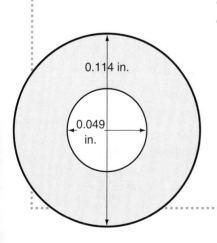

0.114 in.

0.049 in.

14.3 Simplify Rational Expressions

Explore

Use Algeblocks to make the rectangular solids shown below. Determine the surface area and volume of each solid. Then write the ratio of the surface area to the volume.

1.

2.

3.

4.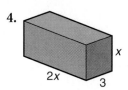

SPOTLIGHT ON LEARNING

WHAT? In this lesson you will learn
- to determine the values of the variable for which a rational expression is undefined.
- to simplify rational expressions.

WHY? Simplifying rational expressions can help you solve problems about interior design, transportation, and farming.

Build Understanding

Recall that a *rational number* is a number that can be expressed in the form $\frac{a}{b}$, where a and b are integers, $b \neq 0$. The denominator b cannot equal zero because division by zero is undefined. Examples of rational numbers include $\frac{0}{8}$, $-\frac{14}{42}$, $\frac{3}{1}$, $0.12121\ldots$, and 23.

> **RATIONAL EXPRESSION**
>
> A rational expression is an expression that can be written in the form $\frac{P}{Q}$ where P and Q are polynomials, $Q \neq 0$.

THINK BACK

A polynomial is a monomial or a sum or difference of monomials.

Examples of rational expressions include $\frac{1}{x}$, $4y + 2$, $\frac{x}{y^2 - 5}$, $\frac{3x^2 - 7}{x + 3}$, and $\frac{17}{77}$. When the denominator of a rational expression is zero, the expression is undefined. You sometimes need to factor the denominator to find the values of the variable that will make the denominator zero.

EXAMPLE 1

For which values of the variable is each expression undefined?

a. $\dfrac{2x^2y}{zw}$

b. $\dfrac{4xt}{t + 3}$

c. $\dfrac{3x}{4x^2 - 9}$

Solution

a. $zw = 0$

  $z = 0 \quad w = 0$

b. $t + 3 = 0$

  $t = -3$

c. $\qquad\qquad 4x^2 - 9 = 0$

  $(2x - 3)(2x + 3) = 0$ Factor.

  $2x - 3 = 0 \qquad\qquad 2x + 3 = 0$

  $x = \dfrac{3}{2} \qquad\qquad x = -\dfrac{3}{2}$ ◀

When you determine the values for a variable for which an expression is undefined, you determine the *restrictions* on that variable. The restrictions on x in Example 1c are $x \neq \dfrac{3}{2}$ and $x \neq -\dfrac{3}{2}$.

When the numerator and denominator of a rational expression have no common factors other than 1, the rational expression is in **simplest form**. As with simplifying rational numbers, to simplify a rational expression, factor the numerator and denominator. Then divide each by the greatest common factor (GCF).

THINK BACK

Recall that usually any real number can be substituted for a variable.

EXAMPLE 2

State any restrictions on the variable x. Then simplify: $\dfrac{4 + x}{2x^2 + 7x - 4}$

Solution

$$\dfrac{4 + x}{2x^2 + 7x - 4} = \dfrac{\overset{1}{\cancel{(4 + x)}}}{(2x - 1)\underset{1}{\cancel{(x + 4)}}} \qquad \begin{array}{l}\text{Factor the denominator and} \\ \text{divide by } (x + 4)\text{, the GCF.}\end{array}$$

$$= \dfrac{1}{2x - 1} \qquad\qquad \text{Simplify.}$$

So, $\dfrac{4 + x}{2x^2 + 7x - 4} = \dfrac{1}{2x - 1}$.

Determine the restrictions on x. Set each factor in the denominator equal to zero.

  $2x - 1 = 0 \qquad\qquad x + 4 = 0$

  $x = \dfrac{1}{2} \qquad\qquad x = -4$

So, the restrictions on x are $x \neq \dfrac{1}{2}, x \neq -4$. ◀

CHECK UNDERSTANDING

In Example 2, if the simplified form of the expression is $\dfrac{1}{2x - 1}$, why must the restriction $x \neq -4$ be included?

Because of the property of -1, $y - x = -1(x - y)$. Therefore,

$$\dfrac{y - x}{x - y} = -1$$

So, you can replace $\dfrac{y - x}{x - y}$ with -1 in a rational expression.

EXAMPLE 3

State any restrictions on the variable x. Then simplify: $\dfrac{12 - 3x}{x^2 - x - 12}$

Solution

$$\frac{12 - 3x}{x^2 - x - 12} = \frac{3(4 - x)}{(x + 3)(x - 4)} \qquad \text{Factor.}$$

$$= \frac{3(-1)(x - 4)}{(x + 3)(x - 4)} \qquad \begin{array}{l}\text{Property of } -1. \\ \text{Divide out common factor.}\end{array}$$

$$= -\frac{3}{x + 3} \qquad \text{Simplify.}$$

So, $\dfrac{12 - 3x}{x^2 - x - 12} = -\dfrac{3}{x + 3}, x \neq -3, x \neq 4.$ ◄

Rational expressions can be used in real world settings.

EXAMPLE 4

INTERIOR DESIGN When the law
firm of C & V had only a few
lawyers, partners and associates
each received an office of
dimensions x ft by x ft. Now that
the firm has grown, they will be
taking on more space in their
building and enlarging the
offices. Each office will be 3 ft
wider, and partners' offices will
also be 6 ft longer. Find a
rational expression for the ratio
of the new area of a partner's

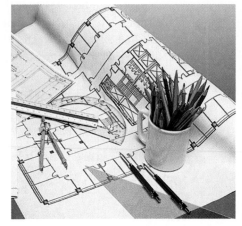

office to the new area of an associate's office. Evaluate the ratio
if the current length of an office is 11 ft.

Solution
The area for a partner's office will be $(x + 6)(x + 3)$.
The area for an associate's office will be $x(x + 3)$.

$$\text{Ratio} = \frac{\text{Area of partner's office}}{\text{Area of associate's office}}$$

$$= \frac{(x + 6)(x + 3)}{x(x + 3)}$$

$$= \frac{(x + 6)}{x}$$

For $x = 11$, the ratio is $\dfrac{17}{11}$, the ratio of the partner's office area to the
associate's office area. ◄

State the values of the variable for which each expression is undefined.

1. $\dfrac{6}{2x}$

2. $\dfrac{x}{3x + 6}$

3. $\dfrac{2x + 6}{(x + 3)(x - 7)}$

4. $\dfrac{x + 7}{x^2 - 2x - 35}$

Simplify each expression. State any restrictions on the variable.

5. $\dfrac{6x}{3x^2}$

6. $\dfrac{5x^2}{25}$

7. $\dfrac{4x - 12}{24}$

8. $\dfrac{6a - 12}{a - 2}$

9. $\dfrac{a - 7}{7a - 49}$

10. $\dfrac{3 - 2b}{6b - 9}$

11. $\dfrac{5 + 2b}{-8b - 20}$

12. $\dfrac{3 + x}{x^2 - 2x - 15}$

13. $\dfrac{4 + x}{x^2 + 10x + 24}$

14. TRANSPORTATION Steve travels x miles each way in his car to visit his grandparents. It takes him y hours to get there and z hours to get home. Express his average speed for the round trip in terms of x, y, and z.

15. WRITING MATHEMATICS Explain why $\dfrac{y^2 - 25}{y - 5}$ is not the same as $y + 5$.

PRACTICE

State the values of the variable for which each expression is undefined.

1. $\dfrac{7}{9mn}$

2. $\dfrac{3}{pqr}$

3. $\dfrac{y}{5y + 15}$

4. $\dfrac{6x}{18x - 12}$

5. $\dfrac{z - 5}{(z - 1)(z + 8)}$

6. $\dfrac{3q + 8}{(q + 6)(q - 4)}$

7. $\dfrac{2x + 13}{x^2 + 2x - 48}$

8. $\dfrac{3y - 11}{y^2 - 10y - 56}$

Simplify each expression. State any restrictions on the variable.

9. $\dfrac{8x}{4x^2}$

10. $\dfrac{10x}{5x^2}$

11. $\dfrac{3y - 12}{15}$

12. $\dfrac{6y + 15}{12}$

13. $\dfrac{4a - 16}{a - 4}$

14. $\dfrac{a - 9}{9a - 81}$

15. $\dfrac{4 - 8b}{40b - 20}$

16. $\dfrac{4 - 3b}{9b - 12}$

17. $\dfrac{2 + x}{x^2 - 5x - 14}$

18. BAKING The baking time for a loaf of bread can be expressed as the ratio of the surface area to the volume. Calculate this ratio for a rectangular loaf of bread with dimensions $4x$ by x by x.

19. MANUFACTURING The efficiency of a machine is the ratio of the work output to the work input. Both are measured in joules. A machine has a work output of $4x^2 + 26x + 12$ joules and a work input of $2x + 1$ joules. Determine the efficiency.

20. WRITING MATHEMATICS Compare simplifying $\dfrac{30x^3}{2x}$ to simplifying $\dfrac{6x^2 - 24x - 30}{4x - 20}$.

EXTEND

Recall that some polynomial expressions must be factored more than once before they are considered to be in simplest form. Be sure to factor completely when factoring rational expressions.

Simplify each expression. State any restrictions on the variable.

21. $\dfrac{x^4 - 1}{x^2 - 2x + 1}$

22. $\dfrac{x^3 - 6x^2 + 9x}{x^3 - 9x}$

23. $\dfrac{x^5 + 10x^4 + 25x^3}{x^3 + 4x^2 - 5x}$

THINK CRITICALLY

24. Write an algebraic expression for which the values $\dfrac{1}{2}, -\dfrac{1}{4}$, and 0 must be excluded.

25. Provide an example of a rational expression, the simplified form of which is $\dfrac{1}{x}$, and which has an x^3 term in the numerator.

26. Simplify: $\dfrac{x^{2n} + 7x^n + 6}{x^{2n} + 9x^n + 8}$

27. Given the equation $\dfrac{(x + y)^q}{(x + y)^p} = x^2 + 2xy + y^2$, write an equation that expresses the relationship between p and q.

MIXED REVIEW

28. STANDARDIZED TESTS Choose the correct product.

$(4 \cdot 10^5)(1.5 \cdot 10^4)$

A. $5.5 \cdot 10^9$ **B.** $6 \cdot 10^{20}$ **C.** $6 \cdot 10^9$ **D.** $6 \cdot 10^{10}$

Write the equation of the line that contains each set of points.

29. $(4, 7)$ and $(-2, -3)$

30. $(3, 6)$ and $(4, -8)$

Tell whether the graph of each parabola opens upward or downward.

31. $y = -6x^2 + 3$

32. $y = 4x^2 - 3$

PROJECT *Connection*   In this activity, you will use the formula for the volume of a cylinder, $V = \pi r^2 h$.

1. In the formula $C = 2\pi r$, C represents the circumference and r represents the radius. Express r in terms of C. Then use your result to express V in terms of C and h.

2. In the Project Connection on page 684 you determined that $h = \dfrac{93.5}{C}$. Substitute this value for h into the new volume formula you wrote in Question 1.

3. Graph the rational function that represents the volume. What are the restrictions on C? Describe and interpret the graph.

4. What are the practical implications of your findings? What dimensions do you recommend for a silo? Justify your response.

Farmers use hay and grain to feed their cattle. The following spreadsheet shows possible mixtures of hay and grain. Note that for each increase of 100 lb of hay, the amount of grain required decreases.

	A	B	C	D	E
1	Hay	Change in	Grain	Change in	Ratio of Change
2	(lb)	Hay (lb)	(lb)	Grain (lb)	in Grain to
3					Change in Hay
4	1000		1316		
5	1100	100	1259	57	0.57
6	1200	100	1208		
7	1300	100	1162		
8	1400	100	1120		
9	1500	100	1081		
10	1600	100	1046		
11	1700	100	1014		

Decision Making

1. Determine the change in the pounds of grain (Column D) for each 100-lb increase in hay. What formula could you use to calculate this in a spreadsheet?

2. Determine the change in pounds of grain per 100-lb change in hay (Column E). What formula could you use to calculate this in a spreadsheet?

The *price ratio* is the ratio of the price of hay to the price of grain. For example, if the price of grain is $0.15/lb and the price of hay is $0.06/lb, the price ratio of hay to grain is 0.06/0.15 = 0.4. The lowest cost combination occurs when the ratio in Column E is equal to or slightly greater than the price ratio.

3. Determine the first number greater than or equal to 0.4 in Column E. How many pounds of hay and grain does that represent?

4. If the price of hay climbs to $0.0725/lb, determine the lowest cost combination.

5. If the price of grain climbs to $0.165/lb and the price of hay is $0.06/lb, determine the lowest cost combination.

6. If the lowest cost combination is 1800 lb of hay and 984 lb of grain and the price of grain is $0.175/lb, determine the price of hay.

Multiply and Divide Rational Expressions

Explore

SPOTLIGHT ON LEARNING

WHAT? In this lesson you will learn
- to multiply and divide rational expressions.
- to find products and quotients of rational expressions in real world settings.

WHY? Multiplying and dividing rational expressions can help you solve problems about physics, geometry, and finance.

Let $x = 5$ for each of the following. Substitute the value for x. Then multiply the two rational numbers. Recall that to multiply two rational numbers, you multiply the numerators and then multiply the denominators. Simplify your answers.

1. $\dfrac{x}{6} \cdot \dfrac{24}{x}$

2. $\dfrac{3x^2}{4} \cdot \dfrac{8}{12x}$

3. $\dfrac{x+6}{x} \cdot \dfrac{2}{x+6}$

4. Multiply $\dfrac{x}{6} \cdot \dfrac{24}{x}$ by multiplying the numerators and then multiplying the denominators as you would if you were multiplying two rational numbers. What is the result? Simplify the expression. How does the simplified expression compare to your answer for Question 1?

5. Multiply $\dfrac{3x^2}{4} \cdot \dfrac{8}{12x}$ by the process you used in Question 4. Simplify the expression. How does the simplified expression compare to your answer for Question 2?

6. Use your answer for Question 2 to predict the expression resulting from multiplying the numerators and denominators of the factors in Question 3 and simplifying. Test your prediction.

Build Understanding

As you saw in Explore, the method for finding the product of rational expressions is similar to the method for finding the product of rational numbers.

EXAMPLE 1

Multiply: $\dfrac{6a^2}{5b} \cdot \dfrac{b}{3d}$

Solution

$$\dfrac{6a^2}{5b} \cdot \dfrac{b}{3d}$$

$$= \dfrac{6a^2 \cdot b}{5b \cdot 3d} \qquad \text{Definition of multiplication.}$$

$$= \dfrac{\overset{2}{\cancel{6}}a^2 \cdot \cancel{b}}{5\cancel{b} \cdot \cancel{3}d} \qquad \text{Divide out common factors.}$$

$$= \dfrac{2a^2}{5d} \qquad \text{Simplify.}$$

◀

When multiplying rational expressions, the polynomials need to be factored completely before you divide out any common factors.

EXAMPLE 2

Multiply: $\dfrac{x}{x+2} \cdot \dfrac{3x+6}{x^2-4x}$

Solution

$$\dfrac{x}{x+2} \cdot \dfrac{3x+6}{x^2-4x}$$

$$= \dfrac{x}{(x+2)} \cdot \dfrac{3(x+2)}{x(x-4)} \qquad \text{Factor.}$$

$$= \dfrac{\cancel{x}}{\cancel{(x+2)}} \cdot \dfrac{3\cancel{(x+2)}}{\cancel{x}(x-4)} \qquad \begin{array}{l}\text{Divide out}\\\text{common factors.}\end{array}$$

$$= \dfrac{3}{(x-4)} \qquad \text{Simplify.} \quad \blacktriangleleft$$

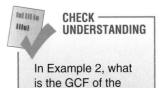

CHECK UNDERSTANDING

In Example 2, what is the GCF of the numerator and denominator?

The method for determining the quotient of rational expressions is similar to the method for determining the quotient of rational numbers. To divide one rational expression by another, multiply the first by the reciprocal of the second. After that, proceed as in multiplication, factoring polynomials completely and dividing out common factors.

EXAMPLE 3

Divide: $\dfrac{(x+2)}{x} \div \dfrac{(x-2)}{x^2}$

Solution

$$\dfrac{(x+2)}{x} \div \dfrac{x-2}{x^2}$$

$$= \dfrac{(x+2)}{x} \cdot \dfrac{x^2}{(x-2)} \qquad \begin{array}{l}\text{Multiply by the reciprocal}\\\text{of the divisor.}\end{array}$$

$$= \dfrac{(x+2)}{\cancel{x}} \cdot \dfrac{\cancel{x}\cdot x}{(x-2)} \qquad \begin{array}{l}\text{Divide out } x,\text{ the}\\\text{common factor.}\end{array}$$

$$= \dfrac{x(x+2)}{(x-2)} \qquad \text{Simplify.} \quad \blacktriangleleft$$

To multiply a rational expression by a polynomial such as $3x+1$, for example, rewrite the polynomial as $\dfrac{3x+1}{1}$. To divide a rational expression by $3x+1$, multiply the rational expression by the reciprocal of $3x+1$, or $\dfrac{1}{3x+1}$.

ALGEBRA: WHO, WHERE, WHEN

In the early 17th century, German astronomer Johannes Kepler discovered that the orbits of the planets were not circular but elliptical. He also discovered the direct relation between the distance between the sun and a planet and the rate at which the planet moves in its orbit. Specifically, the cube of the mean distance d between a planet and the sun divided by the square of its orbital period t is constant $\left(\dfrac{d^3}{t^2}\right)$.

This is critical in understanding the paths of moons, planets, and our satellites launched from Earth.

EXAMPLE 4

PHYSICS Newton's second law states that the force in newtons N is equal to the product of the mass in kg and the acceleration in m/s^2. So, $F = ma$.

If the force on an object is $\left(\dfrac{x^2 - 16}{x - 2}\right)$ newtons and the mass is $(6x - 24)$ kilograms, determine the acceleration of the object.

Solution

$$a = F \div m \qquad \text{Since } F = ma.$$

$$a = \frac{x^2 - 16}{x - 2} \div (6x - 24)$$

$$= \frac{x^2 - 16}{(x - 2)} \cdot \frac{1}{6x - 24} \qquad \text{Multiply by the reciprocal.}$$

$$= \frac{(x + 4)(x - 4)}{(x - 2)} \cdot \frac{1}{6(x - 4)} \qquad \text{Factor.}$$

$$= \frac{(x + 4)(x \cancel{- 4})}{(x - 2)} \cdot \frac{1}{6(x \cancel{- 4})} \qquad \text{Divide out common factors.}$$

$$= \frac{(x + 4)1}{(x - 2)6} \qquad \text{Multiply.}$$

$$= \frac{x + 4}{6x - 12}$$

The acceleration of the object is $\dfrac{x + 4}{6x - 12}$ m/s^2. ◀

TRY THESE

Perform the indicated operation.

1. $\dfrac{3x}{5} \cdot \dfrac{10}{12x}$

2. $\dfrac{4x^2}{7} \cdot \dfrac{14}{5x}$

3. $\dfrac{1}{4n} \div \dfrac{6n}{15}$

4. $\dfrac{7r^2}{5} \div \dfrac{3r}{21}$

5. $\dfrac{ab^2}{c} \cdot \dfrac{3c^2}{b}$

6. $\dfrac{2u^2}{v^2w} \cdot \dfrac{vw^2}{5}$

7. $\dfrac{4d^2}{7e} \div \dfrac{8d}{e^2}$

8. $\dfrac{3q^2r^2}{4} \div \dfrac{9qr}{3}$

9. $\dfrac{m - 3}{6(m + 4)} \cdot \dfrac{3(m + 4)}{m - 3}$

10. $\dfrac{x - 6}{8x + 12} \cdot \dfrac{10x + 15}{3x - 18}$

11. $\dfrac{(3x + 6)(x - 1)}{12x} \div \dfrac{(x + 2)}{8}$

12. $\dfrac{3x^2 - 10x - 8}{6x} \div \dfrac{2x^2 - 32}{-5x - 20}$

13. **WRITING MATHEMATICS** Write a paragraph that compares and contrasts multiplying $\dfrac{2}{3} \cdot \dfrac{1}{4}$ to multiplying $\dfrac{2}{3x} \cdot \dfrac{x}{4}$.

14. **PHYSICS** In physics, the term *work* represents a force
    multiplied by the distance a body moves in the
    direction of the force applied. That is, $W = Fd$,
    where W represents work, or energy, in joules,
    F represents force in newtons, and d is the distance
    through which the force is applied
    (1 joule = 1 newton • 1 meter).

    If a body moves $\dfrac{x^2 + 13x + 42}{(x + 6)}$ meters in the

    direction of a force of $\left(\dfrac{x + 4}{x^2 - 16}\right)$ newtons,

    determine the work.

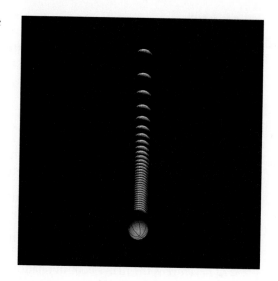

PRACTICE

Perform the indicated operation.

1. $\dfrac{5x}{7} \cdot \dfrac{14}{6x}$

2. $\dfrac{1}{6x} \cdot \dfrac{18x^2}{11}$

3. $\dfrac{1}{5k} \div \dfrac{3}{20k^2}$

4. $\dfrac{8n}{5} \div \dfrac{5}{24n^2}$

5. $\dfrac{a^2b}{c} \cdot \dfrac{2ac^2}{a}$

6. $\dfrac{wu^2}{v^2w} \cdot \dfrac{v^2w}{3u}$

7. $\dfrac{5c^2d}{9ce^2} \div \dfrac{15c^2d^2}{18e}$

8. $\dfrac{8qr^2}{12qr} \div \dfrac{9qrs}{6s}$

9. $\dfrac{m - 5}{2(m + 6)} \cdot \dfrac{4(m + 6)}{8(m - 5)}$

10. $\dfrac{x - 3}{6(2x + 1)} \cdot \dfrac{3(2x + 1)}{4(x - 3)}$

11. $\dfrac{4(x + 3)(x - 2)}{6x} \div \dfrac{(x + 3)(x - 3)}{3(x - 3)}$

12. $\dfrac{(2x + 3)(x - 5)}{4x} \div \dfrac{3(x + 5)(x - 5)}{6(x + 5)}$

13. $\dfrac{7x + 35}{3x^2 - 108} \cdot (6x + 36)$

14. $\dfrac{8x + 32}{4x^2 - 100} \cdot (3x - 15)$

15. $\dfrac{5x^2 - 49}{x^2 - x - 42} \div (3x + 21)$

16. $\dfrac{x^2 - 16}{x^2 - 7x + 12} \div (5x + 20)$

17. $\dfrac{3x^2 - 17x - 6}{3x^2 - 108} \cdot \dfrac{5x}{-21x - 7}$

18. $\dfrac{2x^2 + x - 10}{5x^2 - 20} \cdot \dfrac{5x}{-6x - 15}$

19. $\dfrac{x^2 + x - 6}{x^2 - 9} \div \dfrac{x^2 - 4}{7x - 21}$

20. $\dfrac{x^2 + 5x - 36}{x^2 - 81} \div \dfrac{x^2 - 16}{6x - 54}$

GEOMETRY Find the area of each rectangle.

21. length: $\dfrac{6x - x^2}{2x + 4}$, width: $\dfrac{x^2 - 4}{36 - x^2}$

22. length: $\dfrac{a^2 - 1}{ab^2 - b}$, width: $\dfrac{b}{3 - 3a}$

23. **WRITING MATHEMATICS** Explain or make a flowchart of the steps you would use to divide
    one rational expression by another rational expression.

To simplify rational expressions that involve both multiplication and division, perform operations from left to right unless otherwise indicated by parentheses.

Perform the indicated operation.

24. $\dfrac{18 - 4x}{3x + 2} \div \dfrac{6x - 18}{-(6x + 4)} \cdot \dfrac{3x - 9}{81 - 4x^2}$

25. $\dfrac{t^2 - t}{t^2 - 2t - 3} \cdot \dfrac{t^2 + 2t + 1}{t^2 + 4t} \div \dfrac{t^2 - 3t - 4}{2t^2 - 32}$

FINANCE Before making a loan to a company, a bank conducts a *ratio analysis* to determine how the company is doing in the marketplace.

26. The *current ratio*, the ratio of current assets to current liabilities, is one measure of the liquidity (convertibility of assets to cash) of the company. If the current assets can be represented by $\dfrac{2x^2 + 4x - 30}{2x^2 - 18}$ and the liabilities by $\dfrac{3x + 15}{4x + 12}$, determine the current ratio.

27. *Leverage* is the firm's debt in relation to its equity. If the debt can be represented by $\dfrac{2x^2 - 2x - 4}{4x - 8}$ and the equity by $\dfrac{x^2 - 1}{4x - 4}$, determine the *debt to equity ratio*.

28. *Profitability ratios* measure how efficiently a company is managed. The *profit margin* is the ratio of the gross profit to the net sales. If the net sales can be represented by $10x^2 + 30x + 20$ and the gross profit by $\dfrac{x^3 + 3x^2 + 2x}{x}$, determine the profit margin.

THINK CRITICALLY

29. Find two different pairs of rational expressions whose product is $\dfrac{x^2 + 7x + 10}{3x^2 - 3}$.

30. Find two different pairs of rational expressions whose product is $\dfrac{m^2 - m - 6}{2m^2 + 5m - 3}$.

31. Find two different pairs of rational expressions whose quotient is $\dfrac{3x}{x - 5}$.

32. Find two different pairs of rational expressions whose quotient is $\dfrac{x + 4}{x + 6}$.

MIXED REVIEW

33. STANDARDIZED TESTS The absolute value equation $|3 - 4x| = 12$ can be expressed as

    A. $3 - 4x = 12$ and $3 - 4x = -12$ **B.** $3 - 4x = 12$ or $-3 + 4x = -12$

    C. $3 - 4x = 12$ or $3 - 4x = -12$ **D.** $3 - 4x > 12$ and $3 - 4x < -12$

Find the constant of variation k if y varies directly as x.

34. $y = 8.1, x = 9$

35. $y = 9.9, x = 3.3$

Solve each inequality.

36. $-3x > x - 16$

37. $6x + 12 < 3x + 6$

Perform the indicated operation.

38. $\dfrac{x^2 + 2x - 15}{25 - x^2} \cdot \dfrac{x^2 - 4x - 5}{x^2 - 2x - 3}$

39. $\dfrac{x - 3}{x - 6} \div \dfrac{x^2 - 9}{x^2 - 2x - 24}$

14.5 Divide Polynomials

Explore

● Use Algeblocks and the quadrant mat to divide a polynomial by a monomial .

Divide: $\dfrac{2x^2 - 4x}{2x}$

Place the divisor, $2x$, in the horizontal axis.

Use tiles for $2x^2 - 4x$ to form rectangular areas with $2x$ as their boundaries.

Determine the other dimension of the rectangular area. Read the answer from the mat: $x - 2$.

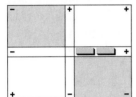
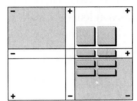
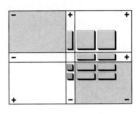

> **SPOTLIGHT ON LEARNING**
>
> **WHAT?** In this lesson you will learn
> • to divide one polynomial by another polynomial.
>
> **WHY?** You can use division of polynomials to solve problems in geometry, automobile depreciation, and biology.

Using Algeblocks, you can divide a polynomial by a binomial.

Divide: $\dfrac{2x^2 - 8x}{x - 4}$

Place the divisor, $x - 4$, in the horizontal axis.

Use tiles for $2x^2 - 8x$ to form rectangular areas with $x - 4$ as their boundaries.

Determine the other dimension of the rectangular area. Read the answer from the mat, $2x$.

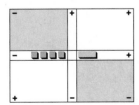
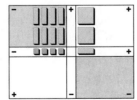
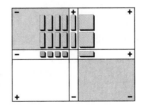

Use Algeblocks and a quadrant mat to divide.

1. $\dfrac{3x^2 - 6x}{3x}$

2. $\dfrac{x^2 + 2x}{x}$

3. $\dfrac{3x^2 + 6x}{x + 2}$

4. $\dfrac{x^2 - x}{x - 1}$

Build Understanding

Recall that to divide a polynomial by a monomial, each term of the polynomial must be divided by the monomial.

EXAMPLE 1

Divide: $(12x^4 + 4x^3 - x^2) \div 4x^3$

Solution

$$\frac{12x^4 + 4x^3 - x^2}{4x^3}$$

$$= \frac{12x^4}{4x^3} + \frac{4x^3}{4x^3} - \frac{x^2}{4x^3} \quad \text{Divide each term.}$$

$$= 3x + 1 - \frac{1}{4x} \qquad \blacktriangleleft$$

To find the quotient of two polynomials with common factors, use the techniques you learned earlier in the chapter. To divide one polynomial by another when they have no common factors, use long division. Note that

$$\frac{\text{dividend}}{\text{divisor}} = \text{quotient} + \frac{\text{remainder}}{\text{divisor}}, \text{ or}$$

$$\text{dividend} = \text{quotient} \cdot \text{divisor} + \text{remainder}$$

EXAMPLE 2

Divide and check: $\dfrac{3a^2 + 2a + 4}{a - 2}$

Solution

$$
\begin{array}{r}
3a + 8 \\
a - 2 \overline{\smash{)}\, 3a^2 + 2a + 4} \\
\underline{3a^2 - 6a} \\
8a + 4 \\
\underline{8a - 16} \\
20
\end{array}
$$

Divide: Think $3a^2 \div a = 3a$.
Multiply: $(a - 2)(3a) = 3a^2 - 6a$.
Subtract: $(3a^2 + 2a) - (3a^2 - 6a) = 8a$.
Bring down the 4.
Divide: Think $8a \div a = 8$.
Multiply: $(a - 2)(8) = 8a - 16$.
Subtract: $(8a + 4) - (8a - 16) = 20$.

So, $\dfrac{3a^2 + 2a - 4}{a - 2} = 3a + 8 + \dfrac{20}{a - 2}.$ Write the remainder as a rational expression.

Check

$$\text{dividend} = \text{quotient} \cdot \text{divisor} + \text{remainder}$$

$$3a^2 + 2a + 4 \stackrel{?}{=} (3a + 8)(a - 2) + 20$$

$$\stackrel{?}{=} 3a^2 + 8a - 6a - 16 + 20$$

$$= 3a^2 + 2a + 4 \checkmark \qquad \blacktriangleleft$$

Before dividing one polynomial by another, arrange the terms of the polynomials in descending order of the exponents of a variable. If the dividend or the divisor has missing terms, insert these terms with zero coefficients.

EXAMPLE 3

Divide: $(-5x - 3 + 4x^3) \div (3 + 2x)$

Solution

Arrange each expression in descending powers of exponents.

$$(4x^3 - 5x - 3) \div (2x + 3)$$

Use $0x^2$ as a place holder for the missing x^2 term.

$$(4x^3 + 0x^2 - 5x - 3) \div (2x + 3)$$

$$
\begin{array}{r}
2x^2 - 3x + 2 \\
2x + 3 \overline{)\, 4x^3 + 0x^2 - 5x - 3} \\
\underline{4x^3 + 6x^2} \\
-6x^2 - 5x \\
\underline{-6x^2 - 9x} \\
4x - 3 \\
\underline{4x + 6} \\
-9
\end{array}
$$

Use long division.

So, $\dfrac{-5x - 3 + 4x^3}{(3 + 2x)} = 2x^2 - 3x + 2 - \dfrac{9}{2x - 3}.$ ◄

CHECK UNDERSTANDING

Check the solution for Example 3. Use the general equation that was given to check the solution for Example 2.

Division of polynomials can be applied to geometry.

EXAMPLE 4

GEOMETRY The volume of a rectangular prism is $(2x^3 + x^2 - 22x + 3)$ in.^{3} Find the surface area of the base of the rectangular prism if the height of the prism is $(x - 3)$ in.

Solution

Find the surface area of the base lw by dividing the volume lwh by the height h.

$$
\begin{array}{r}
2x^2 + 7x - 1 \\
x - 3 \overline{)\, 2x^3 + x^2 - 22x + 3} \\
\underline{2x^3 - 6x^2} \\
7x^2 - 22x \\
\underline{7x^2 - 21x} \\
-x + 3 \\
\underline{-x + 3} \\
0
\end{array}
$$

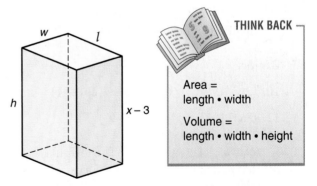

THINK BACK

Area = length • width

Volume = length • width • height

The surface area of the base of the prism is $(2x^2 + 7x - 1)$ in.^{2} ◄

In Exercises 1–2, list the dividend, the divisor, the quotient, and the remainder in that order.

1. $\dfrac{x^2 + 8}{x - 2} = x + 2 + \dfrac{512}{x - 2}$

2. $\dfrac{x^2 + 16}{x + 4} = x - 4 + \dfrac{32}{x + 4}$

Divide.

3. $(6x^3 + 2x^2 + x) \div 2x^2$

4. $(8x^3 + 4x^2 + 2x) \div 4x^2$

5. $(a^2 + 3a - 10) \div (a - 2)$

6. $(4m^2 - 6m - 5) \div (m - 4)$

7. $(2p^3 + 3p^2 - 5p + 1) \div (2p + 1)$

8. $(d^3 + 12) \div (d - 4)$

9. GEOMETRY The area of a rectangle is $8t^2 + 26t + 15$, and the width is $2t + 5$. Find the length.

10. WRITING MATHEMATICS Discuss what it means if you get a remainder of zero when dividing one polynomial by another.

PRACTICE

In Exercises 1–2, list the dividend, the divisor, the quotient, and the remainder in that order.

1. $\dfrac{x^2 + 10}{x - 4} = x + 4 + \dfrac{26}{x - 4}$

2. $\dfrac{x^2 + 12}{x - 3} = x + 3 + \dfrac{21}{x - 3}$

Divide.

3. $(12x^3 + 6x^2 + x) \div 3x^2$

4. $(15x^3 + 10x^2 + x) \div 5x^2$

5. $(a^2 + 4a - 21) \div (a - 3)$

6. $(b^2 + 6b - 16) \div (b + 8)$

7. $(12m^2 - 2m - 2) \div (3m + 1)$

8. $(27y^3 + 27y^2 + 9y + 1) \div (3y + 1)$

9. WRITING MATHEMATICS Explain how dividing a polynomial by a polynomial is similar to dividing an integer by an integer.

Divide.

10. $(2p^3 + 9p^2 - 6p + 2) \div (2p + 1)$

11. $(3q^3 + 5q^2 + 4q - 3) \div (3q - 1)$

12. $(8x^2 + 16 + 4x^4) \div (-4 + x)$

13. $(5 + 10y^2 + 5y^4) \div (-5 + y)$

AUTOMOBILE DEPRECIATION The formula $A = A_0(r^3 + 3r^2 + 3r + 1)$ can be used to determine the value of a car after 3 years, where A_0 represents the original cost of the car and r represents the rate of depreciation (represented by a negative decimal in the formula).

14. If the original price of the car can be represented by $\dfrac{10{,}000}{r + 1}$, find an expression that represents the value of the car after 3 years.

15. Evaluate the expression you found in Exercise 14 if $r = -20\%$.

16. If the original price of the car is $\dfrac{18{,}000}{r + 1}$ and $r = -15\%$, determine the value of the car after 3 years.

EXTEND

When a polynomial contains more than one variable, arrange the terms in descending order of one of the variables. To divide $(15mn^2 + 5m^2n + 10m^3 + 25n^3)$ by $(m - n)$, arrange the terms in descending order of m, the first variable in the divisor. Then,

$$(15mn^2 + 5m^2n + 10m^3 + 25n^3) \div (m - n) = 10m^2 + 15mn + 30n^2 + \frac{55n^3}{m - n}$$

Divide.

17. $(4m^2n + 12mn^2 + 8m^3 + 16n^3) \div (m - n)$

18. $(4m^2n + 12mn^2 + 8m^3 + 16n^3) \div (2n - m)$

BIOLOGY The population of a bacteria colony after 4 days is represented by the formula $P = P_0(r^4 + 4r^3 + 6r^2 + 4r + 1)$, where P_0 represents the initial number of bacteria and r represents the rate of increase in the number of bacteria per day, expressed as a decimal.

19. If the original number of bacteria can be represented by $\frac{336}{1 + r}$, find an expression that represents the number of bacteria after 4 days.

20. Evaluate to the nearest whole number the expression you found in Exercise 19 if $r = 12\%$.

21. If the original number of bacteria is $\frac{460}{1 + r}$ and $r = 15\%$, determine to the nearest whole number the number of bacteria after 4 days.

THINK CRITICALLY

Determine the value of k that will make the divisor a factor of the dividend.

22. $\dfrac{x^3 + 5x^2 + 5x + k}{x + 3}$

23. $\dfrac{x^3 + 2x^2 + kx + 45}{x - 5}$

24. $\dfrac{2x^3 + kx^2 + 18}{2x + 3}$

MIXED REVIEW

25. STANDARDIZED TESTS Select the correct solution for the following set of equations.

$$3x - 5y + 6z = 54$$
$$2x - 5y + 8z = 61$$
$$2z = 10$$

A. $x = 3, y = -3, z = 5$

B. $x = -3, y = 3, z = 5$

C. $x = 9, y = -3, z = 5$

D. $x = 3, y = -1, z = 5$

In Exercises 26–28, match the equation with the graph.

26. $y = 2x^2 + 3x + 1$

27. $y = -\dfrac{1}{2}x^2 - 4x - 2$

28. $y = -\dfrac{1}{4}x^2 + 4x + 2$

a.
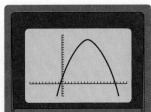

b.
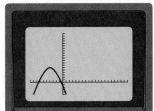

c.

14.5 **Divide Polynomials** **701**

14.6 Add and Subtract Rational Expressions

Explore

● Recall that to add two rational numbers, write equivalent rational numbers having the same denominator and then add the numerators.

Let $x = 10$ for each of the following. Substitute the value for x. Then add the two rational numbers. Simplify your answers and leave them in improper form.

1. $\dfrac{x}{5} + \dfrac{4}{x}$ **2.** $\dfrac{3x}{4} + \dfrac{1}{2x}$ **3.** $\dfrac{(x-6)}{(x-1)} + \dfrac{2}{(x-4)}$

4. When adding rational numbers, how do you determine what the denominator should be?

Determine the denominators you would use to add the following rational numbers.

5. $\dfrac{1}{2} + \dfrac{1}{3}$ **6.** $\dfrac{1}{2} + \dfrac{1}{6}$

7. $\dfrac{1}{x} + \dfrac{1}{x}$ **8.** $\dfrac{1}{4} + \dfrac{1}{x}$

9. $\dfrac{2}{3x} + \dfrac{4}{x}$ **10.** $\dfrac{2}{x^2} + \dfrac{4}{3x}$

11. $\dfrac{1}{(x-5)} + \dfrac{3}{8}$ **12.** $\dfrac{6}{2x} + \dfrac{3}{2(x+1)}$

13. Determine the denominators you would use to add each of the rational expressions in Questions 1–3.

Build Understanding

● Adding or subtracting rational expressions is similar to adding and subtracting rational numbers. When the denominators are the same, add or subtract the numerators.

EXAMPLE 1

Add: $\dfrac{6}{x+3} + \dfrac{4}{x+3}$

Solution

$$\frac{6}{x+3} + \frac{4}{x+3} = \frac{10}{x+3}$$

◀

When the denominators of rational expressions are different, rewrite the expressions as equivalent expressions with a **least common denominator (LCD)** before adding or subtracting.

Use the following steps to determine the LCD.

- Determine the prime factorization for each denominator. Use exponents where needed.
- For each common prime factor, choose the factor with the greatest exponent.
- Multiply these factors.

EXAMPLE 2

Determine the LCD of $\dfrac{5a}{12ab^2}$ and $\dfrac{7b}{15a^3b}$.

Solution

Factor each denominator: $12ab^2 = 2^2 \cdot 3 \cdot a \cdot b^2$
$$15a^3b = 3 \cdot 5 \cdot a^3 \cdot b$$

List the prime factors.	2	3	5	a	b
Determine the greatest exponent for each factor.	2	1	1	3	2

The LCD is $2^2 \cdot 3 \cdot 5 \cdot a^3 \cdot b^2$ or $60a^3b^2$. ◄

To add or subtract rational expressions, use the LCD to rewrite each expression as an equivalent rational expression having the LCD as its denominator. Then add or subtract the numerators.

EXAMPLE 3

Subtract: $\dfrac{5}{6m^2} - \dfrac{7 - 2m^2}{8m}$

Solution

$$\dfrac{5}{6m^2} - \dfrac{7 - 2m^2}{8m}$$
The LCD is $2^3 \cdot 3 \cdot m^2 = 24m^2$.

$$= \dfrac{5}{6m^2}\left(\dfrac{4}{4}\right) - \dfrac{(7 - 2m^2)}{8m}\left(\dfrac{3m}{3m}\right)$$
Write equivalent expressions that have the LCD.

$$= \dfrac{20}{24m^2} - \dfrac{(21m - 6m^3)}{24m^2}$$

$$= \dfrac{20 - (21m - 6m^3)}{24m^2}$$
Subtract.

$$= \dfrac{20 - 21m + 6m^3}{24m^2}$$
Simplify. ◄

You can add and subtract rational expressions to solve many real world applications.

EXAMPLE 4

NAVIGATION A boat takes $\dfrac{3}{t-2}$ hours to travel upstream and $\dfrac{8-t}{t+2}$ hours to travel downstream. Write an expression for the time it takes for the round trip.

Solution

$$\frac{3}{t-2} + \frac{8-t}{t+2}$$
    The LCD is $(t+2)(t-2)$.

$$= \left(\frac{3}{t-2}\right)\left(\frac{t+2}{t+2}\right) + \left(\frac{8-t}{t+2}\right)\left(\frac{t-2}{t-2}\right)$$
    Write equivalent expressions.

$$= \frac{3(t+2) + (8-t)(t-2)}{(t+2)(t-2)}$$

$$= \frac{3t + 6 + 8t - 16 - t^2 + 2t}{(t+2)(t-2)}$$

$$= \frac{-t^2 + 13t - 10}{(t+2)(t-2)}$$
    Combine like terms in the numerator.

The total time is $\dfrac{-t^2 + 13t - 10}{(t+2)(t-2)}$ hours. ◀

TRY THESE

Determine the least common denominator for each set of expressions.

1. $\dfrac{5}{mn^2}, \dfrac{101}{mn}$ **2.** $\dfrac{8}{a^2b}, \dfrac{3}{2ac}$ **3.** $\dfrac{121}{x-4}, \dfrac{199}{x-5}$

Add or subtract. Simplify if possible.

4. $\dfrac{a}{11} + \dfrac{7a}{11}$ **5.** $\dfrac{5}{7y} - \dfrac{9}{7y}$ **6.** $\dfrac{3p}{p-4} - \dfrac{12}{p-4}$

7. $\dfrac{6}{5x} + \dfrac{-2}{15x}$ **8.** $\dfrac{6}{5a^2} + \dfrac{2}{15a} - \dfrac{5}{a}$ **9.** $\dfrac{4}{a+4} + \dfrac{5}{a-9}$

10. $\dfrac{2x+5}{(x+2)(x-2)} + \dfrac{-2}{x+2}$ **11.** $\dfrac{4z+2}{16-z^2} + \dfrac{4}{z-4}$

12. **COOKING** It takes $\dfrac{2}{x+2}$ hours to prepare a specific dinner and $\dfrac{1}{x}$ hours to cook it. If you want to serve the dinner at a certain time, how many hours before that time must you begin preparing it?

13. **WRITING MATHEMATICS** Write a paragraph that explains in your own words how to add two rational expressions with unlike denominators.

Determine the least common denominator.

1. $\dfrac{3}{ab^2}, \dfrac{11}{b^3}$

2. $\dfrac{101}{x-2}, \dfrac{19}{x-3}$

3. $\dfrac{17}{3x}, \dfrac{13}{x(x+1)}$

Add or subtract. Simplify if possible.

4. $\dfrac{12}{5y} - \dfrac{7}{5y}$

5. $\dfrac{22}{7z} + \dfrac{6}{7z}$

6. $\dfrac{3}{p+2} + \dfrac{p-3}{p+2}$

7. $\dfrac{8d}{d-6} - \dfrac{d+42}{d-6}$

8. $\dfrac{13e}{e-12} - \dfrac{e+144}{e-12}$

9. $\dfrac{4}{b-4} - \dfrac{b}{b-4}$

10. $\dfrac{9}{2x} + \dfrac{-5}{8x}$

11. $\dfrac{8}{3x} + \dfrac{-1}{9x}$

12. $\dfrac{5}{4t^2} - \dfrac{3}{8t} - \dfrac{2}{t}$

13. **LIGHT** The *lens formula,* $\dfrac{1}{d_o} + \dfrac{1}{d_i} = \dfrac{1}{f}$, expresses the relationship between the location of an object serving as a light source d_o, the location of the image d_i, and the focal length of the lens f.

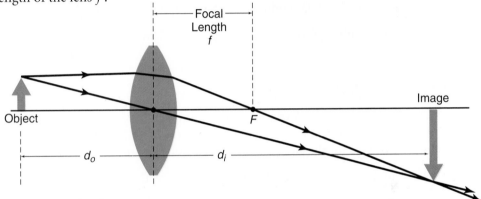

a. Solve the lens equation for d_i .

b. If the focal length of the cornea-lens part of your eye is approximately 1.65 cm and you look at an object 50 m (5000 cm) away, find d_i . Round to two decimal places.

c. If the object is moved so that it is only 30 cm away (reading distance), find d_i. Round to two decimal places.

Add or subtract. Simplify if possible.

14. $\dfrac{3}{a+6} + \dfrac{2}{a-5}$

15. $\dfrac{5}{a+7} + \dfrac{4}{a-4}$

16. $\dfrac{t}{2(t+3)} - \dfrac{2}{3(t+3)}$

17. $\dfrac{r}{3(r+3)} - \dfrac{3}{4(r+3)}$

18. $\dfrac{4y+16}{(y+4)(y-4)} + \dfrac{-4}{y+4}$

19. $\dfrac{5x+3}{(5-x)(5+x)} + \dfrac{5}{x-5}$

20. $\dfrac{6q-1}{81-q^2} - \dfrac{2q}{q+9}$

21. $\dfrac{10v-1}{100-v^2} - \dfrac{5v}{v+10}$

22. $\dfrac{11}{121-u^2} - \dfrac{u^2}{u+11}$

23. $\dfrac{3a}{6a^2+13a+2} + \dfrac{a+1}{a^2+5a+6}$

24. $\dfrac{5b}{3b^2-b-2} + \dfrac{b+2}{b^2+4b-5}$

25. **WRITING MATHEMATICS** When adding or subtracting two rational numbers, such as $\dfrac{5}{6} - \dfrac{1}{2}$, you can find a common denominator by multiplying the denominators instead of using the LCD. Why might you not want to do that for rational expressions?

EXTEND

Combine. Follow the order of operations.

26. $\left(\dfrac{2}{x + 2} - \dfrac{1}{x + 5}\right)\left(\dfrac{2x + 10}{x^2 + 8x}\right)$

27. $\left(\dfrac{3}{x - 5} - \dfrac{1}{x^2 - 25}\right) \div \left(\dfrac{3x + 14}{x - 8}\right)$

GEOMETRY Write an expression for the perimeter of a triangle with sides of the given lengths.

28. $x, \dfrac{3}{x}, \dfrac{4x}{1 - x}$

29. $\dfrac{1}{x}, \dfrac{2}{x + 1}, \dfrac{x}{6x - 1}$

30. $\dfrac{2}{x}, \dfrac{3}{x + 2}, \dfrac{x}{3x - 1}$

31. $x, \dfrac{4}{x}, \dfrac{10x}{1 - 2x}$

THINK CRITICALLY

32. Provide an original example of the addition or subtraction of two rational expressions in which using the LCD produces an answer that is not in simplest form.

33. Provide an original example of the addition or subtraction of two rational expressions in which using the LCD produces an answer that is in simplest form.

34. Write the complete group of rational expressions having k as the numerator and $2x - 6$ as the LCD.

MIXED REVIEW

35. STANDARDIZED TESTS A window washer dropped his ring from a window 72 m above the ground. If the acceleration due to gravity on Earth is approximately -9.8 m/s^2, how long will it take for the ring to reach the ground? Use the formula $d = \dfrac{1}{2}|a|t^2$.

    A. 14.69 s   **B.** 2.74 s   **C.** 7.55 s   **D.** 3.83 s

36. A mini-van dealer estimates that she sold 25 ± 4 green vans last year. Write an absolute value inequality that models the situation.

Simplify each expression.

37. $3(12 - 3^2)^3 - 2(7 - 1)^2$

38. $[1260 \div (4 - 2^8)]^2 - 6 + 5^3$

In Exercises 39–41, match the equation with the graph.

39. $y = 2x^2$

40. $y = x^2 - 2$

41. $y = (x + 2)^2$

a.

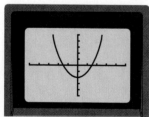

b.

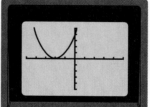

c.

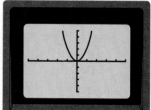

Add or subtract as indicated. Simplify if possible.

42. $\dfrac{7}{6t^2} + \dfrac{12}{12t} - \dfrac{3}{t}$

43. $\dfrac{3}{8r^2} + \dfrac{8}{16r} - \dfrac{2}{r}$

14.7 Complex Rational Expressions

Explore

Recall that a mixed number such as $4\frac{3}{5}$ is the sum of an integer and a fraction $\left(4 + \frac{3}{5}\right)$. You can express a mixed number as an improper fraction. For example,

$$4\frac{3}{5} = 4 + \frac{3}{5}$$

$$= \frac{4}{1} \cdot \left(\frac{5}{5}\right) + \frac{3}{5}$$

$$= \frac{20}{5} + \frac{3}{5}$$

$$= \frac{23}{5}$$

Evaluate each of the following for $x = 4$.

1. $3 - \dfrac{1}{x}$ **2.** $x + 3 + \dfrac{6}{x + 3}$ **3.** $2 + \dfrac{x^2 - 4}{x^2 - 9}$

4. Describe the procedure you would use to convert a mixed expression to a rational expression. Use an example to illustrate your point.

Recall that a fraction bar indicates division. For example, $\dfrac{3}{4} = 3 \div 4$. When you divide 4 into 3, the result is 0.75.

When both the numerator and denominator are rational numbers, you can divide the numerator by the denominator. For example,

$$\frac{\frac{2}{3}}{\frac{3}{4}} = \frac{2}{3} \div \frac{3}{4}$$

Write a division expression for each of the following. Then evaluate the expression.

5. $\dfrac{\frac{5}{3}}{\frac{15}{4}}$ **6.** $\dfrac{\frac{1}{2}}{\frac{2}{3}}$ **7.** $\dfrac{8}{\frac{1}{4}}$

Build Understanding

A **mixed expression** is the sum or difference of a polynomial and a rational expression. Each of the following is a mixed expression.

$$4 + \frac{3}{2x}$$ $$p - 3 - \frac{p}{p - 1}$$

Converting mixed expressions to rational expressions is similar to converting mixed numbers to improper fractions.

EXAMPLE 1

Convert $3 + \dfrac{2}{p}$ to a rational expression.

Solution

$$3 + \frac{2}{p} = \frac{3}{1} \cdot \left(\frac{p}{p}\right) + \frac{2}{p} \qquad \text{Express 3 as a rational number.}$$

$$= \frac{3p}{p} + \frac{2}{p} \qquad \text{Write equivalent expressions with LCD } = p.$$

$$= \frac{3p + 2}{p} \qquad \text{Add.} \qquad \blacktriangleleft$$

A **complex rational expression** contains one or more rational expressions in its numerator or denominator. Examples include

$$\frac{\dfrac{2}{x}}{\dfrac{3}{y}} \qquad \frac{5 + \dfrac{4}{2x}}{x} \qquad \frac{\dfrac{4}{3b - 5}}{2b - \dfrac{3}{b - 5}} \qquad \frac{\dfrac{x - 3}{x}}{\dfrac{x - 2}{x - 3}}$$

To simplify a complex rational expression, determine the LCD of all the rational expressions in both the numerator and the denominator. Then multiply the numerator and the denominator by the LCD.

EXAMPLE 2

Simplify: $\dfrac{\dfrac{3}{x}}{\dfrac{6}{y}}$

Solution

$$\frac{\dfrac{3}{x}}{\dfrac{6}{y}} = \frac{\dfrac{3}{x}}{\dfrac{6}{y}} \cdot \frac{xy}{xy} \qquad \text{The LCD is } xy, \text{ so multiply by } \frac{xy}{xy}.$$

$$= \frac{\dfrac{3}{\cancel{x}} \cdot \dfrac{\cancel{x}y}{1}}{\dfrac{6}{\cancel{y}} \cdot \dfrac{x\cancel{y}}{1}} \qquad \text{Divide out common factors.}$$

$$= \frac{3y}{6x}$$

$$= \frac{y}{2x} \qquad \text{Simplify.} \qquad \blacktriangleleft$$

Some complex rational expressions contain mixed expressions.

EXAMPLE 3

Simplify: $\dfrac{\dfrac{2}{m} + 3}{5 - \dfrac{4}{n}}$

Solution

$\dfrac{\dfrac{2}{m} + 3}{5 - \dfrac{4}{n}} = \dfrac{\dfrac{2}{m} + 3}{5 - \dfrac{4}{n}} \cdot \dfrac{mn}{mn}$ The LCD is mn, so multiply by $\dfrac{mn}{mn}$.

$= \dfrac{\left(\dfrac{2}{m} \cdot mn\right) + (3 \cdot mn)}{(5 \cdot mn) - \left(\dfrac{4}{n} \cdot mn\right)}$ Use the distributive property.

$= \dfrac{2n + 3mn}{5mn - 4m}$ ◀

Complex rational expressions can model real world situations.

EXAMPLE 4

AGRICULTURE In April, a gardener was able to purchase x sacks of fertilizer for $600. In August, the fertilizer was on sale for $10 less per sack. How many sacks at the sale price could the gardener purchase for $600?

Solution

April price: p August price: $p - 10$

Sacks purchased in April: $x = \dfrac{600}{p}$, so $p = \dfrac{600}{x}$

Sacks that can be purchased in August $=$

$\dfrac{600}{p - 10}$

$= \dfrac{600}{\dfrac{600}{x} - 10}$ Substitute $\dfrac{600}{x}$ for p.

$= \dfrac{600}{\dfrac{600}{x} - 10} \cdot \dfrac{x}{x}$ The LCD is x, so multiply by $\dfrac{x}{x}$.

$= \dfrac{600x}{600 - 10x}$ Use the distributive property.

$= \dfrac{10(60x)}{10(60 - x)}$ Simplify.

$= \dfrac{60x}{60 - x}$

The farmer could purchase $\dfrac{60x}{60 - x}$ sacks.   ◀

Convert each mixed expression to a rational expression.

1. $5 + \dfrac{9}{t}$

2. $2x - \dfrac{x + 5}{x}$

3. $3n + \dfrac{2n + 3}{4n + 5}$

4. WRITING MATHEMATICS Explain how converting a mixed expression to a rational expression is similar to converting a mixed number to an improper fraction. Show an example of each and explain the steps.

5. TESTING On a particular true-false examination, there were $\dfrac{100}{4x + 1}$ true statements and $\dfrac{80}{3x + 2}$ false statements. Determine the ratio of true statements to false statements.

Simplify. Write your answer in simplest form.

6. $\dfrac{\frac{5}{x}}{\frac{7}{y}}$

7. $\dfrac{\frac{8}{c}}{\frac{3}{d}}$

8. $\dfrac{\frac{13}{2u}}{\frac{6}{5z}}$

9. $\dfrac{\frac{1}{p} - \frac{3}{q}}{\frac{11}{2q}}$

10. $\dfrac{\frac{1}{r} - \frac{1}{t}}{\frac{7}{3t}}$

11. $\dfrac{\frac{6}{a} - 5}{-2 - \frac{3}{a}}$

12. $\dfrac{7 - \frac{5}{g + 3}}{14 - \frac{10}{g + 3}}$

13. AGRICULTURE Refer to Example 4. Suppose the gardener had purchased 12 sacks of fertilizer in April. How many sacks could be purchased in August?

PRACTICE

Convert each mixed expression to a rational expression.

1. $11 - \dfrac{13}{b}$

2. $3x - \dfrac{x + 2}{x}$

3. $\dfrac{2z - 5}{3z} + 3z - 1$

4. $n + \dfrac{n - 4}{3n + 4}$

5. $t - 3 - \dfrac{5}{t + 1}$

6. $b + 5 + \dfrac{5}{b - 5}$

Simplify. Write your answer in simplest form.

7. $\dfrac{\frac{3}{x}}{\frac{4}{y}}$

8. $\dfrac{\frac{5}{m}}{\frac{6}{n}}$

9. $\dfrac{\frac{1}{p} - \frac{3}{q}}{\frac{4}{q}}$

10. $\dfrac{\frac{2}{r} + \frac{3}{t}}{-\frac{5}{r}}$

11. $\dfrac{\frac{7}{2u}}{\frac{8}{3v}}$

12. $\dfrac{\frac{5}{uv^2}}{\frac{2}{u^2v}}$

13. $\dfrac{\frac{1}{r} - \frac{1}{t}}{-\frac{6}{5t}}$

14. $\dfrac{\frac{1}{m} - \frac{1}{n}}{\frac{3}{2m} - \frac{3}{2n}}$

15. $\dfrac{\frac{4}{p} + 5}{\frac{5}{q}}$

16. $\dfrac{\frac{8}{a} - 9}{\frac{6}{b}}$

17. $\dfrac{\frac{c}{d^2} + \frac{3}{d}}{\frac{c}{d} - 1}$

18. $\dfrac{\frac{4}{e^2} + \frac{f}{e}}{\frac{2}{e} - 12}$

19. ELECTRICITY In a parallel circuit with two light bulbs, the total resistance R can be determined by the formula $R = \dfrac{R_1 \cdot R_2}{R_1 + R_2}$. If $R_1 = \dfrac{5}{q}$ ohm and $R_2 = \dfrac{10}{q^2}$ ohm, determine R.

20. TRAVEL Stacey drives m km at 40 km/h and returns on the same route at 60 km/h. Determine Stacey's average speed for the round trip.

21. VOTING In Urbania, $\dfrac{250,000}{3x - 10}$ people voted yes on a bond issue and $\dfrac{4,000,000}{x^2}$ people voted no. Determine the ratio of yes votes to no votes.

22. WRITING MATHEMATICS Example 2 shows one procedure for simplifying a complex expression. Next to Example 2, Communicating About Algebra shows a different procedure. Choose the procedure you prefer and explain why you prefer it.

EXTEND

When the rational expressions in the numerator and denominator of the complex rational expression have denominators that are polynomials of more than one term, factor first, if possible. Then determine the LCD. For example,

$$\dfrac{\dfrac{4x}{x^2 - 25} + \dfrac{6}{x + 5}}{\dfrac{3}{2x + 10}} = \dfrac{\dfrac{4x}{(x + 5)(x - 5)} - \dfrac{6}{(x + 5)}}{\dfrac{3}{2(x + 5)}}$$

Factor: the LCD is $2(x + 5)(x - 5)$.

23. Simplify the expression above.

Simplify.

24. $\dfrac{\dfrac{3x}{x^2 - 49} + \dfrac{7}{x + 7}}{\dfrac{1}{x - 7} - \dfrac{7}{2x + 14}}$

25. $\dfrac{\dfrac{2b}{b - 4} - \dfrac{3}{b^2}}{\dfrac{5}{5b - 20} + \dfrac{3}{4b^2 - 16b}}$

26. $\dfrac{\dfrac{6}{n^2 + n - 20} + \dfrac{3}{n - 4}}{\dfrac{2}{3n - 12} + \dfrac{3}{n + 5}}$

GEOMETRY Determine the width of each rectangle with the given length and area.

27. $\dfrac{x + 1}{x - 1}$ ft; $\dfrac{x^2 + 2x + 1}{4}$ ft^2

28. $\dfrac{2x + 10}{3x + 18}$ ft; $\dfrac{x^2 + 14x + 45}{x^2 + 6x}$ ft^2

THINK CRITICALLY

29. Is $\dfrac{5}{6}$ a complex rational expression? Explain.

30. Provide an example of a complex rational expression that is its own reciprocal.

Simplify.

31. $\dfrac{\dfrac{1}{a + 1}}{a - \dfrac{1}{a + \dfrac{1}{a}}}$

32. $\dfrac{2}{c + \dfrac{1}{1 + \dfrac{c + 1}{5 - c}}}$

14.8 Solve Rational Equations

Explore

SPOTLIGHT ON LEARNING

WHAT? In this lesson you will learn
- to solve rational equations.
- to set up and solve uniform motion problems and work problems.

WHY? Knowing how to solve rational equations can help you to solve problems in travel, finance, water storage, and agriculture.

1. **a.** Determine the least integer that produces integral coefficients when you multiply both sides of the equation by the integer.

$$3 + \frac{x}{4} = \frac{x}{9}$$

   b. What way besides trial and error can you use to determine this number?

   c. Write and solve the resulting equation.

Solve for x.

2. $\dfrac{x}{2} + \dfrac{x}{5} = 50$ 3. $\dfrac{2x}{7} = \dfrac{3x}{8} - \dfrac{5}{4}$ 4. $\dfrac{3x}{10} = \dfrac{2x}{15} + \dfrac{3}{2}$

5. How would you solve an equation such as $\dfrac{4}{x} + \dfrac{8}{x} = 2$?

6. What method would you use to solve any equation that contains one or more rational expressions?

Build Understanding

A **rational equation** is an equation that contains one or more rational expressions. To eliminate the denominators (other than 1), multiply both sides of the rational equation by the LCD of all the expressions. Then solve the resulting equation. Check each solution *in the original equation* to make sure that it is a solution.

EXAMPLE 1

Solve and check: $\dfrac{3}{x} + \dfrac{1}{2x} = 7$

Solution
Algebraic Method

$$\frac{3}{x} + \frac{1}{2x} = 7 \qquad \text{The LCD is } 2x.$$

$$\frac{3}{x}(2x) + \frac{1}{2x}(2x) = 7(2x) \qquad \text{Multiply each term by the LCD.}$$

$$6 + 1 = 14x$$

$$7 = 14x$$

$$\frac{1}{2} = x \qquad \text{Solve the resulting equation.}$$

Check

$$\frac{3}{\frac{1}{2}} + \frac{1}{2\left(\frac{1}{2}\right)} \stackrel{?}{=} 7$$

$$6 + 1 \stackrel{?}{=} 7$$

$$7 = 7 \checkmark$$

The solution is $\frac{1}{2}$.

Graphing Method
Use a graphing utility to graph each side of the equation. Determine where the graphs of $y_1 = \frac{3}{x} + \frac{1}{2x}$ and $y_2 = 7$ intersect.

Since the value of y_1 and y_2 is 7 when $x = \frac{1}{2}$, the solution is $x = \frac{1}{2}$. ◄

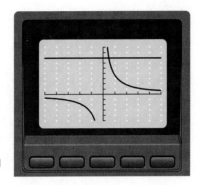

When you multiply both sides of a rational equation by an LCD that contains a variable and then solve the resulting equation, you may find solutions that make the denominator of the original equation equal to zero. Such solutions are called *extraneous solutions*.

EXAMPLE 2

Solve and check: $\dfrac{3}{a - 3} + 2 = \dfrac{a}{a - 3}$

Solution

$$\frac{3}{a - 3} + 2 = \frac{a}{a - 3} \qquad \text{The LCD is } (a - 3).$$

$$\frac{3}{(a - 3)}(a - 3) + 2(a - 3) = \frac{a}{(a - 3)}(a - 3) \qquad \text{Multiply each term by the LCD.}$$

$$3 + 2a - 6 = a$$

$$3 = a \qquad \text{Solve the resulting equation.}$$

Check

$$\frac{3}{3 - 3} + 2 \stackrel{?}{=} \frac{3}{3 - 3}$$

Since the denominator is zero when $a = 3$, 3 is an extraneous solution. The original equation has no solution. ◄

You may get two solutions to a rational equation. However, one or both of them may be extraneous.

COMMUNICATING ABOUT ALGEBRA

Determine whether the solutions of

$$6x = 7x \text{ and}$$

$$\frac{1}{6x} = \frac{1}{7x}$$

are equivalent. Discuss whether the equations are equivalent.

EXAMPLE 3

Solve and check: $\dfrac{c}{c + 3} - \dfrac{4}{c - 3} = \dfrac{c - 27}{c^2 - 9}$

Solution

$$\dfrac{c}{c + 3} - \dfrac{4}{c - 3} = \dfrac{c - 27}{c^2 - 9} \qquad \text{The LCD is } (c + 3)(c - 3).$$

$$\dfrac{c}{(c + 3)}(c + 3)(c - 3) - \dfrac{4}{(c - 3)}(c + 3)(c - 3)$$

$$= \dfrac{c - 27}{(c + 3)(c - 3)}(c + 3)(c - 3)$$

$$c(c - 3) - 4(c + 3) = c - 27$$
$$c^2 - 3c - 4c - 12 = c - 27$$
$$c^2 - 8c + 15 = 0$$
$$(c - 3)(c - 5) = 0 \qquad \text{Solve by factoring.}$$
$$c - 3 = 0 \text{ or } c - 5 = 0$$
$$c = 3 \qquad c = 5$$

Check

Substitute 3 for c in the original equation.

$$\dfrac{3}{3 + 3} - \dfrac{4}{3 - 3} \stackrel{?}{=} \dfrac{3 - 27}{3^2 - 9}$$

Since two denominators in the original equation are 0 when $c = 3$, 3 is an extraneous solution.

Substitute 5 for c in the original equation.

$$\dfrac{5}{5 + 3} - \dfrac{4}{5 - 3} \stackrel{?}{=} \dfrac{5 - 27}{5^2 - 9}$$

$$\dfrac{5}{8} - \dfrac{4}{2} \stackrel{?}{=} \dfrac{-22}{16}$$

$$\dfrac{-11}{8} = \dfrac{-11}{8} \checkmark \quad \blacktriangleleft$$

Rational equations can be used to solve problems involving *uniform (constant) motion.*

EXAMPLE 4

TRAVEL Debbie and Mindy each traveled 300 mi to visit Lamb's Farm. Debbie drove at an average speed of 10 mi/h faster than Mindy drove and arrived 1 h earlier than Mindy. How fast did each woman drive?

Solution

Use the information given in the problem to set up a rational equation. You know that Debbie made the trip in 1 h less than Mindy.

Since distance = rate · time, time = distance ÷ rate. Let r represent the rate at which Debbie traveled. Therefore, $r - 10$ represents the rate at which Mindy traveled.

$$Debbie's\ time\ =\ Mindy's\ time\ -\ 1$$

$$\frac{300}{r} = \frac{300}{r - 10} - 1$$

Solve the equation using an LCD of $r(r - 10)$.

$$\frac{300}{\cancel{r}}(\cancel{r})(r - 10) = \frac{300}{\cancel{(r - 10)}}(r)\cancel{(r - 10)} - 1(r)(r - 10)$$

$$300(r - 10) = 300r - r(r - 10)$$
$$300r - 3000 = 300r - r^2 + 10r$$
$$r^2 - 10r - 3000 = 0$$
$$(r - 60)(r + 50) = 0$$
$$r - 60 = 0 \quad \text{or } r + 50 = 0$$
$$r = 60 \qquad\qquad r = -50$$

Eliminate $r = -50$, since a negative rate of speed does not make sense in this situation. Debbie traveled at 60 mi/h and Mindy traveled at $60 - 10$ or 50 mi/h. ◀

CHECK UNDERSTANDING

In Example 4, check the solutions $r = 60$ and $r = -50$ in the original equation. Is either solution extraneous?

Rational equations can also be used to solve problems involving *wind speed*.

COMMUNICATING ABOUT ALGEBRA

Discuss how to organize the information for Example 5 using a table.

EXAMPLE 5

TRAVEL In calm air, a small airplane can fly at the rate of 140 mi/h. It can fly 700 mi with a tailwind in the same time it can fly 420 mi against the wind. Find the speed of the wind.

Solution

Let w represent the rate of the wind. Then $140 + w$ represents the rate with the wind and $140 - w$ the rate against the wind.

Since the time for the 700-mi trip equals the time for the 420-mi trip, you can write an equation.

$$\frac{700}{140 + w} = \frac{420}{140 - w} \qquad \text{The LCD is}$$
$$\qquad\qquad\qquad\qquad\qquad (140 + w)(140 - w).$$

$$\frac{700}{\cancel{(140 + w)}}\cancel{(140 + w)}(140 - w) = \frac{420}{\cancel{(140 - w)}}(140 + w)\cancel{(140 - w)}$$

$$700(140 - w) = 420(140 + w)$$
$$98,000 - 700w = 58,800 + 420w$$
$$39,200 = 1,120w$$
$$35 = w \qquad \text{Solve the equation.}$$

The speed of the wind was 35 mi/h. ◀

1. **WRITING MATHEMATICS** Explain why it is necessary to check possible solutions of a rational equation in the original equation.

Solve and check. If an equation has no solution, write no solution.

2. $2x + \dfrac{4}{5} = \dfrac{17}{20}$

3. $\dfrac{3t}{4} - \dfrac{5t}{8} = 8$

4. $\dfrac{7t}{3} - \dfrac{8t}{9} = 13$

5. $\dfrac{b}{b-2} - 2 = \dfrac{2}{b-2}$

6. $4 + \dfrac{5}{c} = -11$

7. $\dfrac{3}{4e} - \dfrac{5}{2e} = \dfrac{7}{2}$

8. $\dfrac{h+8}{h} + \dfrac{3}{4h} = 11$

9. $\dfrac{2c+5}{c+10} = \dfrac{c-9}{c+10} + 1$

10. $\dfrac{4d-6}{d-11} = \dfrac{d+8}{d-11} + 2$

11. **BOATING** A canoe can travel 16 mi/h in still water. Going with the current, it can travel 24 mi in the same amount of time that it can travel 12 mi going against the current. Determine the rate of the current.

PRACTICE

Solve and check. If an equation has no solution, write no solution.

1. $2m + \dfrac{2}{3} = \dfrac{23}{10}$

2. $6a - \dfrac{7}{8} = -\dfrac{12}{5}$

3. $\dfrac{3u}{4} + \dfrac{4u}{6} = 17$

4. $\dfrac{4v}{15} + \dfrac{v}{5} = 14$

5. $\dfrac{j}{j+6} = 3 - \dfrac{6}{j+6}$

6. $\dfrac{k}{k+2} - \dfrac{5}{k+2} = 8$

7. $\dfrac{m}{m-5} - \dfrac{13}{m-5} = 7$

8. $4 - \dfrac{6}{n} = 5$

9. $\dfrac{4}{5q} - \dfrac{2}{7q} = \dfrac{1}{35}$

10. $\dfrac{w+3}{w} + \dfrac{2}{3w} = 12$

11. $\dfrac{x-3}{x} + \dfrac{3}{4x} = 16$

12. $\dfrac{3d-8}{d-5} = \dfrac{d+6}{d-5} + 2$

13. **WRITING MATHEMATICS** In your own words, explain the procedure used to rewrite rational equations so they are easier to solve.

14. **TRAVEL** A jet can fly 550 mi/h in calm air. With a tailwind, it can fly 2400 mi in the same time it can fly 2000 mi against the wind. Find the time it takes to make the 2400 mi trip with the tailwind.

15. **FINANCE** Rikuichi's checking account contains $1500 more than Keemo's. If Keemo makes a deposit of $450, the ratio of the amount in Rikuichi's account to the amount in Keemo's account will be 46 : 25. Determine the amount of money in each account.

EXTEND

In Exercises 16–17, factor each denominator before determining the LCD. Then solve and check. Eliminate any extraneous solutions.

16. $\dfrac{2t - 5}{t^2 - t - 2} = \dfrac{t - 5}{t^2 - 4t - 5}$

17. $\dfrac{x + 5}{x^2 + 4x - 5} = \dfrac{2x + 8}{x^2 + 3x - 4}$

18. **BOATING** A motorboat can travel 12 mi downstream in $\dfrac{3}{4}$ of the time it takes to travel the same distance upstream. Determine the rate of the motorboat in still water if the rate of the current is 4 mi/h.

19. **WATER STORAGE** Two pipes can fill a storage tank in 12 h if they run at the same time. If the faster pipe runs 4 times as fast as the slower one, how long would it take for the faster one to fill the tank working alone?

20. **INTERIOR DECORATING** An experienced wallpaper hanger can hang paper twice as fast as his apprentice can. If they work together, they can paper a large room in 4 h. How long would it take each one working alone to do the job?

THINK CRITICALLY

21. Write a rational equation that has an extraneous solution of 4 and one other solution that is not extraneous.

22. Write an original rational equation that has no solution.

23. Write an original rational equation that has zero as its only solution.

Solve and check.

24. $\left(\dfrac{x - 3}{x + 6}\right)^2 \cdot \left(\dfrac{x + 6}{2x - 6}\right)^3 = 1$

25. $\left(\dfrac{x - 2}{x + 4}\right)^2 \div \left(\dfrac{3x - 6}{x + 4}\right)^3 = 1$

PROJECT *Connection*

The table shows the dry matter capacity for standard 12-ft-diameter silos.

Silage, the livestock feed stored in the silo, is usually moist material. Depending on the moisture level, the total tonnage for the silo will vary.

Size, ft (diameter × height)	Volume, ft³	Dry Matter, approximate tons
12 × 30	3390	21
12 × 40	4520	31
12 × 50	5650	42

The total tonnage at different moisture levels is described by the following rational function where the moisture level is expressed as a percent.

$$\dfrac{\text{tons dry matter}}{1.00 - \text{percent moisture level}} = \text{tons in silo}$$

1. Determine how many tons of 50% moisture feed are in a 12-ft × 50-ft silo. How many tons will there be if the moisture level is 65%?

2. At what moisture level will a 12-ft × 40-ft silo hold about 77.5 tons?

3. Use a graphing utility to graph the function for a 12-ft × 30-ft silo. For what moisture levels is the function defined? Why?

The **dimensions** of a quantity are the units in which it is measured. For example, consider the units of a distance formula.

$$\text{distance} = \text{rate} \cdot \text{time} \qquad \text{and} \qquad \text{rate} = \frac{\text{distance}}{\text{time}}$$

If the distance is in inches and the time is in minutes, the rate must be in inches per minute.

You can treat units as though they were variables. Substitute the units into the distance formula to check that both sides represent distance.

$$\text{inches} = \frac{\text{inches}}{\text{minute}} \cdot \text{minutes}$$

Because only identical units divide out, you may need to introduce a **conversion factor** if the units do not match. For example, suppose the rate in the equation above were feet/minute and your answer has to be in inches.

$$\text{inches} = \frac{\text{feet}}{\text{minute}} \cdot \text{minutes}$$

Since inches $\neq$ feet, you need to *convert* feet to inches to get an answer in inches. You can do this by multiplying the rate by 12 inches per foot. Since 12 in. = 1 ft, $\frac{12 \text{ in.}}{\text{ft}} = 1$.

$$\text{inches} = \frac{\text{feet}}{\text{minute}} \cdot \frac{12 \text{ in.}}{\text{foot}} \cdot \text{minutes}$$

This process of using units or dimensions in equations to solve problems is called **dimensional analysis**.

Problem

Each member of a class of 28 students needs 12.5 g of sodium chloride (salt) for a plant-growth experiment. The teacher provides a new 1.5-lb container of the salt. If each student uses the proper amount, what is the mass of the salt that will be left in the container?

Explore the Problem

1. Reread the question in the problem. In what units should the final answer be?

2. First, find the total mass of salt used. The unit "students" is eliminated.

$$\text{mass used} = 28 \text{ students} \cdot 12.5 \frac{g}{\text{student}} = \boxed{} g$$

3. Subtract the mass used from the mass in the container to find the mass of salt left. What additional information do you need?

4. Use 1 lb = 453.6 g as a conversion factor. Write it as a ratio that eliminates pound.

$$\text{mass in container} = 1.5 \text{ lb} \cdot \frac{453.6 \text{ g}}{\text{lb}} = \boxed{} g$$

5. Subtract the mass used from the mass in the container.

6. Suppose there were 24 students, each student used 8.5 g of salt, and the original container held 2.5 lb of salt. Write one equation to solve this problem. Include units and show cancellations.

7. WRITING MATHEMATICS Explain why multiplying one side of an equation by a conversion factor does not upset the equality.

> PROBLEM
> SOLVING PLAN
>
> • Understand
> • Plan
> • Solve
> • Examine

Investigate Further

You can use dimensional analysis to help you set up an equation for a problem. The next problem requires several conversion factors.

The distance light travels through space in one year is called a *light-year*. If light travels 3.0×10^8 meters per second, determine the number of kilometers in a light-year. Assume a 365-day year.

8. Write the conversion factor you can use to change meters to kilometers. Use scientific notation. Then show how to find the number of kilometers light travels per second.

9. To solve this problem, you need to convert seconds to minutes, minutes to hours, and hours to days. Write the conversion factor for each.

10. Write the equation for the number of kilometers in a light-year that combines all the conversions and numerical operations. Show all units. Write the answer in scientific notation.

11. WRITING MATHEMATICS Explain what characteristics of a problem will help you decide whether the strategy of using dimensional analysis is appropriate.

Apply the Strategy

12. MOTION Acceleration requires the dimensions of a length divided by the product of two times, $\dfrac{(\text{length})}{(\text{time})(\text{time})}$. State whether each of the following units can be used to measure acceleration.

    a. feet per second per second
    b. miles per hour per second
    c. meters per kilometer per hour
    d. inches per day per minute

TRAVEL Felipé is planning a car trip with his family through New Mexico, Arizona, and California. He marked their route on a map, measured the segments, and approximated the total length of the route to be 46 cm. He checked the map scale and found that 2 .5 cm = 100 km. The family car averages 22 miles per gallon of gasoline, and gasoline costs $1.16 per gallon.

13. What information do you need to calculate Felipé's family budget for gasoline? Use a reference book to find this information (correct to one decimal place).

14. Show how to set up all the information to compute the answer and check the units by dimensional analysis. Find the answer.

SALES FORCE ORGANIZATION Companies must balance the need for sales personnel to promote products with the goal of cost efficiency. One common method of determining the size of sales force is based on this **workload formula**:

$$S = \frac{C \cdot F \cdot L}{T}$$

where S is the number of salespeople, C is the number of product customers (such as supermarkets or department stores) served, F is the call frequency necessary to service a customer annually, L is the average length of a call, and T is the average amount of selling time a salesperson has per year.

15. Silver Spoon Foods sells its products to 258,000 supermarkets, grocery stores and so on. Assume salespersons call on these customers at least 50 times a year. The average sales call lasts 48 minutes and an average salesperson works 40 hours a week for 50 weeks, but 10 hours a week are taken up by travel and paperwork. Make any necessary conversions and use dimensional analysis to show that the units of the answer are salespeople and determine the number.

REVIEW PROBLEM SOLVING STRATEGIES

READING BETWEEN THE LINES

1. Rectangle *ABCD* is shown. Point *P* is within the rectangle such that the distance from *P* to *A* is 3 units, from *P* to *B* is 4 units, and from *P* to *C* is 5 units. How far is *P* from *D*? Work in pairs to solve the problem.

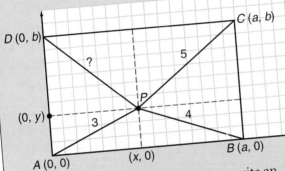

a. Use the known distance *PA* to write an equation for *x* and *y*. Call this Equation I.

b. *PB* is the hypotenuse of a right triangle. Express the length of one leg in terms of *a* and *x* and the other in terms of *y*. Use the Pythagorean theorem to write Equation II.

c. *PC* is the hypotenuse of a right triangle. Express the length of one leg in terms of *a* and *x* and the other in terms of *b* and *y*. Then write Equation III.

d. What do you want to find? What equation can you write? Call it IV.

e. Try to find a way of combining Equations I, II, and III so that the variable expression on one side of the resulting equation is exactly the same as one side of Equation IV.

f. Explain how to complete the problem.

A PERFECT 10

2. What is the smallest ten-digit number you can find that is made up of the digits 0, 0, 1, 1, 2, 2, 3, 3, 4, and 4? The 0s must be separated by zero digits, the 1s by one digit, the 2s by two digits, the 3s by three digits, and the 4s by four digits.

OLD TIMERS

3. Sonja needs to cook onions for exactly one minute. She has two old-fashioned timers. One of her timers takes 5 min to empty its top compartment. The other takes 8 min to empty its top compartment. How can she use her two timers to measure one minute?

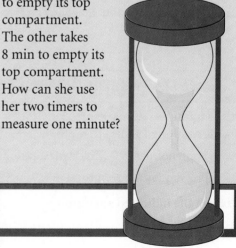

· · · CHAPTER REVIEW · · ·

VOCABULARY

Choose the word from the list that completes each statement.

1. A relation written in the form $y = kxz$, $k \neq 0$, is a(n) __?__.

2. A(n) __?__ is a fraction whose numerator and denominator are polynomials.

3. A(n) __?__ is a relation written in the form $y = \dfrac{k}{x}$ where $k \neq 0$, $x \neq 0$.

a. rational expression

b. joint variation

c. inverse variation

Lessons 14.1 and 14.2   DIRECT, INVERSE, AND JOINT VARIATION   pages 675–685

- Use direct variation, inverse variation, or joint variation to find the value of a given variable or the constant of variation.

Find each variable.

4. If y varies inversely as x and $y = 16$ when $x = 20$, find y when $x = 64$.

5. If y varies jointly as x and z and $y = 60$ when $x = 30$ and $z = 4$, find y when $x = 12$ and $z = 3$.

Lesson 14.3   SIMPLIFY RATIONAL EXPRESSIONS   pages 686–691

- A **rational expression** is an expression that can be written in the form $\dfrac{P}{Q}$ where P and Q are polynomials and $Q \neq 0$. To simplify a rational expression, factor both numerator and denominator and divide out the greatest common factor (GCF), so that the numerator and the denominator have no common factors other than 1.

Simplify each expression. State any restrictions on the variable.

6. $\dfrac{12x}{4x^2}$

7. $\dfrac{6b + 18}{b + 3}$

8. $\dfrac{2y - 1}{4 - 8y}$

Lesson 14.4   MULTIPLY AND DIVIDE RATIONAL EXPRESSIONS   pages 692–696

- To multiply rational expressions, factor and simplify each expression; then multiply the resulting factors. To multiply a rational expression by a polynomial P, rewrite the polynomial as $\dfrac{P}{1}$. To divide a rational expression by a polynomial P, multiply the expression by the reciprocal of P, $\dfrac{1}{P}$.

Perform the indicated operation.

9. $\dfrac{pq^2}{r} \cdot \dfrac{3p^2r}{p}$

10. $\dfrac{4}{3y} \div \dfrac{2}{9y^2}$

11. $\dfrac{n + 7}{2n - 6} \div \dfrac{10(n + 7)}{5(n - 3)}$

12. $\dfrac{6x + 24}{2x^2 - 8} \cdot (3x + 6)$

13. $\dfrac{x^2 - 36}{x^2 + 4x - 12} \div (3x - 18)$

14. $\dfrac{5x^2 + 5x}{x^2 - 3x - 4} \cdot (3x - 12)$

- To divide a polynomial by a monomial, divide each term of the polynomial by the monomial. To divide one polynomial by another when there are no common factors, use long division.

Divide.

15. $(6x^3 - 8x^2 + x) \div 2x^2$ **16.** $(y^2 + y - 56) \div (y - 7)$ **17.** $(5 - 7x^2 + 3x^4) \div (x - 1)$

18. The volume of a rectangular prism is $2x^3 + 5x^2 - x + 2$ and the length is $x + 3$. Find the area of the face bordered by the height and the width.

- To add rational expressions, rewrite the expressions as equivalent expressions with a least common denominator. Then add the numerators.

Add or subtract. Simplify if possible.

19. $\dfrac{7}{g + 2} - \dfrac{g + 6}{g + 2}$ **20.** $\dfrac{8}{3x} - \dfrac{2}{9x}$ **21.** $\dfrac{5}{3z^2} - \dfrac{2}{z} + \dfrac{1}{6z}$ **22.** $\dfrac{t}{2t + 8} - \dfrac{5}{3t + 12}$

- To simplify a complex rational expression, determine the LCD of all the rational expressions in both numerator and denominator and multiply each numerator and denominator by the LCD.

- To solve a rational equation, eliminate the denominators by multiplying both sides of the rational equation by the LCD of all the expressions. Then solve the resulting equation.

Simplify. Write your answer in simplest form.

23. $\dfrac{\frac{4}{a}}{\frac{7}{b}}$ **24.** $\dfrac{\frac{1}{m} - \frac{2}{n}}{\frac{4}{m}}$ **25.** $\dfrac{\frac{2}{x} + \frac{3}{y}}{\frac{1}{2x} - \frac{1}{2y}}$ **26.** $\dfrac{\frac{3}{r} + 2}{\frac{4}{s}}$

Solve and check.

27. $2b + \dfrac{4}{5} = \dfrac{2}{3}$ **28.** $\dfrac{y}{y - 2} - \dfrac{14}{y - 2} = 5$ **29.** $\dfrac{k + 5}{k} - \dfrac{2}{5k} = 10$

- You can use dimensional analysis to solve problems involving different units of measure.

Solve each problem.

30. On a map of the Mississippi, the distance between Biloxi and Jackson measures $6\frac{3}{8}$ in. The map scale is $\frac{3}{4}$ in. = 20 mi. Joyce's car averages 24 mi/gal of gasoline. If gasoline costs $1.29 a gallon, how much, to the nearest cent, will it cost in gasoline for Joyce to drive from Biloxi to Jackson?

CHAPTER ASSESSMENT

CHAPTER TEST

1. **WRITING MATHEMATICS** Write a paragraph explaining what happens to the y-values in the inverse variation $y = \dfrac{24}{x}$ as the x-values increase and decrease. Give examples.

2. **STANDARDIZED TESTS** Which value of k indicates that y varies inversely as x when $y = 14$ and $x = 2$?

   A. 7   **B.** 28   **C.** $\dfrac{1}{7}$ **D.** $\dfrac{1}{28}$

Tell whether each equation represents *direct*, *inverse*, or *joint variation*. State the constant of variation.

3. $y = \dfrac{4}{x}$ 4. $t = 5s$ 5. $A = 2\pi rh$

Determine the LCD for each set of expressions.

6. $\dfrac{15}{2x}, \dfrac{16}{x(x + 2)}$ 7. $\dfrac{3}{8def^2}, \dfrac{8}{6d^2e}, \dfrac{12}{f^2}$

Simplify each expression. State any restrictions on the variable.

8. $\dfrac{5x + 10}{15}$ 9. $\dfrac{x + 3}{x^2 - 10x - 39}$

10. $\dfrac{x^2 + 10x + 25}{x^2 + 11x + 30}$ 11. $\dfrac{c^2 - c - 56}{c^2 - 16c + 64}$

Perform the indicated operation.

12. $\dfrac{3x^2 - 15x}{4x + 2} \cdot \dfrac{8x + 4}{125 - 5x^2}$

13. $\dfrac{4x^2 - 8x}{3x + 1} \cdot \dfrac{9x + 3}{16 - 4x^2}$

14. $\dfrac{x^2 + x - 30}{x^2 - 36} \div \dfrac{x^2 - 25}{3x - 18}$

Add or subtract.

15. $\dfrac{7}{5r^2} + \dfrac{9}{10r} - \dfrac{5}{r}$ 16. $\dfrac{6}{a + 3} + \dfrac{14}{a - 10}$

17. $\dfrac{21}{16a^2} - \dfrac{5}{24ab} - \dfrac{1}{6b^2}$ 18. $\dfrac{7q - 5}{49 - q^2} - \dfrac{7}{q - 7}$

19. **WRITING MATHEMATICS** Write a paragraph explaining when and why you should use long division.

Convert each mixed expression into a rational expression.

20. $2p - \dfrac{4p - 7}{4p - 6}$ 21. $q - 5 - \dfrac{4}{q + 3}$

22. $\dfrac{2k - 1}{k + 10} - k$ 23. $a + 4 + \dfrac{4}{a - 4}$

Simplify. Write your answer in simplest form.

24. $\dfrac{\dfrac{1}{5c^2} - \dfrac{16}{5d^2}}{\dfrac{1}{c^2d} + \dfrac{4}{cd^2}}$ 25. $\dfrac{\dfrac{1}{x} - \dfrac{1}{y}}{\dfrac{3}{2x} - \dfrac{3}{2y}}$

26. $\dfrac{\dfrac{6}{a^2} + \dfrac{6}{a}}{\dfrac{3}{a} - 18}$ 27. $\dfrac{5 - \dfrac{6}{x + 9}}{10 - \dfrac{12}{x + 9}}$

28. For what values of x is the equation $2x^2 + 4x = 3x + 15$ true?

Write and solve an equation for each problem.

29. Determine the area of a rectangle that has a length of $\dfrac{p^2 - 4}{p^2q - 2p}$ cm and a width of $\dfrac{p^2 - p}{p^2 - 3p + 2}$ cm.

30. In calm air an airplane can fly at the rate of 960 mi/h. It can fly 3000 mi with a tailwind in the same time it can fly 2760 mi against the wind. Find the speed of the wind.

31. Earth is about 93,000,000 mi from the sun. If light travels at a speed of about 5.88×10^{12} mi/y, about how many minutes does it take light to travel from the sun to the Earth? Assume a 365-day year and use scientific notation.

PERFORMANCE ASSESSMENT

USE ALGEBLOCKS Use Algeblocks and the Quadrant Mat to divide polynomials by monomials and binomials. Create division examples such that there is no remainder when the polynomial is divided by the monomial or binomial. Have a partner use Algeblocks to find the quotient. Check that each step is correct.

Examples:

$$\frac{(x + 3)(x + 4)}{x + 3} \qquad \frac{x^2 + 7x + 12}{x + 3} \qquad \text{Have a partner find } (x^2 + 7x + 12) \div (x + 3).$$

$$\frac{4x(x - 8)}{4x} \qquad \frac{4x^2 - 32x}{4x} \qquad \text{Have a partner find } (4x^2 - 32x) \div 4x.$$

USE ALGEBLOCKS Use Algeblocks to make different rectangular solids. Have a partner determine the surface area and volume of each solid, write the ratio of the surface area to the volume, and then simplify the ratio if it can be simplified. Check that each step is correct.

USE A GRID Use a grid to draw different rectangles having a certain area and having lengths and widths that are whole numbers. Experiment with different areas and make a table to represent each area showing the lengths and widths which are factor pairs of each area.

PLAN A TRIP Plan a trip you would like to take. On a map, measure the distance of the route you could follow. Note the map scale. Find out the mileage that the car you would ride in gets to a gallon of gasoline. Decide whether you will use regular, mid-range, or high-test gasoline, and find out the cost per gallon. Then calculate how much in gasoline the trip would cost.

PROJECT ASSESSMENT

PROJECT *Connection* — Prepare a report for the farmer summarizing your exploration of silos. Be sure to include an explanation of the methods you used, relevant tables, graphs, mathematical interpretations, and your results and recommendations.

1. Have one student in your group play the role of the farmer, asking questions about building the silo. The other students act as consultants using their reports to respond to questions and justify their recommendations. Both sides should agree on the silo that represents the best choice for the situation.

2. The consultants can propose other investigations that might be useful for the farmer. These investigations may involve geometric questions such as fencing an area or financial questions such as break-even/profit analysis of the farmer's business operations.

•••CUMULATIVE REVIEW•••

1. If y varies jointly as x and z, and $y = 36$ when $x = 2$ and $z = 9$, find x when $y = 40$ and $z = 4$.

Simplify each expression. State any restrictions on the variable.

2. $\dfrac{2x^3 - 18x}{4x - 12}$

3. $\dfrac{x^2 + x - 20}{x^2 + 3x - 10}$

Perform each operation.

4. $(3n^3 + 8n^2 - 23n + 5) + (9n^3 - 8n^2 - 15)$

5. $(4y - 7)(2y^2 + 9y - 2)$

6. $(6t^3 - 3t + 11) - (2t^3 - 5t^2 + 3t - 1)$

7. $(x^4 + 5x^3 - 7x - 2) \div (x + 3)$

8. WRITING MATHEMATICS Explain why $\dfrac{4x}{4y} = \dfrac{x}{y}$ but $\dfrac{x + 4}{y + 4} \neq \dfrac{x}{y}$.

Solve each equation or inequality.

9. $4x + 15 = 2x - 9\left(\dfrac{1}{3}x - 2\right)$

10. $\dfrac{2}{3}x + 5 > 8 - \dfrac{5}{6}x$ 11. $-3m \leq 12$

12. $5n^2 + 27n - 18 = 0$ 13. $2y^2 + 5y = 4$

14. $\sqrt{x + 7} + 9 = x + 4$

Simplify each expression.

15. $\dfrac{x^3 - 16x}{x^2 + 3x - 10} \div \dfrac{x^2 + 2x - 8}{x^2 - 4x + 4}$

16. $\dfrac{y + 4}{3y^2 - 27} + \dfrac{6}{y - 3} - \dfrac{4}{3y + 9}$

17. $4\sqrt{18} + 2\sqrt{72} - 4\sqrt{\dfrac{1}{2}}$

18. $\dfrac{(x^2y^{-4})^6}{(x^{-4}y^8)^{-3}}$ 19. $\dfrac{4ab^5c^{-2}}{12a^3b^{-4}c^{-3}}$

Solve for x.

20. $|x + 2| = 2$ 21. $|2x + 3| + 11 = 12$

22. $2|x - 5| - 6 = 0$ 23. $-3|x + 4| = -12$

The data show how many pitches a baseball pitcher threw in consecutive starts. Use the data for Questions 24–26.

112, 75, 98, 140, 52, 101, 110, 94, 82

24. Find the mean number of pitches thrown.

25. Make a boxplot representing the data.

26. WRITING MATHEMATICS Of bar graph, line graph, or circle graph, which type of graph would you choose to represent the data? Why?

27. Solve for x in the equation $xy - 5 = x + 4$.

28. A chemist has 450 mL of 25% saline solution. How much water should be added to dilute the solution to 20% saline?

29. Donald needs 3 hours more than Mickey to paint a room by himself. Working together, they can paint the room in 2 hours. How long would it take Donald to paint the room himself?

Identify the domain and range of each function.

30. $y = 2x - 3$ 31. $y = x^2 - 2$

32. $y = -|x - 3| + 1$ 33. $y = \sqrt{x + 2} + 1$

34. STANDARDIZED TESTS If $n < 0$, the determinant $\begin{vmatrix} 5 & n \\ 2 & 4 \end{vmatrix}$ has the value

A. $20 - 2n$ B. $20 + 2n$
C. $2n - 20$ D. $5n - 8$
E. $10 - 4n$

35. A concession stand manager adds pompoms and T-shirts to her inventory. There is room to store at most 1500 of the new items. The profit on pompoms is $3.00 and the profit on T-shirts is $5.00, so if x is pompoms and y is T-shirts then her profit is $P = 3x + 5y$. The manager can order no more than 800 pompoms and 1000 T-shirts. How many pompoms should she order? How many T-shirts? What is the maximum profit?

· · · STANDARDIZED TEST · · ·

STUDENT PRODUCED ANSWERS Solve each question and on the answer grid write your answer at the top and fill in the ovals.

Notes: Mixed numbers such as $1\frac{1}{2}$ must be gridded as 1.5 or 3/2. Grid only one answer per question. If your answer is a decimal, enter the most accurate value the grid will accommodate.

1. If $\sin A = \dfrac{3}{5}$, find the value of $\tan A$.

2. Solve the equation $x^2 - 4x - 2 = 0$. Find the product of the two solutions.

3. Pedro has 12 more dimes than nickels. He has a total of $2.70. Find the total number of coins that Pedro has.

4. Find the slope of a line perpendicular to the line that passes through the points $(-3, 4)$ and $(5, 1)$.

5. Find the measure for the perimeter of the figure shown.

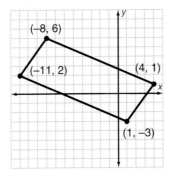

6. Suppose you toss a fair coin once and roll a die twice. Find the number of different outcomes there are in the sample space.

7. If p varies jointly as m and n, and $p = 12$ when $m = 8$ and $n = 6$, find the value of m when $p = 20$ and $n = 5$.

8. Find the remainder when $2x^3 - 54$ is divided by $x - 3$.

9. Sum the solutions to the systems of equations.
$$\begin{cases} x + 2y = 10 \\ 2x + y = 5 \end{cases}$$

10. Find the number of rows in matrix Q.
$$Q = \begin{bmatrix} 7 & 6 \\ 3 & 1 \\ 2 & 9 \\ 4 & 0 \end{bmatrix}$$

11. Simplify the expression. Enter the answer as a fraction.
$$\frac{\dfrac{2}{3} - \dfrac{3}{5}}{\dfrac{5}{6} + \dfrac{1}{4}}$$

12. Find the degree of the monomial $(2x^3y^5)^4$.

13. Every morning Sebastian runs 8 miles, then walks 2 miles. He runs 6 mi/h faster than he walks. If his total time yesterday was 1 hour, find Sebastian's rate running.

14. So far, a survey taken asking people their favorite color has shown that 45 people prefer blue, 38 prefer red, 25 like green, 18 prefer yellow, and 24 people like various other colors. Find the probability, as a decimal, that the next person asked will choose yellow.

15. Evaluate the expression. Enter the answer in decimal notation.
$$\frac{(4.2 \cdot 10^7)(2 \cdot 10^{-9})}{(2.1 \cdot 10^{-4})}$$

16. Find the value of x in the determinant equation.
$$\begin{vmatrix} x & 4 \\ -3 & 5 \end{vmatrix} + \begin{vmatrix} -4 & 2 \\ x & 1 \end{vmatrix} = 11$$

17. If y varies inversely as x and x is 12 when y is 5 then what is x when y is 30?

18. What is the mode of this set of data?

24 22 22 20 25 21 23 23 21 20
22 25 24 20 22 23 22 25 21 25

19. Which of the following is the greatest?
$$\frac{76}{16} \qquad \sqrt{10} \qquad \frac{159}{50} \qquad 3.\overline{15}$$

Geometry
Quick Notes

Geometry Basics

All geometric figures are made up of at least one point.

| point | line | ray | line segment (or segment) | angle |

About Lines

Lines in a plane can be either parallel to each other or they can intersect each other.

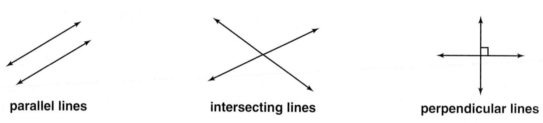

parallel lines   intersecting lines   perpendicular lines

About Angles

Angles are measured in degrees.

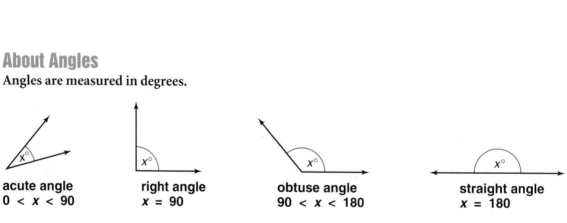

acute angle
$0 < x < 90$

right angle
$x = 90$

obtuse angle
$90 < x < 180$

straight angle
$x = 180$

Complementary and Supplementary Angles

Two angles are complementary if the sum of their measures is exactly 90°.

Two angles are supplementary if the sum of their measures is exactly 180°.

About Triangles

Triangles are three-sided plane figures. They can be classified according to the measures of their sides or their angles.

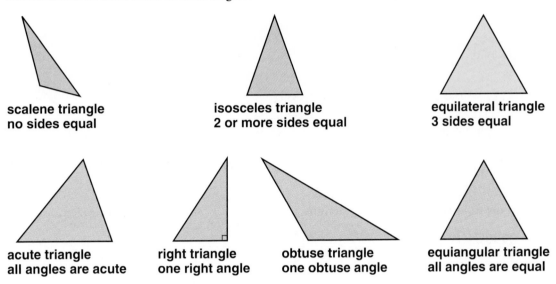

scalene triangle
no sides equal

isosceles triangle
2 or more sides equal

equilateral triangle
3 sides equal

acute triangle
all angles are acute

right triangle
one right angle

obtuse triangle
one obtuse angle

equiangular triangle
all angles are equal

About Quadrilaterals

Quadrilaterals are four-sided plane figures. Each figure in the diagram has all the properties of the figures preceding it, including the properties listed with that figure.

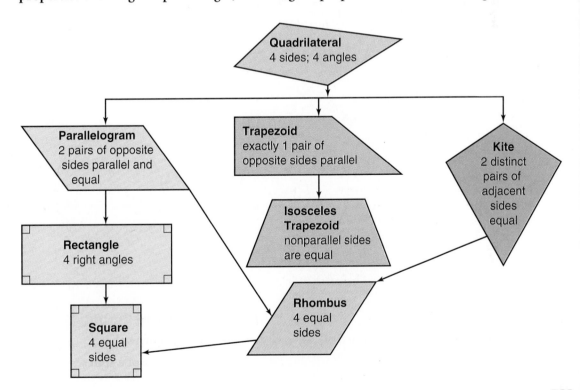

Quadrilateral
4 sides; 4 angles

Parallelogram
2 pairs of opposite sides parallel and equal

Trapezoid
exactly 1 pair of opposite sides parallel

Kite
2 distinct pairs of adjacent sides equal

Rectangle
4 right angles

Isosceles Trapezoid
nonparallel sides are equal

Square
4 equal sides

Rhombus
4 equal sides

About Other Polygons

Polygons are plane figures made up of segments and angles. Triangles and four-sided figures are also polygons.

pentagon

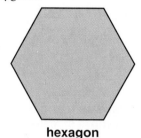

hexagon

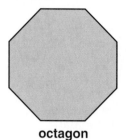

octagon

Perimeter Formulas

In the following formulas, l = length, w = width, s = side, and P = perimeter.

Perimeter of a rectangle $P = 2l + 2w$
Perimeter of a square $P = 4s$

Area Formulas

In the following formulas, b = base, B = long base, h = height, l = length, w = width, s = side, and A = area.

Area of a parallelogram $A = bh$
Area of a rectangle $A = lw$
Area of a square $A = s^2$
Area of a trapezoid $A = \frac{1}{2}(B + b)h$
Area of a triangle $A = \frac{1}{2}bh$

About Circles and Spheres

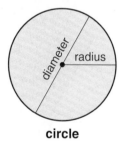

circle

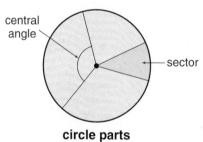

circle parts

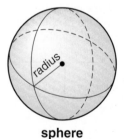

sphere

Circle Formulas

Circumference of a circle $C = 2\pi r$ or $C = \pi d$
Area of a circle $A = \pi r^2$

Sphere Formulas

Area of a sphere $A = 4\pi r^2$
Volume of a sphere $V = \frac{4}{3}\pi r^3$

About Geometric Solid Figures

Geometric solid figures are made up of plane polygons. Below are some geometric solid right figures.

cone

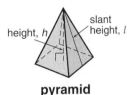

pyramid

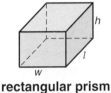

cylinder   rectangular prism

cube

Base

The cone and the pyramid have one base. The cylinder and the prism have two bases and they are parallel. The cone and cylinder have circular bases. The base of a pyramid or prism can be any polygonal shape.

Lateral Surface

The lateral surface is the side or sides of the solid figure other than a base. The cone and cylinder have one lateral surface. The lateral surface of a pyramid is made up of triangles. The lateral surface of a right prism is made up of rectangles.

Slant Height

The slant height of a cone is measured from the vertex of the cone to the edge of its base. The slant height of a pyramid is measured from the vertex to the center of one side of the base.

Formulas

Total surface area of a right circular cone $T = \pi r(l + r)$

Volume of a cone $V = \dfrac{1}{3}\pi r^2 h$

Total surface area of a right cylinder $T = 2\pi r(r + h)$
Volume of a cylinder $V = \pi r^2 h$

Total surface area of a rectangular prism $T = 2(lw + lh + wh)$
Volume of a rectangular prism $V = lwh$

Total surface area of a cube $T = 6s^2$
Volume of a cube $V = s^3$

Technology

Quick Notes

Using the Texas Instruments TI-82 Graphing Calculator

- **For additional features and instructions, consult your user's manual.**

  THE KEYBOARD The feature accessed when a key is pressed is shown in white on the key. To access the features in blue above each key, first press the 2ND key. To access what is in white above the keys, first press ALPHA.

  CALCULATIONS Calculations are performed in the Home Screen. This screen may be returned to at any time by pressing 2ND QUIT. The calculator evaluates according to the order of operations. Press ENTER to calculate. For 3 + 4 × 5 ENTER, the result is 23. You can replay the previous line by pressing 2ND ENTER. Use the arrow keys to edit.

Displaying Graphs

- GRAPH FEATURE To enter an equation, press Y=. Enter an equation such as $y = 2x + 3$ using the X, T, θ key for x. Then press GRAPH to display the graph in the viewing window.

  VIEWING WINDOW To set range values for the viewing window, press WINDOW. Press ▶ to access FORMAT where you can choose features such as Grid Off or Grid On.

  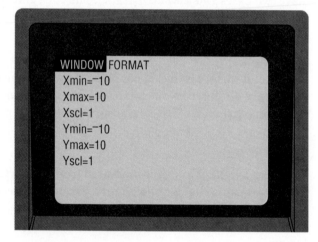

  ZOOM FEATURE Press ZOOM and then 6 (Standard) to set a standard viewing window. Press ZOOM 1 (Box) to highlight a particular area and zoom in on that part of the graph. Press ZOOM 8 (Integer) to set values for a friendly window.

  TRACE FEATURE Pressing TRACE places the cursor directly on the graph and shows the x- and y-coordinates of the point where the cursor is located. You can move the cursor along the graph using the right and left arrow keys.

  TABLE FEATURE Press 2ND TBLSET to set up the table. Then press 2ND TABLE to see a table of values for each equation.

INTERSECTION FEATURE To determine the coordinates of the point of intersection of two graphs, press 2ND CALC, then 5 (Intersect). The calculator will then prompt you to identify the first graph. Use the right and left arrow keys to move the cursor to the first graph, close to the point of intersection. Repeat to identify the second graph and get the coordinates of the point of intersection.

Statistics

- **ENTERING DATA** Enter data into lists by pressing STAT 1 (Edit).

  CALCULATING STATISTICS Return to the Home Screen by pressing 2ND QUIT. To calculate the mean of List 1, press 2ND LIST ▶ (MATH) 3 (Mean) 2ND L1 ENTER. To calculate the median, choose 4 instead of 3. To calculate statistics, press STAT ▶ (CALC) 1 (1 - Var Stats) ENTER. You can also choose 2 to calculate two variable statistics and 5 to calculate linear regression. To see the lower quartile of a boxplot, press VARS 5 (Statistics) ▶ ▶ ▶ (BOX) 1 (Q1). Use 2 for median and 3 for the upper quartile.

  GRAPHING DATA To graph your data, press 2ND STATPLOT and choose a scatter plot, a line graph, a boxplot, or a histogram. Then choose the data list. Then press GRAPH to draw the graph.

Using the Casio CFX-9800G Graphing Calculator

- **For additional features and instructions, consult your user's manual.**

  THE KEYBOARD The feature accessed when a key is pressed is shown in white on the key. To access the features in gold above each key, first press the SHIFT key. To access what is in red above the keys, first press ALPHA.

  THE MAIN MENU This is the screen you see when you first turn the calculator on. Highlight and press EXE or press the number to choose the menu item. You can access the Main Menu at any time by pressing the MENU key.

  PERFORMING CALCULATIONS Calculations are performed by pressing 1 (for COMPutations) in the Main Menu. The calculator evaluates according to the order of operations. Press EXE (for EXEcute) to calculate. For 3 + 4 × 5 EXE, the result is 23. You can replay the previous line by pressing ◀. Use the arrow keys to edit.

Displaying Graphs

- **COMP MODE** Press 1 in the Main Menu. To enter an equation, press GRAPH. Enter an equation such as $y = 2x + 3$ using the X, θ, T key for x. Then press EXE to display the graph in the viewing window.

  GRAPH MODE Press 6 (GRAPH) in the Main Menu. Then press AC. Use the up or down arrows to choose a location to store the equation. Enter the equation as above. Then press F6 (DRW) to display the graph in the viewing window.

VIEWING WINDOW To set range values for the viewing window, press RANGE. You can use F1 (INIT) to set standard values for a viewing window.

ZOOM FEATURE Press SHIFT F2 (ZOOM) to access this feature. Press F1 (BOX) to highlight a particular area and zoom in on that part of the graph. Press F5 (AUT) to set range values for a friendly window.

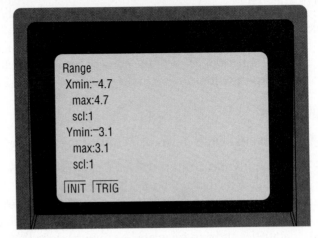

Range
Xmin:⁻4.7
 max:4.7
 scl:1
Ymin:⁻3.1
 max:3.1
 scl:1
INIT TRIG

TRACE FEATURE Pressing SHIFT F1 (TRACE) places the cursor directly on the graph and shows the x- and y-coordinates of the point where the cursor is located. You can move the cursor along the graph using the right arrow keys.

TABLE FEATURE Press 8 (TABLE) in the Main Menu. Press AC to clear the screen. Then press F1 (RANGE FUNC). Select the function. Then press F5 (RNG) to set up the table. Press F6 (TBL) to see the table of values. You can press SHIFT QUIT to return to a previous screen.

INTERSECTION FEATURE To determine the coordinates of the point of intersection of two graphs, press 2ND CALC, then 5 (INTERSECT). The calculator will then prompt you to identify the first graph. Use the right and left arrow keys to move the cursor to the first graph close to the point of intersection. Repeat to identify the second graph and get the coordinates of the point of intersection.

Statistics

ENTERING DATA From the Main Menu, press 3 (SD). Press SHIFT SET UP and select STOre for S-data. Then press EXIT AC to clear the screen. Enter data, pressing F1 after each entry.

CALCULATING STATISTICS To calculate the mean ($\bar{x}$), press F4 (DEV) F1 ($\bar{x}$) EXE. To calculate the median, press F4 (DEV) F4 ▼ and F2 (Med) EXE.

GRAPHING DATA To graph your data, press SHIFT SET UP and select DRAW for S-graph. Press EXIT and then GRAPH EXE.

Using the Hewlett-Packard 38G Graphing Calculator

For additional features and instructions, consult your user's manual.

THE KEYBOARD The feature accessed when a key is pressed is shown in yellow on the key. To access the features in green above each key, first press the green key. To access what is in red below the keys, first press A...Z. The blank keys at the top of the keyboard are used for the menu items at the bottom of the screen. Pressing the Menu key at the far right will return menu items to the screen.

CALCULATIONS Calculations are performed in the Home Screen. This screen may be returned to at any time by pressing HOME. The calculator evaluates according to the order of operations. Press ENTER to calculate. For 3 + 4 × 5 ENTER, the result is 23. You can edit a previous line by using the arrow keys to choose a line you want to edit. Press the Copy key. Use the arrow keys to edit.

Displaying Graphs

- **GRAPH FEATURE** To enter an equation, press LIBrary. Select Function. Press ENTER. Press the Edit key to enter an equation such as $y = 2x + 3$, using the X, T, θ key for x. Then press PLOT to display the graph in the viewing window.

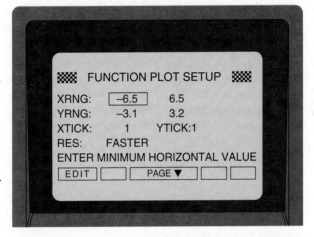

VIEWING WINDOW Press the green key and VIEWS. Select Auto Scale to get a friendly window, or you can press the green key and PLOT to change the range in the viewing window. Next, press PLOT to show the graph.

ZOOM FEATURE Press the Menu key. Press the Zoom key. Then select Box... to highlight a particular area and zoom in on that part of the graph.

TRACE FEATURE Pressing the Trace key puts the cursor directly on the graph and shows the x- and y-coordinates of the point of its location. You can move the cursor along the graph using the right and left arrow keys.

TABLE FEATURE To display a table of values for the graphed equation, press NUMber.

INTERSECTION FEATURE To determine the coordinates of the point of intersection of two graphs, move the cursor close to the point of intersection. Press the Function key and select Intersection. This verifies one of your equations. Press ENTER. Then press the Function key and select Intersection to verify the other equation. Press ENTER.

Statistical Features

- **ENTERING DATA** Data may be entered into a list by pressing LIB and selecting Statistics.

CALCULATING STATISTICS To calculate mean, median, upper, and lower quartiles of the data entered in C1, press the Stats key and use the arrow keys to access all the information.

GRAPHING DATA To graph your data, press the green key and then PLOT. Then press the Choos key. Select BoxWhisker or Histogram and ENTER. Then press the green key and VIEWS. Select Auto Scale.

GLOSSARY

• • A • •

absolute value (p. 68)   The distance of a number from 0 on the number line. The absolute value of x is written as $|x|$. For any real number x, $|x| = x$ if $x > 0$ or $x = 0$ and $|x| = -x$ if $x < 0$.

absolute value function (p. 416)   The function $f(x) = |x|$. The corresponding absolute value equation is $y = |x|$ where x is any real number and y is any number greater than or equal to zero.

absolute value inequality (p. 432)   An inequality of the form $|ax \pm b| < c$ or $|ax \pm b| > c$.

acute triangle (p. 619)   A triangle with three acute angles.

addition property of equality (p. 124)   For all real numbers a, b, and c, if $a = b$, then $a + c = b + c$.

addition property of inequality (p. 230)   For all real numbers a, b, and c, if $a > b$, then $a + c > b + c$ and if $a < b$, then $a + c < b + c$.

additive identity property (p. 80)   For any real number a, $a + 0 = a$.

additive inverse property (p. 80)   For any real number a, $a + (-a) = 0$. (See also **opposite**.)

Algeblocks (p. 56)   Models that physically represent algebraic concepts and operations.

algebraic expression (p. 13, 60)   An expression having at least one number, operation, and variable.

amplitude (of a function) (p. 660)   The distance from the middle line of the graph of a function to the maximum or minimum value of the function.

associative property of addition (p. 80)   For any real numbers a, b, and c, $(a + b) + c = a + (b + c)$.

associative property of multiplication (p. 89)   For any real numbers a, b, and c, $(a \cdot b) \cdot c = a \cdot (b \cdot c)$.

average (p. 32)   The mean of a set of data.

average rate of change (p. 271)   The slope formula can be used to find average rate of change.

axis of symmetry (p. 469)   A line, such as the vertical line passing through the vertex of a parabola, which divides a figure so that when folded, the two halves coincide.

• • B • •

bar graph (p. 19)   A display of statistical information in which horizontal or vertical bars, or rectangles, represent data to be compared.

base (of a power) (p. 59)   The repeating factor in a power. For example, in 4^3 the base is 4.

binomial (p. 541)   A polynomial of two terms.

boundary (of two half-planes) (p. 374)   The line, or graph of a linear equation, that divides the coordinate plane into two half-planes.

boxplot (p. 248)   A graph that uses the three quartiles and the least and greatest values of a set of data to provide a visual display of the data.

• • C • •

cells (p. 14)   The spaces formed by the horizontal rows and vertical columns on a computer spreadsheet, in which formulas or data can be entered.

circle graph (p. 19)   A display of statistical information in which data is represented by sectors.

closed half-plane (p. 375)   A half-plane that includes the boundary line.

cluster sampling (p. 22)   A type of statistical sampling in which a specifically defined portion of a population is chosen at random and interviewed.

coefficient (p. 61)   The numerical, nonvariable portion of a monomial.

coefficient of correlation (p. 299)   The statistical measure r of how closely a set of data approximates a line. The range of the coefficient of correlation is from -1 to 1.

commutative property of addition (p. 80)   For any real numbers a and b, $a + b = b + a$.

commutative property of multiplication (p. 89)   For any real numbers a and b, $a \cdot b = b \cdot a$.

complementary angles (p. 620)   Two angles the sum of whose measures is 90°. Each angle is the *complement* of the other.

complementary events (p. 400) The two possible outcomes in a random experiment in which the event occurs or the event does not occur. The probabilities of two complements total 1.

complex rational expression (p. 708) A rational expression that contains one or more rational expressions in its numerator or denominator.

compound inequality (p. 241, 430) Two inequalities connected by *and* or *or*.

computer spreadsheets (p. 14) Computer software programs that use formulas to calculate and analyze data and display it in columns and rows.

conjugates (p. 629) Two expressions that are the sum and difference of the same two terms, for example, $a + \sqrt{b}$ and $a - \sqrt{b}$.

conjunction (p. 241, 430) Two statements joined by the word *and*. A conjunction is true if and only if both of its statements are true.

consecutive integers (p. 107) Any integers in counting, or successive, order.

consistent (system) (p. 348) A system of equations with at least one solution.

constant (p. 61) In an algebraic expression, a number not multiplied or divided by a variable.

constant of variation (pp. 307, 679, 681) The constant k in the equation of a direct variation $y = kx$, an inverse variation $xy = k$, or a joint variation $y = kwx$.

constraints (p. 388) Conditions that limit business activity. In linear programming, constraints are represented by inequalities.

convenience sampling (p. 22) A type of statistical sampling in which a readily available portion of a population is chosen, and all members of that portion are surveyed.

conversion factor (p. 718) A ratio of two equal quantities used to convert one dimensional unit to another.

coordinate (p. 177) The real number that corresponds to a point on a number line; the first of the two paired numbers corresponding to a given point in the coordinate plane.

coordinate plane (p. 177) A two-dimensional grid consisting of two perpendicular number lines called the x-axis and the y-axis.

corresponding elements (of a matrix) (p. 28) Elements in the same position in each of two matrices having the same dimensions.

cosine (p. 653) For either acute angle in a right triangle, the ratio of the length of the side adjacent to the angle to the length of the hypotenuse.

crest (of a graph) (p. 660) The maximum value of the graph of a periodic function.

• • D • •

data points (p. 169) Points on a coordinate plane that correspond to statistical information.

degree (of an expression) (p. 62) The greatest exponent of the variables in the expression in simplest form.

degree of a monomial (p. 521) The sum of the degrees of all the variable factors in a monomial.

degree of a polynomial (p. 542) The greatest degree of any of the terms of a polynomial.

degree of a variable (in a monomial) (p. 521) The number of times a variable occurs as a factor in a monomial.

dense (p. 446) A characteristic of real numbers because between any two real numbers there exists another real number.

dependent system (p. 349) A system of equations with the same graph.

dependent (variable) (p. 299) The output value of a relation.

determinant (p. 353) A numerical value associated with a square matrix. The determinant of a matrix is symbolized by using vertical bars in place of matrix brackets.

difference of two squares See **product of a sum and a difference**.

dimensional analysis (p. 718) The method of using units or dimensions in the solution of problems.

dimensions (of a matrix) (p. 27) The number of horizontal rows and the number of vertical columns in a matrix.

dimensions (of a quantity) (p. 718) The units in which a quantity is measured.

direct variation (p. 307, 679) When y varies directly as x and k is the constant of variation, an equation can be written in the form $y = kx, k \neq 0$.

discriminant (p. 491, 597)   The expression $b^2 - 4ac$, under the radical symbol in the quadratic formula. The discriminant can be used to determine the number of real number solutions of any quadratic equation.

disjunction (p. 242, 425)   Two statements joined by the word *or*. A disjunction is true if at least one of the statements is true.

distance formula (p. 642)   The formula $d = \sqrt{(x_2 - x_1)^2 + (y_2 - y_1)^2}$ used to determine the distance d between any two points (x_1, y_1) and (x_2, y_2) in the coordinate plane.

distributive property of multiplication over addition (p. 96)   For all real numbers a, b, and c, $a(b + c) = ab + ac$, and $(b + c)a = ba + ca$.

division (p. 90)   For any real numbers a and b, $b \neq 0, a \div b = a \cdot \dfrac{1}{b}$.

division property of equality (p. 130)   For all real numbers a, b, and c, if $a = b$, and $c \neq 0$, then $\dfrac{a}{c} = \dfrac{b}{c}$.

division property of inequality (p. 231)   For all real numbers a, b, and c, if $a > b$, and $c > 0$, then $\dfrac{a}{c} > \dfrac{b}{c}$, and if $a > b$, and $c < 0$, then $\dfrac{a}{c} < \dfrac{b}{c}$.

domain (of a function) (p. 172)   The input values of a function.

• • E • •

element (of a matrix) (p. 27)   Each number or entry in a matrix.

element (of a set) (p. 437)   An item or number that is part of a set. For example, 4 is an element of $\{2, 4, 6, 8\}$.

elimination method (p. 342)   A method for solving a system of linear equations by using multiplication and addition to eliminate one of the variables.

equation (p. 115)   A statement that two numbers or expressions are equal.

equilateral triangle (p. 619)   A triangle having three sides equal in length.

equivalent equations (p. 124)   Equations that have the same solution.

equivalent inequalities (p. 230)   Inequalities that have the same solution.

evaluate (an expression) (p. 60)   To find the value of an algebraic expression by substituting a given number for each variable and simplifying.

evaluate a function (p. 172)   To find the output, $f(x)$, for a given input x.

event (p. 398)   Any one of the possible outcomes of an experiment.

experimental probability (p. 38)   The probability of an event determined by observation or measurement. The experimental probability, $P(E)$, of an event E is given by
$$P(E) = \frac{\text{number of times } E \text{ occurs}}{\text{total number of trials}}$$

exponent (p. 59)   A superscript number showing how many times a base is used as a factor. For example, in 2^5, 5 is the exponent.

extraneous solution (p. 713)   An apparent solution that does not satisy the original equation. Extraneous solutions often result from squaring both sides of an equation.

extremes (of a proportion) (p. 136)   In the proportion $a{:}b = c{:}d$, a and d are the extremes.

• • F • •

factor (p. 88)   Any of two or more numbers multiplied to produce a product. For example, in $3(4) = 12$, 3 and 4 are factors of 12.

factored completely (p. 591)   A polynomial is factored completely when it cannot be factored further.

factored form (of a polynomial) (p. 575)   A polynomial expressed as a product of other polynomials.

feasible region (p. 388)   The intersection of the graphs of a system of constraints; the region of the coordinate plane that includes all possible solutions of the objective function.

finite (set) (p. 437)   A set whose elements can be counted or listed.

FOIL (p. 553)   A method for determining the product of two binomials by finding the product of the First terms, the Outside terms, the Inside terms, and the Last terms, and then simplifying the result.

formula (p. 152)   A literal equation that expresses a relationship between two or more vaariables.

frequency (p. 8)   The number of times an item occurs in a set of data.

frequency table (p. 8) A method of recording and organizing data using tally marks to show how often an item occurs in a set of data.

function (p. 171) A relation having exactly one output value for each input value in a set of paired values. In a function, more than one input value may have the same output value.

function notation (p. 172) The notation for representing a rule that associates an input value (independent variable) with an output value (dependent variable). The most commonly used function notation is the "f of x" notation, written "$f(x)$."

function rule (p. 172) The description of a function.

fundamental counting principle (p. 399) If one event can occur in m ways and, following that, a second event can occur in n ways, then the total number of possible outcomes equals $m \cdot n$.

• • G • •

geometric mean (p. 641) A number b is the geometric mean between two numbers m and n if $b = \sqrt{m \cdot n}$.

• • H • •

half-planes (p. 374) The two regions into which the graph of a linear equation divides the coordinate plane.

horizontal axis (p. 169) The x-axis of the coordinate plane.

horizontal shift (of a graph) (p. 417) The left or right translation of a graph in relation to another graph.

• • I • •

identity element for multiplication (p. 89) The number 1, because the product of any number and 1 is the number itself.

inconsistent (system) (p. 349) A system of equations with no solution.

independent (system) (p. 348) A system of equations that has exactly one solution.

independent (variable) (p. 299) The input value of a relation.

inequality (p. 67, 222) A statement that two numbers or expressions are not equal; a statement containing an inequality symbol.

inequality symbol (p. 67) A symbol used to compare, or order, real numbers. The inequality symbols are $<$, $>$, $\leq$, $\geq$, and $\neq$.

infinite (set) (p. 437) A set whose elements cannot be counted or listed, but go on forever. The set $\{0, 1, 2, 3, \ldots\}$ is an infinite set.

integer(s) (p. 56) All whole numbers and their opposites.

inverse operation (p. 90) An operation "undoing" what another operation does. For example, multiplication and division are inverse operations.

inverse variation (p. 679) When y varies inversely as x and k is the constant of variation, an equation can be written in the form $y = \dfrac{k}{x}, k \neq 0$.

irrational number (p. 65, 438) A number that cannot be expressed as the quotient of two integers, a terminating decimal, or a repeating decimal. For example, π and $\sqrt{2}$ are irrational numbers. (See also **perfect square**.)

isosceles triangle (p. 619) A triangle having at least two sides equal in length.

• • J • •

joint variation (p. 681) When y varies directly as the product of w and x and k is the constant of variation, an equation can be written in the form $y = kwx, k \neq 0$.

• • L • •

leading coefficient (p. 578) The coefficient of the first term in a polynomial in standard form.

least common denominator (LCD) (p. 703) The least positive common multiple of the denominators of two or more fractions or rational expressions.

like radicals (p. 631) Two or more radicals with the same radicand.

like terms (p. 97, 542) Two or more terms in an algebraic expression that have the same variable base and exponent.

linear equation in two variables (p. 183) An equation that can be written in the form $y = ax + b$, representing a linear function.

linear function (p. 183)   A function that can be represented by a straight line or by a linear equation in two variables of the form $y = ax + b$, where x and y are variables and a and b are constants.

linear inequality in two variables (p. 373)   An inequality that results when the equal symbol in a linear equation is replaced by $<$, $>$, $\leq$, $\geq$, or $\neq$.

linear programming (p. 388)   A method used by business and government to maximize or minimize quantities such as profit and cost.

linear system (p. 328)   Two or more linear equations that are considered together.

line graph (p. 19)   A display of statistical information in which points representing data are plotted and then connected with line segments.

line of best fit (p. 299)   The line that approximates a trend for the data on a scatter plot. Also called a *regression line*.

line of reflection (p. 416)   The line over which a figure is reflected, or flipped.

line of symmetry (pp. 413, 417, 469)   A line that divides a figure so that when the figure is folded over that line, the two parts match exactly.

line plot (p. 19)   A display of statistical information in which each item of data is represented by an X with X's stacked vertically in columns.

literal equation (p. 152)   An equation that contains at least two different variables.

• • M • •

matrix *(plural: matrices)* (p. 27)   A rectangular arrangement of numbers in rows and columns and enclosed with brackets.

maximize (an objective function) (p. 390)   To determine the greatest possible value of an objective function. Maximums occur at or near a vertex of the feasible region of a linear programming graph.

maximum (point) (p. 468)   See **vertex**.

mean (p. 32)   The sum of the items in a set of data divided by the number of items. Also called the *average*.

means (of a proportion) (p. 136)   In the proportion $a:b = c:d$, b and c are called the means.

measure of central tendency (p. 32)   A statistical measurement, such as the mean, the median, or the mode, used to describe a set of data.

median (p. 33)   The middle value of a set of data that are arranged in numerical order. When there are two middle values, the median is the average of these two values.

midpoint formula (p. 643)   The point that is equidistant from the endpoints of a line segment. The midpoint M of the line segment whose endpoint are $P_1(x_1, y_1)$ and $P_2(x_2, y_2)$ is $M\left(\dfrac{x_1 + x_2}{2}, \dfrac{y_1 + y_2}{2}\right)$.

minimize (an objective function) (p. 390)   To determine the least possible value of an objective function. Minimums occur at or near a vertex of the feasible region of a linear programming graph.

minimum (point) (p. 468)   See **vertex**.

mixed expression (p. 707)   The sum or difference of a polynomial and a rational expression.

mode (p. 33)   The number or element in a set of data occurring most often.

model (p. 55)   A physical or algebraic representation.

monomial (p. 521)   A term that is a number, a variable, or the product of a number and one or more variables.

multiplication property of equality (p. 129)   For all real numbers a, b, and c, if $a = b$, then $ca = cb$.

multiplication property of inequality (p. 231)   For all real numbers a, b, and c, if $a > b$, and $c > 0$, then $ac > bc$, and if $a > b$ and $c < 0$, then $ac < bc$.

multiplicative identity property (p. 90)   For any real number a, $a \cdot 1 = 1 \cdot a = a$.

multiplicative inverse (p. 90)   The reciprocal of a number. For example, a and $\dfrac{1}{a}$ are multiplicative inverses.

multiplicative inverse property (p. 90)   For any nonzero real number b, there exists a real number $\dfrac{1}{b}$ such that $b \cdot \dfrac{1}{b} = 1$.

multiplicative property of zero (p. 89)   For any real number a, $a \cdot 0 = 0 \cdot a = 0$.

• • N • •

negative integers (p. 56)   Integers less than zero.

negative reciprocal (p. 279)   One of two fractions or ratios whose product is -1.

numerical expression (p. 59)   Two or more numbers connected by operations such as addition, subtraction, multiplication, or division.

objective function (p. 390)   An equation used to represent a quantity such as profit or cost.

obtuse triangle (p. 619)   A triangle with one obtuse angle.

open half-plane (p. 375)   A half-plane that does not include the boundary line.

open sentence (p. 115)   An equation that contains one or more variables. An open sentence may be true or false, depending on what values of the variable are substituted.

opposite (p. 67, 74)   The opposite of any real number x is $-x$. Two opposites have a sum of zero. For example, -27 and 27 are opposites. Also called the *additive inverse* of the number.

order (p. 67)   To arrange numbers according to value, either from least to greatest or greatest to least.

ordered pair (p. 169)   A pair of numbers named in a specific order; a pair of real numbers (x, y) corresponding to a point in the coordinate plane.

order of operations (p. 60, 95)   Rules for evaluating algebraic expressions: Perform operations within grouping symbols first. Then perform all calculations involving exponents. Multiply or divide in order from left to right. Finally, add or subtract in order from left to right.

origin (p. 66, 169)   The point that corresponds to 0 on a number line. The point in the coordinate plane where the horizontal and vertical axes intersect.

outcome (p. 398)   The result of each trial of an experiment.

outlier (p. 248)   A value far to the left of the first quartile or far to the right of the third quartile in a boxplot for a data set.

parabola (p. 463)   The "U-shaped" graph of a quadratic equation of the form $y = ax^2 + bx + c$.

parallel lines (p. 279)   Lines in the same coordinate plane that do not intersect. Two nonvertical lines are parallel if and only if they have the same slope.

percent discount (p. 136)   The percent amount by which a price is reduced, as for a sale.

perfect square (p. 65)   The product of a number or polynomial times itself. For example, 2^2, y^2, $(x + 2)^2$ are perfect squares. The square root of any number that is not a perfect square is an irrational number.

period (of a function) (p. 660)   The horizontal distance from any point on the graph of a function to that point where the graph begins to repeat.

perpendicular lines (p. 279)   Two lines that intersect to form right angles. The slopes of two perpendicular lines are negative reciprocals of each other.

pictograph (p. 18)   A display of statistical information in which data is represented by symbols or pictures.

point of reflection (p. 469)   The reflection of any point on a figure, such as a parabola, across its axis of symmetry. For example, for the graph of $y = x^2$, the point of reflection of $(-3, 9)$ is $(3, 9)$.

point-slope form (of a linear equation) (p. 287, 293)   A linear equation in the form $(y - y_1) = m(x - x_1)$, where m is the slope and (x_1, y_1) are the coordinates of a given point on the line.

polynomial (p. 541)   A monomial or the sum or difference of monomials.

population (p. 21)   The total number of people making up a whole, considered for statistical purposes.

positive integers (p. 56)   Integers that are greater than zero.

power (p. 59)   A number that can be written as the product of equal factors. Powers can be expressed using exponents. The power $2 \cdot 2 \cdot 2$ can be written 2^3, read "two to the third power" or "two cubed."

principal square root (p. 439)   The positive square root of a number k, written $\sqrt{k}$.

principle of squaring (p. 636)   If the equation $a = b$ is true, then the equation $a^2 = b^2$ is also true.

probability (of an event) (p. 38, 398)   The chance or likelihood that an event will occur. The value of a probability ranges from 0 to 1. An impossible event has a probability of 0. A certain event has a probability of 1. (See also **theoretical probability**.)

product (p. 88)   The result when two or more factors are multiplied.

product of a sum and a difference (p. 554)   The product of two binomials that are the sum and difference of the same two terms: $(a + b)(a - b) = a^2 - b^2$.

product property of square roots (p. 439, 625)   For all real numbers a and b, where $a \geq 0$ and $b \geq 0$, $\sqrt{ab} = \sqrt{a} \cdot \sqrt{b}$.

proof (p. 446)   A logical sequence of statements that show another statement to be true.

properties of equality (p. 446)   For all real numbers a, b, and c,

Reflexive Property: $a = a$

Symmetric Property: If $a = b$, then $b = a$.

Transitive Property: If $a = b$ and $b = c$, then $a = c$.

Substitution Property: If $a = b$, then a may replace b or b may replace a in any statement.

property of exponents: power of a power (p. 523)   For any real number a and for all positive integers m and n, $(a^m)^n = a^{mn}$.

property of exponents: power of a product (p. 523)   For any real numbers a and b and for positive integer m, $(ab)^m = a^m b^m$.

property of exponents: power of a quotient rule (p. 529)   For all real numbers a and b, $b \neq 0$, and for any positive integer m, $\left(\dfrac{a}{b}\right)^m = \dfrac{a^m}{b^m}$.

property of exponents: product of powers (p. 522)   For any real number a and for all positive integers m and n, $a^m \cdot a^n = a^{m+n}$.

property of exponents: quotient rule (p. 528)   For any real number a, $a \neq 0$, and for positive integers m and n, $\dfrac{a^m}{b^n} = a^{m-n}$.

property of negative exponents (p. 529)   For any real number a, $a \neq 0$, and for any positive integer n, $a^{-n} = \dfrac{1}{a^n}$.

property of –1 for multiplication (p. 90)   For any real number a, where $a \neq 0$, $a \cdot -1 = -1 \cdot a = -a$.

proportion (p. 136)   An equation that states that two ratios are equal.

Pythagorean relationships (p. 620)   For any triangle the following relationships between the lengths of the sides are true: If $a^2 + b^2 > c^2$, the triangle is acute. If $a^2 + b^2 = c^2$, the triangle is right. If $a^2 + b^2 < c^2$, the triangle is obtuse.

Pythagorean theorem (p. 616)   If a right triangle has lengths a and b and a hypotenuse of length c, then $a^2 + b^2 = c^2$.

Pythagorean triple (p. 617)   A set of three numbers, such as {3, 4, 5} or {5, 12, 13}, that satisfy the Pythagorean theorem.

• • Q • •

quadrants (p. 177)   The four regions into which the x- and y-axes divide a coordinate plane.

quadratic direct variation (p. 683)   A function in which one quantity varies directly as the square of another. For example, the area of a circle varies directly as the square of its radius: $A = \pi r^2$.

quadratic equation (p. 465)   An equation in the form $y = ax^2 + bx + c$, where a, b, and c are real numbers, and $a \neq 0$.

quadratic formula (p. 490)   For a quadratic equation of the form $ax^2 + bx + c = 0$ where a, b, and c are real numbers and $a \neq 0$,
$$x = \frac{-b \pm \sqrt{b^2 - 4ac}}{2a}.$$

quadratic function (p. 468)   A function of the form $f(x) = ax^2 + bx + c$, where a, b, and c are real numbers, and $a \neq 0$.

qualitative graph (p. 202)   A graph that shows general features of a function but does not use precise numerical scales.

quartiles (p. 247)   Three numbers that divide a set of data into four equal parts, or quarters.

quotient property of square roots (p. 439, 625)   For all real numbers a and b, where $a \geq 0$ and $b > 0$, $\sqrt{\dfrac{a}{b}} = \dfrac{\sqrt{a}}{\sqrt{b}}$.

• • R • •

radical (p. 439)   A radical symbol $\sqrt{}$ and its radicand.

radical equation (p. 636)   An equation that contains a radical with a variable in the radicand.

radical expression (p. 625)   An expression containing a radical.

radicand (p. 439, 625)   A number or expression under a radical symbol.

random sampling (p. 22)   A type of statistical sampling in which each member of a population has an equal chance of being selected.

range (of values) (p. 20, 33)   The difference between the greatest and the least values in a set of data.

range (of a function) (p. 172) The output values of a function.

rational equation (p. 712)   An equation that contains one or more rational expressions.

rational expression (p. 686)   An expression that can be written in the form $\frac{P}{Q}$, where P and Q are polynomials, $Q \neq 0$.

rational number (p. 65, 437)   A number that can be expressed in the form $\frac{a}{b}$, where a and b are any integers and $b \neq 0$. A rational number may be expressed as a fraction, a terminating decimal, or a repeating decimal.

rationalizing the denominator (p. 627)   The process of eliminating a radical from the denominator.

real number line (p. 66)   A number line on which every point can be matched with a real number.

real numbers (p. 66, 439)   All rational numbers and irrational numbers.

reciprocal (p. 90)   The multiplicative inverse of a number. The product of a number and its reciprocal is 1.

reflection (p. 416)   A transformation in which a figure is flipped over a line of reflection.

relation (p.171)   A set of ordered pairs of data.

replacement set (p. 61)   The set of numbers that can be substituted for a variable.

right triangle (p. 619)   A triangle with one right angle.

root(s) (of an equation) (p. 477)   The solution of an equation.

• • S • •

sample (p. 21)   The representative portion of a population that is used for a statistical study, as in a survey.

sample space (p. 398)   All the possible outcomes of an event.

scale factor (p. 650)   The number by which the original dimensions of a figure are multiplied in order to enlarge or reduce the original figure proportionally. The scale factor is found by writing the ratio of a pair of corresponding sides.

scalene triangle (p. 619)   A triangle having no sides equal in length.

scatter plot (p. 206)   A graph in which the data are shown as points in a coordinate plane.

scientific notation (p. 532)   A number written in the form $m \times 10^n$ where $1 \leq m < 10$ and n is any integer.

set (of numbers) (p. 437)   Any group of numbers having one or more common attributes.

similar (figures) (p. 648)   Two figures are similar if they have the same shape but not necessarily the same size.

simplest form (of a polynomial) (p. 544)   A polynomial is in simplest form when it contains no like terms.

simplest form (of a radical) (p. 625)   A radical expression in which the radicand has no perfect-square factors other than 1, has no fractions, and no denominator contains a radical.

simplest form (of a rational expression) (p. 687)   A rational expression is in simplest form when the numerator and denominator of the expression have no common factors other than 1.

simplify (an expression) (p. 95)   To change an expression to an equivalent expression with fewer terms.

sine (p. 653)   For either acute angle in a right triangle, the ratio of the length of the side opposite the angle to the length of the hypotenuse.

sine curve (p. 660)   The graph of the sine function.

slope (of a line) (p. 268, 269)   The ratio of the number of units the line rises or falls vertically (the *rise*) to the number of units the line moves horizontally from left to right (the *run*). For a line connecting two points $P(x_1, y_1)$ and $Q(x_2, y_2)$, the slope of the line $= \dfrac{\text{rise}}{\text{run}} = \dfrac{y_2 - y_1}{x_2 - x_1}$

The slope of a horizontal line is 0. The slope of a vertical line is undefined.

slope-intercept form (of a linear equation) (p. 277, 293)   A linear equation in the form $y = mx + b$, where m is the slope and b is the y-intercept of the graph.

solution (p. 115)   A value of a variable that makes an equation true.

solution (of a system of equations) (p. 329)   An ordered pair that is a solution of all the equations in a system.

square matrix (p. 27)   A matrix that has the same number of rows and columns.

GLOSSARY

square of a binomial (p. 554) A trinomial whose first and last terms are the squares of the terms of the binomial and whose middle term is twice the product of the two terms: $(a + b)^2 = a^2 + 2ab + b^2$ and $(a - b)^2 = a^2 - 2ab + b^2$.

square root method (p. 483) A method for solving an equation of the form $ax^2 + b = c$ that involves isolating the squared term on one side of the equation and then taking the square root of both sides.

square root property (p. 483) If $x^2 = k$, then $x = \sqrt{k}$ or $x = -\sqrt{k}$ for any real number k, $k > 0$. If $k = 0$, then $x^2 = 0$ has one solution, 0.

standard form (of a linear equation) (p. 294) An equation of the form $Ax + By = C$, where A, B, and C are integers and A and B are not both 0.

standard form (of a quadratic equation) (p. 477) An equation of the form $d = ax^2 + bx + c$ when $d = 0$.

standard notation (p. 532) The decimal form of any number, for example 12.45, 43, or 0.0087.

subset (p. 431) Set A is a subset of set B if every element of set A is also an element of set B.

substitution method (p. 335) An algebraic method for solving a system of linear equations.

subtraction of real numbers (p. 79) For all real numbers a and b, $a - b = a + (-b)$.

subtraction property of equality (p. 125) For all real numbers a, b, and c, if $a = b$, then $a - c = b - c$.

subtraction property of inequality (p. 230) For all real numbers a, b, and c, if $a > b$, then $a - c > b - c$, and if $a < b$, then $a - c < b - c$.

sum of the angles of a triangle (p. 620) The sum of the measures of the angles of a triangle is 180°.

supplementary angles (p. 621) Two angles whose sum is 180°. Each angle is the *supplement* of the other.

survey (p. 21) A study of the opinions or behavior of a population. Also, the collection of data through questioning or polling a population.

symmetry (p. 413) A figure is said to have symmetry if it has a line of symmetry.

systematic sampling (p. 22) A type of statistical sampling in which members of a population are chosen by a rule or pattern applied to the entire population.

system of linear equations (p. 328) Two or more linear equations that are considered together. Also called a *linear system*.

system of linear inequalities (p. 379) Two or more inequalities that are considered together.

• • T • •

tangent (p. 653) For either acute angle in a right triangle, the ratio of the length of the side opposite the angle to the length of the side adjacent to the angle (not the hypotenuse).

terms (p. 61) The parts of a variable expression that are separated by addition or subtraction signs. For example, the expression $x + y$ has two two terms, x and y.

theoretical probability (p. 398) The probability of an event, $P(E)$, assigned by determining the number of favorable outcomes and the number of possible outcomes in a sample space.
$$P(E) = \frac{\text{number of favorable outcomes}}{\text{number of possible outcomes}}$$

transformation (of a graph) (p. 418) A change in the position or shape of a graph as a result of an operation such as translation or reflection.

translation (of a graph) (p. 417) A slide or shift that moves the graph to a new position horizontally or vertically in the coordinate plane.

trend (p. 207) The general pattern formed by the points in a scatter plot.

trichotomy property (p. 221, 445) For all real numbers a and b, exactly one of the following is true: $a = b$, $a < b$, or $a > b$.

trigonometric ratios (p. 653) Ratios of the lengths of the sides of a right triangle. The three common ratios are the *sine, cosine,* and *tangent*.

trigonometry (p. 653) A word from a Greek term meaning "measure of triangles."

trinomial (p. 541) A polynomial of three terms.

trough (p. 660) The minimum value of the graph of a periodic function.

• • V • •

value (of a variable) (p. 115)   The number assigned to or substituted for a variable.

variable (p. 13, 60)   A letter or other symbol used to represent a number.

variable expression (p. 13, 60)   An expression having at least one number, operation, and variable.

vertex (of a parabola) (p. 468)   The point where a parabola has its maximum (highest) or minimum (lowest) point.

vertical axis (p. 169)   The y-axis of the coordinate plane.

vertical line test (p. 178)   A test used on the graph of a relation to determine whether the relation is also a function. If a vertical line drawn through the graph of a relation intersects the graph in more than one point, the graph does not represent a function.

vertical shift (of a graph) (p. 417)   The up or down translation of a graph in relation to another graph.

• • W • •

whiskers (p. 248)   Horizontal lines in a boxplot showing the range of the data and the first and fourth quartiles.

• • X • •

x-axis (p. 178)   The horizontal number line that divides the coordinate plane.

x-coordinate (p. 178)   The first number in an ordered pair representing a point in the coordinate plane. The x-coordinate determines the horizontal location of the point.

x-intercept (p. 276, 477)   The x-coordinate of the point where a graph crosses the x-axis.

• • Y • •

y-axis (p. 183)   The vertical number line that divides the coordinate plane.

y-coordinate (p. 183)   The second number in an ordered pair representing a point in the coordinate plane, also called the $f(x)$-coordinate for functions. The y-coordinate determines the vertical location of the point.

y-intercept (p. 269, 276)   The y-coordinate of the point where a graph crosses the y-axis.

• • Z • •

zero pair (p. 74, 121)   A pair of integers or algebraic terms whose sum is zero. Zero pairs are opposites.

zero product property (p. 595)   For any real numbers a and b, if $ab = 0$, then $a = 0$ or $b = 0$, or both $a = 0$ and $b = 0$.

zero property of exponents (p. 528)   For any real number a, $a \neq 0$, $a^0 = 1$.

GLOSSARY

• • • SELECTED ANSWERS • • •

Chapter 1 Data and Graphs

Lesson 1.2, pages 8–12

TRY THESE

1.

Number of Hours Health Club Members Exercise per Week

Number of Hours	Tally	Frequency
0	II	2
1	III	3
2	LHT	5
3	III	3
4	II	2
5	II	2
6	I	1
7	II	2

3. $2h$

5.

Number of Books Read During Summer Reading Program

Number of Books	Tally	Frequency
0–4	IIII	4
5–9	LHT II	7
10–14	LHT III	8
15–19	III	3
20–24	III	3

7. 88%

PRACTICE   1. 10   **3.** 11%   **5.** 8 more magazines

EXTEND   9. 0.370   **13.** $6.00–$6.99; 11; 7

THINK CRITICALLY   15. Yes; the exact number of free throws he had made would have been shown as well as the number of boys who had made that number.

Lesson 1.3, pages 13–17

TRY THESE   1. 4   **3.** 3   **5.** 1   **9.** 46, 79

PRACTICE   1. 25   **3.** 35   **5.** 12   **7.** 8   **9.** 6.4
11. Store 1 is 237; Store 2 is 150; Store 3 is 410.
13. 717.5   **15.** Change line 20 to PRINT B * H.
17. 212

EXTEND   23. G is 208; PG is 173; PG-13 is 174; R is 242   **25.** 10 INPUT H, M, N

    20 PRINT 0.5 * H * (M + N)

THINK CRITICALLY   27. Possible answers: 2 * A; B − 2; A * B/C   **29.** Possible answers: B * C; 45 + C   **31.** Possible answers: 2(A * B) + 2C; 9 * B   **33.** no; if X = 2, Y = 3, and Z = 4, X * Y + Z = 10, and X * (Y + Z) = 14

MIXED REVIEW   34. 29%   **35.** C

Lesson 1.5, pages 21–26

TRY THESE   1. Convenience: people stopping in at campaign headquarters are likely to be strong supporters of the mayor.   **3.** Systematic: only people with jobs and cars will be surveyed.   **5.** 8   **7.** 56

PRACTICE   1. Systematic: the only people surveyed will be those who enjoy popcorn.   **3.** Convenience: most of the people surveyed are likely to enjoy popcorn; people taking a tour may feel obligated to answer more positively.   **5.** 2   **7.** 3   **9.** 26 **11.** 1

EXTEND   15. 12.5 million   **17.** 121,200

THINK CRITICALLY   19. 65%; 59%

Lesson 1.6, pages 27–31

TRY THESE   1. 9, 15, 7, 2, 6; 1×5

3. $\begin{bmatrix} 2 & 7 & 7 & 0 \\ -24 & 4 & 15 & 21 \end{bmatrix}$

5. $A = \begin{bmatrix} 117 & 88 \\ 91 & 95 \end{bmatrix}$; $B = \begin{bmatrix} 38 & 35 \\ 25 & 40 \end{bmatrix}$; $C = \begin{bmatrix} 13 & 9 \\ 6 & 10 \end{bmatrix}$

PRACTICE   1. 4, 5, 8, 3

3. $\begin{bmatrix} 3 & 11 \\ 10 & -4 \end{bmatrix}$ **5.** $\begin{bmatrix} 3 & 4 \\ 0 & 6 \\ 4 & 6 \end{bmatrix}$

7. $I = \begin{bmatrix} 317 & 490 & 166 \\ 555 & 207 & 181 \end{bmatrix}$; $N = \begin{bmatrix} 52 & 70 & 48 \\ 88 & 86 & 66 \end{bmatrix}$;

$S = \begin{bmatrix} 61 & 90 & 77 \\ 114 & 98 & 50 \end{bmatrix}$

9. $\begin{bmatrix} 992 & 1008 & 960 \\ 888 & 912 & 920 \\ 1184 & 1128 & 1152 \\ 816 & 872 & 824 \end{bmatrix}$

EXTEND 13. $\begin{bmatrix} 11 & 15 & 24 \\ 19 & 31 & 28 \end{bmatrix}$

THINK CRITICALLY 15. Double each element.

17. $\begin{bmatrix} 12,400 & 12,600 & 12,000 \\ 11,100 & 11,400 & 11,500 \\ 14,800 & 14,100 & 14,400 \\ 10,200 & 10,900 & 10,300 \end{bmatrix}$

MIXED REVIEW 21. B **22.** 2×3

23. $\begin{bmatrix} 265 & 205 & 370 \\ 244 & 144 & 160 \end{bmatrix}$ **24.** $\begin{bmatrix} 37 & 23 & 26 \\ 68 & 14 & 0 \end{bmatrix}$

Lesson 1.7, pages 32–37

TRY THESE 1. 84; 81; 77; 23 **3.** 207; 197; none; 86
5. 188.2 cm; 185 cm; 180 cm; 30 cm **7.** 80.5 s; 80 s;
80 s and 85 s; 23 s

PRACTICE 1. 27.25; 26.5; 23; 15 **3.** 240; 235; 235;
100 **5.** 4394; 4411.5; none; 279 **7.** $13,685;
$13,715; none; $3,020 **11.** 405.3 ft; 403 ft; 400 ft
13. 211.3; 210.8; 210.5; 3.7

EXTEND 17. mode **19.** range **21.** $14 or $15
23. $12, because all values appear the same number of
times

THINK CRITICALLY 25. 15: 6; 16: 96; 17: 120; 18: 18
27. The data item must equal at least one item already
in the set.

MIXED REVIEW 30. 2 **31.** 4 **32.** 8 **33.** 42
34. A

Lesson 1.8, pages 38–41

TRY THESE 1. about 1 **3.** about 0.375 **5.** 80,000
7. $\frac{21}{25} = 0.84$, or 84% **9.** 0

PRACTICE 1. $\frac{251}{365} = 0.688$, or 68.8% **3.** $\frac{6}{25}$, or 24%
5. 0% **7.** about 175,200 **9.** about 51.4%
11. about 21.6%

Lesson 1.9, pages 42–45

APPLY THE STRATEGY 15. 42 **17.** 42% **19.** 32,374
copies/week **21.** 2,343,750 ft^2

Chapter Review, pages 46–47

1. a **2.** d **3.** e **4.** b **5.** c **7.** 40%

8.

	A	B	C	D
1		Regal	Ovenmaster	Total
2	Campus	38	52	90 ⟵ =B2+C2
3	Eastwood	49	70	119 ⟵ =B3+C3
4	Central	25	41	66 ⟵ =B4+C4

9. 2956 **10.** convenience sampling **11.** 18,000

12. $\begin{bmatrix} 10 & 7 & 17 & 10 \\ 4 & 9 & 1 & 10 \end{bmatrix}$ **13.** $\begin{bmatrix} 13 & 6 & 10 & 3 \\ 23 & 30 & 2 & 10 \end{bmatrix}$

14. $\begin{bmatrix} 20 & 10 & 19 & 9 \\ 25 & 39 & 3 & 15 \end{bmatrix}$ **15.** 81.7; 81.5; 80 and 84; 8

16. about 3.3; 3; 3; 5 **17.** 16% **18.** about 26.1%

Chapter 2 Variables, Expressions, and Real Numbers

Lesson 2.2, pages 59–64

TRY THESE 1. 12 **3.** 15 **5.** 13 **7.** 2 **9.** 9
11. 40 **15.** degree 2; 3 terms; −1 **17.** 4 **19.** 1
21. $-\frac{3}{4}$ **23.** $8s - 4$
PRACTICE 1. 35 **3.** 33 **5.** 17 **7.** 19 **9.** 40
11. 1 **13.** 33 **15.** 54 **17.** 73 **19.** 337
21. $3.50 + x$ **23.** 48 **25.** 23 **27.** 7; 10; 13; 16
29. 1; 2; 3; 5 **31.** degree 4; 4 terms; −3 **33.** degree
6; 6 terms; 10 **35.** 9 **37.** −3 **39.** $\frac{y}{8}$

EXTEND 41. 18 **43.** 23 **45.** 37,543 **47.** 15.6
49. 622 **51.** $450 + 0.02x$ **53.** 163.1 cm or about
5 ft 4 in.

THINK CRITICALLY 55. no; possible example: $3^2 \neq 2^3$

Lesson 2.3, pages 65–72

TRY THESE 1. yes; it is the ratio of two integers
3. no; 3 is not a perfect square so $\sqrt{3}$ is irrational
5. no; 11 is not a perfect square, so $\sqrt{11}$ is irrational
7. E **9.** D **11.** G **13.** < **15.** > **17.** 9 **19.** 4.9
23. Fairbanks; $-4 > -20$
PRACTICE 1. yes; $5\frac{1}{4} = \frac{21}{4}$ **3.** no; it is a
nonterminating, nonrepeating decimal **5.** no; 5 is
not a perfect square, so $\sqrt{5}$ is irrational **7.** yes;
$6.2 = \frac{62}{10}$ **9.** yes; $\sqrt{49} = 7$ **11.** yes; it is a
nonterminating, repeating decimal **13.** C **15.** H
17. E **19.** A **21.** G **23.** < **25.** < **27.** <
29. > **31.** 2 **33.** $7\frac{2}{3}$ **35.** −2.3

EXTEND

43. $\sqrt{3}$, 0.51, 0.5, $\frac{9}{20}$, $-\frac{3}{10}$, -0.98, -1.1, $-\sqrt{4}$

45. <   **47.** <   **49.** >

MIXED REVIEW   **54.** $0.15x$ **55.** $\frac{x}{3}$ **56.** B   **57.** <
58. >   **59.** >   **60.** <

Lesson 2.5, pages 77–83

TRY THESE   **1.** -1 **3.** $-\frac{9}{11}$ **5.** 0.7   **7.** $\frac{7}{12}$

11. 7   **13.** $-\frac{1}{8}$ **15.** -9 **17.** 4.5   **19.** $-\$8$ **21.** -10

23. 5.6   **25.** $\frac{17}{20}$

PRACTICE   **3.** 5   **5.** -17 **7.** -0.6 **9.** 1.5

11. -6.3 **13.** -12 **15.** $-\frac{3}{10}$ **17.** $2\frac{2}{3}$ **19.** $-1\frac{1}{6}$

21. $\frac{4}{9}$ **23.** -4 **25.** -11 **27.** -6 **29.** -9.3

31. 0.1   **33.** 0.93   **35.** $1\frac{5}{24}$ **37.** $-1\frac{1}{4}$ **39.** $-\frac{1}{10}$

41. 2748 y   **43.** -7 **45.** 44   **47.** $\frac{2}{5}$ **49.** -0.67
51. at the 28-yd line

EXTEND   **53.** -10 **55.** 2   **57.** -14 **59.** -7 **61.** -3

63. $13\frac{7}{8}$ **65.** 16°F

Lesson 2.7, pages 88–94

TRY THESE   **3.** -3.2 **5.** $25\frac{1}{3}$

7. $-\frac{4}{3}$ **9.** $-\frac{5}{17}$ **11.** 1.55   **13.** $3\frac{1}{2}$

15. $-3(2) = -6$

PRACTICE   **1.** -99 **3.** 105   **5.** -2.04 **7.** 0.0027

9. $-\frac{7}{16}$ **11.** $-\frac{33}{8}$ or $-4\frac{1}{8}$ **13.** $-\frac{1}{21}$ **15.** $-\frac{6}{19}$

19. -25 **21.** 17   **23.** -0.04 **25.** 80   **27.** $-2\frac{1}{7}$

29. $1\frac{1}{4}$ **31. a.** $1125m - 100m - 5m$ **b.** $4080

EXTEND   **33.** 19   **35.** 10   **37.** 13   **39.** -26

THINK CRITICALLY   **45.** no; $-8 \div 3 = -\frac{8}{3}$, which is
not an integer   **47.** $|ab|$ **49.** $\left|\frac{a}{b}\right|$

Lesson 2.8, pages 95–101

TRY THESE   **1.** -3 **3.** 71   **5.** 0.059   **7.** -26 **9.** -4
11. $4b - 12$ **13.** $-2m + 16$ **15.** no   **17.** no
19. $41 - 4x$ **23.** -5°C

PRACTICE   **1.** 27   **3.** -31 **5.** 10.28   **7.** -1.76

9. 0.24   **11.** $\frac{19}{50}$ **13.** 0   **15.** 33   **17.** -13

19. $2x - 7$ **21.** $9 - 24c$ **23.** $17a - 2b + 2c$
25. $2r - s + 6$ **27.** $P = 2(l + w)$

EXTEND   **29.** -5 **31.** -1 **33.** 72   **35.** 30   **37.** 0

39. 8   **41.** $(7 + 1) \cdot 2 + 4$ **43.** 750.2 kilocalories
MIXED REVIEW   **48.** both

49. $A + B$

$$\begin{bmatrix} 5 & 6 & -6 \\ 13 & 0 & 5 \\ 0 & 6 & 5 \end{bmatrix}$$

50. $A - B$

$$\begin{bmatrix} 1 & 4 & -6 \\ 3 & -4 & 1 \\ -2 & 2 & 3 \end{bmatrix}$$

51. D   **52.** F   **53.** C   **54.** G   **55.** B

Lesson 2.9, pages 102–105

APPLY THE STRATEGY   **15. a.** $10a + b$; $10b + a$
b. $(10a + b) + (10b + a)$
c. $(10a + b) + (10b + a)$

$= 10a + b + 10b + a$ associative property

$= 10a + a + 10b + b$ commutative property

$= 11a + 11b$ Combine like terms.

$= 11(a + b)$ distributive property

$11(a + b)$ represents a whole number divisible by 11.

Chapter Review, pages 106–107

1. f   **2.** b   **3.** d   **4.** c   **5.** e   **6.** a   **7.** -2 **8.** $4x$
9. $-2y + 5$ **10.** 7   **11.** 112   **12.** 10
13–15.

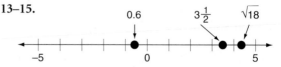

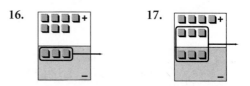

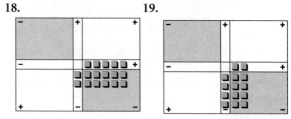

20. $-1\frac{3}{4}$ **21.** 9.8   **22.** -17 **23.** $12\frac{1}{2}$ **24.** -0.4

25. -44 **26.** -140 **27.** $-2\frac{2}{3}$ **28.** 16   **29.** -160

30. $-\frac{1}{6}$ **31.** -340 **32.** 0   **33.** $-8c + 4$

34. $x + 2y - 3z$ **35.** Let n and $n + 1$ represent
consecutive whole numbers; $n + (n + 1) = 2n + 1$;
$2n$ represents an even number, so $2n + 1$ represents an
odd number.

Chapter 3 Linear Equations

Lesson 3.1, pages 115–119

TRY THESE   **1.** 16   **3.** 5   **5.** 28   **7.** 49   **9.** 4
11. –5   **13.** 8   **15.** 3   **17.** 9   **19.** 6   **21.** 2.1
23. 8   **25.** 3   **27.** height = 8 units; methods will vary

PRACTICE   **1.** 40   **3.** 480   **5.** 75   **7.** 14   **9.** 9.1
11. 52   **13.** 10   **15.** 5   **17.** 8   **19.** 38
21. Equations will vary. $13 - t = 12.5$; $t = 0.5$ g

EXTEND   **25.** 38   **27.** 5   **29.** 6   **31.** –81   **35.** less
37. less   **39.** less   **41.** 29   **43.** 4   **45.** no, since
3 • 2.00 = 6.00, almost the entire $6.16   **47.** Possible
answer: $5.75 is approximately $6.00; 15 • 6.00 =
90.00; $89.95 is approximately $90.00.

THINK CRITICALLY   **49.** The variable must be a
negative number.   **51.** 38   **53.** 77

MIXED REVIEW   **54.** mean = 7.2; median = 5;
mode = 2; range = 21   **55.** C

Lesson 3.3, pages 124–128

TRY THESE   **1.** –3   **3.** 17   **5.** 6   **7.** –4
9. $x - 4 = 7$; $x = 11$;

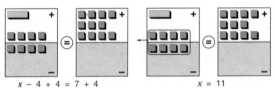

$x - 4 + 4 = 7 + 4$ $x = 11$

11. –5   **13.** –7   **15.** 16   **17.** 47   **19.** $4\frac{1}{4}$ **21.** 20
PRACTICE   **1.** 25   **3.** –20   **5.** 18   **7.** –3   **9.** –14
11. –48   **13.** equations will vary; possible response:
$t + 14 = 71$; $t = 57°$F   **15.** 7   **17.** 0   **19.** –3.6
21. $1\frac{5}{12}$ **23.** 2   **25.** 8
EXTEND   **29.** –12.764   **31.** 5.0928
33. $3000 = C + 300$; 2700 Cal   **35.** $220 = r - 130$;
$r = 350$ Cal/h

THINK CRITICALLY   **37. a.** x decreases   **b.** b increases
39. $7\frac{1}{2}$

Lesson 3.4, pages 129–134

TRY THESE   **1.** divide by 7   **3.** multiply by 3
5. multiply by $\frac{3}{2}$ **7.** divide by 9   **9.** $3x = -9$
11. 15   **13.** 13   **15.** $b = \frac{32.2}{9.2}$ **17.** $d = 7.5(28.5)$

PRACTICE   **1.** divide by 6   **3.** divide by –4   **5.** divide
by –7   **7.** multiply by 5   **9.** multiply by 12

11. multiply by $\frac{4}{3}$ **13.** 120   **15.** 12   **17.** –18
19. –17   **21.** 78   **23.** 3   **25.** –47.6   **27.** 2.5
29. $6x = 1008$; $x = 168$; 168 teams

EXTEND   **33.** 2.5   **35.** –18   **37.** –0.4 or $-\frac{2}{5}$
39. 21.3   **41.** 108.24   **43.** –0.85   **45.** –1.33
47. 0.89   **49.** b; $1242.20   **51.** $4x = 72$; $x = 18$; 18
in.   **53.** $6.875g = 11$; 1.6 km   **55.** $11{,}612m = 36{,}000{,}000$; about 3100 mi

MIXED REVIEW   **57.** D   **58.** 6.5   **59.** 54   **60.** 13.5
61. 12.5   **62.** $8x = (-4)^2$; $x = 2$ **63.** $\frac{72}{2} = x + 15$; $x = 21$

Lesson 3.5, pages 135–139

TRY THESE   **1.** 48   **3.** 8.1   **5.** 130.2   **7.** 29%
9. 167%   **11.** 80   **13.** 104   **15.** $4\frac{1}{2}$%

PRACTICE   **1.** 10.4   **3.** 166.7%   **5.** 62.5%   **7.** 529
9. 63.3%   **11.** 2.6   **13.** 71%   **15.** 3.8 g

EXTEND   **17.** 24.8%   **19.** 2.1   **21.** 51.2   **23.** 74.8
25. $940; find the cost of the car in each state and
subtract, or multiply the base price by the difference in
states' tax rates

THINK CRITICALLY   **27.** true   **29.** false; answers will
vary   **31.** The information is inconsistent. 3 g is 4% of
75 g, but 7 g is 15% of 46.7 g, giving the different RDA
for protein.

MIXED REVIEW   **32.** $\frac{1}{2}$ **33.** $\frac{1}{6}$ **34.** $\frac{5}{12}$ **35.** $\frac{5}{12}$
36. $\frac{7}{12}$ **37.** $\frac{1}{3}$ **38.** B

Lesson 3.6, pages 140–145

TRY THESE   **1.** Add b to both sides.   **3.** Add y to both
sides.   **5.** 9   **7.** 84   **9.** 30   **11.** –8   **13.** 7   **15.** –4
19. $14{,}000 - 1800x = 6800$; $x = 4$ yr

PRACTICE   **1.** 3   **3.** 3   **5.** 3   **7.** 9   **9.** 75   **11.** –10
13. 4   **15.** 11   **17.** –10   **19.** 2.5   **21.** 2   **23.** 56
25. 6   **27.** 9   **29.** 8

EXTEND   **31.** –3   **33.** 3   **35.** $x + 3x = 1$;
$P(\text{even}) = 0.75$; $P(\text{odd}) = 0.25$
37. 70°, 70°, 110°, 110°

THINK CRITICALLY   **39.** P = profit; x = the number of
sweaters sold per month; $19x$ is the store's income;
$1300 is the store's expenses; 162 sweaters   **41.** 27, 29,
31   **43.** 36   **45.** 28

Lesson 3.7, pages 146–151

TRY THESE   **1.** $a = 4$ **3.** $r = 7$ **5.** –3   **7.** 12
9. 5   **11.** 12   **13.** 1.5   **15.** –10

SELECTED ANSWERS

17. The accounts will be equal after 14 weeks.
19. $x = 4$; length $= 17$ mi; area $= 119$ mi²
PRACTICE **1.** −3 **3.** 1 **5.** 4 **7.** 7 **9.** −4 **11.** −9
15. $-\dfrac{1}{2}$ **17.** 2 **19.** −8 **21.** 8
23. $90^2 + 9x = 50 + 13x$; $x = 10$ yd² **25.** b
EXTEND **27.** 9 **29.** 6 **31.** $\dfrac{5}{7}$ **33.** Each board is
12 ft long; he needs 220 ft.
THINK CRITICALLY **35.** The difference between the
coefficients of the variable terms in each equation is 1.

Lesson 3.8, pages 152–155

TRY THESE **1.** no **3.** yes **5.** subtract 7 from both
sides **7.** add m to both sides **9.** divide both sides
by 9 **11.** $w = \dfrac{P - 2l}{2}$ **13.** $N = \dfrac{60H}{15}$
PRACTICE **1.** $m = \dfrac{gs^2}{G}$ **3.** $h = \dfrac{3V}{\pi r^2}$ **5.** $c = \dfrac{C}{1 + r}$

7. $m = \dfrac{E}{c^2}$ **9.** $\dfrac{y - 7}{3}$; $x = 4$ **11.** $\dfrac{V}{lw}$; $h = 3$
13. $\dfrac{E}{tI}$; $V = 6$ **15.** 13 h
17. 701.5 mi **19.** 3 cm **21.** 10.5%
EXTEND **23.** $Q = \dfrac{100A + 10B + C}{A + B + C}$ **25.** $Q = 56$
THINK CRITICALLY **27.** Interest paid is zero.
29. divided by 4 rather than −4
MIXED REVIEW **31.** $m = $ V **32.** $x = -8/-3$
33. $r = 9$ **34.** $b = 4$ **35.** D

Lesson 3.9, pages 156–159

APPLY THE STRATEGY **13.** 50 mL **15.** 60 lb **17.**
121 g **19. a.** 4 qt **b.** 1.6 **c.** 0.8 qt **d.** 4.8 qt

Chapter Review, pages 160–161

1. c **2.** d **3.** e **4.** a **5.** b **6.** 18 **7.** 15 **8.** −6
9. $x + 3 = 8$; $x = 5$ **10.** $x - 3 = 4$; $x = 7$
11. $x - 2 = -5$; $x = -3$ **12.** 9 **13.** −8 **14.** 19
15. −5 **16.** 4 **17.** 18 **18.** 11 **19.** $3\dfrac{2}{3}$ **20.** 6.5
21. 34.5 **22.** 450% **23.** 24% **24.** 6 **25.** −20
26. −15 **27.** 3 **28.** 9 **29.** −40 **30.** 6 **31.** 5
32. 2 **33.** −6 **34.** 1 **35.** $\dfrac{1}{3}$ **36.** $b = \dfrac{a + 3}{2}$
37. $r = \dfrac{12 - 7s}{8}$ **38.** $l = \dfrac{P - 2w}{2}$
39. $1200 at 5%; $900 at 7%

Chapter 4 Functions and Graphs

Lesson 4.2, pages 171–176

TRY THESE **1.** function; domain: −1, −2, −3, −4, −5;
range: 2 **3.** function; domain: 3, 4, 5, 6;
range: 3, 4, 5, 6 **5.** −1; 0; 2 **7.** 1600 ft

PRACTICE **1.** function; domain: $0.50, $0.60, $0.70,
$0.80, $0.90; range: $0.02 **3.** function; domain: 2, 4,
6, 8, 10; range: 0 **5.** not a function **7.** function;
domain: 1, 2, 3, 4, 5; range: 50, 100, 150, 200, 250
9. −2; 0.5; 10 **11.** 7; −1; 7 **13. a.** 4.5 Cal **b.** 18 Cal
c. 153 Cal

EXTEND **15.** not a function **17.** function; domain:
a, b, c, d; range: d, e, f, g
19. a. $f($3,500) = $100 + 0.1($3,500) = 450
b. $f($2,000) = $100 + 0.1($2,000) = 300
c. $f($1,900) = $200 + 0.05($1,900) = 295

THINK CRITICALLY **23.** $f(x) = 2x + 1$ **25.** Possible
answer: cost of a pizza, where $x = $ number of toppings
27. −4 or 4

MIXED REVIEW **28.** 46.2 cm **29.** 41 cm **30.** no
mode **31.** 45 cm **32.** 294 **33.** −5 **34.** C

Lesson 4.3, pages 177–182

TRY THESE **1.** $(-4, -2)$ **3.** $\left(-2\dfrac{1}{2}, 1\right)$ or $(-2.5, 1)$
9. function; domain: all real numbers; range: all real
numbers ≥ 0 **11.** not a function
PRACTICE **1.** x-axis **3.** Quadrant IV **5.** $(-2, -1)$
7. $\left(-3\dfrac{1}{2}, 0\right)$ **9.** $(0, -2.5)$ or $\left(0, -2\dfrac{1}{2}\right)$
15. function; domain: 0–120 in.; range 0–240 in.
19. not a function
EXTEND **21.** $f(x) = 20 - x$ (Each period is $\dfrac{60}{3}$, or
20 min.)
THINK CRITICALLY **27.** (a, b) **29.** $(a, -b)$
31. $(-a, b)$

Lesson 4.4, pages 183–188

TRY THESE
1. **3.**

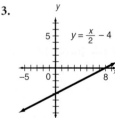

5. No; the point is not on the line. **7.** No; if the line
were extended, the point would still not be on the line.

9. No; the point is not on the line.

PRACTICE

1. 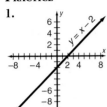 **3.**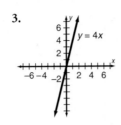

7. No; the point is not on the line.   **9.** Yes; if the line were extended the point would be on the line.   **11.** No; if the line were extended the point would still not be on the line.   **13.** Nonlinear; as each income amount increases by \$15,000, the tax rate increases by a different amount (13%, 0%, 3%, 0%)

EXTEND

15. 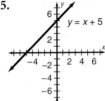 **17.**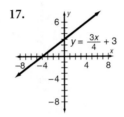

THINK CRITICALLY   21. Not a linear function; this is not a function at all. If this were graphed, it would be a vertical line through the x-axis at 1. This indicates that, for the input 1, there are an infinite number of different outputs.   **23.** Linear function

Lesson 4.6, pages 192–196

TRY THESE   1.

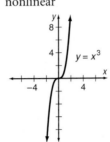

3. 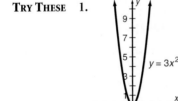 **5.**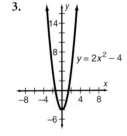

7. Yes; if the curve were extended, the point would be on the curve.   **9.** Yes; the point is on the curve.

PRACTICE

1. 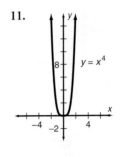   **7.** Accept 1992 or 1993.

9. nonlinear   **11.**

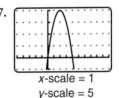

13. 700–800 ft^2 **15.** The charge per ft^2 decreases as the square footage increases.

EXTEND   17.   **19.** 4 s

THINK CRITICALLY   23. One opens up and the other opens down.

MIXED REVIEW   25. \$76,000   **26.** $9a + 7b - 3ab$
27. $11r^2 + 3r - 3$ **28.** B   **29.** Quadrant II
30. Quadrant IV   **31.** x-axis   **32.** Quadrant III

Lesson 4.7, pages 197–201

TRY THESE   1. $y = 9x - 5; y = 22$
3. $y = 3(x - 5); y = x + 7$ **5.** $x = -3$ **9.** $x = 2$

PRACTICE

1. $y = -3x + 5;$ **3.** $y = 2x - 4;$
$y = -10;$ $y = 2(x - 2);$
$x = 5$ $x = $ all real numbers

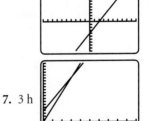

7. 3 h

EXTEND 9. $\$1.99x = 20 + 0.89x$ **11.** Plan A: $\$37.81$; Plan B: $\$36.91$

13. $x = 0$ 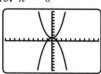 **15.** no solution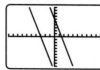

17. two solutions ($x = \pm \sqrt{2}$)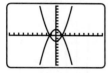

19. The graphs intersect at (6, 15), meaning that 6 min after the hot water was turned on, 15 gal of hot water and 15 gal of cold water, or 30 gal altogether, have run. This is 10 gal more than the sink will hold. Vinje guessed correctly.

THINK CRITICALLY 21. not possible **23.** one line (the two graphs coincide) **25.** a line and a curve that intersect in two points

MIXED REVIEW 26. D **27.** 144 **28.** 48 **29.** $-cde$

30. $\dfrac{mo}{np}$ **31.** -11 **32.** -5 **33.** 4

Lesson 4.8, pages 202–205

APPLY THE STRATEGY 15. a; since the container is wider at the bottom and narrower at the top, the height rises gently at first, then more steeply later.

Lesson 4.9, pages 206–211

TRY THESE
1. Number of Daily Newspapers in U.S.

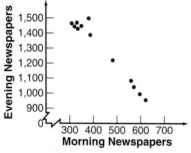

PRACTICE
1. U.S. Energy Production and Imports

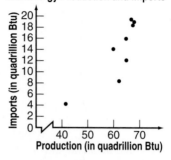

3. World Area and Population, 1992

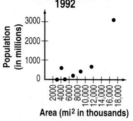

EXTEND 7. As the number of miles increase, so do the number of stations.

THINK CRITICALLY 9. d; the points for these ordered pairs of data form a straight line because the perimeter of a square increases 4 units for every unit the length of a side of a square increases. **11.** b; people's heights should not affect their Science grade, and their Science grade should not affect their height.

MIXED REVIEW 14. 20,000 **15.** 15 **16.** -15
17. -15 **18.** C **19.** A **20.** $y = 45$ **21.** $y = -3$
22. $y = -3$

Chapter Review, pages 212–213

1. b **2.** e **3.** d **4.** c **5.** a **6.** Function; domain: 3, 5, 7, 9, 11; range: 2, 4 **7.** Not a function
8. Function; domain: 5, 10, 15, 20, 25; range: 5, 10, 15, 20, 25 **9.** $(-8, -4)$ **10.** $(-3, 6)$ **11.** $(4, 2)$
12. $(9, 0)$ **13.** $(2, -4)$ **14.** Yes; possible explanations: a vertical line cannot be drawn through any two points; for every different x-value, there is only one $f(x)$-value

15.

$y = -x + 3$

16.

$y = 3x - 2$

17.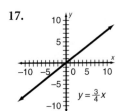

18.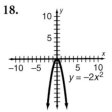

19.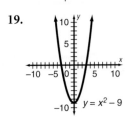

20.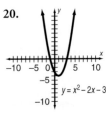

21. $y = 3x - 4$;
$y = 8; x = 4$

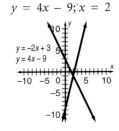

22. $y = -2x + 3$;
$y = 4x - 9; x = 2$

23. $y = 4(x - 1)$;
$y = 4x - 4$;
$x = $ all real numbers

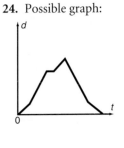

24. Possible graph:

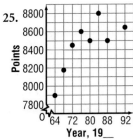

25.

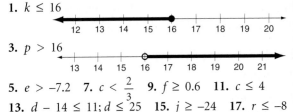

Year, 19__

26. As the years increase, the number of points tend to increase.

Chapter 5 Linear Inequalities

Lesson 5.1, pages 221–226

TRY THESE **1.** $c > -3.5$ **3.** $c < -3.5$

5.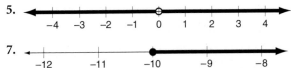

7.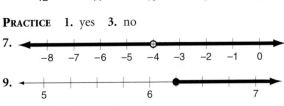

PRACTICE **1.** yes **3.** no

7.

9.

EXTEND **17.** no; accept $g \leq 7.3$ or $g \geq 7.3$ **19.** no; accept $v > 10$, $v \geq 10$, $v \neq 10$, $10 > v$, or $10 \geq v$
21. yes **23.** $x \geq 6$ **25.** $x \neq \frac{1}{2}$ **27.** $x \leq 37.4$
29. a. $s \leq 55$ **b.** No; it is impossible for a car to travel less than 0 mi/h.

THINK CRITICALLY **31.** always true **33.** never true; possible example: $2 < 3$ but 3 is not less than 2
35. sometimes; possible example: if $a = 2$, $b = 1$, then $2 < 3$ but 2 is not less than 1

Lesson 5.3, pages 230–234

TRY THESE
1. $a \leq 7$

3. $c \geq 4.8$

5. $m \leq 4$

PRACTICE
1. $k \leq 16$

3. $p > 16$

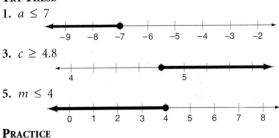

5. $e > -7.2$ **7.** $c < \frac{2}{3}$ **9.** $f \geq 0.6$ **11.** $c \leq 4$
13. $d - 14 \leq 11; d \leq 25$ **15.** $j \geq -24$ **17.** $r \leq -8$
19. $q \geq 8.6$ **21.** $z > 7.2$ **23.** $h > 13.5$
25. $t \leq -\frac{1}{6}$ **27.** $s < 5$ **29.** $\frac{A}{12} \geq 10; A \geq 120$

EXTEND
31. $h \leq 8$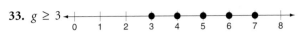

33. $g \geq 3$

35. The graph of $y = x - 4$ is graphed for all values of x when $x < 3$. **37. a.** $0.75w \leq 25$

b. $w \leq 33\frac{1}{3}$; however, only integers from 1 to 33 make sense as solutions to this problem

THINK CRITICALLY **39.** always true

MIXED REVIEW **43.** \$8.06 **44.** \$8.00 **45.** no mode
46. \$2.30 **47.** $e = 1.6$ **48.** $k = -20$ **49.** $n = -36$
50. y-axis **51.** Quadrant II **52.** Quadrant IV **53.** C

Lesson 5.4, pages 235–240

TRY THESE

1. $c \geq 4$

3. $s < -36$

5. $1 \leq x$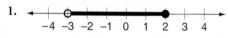

7. $x > -10$

9. $40 + 25h \geq 150$; 4.4 hours or more

PRACTICE **1.** $-5 \leq d$ **3.** $c \geq 12$ **5.** $4 \geq p$
7. $e > -8$ **9.** $x < 7$ **11.** $k > 10$ **15.** $q > 6$
17. $r < 5$ **19.** no solution **21.** all real numbers
23. The solution is $x \geq 5$. This will be graphed as a straight horizontal line beginning at (5, 1) and drawn to the right.

EXTEND **25.** distributed 2 over h but not over 4; $h \geq 2$ **27.** incorrectly thought that if all variables were eliminated, the solution is all real numbers; no solution **29.** $z + 27 < 4z - 6$; $z > 11$
31. $100 + 0.06s \geq \$360$; at least \$4333.34

MIXED REVIEW **37.** \$1000 **38.** 8 **39.** -8 **40.** -8
41. B **42.** $x = 36$ **43.** $y > 1$

Lesson 5.5, pages 241–246

TRY THESE

1.

3.

5.

7. $x < 2$ or $x > 2$ **9.** $0 < x < 4$ **11.** $x > -2$ and $x < 4$, or $-2 < x < 4$ **13.** $x \leq -9$ or $x > 4$

PRACTICE **1.** no **3.** yes **5.** no **7.** $x < 0$ or $x > 0$ **9.** $0 \leq x \leq 5$

11.

13.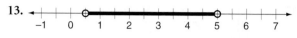

15.

17. $a < -5$ or $a \geq -1$

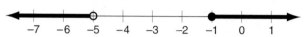

19. $j < 5$ or $j > 7$

21. $24 < b < 32$

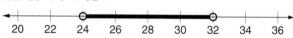

23. $50 \leq 7(n + 12) \leq 68$; $-4 \leq n \leq -3$ **25.** $x < 5$ or $x > 7$ **27.** $x < 8$ or $x > 13$ **29.** $x > 5$ and $x < 7$, or $5 < x < 7$ **31. a.** $t < 55$ or $t > 80$
b. temperatures less than 55°F or greater than 80°F;

EXTEND
33. $2 \leq d < \frac{8}{3}$

35. $x < 8$ or $x > 15$

37. $2 < b < 10$

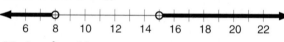

39. a. $155 \leq w - 30 \leq 199$
b. any weight between 185 and 229 lb;

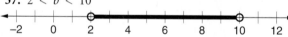

THINK CRITICALLY
41. a. $-1 < x < 6$

b. $x < -8$ or $x > -6$

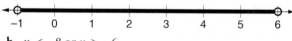

43. all real values of x

MIXED REVIEW   **44.** 5:40   **45.** $b = -2$ **46.** $m = 1$

47.

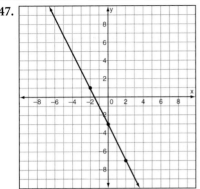

48.

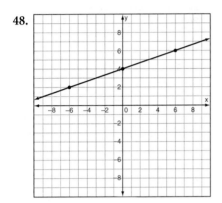

49. D

Lesson 5.6, pages 247–251

TRY THESE   **1.** $500, $562.50, $675
3. $400 \leq c < 500$

PRACTICE   **1.** 24.5 (million), 28 (million), 45 (million)
3. $45 \leq p \leq 65$

5. No outliers. The range is 27.

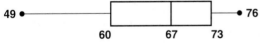

7. Team A; Team A; 130, 198

EXTEND   **11.** 4, 8, 12   **13. a.** $3 \leq v \leq 12$ or
$4 \leq v \leq 54$ **b.** $3 \leq v \leq 8, 4 \leq v \leq 12$, or
$8 \leq v \leq 54$ **c.** $3 \leq v \leq 54$

THINK CRITICALLY   **15.** d; quite consistent
temperatures year round   **17.** c; temperatures vary
from cold to warm but not as much as do the
temperatures for Minneapolis

Lesson 5.7, pages 252–255

APPLY THE STRATEGY   **9.** 5.2 h   **11.** maximum, 216
red; minimum, 144 blue   **13.** $6692°F \leq t \leq 7592°F$
15. $m \geq 7$

Chapter Review, pages 256–257

1. e   **2.** d   **3.** a   **4.** c   **5.** b

6.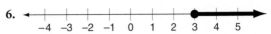

7.

8.

9.

10. $<$ **11.** $\leq$ **12.** $<$ **13.** $>$

14. $a > 5$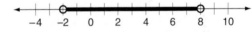

15. $x < 6$

16. $-10 < y$

17. $3 < w$

18. $x > 20$ **19.** $b > -6$ **20.** $x \leq 15$ **21.** $2 > r$
22. $x > 36$ **23.** $x > 12$ **24.** $x < \dfrac{1}{3}$ **25.** $y \leq -4$

26. $-2 < x < 8$

27. $x > 7$ or $x < 3$

28. $y \leq -3$ or $y \geq 5$

29. no solution
30. $x \leq -1$ or $x \geq 5$

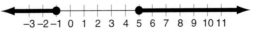

31. $x \leq -3$ or $x \geq 2$

32.

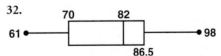

33. $70 \leq s \leq 86.5$ **34.** $w \geq 7$ **35.** length, 39 ft;
width, 8ft

Chapter 6 Linear Functions and Graphs

Lesson 6.2, pages 328–334

TRY THESE **1.** k **3.** j **5.** s **7.** 0.5 **9.** 0
11. undefined **13.** line a **15.** line e

PRACTICE **1.** 0.5 **3.** $-\dfrac{8}{3}$ **5.** 3 **7.** –3
9. undefined **11.** line c **13.** line p **15.** 1

EXTEND **35.** The slope for each year of the graph is positive, except for 1987–88. The negative slope indicates that the number of flights decreased.
37. 540,000 flights per year **39.** 2222.8 ft

MIXED REVIEW **41.** –66 **42.** $-\dfrac{1}{7}$ **43.** 36
44. linear **45.** nonlinear **46.** nonlinear **47.** B

Lesson 6.3, pages 276–284

TRY THESE **1.** 3, –2 **3.** 3, 3

5. 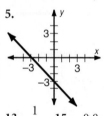 **7.**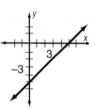

13. $\dfrac{1}{20}$ **15.** –0.8

PRACTICE **1.** 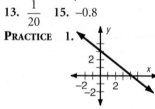 **3.**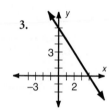

9. c **11.** b **13.** $y = -2.5x$ **15.** $y = -\dfrac{7}{8} + \dfrac{3}{5}$
17. 3 **19.** $\dfrac{1}{3}$ **21.** –1 **23.** –2

EXTEND **27.** $y = 3x$ **29.** $y = -2x - 3$
31. $y = -\dfrac{2}{3}x - 2$

THINK CRITICALLY **39.** $(-4, 2)$ **41.** the slopes are negative reciprocals; –1

MIXED REVIEW **44.** Edgar **45.** approximately 57% (4 out of 7) **46.** $\dfrac{2}{5}$ **47.** $4\dfrac{1}{2}$ h **48.** distributive property of multiplication over addition
49. associative property of addition
50. commutative property of multiplication
51. additive inverse property **52.** D

Lesson 6.4, pages 285–292

TRY THESE **1.** $y = 0.5x + 3.5$ **3.** $y = -2x + 7$

5. $y = -6$ **7.** $y = 1.5x - 3.5$ **9.** $y = \dfrac{3}{2}x + 2$
11. $y - 4 = -3(x - 1)$; $y = -3x + 7$
13. $y - 1 = \dfrac{1}{2}(x + 2)$; $y = \dfrac{1}{2}x + 2$
15. $y = -0.5x + 20.17$

PRACTICE **1.** $b = -2.5$ **3.** $b = 10.5$
5. $y = -4x + 2$ **7.** $y = -1.5x - 2.5$
9. $y = 6x - 18.5$ **11.** $y - 2 = -0.2(x + 3)$;
$y = -0.2x + 1.4$ **13.** $y + 2 = -3(x + 4)$;
$y = -3x - 14$ **15.** $y = 0.5x + 25$
17. $y = 1.4x - 38$

EXTEND **21.** $y = -0.25x - 0.5$ **23.** $y = 8x - 16$
25. $y = 1.25x - 6.25$ **27.** $y = 0.75x - 5.5$
29. $y - 2 = -\dfrac{7}{2}(x + 1)$ **31a.** $y = 10x + 75$

31b. 225 pairs

THINK CRITICALLY **33.** write both in slope-intercept form: $y = x - 1$ **35.** $y = a + 3$; $y = -a - 9$

MIXED REVIEW **36.** 1.75 **37.** 3.61 **38.** 6.29
39. 4.47 **40.** 4 **41.** $\dfrac{1}{3}$ **42.** 0 **43.** 9
44. –2.4 **45.** $2\dfrac{17}{40}$ **46.** $5\dfrac{3}{20}$ **47.** $8\dfrac{29}{30}$ **48.** D

Lesson 6.5, pages 293–298

TRY THESE **1.** $-8x + y = 5$ or $8x - y = -5$
3. $3x + 2y = 0$ **5.** $y = 9$ **7.** $x = -7$
9. $-x + 6y = -4$ **11.** $x + y = 11$ **13.** $x = -3$
15. $x - y = 0$ **17.** $y = -3$
19a. $60x + 90y = 1260$ **19b.** The graph has y-intercept (0, 14), x-intercept (21, 0)

PRACTICE **1.** $-x - y = 0$ or $x + y = 0$
3. $-3x + 2y = 6$ **5.** $2x + y = 18$ **7.** $y = -3$
9. $-2x + y = 1$ **11.** $y = 0$ **13a.** $x + 2y = 18$

EXTEND **15.** $2x + 10y = 15$ **17.** $-18x + 5y = 40$
19. $27x + 20y = 3$ **21.** $-18x + 12y = -7$
23. $-15x + y = -75$ **25.** $10x + 16y = 75$

THINK CRITICALLY **33.** horizontal: $y = -5$; vertical: $x = 2$ **35.** $x = 1, x = -1$ **37.** $x = -8$

MIXED REVIEW **39.** B **40.** 43.8 **41.** 208.3%
42. 920.8 **43.** 2.2 **44.** $x = 13$ **45.** $x = 40$
46. $x = -3$ **47.** 1250%

Lesson 6.6, pages 299–306

TRY THESE **1.** pos **3.** pos **5.** neg **7.** 2005

PRACTICE **1.** negative **3.** nonlinear **5.** positive
7. $y = 4.75x + 4.025$ **9.** a; 0.99; data are positively correlated and each data point lies very close to what would be the line of best fit

EXTEND **11.** negatively correlated; as the number of

cans redeemed increases, the number found in the trash decreases   **13.** –0.997 rounded to nearest thousandth; points for each pair of data lie close to the line of best fit   **15.** 303, to the nearest integer   **17.** The coefficient of correlation is –0.04, showing almost no relationship between the length of the drive and the number of cans collected weekly

THINK CRITICALLY   19. actual number = 149; estimated from the equation = 149.16; percent difference is 0.1%

MIXED REVIEW

21. $\begin{bmatrix} 3 & -2 & 4 \\ 0 & 0 & 4 \\ 4 & 4 & -4 \end{bmatrix}$ **22.** $\begin{bmatrix} 2 & 1 & 4 \\ -1 & 1 & -2 \\ -7 & 10 & -9 \end{bmatrix}$

23. 25   **24.** 267   **25.** 99   **26.** $x > 8$ **27.** $x < 3$
28. $x \geq -5$ **29.** C   **30.** $m = 0.25, b = -25$
31. $m = 4.2, b = 0.3$ **32.** $m = 1, b = 0$
33. $m = -1, b = -1$ **34.** parallel: $y = 1.2x + 19$;
perpendicular: $y = -\dfrac{5}{6}x + 2\dfrac{14}{15}$

Lesson 6.7, pages 307–311

TRY THESE   1. $y = 50$ **3.** $y = 23.8$ **5.** $k = 0.8$;
$y = 0.8x$ **7.** no   **9.** yes   **11.** yes   **13.** no
15. 170

PRACTICE   1. $k = 1.75, y = 1.75x$ **3.** $k = 2.4$,
$y = 2.4x$ **5.** $k = 4; y = 4x$ **7.** $k = -0.2, y = -0.2x$
9. $k = -0.75, y = -0.75x$ **11.** 11.25   **13.** –35
15. the value of π, approximately 3.14   **17.** no

EXTEND   21. 0.38

THINK CRITICALLY   25. No, an unidentified slope indicates a vertical line and means that k, the constant of variation, will be undefined.

Chapter Review, pages 316–317

1. b   **2.** c   **3.** a   **4.** c   **5.** $\dfrac{2}{5}$ **6.** 2
7. –3   **8.** $y = \dfrac{3}{4}x + 8$ **9.** $y = -2x + 3$
10. $y = -\dfrac{1}{3}x - 2$ **11.** perpedicular
12. parallel   **13.** neither
14. $y - 2 = -9(x - 6); y = -9x + 56$
15. $m = 6; y - 3 = 6(x - 3); y = 6x - 15$
16. $y - 2 = -\dfrac{1}{2}(x - 9); y = \dfrac{1}{2}x + 6.5$
17. $-2x + y = 1$ **18.** $2x + y = 11$
19. $-3x + 8y = 17$ **20.** $3x + 2y = 8$ **21.** zero
22. positive   **23.** negative   **24.** $k = \dfrac{2}{3}; y = \dfrac{2}{3}x$
25. $k = 3.2; y = 3.2x$ **26.** $k = -1.75; y = -1.75x$

27.

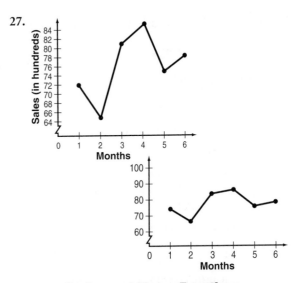

Chapter 7 Systems of Linear Equations

Lesson 7.2, pages 328–334

TRY THESE   1. yes   **3.** yes
5.

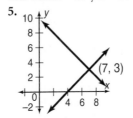

7.

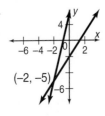

9. (3, $17)

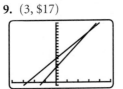

PRACTICE   1. yes   **3.** no   **5.** $3x + 2y = 28$,
$2x + y = 17$ **7.** $6/yd for cotton; $5 yd for rayon
9.   **11.**

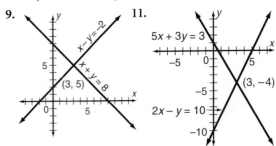

17. $y = 0.06x + 44, y = 0.09x + 35$
21. $y = 15x + 250, y = 20x + 200$ **23.** 10 weeks; $400

EXTEND   25. (4, 3)   **27.** 24   **29.** $(-1, -5)$

SELECTED ANSWERS

THINK CRITICALLY **31.** no; because two distinct lines may intersect in, at most, one point; infinitely many **33.** $(0, b)$

Lesson 7.3, pages 335–341
TRY THESE **1.** $(15, 3)$ **3.** $(3, 2)$ **5.** $\left(\frac{1}{3}, 2\right)$
7. $x + y = 8000, 0.06x + 0.08y = 530$
PRACTICE **1.** $(2, -7)$ **3.** $(2, -3)$ **5.** $(0, 2)$
7. $(7, -1)$ **9.** $\left(-1, \frac{2}{7}\right)$ **11.** $(10, 1)$
13. $x + y = 90, x = 5y$ **15.** $y = 50 + 40x$, $y = 30 + 45x$ **17.** For a job lasting less than 4 h, Paula's is cheaper. For a job lasting more than 4 h, Reliable would be cheaper.
EXTEND **19.** $(0.5, 3)$ **23.** 35 **25.** rate of boat in still water is 3 mi/h; rate of current is 1 mi/h
27. $(3, 2, 1)$
THINK CRITICALLY **31.** $(4, 2)$ **33.** $A = 2, B = 5$

Lesson 7.4, pages 342–347
TRY THESE **1.** Multiply equation 1 by 7 and equation 2 by 4, or multiply equation 1 by 3 and equation 2 by 5.
3. $(8, 3)$ **5.** $(-2, 5)$ **7.** $(-6, -12)$ **9.** 3 at $200, 13 at $165
PRACTICE **1.** Multiply equation 1 by 3 and equation 2 by 5, or multiply equation 1 by 7 and equation 2 by 6.
3. $(1, -2)$ **5.** $(-1, 4)$ **7.** $(-3, -2)$ **9.** $(-1.5, 0.25)$
11. $(-0.5, 1)$ **13.** $(9, -7)$ **15.** 36 Mercurys, 53 Whirlwinds
EXTEND **19.** $(6, 10)$ **21.** $\left(\frac{58}{27}, -\frac{7}{9}\right)$ **23.** 1986
25. The first equation, $N = 1200 - 10t$
27. $N \approx 1010, t \approx 19.0$; the equilibrium point of about 1010 passengers per hour is reached when a bus completes the route in about 19 min.
THINK CRITICALLY **29.** $(a + b, a - b)$ **31.** 42
33. $(2, 4), (-6, 0), (6, -2)$
MIXED REVIEW **35.** $5a - 35$ **36.** $-2x + 12$
37. $2b + 6b^2$ **38.** $8m - 4$ **39.** 74,800 mi **40.** D
41. $(6, 11)$

Lesson 7.5, pages 348–352
TRY THESE **1.** $(9, 4)$ **3.** inconsistent
5. independent **7.** Consolidated: $y = 5.6x - 35$; Co-op: $y = 5.6x - 25$
PRACTICE **1.** inconsistent **3.** inconsistent
5. dependent **7.** inconsistent **9.** independent
EXTEND **13.** 15; −25; because then it would be a

dependent, consistent system **15.** −3
THINK CRITICALLY **17.** The slopes are equal. The y-intercepts are unequal. **19.** A: 3,600; B: 4,100; C: 4,200

Lesson 7.6, pages 353–359
TRY THESE **1.** −2 **3.** −22
5. 32 **7.** $(5, -4)$ **9.** $(2, -1)$
PRACTICE **1.** −3 **3.** 0 **5.** −4 **7.** 2 **9.** −171
11. $-8k$ **15.** $(-1, 4)$ **17.** $(-1, 2)$ **19.** $\left(\frac{-28}{3}, \frac{-49}{9}\right)$
21. $(6, 9)$ **23.** $(4, 6)$ **25.** $(3, -7)$ **27.** 15 at $30, 25 at $22 **29.** 4 g each

EXTEND **31.** −63 **33.** André, $18; Franco, $12
35. 0 **37.** $(6, 10)$
THINK CRITICALLY
39. $\begin{vmatrix} a & b \\ c & d \end{vmatrix} = ad - bc = 4$
The value of $\begin{vmatrix} c & d \\ a & b \end{vmatrix}$ is $cb - ad$.
And $cb - ad = bc - ad$, which is the additive inverse of $ad - bc$. So $bc - ad = -(ad - bc)$.
$$= -(4)$$
41. $\begin{vmatrix} d & c \\ b & a \end{vmatrix} = da - bc$; since $da - bc = ad - bc$, and $ad - bc = 4, da - bc = 4$
MIXED REVIEW **43.** 39% or 0.39 **44.** $2800
45–48. **49.** B **50.** $(5, -2)$

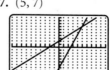

Lesson 7.7, pages 360–365
APPLY THE STRATEGY **19.** 56 dimes, 24 quarters
21. 35, 69 **23.** $(29)(14) = 406$ **25.** $26,500
27. $w = 5$ ft, $l = 14$ ft **29.** 24 $10 bills, 34 $5 bills, and 36 $1 bills

Chapter Review, pages 364–365
1. c **2.** a **3.** d **4.** b **5.** yes **6.** no
7. $(5, 7)$ **8.** $(4, 13)$

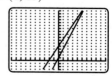

9. $(1, 0)$

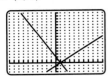

10. $(-6, -5)$

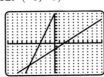

11. $(1, -2)$

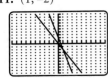

12. $(-6, -8)$

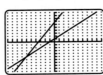

13. $(2, -3)$ **14.** $\left(5, \dfrac{1}{2}\right)$ **15.** $(3, -2)$ **16.** $(2, 1)$

17. 18 **18.** $(3, 4.5)$ **19.** $(8, -5)$ **20.** $(-3, -1)$

21. $(-17, -14)$ **22.** Chen, \$53; Lisa, \$37 **23.** $(5, -2)$

24. inconsistent **25.** dependent **26.** $(-1, -1)$

27. $(2, 1)$ **28.** $(3, 4)$ **29.** $w = 18$ in., $l = 72$ in.

30. 16, 22

Chapter 8 Systems of Linear Inequalities

Lesson 8.1, pages 373–378

TRY THESE **1.** no **3.** no

5.

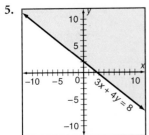

7. $y \geq \dfrac{1}{3}x + 3$ or $3y \geq x + 9$

PRACTICE **1.** no **3.** yes

5.

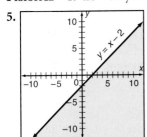

7.

11. $y < x - 4$ **13.** $y \geq -2x + 1$

EXTEND **17.** not equivalent **19.** equivalent

THINK CRITICALLY **25.** dashed line; half-plane above the line **27.** solid line; half-plane above the line

29. $x > 0$

33.

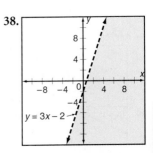

-1 1

34.

-2.3 6

35. $(-3, 2)$ **36.** $(1.5, 2.5)$

37.

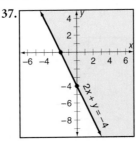

$2x + y = -4$

38.

$y = 3x - 2$

Lesson 8.2, pages 379–385

TRY THESE

1.

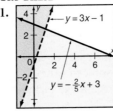

$y = 3x - 1$

$y = -\dfrac{2}{5}x + 3$

3.

$y = -x + 1$

$y = x$

5.

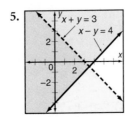

$x + y = 3$

$x - y = 4$

7.

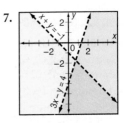

$x + y = 2$

$3x - y = 4$

11a. no **11b.** no **11c.** yes **11d.** yes

PRACTICE **1.** no **3.** no

5.

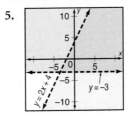

$y = 2x + 4$

$y = -3$

7.

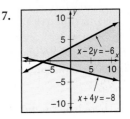

$x - 2y = -6$

$x + 4y = -8$

9.

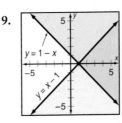

$y = 1 - x$

$y = x - 1$

11.

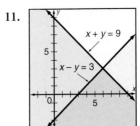

$x + y = 9$

$x - y = 3$

17. $\begin{cases} y < -2x + 4 \\ y > 2x - 1 \end{cases}$ **19.** $\begin{cases} y \geq x + 2 \\ y \leq \dfrac{3}{5}\,x + 3 \end{cases}$

21.b. possible combinations: 0 chairs and 50 tables, 10 chairs and 35 tables, 50 chairs and 10 tables

EXTEND   23. $2x + y \leq 3$ **25.** both

27. $\begin{cases} x > 0 \\ y > 0 \end{cases}$ **29.** $\begin{cases} x \geq 0 \\ y < 0 \end{cases}$

31. 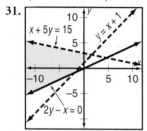 **33.**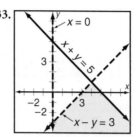

b. possible combinations: 5000 in mutual funds and 4000 in CD's, 6000

THINK CRITICALLY

39. Not possible; because coordinate plane extends to infinity in all directions, it is impossible to write equations for boundary lines of a region that includes the entire plane.

Lesson 8.4, pages 388–393

TRY THESE   1. The point is not within the region.
3. minimum 0 at $(0, 0)$; maximum 21 at $(0, 7)$
5. minimum 2 at $(0, 1)$; maximum 38 at $(9, 1)$
7. minimum 0 at $(0, 0)$; maximum 15 at $(4, 3)$ and $(5, 0)$ **9.** maximum is 16.5 at $(3.5, 9.5)$

PRACTICE   1. The point lies within the region.
3. The point lies within the region.   **5.** minimum -15 at $(0, 5)$; maximum 3 at $(3, 1)$ **7.** minimum -15 at $(-3, 3)$; maximum 0 at $(0, 0)$ and $(-2, -3)$
9. maximum is 12 at $(0, 3)$ **11.** maximum is 81 at $(16, 5)$ **13.** She should choose $C = 3x + 2y$. At the point $(0, 2)$, this function has a minimum value of 4.

EXTEND

17. There is a minimum value but no maximum value because the graph of the constraints is unbounded.

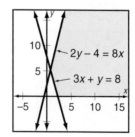

19. There is a minimum value but no maximum value, because the graph of the constraints is unbounded.

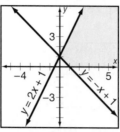

21. For any value less than M, the graph misses the feasible region.

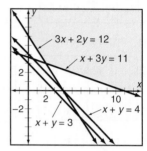

THINK CRITICALLY   23. It represents that profit P is $1.00 on each bracelet and $2.00 on each necklace.

Lesson 8.5, pages 394–397

APPLY THE STRATEGY   15 a. Let $x =$ number of trick skis. Let $y =$ number of slalom skis.

$6x + 4y \leq 96$

$x + y \leq 20$

$x \geq 0$

$y \geq 0$

15b.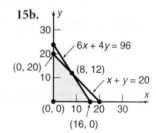

15c. $P = 45x + 30y$

vertex	$P = 45x + 30y$	Profit, P
$(0, 0)$	$45(0) + 30(0)$	$0
$(0, 20)$	$45(0) + 30(20)$	$600
$(16, 0)$	$45(16) + 30(0)$	$720
$(8, 12)$	$45(8) + 30(12)$	$720

15d. 16 pairs of trick skis and 0 pairs of slalom skis or 8 pairs of trick skis and 12 pairs of slalom skis; $720
17. 75 bicycles and 35 rowers;
$P = 50(75) + 75(35) = \$6375$

Lesson 8.6, pages 398–403

TRY THESE   1. 104   **3.** $\dfrac{26}{104}$ or $\dfrac{1}{4}$ **5.** 360 shirts
7. $\dfrac{1}{18}$
PRACTICE   1. 26^3, which is 17,576   **3.** $90 \cdot 26^3$, which is 1,581,840   **5.** $26 \cdot 90$, which is 2340
7. $9 \cdot 2 \cdot 10$, which is 180   **9.** $6 \cdot 8 \cdot 5 = 240$
11. $\left(\dfrac{1}{2}\right)^5 = \dfrac{1}{32}$

EXTEND   13. The game is unfair;
$P(A \text{ wins}) = \dfrac{27}{36}$ or $\dfrac{3}{4}$ of the time;

$P(B \ wins) = \dfrac{9}{36}$ or $\dfrac{1}{4}$ of the time

THINK CRITICALLY **15.** $\dfrac{1}{16}$ **17.** $\dfrac{1}{4}$; $\dfrac{1}{4}$

19. Exercises 16–18 include the sample space of all possible outcomes for a family with four children. The sum of all possible outcomes in a probability experiment is exactly 1.

Chapter Review, pages 404–405

1. d **2.** c **3.** b **4.** a **5.** no **6.** no **7.** yes

8. $y = x + 3$

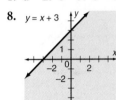

9. $3x + 6y = 12$

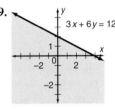

10. $3x - 2y = 6$

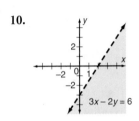

11. $y = x + 1$ $y = 2x - 2$

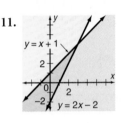

12. $y = -2x + 3$ $3y = x + 6$

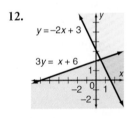

13. $3x - 2y = 0$ $x - 4y = 4$

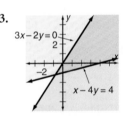

14.

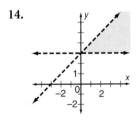

15.

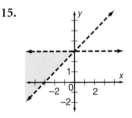

16.

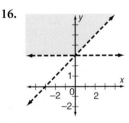

17. maximum is 24 at $(0, 8)$ **18. a.** $P = 3x + 2y$
18b. $x + y \geq 20, x + y \leq 40, y \geq x, y \leq 20$

18c. The maximum profit given the constraints is $100 earned by selling 20 apple pies and 20 blueberry pies.

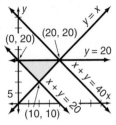

19. 216

Chapter 9 Absolute Value and the Real Number System

Lesson 9.2, pages 416–423

TRY THESE **1.** $(2.5, 0); x = 2.5$ **3.** $(0, -1.5); x = 0$
5. $(-4, -9); x = -4$ **7.** $(-3, -1); x = -3$ **9.** The rays that form the V in the graph of $y = 4|x|$ are farther from the x-axis than those of $y = |x|$. **11.** The rays that form the V in the graph of $y = -2|x|$ are farther from the x-axis than those of $y = |x|$. The graph opens downward, unlike the graph of $y = |x|$.
13. d **15.** a

17.

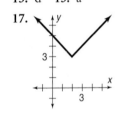

19.

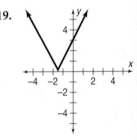

PRACTICE **1.** $(0, -4); x = 0$ **3.** $(1.5, 0); x = 1.5$
5. $(4, -3); x = 4$ **7.** $(-1, -2); x = -1$ **9.** The rays that form the V in the graph of $y = 3|x|$ are farther from the x-axis than those of $y = |x|$. **11.** The rays that form the V in the graph of $y = -\dfrac{1}{5}|x|$ are closer to the x-axis than those of $y = |x|$. The graph opens downward, unlike the graph of $y = |x|$. **13.** b
15. c **17.** d **19.** a

21.

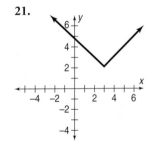

23.

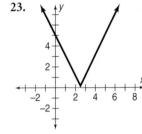

31. $y = -|x - 3| + 3$ **33.** A: $y = |x| - 3$, B: $y = 2|x| - 3$, C: $y = 3|x| - 3$

EXTEND **35.** $(-1, 1); (1, 1)$ **37.** x-axis, y-axis
39. x-axis, y-axis, $y = x$, $y = -x$ **41.** x-axis, y-axis
43. any line passing through the center of the circle
45.

THINK CRITICALLY **47.** Quadrant IV is reflected into Quadrant I; x-axis **49.** no

Lesson 9.3, pages 424–429

TRY THESE **1.** $2x - 3 = 12$ or $2x - 3 = -12$
3. $2x + 5 = 17$ or $2x + 5 = -17$ **5.** $-6, 12$

7. $6, 10$ **9.** $-1, 10$ **11.** $-1, 7$ **13.** $-1, \dfrac{3}{2}$

15. no solution **17.** $-5, -1$ **19.** $|x - 30| = 1.5$

PRACTICE **1.** $4x = 12$ or $4x = -12$
3. $3x + 4 = 11$ or $3x + 4 = -11$
5. $5x - 1.5 = 8.5$ or $5x - 1.5 = -8.5$ **7.** $-8, 8$
9. no solution **11.** $3, 13$ **13.** $-1, 6$ **15.** $-\dfrac{4}{3}, \dfrac{2}{3}$

17. $-13, 5$ **19.** $-6, 0$ **21.** $-\dfrac{5}{2}, \dfrac{7}{2}$ **23.** $-5, -1$

25. $-2, 12$ **27.** $-\dfrac{8}{3}, \dfrac{16}{3}$ **29.** $|x - 9| = 3$; max 12
mo; min 6 mo

EXTEND **31.** $-\dfrac{5}{4}, \dfrac{5}{6}; -\dfrac{5}{4}; \dfrac{5}{6}$ **33.** $\dfrac{3}{5}$ **35.** $-2, 6$

37. $3, 13$ **39.** $|x - 3.5| = 0.5$; max average 4.0, min average 3.0

THINK CRITICALLY **43.** y_1 would be V-shaped and y_2 would be a line that intersects y_1 in exactly one point.
45. $\dfrac{d - e}{c}, \dfrac{d + e}{c}$ **47.** $\dfrac{4 - c}{3}, -c - 4$

Lesson 9.4, pages 430–436

TRY THESE **1.** $-6 < x + 3$ and $x + 3 < 6$
3. $2x - 4 \le -6$ or $2x - 4 \ge 6$ **5.** $-8 < z < 2$
7. $t \le 3$ or $t \ge 7$ **9.** c **11.** d **13.** $q \le 3$ or $q \ge 4$
15. $-4 < t < \dfrac{2}{3}$ **17.** all reals **19.** $|x - 85| \le 3$

PRACTICE **1.** $-5 < x + 1$ and $x + 1 < 5$
3. $x - 4 < -6$ or $x - 4 > 6$ **5.** $-6 \le 2x - 2$ and $2x - 2 \le 6$ **7.** $2x + 5 \le -3$ or $2x + 5 > 3$

9. $-4 < z < 2$

11. $x < 3$ or $x > 7$

13. $-3 \le x \le 4$

15. $w \le -6$ or $w \ge 1$

17. b **19.** c

21. $q \le -1$ or $q \ge 5$

23. $-5 < t < \dfrac{5}{3}$

25. $z < -2$ or $z > 6$

27. $-7 \le q \le 2$

29. (all reals except $x = 0$) **31.** no solution
33. (all x except $x = 2$) **35.** all reals
37. $-3 < x < 9$ **39.** $x = -2$ **41.** $-\dfrac{3}{2} \le z \le \dfrac{1}{2}$
43. $2 \le t \le 8$ **45.** $\dfrac{1}{2} < x < \dfrac{7}{2}$
47. $|x - 21| \le 10$

EXTEND **51.** $x > \dfrac{1}{3}$ **53.** $x \le \dfrac{5}{4}$ **55.** $-7 < x \le 11$
57. $x \le 2$ or $x \ge 8$ **59.** $|x - 1.05| \le 0.35$

THINK CRITICALLY **61.** $|x| > 0$ **63.** Since $<$ includes all numbers that $\ge$ does not, any real number that does not satisfy $|x| < c$ must satisfy $|x| \ge c$.

MIXED REVIEW **65.** C **66.** $y = \dfrac{3}{4}x - \dfrac{5}{2}$
67. $y = \dfrac{2}{5}x + \dfrac{16}{5}$ **68.** $(6, 3)$ **69.** $(2, 4)$

70. $x < -10.5$ or $x > 4.5$ **71.** $-\dfrac{14}{3} < x < \dfrac{28}{3}$

Lesson 9.5, pages 437–444

TRY THESE

1.

3. irrationals, reals **5.** wholes, integers, rationals, reals **7.** $\dfrac{-3}{1}$ **9.** $\dfrac{38}{7}$ **11.** $\dfrac{1}{100}$ **13.** $\dfrac{35}{99}$ **15.** $\dfrac{41}{333}$

17. 0.45 **19.** -1.4 **21.** $0.\overline{81}$ **23.** not a real number

25. 52 **27.** 9.06 **29.** -6.86 **31.** 0.94 **33.** $\dfrac{263}{100}, 162$

PRACTICE

1.

3. wholes, integers, rationals, reals **5.** rationals, reals
7. rationals, reals **9.** irrationals, reals **11.** rationals, reals **13.** $\dfrac{522}{100}$ **15.** $\dfrac{78}{1000}$ **17.** $\dfrac{4}{33}$ **19.** $\dfrac{167}{999}$ **21.** $\dfrac{157}{999}$

23. 1.89 **25.** 52 **27.** $\dfrac{30}{23}$ **29.** -0.8 **31.** 1.7

33. 12.29 **35.** −12.88 **37.** 0.75 **39.** 0.3
41. −47.25 **43.** about 0.45 s

EXTEND   45. 3.00 **47.** −0.30 **49.** 4.22
51. 6.23 cm **53.** 81, 100, 121

THINK CRITICALLY   55. $\sqrt{\dfrac{1}{7}}$ **57.** $\sqrt[3]{0.064}$

59. false; answers will vary **61.** A rational number k
is a perfect square if there is a number x such that
$x^2 = k$. **63.** $x \geq -3$

Lesson 9.6, pages 445–451

TRY THESE   1. $-3, -1, 0, 2, \sqrt{5}, 4$

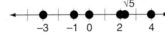

3. $\dfrac{3}{4}$, 1.2, $\dfrac{5}{4}$, $\dfrac{3}{2}$, 1.8

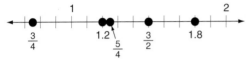

7. symmetric **9.** reflexive **11.** reflexive property;
substitution property **13.** carpool

PRACTICE   1. $-4, -\sqrt{6}, -2, 0, 3, 5$

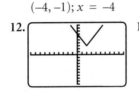

3. $\dfrac{6}{5}$, 1.3, $\dfrac{8}{5}$, .8, $\dfrac{19}{10}$ **5.** $\dfrac{4}{7}$, $\dfrac{5}{8}$, $\dfrac{6}{9}$ **7.** $1\dfrac{6}{7}$, 2.1, $\dfrac{15}{7}$
9. −0.1511, −0.151, −0.1501 **15.** reflexive
17. symmetric **19.** reflexive property; substitution
property **21.** Flora **23.** Flora, Vesta, Iris, Metis,
Hebe, Astraea, Juno, Ceres, Pallas, Hygeia

EXTEND   27. No; a number is not less than itself;
$3 < 3$ **29.** No; if $2 < 3$ and $2 + 4 < 7$, then
$3 + 4 < 7$ is a false statement. **31.** false **33.** true

THINK CRITICALLY   35. For all real numbers a and b,
one and only one of the following properties is true:
$a = b$; $a \neq b$. **37.** Yes; if $a \neq b$, then $b \neq a$.
39. No; if $3 \neq 4$ and $3 + 5 \neq 9$, then $4 + 5 \neq 9$ is a
false statement.

MIXED REVIEW   41. 85.5 **42.** 92 **43.** 88.5 **44.** 29
45. B **46.** $\dfrac{17}{21}$, $\dfrac{13}{16}$, $\dfrac{17}{19}$ **47.** −0.1611, −0.161, 0.1601

Lesson 9.7, pages 452–455

APPLY THE STRATEGY   13. a. The slope of the
transportation workers' graph would rise more slowly
because the average yearly increase is less. The graph

for Exercise 1 supports the prediction.
b. $15.45; $y = 15.45x - 948.5$ **c.** $596.50

Chapter Review, pages 456–457

1. e **2.** c **3.** a **4.** d **5.** b

6. 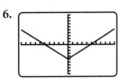 **7.**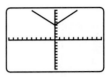

8. 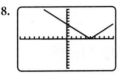 **9.**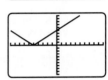

10.   **11.**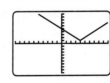

$(-4, -1); x = -4$ $(4, 1); x = 4$

12.   **13.**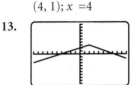

$(2, 3); x = 2$ $(2, 3); x = 2$

14. −6, 6 **15.** −3, 13 **16.** −12, 6 **17.** no solution
18. $-5 < x < 9$

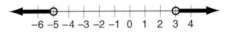

19. $z < -5$ or $z > 3$

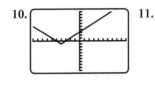

20. $-6 \leq w \leq 2$

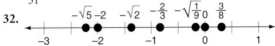

21. all reals; graph is the number line
22. $|x - 97| \leq 3$ **23.** 0.18 **24.** 0.5 **25.** 0.55
26. 17 **27.** 0.27 **28.** −0.7 **29.** not a real number
30. $\dfrac{25}{31}$ **31.** 64

32.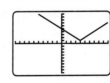

33.

Statement	Reason
1. $a = b$	given
2. $a - c = a - c$	reflexive property
3. $a - c = b - c$	substitution property

34. 2800 million metric tons

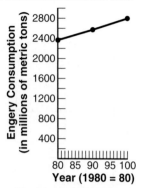

Chapter 10 Graphs of Quadratic Functions

Lesson 10.2, pages 468–476

TRY THESE   **1.** downward   **3.** upward
5. $(0, 0), x = 0$ **7.** $(0, 15), x = 0$
9. $\left(-\frac{5}{6}, -\frac{83}{12}\right), x = -\frac{5}{6}$ **11.** 6 units up   **13.** wider
15. 8 units left   **17.** a
19.

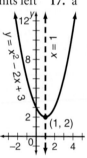

21.

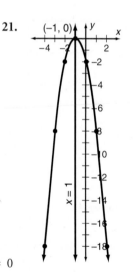

PRACTICE   **1.** upward
3. upward   **5.** $(0, 0), x = 0$
7. $(0, 14), x = 0$ **9.** $(-1, -10), x = -1$
11. $(-4, 22), x = -4$ **13.** 10 units down
15. narrower   **17.** 25 units left   **19.** c   **21.** a   **23.** a
25.

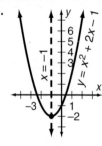

27.

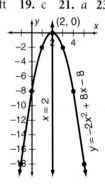

29. (312.50, 103.91); the high point will be 3.9 ft above point A and 312.5 ft from point A.
EXTEND   **31.** $y = 3x^2 - 2x + 4$
33. $y = 2x^2 - x - 2$ **35.** $h = -16t^2 + 25t + 6$
37. about 1,093 million, or about 1.093 billion
39. $(-2.29, 0.25); (2.29, 0.25)$
THINK CRITICALLY   **41.** $(0, c)$ **43.** c moves each graph up c units if c is positive and down c units if c is negative
MIXED REVIEW   **46.** $x = -13.5$ **47.** $x = -\frac{22}{5}$
48. C   **49.** A   **50.** It is shifted up 3 units and to the right by 1 unit. The rays that form the V are closer to the y-axis.   **51.** It is shifted down 1 unit and to the left by 3 units. The rays that form the V are farther away from the y-axis.   **52.** $(2, -25); x = 2$
53. $\left(\frac{3}{2}, -13\right); x = \frac{3}{2}$

Lesson 10.3, pages 477–482

TRY THESE   **1.** $0 = 3x^2 - 4x + 8$ **3.** $0 = \frac{1}{2}x^2 - 3x$
5. equivalent   **7.** $-5, 3$ **9.** 5
11. $0 = -x^2 + 6x - 7; 1.59, 4.41$ **13.** $-3, 5$
15. $-1.68, 2.68$ **17.** 0   **19.** 1.29 and 43.83
PRACTICE   **1.** $0 = 2x^2 - 5x + 3$ **3.** $0 = \frac{1}{4}x^2 - 5x$
5. equivalent   **7.** $-4, 6$ **9.** $-5, 2$ **11.** 6
13. $0.31, 3.19$ **15.** $-0.78, 1.28$
17. $0 = x^2 + 8x - 65; 5, -13$
19. $0 = -x^2 + 2x + 4; -1.24, 3.24$
21. $0 = 4x^2 - 2x - 4; -0.78, 1.28$ **23.** $-2.27, 5.27$
25. 1   **27.** 2   **29.** 2   **31.** April 1994
EXTEND   **35.** 16 or 44   **37.** 60
MIXED REVIEW   **43.** C   **44.** infinitely many
45. one   **46.** $\frac{-8}{1}$ **47.** $\frac{533}{100}$ **48.** $\frac{-19}{2}$ **49.** $\frac{19}{200}$
50. $0 = x^2 + 5x + 6; -2, -3$
51. $0 = -x^2 + 3x + 3; -0.79, 3.79$

Lesson 10.4, pages 483–489

TRY THESE   **1.** $-9, 9$ **3.** $\pm 4\sqrt{3}$ **5.** $\pm 2\sqrt{10}$
7. $-4, 6$ **9.** $8, -10$ **11.** $-1, -\frac{1}{3}$
13. no real number solution   **15.** ± 2.29 **17.** ± 2.58
19. $0.45, 3.55$ **21.** $-1.72, 0.72$
PRACTICE   **1.** $-7, 7$ **3.** $\pm \frac{5}{9}$ **5.** $\pm \frac{19}{22}$
7. no real number solution   **9.** $\pm 4\sqrt{5}$ **11.** $\pm 7\sqrt{2}$
13. $-11, 19$ **15.** $\pm 7\sqrt{10}$ **17.** $-2, -10$

19. $-4.5, \dfrac{11}{6}$ **21.** $-1.2, 0.6$ **23.** ± 2.65 **25.** ± 2.41
27. $\pm \sqrt{8.54}$ **29.** $1.95, 10.05$ **31.** $-0.75, 2.35$

EXTEND **35.** $x < -12$ or $x > 12$ **37.** $x < -6$ or
$x > 6$ **39.** $0 < x < 4$ **41.** 2.82 m **43.** 7.64 m

THINK CRITICALLY **45.** 4 **47.** $A = P\left(1 + \dfrac{r}{100}\right)^4$

49. less; in the simple interest account, you would
have \$929.60; in the compounded account, you would
have \$933.12

MIXED REVIEW **51.** D **52.** multiplicative inverse
53. multiplication property of zero **54.** $(5, 1); x = 5$
55. $(-4, -2); x = -4$ **56.** $-5.45, -0.55$
57. $2.31, 7.69$

Lesson 10.5, pages 490–497

TRY THESE **1.** -56 **3.** 0 **5.** 0 **7.** 1 **9.** 0
11. $-1, 8$ **13.** $-5, 2$ **15.** $-3.12, 1.12$
17. $0.88, -0.68$ **19.** 36 **21.** ± 8

PRACTICE **1.** -87 **3.** 0 **5.** 2 **7.** 1 **9.** 0
11. $-1, 7$ **13.** $5, 6$ **15.** $-3, 10$ **17.** $-7, 3$
19. \$37.00 **21.** $-\dfrac{1}{2}, \dfrac{5}{7}$ **23.** $-5.37, 0.37$
25. $-1.65, 3.65$ **27.** $-2.87, -0.46$ **29.** 49 **31.** ± 40

EXTEND **35.** $x^2 - 4x - 21 = 0$
37. $2x^2 - 13x - 7 = 0$ **39.** $x^2 - c^2 = 0$
41. $x^2 + 37x - 650 = 0; 13, -50$

THINK CRITICALLY **43.** $k < 9$ **45.** $\dfrac{2a + 1 \pm \sqrt{21}}{2}$

47. The solutions are all rational numbers.

MIXED REVIEW **49.** C **50.** function **51.** not a
function **52.** $4 - 3x = 21$ or $4 - 3x = -21$
53. $2x - \dfrac{1}{2} = \dfrac{1}{4}$ or $2x - \dfrac{1}{2} = -\dfrac{1}{4}$ **54.** $\dfrac{7}{2}, 15$
55. $-\dfrac{1}{2}, \dfrac{1}{3}$

Lesson 10.6, pages 498–501

APPLY THE STRATEGY **13a.** $l = 120 - 2w$
13b. $A(w) = w(120 - 2w)$ or $120 - 2w^2$
13c. $l = 60$ ft, $w = 30$ ft, $A = 1800$ ft^2
13d. To make a square, divide the total amount of
fencing by 3. Each side of fencing must be 40 ft; for a
40-ft square $A = 1600$ ft^2

Lesson 10.7, pages 502–507

TRY THESE

1. quadratic

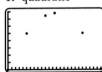

3. neither

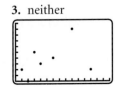

PRACTICE **3.** $74.121; 179.721$
5. $y = 0.783x + 128.857$

EXTEND **9.** $c = 179; 1600a + 40b + c = 238$;
$6400a + 80b + c = 362$ **11.** $291.875, 448.375$; the
equation does not seem to be a good fit at higher
temperatures

THINK CRITICALLY **13.** false

MIXED REVIEW **15.** D

16. $m < 4$;

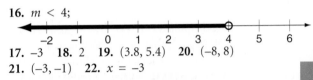

17. -3 **18.** 2 **19.** $(3.8, 5.4)$ **20.** $(-8, 8)$
21. $(-3, -1)$ **22.** $x = -3$

Chapter Review, pages 508–509

1. b **2.** d **3.** c **4.** a **5.** c **6.** a **7.** b
8. opens downward **9.** shifted 3 units left
10. wider **11.** shifted down 3 units
12. $(0, 5); x = 0$ **13.** $(1, 9); x = 1$
14. $(-3, -10); x = -3$ **15.** $1, 2$ **16.** $-4, 5$
17. $1.32, -5.32$ **18.** ± 7 **19.** $\pm 4\sqrt{2}$ **20.** $\pm 4\sqrt{3}$
21. $-2, -5$ **22.** 1.5 **23.** $2.79, -1.79$ **24.** two
25. zero **26.** two **27.** 3 yards by 3 yards
28. 10%

Chapter 11 Polynomials and Exponents

Lesson 11.2, pages 521–526

TRY THESE **1.** 10 **3.** 1 **5.** $-6a^4$ **7.** $-18c^{11}d^3$
9. 729 **11.** z^{28} **13.** $-27b^3$ **15.** $-64m^6$

PRACTICE **1.** 2 **3.** 8 **5.** 1 **7.** 6 **9.** a^7
11. $-10x^5$ **13.** $-20g^{10}$ **15.** $18a^3b^6$

EXTEND **35.** $-36x^9y^5$ **37.** $-8p^9q^{19}$ **39.** $W = I^2R$

THINK CRITICALLY **41.** 3 **43.** 5 **45.** $q = p + 1$

Lesson 11.3, pages 527–531

TRY THESE **1.** $3x^2$ **3.** $3a^8b^2$ **5.** 1 **7.** $\dfrac{1}{16}$ **9.** $\dfrac{1}{3g^2h^2}$
11. $2q$ **13.** $\dfrac{64}{125}$ **15.** $\dfrac{32x^{20}}{y^{15}}$ **17.** $4x^2 \div -2x = -2x$

PRACTICE **1.** $8a$ **3.** $-3ab^3$ **5.** $6y$ **7.** $\dfrac{31}{7y^4}$ **9.** 79
11. $\dfrac{3}{4}$ **13.** $\dfrac{13x^4}{y^2}$ **15.** $\dfrac{-3s}{t}$ **17.** $\dfrac{16}{81}$ **19.** $\dfrac{4c^8}{25d^2}$ **21.** x

EXTEND **23.** $\dfrac{64a^{13}b^5}{9}$ **25.** $\dfrac{-gh}{2}$ **27.** $\dfrac{25w^6x^4}{4}$

31. $7,467,802$

THINK CRITICALLY **33.** 3 **35.** 2 **37.** If p and q are
positive and $n(p - q)$ is negative, then p must be less
than q.

41.

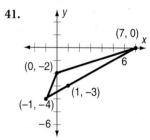

42. Connecting the points (–1, –4), (1, –3), and (7, 0) appears to result in a line that forms the side of a triangle. To justify the conclusion that all the points are in a line, you must use the point slope form and points (–1, –4) and (7, 0) to find the equation of the lines connecting them and then substitute the values (1, –3) in that equation to see whether they are a solution of the equation. 43. $x > -3$ 44. $x \leq 7$ 45. $x < -\dfrac{1}{2}$

46. $x < 1$ 47. C 48. $2b^2$ 49. $\dfrac{-5}{x^2y}$ 50. $\dfrac{64a^6}{27b^9}$

51. 81

Lesson 11.4, pages 532–537

TRY THESE 1. 3.25×10^9 3. 7.2×10^7
5. 3.15×10^{-3} 7. 5.4×10^{-6} 9. 7.35×10^9
11. 8×10^7 13. 5.776×10^9 15. 6.72×10^7

PRACTICE 1. 512,000,000 3. 0.000012
5. 3.14×10^5 7. 2.3×10^7 9. 4.15×10^{-4}
11. 8×10^{-6} 13. 3.12×10^{14} 15. 3×10^3

EXTEND 27. 1×10^{36} 29. 3×10^5 31. 4×10^{17}
33. $(3.7 \times 10^{-1}) \dfrac{\text{m/s}}{°\text{C}}$

THINK CRITICALLY 35. 4 37. 53 39. $ab > 10$

Lesson 11.6, pages 541–547

TRY THESE 1. 6 3. 3
5. $-2x^4y + 3x^2y^2 + 6xy^3 + 9$ 7. $2x^2 + 3x - 5$
9. $9x^2y + xy + 8y$ 11. $-8a^2 - 2a + 7$
13. $-a - 2b - c$ 15. $5b^2 + 9b - 5$
17. $9p^2 + 5p - 4q - 6q^2$

PRACTICE 1. 2 3. 4 5. 7 7. $12a - 2c$
9. $-3p + 6q + 9$ 11. $-4a + b - 6c$
13. $6y^4 + 8y^2 - 10$ 15. $5x + 10y - 24$
17. $4a + 3b + 4c$ 19. $5d^3 - 9d^2 + 4d$
21. $2x^2 + 8xy - 2y^2$ 23. $32x^2$

EXTEND 25. $5x$ 27. 1.72 29. $4g^2 - 4g - 3$
31. W_{st} and W_e 33. 120 ft

THINK CRITICALLY 35. $-7x$ 37. It is true;
$(a - b) + (b - a) = 0$; therefore $b - a$ is the

opposite of $a - b$.
41. 10.15 42. $11,250 43. $19,500 44. 3,400,000
45. 102,000,000 46. 0.00000048 47. 0.0000303

Lesson 11.8, pages 551–557

TRY THESE 1. $3x^2 + 5x$ 3. $-4y - 8$
5. $x^2 + 15x + 56$ 7. $2x^2 - 7x - 4$
9. $a^3 - 5a^2 + 8a - 4$ 11. $a^2 + 4a + 4$
13. $x^2 - 25$ 15. $(x + 2)(2x + 1) = 2x^2 + 5x + 2$

PRACTICE 1. $x^2 - 5x$ 3. $-10z^2 - 15z$
5. $-6m^2 + 12m$ 7. $2x^3y - 2xy^3$ 9. $x^2 + 7x + 10$
11. $a^2 - 10a + 24$ 13. $c^2 - 2c - 15$
15. $g^4 + 5g^2 - 14$

EXTEND 35. $q^3 - 6q^2 + 12q - 8$
37. $g^3 + g^2 - 5g + 3$ 39. 4.5 m 41. 6 ft
43. 21,005,820

THINK CRITICALLY 45. $n = 7$
47. Since $(a + b)^2 = a^2 + 2ab + b^2$, it is $2ab$ greater than $a^2 + b^2$.

Lesson 11.9, pages 558–563

APPLY THE STRATEGY 13. same direction 15. Their distances are equal; $4(t + 0.5) = 6t$; $t = 1$ h; Ivan hikes for 1 h and covers 6 km; Darrell hikes for 1.5 h and also covers 6 km. 17. 4:06 P.M.

19. 9.75 or $9\dfrac{3}{4}$ mi

Chapter Review, pages 562–567

1. b 2. c 3. a 4. $-3y^2$ 5. $3x$ 6. $6xy$ 7. x
8. z^6 9. $-6a^7$ 10. $-12x^3y^7$ 11. $6a^4b^4c^3$ 12. $3x^2$
13. $-6a^3$ 14. $\dfrac{p^5}{4q^2}$ 15. $-5x$ 16. $\dfrac{27g^6}{8h^{12}}$
17. $4.5 \cdot 10^{10}$ 18. $3.15 \cdot 10^3$ 19. $2.4 \cdot 10^9$
20. $6.9 \cdot 10^{-8}$ 21. $2x + y$ 22. $-2x^2 + 3y$
23. $3x - 5$ 24. $4x - y$ 25. $11r - s + t$
26. $3x^2 + 7x - 10$ 27. $2a - 2b + 6c$
28. $3x^2 + 3xy + 3$ 29. $2x^2 + 5x + 2$
30. $-x^2 + 3x - 2$ 31. $x^2 + 2x - 3$ 32. $k^2 + 4k$
33. $-15x^2 + 5xy$ 34. $6a^2b - 8ab^3$
35. $a^2 + 4a + 3$ 36. $x^2 + 2x - 8$
37. $2r^2 - rs - s^2$ 38. $y^2 - 36$ 39. $b^2 + 4b + 4$
40. $y^3 + y^2 - 10y + 8$ 41. 1:50 P.M.

Chapter 12 Polynomials and Factoring

Lesson 12.2, pages 575–581

TRY THESE 1. $3x(3x + 1)$ 3. $2y^2(y - 8)$
5. $(x + 3)(x + 2)$ 7. $(x - 3)(x - 1)$

9. $(x - 4)(x + 2)$ **11.** $(x + 7)(x - 4)$
13. $(2x + 3)(x + 1)$
PRACTICE   **1.** $2a(a^3 + 4)$ **3.** $4ab^2(2 - 3ab)$
5. $3(2c^2 - 3d^2)$ **7.** $2x(2x^2 - x + 7)$
15. $(x + 1)(x + 4)$ **17.** $(z + 5)(z + 3)$
19. $(x - 3)(x - 3)$ **21.** $(z - 9)(z - 1)$
23. $(y + 8)(y - 1)$ **25.** $(y + 12)(y - 1)$
EXTEND   **39.** $3b^{2n}$ **41.** $\left(\dfrac{1}{2}\right)(Z - N)$

THINK CRITICALLY   **43.** $7, -7, 13, -13, 8, -8$

Lesson 12.3, pages 582–585

TRY THESE   **1.** $(a + 4)(a - 4)$ **3.** not factorable
5. $(3e^2 + 4f)(3e^2 - 4f)$ **7.** yes; $(a - 5)^2$ **9.** yes;
$(c^2 - 2)^2$ **11.** $x^2 - 6x + 9 = (x - 3)^2$

PRACTICE   **1.** $(a - 8)^2$ **3.** $(c + 3)^2$ **5.** $(e - 10)^2$
7. $(9h + 2)^2$ **9.** $(10p + 3a)^2$
11. $(v + 11)(v - 11)$ **13.** $(y + 1)(y - 1)$
15. $3(3a^2 - 4)$ **17.** $(6c^2 + d)(6c^2 - d)$
19. $2x + 5$ and $2x - 5$

EXTEND   **21.** $(4r^2 - s)^2$ **23.** not factorable
25. $(11x + 17)(11x - 17)$ **27a.** 7 and 5
$(49 - 25 = 24)$; 5 and 1 $(25 - 1) = 24$ **b.** 12 and 2
$(7 + 5, 7 - 5)$; 6 and 4 $(5 + 1, 5 - 1)$

THINK CRITICALLY
29. $(a + b + c + d)(a + b - c - d)$ **31.** 27

MIXED REVIEW

33.

34.

35.

36. B   **37.** $(8, 16)$ **38.** $(5, -5)$ **39.** $(1, 3)$

Lesson 12.4, pages 586–590

TRY THESE   **1.** $12(c + d)$ **3.** $(g - 4)(b + 3)$
5. $(2 + g)(h - k)$ **7.** $(y^3 + 3)(y - 2)$
9. $(w - 3)(2z - 1)$ **11.** $(x + 1)(y - 2)$

PRACTICE   **1.** $(5 + w)(x + 1)$ **3.** $(x + 2)(y + 5)$
5. $(x - 7)(y + 2)$ **7.** $(p + q)(s - 2t)$
9. $(3a + 1)(4b - 5)$ **11.** $(3w - 1)(z + 4)$

EXTEND   **21.** $(a + b + c)(x + 2)$
23. $(p + q - r)(a - b)$ **25.** $2x + 2y - 2$
27a. $(x + 2)(y + 4)$ **b.** *ya ka* 3 *bha ka* 9

Lesson 12.5, pages 591–594

TRY THESE   **1.** $4(x - 2)(x + 4)$ **3.** $5(z + 2)(z - 2)$
5. $(x + 2)(y + 4)(y - 4)$ **7.** $a, b + 5, b + 8$
PRACTICE   **1.** $2(x + 5)(x + 7)$ **3.** $5(z - 6)(z + 3)$
5. $4b(c - 2)(c + 5)$ **7.** $3(x + 5)(x - 5)$

EXTEND   **19.** $(a + 3)(b + 6)(b + 2)$
21. $6(p - 2)(q + 3)(q - 3)$
23. $(a + 1)(b + 2)(b + 5)$
25. $(m + 2)(m - 2)(n + 5)(n + 4)$
27a. $V = \pi h(R + r)(R - r)$ **b.** 5820.0 lb
29. $(x + 1)(x^2 + x + 1)$
31. $(y - 2)(y^2 + 2y + 4)(y + 2)(y^2 - 2y + 4)$
THINK CRITICALLY   **33.** $2(2x + 3y)(4x^2 - 6xy + 9y^2)$
35. 7

MIXED REVIEW   **36.** B   **37.** $(x + 2)(y - 5)$
38. $(2a + 3)(b - 1)$ **39.** $(2y + 3)(2y - 3)(z + 1)$

Lesson 12.6, pages 595–601

TRY THESE   **1.** $x = 0, x = -3$ **3.** $x = -5, x = 1$
5. $x = 0, x = 6$ **7.** $x = -7, x = 2$
9. $z = 4, z = -3$ **11.** $y = -4, y = -2$
13. 5 yd, 11 yd   **15.** not factorable
PRACTICE   **1.** $0, -1$ **3.** $-3, 8$ **5.** $0, 5$ **7.** $10, -7$
9. $-4, -7$ **11.** $-6, 3$ **13.** $-2, 7$ **15.** $-\dfrac{1}{2}, -3$

17. $-2, 6$ **19.** 2 m, 9 m   **21.** yes; $(x - 8)(x - 4)$
23. yes; $(x - 10)(x + 9)$ **25.** not factorable
27. not factorable   **29.** $11, 12; -12, -11$

EXTEND   **31.** $1, 3$ **33.** 12 m, 8 m   **35.** 4 units by
7 units

THINK CRITICALLY   **37.** $x^2 + 12x = 0$
39. $3x^2 + x - 2 = 0$ **41.** $0, 2, 6$
MIXED REVIEW   **42.** -7 **43.** 23   **44.** $\dfrac{4}{3}$

45. $\{6, 0, -9\}$ **46.** $\{3, 5, 8\}$ **47.** $\{4, 0, 9\}$ **48.** D
49. $7x^2 + 4x - 5$ **50.** $2ab - 4a + 7b$

Lesson 12.7, pages 602–605

APPLY THE STRATEGY   **15a.** $s^2 + 4s - 60 = 0$
b. $s = 6$ ft   **17a.** 7 in. × 9 in.   **b.** 10 in. × 12 in.

19. 39, 40 and $-40, -39$ **21a.** 3 in
b. 18 in. × 25 in.   **23.** 6 ft

Chapter Review, pages 606–607

1. c   **2.** a   **3.** d   **4.** b   **5.** $(x + 1)(x + 3)$
6. $(x - 2)(x - 3)$ **7.** $(x + 4)(x - 2)$
8. $(x + 1)(x - 5)$ **9.** $5d(3 + 5d)$
10. $3yz(3xz - y + 2x^2)$ **11.** $(z + 8)(z + 3)$
12. $(y + 9)(y - 1)$ **13.** $2x - 5, x + 3$
14. $(c + 4)^2$ **15.** not factorable   **16.** $(r - 3)^2$

17. $(w + 10)(w - 10)$ **18.** $3 + 2x, 3 - 2x$
19. $(4 + y)(x - 3)$ **20.** $(a + 4)(b + 2)$
21. $(p - q)(r - s)$ **22.** $(c - 1)(d + 5)$
23. $a + 3, b - 5$ **24.** $3(x + 2)(x + 7)$
25. $2a(b^2 + 5)(b + 3)(b - 3)$
26. $6(y - 3)(y - 3)$ **27.** $3x(y + 2)(y - 2)$
28. $4, x + 6, x - 6$ **29.** $0, -5$ **30.** $0, 9$ **31.** $-7, 10$
32. $-6, -3$ **33.** 14 and 16 or -14 and -16 **34.** 6 in.
35. 7 and 11

Chapter 13 Geometry and Radical Expressions

Lesson 13.2, pages 618–624

TRY THESE **1.** obtuse, scalene **3.** acute, scalene
5. $67°$ **7.** $90°$ **9.** $90°, 69°, 21°$ **11.** obtuse
13. right **15.** 11.3 ft

PRACTICE **1.** acute, isosceles **3.** right, scalene
5. $35°$ and $110°$

EXTEND **9.** 29 ft **11.** yes; the longest diagonal is
66.52.; find this length by squaring 65, adding 200 (the
square of the diagonal of the square face) and finding
the square root of the sum.

THINK CRITICALLY **15.** two lengths equal, with $a^2 + b^2 > c^2$; c can be either the longest side or one of two
equal longer sides.

MIXED REVIEW **16.** 12 **17.** -17 **18.** 41 **19.** C

20. 2, 4 **21.** 7, 6 **22.** $\frac{1}{2}, \frac{1}{3}$ **23.** $(x + 4)(x - 4)$

24. $4y(x - 6)(x + 2)$ **25.** $(x^2 + 9)(x + 3)(x - 3)$

Lesson 13.3, pages 625–630

TRY THESE **1.** $5\sqrt{2}$ **3.** $8\sqrt{3}$ **5.** $2\sqrt{3}$

7. $\frac{\sqrt{30}}{6}$ **9.** $30\sqrt{5}$ km

PRACTICE **1.** $3\sqrt{2}$ **3.** $10\sqrt{3}$ **5.** $3n$
7. $12c\sqrt{2c}$ **9.** 3 **11.** 50 **13.** $12 - 6\sqrt{3}$

15. $12 - 2\sqrt{15}$ **17.** 3 **19.** $\frac{2\sqrt{5}}{5}$ **21.** 52

23. 36

EXTEND **27.** $\frac{s\sqrt{3}}{2}$ **29.** $\sqrt{5} + \sqrt{3}; 2$

31. $\sqrt{10} - 4; -6$ **33.** $\frac{3 + \sqrt{3} + 3\sqrt{2} + \sqrt{6}}{6}$

THINK CRITICALLY **35.** 64, 36
37. $(m + \sqrt{n})(m - \sqrt{n}) = m^2 - n$. Since both m^2
and n are rational numbers, their difference is also a

rational number. Therefore, the product is rational.

MIXED REVIEW **38.** 9 **39.** -2 **40.** 3 **41.** $-4; 3$
42. $-3; \frac{11}{2}$ **43.** $1; 4$ **44.** D

Lesson 13.4, pages 631–635

TRY THESE **1.** $7\sqrt{7}$ **3.** $19\sqrt{2}$
5. $-4\sqrt{10} + 4\sqrt{5}$ **7.** $27\sqrt{5}$ **9.** $\frac{\sqrt{2}}{4}$

PRACTICE **1.** $9\sqrt{3}$ **3.** $8\sqrt{5}$ **5.** $3\sqrt{3} + 6\sqrt{2}$
7. $16\sqrt{h}$ **9.** $19\sqrt{2x}$ **11.** $9\sqrt{2}$ **13.** $3\sqrt{2n + 2}$

EXTEND **23a.** $14\sqrt{2}$ mi **23b.** $28 - 14\sqrt{2}$ mi
25. yes **27.** yes **29a.** The expressions are equal.

29b. $(\sqrt{5} - 1)^2 = 6 - 2\sqrt{5}$ **29c.** $\sqrt{6} + 2$

29d. $\sqrt{3} + 2$

MIXED REVIEW **30.** -26 **31.** 24 **32.** 36 **33.** -9
34. 7 **35.** -27 **36.** -3 **37.** $4x - 1$
38. $-3 < x < 3$ **39.** $x \leq -6$ or $x \geq 6$ **40.** C
41. $6\sqrt{7}$ **42.** $\sqrt{2}$ **43.** $8\sqrt{3}$ **44.** $11\sqrt{x}$

Lesson 13.5, pages 636–641

TRY THESE **1.** 9 **3.** 28 **5.** 12 **7.** 13 **9.** 9

11. 5 **13.** $11 = \sqrt{\frac{A}{\pi}}; 121\pi \text{cm}^2$

PRACTICE **1.** 81 **3.** 1.44 **5.** 6 **7.** $x - 5$
9. 4 **11.** 81 **13.** 36 **15.** 9 **17.** 58 **19.** 72

EXTEND **39.** 2 **41.** $\frac{9}{16}$ **43.** 218 **45.** 12 **47.** 45

51. 10.5

THINK CRITICALLY **49.** Graph of
$y_2 = \sqrt{x} + 2$ same as $y_1 = \sqrt{x}$
but is shifted upward 2 units.

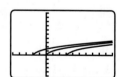

Graph of $y_3 = \sqrt{x - 3}$ is the
same as $y_1 = \sqrt{x}$ but is shifted
right 3 units **51.** 10.5

MIXED REVIEW **52.** 7, 8, 9 **53.** 41.5, 39.5, 37
54. $x \leq 7$ **55.** $x < 11$ **56.** $x > -2$ **57.** 2, 3
58. $-1, 4$ **59.** 9 **60.** 2 **61.** 7

Lesson 13.6, pages 642–647

TRY THESE **1.** 10 **3.** 13 **5.** 17 **7.** $(4, 3)$

9. $\left(\frac{17}{2}, \frac{27}{2}\right)$

PRACTICE 1. 10 3. $2\sqrt{5}$ 5. $5\sqrt{2}$ 7. 25
9. 41 11. $(3, -7)$ 13. $(-2.5, -11)$ 15. $(m, 0)$
17. equilateral

EXTEND 21. $AM = MB = \sqrt{34}$ 23. 6 or 12
THINK CRITICALLY 25a. 2 25b. $y = 2x + 2$
25c. $(-2, -2)$ 25d. $3\sqrt{5}$

MIXED REVIEW 27. 12 28. 1 29. 18 30. 2
31. 8 32. 9 33. D 34. $5 \pm \sqrt{3}$ 35. $\dfrac{2 \pm \sqrt{10}}{3}$
36. $\dfrac{-5 \pm \sqrt{2}}{2}$ 37. $5\sqrt{2}$ 38. 20 39. $\sqrt{29}$

Lesson 13.8, pages 652–658

TRY THESE 1. $\cos B$ or $\sin C$ 3. $\tan C$ 5. $\dfrac{5}{12}$
7. $\dfrac{12}{13}$ 9. $\dfrac{12}{5}$ 11. 0.8333 13. 0.7879

PRACTICE 1. 0.5526 3. 0.7778
5. $\angle G = 28°, \angle H = 62°$ 7. $\angle S = 53°, \angle T = 37°$
9. $\angle X = 32°, \angle Y = 58°$ 11. 35° 13. 87°
15. 12° 17. 2.3 m or less
EXTEND 21. $\text{Tan } C = \dfrac{10}{100} = 0.10; m\angle C = 5.7°$ to
the nearest tenth of a degree; Find side DE of $\triangle CDE$:
$\sin 5.7° = \dfrac{DE}{CD}; \sin 5.7° = \dfrac{DE}{2,640 \text{ ft}};$
$DE = \sin 5.7°(2,640) = 262.2$ ft to the nearest tenth.
THINK CRITICALLY 23. $\tan x = \dfrac{\sin x}{\cos x};$
$\tan x = \dfrac{\dfrac{opp.}{hyp.}}{\dfrac{adj.}{hyp.}} = \dfrac{opp.}{hyp.} \times \dfrac{hyp.}{adj.} = \dfrac{opp.}{adj.},$
the tangent ratio
MIXED REVIEW 24. 1 25. 3 26. 3 27. 6
28. 2 29. 4 30. $x = 7; x = \dfrac{-3}{2}$
31. $x = 5; x = -5$ 32. $x = \sqrt{11}; x = -\sqrt{11}$
33. $x = -1; x = -6$ 34. 35, 28 and $-35, -28$
35. B

Lesson 13.10, pages 662–665

APPLY THE STRATEGY 15. 128 mi 17a. 12.3 ft
17b. 56.2 ft^2 19. 8.6 to 8.7 in., depending on
whether the method of similar triangles or
trigonometric ratios is used

Chapter Review, pages 666–667

1. b 2. c 3. a 4. obtuse 5. right 6. acute
7. $m\angle Z = 18°, m\angle P = 33°, m\angle Q = 129°,$

$m\angle R = 18°$, XY is 20, YZ is 47.5
8. $m\angle C = 41°, m\angle D = 80°, m\angle E = 59°,$
$m\angle F = 41°$, DE is 20, EF is 22.5
9. $2\sqrt{7}$ 10. $5|y|$ 11. 16 12. $4\sqrt{3} + 3\sqrt{5}$
13. $2\sqrt{2}$ 14. 5 15. $6\sqrt{5} - 3\sqrt{2}$ 16. $14\sqrt{2a}$
17. $6|r|\sqrt{3}$ 18. 25 19. 2 20. 6 21. 10
22. 2; $(-5, 4)$ 23. 5; $(7.5, 0)$ 24. 0.4103
25. $m\angle G = 48°, m\angle I = 42°$ 26. c 27. b 28. a
29. 12 ft; 13 ft

Chapter 14 Rational Expressions

Lesson 14.2, pages 679–685

TRY THESE 1. joint; $k = 2$ 3. inverse; $k = 10$
5. yes; $k = 60$ 7. $c = \dfrac{k}{d^2}$ 9. 25 11. 12
13. 15 N

PRACTICE 1. direct; $k = 9.5$ 3. inverse; $k = 72$
5. direct; $k = 0.14$ 7. 48 9. 7 11. 50 13. 2100
15. $d = kr$ 17. $w = \dfrac{k}{l}$
EXTEND 19. 605 21. about 35 min
THINK CRITICALLY 23. $y = x, y = -x$ 25. $(2, 2)$
MIXED REVIEW 27. B 28. $y = -2x + 6$
29. $y = \dfrac{2}{3}x + 2$ 30. $x < -3$ or $x > 9$
31. $-6 < x < 2$ 32. 3 33. 25

Lesson 14.3, pages 686–691

TRY THESE 1. $x = 0$ 3. $x = -3, 7$ 5. $\dfrac{2}{x}, x \neq 0$
7. $\dfrac{x - 3}{6}$ 9. $\dfrac{1}{7}; a \neq 7$

PRACTICE 1. $m = 0; n = 0$ 3. $y = -3$
5. $z = 1, -8$ 7. $x = -8, 6$ 9. $\dfrac{2}{x}; x \neq 0$
EXTEND 21. $\dfrac{(x^2 + 1)(x + 1)}{(x - 1)}; x \neq 1$
23. $\dfrac{x^2(x + 5)}{(x - 1)}; x \neq 0, 1, -5$
THINK CRITICALLY 27. $q = p + 2$
MIXED REVIEW 28. C 29. $y = \dfrac{5}{3}x + \dfrac{1}{3}$
30. $y = -14x + 48$ 31. downward 32. upward

Lesson 14.4, pages 692–696

TRY THESE 1. $\dfrac{1}{2}$ 3. $\dfrac{5}{8n^2}$ 5. $3abc$ 7. $\dfrac{de}{14}$
9. $\dfrac{1}{2}$ 11. $\dfrac{2x - 2}{x}$

PRACTICE 1. $\dfrac{5}{3}$ 3. $\dfrac{4}{3k}$ 5. $2a^2bc$ 7. $\dfrac{2}{3cde}$

9. $\dfrac{1}{4}$ 11. $\dfrac{2x-4}{x}$ 13. $\dfrac{14x+70}{x-6}$ 15. $\dfrac{1}{3x+18}$

17. $-\dfrac{5x}{21x+126}$ 19. $\dfrac{7}{x+2}$ 21. $\dfrac{x^2-2x}{12+2x}$

EXTEND 25. $\dfrac{2t-2}{t-3}$ 27. 2 or 2:1

MIXED REVIEW 33. C 34. 0.9 35. 3 36. $x<4$

37. $x<-2$ 38. -1 39. $\dfrac{x+4}{x+3}$

Lesson 14.5, pages 697–701

TRY THESE 1. $x^2+8, x-2, x+2, 12$

3. $3x+1+\dfrac{1}{2x}$ 5. $a+5$

7. $p^2+p-3+\dfrac{4}{2p+1}$ 9. $4t+3$

PRACTICE 1. $x^2+10, x-4, x+4, 26$

3. $4x+2+\dfrac{1}{3x}$ 5. $a+7$ 7. $4m-2$

EXTEND 17. $8m^2+12mn+24n^2+\dfrac{40n^2}{m-n}$

19. $336r^2+1008r^2+1008r+336$ 21. 700

THINK CRITICALLY 23. -44

MIXED REVIEW 25. A 26. c 27. b 28. a

Lesson 14.6, pages 702–706

TRY THESE 1. mn^2 3. $(x-4), (x-5)$ 5. $-\dfrac{4}{7y}$

7. $-\dfrac{16}{15x}$ 9. $\dfrac{9a-16}{(a+4)(a-9)}$ 11. $\dfrac{-14}{16-x^2}$

PRACTICE 1. ab^3 3. $3x(x+1)$ 5. $\dfrac{4}{z}$ 7. 7

9. -1 11. $\dfrac{23}{9x}$

EXTEND 27. $\dfrac{x-8}{x^2-25}$ 29. $\dfrac{x^3+19x+3x-1}{x(x+1)(6x-1)}$

31. $\dfrac{-2x^3+11x^2-8x+4}{x(1-2x)}$

MIXED REVIEW 35. D 36. $|x-25|\le4$ 37. 9

38. 144 39. c 40. a 41. b 42. $\dfrac{-15t+7}{6t^2}$

43. $\dfrac{-12r+3}{8r^2}$

Lesson 14.7, pages 707–711

TRY THESE 1. $\dfrac{5t+9}{t}$ 3. $\dfrac{12n^2+17n+3}{4n+5}$

5. $\dfrac{15x+10}{16x+4}$ 7. $\dfrac{8d}{3c}$ 9. $\dfrac{2q-6p}{11p}$

PRACTICE 1. $\dfrac{11b-13}{b}$ 3. $\dfrac{9z^2-z-5}{3z}$

5. $\dfrac{t^2-2t-8}{t+1}$ 7. $\dfrac{3y}{4x}$ 9. $\dfrac{q-3p}{4p}$

EXTEND 23. $\dfrac{20x-60}{3x-15}$ 25. $\dfrac{8b^3-12b+48}{4b^2+3b}$

27. $\dfrac{x^2-1}{4}$ ft

THINK CRITICALLY 29. yes; 5 and 6 are both rational

expressions 31. $\dfrac{a^2+1}{a^4+a^3}$

Lesson 14.8, pages 712–717

TRY THESE 3. 64 5. no solution 7. $-\dfrac{1}{2}$

9. no solution 11. $5\dfrac{1}{2}$ mi/h

PRACTICE 1. $\dfrac{49}{60}$ 3. 12 5. no solution 7. $\dfrac{11}{3}$

9. 18 11. $-\dfrac{3}{20}$ 15. Rikuichi = \$2300,

Keemo = \$1250

EXTEND 17. no solution 19. 15 h

THINK CRITICALLY 25. $\dfrac{29}{13}$

Lesson 14.9, pages 718–721

APPLY THE STRATEGY 13. need number of km per mi;
use 1 mi = 1.6 km

Chapter Review, pages 722–723

1. b 2. a 3. c 4. 5 5. 18 6. $\dfrac{3}{x}$; $x\ne0$

7. 6; $b\ne-3$ 8. $-\dfrac{1}{4}$; $y\ne\dfrac{1}{2}$ 9. $3p^2q^2$ 10. $6y$

11. $\dfrac{1}{4}$ 12. $\dfrac{9x+36}{x-2}$ 13. $\dfrac{1}{3x-6}$ 14. $15x$

15. $3x-4+\dfrac{1}{2x}$ 16. $y+8$

17. $3x^3+3x^2-4x-4+\dfrac{1}{x-1}$

18. $2x^2-x+2-\dfrac{4}{x+3}$ 19. $\dfrac{1-g}{g+2}$ 20. $\dfrac{22}{9x}$

21. $\dfrac{10-11z}{6z^2}$ 22. $\dfrac{3t-10}{6t+24}$ 23. $\dfrac{4b}{7a}$ 24. $\dfrac{n-2m}{4n}$

25. $\dfrac{4y+6x}{y-x}$ 26. $\dfrac{3s+2rs}{4r}$ 27. $-\dfrac{1}{15}$ 28. -1

29. $\dfrac{23}{45}$ 30. \$9.14

INDEX

INDEX

Photo Credits

CONTENTS
p. vi: Lillian Gee/Picture It Corp.; p. vii: Terry Qing/FPG International Corp.; p. viii: Michael Hart/FPG International Corp.; p. ix: Jacob Taposchaner/FPG International Corp.; p. x: Lillian Gee/Picture It Corp.; p. xi: DL-FC/FPG International Corp.; p. xii: Lillian Gee/Picture It Corp.; p. xiii: Comstock; p. xiv: John Terence Turner/FPG International Corp.; p. xv: Lillian Gee/Picture It Corp.; p. xvi: Lillian Gee/Picture It Corp.; p. xvii: Lillian Gee/Picture It Corp.; p. xviii: Suzanne Murphy/FPG International Corp.; p. xix: Lillian Gee/Picture It Corp.

CHAPTER 1
p. 2: Ken Chernus/FPG International Corp. (top); David Hamilton/The Image Bank (bottom); p. 3: Bruce Forster/Tony Stone Images; p. 4: Grandadam/Photo Researchers, Inc.; p. 5: Jeff Greenberg/Photo Researchers; p. 6: Lawrence Fried/The Image Bank; p. 7: Steven W. Jones/FPG International Inc. (top); Richard Laird/FPG International Corp. (bottom); p. 8: Blair Seitz/Photo Researchers, Inc.; p. 10: Marc Romanelli/The Image Bank; p. 11: Michael Melford/The Image Bank (top); Telegraph Colour Library/FPG International Corp. (bottom); p. 12: Porterfield/Chickering/Photo Researchers, Inc.; p. 14: Ira Block/The Image Bank; p. 15: Lillian Gee/Picture It Corp.; p. 16: Gerard Champlong/The Image Bank; p. 18: The Image Bank; p. 19: John Lewis Stage/The Image Bank; p. 20: Jeff Isaac Greenberg/Photo Researchers, Inc.; p. 21: Skip Hine; p. 23: Roger Miller Photo, Ltd. (top); F. Roiter/The Image Bank (bottom); p. 24: Ken Huang/The Image Bank; p. 26: McDonald Studios/FPG International Corp. (left); P. Harris/Shooting Star (right); p. 27: Duomo Photo/The Image Bank; p. 28: David De Lossy/The Image Bank; p. 29: Anthony Meshkinyar/Tony Stone Images; p. 30: Sauzereau O/Explorer/Photo Researchers, Inc.; p. 32: Kevin Forest/The Image Bank; p. 34: Lou Jones/The Image Bank; p. 39: Grant Faint/The Image Bank; p. 42: Richard Mackson/FPG International Corp.; p. 44: Steve Dunwell/The Image Bank

CHAPTER 2
p. 52: Stim/Photri, Inc. (top); NSS/LAB/The Image Bank (bottom); p. 53: Mark Burnett/Photo Researchers, Inc.; p. 54: Mark C. Burnett/Photo Researchers, Inc.; p. 55: Ted Kawalerski/The Image Bank; p. 56: Erik von Fisher/Photonics (top); Les Riess/Photri, Inc. (bottom); p. 58 Keith Kent/Science Photo Library/Photo Researchers, Inc.; p. 59: J. P. Pieuchot/The Image Bank; p. 60: Terje Rakke/The Image Bank; p. 63: Romilly Lockyer/The Image Bank; p. 64: Alvis Upitis/The Image Bank (bottom); p. 66: Bachmann/Photo Researchers, Inc.; p. 69: Farrell Grehan/Photo Researchers, Inc.; p. 71: Syd Greenberg/Photo Researchers, Inc. (top); Photo Researchers, Inc. (top middle); Tom Burnside/Photo Researchers, Inc. (bottom middle); Courtesy of Oldsmobile (bottom); p. 72: Science Photo Library/Photo Researchers, Inc.; p. 73: Lillian Gee/Picture It Corp.; p. 81: Harald Sund/The Image Bank; p. 82: Bill Billingham/Photri, Inc.; p. 87: Roger Miller Photo, Ltd.; p. 91: Rosenthal/Superstock; p. 92: Photri, Inc.; p. 93: Photri, Inc.; p. 94: David R. Frazier Photolibrary/Photo Researchers, Inc.; p. 96: Stephen J. Krasemann/Photo Researchers, Inc.; p. 99: Tom Wilson/FPG International Corp.; p. 101: Scott Markewitz/FPG International Corp.; p. 102: Art Stein/Photo Researchers, Inc. (top and bottom); p. 104: Bruce Roberts/Photo Researchers, Inc.

CHAPTER 3
p. 112: Eunice Harris/Science Source/Photo Researchers, Inc. (top); Ken Lax/Photo Researchers, Inc. (bottom); p. 113: Will & Deni McIntyre/Photo Researchers, Inc.; p. 114: Ron Chapple/FPG International Corp.; p. 115: Lillian Gee/Picture It Corp.; p. 116: Lillian Gee/Picture It Corp.; p. 118: Schneps/The Image Bank; p. 119: Courtesy of International Business Machines Corp.; p. 120: Benn Mitchell/The Image Bank; p. 124: Photri, Inc.; p. 127: Renee Lynn/Photo Researchers, Inc.; p. 128: Charles D. Winters/Photo Researchers, Inc.; p. 130: Lillian Gee/Picture It Corp; p. 133: Photri, Inc.; p. 134: Marc Romanelli/The Image Bank; p. 135: Lillian Gee/Picture It; p. 136: Gilda Schiff/Photo Researchers, Inc.; p. 139: John Clark/Photri, Inc.; p. 142: Robert E Daemmich/Tony Stone Images; p. 143: Dean Siracusa/FPG International Corp.; p. 144:

Jeff Greenberg; p. 145: Charles D. Winters/Photo Researchers, Inc.; p. 147: Will & Deni McIntyre/The Image Bank; p. 149: Nancy Brown/The Image Bank; p. 151: Richard Anderson (bottom); p. 152: Lillian Gee/Picture It Corp.; p. 153: Lillian Gee/Picture It Corp.; p. 156: Diane Padys/FPG International Corp.; p. 157: Lillian Gee/Picture It Corp.; p. 158: Bill Varie/The Image Bank (top); Thomas Digory/The Image Bank (bottom); p. 159: Obremski/The Image Bank

CHAPTER 4

p. 166: Robert E. Daemmrich/Tony Stone Images (top); Jim McNee/FPG International Corp. (bottom); p. 167: Tetrel/Photo Researchers, Inc.; p. 168: Robert E. Daemmrich/Tony Stone Images; p. 169: Duomo Photography/The Image Bank; p. 170: Larry Dale Gordon Studio/The Image Bank; p. 171: Wachter/Photri, Inc.; p. 175: Al Hamdan/The Image Bank; p. 177: Joseph Devenney/The Image Bank; p. 179: Anne-Marie Weber/FPG International Corp.; p. 184: Lillian Gee/Picture It Corp.; p. 186: Maria Taglienti/The Image Bank; p.187: Nick Sebastian/Photri, Inc.; p. 191: Lillian Gee/Picture It Corp.; p. 195: David Madison/Tony Stone Images; p. 196: Jeff Greenberg; p. 198: Mel Digiacomo/The Image Bank; p. 200: Peter Gridley/FPG International Corp.; p. 201: Stephen Simpson/FPG International Corp.; p. 202: Aram Gesar/The Image Bank; p. 204: Ed Braverman/FPG International Corp.; p. 206: Lillian Gee/Picture It Corp.; p. 207: Telegraph Colour Library/FPG International Corp.; p. 208: Vince Streano/Tony Stone Images (top); Art Montes De Oca/FPG International Corp. (bottom); p. 209: Brett Froomer/The Image Bank; p. 210: Lonnie Duka/Tony Stone Images; p. 211: Lou Jones/The Image Bank

CHAPTER 5

p. 218: Joseph Nettis/Photo Researchers, Inc. (top); Michael Salas/The Image Bank (middle); Bruce Ayres/Tony Stone Images (bottom); p. 219: William McCoy/Rainbow; p. 220: Will & Deni McIntyre/Photo Researchers, Inc.; p. 221: Obremski/The Image Bank; p. 223: Takeshi Takahara/Photo Researchers Inc.; p. 224: Robert E. Daemmrich/Tony Stone Images (top); Alex Stewart/The Image Bank (bottom); p. 225: Lani Novak Howe/Photri Inc., p. 226: Will McIntrye/Photo Reasearchers Inc.; p. 228: Lillian Gee/Picture It Corp.; p. 233: Margot Granitsas/Photo Researchers, Inc.; p. 235: Blair Seitz/Photo Reasearchers, Inc.; p. 237: Gatzen/Photri, Inc., p. 238: Rafael Macia/Photo Researchers, Inc.; p. 240: Blair Seitz/Photo Researchers Inc.; p. 241: Alan Carruthers/Photo Researchers Inc.; p. 243: Ken Cavanagh/Photo Researchers Inc.; p. 246: Philippe Sion/The Image Bank; p. 247: Courtesy of International Business Machines Corp.; p. 251: Jay Freis/The Image Bank; p. 252: Lillian Gee/Picture It Corp., p. 253: Dick Luria/FPG International Corp.; p. 254: Michael Ventura/Tony Stone Images

CHAPTER 6

p. 262: Patti McConville/The Image Bank (top); Yellow Dog Prods./The Image Bank (bottom); p. 263: Ron Chapple/FPG International Corp.; p. 264: Renee Lynn/Photo Researchers, Inc.; p. 265: Jeff Greenberg; p. 268: David Brownell/The Image Bank; p. 270: Luis Castaneda/The Image Bank (top); Grant Faint/The Image Bank (bottom); p. 271: Jeffrey Sylvester/FPG International Corp., p. 275: Mosallem/ FPG International Corp.; p. 277: Andre Gallant/The Image Bank; p. 278: Jurgen Vogt/The Image Bank (top); Robert Kristofik/The Image Bank (bottom); p. 280: Roger Miller Photo, Ltd.; p. 281: Maria Taglienti/The Image Bank; p. 284: Walter Bibikow/The Image Bank; p. 287: Paul Simcock/The Image Bank; p. 288: Greater Houston Partnership; p. 289: U. S. Department of Agriculture; p. 290: Dr. C. W. Biedel/Photri, Inc.; p. 292: Aluminum Company of America (top); Veina Brainard/Photri, Inc. (bottom); p. 295: Robert E. Daemmrich/Tony Stone Images; p. 302: Larry Fried/The Image Bank; p. 304: Jeff Cadge/The Image Bank; p. 305: Kaz Mori/The Image Bank; p. 308: Alberto Incrocci/The Image Bank; p. 310: U.S. Department of Commerce/American Petroleum Institute; p. 313: Holiday Inn, Inc. (top); Courtesy of Hewlett-Packard Co. (bottom); p. 314: HMS Images/The Image Bank

CHAPTER 7

p. 322: Alcoa (top); Telegraph Colour Library/FPG International Corp. (bottom); p. 323: Charles Thatcher/Tony Stone Images (top); Lawrence Migdale/Photo Researchers, Inc. (bottom); p. 324: Tourism British Columbia; p. 326: Lillian Gee/Picture It Corp.; p. 327: Photri, Inc.; p. 328: Roger Miller Photo, Ltd.; p. 330: Comstock; p. 331: Theodore Anderson/The Image Bank; p. 332: Ron Chapple/FPG International Corp.; p. 333:

Sebastian/Photri, Inc.; p. 334: Photri, Inc.; p. 335: Lori Adamski Peek/Tony Stone Images; p. 337: Lou Jones/The Image Bank; p. 338: Michael Krasowitz/FPG International Corp.; p. 339: B. Kulik/Photri, Inc.; p. 340: Comstock; p. 341: InnerLight/The Image Bank; p. 343: Lani Novak Howe/Photri, Inc.; p. 345: Photri, Inc. (top); Dick Luria/FPG International Corp. (bottom); p. 347: Photri, Inc.; p. 348: Photri, Inc.; p. 351: Comstock; p. 355: Ken Kaminsky/Photri, Inc.; p. 356: Blair Seitz/Photo Researchers, Inc.; p. 358: Dick Luria/FPG International Corp.; p. 359: Photri, Inc.; p. 360: Bennett/Photri, Inc.; p. 361: Jack Novak/Photri, Inc.; p. 362: Michael Melford/The Image Bank

CHAPTER 8

p. 370: Don Smeltzer/Tony Stone Images (top); Michael Krasowitz/FPG International Corp. (bottom); p. 371: Jon Riley/Tony Stone Images (left); p. 372: David Young Wolff/Tony Stone Images; p. 373: David R. Frazier Photolibrary/Photo Researchers, Inc.; p. 377: Ken Lax/Photo Researchers, Inc.; p. 378: Jeff Isaac Greenberg/Photo Researchers, Inc.; p. 379: Alvis Upitis/The Image Bank; p. 380: Superstock; p. 381: Burton McNeely/The Image Bank; p. 382: Steve Niedorf/The Image Bank; p. 383: Photri, Inc.; p. 384: Alan Brown/Photonics; p. 385: Gregory Heisler/The Image Bank (top and top center); Jay Brousseau/The Image Bank (bottom center); Larry Gatz/The Image Bank (bottom); p. 386: Lillian Gee/Picture It Corp.; p. 389: Alvis Upitis/The Image Bank; p. 392: Dick Luria/FPG International Corp.; p. 393: Rafael Macia/Photo Researchers, Inc.; p. 396: Jean-Marc Barey/Agence Vandystadt/Photo Researchers, Inc. (top); Jeff Greenberg/Photo Researchers, Inc. (bottom); p. 398: Lillian Gee/Picture It Corp.; p. 401: Richard Nowitz/Photo Researchers, Inc.; p. 402: Image Makers/The Image Bank

CHAPTER 9

p. 410: Baron Wolman/Tony Stone Images (top); Superstock (bottom); p. 411: Steve Dunwell/The Image Bank; p. 412: R. W. Jones/Westlight (top); Peter Gridley/FPG International Corp. (bottom); p. 415: Andy Caulfield/The Image Bank; p. 416: Lani Novak Howe/Photri, Inc.; p. 420: Lani Novak Howe/Photri, Inc.; p. 423: Elaine Sulle/The Image Bank; p. 425: Art Stein/Photo Researchers, Inc. (top); Harald Sund/The Image Bank (center); Art Stein/The Image Bank (bottom); p. 427: Gary Bistram/The Image Bank; p. 428: Comstock (top); Rafael Macia/Photo Researchers, Inc. (bottom); p. 432: Geoff Gove/The Image Bank; p. 433: Blair Seitz/Photo Researchers, Inc.; p. 435: Donal Philby/FPG International Corp.; p. 436: Michael Hart/FPG International Corp.; p. 438: Alberto Incrocci/The Image Bank; p. 441: Patti McConville/The Image Bank; p. 442: Sergio Duarte/The Image Bank; p. 443: David Doody/FPG International Corp.; p. 444: Robert J. Bennett/Photri, Inc. (top); Mark Scott/FPG International Corp. (bottom); p. 447: Frank Cezus/Tony Stone Images; p. 448: Grant V. Faint/The Image Bank; p. 449: NASA/Science Photo Library/Photo Researchers, Inc.; p. 451: Tom Wilson/FPG International Corp.; p. 453: Comstock

CHAPTER 10

p. 462: Telegraph Colour Library/FPG International Corp. (top); Andy Sacks/Tony Stone Images (bottom); p. 463: John Banagan/The Image Bank; p. 464: Gerard Loucel/Tony Stone Images; p. 465: Kathleen Campbell/Tony Stone Images; p. 470: Donovan Reese/Tony Stone Images; p. 471: Tom Wilson/FPG International Corp.; p. 473: Bernard Roussel/The Image Bank; p. 474: Courtesy of Johnson Controls, Inc.; p; 475: Comstock; p. 476: Richard Hutchings/Photo Researchers, Inc.; p. 477: Lillian Gee/Picture It Corp.; p. 480: Photri, Inc.; p. 481: Philip & Karen Smith/Tony Stone Images; p. 485: B. Kulik/Photri, Inc.; p. 489: Chuck Kuhn/The Image Bank; p. 491: F. Roiter/The Image Bank; p. 493: Keith Wood/Tony Stone Images; p. 497: Bill Howe/Photri, Inc.; p. 498: James Kirby/Photri, Inc.; p. 499: Hans Reinhard/ OKAPIA/Photo Researchers, Inc.; p. 500: Michael P. Gadomski/Photo Researchers, Inc.; p. 502: G. Randall/FPG International Corp. (top); Stan Osolinski/FPG International Corp. (bottom); p. 506: Blair Seitz/Photo Researchers, Inc.

CHAPTER 11

p. 514: NASA/Science Photo Library/Photo Researchers, Inc. (top); Kay Chernish/The Image Bank (bottom); p. 515: Stephen Derr/The Image Bank; p. 516: Stephen Dalton/Photo Researchers, Inc. (top); Frank Whitney/The Image Bank (bottom); p. 518: Lillian Gee/Picture It Corp.; p. 523: Lillian Gee/Picture It Corp.; p. 526: P. & G. Bowater/The Image Bank; p. 527: Marvin E. Newman/The Image Bank; p. 531:

Superstock; p. 533: Steven Hunt/The Image Bank; p. 535: NASA/Science Source/Photo Reseachers, Inc. (top); Rich LaSalle/Tony Stone Images (bottom); p. 536: Gerard Christopher/Viacom/Shooting Star; p. 537: Wendt Worldwide; p. 539: Photri, Inc.; p. 540: M. Keller/Superstock; p. 541: Erik Simonsen/The Image Bank; p. 542: Trevor Bonderud/Westlight; p. 543: Lillian Gee/Picture It Corp.; p. 546: Melissa Grimes-Guy/Photo Researchers, Inc.; p. 547: Guy Sauvage/Agence Vandystadt/Photo Researchers, Inc.; p. 548: Lillian Gee/Picture It Corp.; p. 551: M. Rutherford/ Superstock; p. 552: Walter Hodges/Westlight; p. 557: Superstock; p. 558: Frank Whitney/The Image Bank; p. 559: Hugh Sitton/Tony Stone Images (top); Randy Faris/Westlight (bottom); p. 560: Jim Zuckerman/Westlight; p. 561: C. Aurness/Westlight

CHAPTER 12

p. 570: Keith Wood/Tony Stone Images; p. 571: Luis Casteneda/The Image Bank; p. 572: Michael O'Leary/Tony Stone Images; p. 573: Lillian Gee/Picture It Corp; p. 575: Renee Lynn/Photo Researchers, Inc.; p. 579: Dr. David Wexler, colored by Dr. Jeremy Burgess/Science Photo Library/Photo Researchers, Inc.; p. 581: Colin Molyneux/The Image Bank (left); Photri, Inc. (right); p. 584: Photri, Inc.; p. 586: Tim Bieber/The Image Bank; p. 587: D. & I. McDonald/Photri, Inc.; p. 589: Photri, Inc.; p. 590: Photri, Inc.; p. 592: Comstock; p. 594: Tim Davis/Photo Reseachers, Inc.; p. 595: Bachmann/Photo Researchers, Inc.; p. 597: Roger Miller Photo, Ltd.; p. 600: Obremski/The Image Bank; p. 601: Photri, Inc.; p. 603: Garry McMichael/Photo Researchers, Inc. (top); Blair Seitz/Photo Researchers, Inc. (bottom); p. 604: Harald Sund/The Image Bank (bottom)

CHAPTER 13

p. 612: Maria Taglienti/The Image Bank; p. 613: Tim Bieber/The Image Bank; p. 614: Blair Seitz/Photo Researchers, Inc. (top); Marc Romanelli/The Image Bank (bottom); p. 615: Lillian Gee/Picture It Corp.; p. 617: Peter Miller/The Image Bank; p. 619: John Grant/Photri, Inc. (top); Lani Novak Howe/Photri, Inc. (bottom); p. 624: Mark Burnett/Photo Researchers, Inc. (top); Laima Druskis/Photo Researchers, Inc. (bottom); p. 628: Photri, Inc.; p. 630: Derek Redfearn/The Image Bank (top); Marc Romanelli/The Image Bank (bottom); p. 632: Harald Sund/The Image Bank; p. 634: Photri, Inc.; p. 639: Marcel Isy-Schwart/The Image Bank; p. 640: Francesco Ruggeri/The Image Bank (top); Weinberg/Clark/The Image Bank (bottom); p. 644: Marc Romanelli/The Image Bank; p. 647: Photri, Inc.; p. 648: Lillian Gee/Picture It Corp; p. 654: Photri, Inc.; p. 655: Steve Krongard/The Image Bank; p. 658: Grant V. Faint/The Image Bank; p. 661: Lillian Gee/Picture It Corp.; p. 662: Anne Rippy/The Image Bank; p. 663: Photri, Inc.

CHAPTER 14

p. 672: Comstock; p. 673: U. S. Department of Agriculture; p. 674: David W. Hamilton/The Image Bank; p. 676: Lillian Gee/Picture It Corp., p. 677: Bob Thomason/Tony Stone Images; p. 678: Courtesy of New Idea, Coldwater, OH; p. 680: Marc Grimberg/The Image Bank; p. 681: U.S. Dept. of Housing & Urban Development; p. 682: Kristin Finnegan/Tony Stone Images; p. 688: David Jeffrey/The Image Bank; p. 689: Michael Neveux/Westlight; p. 691: Photri, Inc. (top); Alvis Upitis/The Image Bank (bottom); p. 695: Ben Rose/The Image Bank; p. 704: Ron Chapple/FPG International Corp.; p. 709: U. S. Dept. of Agriculture; p. 711: Lee Balterman/FPG International Corp.; p. 714: Grant Faint/The Image Bank; p. 713: Wendt Worldwide; p. 716: Burton McNeely/The Image Bank; p. 718: Lillian Gee/Picture It Corp.; p. 719: Lou Jones/The Image Bank; p. 720: E. Burciaga/Photri, Inc.